Means Residential Detailed Costs

Contractor's Pricing Guide 2007

Senior Editor
Robert W. Mewis, CCC

Contributing Editors
Christopher Babbit
Ted Baker
Barbara Balboni
Robert A. Bastoni
John H. Chiang, PE
Gary W. Christensen
Cheryl Elsmore
Robert J. Kuchta
Robert C. McNichols
Melville J. Mossman, PE
Jeannene D. Murphy
Stephen C. Plotner
Eugene R. Spencer
Marshall J. Stetson
Phillip R. Waier, PE

Senior Engineering Operations Manager
John H. Ferguson, PE

Senior Vice President & General Manager
John Ware

Vice President of Direct Response
John M. Shea

Director of Product Development
Thomas J. Dion

Production Manager
Michael Kokernak

Production Coordinator
Wayne D. Anderson

Technical Support
Jonathan Forgit
Mary Lou Geary
Jill Goodman
Gary L. Hoitt
Genevieve Medeiros
Paula Reale-Camelio
Kathryn S. Rodriguez
Sheryl A. Rose

Book & Cover Design
Norman R. Forgit

RSMeans

Means Residential Detailed Costs

Contractor's Pricing Guide 2007

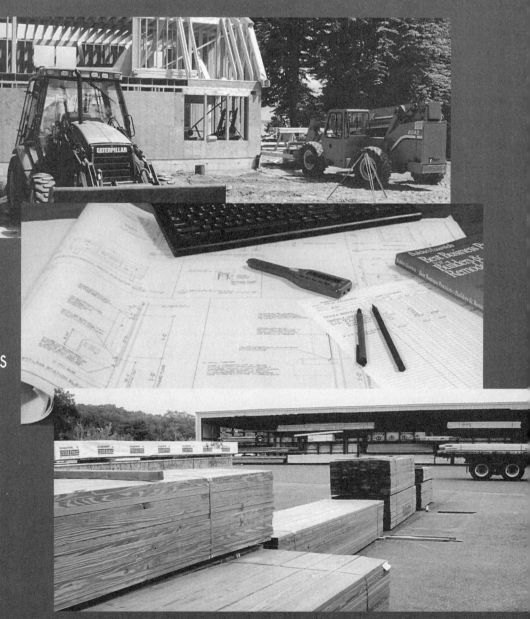

- Unit Costs for Thousands of Residential Building Components
- Cost Adjustment Factors for Your Location
- Daily Productivities & Standard Crews
- Overhead & Profit Guidance

$39.95 per copy (in United States).
Price subject to change without prior notice.

Copyright © 2006

RSMeans
Construction Publishers & Consultants
63 Smiths Lane
Kingston, MA 02364-0800
(781) 422-5000

The authors, editors and engineers of Reed Construction Data's RSMeans product line apply diligence and judgment in locating and using reliable sources for the information published. **However, RSMeans makes no express or implied warranty or guarantee in connection with the content of the information contained herein, including the accuracy, correctness, value, sufficiency, or completeness of the data, methods and other information contained herein. RSMeans makes no express or implied warranty of merchantability or fitness for a particular purpose.** RSMeans shall have no liability to any customer or third party for any loss, expense, or damage, including consequential, incidental, special or punitive damage, including lost profits or lost revenue, caused directly or indirectly by any error or omission, or arising out of, or in connection with, the information contained herein.

No part of this publication may be reproduced, stored in a retrieval system, or transmitted in any form or by any means without prior written permission of Reed Construction Data.

Printed in the United States of America.

10 9 8 7 6 5 4 3 2 1

ISSN 1074-0481

ISBN 0-87629-872-2

Foreword

Reed Construction Data's portfolio of project information products and services includes national, regional, and local construction data, project leads, and project plans, specifications, and addenda available online or in print. Reed Bulletin (www.reedbulletin.com) and Reed CONNECT™ (www.reedconnect.com) deliver the most comprehensive, timely, and reliable project information to support contractors, distributers, and building product manufacturers in identifying, bidding, and tracking projects. The Reed First Source (www.reedfirstsource.com) suite of products used by design professionals for the search, selection, and specification of nationally available building projects consists of the First Source™ annual, SPEC-DATA™, MANU-SPEC™, First Source CAD, and manufacturer catalogs. Reed Design Registry, a database of more than 30,000 U.S. architectural firms, is also published by Reed Construction Data. RSMeans (www.rsmeans.com) provides construction cost data, training, and consulting services in print, CD-ROM, and online. Associated Construction Publications (www.reedpubs.com) reports on heavy, highway, and non-residential construction through a network of 14 regional construction magazines. Reed Construction Data (www.reedconstructiondata.com), headquartered in Atlanta, is a subsidiary of Reed Business Information (www.reedbusinessinformation.com), North America's largest business-to-business information provider. With more than 80 market-leading publications and 55 websites, Reed Business Information's wide range of services also includes research, business development, direct marketing lists, training and development programs, and technology solutions. Reed Business Information is a member of the Reed Elsevier plc group (NYSE: RUK and ENL)—a leading provider of global information-driven services and solutions in the science and medical, legal, education, and business-to-business industry sectors.

Our Mission

Since 1942, RSMeans has been actively engaged in construction cost publishing and consulting throughout North America.

Today, over 60 years after RSMeans began, our primary objective remains the same: to provide you, the construction and facilities professional, with the most current and comprehensive construction cost data possible.

Whether you are a contractor, an owner, an architect, an engineer, a facilities manager, or anyone else who needs a reliable construction cost estimate, you'll find this publication to be a highly useful and necessary tool.

Today, with the constant flow of new construction methods and materials, it's difficult to find the time to look at and evaluate all the different construction cost possibilities. In addition, because labor and material costs keep changing, last year's cost information is not a reliable basis for today's estimate or budget.

That's why so many construction professionals turn to RSMeans. We keep track of the costs for you, along with a wide range of other key information, from city cost indexes . . . to productivity rates . . . to crew composition . . . to contractor's overhead and profit rates.

RSMeans performs these functions by collecting data from all facets of the industry and organizing it in a format that is instantly accessible to you. From the preliminary budget to the detailed unit price estimate, you'll find the data in this book useful for all phases of construction cost determination.

The Staff, the Organization, and Our Services

When you purchase one of RSMeans' publications, you are, in effect, hiring the services of a full-time staff of construction and engineering professionals.

Our thoroughly experienced and highly qualified staff works daily at collecting, analyzing, and disseminating comprehensive cost information for your needs. These staff members have years of practical construction experience and engineering training prior to joining the firm. As a result, you can count on them not only for the cost figures, but also for additional background reference information that will help you create a realistic estimate.

The RSMeans organization is always prepared to help you solve construction problems through its five major divisions: Construction and Cost Data Publishing, Electronic Products and Services, Consulting and Research Services, Insurance Services, and Professional Development Services.

Besides a full array of construction cost estimating books, RSMeans also publishes a number of other reference works for the construction industry. Subjects include construction estimating and project and business management; special topics such as HVAC, roofing, plumbing, and hazardous waste remediation; and a library of facility management references.

In addition, you can access all of our construction cost data electronically using *Means CostWorks®* CD or on the Web.

What's more, you can increase your knowledge and improve your construction estimating and management performance with an RSMeans Construction Seminar or In-House Training Program. These two-day seminar programs offer unparalleled opportunities for everyone in your organization to get updated on a wide variety of construction-related issues.

RSMeans is also a worldwide provider of construction cost management and analysis services for commercial and government owners, and of claims and valuation services for insurers.

In short, RSMeans can provide you with the tools and expertise for constructing accurate and dependable construction estimates and budgets in a variety of ways.

Robert Snow Means Established a Tradition of Quality That Continues Today

Robert Snow Means spent years building RSMeans, making certain he always delivered a quality product.

Today, at RSMeans, we do more than talk about the quality of our data and the usefulness of our books. We stand behind all of our data, from historical cost indexes to construction materials and techniques to current costs.

If you have any questions about our products or services, please call us toll-free at 1-800-334-3509. Our customer service representatives will be happy to assist you. You can also visit our Web site at www.rsmeans.com.

Table of Contents

Foreword	v
How the Book Is Built: An Overview	ix
How To Use the Book: The Details	x
Unit Price Section	1
Reference Section	291
Construction Equipment Rental Costs	293
Crew Listings	305
Location Factors	332
Reference Tables	337
Abbreviations	356
Index	359
Other RSMeans Products and Services	372
Labor Trade Rates including Overhead & Profit	Inside Back Cover

How the Book is Built: An Overview

NEW In 2007...

The Construction Specifications Institute (CSI) and Construction Specifications Canada (CSC) have produced the 2004 edition of MasterFormat, a new and updated system of titles and numbers used extensively to organize construction information.

All unit price data in the RSMeans cost data books is now arranged in the 50-division MasterFormat 2004 system.

A Powerful Construction Tool

You have in your hands one of the most powerful construction tools available today. A successful project is built on the foundation of an accurate and dependable estimate. This book will enable you to construct just such an estimate.

For the casual user the book is designed to be:

- quickly and easily understood so you can get right to your estimate.
- filled with valuable information so you can understand the necessary factors that go into the cost estimate.

For the regular user, the book is designed to be:

- a handy desk reference that can be quickly referred to for key costs.
- a comprehensive, fully reliable source of current construction costs and productivity rates, so you'll be prepared to estimate any project.
- a source book for preliminary project cost, product selections, and alternate materials and methods.

To meet all of these requirements we have organized the book into the following clearly defined sections.

Unit Price Section

All cost data has been divided into the 50 divisions according to the MasterFormat system of classification and numbering For a listing of these divisions and an outline of their subdivisions, see the Unit Price Section Table of Contents.

Reference Section

This section includes information on Equipment Rental Costs, Crew Listings, Location Factors, Reference Tables, and a listing of Abbreviations.

Equipment Rental Costs: This section contains the average costs to rent and operate hundreds of pieces of construction equipment.

Crew Listings: This section lists all the crews referenced in the book. For the purposes of this book, a crew is composed of more than one trade classification and/or the addition of power equipment to any trade classification. Power equipment is included in the cost of the crew. Costs are shown both with bare labor rates and with the installing contractor's overhead and profit added. For each, the total crew cost per eight-hour day and the composite cost per labor-hour are listed.

Location Factors: You can adjust total project costs to over 900 locations throughout the U.S. and Canada by using the data in this section.

Reference Tables: At the beginning of selected major classifications in the Unit Price Section are reference numbers shown in a shaded box. These numbers refer you to related information in the Reference Section. In this section, you'll find reference tables, explanations, and estimating information that support how we develop the unit price data, technical data, and estimating procedures.

Abbreviations: A listing of the abbreviations used throughout this book, along with the terms they represent, is included in this section.

Index

A comprehensive listing of all terms and subjects in this book will help you quickly find what you need when you are not sure where it falls in MasterFormat.

The Scope of This Book

This book is designed to be as comprehensive and as easy to use as possible. To that end we have made certain assumptions and limited its scope in two key ways:

1. We have established material prices based on a national average.
2. We have computed labor costs based on a 7 major-region average of residential wage rates.

Project Size

This book is intended for use by those involved primarily in Residential construction costing less than $850,000. This includes the construction of homes, row houses, townhouses, condominiums, and apartments.

With reasonable exercise of judgment the figures can be used for any building work. For other types of projects, such as repair and remodeling or commercial buildings, consult the appropriate RSMeans publication for more information.

How to Use the Book: The Details

What's Behind the Numbers? The Development of Cost Data

The staff at RSMeans continuously monitors developments in the construction industry in order to ensure reliable, thorough and up-to-date cost information.

While *overall* construction costs may vary relative to general economic conditions, price fluctuations within the industry are dependent upon many factors. Individual price variations may, in fact, be opposite to overall economic trends. Therefore, costs are continually monitored and complete updates are published yearly. Also, new items are frequently added in response to changes in materials and methods.

Costs–$ (U.S.)

All costs represent U.S. national averages and are given in U.S. dollars. The RSMeans Location Factors can be used to adjust costs to a particular location. The Location Factors for Canada can be used to adjust U.S. national averages to local costs in Canadian dollars. No exchange rate conversion is necessary.

Material Costs

The RSMeans staff contacts manufacturers, dealers, distributors, and contractors all across the U.S. and Canada to determine national average material costs. If you have access to current material costs for your specific location, you may wish to make adjustments to reflect differences from the national average.

Included within material costs are fasteners for a normal installation. RSMeans engineers use manufacturers' recommendations, written specifications, and/or standard construction practice for size and spacing of fasteners. Adjustments to material costs may be required for your specific application or location. Material costs do not include sales tax.

Labor Costs

Labor costs are based on the average of residential wages from across the U.S. for the current year. Rates, along with overhead and profit markups, are listed on the inside back cover of this book.

- If wage rates in your area vary from those used in this book, or if rate increases are expected within a given year, labor costs should be adjusted accordingly.

Labor costs reflect productivity based on actual working conditions. These figures include time spent during a normal workday on tasks other than actual installation, such as material receiving and handling, mobilization at site, site movement, breaks, and cleanup.

Productivity data is developed over an extended period so as not to be influenced by abnormal variations and reflects a typical average.

Equipment Costs

Equipment costs include not only rental costs, but also operating costs for equipment under normal use. Equipment and rental rates are obtained from industry sources throughout North America—contractors, suppliers, dealers, manufacturers, and distributors.

Factors Affecting Costs

Costs can vary depending upon a number of variables. Here's how we have handled the main factors affecting costs.

Quality—The prices for materials and the workmanship upon which productivity is based represent sound construction work. They are also in line with U.S. government specifications.

Overtime—We have made no allowance for overtime. If you anticipate premium time or work beyond normal working hours, be sure to make an appropriate adjustment to your labor costs.

Productivity—The productivity, daily output, and labor-hour figures for each line item are based on working an eight-hour day in daylight hours in moderate temperatures. For work that extends beyond normal work hours or is performed under adverse conditions, productivity may decrease. (See the section in "How To Use the Unit Price Pages" for more on productivity.)

Size of Project—The size, scope of work, and type of construction project will have a significant impact on cost. Economies of scale can reduce costs for large projects. Unit costs can often run higher for small projects. Costs in this book are intended for the size and type of project as previously described in "How the Book Is Built: An Overview." Costs for projects of a significantly different size or type should be adjusted accordingly.

Location—Material prices in this book are for metropolitan areas. However, in dense urban areas, traffic and site storage limitations may increase costs.

Beyond a 20-mile radius of large cities, extra trucking or transportation charges may also increase the material costs slightly. On the other hand, lower wage rates may be in effect. Be sure to consider both of these factors when preparing an estimate, particularly if the job site is located in a central city or remote rural location.

In addition, highly specialized subcontract items may require travel and per-diem expenses for mechanics.

Other Factors —
- season of year
- contractor management
- weather conditions
- local union restrictions
- building code requirements
- availability of:
  - adequate energy
  - skilled labor
  - building materials
- owner's special requirements/restrictions
- safety requirements
- environmental considerations

General Conditions—The extreme right-hand column of each chart gives the "Total Including O&P." These figures contain the original installing subcontractor's O&P. Therefore, it is necessary for a general contractor to add a percentage of all subcontracted items. For a detailed breakdown of O&P see the inside back cover of this book.

Overhead & Profit—Tables in the reference section give details on overhead. For Unit Price costs, these tables can be used by the general contractor as a guide to determine the appropriate overhead and profit markups.

Unpredictable Factors—General business conditions influence "in-place" costs of all items. Substitute materials and construction methods may have to be employed. These may affect the installed cost and/or life cycle costs. Such factors may be difficult to evaluate and cannot necessarily be predicted on the basis of the job's location in a particular section of the country. Thus, where these factors apply, you may find significant but unavoidable cost variations for which you will have to apply a measure of judgment to your estimate.

Rounding of Costs

In general, all unit prices in excess of $5.00 have been rounded to make them easier to use and still maintain adequate precision of the results. The rounding rules we have chosen are in the following table.

Prices from . . .	Rounded to the nearest . . .
$.01 to $5.00	$.01
$5.01 to $20.00	$.05
$20.01 to $100.00	$.50
$100.01 to $300.00	$1.00
$300.01 to $1,000.00	$5.00
$1,000.01 to $10,000.00	$25.00
$10,000.01 to $50,000.00	$100.00
$50,000.01 and above	$500.00

Final Checklist

Estimating can be a straightforward process provided you remember the basics. Here's a checklist of some of the steps you should remember to complete before finalizing your estimate.

Did you remember to . . .

- factor in the Location Factor for your locale?
- take into consideration which items have been marked up and by how much?
- mark up the entire estimate sufficiently for your purposes?
- read the background information on techniques and technical matters that could impact your project time span and cost?
- include all components of your project in the final estimate?
- double check your figures for accuracy?
- call RSMeans if you have any questions about your estimate or the data you've found in our publications?

Remember, RSMeans stands behind its publications. If you have any questions about your estimate . . . about the costs you've used from our books . . . or even about the technical aspects of the job that may affect your estimate, feel free to call the RSMeans editors at 1-800-334-3509.

Unit Price Section

Table of Contents

Div. No.		Page
	General Requirements	
01 11	Summary of Work	6
01 21	Allowances	6
01 31	Project Management and Coordination	6
01 41	Regulatory Requirements	7
01 54	Construction Aids	7
01 56	Temporary Barriers and Enclosures	10
01 71	Examination and Preparation	10
01 74	Cleaning and Waste Management	10
	Existing Conditions	
02 21	Surveys	12
02 32	Geotechnical Investigations	12
02 41	Demolition	12
02 83	Lead Remediation	16
	Concrete	
03 01	Maintenance of Concrete	20
03 05	Common Work Results for Concrete	20
03 11	Concrete Forming	20
03 15	Concrete Accessories	22
03 21	Reinforcing Steel	23
03 22	Welded Wire Fabric Reinforc.	24
03 24	Fibrous Reinforcing	24
03 30	Cast-In-Place Concrete	24
03 31	Structural Concrete	25
03 35	Concrete Finishing	26
03 39	Concrete Curing	27
03 41	Precast Structural Concrete	27
03 48	Precast Concrete Specialties	27
03 63	Epoxy Grouting	28
03 82	Concrete Boring	28
	Masonry	
04 01	Maintenance of Masonry	30
04 05	Common Work Results for Masonry	30
04 21	Clay Unit Masonry	33
04 22	Concrete Unit Masonry	34
04 23	Glass Unit Masonry	36
04 24	Adobe Unit Masonry	36
04 27	Multiple-Wythe Unit Masonry	37
04 41	Dry-Placed Stone	37
04 43	Stone Masonry	37
04 51	Flue Liner Masonry	39
04 57	Masonry Fireplaces	39
04 72	Cast Stone Masonry	39
	Metals	
05 01	Maintenance of Metals	43
05 12	Structural Steel Framing	46
05 31	Steel Decking	47
05 41	Structural Metal Stud Framing	48
05 42	Cold-Formed Metal Joist Framing	51
05 44	Cold-Formed Metal Trusses	57
05 51	Metal Stairs	58
05 52	Metal Railings	58
05 58	Formed Metal Fabrications	59
05 71	Decorative Metal Stairs	59
	Wood, Plastics & Composites	
06 05	Common Work Results for Wood, Plastics & Composites	62
06 11	Wood Framing	71
06 12	Structural Panels	81
06 13	Heavy Timber	82
06 15	Wood Decking	83
06 16	Sheathing	83

Div. No.		Page
06 17	Shop-Fabricated Structural Wood	85
06 18	Glued-Lamin. Construction	86
06 22	Millwork	87
06 25	Prefinished Paneling	90
06 26	Board Paneling	92
06 43	Wood Stairs and Railings	92
06 44	Ornamental Woodwork	93
06 48	Wood Frames	94
06 49	Wood Screens and Exterior Wood Shutters	95
06 65	Plastic Simulated Wood Trim	96
	Thermal & Moisture Protection	
07 01	Operation and Maint. of Thermal and Moisture Protection	100
07 05	Common Work Results for Thermal and Moisture Protection	100
07 11	Dampproofing	101
07 19	Water Repellents	101
07 21	Thermal Insulation	101
07 22	Roof and Deck Insulation	104
07 24	Exterior Insulation and Finish Systems	105
07 26	Vapor Retarders	105
07 31	Shingles and Shakes	106
07 32	Roof Tiles	108
07 41	Roof Panels	108
07 42	Wall Panels	109
07 46	Siding	110
07 51	Built-Up Bituminous Roofing	112
07 52	Modified Bituminous Membrane Roofing	113
07 58	Roll Roofing	113
07 61	Sheet Metal Roofing	114
07 62	Sheet Metal Flashing and Trim	114
07 65	Flexible Flashing	115
07 71	Roof Specialties	116
07 72	Roof Accessories	118
07 92	Joint Sealants	119
	Openings	
08 01	Operation and Maintenance of Openings	122
08 05	Common Work Results for Openings	123
08 11	Metal Doors and Frames	124
08 12	Metal Frames	124
08 13	Metal Doors	125
08 14	Wood Doors	127
08 16	Composite Doors	130
08 17	Integrated Door Opening Assemblies	130
08 31	Access Doors and Panels	131
08 32	Sliding Glass Doors	132
08 36	Panel Doors	132
08 51	Metal Windows	133
08 52	Wood Windows	134
08 53	Plastic Windows	142
08 61	Roof Windows	145
08 62	Unit Skylights	146
08 71	Door Hardware	147
08 75	Window Hardware	149
08 79	Hardware Accessories	150
08 81	Glass Glazing	150
08 83	Mirrors	151
08 87	Glazing Surface Films	151
08 91	Louvers	152
08 95	Vents	152

Div. No.		Page
	Finishes	
09 01	Maintenance of Finishes	154
09 05	Common Work Results for Finishes	154
09 22	Supports for Plaster and Gypsum Board	155
09 23	Gypsum Plastering	158
09 24	Portland Cement Plastering	158
09 26	Veneer Plastering	159
09 28	Backing Boards and Underlayments	159
09 29	Gypsum Board	160
09 30	Tiling	162
09 51	Acoustical Ceilings	164
09 53	Acoustical Ceiling Suspension Assemblies	165
09 63	Masonry Flooring	165
09 64	Wood Flooring	166
09 65	Resilient Flooring	167
09 66	Terrazzo Flooring	168
09 68	Carpeting	169
09 72	Wall Coverings	170
09 91	Painting	171
09 93	Staining and Transparent Finishing	184
09 96	High-Performance Coatings	184
	Specialties	
10 21	Compartments and Cubicles	186
10 28	Toilet, Bath, and Laundry Accessories	187
10 31	Manufactured Fireplaces	187
10 32	Fireplace Specialties	188
10 35	Stoves	189
10 44	Fire Protection Specialties	189
10 55	Postal Specialties	189
10 56	Storage Assemblies	189
10 57	Wardrobe and Closet Specialties	190
10 73	Protective Covers	190
10 74	Manufactured Exterior Specialties	191
10 75	Flagpoles	191
	Equipment	
11 24	Maintenance Equipment	194
11 26	Unit Kitchens	194
11 31	Residential Appliances	194
11 33	Retractable Stairs	196
11 41	Food Storage Equipment	197
	Furnishings	
12 21	Window Blinds	200
12 22	Curtains and Drapes	200
12 23	Interior Shutters	201
12 24	Window Shades	201
12 32	Manufactured Wood Casework	201
12 34	Manufactured Plastic Casework	205
12 36	Countertops	205
12 93	Site Furnishings	207
	Special Construction	
13 11	Swimming Pools	210
13 12	Fountains	210
13 17	Tubs and Pools	211
13 24	Special Activity Rooms	211
13 34	Fabricated Engineered Structures	212

Table of Contents (cont.)

Div. No.		Page
	Conveying Equipment	
14 21	Electric Traction Elevators	214
14 42	Wheelchair Lifts	214
	Plumbing	
22 05	Common Work Results for Plumbing	216
22 07	Plumbing Insulation	217
22 11	Facility Water Distribution	218
22 13	Facility Sanitary Sewerage	224
22 14	Facility Storm Drainage	226
22 31	Domestic Water Softeners	226
22 33	Electric Domestic Water Heaters	226
22 34	Fuel-Fired Domestic Water Heaters	227
22 41	Residential Plumbing Fixtures	227
22 42	Commercial Plumbing Fixt.	230
22 51	Swimming Pool Plumbing Systems	230
	Heating, Ventilating, and Air Conditioning	
23 05	Common Work Results for HVAC	232
23 07	HVAC Insulation	232
23 09	Instrumentation and Control for HVAC	233
23 13	Facility Fuel-Storage Tanks	233
23 21	Hydronic Piping and Pumps	234
23 31	HVAC Ducts and Casings	234
23 33	Air Duct Accessories	234
23 34	HVAC Fans	235
23 37	Air Outlets and Inlets	236
23 41	Particulate Air Filtration	237
23 42	Gas-Phase Air Filtration	238
23 43	Electronic Air Cleaners	238
23 51	Breechings, Chimneys, and Stacks	238
23 52	Heating Boilers	238
23 54	Furnaces	239

Div. No.		Page
23 62	Packaged Compressor and Condenser Units	241
23 74	Packaged Outdoor HVAC Equipment	241
23 81	Decentralized Unitary HVAC Equipment	242
23 82	Convection Heating and Cooling Units	243
23 83	Radiant Heating Units	243
	Electrical	
26 05	Common Work Results for Electrical	246
26 24	Switchboards and Panelboards	255
26 27	Low-Voltage Distribution Equipment	255
26 28	Low-Voltage Circuit Protective Devices	257
26 32	Packaged Generator Assemblies	258
26 36	Transfer Switches	258
26 41	Facility Lightning Protection	258
26 51	Interior Lighting	259
26 55	Special Purpose Lighting	259
26 56	Exterior Lighting	260
26 61	Lighting Systems and Accessories	260
	Communications	
27 41	Audio-Video Systems	262
	Electronic Safety & Security	
28 16	Intrusion Detection	264
28 31	Fire Detection and Alarm	264
	Earthwork	
31 05	Common Work Results for Earthwork	266
31 11	Clearing and Grubbing	266
31 13	Selective Tree and Shrub Removal and Trimming	266
31 14	Earth Stripping & Stockpiling	266
31 22	Grading	267

Div. No.		Page
31 23	Excavation and Fill	267
31 25	Erosion and Sedimentation Controls	269
31 31	Soil Treatment	269
	Exterior Improvements	
32 01	Operation and Maintenance of Exterior Improvements	272
32 06	Schedules for Exterior Improvements	272
32 11	Base Courses	272
32 12	Flexible Paving	273
32 13	Rigid Paving	273
32 14	Unit Paving	274
32 16	Curbs and Gutters	274
32 31	Fences and Gates	275
32 32	Retaining Walls	278
32 84	Planting Irrigation	279
32 91	Planting Preparation	279
32 92	Turf and Grasses	280
32 93	Plants	281
32 94	Planting Accessories	282
32 96	Transplanting	283
	Utilities	
33 05	Common Work Results for Utilities	287
33 11	Water Utility Distribution Piping	287
33 12	Water Utility Distribution Equipment	288
33 21	Water Supply Wells	288
33 31	Sanitary Utility Sewerage Piping	288
33 36	Utility Septic Tanks	289
33 41	Storm Utility Drainage Piping	289
33 44	Storm Utility Water Drains	289
33 46	Subdrainage	289
33 49	Storm Drainage Structures	290
33 51	Natural-Gas Distribution	290

How to Use the Unit Price Pages

Important
Prices in this section are listed in two ways: as bare costs and as costs including overhead and profit of the installing contractor. In most cases, if the work is to be subcontracted, it is best for a general contractor to add an additional 10% to the figures found in the column titled "TOTAL INCL. O&P."

Unit
The unit of measure listed here reflects the material being used in the line item. For example: roof trusses are priced per each (Ea.).

Productivity
The daily output represents typical total daily amount of work that the designated crew will produce. Labor-hours are a unit of measure for the labor involved in performing a task. To derive the total labor-hours for a task, multiply the quantity of the item involved times the labor-hour figure shown.

Line Number Determination
Each line item is identified by a unique twelve-digit number.

MasterFormat
Division (06)
06 17 53.10 5050

MasterFormat Level 2 (06 17 00)
06 17 53.10 5050

MasterFormat Level 3
06 17 53.10 5050

06 17 53.10 5050
RSMeans 12-Digit
Line Number

Description
This line item describes a common wood truss that will span 20'. It will be installed by an F-6 Crew at the rate of 62 per day, or .645 labor hours each.

Reference Number [R061636-20]
These reference numbers refer to charts, tables, estimating data, cost derivations and other information which may be useful to the user of this book. This information is located in the Reference Section of this book.

Bare Costs are developed as follows for line no. 06 17 53.10 5050
Mat. is Bare Material Cost ($64.50)
Labor for Crew F-6 = Labor-hour Cost ($23.25) × Labor-hour Units (.645) = $15.00 (Rounded)
Equipment = Equipment Cost per Labor-hour ($18.10) × Labor-hour Units (.645) = $11.70
Total = Mat. Cost ($64.50) + Labor Cost ($15.00) + Equip. Cost ($11.70) = $91.20 per EA.

(Note: Where the crew is indicated, Equipment and Labor cost are derived from the Crew Tables. See example above.)

Total Costs Including O&P are developed as follows:
Mat. is Bare Material Cost + 10% = 64.50 + $6.45 = $70.95
Labor for Crew F-6 = Labor-hour Cost ($39.18) × Labor-Hour Units (.645) = $25.27 (Rounded)
Equipment = Equipment Cost per Labor-hour ($19.91) × Labor-hour Units (.645) = $12.80 (Rounded)
Total = Mat. Cost ($70.95) + Labor Cost ($25.25) + Equip. Cost ($12.80) = $109.00

(Note: Where a crew is indicated, Equipment and Labor costs are derived from the Crew Tables. See example above. "Total incl. O&P" costs may be rounded.)

Crew F-6

Crew No.	Bare Costs		Incl. Subs O & P		Cost Per Labor-Hour	
Crew F-6	Hr.	Daily	Hr.	Daily	Bare Costs	Incl. O&P
2 Carpenters	$25.70	$411.20	$43.60	$697.60	$23.25	$39.18
2 Building Laborers	18.70	299.20	31.75	508.00		
1 Equip. Oper. (crane)	27.45	219.60	45.20	361.60		
1 Hyd. Crane, 12 Ton		724.00		796.40	18.10	19.91
40 L.H., Daily Totals		$1654.00		$2363.60	$41.35	$59.09

06 17 Shop-Fabricated Structural Wood

06 17 53 – Shop-Fabricated Wood Trusses

06 17 53.10 Roof Trusses	Crew	Daily Output	Labor-Hours	Unit	Material	2007 Bare Costs Labor	Equipment	Total	Total Incl O&P
0010 ROOF TRUSSES									
0020 For timber connectors, see div. 06 05 23.60									
5000 Common wood, 2" x 4" metal plate connected, 24" O.C., 4/12 slope									
5010 1' overhang, 12' span	F-5	55	.582	Ea.	42	12.95		54.95	68
5050 20' span	F-6	62	.645		64.50	15	11.70	91.20	109
5100 24' span		60	.667		75	15.50	12.05	102.55	122
5150 26' span		57	.702		77.50	16.30	12.70	106.50	127
5200 28' span		53	.755		69	17.55	13.65	100.20	121
5240 30' span		51	.784		96.50	18.25	14.20	128.95	152
5250 32' span		50	.800		100	18.60	14.50	133.10	157
5280 34' span		48	.833		122	19.35	15.10	156.45	183
5350 8/12 pitch, 1' overhang, 20' span		57	.702		68	16.30	12.70	97	116
5400 24' span		55	.727		86.50	16.90	13.15	116.55	138
5450 26' span		52	.769		90	17.90	13.90	121.80	144
5500 28' span		49	.816		99.50	19	14.80	133.30	157
5550 32' span		45	.889		119	20.50	16.10	155.60	184
5600 36' span		41	.976		149	22.50	17.65	189.15	221

Division 1
General Requirements

01 11 Summary of Work

01 11 31 – Professional Consultants

01 11 31.10 Architectural Fees

		Crew	Daily Output	Labor-Hours	Unit	Material	2007 Bare Costs Labor	Equipment	Total	Total Incl O&P
0010	**ARCHITECTURAL FEES** R011110-10									
0020	For new construction									
0060	Minimum				Project					4.90%
0090	Maximum									16%
0100	For alteration work, to $500,000, add to fee									50%
0150	Over $500,000, add to fee				↓					25%

01 11 31.20 Construction Management Fees

0010	**CONSTRUCTION MANAGEMENT FEES**									
0060	For work to $10,000				Project					10%
0070	To $25,000									9%
0090	To $100,000				↓					6%

01 11 31.75 Renderings

0010	**RENDERINGS** Color, matted, 20" x 30", eye level,									
0050	Average				Ea.	2,650			2,650	2,925

01 21 Allowances

01 21 16 – Contingency Allowances

01 21 16.50 Contingencies

0010	**CONTINGENCIES**									
0020	For estimate at conceptual stage				Project					20%
0150	Final working drawing stage				"					3%

01 21 63 – Taxes

01 21 63.10 Taxes

0010	**TAXES** R012909-80									
0020	Sales tax, State, average				%	4.65%				
0050	Maximum R012909-85					7%				
0200	Social Security, on first $94,200 of wages						7.65%			
0300	Unemployment, combined Federal and State, minimum						2.10%			
0350	Average						6.20%			
0400	Maximum				↓		8%			

01 31 Project Management and Coordination

01 31 13 – Project Coordination

01 31 13.30 Insurance

0010	**INSURANCE** R013113-40									
0020	Builders risk, standard, minimum				Job					.22%
0050	Maximum R013113-60									.59%
0200	All-risk type, minimum									.25%
0250	Maximum				↓					.62%
0400	Contractor's equipment floater, minimum				Value					.50%
0450	Maximum				"					1.50%
0600	Public liability, average				Job					1.55%
0800	Workers' compensation & employer's liability, average									
0850	by trade, carpentry, general				Payroll		18.51%			
0900	Clerical						.60%			
0950	Concrete						15.79%			
1000	Electrical						6.40%			
1050	Excavation				↓		10.34%			

01 31 Project Management and Coordination

01 31 13 – Project Coordination

01 31 13.30 Insurance		Crew	Daily Output	Labor-Hours	Unit	Material	2007 Bare Costs Labor	2007 Bare Costs Equipment	Total	Total Incl O&P
1100	Glazing				Payroll		13.82%			
1150	Insulation						15.24%			
1200	Lathing						10.71%			
1250	Masonry						14.97%			
1300	Painting & decorating						12.89%			
1350	Pile driving						22.94%			
1400	Plastering						14.62%			
1450	Plumbing						7.78%			
1500	Roofing						31.75%			
1550	Sheet metal work (HVAC)						11.09%			
1600	Steel erection, structural						38.86%			
1650	Tile work, interior ceramic						9.63%			
1700	Waterproofing, brush or hand caulking						7.27%			
1800	Wrecking						40.51%			
2000	Range of 35 trades in 50 states, excl. wrecking, min.						2.50%			
2100	Average						16.20%			
2200	Maximum						110.10%			

01 41 Regulatory Requirements

01 41 26 – Permits

01 41 26.50 Permits		Crew	Daily Output	Labor-Hours	Unit	Material	Labor	Equipment	Total	Total Incl O&P
0010	**PERMITS**									
0020	Rule of thumb, most cities, minimum				Job					.50%
0100	Maximum				"					2%

01 54 Construction Aids

01 54 23 – Temporary Scaffolding and Platforms

01 54 23.60 Pump Staging		Crew	Daily Output	Labor-Hours	Unit	Material	Labor	Equipment	Total	Total Incl O&P
0010	**PUMP STAGING**, Aluminum									
1300	System in place, 50' working height, per use based on 50 uses	2 Carp	84.80	.189	C.S.F.	5.65	4.85		10.50	14.50
1400	100 uses		84.80	.189		2.83	4.85		7.68	11.35
1500	150 uses		84.80	.189		1.90	4.85		6.75	10.35

01 54 23.70 Scaffolding		Crew	Daily Output	Labor-Hours	Unit	Material	Labor	Equipment	Total	Total Incl O&P
0010	**SCAFFOLDING**									
0015	Steel tube, regular, no plank, labor only to erect & dismantle									
0090	Building exterior, wall face, 1 to 5 stories, 6'-4" x 5' frames	3 Carp	8	3	C.S.F.		77		77	131
0200	6 to 12 stories	4 Carp	8	4			103		103	174
0310	13 to 20 stories	5 Carp	8	5			129		129	218
0460	Building interior, wall face area, up to 16' high	3 Carp	12	2			51.50		51.50	87
0560	16' to 40' high		10	2.400			61.50		61.50	105
0800	Building interior floor area, up to 30' high		150	.160	C.C.F.		4.11		4.11	7
0900	Over 30' high	4 Carp	160	.200	"		5.15		5.15	8.70
0906	Complete system for face of walls, no plank, material only rent/mo				C.S.F.	34.50			34.50	37.50
0908	Interior spaces, no plank, material only rent/mo				"	3.30			3.30	3.63
0910	Steel tubular, heavy duty shoring, buy									
0920	Frames 5' high 2' wide				Ea.	82.50			82.50	91
0925	5' high 4' wide					93.50			93.50	103
0930	6' high 2' wide					94.50			94.50	104
0935	6' high 4' wide					110			110	121

01 54 Construction Aids

01 54 23 – Temporary Scaffolding and Platforms

01 54 23.70 Scaffolding

		Crew	Daily Output	Labor-Hours	Unit	Material	2007 Bare Costs Labor	2007 Bare Costs Equipment	Total	Total Incl O&P
0940	Accessories									
0945	Cross braces				Ea.	16			16	17.60
0950	U-head, 8" x 8"					19.30			19.30	21
0955	J-head, 4" x 8"					14.10			14.10	15.50
0960	Base plate, 8" x 8"					15.70			15.70	17.25
0965	Leveling jack				↓	33.50			33.50	37
1000	Steel tubular, regular, buy									
1100	Frames 3' high 5' wide				Ea.	64			64	70
1150	5' high 5' wide					73.50			73.50	81
1200	6'-4" high 5' wide					92.50			92.50	102
1350	7'-6" high 6' wide					160			160	175
1500	Accessories cross braces					18.15			18.15	19.95
1550	Guardrail post					16.50			16.50	18.15
1600	Guardrail 7' section					8.80			8.80	9.70
1650	Screw jacks & plates					24			24	26.50
1700	Sidearm brackets					31			31	34
1750	8" casters					33			33	36.50
1800	Plank 2" x 10" x 16'-0"					44			44	48.50
1900	Stairway section					270			270	296
1910	Stairway starter bar					32			32	35
1920	Stairway inside handrail					58.50			58.50	64
1930	Stairway outside handrail					80.50			80.50	88.50
1940	Walk-thru frame guardrail				↓	40.50			40.50	44.50
2000	Steel tubular, regular, rent/mo.									
2100	Frames 3' high 5' wide				Ea.	5			5	5.50
2150	5' high 5' wide					5			5	5.50
2200	6'-4" high 5' wide					4.50			4.50	4.95
2250	7'-6" high 6' wide					7			7	7.70
2500	Accessories, cross braces					1			1	1.10
2550	Guardrail post					1			1	1.10
2600	Guardrail 7' section					1			1	1.10
2650	Screw jacks & plates					2			2	2.20
2700	Sidearm brackets					2			2	2.20
2750	8" casters					8			8	8.80
2800	Outrigger for rolling tower					3			3	3.30
2850	Plank 2" x 10" x 16'-0"					6			6	6.60
2900	Stairway section					40			40	44
2940	Walk-thru frame guardrail				↓	2.50			2.50	2.75
3000	Steel tubular, heavy duty shoring, rent/mo.									
3250	5' high 2' & 4' wide				Ea.	5			5	5.50
3300	6' high 2' & 4' wide					5			5	5.50
3500	Accessories, cross braces					1			1	1.10
3600	U - head, 8" x 8"					1			1	1.10
3650	J - head, 4" x 8"					1			1	1.10
3700	Base plate, 8" x 8"					1			1	1.10
3750	Leveling jack					2			2	2.20
5700	Planks, 2x10x16'-0", labor only to erect & remove to 50' H	3 Carp	72	.333			8.55		8.55	14.55
5800	Over 50' high	4 Carp	80	.400	↓		10.30		10.30	17.45

01 54 23.80 Staging Aids

		Crew	Daily Output	Labor-Hours	Unit	Material	Labor	Equipment	Total	Total Incl O&P
0010	**STAGING AIDS** and fall protection equipment									
0100	Sidewall staging bracket, tubular, buy				Ea.	35			35	38.50
0110	Cost each per day, based on 250 days use				Day	.14			.14	.15

01 54 Construction Aids

01 54 23 – Temporary Scaffolding and Platforms

01 54 23.80 Staging Aids

		Crew	Daily Output	Labor-Hours	Unit	Material	2007 Bare Costs Labor	2007 Bare Costs Equipment	Total	Total Incl O&P
0200	Guard post, buy				Ea.	18.20			18.20	20
0210	Cost each per day, based on 250 days use				Day	.07			.07	.08
0300	End guard chains, buy per pair				Pair	27.50			27.50	30.50
0310	Cost per set per day, based on 250 days use				Day	.14			.14	.15
1010	Cost each per day, based on 250 days use				"	.03			.03	.03
1100	Wood bracket, buy				Ea.	14.85			14.85	16.35
1110	Cost each per day, based on 250 days use				Day	.06			.06	.07
2010	Cost per pair per day, based on 250 days use				"	.41			.41	.45
2100	Steel siderail jack, buy per pair				Pair	73			73	80.50
2110	Cost per pair per day, based on 250 days use				Day	.29			.29	.32
3010	Cost each per day, based on 250 days use				"	.18			.18	.19
3100	Aluminum scaffolding plank, 20" wide x 24' long, buy				Ea.	720			720	790
3110	Cost each per day, based on 250 days use				Day	2.87			2.87	3.16
4010	Cost each per day, based on 250 days use				"	.82			.82	.90
4100	Rope for safety line, 5/8" x 100' nylon, buy				Ea.	55			55	60.50
4110	Cost each per day, based on 250 days use				Day	.22			.22	.24
4200	Permanent U-Bolt roof anchor, buy				Ea.	32.50			32.50	35.50
4300	Temporary (one use) roof ridge anchor, buy				"	26			26	28.50
5000	Installation (setup and removal) of staging aids									
5010	Sidewall staging bracket	2 Carp	64	.250	Ea.		6.45		6.45	10.90
5020	Guard post with 2 wood rails	"	64	.250			6.45		6.45	10.90
5030	End guard chains, set	1 Carp	64	.125			3.21		3.21	5.45
5100	Roof shingling bracket		96	.083			2.14		2.14	3.63
5200	Ladder jack		64	.125			3.21		3.21	5.45
5300	Wood plank, 2x10x16'	2 Carp	80	.200			5.15		5.15	8.70
5310	Aluminum scaffold plank, 20" x 24'	"	40	.400			10.30		10.30	17.45
5410	Safety rope	1 Carp	40	.200			5.15		5.15	8.70
5420	Permanent U-Bolt roof anchor (install only)	2 Carp	40	.400			10.30		10.30	17.45
5430	Temporary roof ridge anchor (install only)	1 Carp	64	.125			3.21		3.21	5.45

01 54 36 – Equipment Mobilization

01 54 36.50 Mobilization or Demob.

		Crew	Daily Output	Labor-Hours	Unit	Material	2007 Bare Costs Labor	2007 Bare Costs Equipment	Total	Total Incl O&P
0010	**MOBILIZATION OR DEMOB.** (One or the other, unless noted)									
0015	Up to 25 mi haul dist (50 mi RT for mob/demob crew)									
0020	Dozer, loader, backhoe, excav., grader, paver, roller, 70 to 150 H.P.	B-34N	4	2	Ea.		42	124	166	207
0900	Shovel or dragline, 3/4 C.Y.	B-34K	3.60	2.222			46.50	160	206.50	255
1100	Small equipment, placed in rear of, or towed by pickup truck	A-3A	8	1			20.50	11.40	31.90	46.50
1150	Equip up to 70 HP, on flatbed trailer behind pickup truck	A-3D	4	2			40.50	46	86.50	119
2000	Mob & demob truck-mounted crane up to 75 ton, driver only	1 Eqhv	3.60	2.222			61		61	100
2200	Crawler-mounted, up to 75 ton	A-3F	2	8			194	305	499	655
2500	For each additional 5 miles haul distance, add						10%	10%		
3000	For large pieces of equipment, allow for assembly/knockdown									
3100	For mob/demob of micro-tunneling equip, see section 33 05 23.19									

01 54 39 – Construction Equipment

01 54 39.70 Small Tools

		Crew	Daily Output	Labor-Hours	Unit	Material	2007 Bare Costs Labor	2007 Bare Costs Equipment	Total	Total Incl O&P
0010	**SMALL TOOLS**									
0020	As % of contractor's work, minimum				Total					.50%
0100	Maximum				"					2%

01 56 Temporary Barriers and Enclosures

01 56 13 – Temporary Air Barriers

01 56 13.60 Tarpaulins		Crew	Daily Output	Labor-Hours	Unit	Material	2007 Bare Costs Labor	Equipment	Total	Total Incl O&P
0010	**TARPAULINS**									
0020	Cotton duck, 10 oz. to 13.13 oz. per S.Y., minimum				S.F.	.48			.48	.53
0050	Maximum					.62			.62	.68
0200	Reinforced polyethylene 3 mils thick, white					.15			.15	.17
0300	4 mils thick, white, clear or black					.20			.20	.22
0730	Polyester reinforced w/ integral fastening system 11 mils thick					1.07			1.07	1.18

01 71 Examination and Preparation

01 71 23 – Field Engineering

01 71 23.13 Construction Layout

		Crew	Daily Output	Labor-Hours	Unit	Material	Labor	Equipment	Total	Total Incl O&P
0010	**CONSTRUCTION LAYOUT**									
1100	Crew for layout of building, trenching or pipe laying, 2 person crew	A-6	1	16	Day		415	58	473	760
1200	3 person crew	A-7	1	24	"		660	58	718	1,200

01 74 Cleaning and Waste Management

01 74 13 – Progress Cleaning

01 74 13.20 Cleaning Up

		Crew	Daily Output	Labor-Hours	Unit	Material	Labor	Equipment	Total	Total Incl O&P
0010	**CLEANING UP**									
0020	After job completion, allow, minimum				Job					.30%
0040	Maximum				"					1%

Division 2
Existing Conditions

02 21 Surveys

02 21 13 – Site Surveys

02 21 13.09 Topographical Surveys

		Crew	Daily Output	Labor-Hours	Unit	Material	2007 Bare Costs Labor	Equipment	Total	Total Incl O&P
0010	**TOPOGRAPHICAL SURVEYS**									
0020	Minimum	A-7	3.30	7.273	Acre	17	201	17.60	235.60	375
0100	Maximum	A-8	.60	53.333	"	52	1,450	97	1,599	2,600

02 21 13.13 Boundary and Survey Markers

		Crew	Daily Output	Labor-Hours	Unit	Material	Labor	Equipment	Total	Total Incl O&P
0010	**BOUNDARY AND SURVEY MARKERS**									
0300	Lot location and lines, minimum, for large quantities	A-7	2	12	Acre	30	330	29	389	620
0320	Average	"	1.25	19.200		48	530	46.50	624.50	995
0400	Maximum, for small quantities	A-8	1	32	↓	64	865	58	987	1,575
0600	Monuments, 3' long	A-7	10	2.400	Ea.	30	66	5.80	101.80	150
0800	Property lines, perimeter, cleared land	"	1000	.024	L.F.	.03	.66	.06	.75	1.20
0900	Wooded land	A-8	875	.037	"	.05	.99	.07	1.11	1.79

02 32 Geotechnical Investigations

02 32 13 – Subsurface Drilling and Sampling

02 32 13.10 Boring and Exploratory Drilling

		Crew	Daily Output	Labor-Hours	Unit	Material	Labor	Equipment	Total	Total Incl O&P
0010	**BORING AND EXPLORATORY DRILLING**									
0020	Borings, initial field stake out & determination of elevations	A-6	1	16	Day		415	58	473	760
0100	Drawings showing boring details				Total		185		185	270
0200	Report and recommendations from P.E.						415		415	595
0300	Mobilization and demobilization, minimum	B-55	4	4			78	215	293	370
0350	For over 100 miles, per added mile		450	.036	Mile		.69	1.91	2.60	3.27
0600	Auger holes in earth, no samples, 2-1/2" diameter		78.60	.204	L.F.		3.97	10.95	14.92	18.75
0800	Cased borings in earth, with samples, 2-1/2" diameter		55.50	.288	"	16.55	5.60	15.50	37.65	45
1400	Drill rig and crew with truck mounted auger	↓	1	16	Day		310	860	1,170	1,475
1500	For inner city borings add, minimum									10%
1510	Maximum									20%

02 41 Demolition

02 41 13 – Selective Site Demolition

02 41 13.17 Demolish, Remove Pavement and Curb

		Crew	Daily Output	Labor-Hours	Unit	Material	Labor	Equipment	Total	Total Incl O&P
0010	**DEMOLISH, REMOVE PAVEMENT AND CURB** R024119-10									
5010	Pavement removal, bituminous roads, 3" thick	B-38	690	.035	S.Y.		.73	.52	1.25	1.79
5050	4" to 6" thick		420	.057			1.20	.85	2.05	2.95
5100	Bituminous driveways		640	.038			.79	.56	1.35	1.94
5200	Concrete to 6" thick, hydraulic hammer, mesh reinforced		255	.094			1.97	1.41	3.38	4.86
5300	Rod reinforced	↓	200	.120	↓		2.51	1.79	4.30	6.20
5600	With hand held air equipment, bituminous, to 6" thick	B-39	1900	.025	S.F.		.48	.09	.57	.92
5700	Concrete to 6" thick, no reinforcing		1600	.030			.57	.11	.68	1.09
5800	Mesh reinforced		1400	.034			.65	.13	.78	1.25
5900	Rod reinforced	↓	765	.063	↓		1.19	.24	1.43	2.29
6000	Curbs, concrete, plain	B-6	360	.067	L.F.		1.40	.68	2.08	3.08
6100	Reinforced		275	.087			1.83	.88	2.71	4.04
6200	Granite		360	.067			1.40	.68	2.08	3.08
6300	Bituminous	↓	528	.045	↓		.95	.46	1.41	2.11

02 41 13.33 Minor Site Demolition

		Crew	Daily Output	Labor-Hours	Unit	Material	Labor	Equipment	Total	Total Incl O&P
0010	**MINOR SITE DEMOLITION** R024119-10									
0015	No hauling, abandon catch basin or manhole	B-6	7	3.429	Ea.		72	35	107	159
0020	Remove existing catch basin or manhole, masonry		4	6			126	61	187	278
0030	Catch basin or manhole frames and covers, stored	↓	13	1.846	↓		38.50	18.70	57.20	85.50

02 41 Demolition

02 41 13 – Selective Site Demolition

02 41 13.33 Minor Site Demolition

		Crew	Daily Output	Labor-Hours	Unit	Material	2007 Bare Costs Labor	Equipment	Total	Total Incl O&P
0040	Remove and reset	B-6	7	3.429	Ea.		72	35	107	159
1000	Masonry walls, block or tile, solid, remove	B-5	1800	.022	C.F.		.46	.56	1.02	1.39
1100	Cavity wall		2200	.018			.38	.46	.84	1.14
1200	Brick, solid		900	.044			.92	1.13	2.05	2.79
1300	With block back-up		1130	.035			.73	.90	1.63	2.22
1400	Stone, with mortar		900	.044			.92	1.13	2.05	2.79
1500	Dry set		1500	.027			.55	.68	1.23	1.67
2900	Pipe removal, sewer/water, no excavation, 12" diameter	B-6	175	.137	L.F.		2.87	1.39	4.26	6.35
2960	21"-24" diameter		120	.200	"		4.19	2.03	6.22	9.30
4000	Sidewalk removal, bituminous, 2-1/2" thick		325	.074	S.Y.		1.55	.75	2.30	3.41
4050	Brick, set in mortar		185	.130			2.72	1.32	4.04	6
4100	Concrete, plain, 4"		160	.150			3.14	1.52	4.66	6.90
4200	Mesh reinforced		150	.160			3.35	1.62	4.97	7.40
4300	Slab on grade removal, plain	B-5	45	.889	C.Y.		18.35	22.50	40.85	56
4310	Mesh reinforced		33	1.212			25	30.50	55.50	76
4320	Rod reinforced		25	1.600			33	40.50	73.50	100
4400	For congested sites or small quantities, add up to								200%	200%
4450	For disposal on site, add	B-11A	232	.069			1.56	4.26	5.82	7.30
4500	To 5 miles, add	B-34D	76	.105			2.21	6	8.21	10.30

02 41 13.60 Fencing Demolition

		Crew	Daily Output	Labor-Hours	Unit	Material	Labor	Equipment	Total	Total Incl O&P
0010	**FENCING DEMOLITION** R024119-10									
1600	Fencing, barbed wire, 3 strand	2 Clab	430	.037	L.F.		.70		.70	1.18
1650	5 strand	"	280	.057			1.07		1.07	1.81
1700	Chain link, posts & fabric, remove only, 8' to 10' high	B-6	445	.054			1.13	.55	1.68	2.49
1750	Remove and reset	"	70	.343			7.20	3.48	10.68	15.90

02 41 16 – Structure Demolition

02 41 16.13 Building Demolition

		Crew	Daily Output	Labor-Hours	Unit	Material	Labor	Equipment	Total	Total Incl O&P
0010	**BUILDING DEMOLITION** Large urban projects, incl. 20 mi. haul R024119-10									
0500	Small bldgs, or single bldgs, no salvage included, steel	B-3	14800	.003	C.F.		.07	.13	.20	.26
0600	Concrete	"	11300	.004	"		.09	.17	.26	.33
0605	Concrete, plain	B-5	33	1.212	C.Y.		25	30.50	55.50	76
0610	Reinforced		25	1.600			33	40.50	73.50	100
0615	Concrete walls		34	1.176			24.50	30	54.50	74
0620	Elevated slabs		26	1.538			32	39	71	96.50
0650	Masonry	B-3	14800	.003	C.F.		.07	.13	.20	.26
0700	Wood	"	14800	.003	"		.07	.13	.20	.26
1000	Single family, one story house, wood, minimum				Ea.				2,525	2,975
1020	Maximum								4,400	5,275
1200	Two family, two story house, wood, minimum								3,300	3,950
1220	Maximum								6,375	7,700
1300	Three family, three story house, wood, minimum								4,400	5,275
1320	Maximum								7,700	9,250

02 41 16.17 Bldg. Footings and Foundations Demolition

		Crew	Daily Output	Labor-Hours	Unit	Material	Labor	Equipment	Total	Total Incl O&P
0010	**BLDG. FOOTINGS AND FOUNDATIONS DEMOLITION** R024119-10									
0200	Floors, concrete slab on grade,									
0240	4" thick, plain concrete	B-9C	500	.080	S.F.		1.53	.36	1.89	2.99
0280	Reinforced, wire mesh		470	.085			1.63	.38	2.01	3.18
0300	Rods		400	.100			1.91	.45	2.36	3.73
0400	6" thick, plain concrete		375	.107			2.04	.48	2.52	3.99
0420	Reinforced, wire mesh		340	.118			2.25	.53	2.78	4.40
0440	Rods		300	.133			2.55	.60	3.15	4.98
1000	Footings, concrete, 1' thick, 2' wide	B-5	300	.133	L.F.		2.75	3.38	6.13	8.35

02 41 Demolition

02 41 16 – Structure Demolition

02 41 16.17 Bldg. Footings and Foundations Demolition		Crew	Daily Output	Labor-Hours	Unit	Material	2007 Bare Costs Labor	Equipment	Total	Total Incl O&P
1080	1'-6" thick, 2' wide	B-5	250	.160	L.F.		3.31	4.06	7.37	10
1120	3' wide	↓	200	.200			4.13	5.05	9.18	12.55
1200	Average reinforcing, add				↓				10%	10%
2000	Walls, block, 4" thick	1 Clab	180	.044	S.F.		.83		.83	1.41
2040	6" thick		170	.047			.88		.88	1.49
2080	8" thick		150	.053			1		1	1.69
2100	12" thick	↓	150	.053			1		1	1.69
2400	Concrete, plain concrete, 6" thick	B-9	160	.250			4.78	1.13	5.91	9.35
2420	8" thick		140	.286			5.45	1.29	6.74	10.65
2440	10" thick		120	.333			6.35	1.50	7.85	12.45
2500	12" thick	↓	100	.400			7.65	1.80	9.45	14.95
2600	For average reinforcing, add								10%	10%
4000	For congested sites or small quantities, add up to				↓				200%	200%
4200	Add for disposal, on site	B-11A	232	.069	C.Y.		1.56	4.26	5.82	7.30
4250	To five miles	B-30	220	.109	"		2.49	8.35	10.84	13.35

02 41 19 – Selective Structure Demolition

02 41 19.13 Selective Building Demolition

0010	**SELECTIVE BUILDING DEMOLITION**
0020	Costs related to selective demolition of specific building components
0025	are included under Common Work Results (XX 05 00)
0030	in the component's appropriate division.

02 41 19.16 Selective Demolition, Cutout

		Crew	Daily Output	Labor-Hours	Unit	Material	Labor	Equipment	Total	Total Incl O&P
0010	**SELECTIVE DEMOLITION, CUTOUT** R024119-10									
0020	Concrete, elev. slab, light reinforcement, under 6 CF	B-9C	65	.615	C.F.		11.75	2.77	14.52	23
0050	Light reinforcing, over 6 C.F.	"	75	.533	"		10.20	2.40	12.60	19.95
0200	Slab on grade to 6" thick, not reinforced, under 8 S.F.	B-9	85	.471	S.F.		9	2.12	11.12	17.55
0250	8 - 16 S.F.		175	.229	"		4.37	1.03	5.40	8.55
0600	Walls, not reinforced, under 6 C.F.		60	.667	C.F.		12.75	3	15.75	25
0650	6 - 12 C.F.	↓	80	.500			9.55	2.25	11.80	18.65
1000	Concrete, elevated slab, bar reinforced, under 6 C.F.	B-9C	45	.889			17	4	21	33.50
1050	Bar reinforced, over 6 C.F.	"	50	.800	↓		15.30	3.60	18.90	30
1200	Slab on grade to 6" thick, bar reinforced, under 8 S.F.	B-9	75	.533	S.F.		10.20	2.40	12.60	19.95
1250	8 - 16 S.F.	"	150	.267	"		5.10	1.20	6.30	9.95
1400	Walls, bar reinforced, under 6 C.F.	B-9C	50	.800	C.F.		15.30	3.60	18.90	30
1450	6 - 12 CF	"	70	.571	"		10.90	2.57	13.47	21.50
2000	Brick, to 4 S.F. opening, not including toothing									
2040	4" thick	B-9C	30	1.333	Ea.		25.50	6	31.50	49.50
2060	8" thick		18	2.222			42.50	10	52.50	83
2080	12" thick		10	4			76.50	18	94.50	150
2400	Concrete block, to 4 S.F. opening, 2" thick		35	1.143			22	5.15	27.15	42.50
2420	4" thick		30	1.333			25.50	6	31.50	49.50
2440	8" thick		27	1.481			28.50	6.65	35.15	55.50
2460	12" thick	↓	24	1.667			32	7.50	39.50	62.50
2600	Gypsum block, to 4 S.F. opening, 2" thick	B-9	80	.500			9.55	2.25	11.80	18.65
2620	4" thick		70	.571			10.90	2.57	13.47	21.50
2640	8" thick		55	.727			13.90	3.27	17.17	27
2800	Terra cotta, to 4 S.F. opening, 4" thick		70	.571			10.90	2.57	13.47	21.50
2840	8" thick		65	.615			11.75	2.77	14.52	23
2880	12" thick	↓	50	.800	↓		15.30	3.60	18.90	30
3000	Toothing masonry cutouts, brick, soft old mortar	1 Brhe	40	.200	V.L.F.		4.01		4.01	6.65
3100	Hard mortar		30	.267			5.35		5.35	8.90
3200	Block, soft old mortar	↓	70	.114			2.29		2.29	3.81

02 41 Demolition

02 41 19 – Selective Structure Demolition

02 41 19.16 Selective Demolition, Cutout

		Crew	Daily Output	Labor-Hours	Unit	Material	2007 Bare Costs Labor	2007 Bare Costs Equipment	Total	Total Incl O&P
3400	Hard mortar	1 Brhe	50	.160	V.L.F.		3.21		3.21	5.35
6000	Walls, interior, not including re-framing,									
6010	openings to 5 S.F.									
6100	Drywall to 5/8" thick	1 Clab	24	.333	Ea.		6.25		6.25	10.60
6200	Paneling to 3/4" thick		20	.400			7.50		7.50	12.70
6300	Plaster, on gypsum lath		20	.400			7.50		7.50	12.70
6340	On wire lath		14	.571			10.70		10.70	18.15
7000	Wood frame, not including re-framing, openings to 5 S.F.									
7200	Floors, sheathing and flooring to 2" thick	1 Clab	5	1.600	Ea.		30		30	51
7310	Roofs, sheathing to 1" thick, not including roofing		6	1.333			25		25	42.50
7410	Walls, sheathing to 1" thick, not including siding		7	1.143			21.50		21.50	36.50

02 41 19.19 Selective Demolition, Dump Charges

		Crew	Daily Output	Labor-Hours	Unit	Material	Labor	Equipment	Total	Total Incl O&P
0010	**SELECTIVE DEMOLITION, DUMP CHARGES** R024119-10									
0020	Dump charges, typical urban city, tipping fees only									
0100	Building construction materials				Ton					70
0200	Trees, brush, lumber									50
0300	Rubbish only									60
0500	Reclamation station, usual charge									85

02 41 19.21 Selective Demolition, Gutting

		Crew	Daily Output	Labor-Hours	Unit	Material	Labor	Equipment	Total	Total Incl O&P
0010	**SELECTIVE DEMOLITION, GUTTING** R024119-10									
0020	Building interior, including disposal, dumpster fees not included									
0500	Residential building									
0560	Minimum	B-16	400	.080	SF Flr.		1.58	1.32	2.90	4.14
0580	Maximum	"	360	.089	"		1.76	1.47	3.23	4.60
0900	Commercial building									
1000	Minimum	B-16	350	.091	SF Flr.		1.81	1.51	3.32	4.73
1020	Maximum	"	250	.128	"		2.53	2.12	4.65	6.60

02 41 19.23 Selective Demolition, Rubbish Handling

		Crew	Daily Output	Labor-Hours	Unit	Material	Labor	Equipment	Total	Total Incl O&P
0010	**SELECTIVE DEMOLITION, RUBBISH HANDLING** R024119-10									
0020	The following are to be added to the demolition prices									
0400	Chute, circular, prefabricated steel, 18" diameter	B-1	40	.600	L.F.	37	11.60		48.60	60.50
0440	30" diameter	"	30	.800	"	44	15.50		59.50	75
0700	10 C.Y. capacity (4 Tons)				Week					375
0725	Dumpster, weekly rental, 1 dump/week, 20 C.Y. capacity (8 Tons)									440
0800	30 C.Y. capacity (10 Tons)									665
0840	40 C.Y. capacity (13 Tons)									805
1000	Dust partition, 6 mil polyethylene, 1" x 3" frame	2 Carp	2000	.008	S.F.	.28	.21		.49	.66
1080	2" x 4" frame	"	2000	.008	"	.33	.21		.54	.71
2000	Load, haul, and dump, 50' haul	2 Clab	24	.667	C.Y.		12.45		12.45	21
2040	100' haul		16.50	.970			18.15		18.15	31
2080	Over 100' haul, add per 100 L.F.		35.50	.451			8.45		8.45	14.30
2120	In elevators, per 10 floors, add		140	.114			2.14		2.14	3.63
3000	Loading & trucking, including 2 mile haul, chute loaded	B-16	45	.711			14.05	11.80	25.85	37
3040	Hand loading truck, 50' haul	"	48	.667			13.20	11.05	24.25	34.50
3080	Machine loading truck	B-17	120	.267			5.60	5.10	10.70	15
5000	Haul, per mile, up to 8 C.Y. truck	B-34B	1165	.007			.14	.46	.60	.74
5100	Over 8 C.Y. truck	"	1550	.005			.11	.34	.45	.56

02 83 Lead Remediation

02 83 19 – Lead-Based Paint Remediation

02 83 19.23 Encapsulation of Lead-Based Paint

		Crew	Daily Output	Labor-Hours	Unit	Material	2007 Bare Costs Labor	Equipment	Total	Total Incl O&P
0010	**ENCAPSULATION OF LEAD-BASED PAINT**, water based polymer									
0020	Interior, brushwork, trim, under 6"	1 Pord	240	.033	L.F.	2.46	.77		3.23	3.98
0030	6" to 12" wide		180	.044		3.28	1.03		4.31	5.30
0040	Balustrades		300	.027		1.98	.62		2.60	3.20
0050	Pipe to 4" diameter		500	.016		1.19	.37		1.56	1.92
0060	To 8" diameter		375	.021		1.57	.49		2.06	2.54
0070	To 12" diameter		250	.032		2.36	.74		3.10	3.82
0080	To 16" diameter		170	.047		3.47	1.09		4.56	5.60
0090	Cabinets, ornate design		200	.040	S.F.	2.97	.93		3.90	4.80
0100	Simple design		250	.032	"	2.36	.74		3.10	3.82
0110	Doors, 3'x 7', both sides, incl. frame & trim									
0120	Flush	1 Pord	6	1.333	Ea.	30.50	31		61.50	84.50
0130	French, 10-15 lite		3	2.667		6.05	62		68.05	109
0140	Panel		4	2		36.50	46.50		83	117
0150	Louvered		2.75	2.909		33.50	67.50		101	148
0160	Windows, per interior side, per 15 S.F.									
0170	1 to 6 lite	1 Pord	14	.571	Ea.	21	13.25		34.25	45
0180	7 to 10 lite		7.50	1.067		23	25		48	66
0190	12 lite		5.75	1.391		31	32.50		63.50	87
0200	Radiators		8	1		73.50	23		96.50	119
0210	Grilles, vents		275	.029	S.F.	2.15	.67		2.82	3.48
0220	Walls, roller, drywall or plaster		1000	.008		.59	.19		.78	.96
0230	With spunbonded reinforcing fabric		720	.011		.68	.26		.94	1.17
0240	Wood		800	.010		.74	.23		.97	1.19
0250	Ceilings, roller, drywall or plaster		900	.009		.68	.21		.89	1.09
0260	Wood		700	.011		.84	.27		1.11	1.36
0270	Exterior, brushwork, gutters and downspouts		300	.027	L.F.	1.98	.62		2.60	3.20
0280	Columns		400	.020	S.F.	1.47	.46		1.93	2.38
0290	Spray, siding		600	.013	"	.99	.31		1.30	1.60
0300	Miscellaneous									
0310	Electrical conduit, brushwork, to 2" diameter	1 Pord	500	.016	L.F.	1.19	.37		1.56	1.92
0320	Brick, block or concrete, spray		500	.016	S.F.	1.19	.37		1.56	1.92
0330	Steel, flat surfaces and tanks to 12"		500	.016		1.19	.37		1.56	1.92
0340	Beams, brushwork		400	.020		1.47	.46		1.93	2.38
0350	Trusses		400	.020		1.47	.46		1.93	2.38

02 83 19.26 Removal of Lead-Based Paint

		Crew	Daily Output	Labor-Hours	Unit	Material	2007 Bare Costs Labor	Equipment	Total	Total Incl O&P
0010	**REMOVAL OF LEAD-BASED PAINT**, by chemicals, per application									
0050	Baseboard, to 6" wide	1 Pord	64	.125	L.F.	1.62	2.90		4.52	6.55
0070	To 12" wide		32	.250	"	3.20	5.80		9	13.05
0200	Balustrades, one side		28	.286	S.F.	3.62	6.65		10.27	14.90
1400	Cabinets, simple design		32	.250		3.17	5.80		8.97	13.05
1420	Ornate design		25	.320		4.08	7.40		11.48	16.70
1600	Cornice, simple design		60	.133		1.71	3.09		4.80	7
1620	Ornate design		20	.400		5.05	9.30		14.35	21
2800	Doors, one side, flush		84	.095		1.23	2.21		3.44	4.98
2820	Two panel		80	.100		1.27	2.32		3.59	5.20
2840	Four panel		45	.178		2.25	4.12		6.37	9.30
2880	For trim, one side, add		64	.125	L.F.	1.62	2.90		4.52	6.55
3000	Fence, picket, one side		30	.267	S.F.	3.41	6.20		9.61	13.90
3200	Grilles, one side, simple design		30	.267		3.41	6.20		9.61	13.90
3220	Ornate design		25	.320		4.08	7.40		11.48	16.70
4400	Pipes, to 4" diameter		90	.089	L.F.	1.17	2.06		3.23	4.68

02 83 Lead Remediation

02 83 19 – Lead-Based Paint Remediation

02 83 19.26 Removal of Lead-Based Paint

		Crew	Daily Output	Labor-Hours	Unit	Material	2007 Bare Costs Labor	2007 Bare Costs Equipment	Total	Total Incl O&P
4420	To 8" diameter	1 Pord	50	.160	L.F.	2.03	3.71		5.74	8.35
4440	To 12" diameter		36	.222		2.85	5.15		8	11.65
4460	To 16" diameter		20	.400		5.05	9.30		14.35	21
4500	For hangers, add		40	.200	Ea.	2.54	4.64		7.18	10.45
4800	Siding		90	.089	S.F.	1.17	2.06		3.23	4.68
5000	Trusses, open		55	.145	SF Face	1.85	3.37		5.22	7.60
6200	Windows, one side only, double hung, 1/1 light, 24" x 48" high		4	2	Ea.	25.50	46.50		72	105
6220	30" x 60" high		3	2.667		34	62		96	140
6240	36" x 72" high		2.50	3.200		41	74		115	167
6280	40" x 80" high		2	4		51	93		144	209
6400	Colonial window, 6/6 light, 24" x 48" high		2	4		51	93		144	210
6420	30" x 60" high		1.50	5.333		68	124		192	278
6440	36" x 72" high		1	8		102	186		288	415
6480	40" x 80" high		1	8		102	186		288	415
6600	8/8 light, 24" x 48" high		2	4		51	93		144	210
6620	40" x 80" high		1	8		102	186		288	415
6800	12/12 light, 24" x 48" high		1	8		102	186		288	415
6820	40" x 80" high		.75	10.667		136	247		383	555
6840	Window frame & trim items, included in pricing above									

Division 3
Concrete

03 01 Maintenance of Concrete

03 01 30 – Maintenance of Cast-In-Place Concrete

03 01 30.64 Floor Patching

		Crew	Daily Output	Labor-Hours	Unit	Material	2007 Bare Costs Labor	2007 Bare Costs Equipment	Total	Total Incl O&P
0010	**FLOOR PATCHING**									
0012	Floor patching, 1/4" thick, small areas, regular	1 Cefi	170	.047	S.F.	2.02	1.17		3.19	4.10
0100	Epoxy	"	100	.080	"	3.87	1.99		5.86	7.45

03 05 Common Work Results for Concrete

03 05 13 – Basic Concrete Materials

03 05 13.85 Winter Protection

		Crew	Daily Output	Labor-Hours	Unit	Material	2007 Bare Costs Labor	2007 Bare Costs Equipment	Total	Total Incl O&P
0010	**WINTER PROTECTION**									
0012	For heated ready mix, add, minimum				C.Y.	5.50			5.50	6.05
0050	Maximum				"	6.90			6.90	7.55
0100	Temporary heat to protect concrete, 24 hours, minimum	2 Clab	50	.320	M.S.F.	200	6		206	230
0200	Temporary shelter for slab on grade, wood frame/polyethylene sheeting									
0201	Build or remove, minimum	2 Carp	10	1.600	M.S.F.	291	41		332	390
0210	Maximum	"	3	5.333	"	345	137		482	615
0710	Electrically, heated pads, 15 watts/S.F., 20 uses, minimum				S.F.	.21			.21	.23
0800	Maximum				"	.35			.35	.39

03 11 Concrete Forming

03 11 13 – Structural Cast-In-Place Concrete Forming

03 11 13.25 Forms In Place, Columns

		Crew	Daily Output	Labor-Hours	Unit	Material	2007 Bare Costs Labor	2007 Bare Costs Equipment	Total	Total Incl O&P
0010	**FORMS IN PLACE, COLUMNS**									
1500	Round fiber tube, 1 use, 8" diameter	C-1	155	.206	L.F.	1.56	4.59		6.15	9.50
1550	10" diameter		155	.206		2.04	4.59		6.63	10.05
1600	12" diameter		150	.213		2.35	4.74		7.09	10.65
1700	16" diameter		140	.229		3.98	5.10		9.08	13
5000	Job-built plywood, 8" x 8" columns, 1 use		165	.194	SFCA	2.24	4.31		6.55	9.75
5500	12" x 12" columns, 1 use		180	.178		2.23	3.95		6.18	9.15
7500	Steel framed plywood, 4 use per mo., rent, 8" x 8"		340	.094		3.48	2.09		5.57	7.40
7550	10" x 10"		350	.091		2.45	2.03		4.48	6.15
7600	12" x 12"		370	.086		3.24	1.92		5.16	6.80

03 11 13.45 Forms In Place, Footings

		Crew	Daily Output	Labor-Hours	Unit	Material	2007 Bare Costs Labor	2007 Bare Costs Equipment	Total	Total Incl O&P
0010	**FORMS IN PLACE, FOOTINGS**									
0020	Continuous wall, plywood, 1 use	C-1	375	.085	SFCA	2.64	1.90		4.54	6.15
0150	4 use	"	485	.066	"	.86	1.47		2.33	3.44
1500	Keyway, 4 use, tapered wood, 2" x 4"	1 Carp	530	.015	L.F.	.20	.39		.59	.88
1550	2" x 6"	"	500	.016	"	.29	.41		.70	1.02
5000	Spread footings, job-built lumber, 1 use	C-1	305	.105	SFCA	1.81	2.33		4.14	5.95
5150	4 use	"	414	.077	"	.59	1.72		2.31	3.56

03 11 13.50 Forms In Place, Grade Beam

		Crew	Daily Output	Labor-Hours	Unit	Material	2007 Bare Costs Labor	2007 Bare Costs Equipment	Total	Total Incl O&P
0010	**FORMS IN PLACE, GRADE BEAM**									
0020	Job-built plywood, 1 use	C-2	530	.091	SFCA	1.73	2.04		3.77	5.40
0150	4 use	"	605	.079	"	.56	1.79		2.35	3.66

03 11 13.65 Forms In Place, Slab On Grade

		Crew	Daily Output	Labor-Hours	Unit	Material	2007 Bare Costs Labor	2007 Bare Costs Equipment	Total	Total Incl O&P
0010	**FORMS IN PLACE, SLAB ON GRADE**									
1000	Bulkhead forms w/keyway, wood, 6" high, 1 use	C-1	510	.063	L.F.	.87	1.39		2.26	3.33
1400	Bulkhead form for slab, 4-1/2" high, exp metal, incl keyway & stakes		1200	.027		.84	.59		1.43	1.93
1410	5-1/2" high		1100	.029		.90	.65		1.55	2.09
1420	7-1/2" high		960	.033		1.08	.74		1.82	2.45

03 11 Concrete Forming

03 11 13 – Structural Cast-In-Place Concrete Forming

03 11 13.65 Forms In Place, Slab On Grade

		Crew	Daily Output	Labor-Hours	Unit	Material	2007 Bare Costs Labor	Equipment	Total	Total Incl O&P
1430	9-1/2" high	C-1	840	.038	L.F.	1.88	.85		2.73	3.51
2000	Curb forms, wood, 6" to 12" high, on grade, 1 use		215	.149	SFCA	1.95	3.31		5.26	7.75
2150	4 use		275	.116	"	.63	2.59		3.22	5.10
3000	Edge forms, wood, 4 use, on grade, to 6" high		600	.053	L.F.	.29	1.19		1.48	2.33
3050	7" to 12" high		435	.074	SFCA	.65	1.64		2.29	3.48
4000	For slab blockouts, to 12" high, 1 use		200	.160	L.F.	.68	3.56		4.24	6.80
4100	Plastic (extruded), to 6" high, multiple use, on grade		800	.040	"	.40	.89		1.29	1.95
8760	Void form, corrugated fiberboard, 6" x 12", 10' long		240	.133	S.F.	.91	2.96		3.87	6.05

03 11 13.85 Forms In Place, Walls

		Crew	Daily Output	Labor-Hours	Unit	Material	Labor	Equipment	Total	Total Incl O&P
0010	**FORMS IN PLACE, WALLS**									
0100	Box out for wall openings, to 16" thick, to 10 S.F.	C-2	24	2	Ea.	22	45		67	101
0150	Over 10 S.F. (use perimeter)	"	280	.171	L.F.	1.93	3.87		5.80	8.70
0250	Brick shelf, 4" w, add to wall forms, use wall area abv shelf									
0260	1 use	C-2	240	.200	SFCA	2.14	4.51		6.65	10
0350	4 use		300	.160	"	.85	3.61		4.46	7.05
0500	Bulkhead, with keyway, 1 use, 2 piece		265	.181	L.F.	3.49	4.09		7.58	10.80
0550	3 piece		175	.274		4.40	6.20		10.60	15.35
0600	Bulkhead forms with keyway, 1 piece expanded metal, 8" wall	C-1	1000	.032		1.08	.71		1.79	2.40
0610	10" wall	"	800	.040		1.88	.89		2.77	3.58
2000	Wall, below grade, job-built plywood, to 8' high, 1 use	C-2	300	.160	SFCA	2.49	3.61		6.10	8.85
2150	4 use		435	.110		.94	2.49		3.43	5.25
2400	Over 8' to 16' high, 1 use		280	.171		5.20	3.87		9.07	12.25
2420	2 use		345	.139		1.98	3.14		5.12	7.50
2430	3 use		375	.128		1.68	2.89		4.57	6.75
2440	4 use		395	.122		1.53	2.74		4.27	6.35
2445	Exterior wall, 8' to 16' high, 1 use		280	.171		2.06	3.87		5.93	8.80
2550	4 use		395	.122		.66	2.74		3.40	5.40
3000	For architectural finish, add		1820	.026		5.45	.60		6.05	7
7800	Modular prefabricated plywood, to 8' high, 1 use		1180	.041		2.01	.92		2.93	3.77
7860	4 use		1260	.038		.66	.86		1.52	2.19
8000	To 16' high, 1 use		715	.067		2.65	1.52		4.17	5.50
8060	4 use		790	.061		.87	1.37		2.24	3.28
8100	Over 16' high, 1 use		715	.067		3.18	1.52		4.70	6.05
8160	4 use		790	.061		1.06	1.37		2.43	3.49

03 11 19 – Insulating Concrete Forming

03 11 19.10 Insulating Forms, Left In Place

		Crew	Daily Output	Labor-Hours	Unit	Material	Labor	Equipment	Total	Total Incl O&P
0010	**INSULATING FORMS, LEFT IN PLACE**									
0020	Forms left in place, S.F. is for one face, but incl. forms for both faces									
1000	Panel system, flat cavity, minimum	2 Carp	960	.017	S.F.	2.24	.43		2.67	3.19
1010	Maximum		960	.017		3.36	.43		3.79	4.43
1020	Grid cavity, minimum		960	.017		2.32	.43		2.75	3.28
1030	Maximum		960	.017		3.48	.43		3.91	4.56
1040	Post and beam cavity, minimum		960	.017		2.36	.43		2.79	3.33
1050	Maximum		960	.017		3.54	.43		3.97	4.62
1060	Plank system, flat cavity, minimum		1920	.008		2.60	.21		2.81	3.22
1070	Maximum		1920	.008		3.90	.21		4.11	4.65
1120	Block system, flat cavity, minimum		480	.033		2.48	.86		3.34	4.18
1130	Maximum		480	.033		3.72	.86		4.58	5.55
1140	Grid cavity, minimum		480	.033		2.48	.86		3.34	4.18
1150	Maximum		480	.033		3.72	.86		4.58	5.55
1160	Post and beam cavity, minimum		480	.033		2.48	.86		3.34	4.18
1170	Maximum		480	.033		3.72	.86		4.58	5.55

03 11 Concrete Forming

03 11 23 – Permanent Stair Forming

03 11 23.75 Forms In Place, Stairs	Crew	Daily Output	Labor-Hours	Unit	Material	2007 Bare Costs Labor	Equipment	Total	Total Incl O&P
0010 **FORMS IN PLACE, STAIRS**									
0015 (Slant length x width), 1 use	C-2	165	.291	S.F.	5.50	6.55		12.05	17.20
0150 4 use		190	.253		2.18	5.70		7.88	12.05
2000 Stairs, cast on sloping ground (length x width), 1 use		220	.218		2.11	4.92		7.03	10.65
2100 4 use	↓	240	.200	↓	.69	4.51		5.20	8.40

03 15 Concrete Accessories

03 15 05 – Concrete Forming Accessories

03 15 05.02 Anchor Bolts

	Crew	Daily Output	Labor-Hours	Unit	Material	Labor	Equipment	Total	Total Incl O&P
0010 **ANCHOR BOLTS**									
0015 J-type, plain, incl. nut and washer									
0020 1/2" diameter, 6" long	1 Carp	90	.089	Ea.	1.09	2.28		3.37	5.10
0050 10" long		85	.094		1.24	2.42		3.66	5.45
0100 12" long		85	.094		1.36	2.42		3.78	5.60
0200 5/8" diameter, 12" long		80	.100		1.37	2.57		3.94	5.85
0250 18" long		70	.114		1.61	2.94		4.55	6.75
0300 24" long		60	.133		1.85	3.43		5.28	7.85
0350 3/4" diameter, 8" long		80	.100		1.61	2.57		4.18	6.15
0400 12" long		70	.114		2.01	2.94		4.95	7.20
0450 18" long		60	.133		2.61	3.43		6.04	8.65
0500 24" long	↓	50	.160	↓	3.42	4.11		7.53	10.75

03 15 05.12 Chamfer Strips

	Crew	Daily Output	Labor-Hours	Unit	Material	Labor	Equipment	Total	Total Incl O&P
0010 **CHAMFER STRIPS**									
5000 Wood, 1/2" wide	1 Carp	535	.015	L.F.	.11	.38		.49	.77
5200 3/4" wide		525	.015		.12	.39		.51	.79
5400 1" wide	↓	515	.016	↓	.17	.40		.57	.87

03 15 05.15 Column Form Accessories

	Crew	Daily Output	Labor-Hours	Unit	Material	Labor	Equipment	Total	Total Incl O&P
0010 **COLUMN FORM ACCESSORIES**									
1000 Column clamps, adjustable to 24" x 24", buy				Set	97			97	107
1400 Rent per month				"	11.20			11.20	12.35

03 15 05.25 Expansion Joints

	Crew	Daily Output	Labor-Hours	Unit	Material	Labor	Equipment	Total	Total Incl O&P
0010 **EXPANSION JOINTS**									
0020 Keyed, cold, 24 ga, incl. stakes, 3-1/2" high	1 Carp	200	.040	L.F.	.69	1.03		1.72	2.50
0050 4-1/2" high		200	.040		.84	1.03		1.87	2.66
0100 5-1/2" high		195	.041		.90	1.05		1.95	2.78
2000 Premolded, bituminous fiber, 1/2" x 6"		375	.021		.41	.55		.96	1.38
2050 1" x 12"		300	.027		1.76	.69		2.45	3.10
2500 Neoprene sponge, closed cell, 1/2" x 6"		375	.021		1.56	.55		2.11	2.65
2550 1" x 12"	↓	300	.027	↓	7.15	.69		7.84	9.05
5000 For installation in walls, add						75%			
5250 For installation in boxouts, add						25%			

03 15 05.35 Inserts

	Crew	Daily Output	Labor-Hours	Unit	Material	Labor	Equipment	Total	Total Incl O&P
0010 **INSERTS**									
1000 All size nut insert, 5/8" & 3/4", incl. nut	1 Carp	84	.095	Ea.	4.37	2.45		6.82	8.95
2000 Continuous slotted, 1-5/8" x 1-3/8"									
2100 12 ga., 3" long	1 Carp	65	.123	Ea.	3.96	3.16		7.12	9.70
2150 6" long		65	.123		5.15	3.16		8.31	11
2200 8 ga., 12" long		65	.123		12.30	3.16		15.46	18.90
2300 36" long	↓	60	.133	↓	28	3.43		31.43	37

03 15 Concrete Accessories

03 15 05 – Concrete Forming Accessories

03 15 05.75 Sleeves and Chases

		Crew	Daily Output	Labor-Hours	Unit	Material	2007 Bare Costs Labor	2007 Bare Costs Equipment	Total	Total Incl O&P
0010	**SLEEVES AND CHASES**									
0100	Plastic, 1 use, 9" long, 2" diameter	1 Carp	100	.080	Ea.	1.55	2.06		3.61	5.20
0150	4" diameter		90	.089		4.14	2.28		6.42	8.45
0200	6" diameter		75	.107		7.30	2.74		10.04	12.70

03 15 05.80 Snap Ties, Flat Washer

		Crew	Daily Output	Labor-Hours	Unit	Material	Labor	Equipment	Total	Total Incl O&P
0010	**SNAP TIES, FLAT WASHER**									
0100	3000 lb., to 8"				C	121			121	133
0250	16"					154			154	170
0300	18"					153			153	169
0500	With plastic cone, to 8"					112			112	123
0600	11" & 12"					133			133	146
0650	16"					140			140	154
0700	18"					146			146	161

03 15 05.95 Wall and Foundation Form Accessories

		Crew	Daily Output	Labor-Hours	Unit	Material	Labor	Equipment	Total	Total Incl O&P
0010	**WALL AND FOUNDATION FORM ACCESSORIES**									
0020	Coil tie system									
0700	1-1/4", 36,000 lb., to 8"				C	1,050			1,050	1,150
1200	1-1/4" diameter x 3" long				"	2,025			2,025	2,250
4200	30" long				Ea.	3.47			3.47	3.82
4250	36" long				"	4.01			4.01	4.41

03 15 13 – Waterstops

03 15 13.50 Waterstops

		Crew	Daily Output	Labor-Hours	Unit	Material	Labor	Equipment	Total	Total Incl O&P
0010	**WATERSTOPS**									
0020	PVC, ribbed 3/16" thick, 4" wide	1 Carp	155	.052	L.F.	.91	1.33		2.24	3.25
0050	6" wide		145	.055		1.39	1.42		2.81	3.94
0500	Ribbed, PVC, with center bulb, 6" wide, 3/16" thick		135	.059		1.36	1.52		2.88	4.08
0550	3/8" thick		130	.062		2.12	1.58		3.70	5

03 21 Reinforcing Steel

03 21 10 – Uncoated Reinforcing Steel

03 21 10.60 Reinforcing In Place

		Crew	Daily Output	Labor-Hours	Unit	Material	Labor	Equipment	Total	Total Incl O&P
0015	**REINFORCING IN PLACE** A615 Grade 60, incl. access. labor									
0502	Footings, #4 to #7	4 Rodm	4200	.008	Lb.	.47	.21		.68	.89
0550	#8 to #18		3.60	8.889	Ton	805	246		1,051	1,325
0702	Walls, #3 to #7		6000	.005	Lb.	.47	.15		.62	.77
0750	#8 to #18		4	8	Ton	850	221		1,071	1,325
2400	Dowels, 2 feet long, deformed, #3	2 Rodm	520	.031	Ea.	.37	.85		1.22	1.93
2410	#4		480	.033		.66	.92		1.58	2.36
2420	#5		435	.037		1.03	1.02		2.05	2.94
2430	#6		360	.044		1.48	1.23		2.71	3.82
2600	Dowel sleeves for CIP concrete, 2-part system									
2610	Sleeve base, plastic, for #5 bar, fasten to edge form	1 Rodm	200	.040	Ea.	.40	1.11		1.51	2.41
2615	Sleeve, plastic, for #5 bar x 9" long, snap onto base		400	.020		1.14	.55		1.69	2.24
2620	Sleeve base, for #6 bar		175	.046		.40	1.26		1.66	2.69
2625	Sleeve, for #6 bar		350	.023		1.17	.63		1.80	2.42
2630	Sleeve base, for #8 bar		150	.053		.49	1.47		1.96	3.17
2635	Sleeve, for #8 bar		300	.027		1.30	.74		2.04	2.74
2700	Dowel caps, visual warning only, plastic, #3 to #8	2 Rodm	800	.020		.28	.55		.83	1.30
2720	#7 to #14		750	.021		.50	.59		1.09	1.60

03 21 Reinforcing Steel

03 21 10 – Uncoated Reinforcing Steel

03 21 10.60 Reinforcing In Place

		Crew	Daily Output	Labor-Hours	Unit	Material	2007 Bare Costs Labor	Equipment	Total	Total Incl O&P
2750	Impalement protective, plastic, #3 to #7	2 Rodm	800	.020	Ea.	1.60	.55		2.15	2.75
2760	#7 to #11		775	.021		2.18	.57		2.75	3.42
2770	#11 to #16	↓	750	.021	↓	2.22	.59		2.81	3.49

03 22 Welded Wire Fabric Reinforcing

03 22 05 – Uncoated Welded Wire Fabric

03 22 05.50 Welded Wire Fabric

		Crew	Daily Output	Labor-Hours	Unit	Material	Labor	Equipment	Total	Total Incl O&P
0011	**Welded wire fabric**, 6 x 6 - W1.4 x W1.4 (10 x 10)		3500	.005		.13	.13		.26	.37
0301	6 x 6 - W2.9 x W2.9 (6 x 6) 42 lb. per C.S.F.		2900	.006		.20	.15		.35	.49
0501	4 x 4 - W1.4 x W1.4 (10 x 10) 31 lb. per C.S.F.	↓	3100	.005	↓	.18	.14		.32	.45
0750	Rolls									
0901	2 x 2 - #12 galv. for gunite reinforcing	2 Rodm	650	.025	S.F.	.38	.68		1.06	1.62

03 24 Fibrous Reinforcing

03 24 05 – Fibrous Reinforcing

03 24 05.30 Synthetic Fibers

		Crew	Daily Output	Labor-Hours	Unit	Material	Labor	Equipment	Total	Total Incl O&P
0010	**SYNTHETIC FIBERS**									
0100	Synthetic fibers, add to concrete				Lb.	3.98			3.98	4.38
0110	1-1/2 lb. per C.Y.				C.Y.	6.15			6.15	6.75

03 24 05.70 Steel Fibers

		Crew	Daily Output	Labor-Hours	Unit	Material	Labor	Equipment	Total	Total Incl O&P
0010	**STEEL FIBERS**									
0150	Steel fibers, add to concrete				Lb.	.46			.46	.51
0155	25 lb. per C.Y.				C.Y.	11.50			11.50	12.65
0160	50 lb. per C.Y.					23			23	25.50
0170	75 lb. per C.Y.					35.50			35.50	39
0180	100 lb. per C.Y.				↓	46			46	50.50

03 30 Cast-In-Place Concrete

03 30 53 – Miscellaneous Cast-In-Place Concrete

03 30 53.40 Concrete In Place

		Crew	Daily Output	Labor-Hours	Unit	Material	Labor	Equipment	Total	Total Incl O&P
0010	**CONCRETE IN PLACE**									
0020	Including forms (4 uses), concrete, placement, reinforcing									
0050	steel and finishing unless otherwise indicated									
0500	Chimney foundations, industrial, minimum	C-14C	32.22	3.476	C.Y.	148	83.50	.66	232.16	305
0510	Maximum	"	23.71	4.724	"	172	114	.90	286.90	385
3540	Equipment pad, 3' x 3' x 6" thick	C-14H	45	1.067	Ea.	42.50	26.50	.48	69.48	93
3550	4' x 4' x 6" thick		30	1.600		64.50	40	.72	105.22	140
3560	5' x 5' x 8" thick		18	2.667		115	67	1.20	183.20	241
3570	6' x 6' x 8" thick		14	3.429		157	86	1.54	244.54	320
3580	8' x 8' x 10" thick		8	6		330	150	2.70	482.70	620
3590	10' x 10' x 12" thick	↓	5	9.600	↓	565	241	4.32	810.32	1,025
3800	Footings, spread under 1 C.Y.	C-14C	38.07	2.942	C.Y.	195	71	.56	266.56	335
3850	Over 5 C.Y.		81.04	1.382		266	33.50	.26	299.76	350
3900	Footings, strip, 18" x 9", unreinforced		40	2.800		126	67.50	.53	194.03	254
3920	18" x 9", reinforced		35	3.200		147	77	.61	224.61	294
3925	20" x 10", unreinforced		45	2.489		122	60	.47	182.47	238
3930	20" x 10", reinforced		40	2.800		140	67.50	.53	208.03	270

03 30 Cast-In-Place Concrete

03 30 53 – Miscellaneous Cast-In-Place Concrete

03 30 53.40 Concrete In Place

		Crew	Daily Output	Labor-Hours	Unit	Material	2007 Bare Costs Labor	Equipment	Total	Total Incl O&P
3935	24" x 12", unreinforced	C-14C	55	2.036	C.Y.	120	49	.39	169.39	216
3940	24" x 12", reinforced		48	2.333		138	56	.44	194.44	248
3945	36" x 12", unreinforced		70	1.600		117	38.50	.30	155.80	194
3950	36" x 12", reinforced		60	1.867		133	45	.35	178.35	223
4000	Foundation mat, under 10 C.Y.		38.67	2.896		197	69.50	.55	267.05	335
4050	Over 20 C.Y.		56.40	1.986		174	48	.38	222.38	274
4520	Handicap access ramp, railing both sides, 3' wide	C-14H	14.58	3.292	L.F.	238	82.50	1.48	321.98	405
4525	5' wide		12.22	3.928		247	98.50	1.77	347.27	440
4530	With 6" curb and rails both sides, 3' wide		8.55	5.614		246	141	2.53	389.53	515
4535	5' wide		7.31	6.566		251	165	2.95	418.95	560
4650	Slab on grade, not including finish, 4" thick	C-14E	60.75	1.449	C.Y.	124	35.50	.35	159.85	198
4700	6" thick	"	92	.957	"	120	23.50	.23	143.73	173
4751	Slab on grade, incl. troweled finish, not incl. forms									
4760	or reinforcing, over 10,000 S.F., 4" thick	C-14F	3425	.021	S.F.	1.37	.48	.01	1.86	2.31
4820	6" thick	"	3350	.021	"	2.01	.50	.01	2.52	3.03
5000	Slab on grade, incl. textured finish, not incl. forms									
5001	or reinforcing, 4" thick	C-14G	2873	.019	S.F.	1.35	.44	.01	1.80	2.22
5010	6" thick		2590	.022		2.11	.49	.01	2.61	3.14
5020	8" thick		2320	.024		2.76	.54	.01	3.31	3.93
6203	Retaining walls, gravity, 4' high				C.Y.	143			143	158
6800	Stairs, not including safety treads, free standing, 3'-6" wide	C-14H	83	.578	LF Nose	5.70	14.50	.26	20.46	31
6850	Cast on ground		125	.384	"	4.52	9.60	.17	14.29	21.50
7000	Stair landings, free standing		200	.240	S.F.	4.63	6	.11	10.74	15.40
7050	Cast on ground		475	.101	"	3.45	2.53	.05	6.03	8.15

03 31 Structural Concrete

03 31 05 – Normal Weight Structural Concrete

03 31 05.35 Normal Weight Concrete, Ready Mix

		Crew	Daily Output	Labor-Hours	Unit	Material	2007 Bare Costs Labor	Equipment	Total	Total Incl O&P
0010	**NORMAL WEIGHT CONCRETE, READY MIX**									
0012	Includes local aggregate, sand, portland cement, and water									
0015	Excludes all additives and treatments									
0020	2000 psi				C.Y.	99.50			99.50	110
0100	2500 psi					101			101	111
0150	3000 psi					104			104	114
0200	3500 psi					106			106	116
0300	4000 psi					108			108	119
0350	4500 psi					110			110	121
0400	5000 psi					114			114	125
0411	6000 psi					130			130	143
0412	8000 psi					212			212	233
0413	10,000 psi					300			300	330
0414	12,000 psi					365			365	400
1000	For high early strength cement, add					10%				
2000	For all lightweight aggregate, add					45%				

03 31 05.70 Placing Concrete

		Crew	Daily Output	Labor-Hours	Unit	Material	2007 Bare Costs Labor	Equipment	Total	Total Incl O&P
0010	**PLACING CONCRETE**									
0020	Includes labor and equipment to place and vibrate									
1900	Footings, continuous, shallow, direct chute	C-6	120	.400	C.Y.		8.05	.36	8.41	13.90
1950	Pumped	C-20	150	.427			8.85	5	13.85	20.50
2000	With crane and bucket	C-7	90	.800			16.80	12.45	29.25	41.50
2400	Footings, spread, under 1 C.Y., direct chute	C-6	55	.873			17.50	.78	18.28	30.50

03 31 Structural Concrete

03 31 05 – Normal Weight Structural Concrete

03 31 05.70 Placing Concrete

		Crew	Daily Output	Labor-Hours	Unit	Material	2007 Bare Costs Labor	Equipment	Total	Total Incl O&P
2600	Over 5 C.Y., direct chute	C-6	120	.400	C.Y.		8.05	.36	8.41	13.90
2900	Foundation mats, over 20 C.Y., direct chute		350	.137			2.75	.12	2.87	4.75
4300	Slab on grade, 4" thick, direct chute	↓	110	.436			8.75	.39	9.14	15.15
4350	Pumped	C-20	130	.492			10.20	5.80	16	23.50
4400	With crane and bucket	C-7	110	.655			13.75	10.20	23.95	34
4900	Walls, 8" thick, direct chute	C-6	90	.533			10.70	.47	11.17	18.45
4950	Pumped	C-20	100	.640			13.25	7.50	20.75	30.50
5000	With crane and bucket	C-7	80	.900			18.90	14	32.90	47
5050	12" thick, direct chute	C-6	100	.480			9.65	.43	10.08	16.65
5100	Pumped	C-20	110	.582			12.05	6.85	18.90	27.50
5200	With crane and bucket	C-7	90	.800	↓		16.80	12.45	29.25	41.50
5600	Wheeled concrete dumping, add to placing costs above									
5610	Walking cart, 50' haul, add	C-18	32	.281	C.Y.		5.30	1.67	6.97	10.90
5620	150' haul, add		24	.375			7.10	2.22	9.32	14.50
5700	250' haul, add	↓	18	.500			9.45	2.97	12.42	19.30
5800	Riding cart, 50' haul, add	C-19	80	.113			2.13	.99	3.12	4.69
5810	150' haul, add		60	.150			2.84	1.31	4.15	6.25
5900	250' haul, add	↓	45	.200	↓		3.78	1.75	5.53	8.40

03 35 Concrete Finishing

03 35 29 – Tooled Concrete Finishing

03 35 29.30 Finishing Floors

		Crew	Daily Output	Labor-Hours	Unit	Material	2007 Bare Costs Labor	Equipment	Total	Total Incl O&P
0010	**FINISHING FLOORS**									
0020	Monolithic, screed finish	1 Cefi	900	.009	S.F.		.22		.22	.36
0100	Screed and bull float (darby) finish		725	.011			.27		.27	.44
0150	Screed, float, and broom finish		630	.013			.32		.32	.51
0200	Screed, float, and hand trowel		600	.013			.33		.33	.53
0250	Machine trowel		550	.015			.36		.36	.58
1600	Exposed local aggregate finish, minimum		625	.013		.18	.32		.50	.71
1650	Maximum	↓	465	.017	↓	.26	.43		.69	.98

03 35 29.35 Control Joints, Saw Cut

		Crew	Daily Output	Labor-Hours	Unit	Material	2007 Bare Costs Labor	Equipment	Total	Total Incl O&P
0010	**CONTROL JOINTS, SAW CUT**									
0100	Sawcut in green concrete									
0120	1" depth	C-27	2000	.008	L.F.	.09	.20	.06	.35	.49
0140	1-1/2" depth		1800	.009		.14	.22	.07	.43	.58
0160	2" depth	↓	1600	.010		.18	.25	.07	.50	.68
0200	Clean out control joint of debris	C-28	6000	.001	↓		.03		.03	.05
0300	Joint sealant									
0320	Backer rod, polyethylene, 1/4" diameter	1 Cefi	460	.017	L.F.	.04	.43		.47	.75
0340	Sealant, polyurethane									
0360	1/4" x 1/4" (308 LF/Gal)	1 Cefi	270	.030	L.F.	.17	.74		.91	1.38
0380	1/4" x 1/2" (154 LF/Gal)	"	255	.031	"	.34	.78		1.12	1.63

03 35 29.60 Finishing Walls

		Crew	Daily Output	Labor-Hours	Unit	Material	2007 Bare Costs Labor	Equipment	Total	Total Incl O&P
0010	**FINISHING WALLS**									
0020	Break ties and patch voids	1 Cefi	540	.015	S.F.	.03	.37		.40	.62
0050	Burlap rub with grout	"	450	.018		.03	.44		.47	.74
0300	Bush hammer, green concrete	B-39	1000	.048			.91	.18	1.09	1.75
0350	Cured concrete	"	650	.074	↓		1.41	.28	1.69	2.69

03 35 Concrete Finishing

03 35 33 – Stamped Concrete Finishing

03 35 33.50 Slab Texture Stamping		Crew	Daily Output	Labor-Hours	Unit	Material	2007 Bare Costs Labor	2007 Bare Costs Equipment	Total	Total Incl O&P
0010	**SLAB TEXTURE STAMPING**									
0020	Approx. 3 S.F.- 5 S.F. each, buy, minimum				Ea.	75			75	82.50
0030	Average					138			138	152
0120	Maximum					200			200	220
0200	Commonly used chemicals for texture systems									
0210	Hardener, colored powder				S.F.	.55			.55	.61
0220	Release agent, colored powder					.09			.09	.10
0230	Curing & sealing compound, solvent based					.05			.05	.06
0300	Broadcasting hardener & release agent, stamping	2 Cefi	1000	.016			.40		.40	.64

03 39 Concrete Curing

03 39 13 – Water Concrete Curing

03 39 13.50 Water Curing		Crew	Daily Output	Labor-Hours	Unit	Material	Labor	Equipment	Total	Total Incl O&P
0011	**WATER CURING**									
0020	With burlap, 4 uses assumed, 7.5 oz.	2 Clab	5500	.003	S.F.	.07	.05		.12	.17
0101	10 oz.	"	5500	.003	"	.13	.05		.18	.23

03 39 23 – Membrane Concrete Curing

03 39 23.13 Chemical Compound Membrane Concrete Curing

		Crew	Daily Output	Labor-Hours	Unit	Material	Labor	Equipment	Total	Total Incl O&P
0010	**CHEMICAL COMPOUND MEMBRANE CONCRETE CURING**									
0301	With sprayed membrane curing compound	2 Clab	9500	.002	S.F.	.05	.03		.08	.10

03 39 23.23 Sheet Membrane Concrete Curing

		Crew	Daily Output	Labor-Hours	Unit	Material	Labor	Equipment	Total	Total Incl O&P
0010	**SHEET MEMBRANE CONCRETE CURING**									
0201	Curing blanket, burlap/poly, 2-ply	2 Clab	7000	.002	S.F.	.17	.04		.21	.25

03 41 Precast Structural Concrete

03 41 23 – Precast Concrete Stairs

03 41 23.50 Precast Stairs		Crew	Daily Output	Labor-Hours	Unit	Material	Labor	Equipment	Total	Total Incl O&P
0010	**PRECAST STAIRS**									
0020	Precast concrete treads on steel stringers, 3' wide	C-12	75	.640	Riser	107	16.10	9.65	132.75	155
0300	Front entrance, 5' wide with 48" platform, 2 risers		16	3	Flight	340	75.50	45	460.50	550
0350	5 risers		12	4		535	101	60.50	696.50	825
0500	6' wide, 2 risers		15	3.200		365	80.50	48.50	494	590
1200	Basement entrance stairs, steel bulkhead doors, minimum	B-51	22	2.182		1,350	42	6.45	1,398.45	1,550
1250	Maximum	"	11	4.364		2,175	84	12.85	2,271.85	2,550

03 48 Precast Concrete Specialties

03 48 43 – Precast Concrete Trim

03 48 43.40 Precast Lintels

		Crew	Daily Output	Labor-Hours	Unit	Material	Labor	Equipment	Total	Total Incl O&P
0010	**PRECAST LINTELS**									
0800	Precast concrete, 4" wide, 8" high, to 5' long	D-10	28	1.143	Ea.	29	29.50	20.50	79	103
0850	5'-12' long		24	1.333		74	34.50	24	132.50	165
1000	6" wide, 8" high, to 5' long		26	1.231		40.50	31.50	22	94	122
1050	5'-12' long		22	1.455		98.50	37.50	26	162	199

03 48 43.90 Precast Window Sills

		Crew	Daily Output	Labor-Hours	Unit	Material	Labor	Equipment	Total	Total Incl O&P
0010	**PRECAST WINDOW SILLS**									
0600	Precast concrete, 4" tapers to 3", 9" wide	D-1	70	.229	L.F.	11.75	5.35		17.10	22

03 48 Precast Concrete Specialties

03 48 43 – Precast Concrete Trim

03 48 43.90 Precast Window Sills

		Crew	Daily Output	Labor-Hours	Unit	Material	2007 Bare Costs Labor	2007 Bare Costs Equipment	Total	Total Incl O&P
0650	11" wide	D-1	60	.267	L.F.	14.25	6.25		20.50	26
0700	13" wide, 3 1/2" tapers to 2 1/2", 12" wall	↓	50	.320	↓	15	7.45		22.45	29

03 63 Epoxy Grouting

03 63 05 – Grouting of Dowels and Fasteners

03 63 05.10 Epoxy Only

		Crew	Daily Output	Labor-Hours	Unit	Material	2007 Bare Costs Labor	2007 Bare Costs Equipment	Total	Total Incl O&P
0010	**EPOXY ONLY**									
1500	Chemical anchoring, epoxy cartridge, excludes layout, drilling, fastener									
1530	For fastener 3/4" dia x 6" embedment	B-89A	27	.593	Ea.	5.55	13.25	3.89	22.69	33
1535	1" dia x 8" embedment		24	.667		8.35	14.95	4.37	27.67	39.50
1540	1-1/4" dia x 10" embedment		21	.762		16.65	17.05	5	38.70	53
1545	1-3/4" dia x 12" embedment		20	.800		28	17.90	5.25	51.15	67
1550	14" embedment		17	.941		33.50	21	6.15	60.65	79.50
1555	2" dia x 12" embedment		16	1		44.50	22.50	6.55	73.55	94
1560	18" embedment	↓	15	1.067	↓	55.50	24	7	86.50	109

03 82 Concrete Boring

03 82 16 – Concrete Drilling

03 82 16.10 Concrete Drilling

		Crew	Daily Output	Labor-Hours	Unit	Material	2007 Bare Costs Labor	2007 Bare Costs Equipment	Total	Total Incl O&P
0010	**CONCRETE DRILLING**									
0050	Up to 4" deep in conc/brick floor/wall, incl. bit & layout, no anchor									
0100	Holes, 1/4" diameter	1 Carp	75	.107	Ea.	.10	2.74		2.84	4.76
0150	For each additional inch of depth, add		430	.019		.02	.48		.50	.84
0200	3/8" diameter		63	.127		.09	3.26		3.35	5.65
0250	For each additional inch of depth, add		340	.024		.02	.60		.62	1.05
0300	1/2" diameter		50	.160		.09	4.11		4.20	7.10
0350	For each additional inch of depth, add		250	.032		.02	.82		.84	1.43
0400	5/8" diameter		48	.167		.17	4.28		4.45	7.45
0450	For each additional inch of depth, add		240	.033		.04	.86		.90	1.50
0500	3/4" diameter		45	.178		.20	4.57		4.77	7.95
0550	For each additional inch of depth, add		220	.036		.05	.93		.98	1.65
0600	7/8" diameter		43	.186		.25	4.78		5.03	8.35
0650	For each additional inch of depth, add		210	.038		.06	.98		1.04	1.73
0700	1" diameter		40	.200		.28	5.15		5.43	9
0750	For each additional inch of depth, add		190	.042		.07	1.08		1.15	1.92
0800	1-1/4" diameter		38	.211		.40	5.40		5.80	9.65
0850	For each additional inch of depth, add		180	.044		.10	1.14		1.24	2.05
0900	1-1/2" diameter		35	.229		.61	5.85		6.46	10.60
0950	For each additional inch of depth, add	↓	165	.048	↓	.15	1.25		1.40	2.28
1000	For ceiling installations, add						40%			

Division 4
Masonry

04 01 Maintenance of Masonry

04 01 20 – Maintenance of Unit Masonry

04 01 20.20 Pointing Masonry

		Crew	Daily Output	Labor-Hours	Unit	Material	2007 Bare Costs Labor	Equipment	Total	Total Incl O&P
0010	**POINTING MASONRY**									
0300	Cut and repoint brick, hard mortar, running bond	1 Bric	80	.100	S.F.	.47	2.67		3.14	4.95
0320	Common bond		77	.104		.47	2.77		3.24	5.10
0360	Flemish bond		70	.114		.49	3.05		3.54	5.60
0400	English bond		65	.123		.49	3.28		3.77	6
0600	Soft old mortar, running bond		100	.080		.48	2.13		2.61	4.07
0620	Common bond		96	.083		.47	2.22		2.69	4.21
0640	Flemish bond		90	.089		.49	2.37		2.86	4.48
0680	English bond		82	.098		.49	2.60		3.09	4.86
0700	Stonework, hard mortar		140	.057	L.F.	.62	1.52		2.14	3.22
0720	Soft old mortar		160	.050	"	.62	1.33		1.95	2.91
1000	Repoint, mask and grout method, running bond		95	.084	S.F.	.62	2.24		2.86	4.42
1020	Common bond		90	.089		.62	2.37		2.99	4.63
1040	Flemish bond		86	.093		.62	2.48		3.10	4.81
1060	English bond		77	.104		.62	2.77		3.39	5.30
2000	Scrub coat, sand grout on walls, minimum		120	.067		3.05	1.78		4.83	6.30
2020	Maximum		98	.082		2.19	2.18		4.37	6.05

04 01 30 – Unit Masonry Cleaning

04 01 30.60 Brick Washing

		Crew	Daily Output	Labor-Hours	Unit	Material	2007 Bare Costs Labor	Equipment	Total	Total Incl O&P
0010	**BRICK WASHING** R040130-10									
0012	Acid cleanser, smooth brick surface	1 Bric	560	.014	S.F.	.03	.38		.41	.66
0050	Rough brick		400	.020		.03	.53		.56	.93
0060	Stone, acid wash		600	.013		.04	.36		.40	.63
1000	Muriatic acid, price per gallon in 5 gallon lots				Gal.	5.10			5.10	5.60

04 05 Common Work Results for Masonry

04 05 05 – Selective Masonry Demolition

04 05 05.10 Selective Demolition

		Crew	Daily Output	Labor-Hours	Unit	Material	2007 Bare Costs Labor	Equipment	Total	Total Incl O&P
0010	**SELECTIVE DEMOLITION** R024119-10									
0300	Concrete block walls, unreinforced, 2" thick	2 Clab	1200	.013	S.F.		.25		.25	.42
0310	4" thick		1150	.014			.26		.26	.44
0320	6" thick		1100	.015			.27		.27	.46
0330	8" thick		1050	.015			.29		.29	.48
0340	10" thick		1000	.016			.30		.30	.51
0360	12" thick		950	.017			.31		.31	.53
0380	Reinforced alternate courses, 2" thick		1130	.014			.26		.26	.45
0390	4" thick		1080	.015			.28		.28	.47
0400	6" thick		1035	.015			.29		.29	.49
0410	8" thick		990	.016			.30		.30	.51
0420	10" thick		940	.017			.32		.32	.54
0430	12" thick		890	.018			.34		.34	.57
0440	Reinforced alternate courses & vertically 48" OC, 4" thick		900	.018			.33		.33	.56
0450	6" thick		850	.019			.35		.35	.60
0460	8" thick		800	.020			.37		.37	.64
0480	10" thick		750	.021			.40		.40	.68
0490	12" thick		700	.023			.43		.43	.73
1000	Chimney, 16" x 16", soft old mortar	1 Clab	55	.145	C.F.		2.72		2.72	4.62
1020	Hard mortar		40	.200			3.74		3.74	6.35
1030	16" x 20", soft old mortar		55	.145			2.72		2.72	4.62
1040	Hard mortar		40	.200			3.74		3.74	6.35

04 05 Common Work Results for Masonry

04 05 05 – Selective Masonry Demolition

04 05 05.10 Selective Demolition

		Crew	Daily Output	Labor-Hours	Unit	Material	2007 Bare Costs Labor	Equipment	Total	Total Incl O&P
1050	16" x 24", soft old mortar	1 Clab	55	.145	C.F.		2.72		2.72	4.62
1060	Hard mortar		40	.200			3.74		3.74	6.35
1080	20" x 20", soft old mortar		55	.145			2.72		2.72	4.62
1100	Hard mortar		40	.200			3.74		3.74	6.35
1110	20" x 24", soft old mortar		55	.145			2.72		2.72	4.62
1120	Hard mortar		40	.200			3.74		3.74	6.35
1140	20" x 32", soft old mortar		55	.145			2.72		2.72	4.62
1160	Hard mortar		40	.200			3.74		3.74	6.35
1200	48" x 48", soft old mortar		55	.145			2.72		2.72	4.62
1220	Hard mortar		40	.200			3.74		3.74	6.35
1250	Metal, high temp steel jacket, 24" diameter	E-2	130	.369	V.L.F.		10.35	11.90	22.25	33
1260	60" diameter	"	60	.800			22.50	26	48.50	71
1280	Flue lining, up to 12" x 12"	1 Clab	200	.040			.75		.75	1.27
1282	Up to 24" x 24"		150	.053			1		1	1.69
2000	Columns, 8" x 8", soft old mortar		48	.167			3.12		3.12	5.30
2020	Hard mortar		40	.200			3.74		3.74	6.35
2060	16" x 16", soft old mortar		16	.500			9.35		9.35	15.90
2100	Hard mortar		14	.571			10.70		10.70	18.15
2140	24" x 24", soft old mortar		8	1			18.70		18.70	32
2160	Hard mortar		6	1.333			25		25	42.50
2200	36" x 36", soft old mortar		4	2			37.50		37.50	63.50
2220	Hard mortar		3	2.667			50		50	84.50
2230	Alternate pricing method, soft old mortar		30	.267	C.F.		4.99		4.99	8.45
2240	Hard mortar		23	.348	"		6.50		6.50	11.05
3000	Copings, precast or masonry, to 8" wide									
3020	Soft old mortar	1 Clab	180	.044	L.F.		.83		.83	1.41
3040	Hard mortar	"	160	.050	"		.94		.94	1.59
3100	To 12" wide									
3120	Soft old mortar	1 Clab	160	.050	L.F.		.94		.94	1.59
3140	Hard mortar	"	140	.057	"		1.07		1.07	1.81
4000	Fireplace, brick, 30" x 24" opening									
4020	Soft old mortar	1 Clab	2	4	Ea.		75		75	127
4040	Hard mortar		1.25	6.400			120		120	203
4100	Stone, soft old mortar		1.50	5.333			99.50		99.50	169
4120	Hard mortar		1	8	S.F.		150		150	254
5000	Veneers, brick, soft old mortar		140	.057	S.F.		1.07		1.07	1.81
5020	Hard mortar		125	.064			1.20		1.20	2.03
5100	Granite and marble, 2" thick		180	.044			.83		.83	1.41
5120	4" thick		170	.047			.88		.88	1.49
5140	Stone, 4" thick		180	.044			.83		.83	1.41
5160	8" thick		175	.046			.85		.85	1.45
5400	Alternate pricing method, stone, 4" thick		60	.133	C.F.		2.49		2.49	4.23
5420	8" thick		85	.094	"		1.76		1.76	2.99

04 05 13 – Masonry Mortaring

04 05 13.10 Cement

					Unit	Material			Total	Total Incl O&P
0010	**CEMENT** R040513-10									
0100	Masonry, 70 lb. bag, T.L. lots				Bag	8.55			8.55	9.40
0150	L.T.L. lots					9.05			9.05	9.95
0200	White, 70 lb. bag, T.L. lots					15.40			15.40	16.90
0250	L.T.L. lots					16.50			16.50	18.15

04 05 Common Work Results for Masonry

04 05 16 – Masonry Grouting

04 05 16.30 Grouting

		Crew	Daily Output	Labor-Hours	Unit	Material	2007 Bare Costs Labor	Equipment	Total	Total Incl O&P
0010	**GROUTING**									
0200	Concrete block cores, solid, 4" thk., by hand, 0.067 C.F./S.F. of wall	D-8	1100	.036	S.F.	.28	.87		1.15	1.76
0210	6" thick, pumped, 0.175 C.F. per S.F.	D-4	720	.056		.74	1.17	.21	2.12	3
0250	8" thick, pumped, 0.258 C.F. per S.F.		680	.059		1.09	1.24	.22	2.55	3.51
0300	10" thick, pumped, 0.340 C.F. per S.F.		660	.061		1.44	1.28	.23	2.95	3.96
0350	12" thick, pumped, 0.422 C.F. per S.F.		640	.063		1.78	1.32	.23	3.33	4.42

04 05 19 – Masonry Anchorage and Reinforcing

04 05 19.05 Anchor Bolts

		Crew	Daily Output	Labor-Hours	Unit	Material	Labor	Equipment	Total	Total Incl O&P
0010	**ANCHOR BOLTS**									
0020	Hooked, with nut and washer, 1/2" diam., 8" long	1 Bric	200	.040	Ea.	.74	1.07		1.81	2.58
0030	12" long		190	.042		1.36	1.12		2.48	3.37
0060	3/4" diameter, 8" long		160	.050		1.61	1.33		2.94	3.99
0070	12" long		150	.053		2.01	1.42		3.43	4.57

04 05 19.16 Masonry Anchors

		Crew	Daily Output	Labor-Hours	Unit	Material	Labor	Equipment	Total	Total Incl O&P
0010	**MASONRY ANCHORS**									
0020	For brick veneer, galv., corrugated, 7/8" x 7", 22 Ga.	1 Bric	10.50	.762	C	6.40	20.50		26.90	41
0100	24 Ga.		10.50	.762		5.95	20.50		26.45	40.50
0150	16 Ga.		10.50	.762		22.50	20.50		43	59
0200	Buck anchors, galv., corrugated, 16 gauge, 2" bend, 8" x 2"		10.50	.762		124	20.50		144.50	171
0250	8" x 3"		10.50	.762		128	20.50		148.50	175
0660	Cavity wall, Z-type, galvanized, 6" long, 1/8" diam.		10.50	.762		17.30	20.50		37.80	53
0670	3/16" diameter		10.50	.762		18.10	20.50		38.60	54
0680	1/4" diameter		10.50	.762		36.50	20.50		57	74
0850	8" long, 3/16" diameter		10.50	.762		20	20.50		40.50	56
0855	1/4" diameter		10.50	.762		43	20.50		63.50	81.50
1000	Rectangular type, galvanized, 1/4" diameter, 2" x 6"		10.50	.762		41.50	20.50		62	79.50
1050	4" x 6"		10.50	.762		46.50	20.50		67	85
1100	3/16" diameter, 2" x 6"		10.50	.762		31	20.50		51.50	68
1150	4" x 6"		10.50	.762		27.50	20.50		48	64.50
1500	Rigid partition anchors, plain, 8" long, 1" x 1/8"		10.50	.762		79.50	20.50		100	122
1550	1" x 1/4"		10.50	.762		131	20.50		151.50	178
1580	1-1/2" x 1/8"		10.50	.762		106	20.50		126.50	150
1600	1-1/2" x 1/4"		10.50	.762		215	20.50		235.50	271
1650	2" x 1/8"		10.50	.762		137	20.50		157.50	185
1700	2" x 1/4"		10.50	.762		252	20.50		272.50	310

04 05 19.26 Masonry Reinforcing Bars

		Crew	Daily Output	Labor-Hours	Unit	Material	Labor	Equipment	Total	Total Incl O&P
0010	**MASONRY REINFORCING BARS** R040519-50									
0015	Steel bars A615, placed horiz., #3 & #4 bars	1 Bric	450	.018	Lb.	.45	.47		.92	1.28
0050	Placed vertical, #3 & #4 bars		350	.023		.45	.61		1.06	1.50
0060	#5 & #6 bars		650	.012		.45	.33		.78	1.04
0200	Joint reinforcing, regular truss, to 6" wide, mill std galvanized		30	.267	C.L.F.	13.95	7.10		21.05	27
0250	12" wide		20	.400		15.35	10.65		26	34.50
0400	Cavity truss with drip section, to 6" wide		30	.267		13.45	7.10		20.55	26.50
0450	12" wide		20	.400		13.80	10.65		24.45	33

04 21 Clay Unit Masonry

04 21 13 – Brick Masonry

04 21 13.13 Brick Veneer Masonry

		Crew	Daily Output	Labor-Hours	Unit	Material	2007 Bare Costs Labor	Equipment	Total	Total Incl O&P
0010	**BRICK VENEER MASONRY**	R042110-10								
0015	Material costs incl. 3% brick and 25% mortar waste									
2000	Standard, sel. common, 4" x 2-2/3" x 8", (6.75/S.F.)	R042110-20	D-8	230	.174	S.F.	3.85	4.18		8.03	11.20
2020	Standard, red, 4" x 2-2/3" x 8", running bond (6.75/SF)			220	.182		3.85	4.37		8.22	11.50
2050	Full header every 6th course (7.88/S.F.)	R042110-50		185	.216		4.49	5.20		9.69	13.60
2100	English, full header every 2nd course (10.13/S.F.)			140	.286		5.75	6.85		12.60	17.75
2150	Flemish, alternate header every course (9.00/S.F.)			150	.267		5.10	6.40		11.50	16.30
2200	Flemish, alt. header every 6th course (7.13/S.F.)			205	.195		4.07	4.68		8.75	12.25
2250	Full headers throughout (13.50/S.F.)			105	.381		7.65	9.15		16.80	23.50
2300	Rowlock course (13.50/S.F.)			100	.400		7.65	9.60		17.25	24.50
2350	Rowlock stretcher (4.50/S.F.)			310	.129		2.59	3.10		5.69	8
2400	Soldier course (6.75/S.F.)			200	.200		3.85	4.80		8.65	12.25
2450	Sailor course (4.50/S.F.)			290	.138		2.59	3.31		5.90	8.35
2600	Buff or gray face, running bond, (6.75/S.F.)			220	.182		4.08	4.37		8.45	11.75
2700	Glazed face brick, running bond			210	.190		9.70	4.57		14.27	18.25
2750	Full header every 6th course (7.88/S.F.)			170	.235		11.30	5.65		16.95	22
3000	Jumbo, 6" x 4" x 12" running bond (3.00/S.F.)			435	.092		4.28	2.21		6.49	8.40
3050	Norman, 4" x 2-2/3" x 12" running bond, (4.5/S.F.)			320	.125		4.65	3		7.65	10.10
3100	Norwegian, 4" x 3-1/5" x 12" (3.75/S.F.)			375	.107		3.44	2.56		6	8.05
3150	Economy, 4" x 4" x 8" (4.50/S.F.)			310	.129		4.09	3.10		7.19	9.65
3200	Engineer, 4" x 3-1/5" x 8" (5.63/S.F.)			260	.154		3.26	3.69		6.95	9.75
3250	Roman, 4" x 2" x 12" (6.00/S.F.)			250	.160		5.40	3.84		9.24	12.35
3300	SCR, 6" x 2-2/3" x 12" (4.50/S.F.)			310	.129		4.92	3.10		8.02	10.55
3350	Utility, 4" x 4" x 12" (3.00/S.F.)			450	.089		3.59	2.13		5.72	7.50
3400	For cavity wall construction, add						15%				
3450	For stacked bond, add						10%				
3500	For interior veneer construction, add						15%				
3550	For curved walls, add						30%				

04 21 13.15 Chimney

		Crew	Daily Output	Labor-Hours	Unit	Material	Labor	Equipment	Total	Total Incl O&P
0010	**CHIMNEY**									
0100	Brick, 16" x 16", 8" flue, scaff. not incl.	D-1	18.20	.879	V.L.F.	18.90	20.50		39.40	55
0150	16" x 20" with one 8" x 12" flue		16	1		29	23.50		52.50	71
0200	16" x 24" with two 8" x 8" flues		14	1.143		41.50	26.50		68	90.50
0250	20" x 20" with one 12" x 12" flue		13.70	1.168		34.50	27.50		62	83.50
0300	20" x 24" with two 8" x 12" flues		12	1.333		47.50	31		78.50	104
0350	20" x 32" with two 12" x 12" flues		10	1.600		60.50	37.50		98	129

04 21 13.18 Columns

		Crew	Daily Output	Labor-Hours	Unit	Material	Labor	Equipment	Total	Total Incl O&P
0010	**COLUMNS**									
0050	Brick, 8" x 8", 9 brick per course	D-1	56	.286	V.L.F.	4.94	6.65		11.59	16.55
0100	12" x 8", 13.5 brick		37	.432		7.40	10.10		17.50	25
0200	12" x 12", 20 brick		25	.640		11	14.95		25.95	37
0300	16" x 12", 27 brick		19	.842		14.80	19.65		34.45	49
0400	16" x 16", 36 brick		14	1.143		19.75	26.50		46.25	66
0500	20" x 16", 45 brick		11	1.455		24.50	34		58.50	83.50
0600	20" x 20", 56 brick		9	1.778		31	41.50		72.50	103

04 21 13.30 Oversized Brick

		Crew	Daily Output	Labor-Hours	Unit	Material	Labor	Equipment	Total	Total Incl O&P
0010	**OVERSIZED BRICK**									
0100	Veneer, 4" x 2.25" x 16"	D-8	387	.103	S.F.	4.79	2.48		7.27	9.35
0105	4" x 2.75" x 16"		412	.097		4.33	2.33		6.66	8.65
0110	4" x 4" x 16"		460	.087		4.26	2.09		6.35	8.15
0120	4" x 8" x 16"		533	.075		5.20	1.80		7	8.75
0125	Loadbearing, 6" x 4" x 16", grouted and reinforced		387	.103		7.55	2.48		10.03	12.40

04 21 Clay Unit Masonry

04 21 13 – Brick Masonry

04 21 13.30 Oversized Brick

		Crew	Daily Output	Labor-Hours	Unit	Material	2007 Bare Costs Labor	Equipment	Total	Total Incl O&P
0130	8" x 4" x 16", grouted and reinforced	D-8	327	.122	S.F.	8	2.94		10.94	13.70
0135	6" x 8" x 16", grouted and reinforced		440	.091		7.20	2.18		9.38	11.60
0140	8" x 8" x 16", grouted and reinforced		400	.100		8.10	2.40		10.50	12.90
0145	Curtainwall / reinforced veneer, 6" x 4" x 16"		387	.103		11.75	2.48		14.23	17
0150	8" x 4" x 16"		327	.122		14.25	2.94		17.19	20.50
0155	6" x 8" x 16"		440	.091		11.85	2.18		14.03	16.70
0160	8" x 8" x 16"		400	.100		14.35	2.40		16.75	19.75
0200	For 1 to 3 slots in face, add					25%				
0210	For 4 to 7 slots in face, add					15%				
0220	For bond beams, add					20%				
0230	For bullnose shapes, add					20%				
0240	For open end knockout, add					10%				
0250	For white or gray color group, add					10%				
0260	For 135 degree corner, add					250%				

04 21 13.35 Common Building Brick

			Crew	Daily Output	Labor-Hours	Unit	Material	Labor	Equipment	Total	Total Incl O&P
0010	**COMMON BUILDING BRICK**, Material Only	R042110-20									
0020	Standard, minimum						M	335			335	365
0050	Average (select)						"	405			405	450

04 21 13.45 Face Brick

			Crew	Daily Output	Labor-Hours	Unit	Material	Labor	Equipment	Total	Total Incl O&P
0010	**FACE BRICK** Material Only	R042110-20									
0300	Standard modular, 4" x 2-2/3" x 8", minimum						M	475			475	525
0350	Maximum							545			545	600
2170	For less than truck load lots, add							10			10	11
2180	For buff or gray brick, add							15			15	16.50

04 22 Concrete Unit Masonry

04 22 10 – Concrete Masonry Units

04 22 10.11 Autoclave Aerated Concrete Block

		Crew	Daily Output	Labor-Hours	Unit	Material	Labor	Equipment	Total	Total Incl O&P
0010	**AUTOCLAVE AERATED CONCRETE BLOCK**									
0050	Solid, 4" x 12" x 24", incl mortar	D-8	600	.067	S.F.	1.87	1.60		3.47	4.72
0060	6" x 12" x 24"		600	.067		2.52	1.60		4.12	5.45
0070	8" x 8" x 24"		575	.070		3.16	1.67		4.83	6.25
0080	10" x 12" x 24"		575	.070		4.12	1.67		5.79	7.30
0090	12" x 12" x 24"		550	.073		4.80	1.75		6.55	8.20

04 22 10.14 Concrete Block, Back-Up

			Crew	Daily Output	Labor-Hours	Unit	Material	Labor	Equipment	Total	Total Incl O&P
0010	**CONCRETE BLOCK, BACK-UP**	R042210-20									
0020	Normal weight, 8" x 16" units, tooled joint 1 side										
0050	Not-reinforced, 2000 psi, 2" thick		D-8	475	.084	S.F.	1.17	2.02		3.19	4.64
0200	4" thick			460	.087		1.29	2.09		3.38	4.89
0300	6" thick			440	.091		1.86	2.18		4.04	5.65
0350	8" thick			400	.100		2.02	2.40		4.42	6.20
0400	10" thick			330	.121		2.74	2.91		5.65	7.85
0450	12" thick		D-9	310	.155		2.97	3.62		6.59	9.25
1000	Reinforced, alternate courses, 4" thick		D-8	450	.089		1.41	2.13		3.54	5.10
1100	6" thick			430	.093		1.98	2.23		4.21	5.90
1150	8" thick			395	.101		2.15	2.43		4.58	6.40
1200	10" thick			320	.125		2.87	3		5.87	8.15
1250	12" thick		D-9	300	.160		3.11	3.74		6.85	9.60

04 22 10.16 Concrete Block, Bond Beam

0010	**CONCRETE BLOCK, BOND BEAM**										

04 22 Concrete Unit Masonry

04 22 10 – Concrete Masonry Units

04 22 10.16 Concrete Block, Bond Beam		Crew	Daily Output	Labor-Hours	Unit	Material	2007 Bare Costs Labor	Equipment	Total	Total Incl O&P
0020	Not including grout or reinforcing									
0130	8" high, 8" thick	D-8	565	.071	L.F.	2.19	1.70		3.89	5.25
0150	12" thick	D-9	510	.094		3.05	2.20		5.25	7
0525	Lightweight, 6" thick	D-8	592	.068		2.20	1.62		3.82	5.10
04 22 10.19 Concrete Block, Insulation Inserts										
0010	**CONCRETE BLOCK, INSULATION INSERTS**									
0100	Inserts, styrofoam, plant installed, add to block prices									
0200	8" x 16" units, 6" thick				S.F.	1.13			1.13	1.24
0250	8" thick					1.13			1.13	1.24
0300	10" thick					1.33			1.33	1.46
0350	12" thick					1.40			1.40	1.54
04 22 10.23 Concrete Block, Decorative										
0010	**CONCRETE BLOCK, DECORATIVE**									
5000	Split rib profile units, 1" deep ribs, 8 ribs									
5100	8" x 16" x 4" thick	D-8	345	.116	S.F.	2.68	2.78		5.46	7.55
5150	6" thick		325	.123		3.10	2.96		6.06	8.30
5200	8" thick		300	.133		3.56	3.20		6.76	9.20
5250	12" thick	D-9	275	.175		4.27	4.08		8.35	11.45
5400	For special deeper colors, 4" thick, add					1.02			1.02	1.12
5450	12" thick, add					.87			.87	.96
5600	For white, 4" thick, add					1.02			1.02	1.12
5650	6" thick, add					1.02			1.02	1.12
5700	8" thick, add					.95			.95	1.04
5750	12" thick, add					.87			.87	.96
04 22 10.24 Concrete Block, Exterior										
0010	**CONCRETE BLOCK, EXTERIOR**									
0020	Reinforced alt courses, tooled joints 2 sides									
0100	Normal weight, 8" x 16" x 6" thick	D-8	395	.101	S.F.	2.20	2.43		4.63	6.45
0200	8" thick		360	.111		3.23	2.67		5.90	8
0250	10" thick		290	.138		3.86	3.31		7.17	9.75
0300	12" thick	D-9	250	.192		4	4.48		8.48	11.85
04 22 10.26 Concrete Block Foundation Wall										
0010	**CONCRETE BLOCK FOUNDATION WALL**									
0050	Normal-weight, cut joints, horiz joint reinf, no vert reinf									
0200	Hollow, 8" x 16" x 6" thick	D-8	455	.088	S.F.	2.28	2.11		4.39	6
0250	8" thick		425	.094		2.45	2.26		4.71	6.45
0300	10" thick		350	.114		3.19	2.74		5.93	8.05
0350	12" thick	D-9	300	.160		3.43	3.74		7.17	9.95
0500	Solid, 8" x 16" block, 6" thick	D-8	440	.091		2.30	2.18		4.48	6.15
0550	8" thick	"	415	.096		3.38	2.31		5.69	7.55
0600	12" thick	D-9	350	.137		5	3.20		8.20	10.80
04 22 10.32 Concrete Block, Lintels										
0010	**CONCRETE BLOCK, LINTELS**									
0100	Including grout and horizontal reinforcing									
0200	8" x 8" x 8", 1 #4 bar	D-4	300	.133	L.F.	4.29	2.81	.50	7.60	9.95
0250	2 #4 bars		295	.136		4.49	2.86	.51	7.86	10.25
1000	12" x 8" x 8", 1 #4 bar		275	.145		6	3.07	.55	9.62	12.30
1150	2 #5 bars		270	.148		6.45	3.13	.56	10.14	12.90
04 22 10.34 Concrete Block, Partitions										
0010	**CONCRETE BLOCK, PARTITIONS**									
1000	Lightweight block, tooled joints, 2 sides, hollow									

04 22 Concrete Unit Masonry

04 22 10 – Concrete Masonry Units

04 22 10.34 Concrete Block, Partitions

		Crew	Daily Output	Labor-Hours	Unit	Material	2007 Bare Costs Labor	Equipment	Total	Total Incl O&P
1100	Not reinforced, 8" x 16" x 4" thick	D-8	440	.091	S.F.	1.49	2.18		3.67	5.25
1150	6" thick		410	.098		2.03	2.34		4.37	6.10
1200	8" thick		385	.104		2.48	2.49		4.97	6.90
1250	10" thick		370	.108		3.21	2.60		5.81	7.85
1300	12" thick	D-9	350	.137		3.51	3.20		6.71	9.15
4000	Regular block, tooled joints, 2 sides, hollow									
4100	Not reinforced, 8" x 16" x 4" thick	D-8	430	.093	S.F.	1.22	2.23		3.45	5.05
4150	6" thick		400	.100		1.78	2.40		4.18	5.95
4200	8" thick		375	.107		1.94	2.56		4.50	6.40
4250	10" thick		360	.111		2.66	2.67		5.33	7.35
4300	12" thick	D-9	340	.141		2.89	3.30		6.19	8.70

04 23 Glass Unit Masonry

04 23 13 – Vertical Glass Unit Masonry

04 23 13.10 Glass Block

		Crew	Daily Output	Labor-Hours	Unit	Material	2007 Bare Costs Labor	Equipment	Total	Total Incl O&P
0010	**GLASS BLOCK**									
0150	8" x 8"	D-8	160	.250	S.F.	10.60	6		16.60	21.50
0160	end block		160	.250		35	6		41	48.50
0170	90 deg corner		160	.250		34	6		40	47.50
0180	45 deg corner		160	.250		15.75	6		21.75	27.50
0200	12" x 12"		175	.229		13.55	5.50		19.05	24
0210	4" x 8"		160	.250		8.55	6		14.55	19.45
0220	6" x 8"		160	.250		10.10	6		16.10	21
0700	For solar reflective blocks, add					100%				
1000	Thinline, plain, 3-1/8" thick, under 1,000 S.F., 6" x 6"	D-8	115	.348	S.F.	13.25	8.35		21.60	28.50
1050	8" x 8"		160	.250		10.60	6		16.60	21.50
1400	For cleaning block after installation (both sides), add		1000	.040		.10	.96		1.06	1.71

04 24 Adobe Unit Masonry

04 24 16 – Manufactured Adobe Unit Masonry

04 24 16.06 Adobe Brick

		Crew	Daily Output	Labor-Hours	Unit	Material	2007 Bare Costs Labor	Equipment	Total	Total Incl O&P
0010	**ADOBE BRICK**									
0060	Brick, 10" x 4" x 14", 2.6/S.F.	D-8	560	.071	S.F.	2.77	1.72		4.49	5.90
0080	12" x 4" x 16", 2.3/S.F.		580	.069		4.28	1.66		5.94	7.45
0100	10" x 4" x 16", 2.3/S.F.		590	.068		4.07	1.63		5.70	7.20
0120	8" x 4" x 16", 2.3/S.F.		560	.071		3.02	1.72		4.74	6.15
0140	4" x 4" x 16", 2.3/S.F.		540	.074		2.50	1.78		4.28	5.70
0160	6" x 4" x 16", 2.3/S.F.		540	.074		2.45	1.78		4.23	5.65
0180	4" x 4" x 12", 3.0/S.F.		520	.077		2.27	1.85		4.12	5.55
0200	8" x 4" x 12", 3.0/S.F.		520	.077		2.60	1.85		4.45	5.95

04 27 Multiple-Wythe Unit Masonry

04 27 10 – Multiple-Wythe Unit Masonry

04 27 10.10 Cornices

		Crew	Daily Output	Labor-Hours	Unit	Material	2007 Bare Costs Labor	2007 Bare Costs Equipment	Total	Total Incl O&P
0010	**CORNICES**									
0110	Face bricks, 12 brick/S.F., minimum	D-1	30	.533	SF Face	5.30	12.45		17.75	26.50
0150	15 brick/S.F., maximum	"	23	.696	"	6.35	16.25		22.60	34

04 27 10.30 Brick Walls

		Crew	Daily Output	Labor-Hours	Unit	Material	Labor	Equipment	Total	Total Incl O&P
0016	**BRICK WALLS** R042110-50									
0800	4" wall, face, 4" x 2-2/3" x 8"	D-8	215	.186	S.F.	3.79	4.47		8.26	11.55
0850	4" thick, as back up, 6.75 bricks per S.F.		240	.167		2.81	4		6.81	9.75
0900	8" thick wall, 13.50 brick per S.F.		135	.296		5.85	7.10		12.95	18.20
1000	12" thick wall, 20.25 bricks per S.F.		95	.421		8.80	10.10		18.90	26.50
1050	16" thick wall, 27.00 bricks per S.F.		75	.533		11.95	12.80		24.75	34.50
1200	Reinforced, 4" x 2-2/3" x 8", 4" wall		205	.195		2.81	4.68		7.49	10.90
1250	8" thick wall, 13.50 brick per S.F.		130	.308		5.85	7.40		13.25	18.70
1300	12" thick wall, 20.25 bricks per S.F.		90	.444		8.80	10.65		19.45	27.50
1350	16" thick wall, 27.00 bricks per S.F.		70	.571		11.95	13.70		25.65	36

04 41 Dry-Placed Stone

04 41 10 – Dry Placed Stone

04 41 10.10 Rough Stone Wall

		Crew	Daily Output	Labor-Hours	Unit	Material	Labor	Equipment	Total	Total Incl O&P
0011	**ROUGH STONE WALL**, Dry									
0100	Random fieldstone, under 18" thick	D-12	60	.533	C.F.	8.20	12.45		20.65	29.50
0150	Over 18" thick	"	63	.508	"	9.85	11.85		21.70	30.50

04 43 Stone Masonry

04 43 10 – Stone Masonry

04 43 10.45 Granite

		Crew	Daily Output	Labor-Hours	Unit	Material	Labor	Equipment	Total	Total Incl O&P
0010	**GRANITE**									
2450	For radius under 5', add				L.F.	100%				
2500	Steps, copings, etc., finished on more than one surface									
2550	Minimum	D-10	50	.640	C.F.	87.50	16.45	11.50	115.45	137
2600	Maximum	"	50	.640	"	140	16.45	11.50	167.95	194
2800	Pavers, 4" x 4" x 4" blocks, split face and joints									
2850	Minimum	D-11	80	.300	S.F.	12.90	7.35		20.25	26.50
2900	Maximum	"	80	.300	"	25.50	7.35		32.85	40
3500	Curbing, city street type, See Division 32 16 13.13									

04 43 10.55 Limestone

		Crew	Daily Output	Labor-Hours	Unit	Material	Labor	Equipment	Total	Total Incl O&P
0010	**LIMESTONE**									
0020	Veneer facing panels									
0500	Texture finish, light stick, 4-1/2" thick, 5' x 12'	D-4	300	.133	S.F.	46.50	2.81	.50	49.81	56
0750	5" thick, 5' x 14' panels	D-10	275	.116		50.50	2.99	2.09	55.58	63
1000	Sugarcube finish, 2" Thick, 3' x 5' panels		275	.116		25	2.99	2.09	30.08	35
1050	3" Thick, 4' x 9' panels		275	.116		37.50	2.99	2.09	42.58	49
1200	4" Thick, 5' x 11' panels		275	.116		39.50	2.99	2.09	44.58	51
1400	Sugarcube, textured finish, 4-1/2" thick, 5' x 12'		275	.116		43.50	2.99	2.09	48.58	55.50
1450	5" thick, 5' x 14' panels		275	.116		51.50	2.99	2.09	56.58	64
2000	Coping, sugarcube finish, top & 2 sides		30	1.067	C.F.	44.50	27.50	19.15	91.15	116
2100	Sills, lintels, jambs, trim, stops, sugarcube finish, average		20	1.600		66	41	29	136	172
2150	Detailed		20	1.600		66	41	29	136	172
2300	Steps, extra hard, 14" wide, 6" rise		50	.640	L.F.	22.50	16.45	11.50	50.45	65
3000	Quoins, plain finish, 6"x12"x12"	D-12	25	1.280	Ea.	60	30		90	116

04 43 Stone Masonry

04 43 10 – Stone Masonry

04 43 10.55 Limestone		Crew	Daily Output	Labor-Hours	Unit	Material	2007 Bare Costs Labor	Equipment	Total	Total Incl O&P
3050	6"x16"x24"	D-12	25	1.280	Ea.	80	30		110	138
04 43 10.60 Marble										
0011	**MARBLE**, ashlar, split face, 4" + or – thick, random									
0040	Lengths 1' to 4' & heights 2" to 7-1/2", average	D-8	175	.229	S.F.	14.55	5.50		20.05	25
0100	Base, polished, 3/4" or 7/8" thick, polished, 6" high	D-10	65	.492	L.F.	13.25	12.65	8.85	34.75	45.50
1000	Facing, polished finish, cut to size, 3/4" to 7/8" thick									
1050	Average	D-10	130	.246	S.F.	19.45	6.35	4.42	30.22	37
1100	Maximum	"	130	.246	"	45	6.35	4.42	55.77	65
2200	Window sills, 6" x 3/4" thick	D-1	85	.188	L.F.	7.30	4.40		11.70	15.35
2500	Flooring, polished tiles, 12" x 12" x 3/8" thick									
2510	Thin set, average	D-11	90	.267	S.F.	9.75	6.50		16.25	21.50
2600	Maximum		90	.267		150	6.50		156.50	176
2700	Mortar bed, average		65	.369		9.90	9.05		18.95	26
2740	Maximum		65	.369		150	9.05		159.05	180
2780	Travertine, 3/8" thick, average	D-10	130	.246		12.40	6.35	4.42	23.17	29
2790	Maximum	"	130	.246		30.50	6.35	4.42	41.27	49
3500	Thresholds, 3' long, 7/8" thick, 4" to 5" wide, plain	D-12	24	1.333	Ea.	14.05	31		45.05	67
3550	Beveled		24	1.333	"	16.30	31		47.30	69.50
3700	Window stools, polished, 7/8" thick, 5" wide		85	.376	L.F.	13.10	8.80		21.90	29
04 43 10.75 Sandstone or Brownstone										
0011	**SANDSTONE OR BROWNSTONE**									
0100	Sawed face veneer, 2-1/2" thick, to 2' x 4' panels	D-10	130	.246	S.F.	17.25	6.35	4.42	28.02	34.50
0150	4" thick, to 3'-6" x 8' panels		100	.320		17.25	8.20	5.75	31.20	39
0300	Split face, random sizes		100	.320		12.40	8.20	5.75	26.35	33.50
0350	Cut stone trim (limestone)									
0360	Ribbon stone, 4" thick, 5' pieces	D-8	120	.333	Ea.	152	8		160	180
0370	Cove stone, 4" thick, 5' pieces		105	.381		153	9.15		162.15	183
0380	Cornice stone, 10" to 12" wide		90	.444		189	10.65		199.65	225
0390	Band stone, 4" thick, 5' pieces		145	.276		97.50	6.60		104.10	118
0410	Window and door trim, 3" to 4" wide		160	.250		82.50	6		88.50	101
0420	Key stone, 18" long		60	.667		87	16		103	122
04 43 10.80 Slate										
0010	**SLATE**									
3500	Stair treads, sand finish, 1" thick x 12" wide									
3600	3 L.F. to 6 L.F.	D-10	120	.267	L.F.	21.50	6.85	4.79	33.14	40
3700	Ribbon, sand finish, 1" thick x 12" wide									
3750	To 6 L.F.	D-10	120	.267	L.F.	15.65	6.85	4.79	27.29	34
04 43 10.85 Window Sill										
0010	**WINDOW SILL**									
0020	Bluestone, thermal top, 10" wide, 1-1/2" thick	D-1	85	.188	S.F.	15.45	4.40		19.85	24.50
0050	2" thick		75	.213	"	18	4.98		22.98	28
0100	Cut stone, 5" x 8" plain		48	.333	L.F.	11.40	7.80		19.20	25.50
0200	Face brick on edge, brick, 8" wide		80	.200		2.41	4.67		7.08	10.40
0400	Marble, 9" wide, 1" thick		85	.188		8.40	4.40		12.80	16.55
0900	Slate, colored, unfading, honed, 12" wide, 1" thick		85	.188		17.10	4.40		21.50	26
0950	2" thick		70	.229		24	5.35		29.35	35

04 51 Flue Liner Masonry

04 51 10 – Flue Liner Masonry

04 51 10.10 Flue Lining

		Crew	Daily Output	Labor-Hours	Unit	Material	2007 Bare Costs Labor	Equipment	Total	Total Incl O&P
0010	**FLUE LINING**									
0020	8" x 8"	D-1	125	.128	V.L.F.	3.92	2.99		6.91	9.30
0100	8" x 12"		103	.155		5.85	3.63		9.48	12.50
0200	12" x 12"		93	.172		7.65	4.02		11.67	15.10
0300	12" x 18"		84	.190		13.15	4.45		17.60	22
0400	18" x 18"		75	.213		18.80	4.98		23.78	29
0500	20" x 20"		66	.242		30.50	5.65		36.15	43
0600	24" x 24"		56	.286		39	6.65		45.65	54
1000	Round, 18" diameter		66	.242		28.50	5.65		34.15	40.50
1100	24" diameter	↓	47	.340	↓	59	7.95		66.95	78

04 57 Masonry Fireplaces

04 57 10 – Masonry Fireplaces

04 57 10.10 Fireplace

		Crew	Daily Output	Labor-Hours	Unit	Material	2007 Bare Costs Labor	Equipment	Total	Total Incl O&P
0010	**FIREPLACE**									
0100	Brick fireplace, not incl. foundations or chimneys									
0110	30" x 29" opening, incl. chamber, plain brickwork	D-1	.40	40	Ea.	480	935		1,415	2,075
0200	Fireplace box only (110 brick)	"	2	8	"	158	187		345	485
0300	For elaborate brickwork and details, add					35%	35%			
0400	For hearth, brick & stone, add	D-1	2	8	Ea.	177	187		364	505
0410	For steel angle, damper, cleanouts, add		4	4		124	93.50		217.50	291
0600	Plain brickwork, incl. metal circulator		.50	32	↓	915	745		1,660	2,250
0800	Face brick only, standard size, 8" x 2-2/3" x 4"	↓	.30	53.333	M	480	1,250		1,730	2,600
0900	Stone fireplace, fieldstone, add				SF Face	12.65			12.65	13.90
1000	Cut stone, add				"	13.90			13.90	15.30

04 72 Cast Stone Masonry

04 72 10 – Cast Stone Masonry Features

04 72 10.10 Coping

		Crew	Daily Output	Labor-Hours	Unit	Material	2007 Bare Costs Labor	Equipment	Total	Total Incl O&P
0010	**COPING**									
0050	Precast concrete, 10" wide, 4" tapers to 3-1/2", 8" wall	D-1	75	.213	L.F.	17.10	4.98		22.08	27
0100	12" wide, 3-1/2" tapers to 3", 10" wall		70	.229		11.15	5.35		16.50	21
0150	16" wide, 4" tapers to 3-1/2", 14" wall		60	.267		15.35	6.25		21.60	27.50
0300	Limestone for 12" wall, 4" thick		90	.178		13.50	4.15		17.65	22
0350	6" thick		80	.200		15.75	4.67		20.42	25
0500	Marble, to 4" thick, no wash, 9" wide		90	.178		20.50	4.15		24.65	29.50
0550	12" wide		80	.200		30.50	4.67		35.17	42
0700	Terra cotta, 9" wide		90	.178		5.05	4.15		9.20	12.45
0800	Aluminum, for 12" wall	↓	80	.200	↓	11.95	4.67		16.62	21

04 72 20 – Cultured Stone Veneer

04 72 20.10 Cultured Stone Veneer

		Crew	Daily Output	Labor-Hours	Unit	Material	2007 Bare Costs Labor	Equipment	Total	Total Incl O&P
0010	**CULTURED STONE VENEER**									
0110	On wood frame and sheathing substrate, random sized cobbles, corner stones	D-8	70	.571	V.L.F.	13.05	13.70		26.75	37.50
0120	Field stones		140	.286	S.F.	9	6.85		15.85	21.50
0130	Random sized flats, corner stones		70	.571	V.L.F.	12.05	13.70		25.75	36.50
0140	Field stones		140	.286	S.F.	9.70	6.85		16.55	22
0150	Horizontal lined ledgestones, corner stones		75	.533	V.L.F.	13.05	12.80		25.85	36
0160	Field stones		150	.267	S.F.	9	6.40		15.40	20.50
0170	Random shaped flats, corner stones	↓	65	.615	V.L.F.	13.05	14.80		27.85	39

04 72 Cast Stone Masonry

04 72 20 – Cultured Stone Veneer

04 72 20.10 Cultured Stone Veneer		Crew	Daily Output	Labor-Hours	Unit	Material	2007 Bare Costs Labor	2007 Bare Costs Equipment	Total	Total Incl O&P
0180	Field stones	D-8	150	.267	S.F.	9	6.40		15.40	20.50
0190	Random shaped / textured face, corner stones		65	.615	V.L.F.	13.05	14.80		27.85	39
0200	Field stones		130	.308	S.F.	9	7.40		16.40	22
0210	Random shaped river rock, corner stones		65	.615	V.L.F.	13.05	14.80		27.85	39
0220	Field stones		130	.308	S.F.	9	7.40		16.40	22
0240	On concrete or CMU substrate, random sized cobbles, corner stones		70	.571	V.L.F.	12.50	13.70		26.20	37
0250	Field stones		140	.286	S.F.	8.70	6.85		15.55	21
0260	Random sized flats, corner stones		70	.571	V.L.F.	11.45	13.70		25.15	35.50
0270	Field stones		140	.286	S.F.	9.45	6.85		16.30	22
0280	Horizontal lined ledgestones, corner stones		75	.533	V.L.F.	12.50	12.80		25.30	35.50
0290	Field stones		150	.267	S.F.	8.70	6.40		15.10	20.50
0300	Random shaped flats, corner stones		70	.571	V.L.F.	12.50	13.70		26.20	37
0310	Field stones		140	.286	S.F.	8.70	6.85		15.55	21
0320	Random shaped / textured face, corner stones		65	.615	V.L.F.	12.50	14.80		27.30	38.50
0330	Field stones		130	.308	S.F.	8.70	7.40		16.10	22
0340	Random shaped river rock, corner stones		65	.615	V.L.F.	12.50	14.80		27.30	38.50
0350	Field stones		130	.308	S.F.	8.70	7.40		16.10	22
0360	Cultured stone veneer, #15 felt weather resistant barrier	1 Clab	3700	.002	Sq.	4.23	.04		4.27	4.72
0390	Water table or window sill, 18" long	1 Bric	80	.100	Ea.	12.50	2.67		15.17	18.20

Division 5
Metals

05 01 Maintenance of Metals

05 05 21 – Fastening Methods for Metal

05 05 21.15 Drilling Steel

		Crew	Daily Output	Labor-Hours	Unit	Material	2007 Bare Costs Labor	2007 Bare Costs Equipment	Total	Total Incl O&P
0010	**DRILLING STEEL**									
1910	Drilling & layout for steel, up to 1/4" deep, no anchor									
1920	Holes, 1/4" diameter	1 Sswk	112	.071	Ea.	.13	1.98		2.11	4
1925	For each additional 1/4" depth, add		336	.024		.13	.66		.79	1.43
1930	3/8" diameter		104	.077		.15	2.13		2.28	4.32
1935	For each additional 1/4" depth, add		312	.026		.15	.71		.86	1.55
1940	1/2" diameter		96	.083		.17	2.31		2.48	4.68
1945	For each additional 1/4" depth, add		288	.028		.17	.77		.94	1.68
1950	5/8" diameter		88	.091		.27	2.52		2.79	5.20
1955	For each additional 1/4" depth, add		264	.030		.27	.84		1.11	1.94
1960	3/4" diameter		80	.100		.30	2.77		3.07	5.75
1965	For each additional 1/4" depth, add		240	.033		.30	.92		1.22	2.13
1970	7/8" diameter		72	.111		.36	3.08		3.44	6.40
1975	For each additional 1/4" depth, add		216	.037		.36	1.03		1.39	2.40
1980	1" diameter		64	.125		.41	3.46		3.87	7.20
1985	For each additional 1/4" depth, add		192	.042		.41	1.15		1.56	2.70
1990	For drilling up, add						40%			

05 05 23 – Metal Fastenings

05 05 23.10 Bolts and Hex Nuts

		Crew	Daily Output	Labor-Hours	Unit	Material	2007 Bare Costs Labor	2007 Bare Costs Equipment	Total	Total Incl O&P
0010	**BOLTS & HEX NUTS**, Steel, A307									
0100	1/4" diameter, 1/2" long	1 Sswk	140	.057	Ea.	.07	1.58		1.65	3.17
0200	1" long		140	.057		.08	1.58		1.66	3.18
0300	2" long		130	.062		.11	1.70		1.81	3.45
0400	3" long		130	.062		.16	1.70		1.86	3.51
0500	4" long		120	.067		.18	1.85		2.03	3.79
0600	3/8" diameter, 1" long		130	.062		.12	1.70		1.82	3.46
0700	2" long		130	.062		.15	1.70		1.85	3.50
0800	3" long		120	.067		.20	1.85		2.05	3.82
0900	4" long		120	.067		.25	1.85		2.10	3.88
1000	5" long		115	.070		.31	1.93		2.24	4.10
1100	1/2" diameter, 1-1/2" long		120	.067		.24	1.85		2.09	3.86
1200	2" long		120	.067		.27	1.85		2.12	3.89
1300	4" long		115	.070		.41	1.93		2.34	4.21
1400	6" long		110	.073		.56	2.01		2.57	4.54
1500	8" long		105	.076		.73	2.11		2.84	4.92
1600	5/8" diameter, 1-1/2" long		120	.067		.47	1.85		2.32	4.12
1700	2" long		120	.067		.51	1.85		2.36	4.16
1800	4" long		115	.070		.71	1.93		2.64	4.54
1900	6" long		110	.073		.89	2.01		2.90	4.91
2000	8" long		105	.076		1.29	2.11		3.40	5.55
2100	10" long		100	.080		1.60	2.22		3.82	6.10
2200	3/4" diameter, 2" long		120	.067		.74	1.85		2.59	4.41
2300	4" long		110	.073		1.02	2.01		3.03	5.05
2400	6" long		105	.076		1.30	2.11		3.41	5.55
2500	8" long		95	.084		1.92	2.33		4.25	6.65
2600	10" long		85	.094		2.50	2.61		5.11	7.85
2700	12" long		80	.100		2.91	2.77		5.68	8.60
2800	1" diameter, 3" long		105	.076		1.90	2.11		4.01	6.20
2900	6" long		90	.089		2.93	2.46		5.39	8
3000	12" long		75	.107		5.50	2.95		8.45	11.80
3100	For galvanized, add					75%				
3200	For stainless, add					350%				

05 01 Maintenance of Metals

05 05 23 – Metal Fastenings

05 05 23.15 Chemical Anchors

		Crew	Daily Output	Labor-Hours	Unit	Material	2007 Bare Costs Labor	Equipment	Total	Total Incl O&P
0010	**CHEMICAL ANCHORS**									
0020	Includes layout & drilling									
1430	Chemical anchor, w/rod & epoxy cartridge, 3/4" diam. x 9-1/2" long	B-89A	27	.593	Ea.	13.70	13.25	3.89	30.84	42
1435	1" diameter x 11-3/4" long		24	.667		26.50	14.95	4.37	45.82	59.50
1440	1-1/4" diameter x 14" long		21	.762		50.50	17.05	5	72.55	90
1445	1-3/4" diameter x 15" long		20	.800		95	17.90	5.25	118.15	141
1450	18" long		17	.941		114	21	6.15	141.15	169
1455	2" diameter x 18" long		16	1		145	22.50	6.55	174.05	205
1460	24" long		15	1.067		190	24	7	221	257

05 05 23.20 Expansion Anchors

		Crew	Daily Output	Labor-Hours	Unit	Material	Labor	Equipment	Total	Total Incl O&P
0010	**EXPANSION ANCHORS**									
0100	Anchors for concrete, brick or stone, no layout and drilling									
0200	Expansion shields, zinc, 1/4" diameter, 1-5/16" long, single	1 Carp	90	.089	Ea.	1.11	2.28		3.39	5.10
0300	1-3/8" long, double		85	.094		1.22	2.42		3.64	5.45
0500	2" long, double		80	.100		2.26	2.57		4.83	6.85
0700	2-1/2" long, double		75	.107		2.92	2.74		5.66	7.85
0900	2-3/4" long, double		70	.114		4.33	2.94		7.27	9.75
1100	3-15/16" long, double		65	.123		8.60	3.16		11.76	14.80
2100	Hollow wall anchors for gypsum wall board, plaster or tile									
2500	3/16" diameter, short	1 Carp	150	.053	Ea.	.64	1.37		2.01	3.03
3000	Toggle bolts, bright steel, 1/8" diameter, 2" long		85	.094		.28	2.42		2.70	4.41
3100	4" long		80	.100		.43	2.57		3	4.83
3200	3/16" diameter, 3" long		80	.100		.48	2.57		3.05	4.89
3300	6" long		75	.107		.67	2.74		3.41	5.40
3400	1/4" diameter, 3" long		75	.107		.54	2.74		3.28	5.25
3500	6" long		70	.114		.76	2.94		3.70	5.80
3600	3/8" diameter, 3" long		70	.114		1.04	2.94		3.98	6.10
3700	6" long		60	.133		1.82	3.43		5.25	7.80
3800	1/2" diameter, 4" long		60	.133		2.70	3.43		6.13	8.75
3900	6" long		50	.160		4.45	4.11		8.56	11.90
4000	Nailing anchors									
4100	Nylon nailing anchor, 1/4" diameter, 1" long	1 Carp	3.20	2.500	C	21	64.50		85.50	132
4200	1-1/2" long		2.80	2.857		26.50	73.50		100	155
4300	2" long		2.40	3.333		44.50	85.50		130	194
4400	Metal nailing anchor, 1/4" diameter, 1" long		3.20	2.500		32	64.50		96.50	144
4500	1-1/2" long		2.80	2.857		43.50	73.50		117	173
4600	2" long		2.40	3.333		55.50	85.50		141	206
5000	Screw anchors for concrete, masonry,									
5100	stone & tile, no layout or drilling included									
5200	Jute fiber, #6, #8, & #10, 1" long	1 Carp	240	.033	Ea.	.26	.86		1.12	1.74
5300	#12, 1-1/2" long		200	.040		.38	1.03		1.41	2.16
5400	#14, 2" long		160	.050		.60	1.29		1.89	2.84
5500	#16, 2" long		150	.053		.63	1.37		2	3.02
5600	#20, 2" long		140	.057		1.01	1.47		2.48	3.60
5700	Lag screw shields, 1/4" diameter, short		90	.089		.48	2.28		2.76	4.41
5800	Long		85	.094		.57	2.42		2.99	4.73
5900	3/8" diameter, short		85	.094		.88	2.42		3.30	5.05
6000	Long		80	.100		1.04	2.57		3.61	5.50
6100	1/2" diameter, short		80	.100		1.22	2.57		3.79	5.70
6200	Long		75	.107		1.54	2.74		4.28	6.35
6300	3/4" diameter, short		70	.114		3.44	2.94		6.38	8.75
6400	Long		65	.123		4.16	3.16		7.32	9.95

05 01 Maintenance of Metals

05 05 23 – Metal Fastenings

05 05 23.20 Expansion Anchors		Crew	Daily Output	Labor-Hours	Unit	Material	2007 Bare Costs Labor	Equipment	Total	Total Incl O&P
6600	Lead, #6 & #8, 3/4" long	1 Carp	260	.031	Ea.	.17	.79		.96	1.53
6700	#10 - #14, 1-1/2" long		200	.040		.25	1.03		1.28	2.02
6800	#16 & #18, 1-1/2" long		160	.050		.34	1.29		1.63	2.55
6900	Plastic, #6 & #8, 3/4" long		260	.031		.12	.79		.91	1.47
7000	#8 & #10, 7/8" long		240	.033		.05	.86		.91	1.51
7100	#10 & #12, 1" long		220	.036		.16	.93		1.09	1.77
7200	#14 & #16, 1-1/2" long		160	.050		.09	1.29		1.38	2.28
8950	Self-drilling concrete screw, hex washer head, 3/16" dia x 1-3/4" long		300	.027		.37	.69		1.06	1.57
8960	2-1/4" long		250	.032		.56	.82		1.38	2.02
8970	Phillips flat head, 3/16" dia x 1-3/4" long		300	.027		.38	.69		1.07	1.58
8980	2-1/4" long		250	.032		.56	.82		1.38	2.02

05 05 23.30 Lag Screws		Crew	Daily Output	Labor-Hours	Unit	Material	Labor	Equipment	Total	Total Incl O&P
0010	**LAG SCREWS**									
0020	Steel, 1/4" diameter, 2" long	1 Carp	200	.040	Ea.	.10	1.03		1.13	1.85
0100	3/8" diameter, 3" long		150	.053		.27	1.37		1.64	2.63
0200	1/2" diameter, 3" long		130	.062		.44	1.58		2.02	3.16
0300	5/8" diameter, 3" long		120	.067		.86	1.71		2.57	3.86

05 05 23.50 Powder Actuated Tools and Fasteners		Crew	Daily Output	Labor-Hours	Unit	Material	Labor	Equipment	Total	Total Incl O&P
0010	**POWDER ACTUATED TOOLS & FASTENERS**									
0020	Stud driver, .22 caliber, buy, minimum				Ea.	355			355	390
0100	Maximum				"	575			575	630
0300	Powder charges for above, low velocity				C	18.30			18.30	20
0400	Standard velocity					26			26	28.50
0600	Drive pins & studs, 1/4" & 3/8" diam., to 3" long, minimum	1 Carp	4.80	1.667		13.70	43		56.70	87.50
0700	Maximum	"	4	2		53.50	51.50		105	146

05 05 23.55 Rivets		Crew	Daily Output	Labor-Hours	Unit	Material	Labor	Equipment	Total	Total Incl O&P
0010	**RIVETS**									
0100	Aluminum rivet & mandrel, 1/2" grip length x 1/8" diameter	1 Carp	4.80	1.667	C	5.50	43		48.50	78.50
0200	3/16" diameter		4	2		8.50	51.50		60	96.50
0300	Aluminum rivet, steel mandrel, 1/8" diameter		4.80	1.667		8.45	43		51.45	82
0400	3/16" diameter		4	2		7.75	51.50		59.25	95.50
0500	Copper rivet, steel mandrel, 1/8" diameter		4.80	1.667		9.50	43		52.50	83
0600	Monel rivet, steel mandrel, 1/8" diameter		4.80	1.667		27.50	43		70.50	103
0700	3/16" diameter		4	2		78.50	51.50		130	174
0800	Stainless rivet & mandrel, 1/8" diameter		4.80	1.667		13.70	43		56.70	87.50
0900	3/16" diameter		4	2		25.50	51.50		77	116
1000	Stainless rivet, steel mandrel, 1/8" diameter		4.80	1.667		10.70	43		53.70	84.50
1100	3/16" diameter		4	2		19.35	51.50		70.85	109
1200	Steel rivet and mandrel, 1/8" diameter		4.80	1.667		6.60	43		49.60	80
1300	3/16" diameter		4	2		9.95	51.50		61.45	98
1400	Hand riveting tool, minimum				Ea.	128			128	141
1500	Maximum					245			245	269
1600	Power riveting tool, minimum					925			925	1,025
1700	Maximum					2,325			2,325	2,575

05 12 Structural Steel Framing

05 12 23 – Structural Steel for Buildings

05 12 23.10 Ceiling Supports		Crew	Daily Output	Labor-Hours	Unit	Material	2007 Bare Costs Labor	Equipment	Total	Total Incl O&P
0010	**CEILING SUPPORTS**									
1000	Entrance door/folding partition supports	E-4	60	.533	L.F.	20.50	15.05	1.92	37.47	54
1100	Linear accelerator door supports		14	2.286		93.50	64.50	8.25	166.25	238
1200	Lintels or shelf angles, hung, exterior hot dipped galv.		267	.120		14	3.38	.43	17.81	22.50
1250	Two coats primer paint instead of galv.		267	.120	↓	12.15	3.38	.43	15.96	20.50
1400	Monitor support, ceiling hung, expansion bolted		4	8	Ea.	325	226	29	580	825
1450	Hung from pre-set inserts		6	5.333		350	150	19.20	519.20	700
1600	Motor supports for overhead doors		4	8	↓	165	226	29	420	655
1700	Partition support for heavy folding partitions, without pocket		24	1.333	L.F.	46.50	37.50	4.80	88.80	130
1750	Supports at pocket only		12	2.667		93.50	75	9.60	178.10	261
2000	Rolling grilles & fire door supports		34	.941	↓	40	26.50	3.39	69.89	99.50
2100	Spider-leg light supports, expansion bolted to ceiling slab		8	4	Ea.	133	113	14.40	260.40	385
2150	Hung from pre-set inserts		12	2.667	"	144	75	9.60	228.60	315
2400	Toilet partition support		36	.889	L.F.	46.50	25	3.20	74.70	104
2500	X-ray travel gantry support	↓	12	2.667	"	160	75	9.60	244.60	335

05 12 23.15 Columns, Lightweight										
0010	**COLUMNS, LIGHTWEIGHT**									
1000	Lightweight units (lally), 3-1/2" diameter	E-2	780	.062	L.F.	3.14	1.72	1.98	6.84	8.90
1050	4" diameter	"	900	.053	"	4.62	1.49	1.72	7.83	9.85
8000	Lally columns, to 8', 3-1/2" diameter	2 Carp	24	.667	Ea.	25	17.15		42.15	56.50
8080	4" diameter	"	20	.800	"	37	20.50		57.50	75.50

05 12 23.17 Columns, Structural										
0010	**COLUMNS, STRUCTURAL**									
0020	Shop fab'd for 100-ton, 1-2 story project, bolted conn's.									
0800	Steel, concrete filled, extra strong pipe, 3-1/2" diameter	E-2	660	.073	L.F.	34	2.04	2.34	38.38	43.50
0830	4" diameter		780	.062		37.50	1.72	1.98	41.20	47
0890	5" diameter		1020	.047		45	1.32	1.52	47.84	53.50
0930	6" diameter		1200	.040		59.50	1.12	1.29	61.91	69
0940	8" diameter	↓	1100	.044	↓	59.50	1.22	1.41	62.13	69.50
1100	For galvanizing, add				Lb.	.25			.25	.27
1300	For web ties, angles, etc., add per added lb.	1 Sswk	945	.008		1.03	.23		1.26	1.59
1500	Steel pipe, extra strong, no concrete, 3" to 5" diameter	E-2	16000	.003		1.03	.08	.10	1.21	1.40
1600	6" to 12" diameter		14000	.003		1.03	.10	.11	1.24	1.43
2400	Structural tubing, rect, 5" to 6" wide, light section		11200	.004		1.03	.12	.14	1.29	1.51
2700	12" x 8" x 1/2" thk wall		24000	.002		1.03	.06	.06	1.15	1.31
2800	Heavy section	↓	32000	.002	↓	1.03	.04	.05	1.12	1.26
8090	For projects 75 to 99 tons, add				All	10%				
8092	50 to 74 tons, add					20%				
8094	25 to 49 tons, add					30%	10%			
8096	10 to 24 tons, add					50%	25%			
8098	2 to 9 tons, add					75%	50%			
8099	Less than 2 tons, add				↓	100%	100%			

05 12 23.45 Lintels										
0010	**LINTELS**									
0020	Plain steel angles, under 500 lb.	1 Bric	550	.015	Lb.	.79	.39		1.18	1.51
0100	500 to 1000 lb.		640	.013	"	.77	.33		1.10	1.40
2000	Steel angles, 3-1/2" x 3", 1/4" thick, 2'-6" long		47	.170	Ea.	11.05	4.54		15.59	19.75
2100	4'-6" long		26	.308		19.95	8.20		28.15	35.50
2600	4" x 3-1/2", 1/4" thick, 5'-0" long		21	.381		25.50	10.15		35.65	45
2700	9'-0" long	↓	12	.667	↓	46	17.75		63.75	80

05 12 23.65 Plates										
0010	**PLATES**									

05 12 Structural Steel Framing

05 12 23 – Structural Steel for Buildings

05 12 23.65 Plates

		Crew	Daily Output	Labor-Hours	Unit	Material	2007 Bare Costs Labor	Equipment	Total	Total Incl O&P
0015	Made from recycled materials									
0020	For connections & stiffener plates, shop fabricated									
0050	1/8" thick (5.1 Lb./S.F.)				S.F.	5.25			5.25	5.75
0100	1/4" thick (10.2 Lb./S.F.)					10.45			10.45	11.50
0300	3/8" thick (15.3 Lb./S.F.)					15.70			15.70	17.25
0400	1/2" thick (20.4 Lb./S.F.)					21			21	23
0450	3/4" thick (30.6 Lb./S.F.)					31.50			31.50	34.50
0500	1" thick (40.8 Lb.S.F.)					42			42	46
2000	Steel plate, warehouse prices, no fabrication									
2100	1/4" thick (10.2 Lb./S.F.)				S.F.	4.49			4.49	4.94

05 12 23.79 Structural Steel

		Crew	Daily Output	Labor-Hours	Unit	Material	Labor	Equipment	Total	Total Incl O&P
0010	**STRUCTURAL STEEL**									
0020	Shop fab'd for 100-ton, 1-2 story project, bolted conn's.									
0050	Beams, W 6 x 9	E-2	720	.067	L.F.	11.05	1.87	2.15	15.07	18.10
0100	W 8 x 10		720	.067		12.30	1.87	2.15	16.32	19.45
0200	Columns, W 6 x 15		540	.089		20	2.49	2.86	25.35	30
0250	W 8 x 31		540	.089		41.50	2.49	2.86	46.85	53.50
7990	For projects 75 to 99 tons, add				All	10%				
7992	50 to 75 tons, add					20%				
7994	25 to 49 tons, add					30%	10%			
7996	10 to 24 tons, add					50%	25%			
7998	2 to 9 tons, add					75%	50%			
7999	Less than 2 tons, add					100%	100%			

05 31 Steel Decking

05 31 13 – Steel Floor Decking

05 31 13.50 Floor Decking

		Crew	Daily Output	Labor-Hours	Unit	Material	Labor	Equipment	Total	Total Incl O&P
0010	**FLOOR DECKING** R053100-10									

05 31 23 – Steel Roof Decking

05 31 23.50 Roof Decking

		Crew	Daily Output	Labor-Hours	Unit	Material	Labor	Equipment	Total	Total Incl O&P
0010	**ROOF DECKING**									
2100	Open type, galv., 1-1/2" deep wide rib, 22 gauge, under 50 squares	E-4	4500	.007	S.F.	1.55	.20	.03	1.78	2.12
2600	20 gauge, under 50 squares		3865	.008		1.82	.23	.03	2.08	2.49
2900	18 gauge, under 50 squares		3800	.008		2.36	.24	.03	2.63	3.08
3050	16 gauge, under 50 squares		3700	.009		3.17	.24	.03	3.44	3.99

05 31 33 – Steel Form Decking

05 31 33.50 Form Decking

		Crew	Daily Output	Labor-Hours	Unit	Material	Labor	Equipment	Total	Total Incl O&P
0010	**FORM DECKING**									
6100	Slab form, steel, 28 gauge, 9/16" deep, uncoated	E-4	4000	.008	S.F.	1.03	.23	.03	1.29	1.60
6200	Galvanized		4000	.008		.91	.23	.03	1.17	1.47
6220	24 gauge, 1" deep, uncoated		3900	.008		1.12	.23	.03	1.38	1.71
6240	Galvanized		3900	.008		1.32	.23	.03	1.58	1.93
6300	24 gauge, 1-5/16" deep, uncoated		3800	.008		1.20	.24	.03	1.47	1.81
6400	Galvanized		3800	.008		1.41	.24	.03	1.68	2.04
6500	22 gauge, 1-5/16" deep, uncoated		3700	.009		1.50	.24	.03	1.77	2.16
6600	Galvanized		3700	.009		1.53	.24	.03	1.80	2.19
6700	22 gauge, 2" deep uncoated		3600	.009		1.98	.25	.03	2.26	2.71
6800	Galvanized		3600	.009		1.94	.25	.03	2.22	2.66

05 41 Structural Metal Stud Framing

05 41 13 – Load-Bearing Metal Stud Framing

05 41 13.05 Bracing

		Crew	Daily Output	Labor-Hours	Unit	Material	2007 Bare Costs Labor	Equipment	Total	Total Incl O&P
0010	**BRACING**, shear wall X-bracing, per 10' x 10' bay, one face									
0120	Metal strap, 20 ga x 4" wide	2 Carp	18	.889	Ea.	21.50	23		44.50	63
0130	6" wide		18	.889		34	23		57	76
0160	18 ga x 4" wide		16	1		31	25.50		56.50	77.50
0170	6" wide		16	1		46	25.50		71.50	94
0410	Continuous strap bracing, per horizontal row on both faces									
0420	Metal strap, 20 ga x 2" wide, studs 12" O.C.	1 Carp	7	1.143	C.L.F.	55	29.50		84.50	111
0430	16" O.C.		8	1		55	25.50		80.50	104
0440	24" O.C.		10	.800		55	20.50		75.50	95.50
0450	18 ga x 2" wide, studs 12" O.C.		6	1.333		75	34.50		109.50	141
0460	16" O.C.		7	1.143		75	29.50		104.50	133
0470	24" O.C.		8	1		75	25.50		100.50	126

05 41 13.10 Bridging

		Crew	Daily Output	Labor-Hours	Unit	Material	2007 Bare Costs Labor	Equipment	Total	Total Incl O&P
0010	**BRIDGING**, solid between studs w/ 1-1/4" leg track, per stud bay									
0200	Studs 12" O.C., 18 ga x 2-1/2" wide	1 Carp	125	.064	Ea.	.93	1.64		2.57	3.81
0210	3-5/8" wide		120	.067		1.12	1.71		2.83	4.14
0220	4" wide		120	.067		1.18	1.71		2.89	4.21
0230	6" wide		115	.070		1.54	1.79		3.33	4.73
0240	8" wide		110	.073		1.96	1.87		3.83	5.30
0300	16 ga x 2-1/2" wide		115	.070		1.16	1.79		2.95	4.30
0310	3-5/8" wide		110	.073		1.42	1.87		3.29	4.73
0320	4" wide		110	.073		1.51	1.87		3.38	4.83
0330	6" wide		105	.076		1.93	1.96		3.89	5.45
0340	8" wide		100	.080		2.47	2.06		4.53	6.20
1200	Studs 16" O.C., 18 ga x 2-1/2" wide		125	.064		1.19	1.64		2.83	4.10
1210	3-5/8" wide		120	.067		1.44	1.71		3.15	4.49
1220	4" wide		120	.067		1.52	1.71		3.23	4.58
1230	6" wide		115	.070		1.98	1.79		3.77	5.20
1240	8" wide		110	.073		2.51	1.87		4.38	5.95
1300	16 ga x 2-1/2" wide		115	.070		1.49	1.79		3.28	4.66
1310	3-5/8" wide		110	.073		1.82	1.87		3.69	5.15
1320	4" wide		110	.073		1.93	1.87		3.80	5.30
1330	6" wide		105	.076		2.48	1.96		4.44	6.05
1340	8" wide		100	.080		3.17	2.06		5.23	6.95
2200	Studs 24" O.C., 18 ga x 2-1/2" wide		125	.064		1.72	1.64		3.36	4.68
2210	3-5/8" wide		120	.067		2.08	1.71		3.79	5.20
2220	4" wide		120	.067		2.20	1.71		3.91	5.35
2230	6" wide		115	.070		2.86	1.79		4.65	6.20
2240	8" wide		110	.073		3.63	1.87		5.50	7.15
2300	16 ga x 2-1/2" wide		115	.070		2.15	1.79		3.94	5.40
2310	3-5/8" wide		110	.073		2.63	1.87		4.50	6.05
2320	4" wide		110	.073		2.79	1.87		4.66	6.25
2330	6" wide		105	.076		3.58	1.96		5.54	7.25
2340	8" wide		100	.080		4.58	2.06		6.64	8.55
3000	Continuous bridging, per row									
3100	16 ga x 1-1/2" channel thru studs 12" O.C.	1 Carp	6	1.333	C.L.F.	51.50	34.50		86	115
3110	16" O.C.		7	1.143		51.50	29.50		81	107
3120	24" O.C.		8.80	.909		51.50	23.50		75	96.50
4100	2" x 2" angle x 18 ga, studs 12" O.C.		7	1.143		76	29.50		105.50	134
4110	16" O.C.		9	.889		76	23		99	123
4120	24" O.C.		12	.667		76	17.15		93.15	113
4200	16 ga, studs 12" O.C.		5	1.600		97	41		138	176

05 41 Structural Metal Stud Framing

05 41 13 – Load-Bearing Metal Stud Framing

05 41 13.10 Bridging	Crew	Daily Output	Labor-Hours	Unit	Material	2007 Bare Costs Labor	Equipment	Total	Total Incl O&P
4210 16" O.C.	1 Carp	7	1.143	C.L.F.	97	29.50		126.50	156
4220 24" O.C.	↓	10	.800	↓	97	20.50		117.50	141

05 41 13.25 Framing, Boxed Headers/Beams

	Crew	Daily Output	Labor-Hours	Unit	Material	Labor	Equipment	Total	Total Incl O&P
0010 **FRAMING, BOXED HEADERS/BEAMS**									
0200 Double, 18 ga x 6" deep	2 Carp	220	.073	L.F.	5.35	1.87		7.22	9.05
0210 8" deep		210	.076		6	1.96		7.96	9.90
0220 10" deep		200	.080		7.25	2.06		9.31	11.45
0230 12" deep		190	.084		7.95	2.16		10.11	12.40
0300 16 ga x 8" deep		180	.089		6.85	2.28		9.13	11.45
0310 10" deep		170	.094		8.25	2.42		10.67	13.15
0320 12" deep		160	.100		8.95	2.57		11.52	14.20
0400 14 ga x 10" deep		140	.114		9.55	2.94		12.49	15.50
0410 12" deep		130	.123		10.50	3.16		13.66	16.90
1210 Triple, 18 ga x 8" deep		170	.094		8.65	2.42		11.07	13.65
1220 10" deep		165	.097		10.40	2.49		12.89	15.65
1230 12" deep		160	.100		11.45	2.57		14.02	16.95
1300 16 ga x 8" deep		145	.110		10	2.84		12.84	15.80
1310 10" deep		140	.114		11.90	2.94		14.84	18.05
1320 12" deep		135	.119		13	3.05		16.05	19.45
1400 14 ga x 10" deep		115	.139		13.05	3.58		16.63	20.50
1410 12" deep	↓	110	.145	↓	14.50	3.74		18.24	22.50

05 41 13.30 Framing, Stud Walls

	Crew	Daily Output	Labor-Hours	Unit	Material	Labor	Equipment	Total	Total Incl O&P
0010 **FRAMING, STUD WALLS** w/ top & bottom track, no openings,									
0020 Headers, beams, bridging or bracing									
4100 8' high walls, 18 ga x 2-1/2" wide, studs 12" O.C.	2 Carp	54	.296	L.F.	8.95	7.60		16.55	23
4110 16" O.C.		77	.208		7.15	5.35		12.50	16.95
4120 24" O.C.		107	.150		5.35	3.84		9.19	12.40
4130 3-5/8" wide, studs 12" O.C.		53	.302		10.75	7.75		18.50	25
4140 16" O.C.		76	.211		8.60	5.40		14	18.65
4150 24" O.C.		105	.152		6.45	3.92		10.37	13.75
4160 4" wide, studs 12" O.C.		52	.308		11.30	7.90		19.20	26
4170 16" O.C.		74	.216		9.05	5.55		14.60	19.40
4180 24" O.C.		103	.155		6.75	3.99		10.74	14.20
4190 6" wide, studs 12" O.C.		51	.314		14.30	8.05		22.35	29.50
4200 16" O.C.		73	.219		11.45	5.65		17.10	22
4210 24" O.C.		101	.158		8.60	4.07		12.67	16.40
4220 8" wide, studs 12" O.C.		50	.320		17.50	8.20		25.70	33
4230 16" O.C.		72	.222		14.05	5.70		19.75	25
4240 24" O.C.		100	.160		10.60	4.11		14.71	18.65
4300 16 ga x 2-1/2" wide, studs 12" O.C.		47	.340		10.55	8.75		19.30	26.50
4310 16" O.C.		68	.235		8.35	6.05		14.40	19.45
4320 24" O.C.		94	.170		6.15	4.37		10.52	14.20
4330 3-5/8" wide, studs 12" O.C.		46	.348		12.70	8.95		21.65	29
4340 16" O.C.		66	.242		10.05	6.25		16.30	21.50
4350 24" O.C.		92	.174		7.40	4.47		11.87	15.75
4360 4" wide, studs 12" O.C.		45	.356		13.35	9.15		22.50	30
4370 16" O.C.		65	.246		10.60	6.35		16.95	22.50
4380 24" O.C.		90	.178		7.80	4.57		12.37	16.35
4390 6" wide, studs 12" O.C.		44	.364		16.85	9.35		26.20	34.50
4400 16" O.C.		64	.250		13.40	6.45		19.85	25.50
4410 24" O.C.		88	.182		9.90	4.67		14.57	18.85
4420 8" wide, studs 12" O.C.	↓	43	.372		21	9.55		30.55	39

05 41 Structural Metal Stud Framing

05 41 13 – Load-Bearing Metal Stud Framing

	05 41 13.30 Framing, Stud Walls	Crew	Daily Output	Labor-Hours	Unit	Material	2007 Bare Costs Labor	Equipment	Total	Total Incl O&P
4430	16" O.C.	2 Carp	63	.254	L.F.	16.50	6.55		23.05	29
4440	24" O.C.		86	.186		12.25	4.78		17.03	21.50
5100	10' high walls, 18 ga x 2-1/2" wide, studs 12" O.C.		54	.296		10.75	7.60		18.35	25
5110	16" O.C.		77	.208		8.50	5.35		13.85	18.40
5120	24" O.C.		107	.150		6.25	3.84		10.09	13.40
5130	3-5/8" wide, studs 12" O.C.		53	.302		12.95	7.75		20.70	27.50
5140	16" O.C.		76	.211		10.25	5.40		15.65	20.50
5150	24" O.C.		105	.152		7.55	3.92		11.47	14.95
5160	4" wide, studs 12" O.C.		52	.308		13.55	7.90		21.45	28.50
5170	16" O.C.		74	.216		10.75	5.55		16.30	21.50
5180	24" O.C.		103	.155		7.90	3.99		11.89	15.45
5190	6" wide, studs 12" O.C.		51	.314		17.15	8.05		25.20	32.50
5200	16" O.C.		73	.219		13.60	5.65		19.25	24.50
5210	24" O.C.		101	.158		10.05	4.07		14.12	17.95
5220	8" wide, studs 12" O.C.		50	.320		21	8.20		29.20	37
5230	16" O.C.		72	.222		16.60	5.70		22.30	28
5240	24" O.C.		100	.160		12.30	4.11		16.41	20.50
5300	16 ga x 2-1/2" wide, studs 12" O.C.		47	.340		12.75	8.75		21.50	29
5310	16" O.C.		68	.235		10	6.05		16.05	21.50
5320	24" O.C.		94	.170		7.25	4.37		11.62	15.40
5330	3-5/8" wide, studs 12" O.C.		46	.348		15.35	8.95		24.30	32
5340	16" O.C.		66	.242		12.05	6.25		18.30	24
5350	24" O.C.		92	.174		8.75	4.47		13.22	17.20
5360	4" wide, studs 12" O.C.		45	.356		16.15	9.15		25.30	33.50
5370	16" O.C.		65	.246		12.70	6.35		19.05	24.50
5380	24" O.C.		90	.178		9.20	4.57		13.77	17.85
5390	6" wide, studs 12" O.C.		44	.364		20.50	9.35		29.85	38.50
5400	16" O.C.		64	.250		16	6.45		22.45	28.50
5410	24" O.C.		88	.182		11.65	4.67		16.32	21
5420	8" wide, studs 12" O.C.		43	.372		25	9.55		34.55	43.50
5430	16" O.C.		63	.254		19.70	6.55		26.25	32.50
5440	24" O.C.		86	.186		14.35	4.78		19.13	24
6190	12' high walls, 18 ga x 6" wide, studs 12" O.C.		41	.390		20	10.05		30.05	39
6200	16" O.C.		58	.276		15.70	7.10		22.80	29.50
6210	24" O.C.		81	.198		11.45	5.10		16.55	21
6220	8" wide, studs 12" O.C.		40	.400		24.50	10.30		34.80	44.50
6230	16" O.C.		57	.281		19.20	7.20		26.40	33.50
6240	24" O.C.		80	.200		14.05	5.15		19.20	24
6390	16 ga x 6" wide, studs 12" O.C.		35	.457		24	11.75		35.75	46
6400	16" O.C.		51	.314		18.60	8.05		26.65	34
6410	24" O.C.		70	.229		13.40	5.85		19.25	24.50
6420	8" wide, studs 12" O.C.		34	.471		29.50	12.10		41.60	52.50
6430	16" O.C.		50	.320		23	8.20		31.20	39
6440	24" O.C.		69	.232		16.50	5.95		22.45	28.50
6530	14 ga x 3-5/8" wide, studs 12" O.C.		34	.471		22.50	12.10		34.60	45.50
6540	16" O.C.		48	.333		17.65	8.55		26.20	34
6550	24" O.C.		65	.246		12.65	6.35		19	24.50
6560	4" wide, studs 12" O.C.		33	.485		24	12.45		36.45	47.50
6570	16" O.C.		47	.340		18.60	8.75		27.35	35.50
6580	24" O.C.		64	.250		13.35	6.45		19.80	25.50
6730	12 ga x 3-5/8" wide, studs 12" O.C.		31	.516		31.50	13.25		44.75	57
6740	16" O.C.		43	.372		24	9.55		33.55	42.50
6750	24" O.C.		59	.271		17	6.95		23.95	30.50

05 41 Structural Metal Stud Framing

05 41 13 – Load-Bearing Metal Stud Framing

05 41 13.30 Framing, Stud Walls

		Crew	Daily Output	Labor-Hours	Unit	Material	2007 Bare Costs Labor	Equipment	Total	Total Incl O&P
6760	4" wide, studs 12" O.C.	2 Carp	30	.533	L.F.	33.50	13.70		47.20	60.50
6770	16" O.C.		42	.381		26	9.80		35.80	45
6780	24" O.C.		58	.276		18.15	7.10		25.25	32
7390	16' high walls, 16 ga x 6" wide, studs 12" O.C.		33	.485		31	12.45		43.45	55
7400	16" O.C.		48	.333		24	8.55		32.55	40.50
7410	24" O.C.		67	.239		16.85	6.15		23	29
7420	8" wide, studs 12" O.C.		32	.500		38	12.85		50.85	63.50
7430	16" O.C.		47	.340		29.50	8.75		38.25	47
7440	24" O.C.		66	.242		21	6.25		27.25	33.50
7560	14 ga x 4" wide, studs 12" O.C.		31	.516		31	13.25		44.25	56.50
7570	16" O.C.		45	.356		24	9.15		33.15	42
7580	24" O.C.		61	.262		16.85	6.75		23.60	30
7590	6" wide, studs 12" O.C.		30	.533		39	13.70		52.70	66.50
7600	16" O.C.		44	.364		30	9.35		39.35	49
7610	24" O.C.		60	.267		21.50	6.85		28.35	35
7760	12 ga x 4" wide, studs 12" O.C.		29	.552		43.50	14.20		57.70	72
7770	16" O.C.		40	.400		33.50	10.30		43.80	54.50
7780	24" O.C.		55	.291		23.50	7.50		31	38
7790	6" wide, studs 12" O.C.		28	.571		55	14.70		69.70	85.50
7800	16" O.C.		39	.410		42.50	10.55		53.05	64.50
7810	24" O.C.		54	.296		29.50	7.60		37.10	45.50
8590	20' high walls, 14 ga x 6" wide, studs 12" O.C.		29	.552		47.50	14.20		61.70	76.50
8600	16" O.C.		42	.381		36.50	9.80		46.30	57
8610	24" O.C.		57	.281		25.50	7.20		32.70	40.50
8620	8" wide, studs 12" O.C.		28	.571		58.50	14.70		73.20	89
8630	16" O.C.		41	.390		45	10.05		55.05	66.50
8640	24" O.C.		56	.286		31.50	7.35		38.85	47
8790	12 ga x 6" wide, studs 12" O.C.		27	.593		68	15.25		83.25	101
8800	16" O.C.		37	.432		52	11.10		63.10	76
8810	24" O.C.		51	.314		36	8.05		44.05	53
8820	8" wide, studs 12" O.C.		26	.615		83	15.80		98.80	118
8830	16" O.C.		36	.444		63.50	11.40		74.90	89
8840	24" O.C.		50	.320		44	8.20		52.20	62

05 42 Cold-Formed Metal Joist Framing

05 42 13 – Cold-Formed Metal Floor Joist Framing

05 42 13.05 Bracing

		Crew	Daily Output	Labor-Hours	Unit	Material	Labor	Equipment	Total	Total Incl O&P
0010	**BRACING**, continuous, per row, top & bottom									
0120	Flat strap, 20 ga x 2" wide, joists at 12" O.C.	1 Carp	4.67	1.713	C.L.F.	57.50	44		101.50	138
0130	16" O.C.		5.33	1.501		55.50	38.50		94	127
0140	24" O.C.		6.66	1.201		53.50	31		84.50	112
0150	18 ga x 2" wide, joists at 12" O.C.		4	2		74	51.50		125.50	169
0160	16" O.C.		4.67	1.713		73	44		117	155
0170	24" O.C.		5.33	1.501		71.50	38.50		110	144

05 42 13.10 Bridging

		Crew	Daily Output	Labor-Hours	Unit	Material	Labor	Equipment	Total	Total Incl O&P
0010	**BRIDGING**, solid between joists w/ 1-1/4" leg track, per joist bay									
0230	Joists 12" O.C., 18 ga track x 6" wide	1 Carp	80	.100	Ea.	1.54	2.57		4.11	6.05
0240	8" wide		75	.107		1.96	2.74		4.70	6.80
0250	10" wide		70	.114		2.42	2.94		5.36	7.65
0260	12" wide		65	.123		2.81	3.16		5.97	8.45
0330	16 ga track x 6" wide		70	.114		1.93	2.94		4.87	7.10

05 42 Cold-Formed Metal Joist Framing

05 42 13 – Cold-Formed Metal Floor Joist Framing

05 42 13.10 Bridging		Crew	Daily Output	Labor-Hours	Unit	Material	2007 Bare Costs Labor	Equipment	Total	Total Incl O&P
0340	8" wide	1 Carp	65	.123	Ea.	2.47	3.16		5.63	8.05
0350	10" wide		60	.133		3.04	3.43		6.47	9.15
0360	12" wide		55	.145		3.50	3.74		7.24	10.20
0440	14 ga track x 8" wide		60	.133		3.11	3.43		6.54	9.25
0450	10" wide		55	.145		3.82	3.74		7.56	10.55
0460	12" wide		50	.160		4.41	4.11		8.52	11.85
0550	12 ga track x 10" wide		45	.178		5.60	4.57		10.17	13.90
0560	12" wide		40	.200		6.35	5.15		11.50	15.70
1230	16" O.C., 18 ga track x 6" wide		80	.100		1.98	2.57		4.55	6.55
1240	8" wide		75	.107		2.51	2.74		5.25	7.40
1250	10" wide		70	.114		3.10	2.94		6.04	8.40
1260	12" wide		65	.123		3.60	3.16		6.76	9.30
1330	16 ga track x 6" wide		70	.114		2.48	2.94		5.42	7.70
1340	8" wide		65	.123		3.17	3.16		6.33	8.85
1350	10" wide		60	.133		3.89	3.43		7.32	10.10
1360	12" wide		55	.145		4.49	3.74		8.23	11.30
1440	14 ga track x 8" wide		60	.133		3.99	3.43		7.42	10.20
1450	10" wide		55	.145		4.90	3.74		8.64	11.75
1460	12" wide		50	.160		5.65	4.11		9.76	13.25
1550	12 ga track x 10" wide		45	.178		7.20	4.57		11.77	15.65
1560	12" wide		40	.200		8.15	5.15		13.30	17.65
2230	24" O.C., 18 ga track x 6" wide		80	.100		2.86	2.57		5.43	7.50
2240	8" wide		75	.107		3.63	2.74		6.37	8.65
2250	10" wide		70	.114		4.49	2.94		7.43	9.90
2260	12" wide		65	.123		5.20	3.16		8.36	11.05
2330	16 ga track x 6" wide		70	.114		3.58	2.94		6.52	8.90
2340	8" wide		65	.123		4.58	3.16		7.74	10.40
2350	10" wide		60	.133		5.65	3.43		9.08	12
2360	12" wide		55	.145		6.50	3.74		10.24	13.50
2440	14 ga track x 8" wide		60	.133		5.80	3.43		9.23	12.15
2450	10" wide		55	.145		7.10	3.74		10.84	14.15
2460	12" wide		50	.160		8.20	4.11		12.31	16
2550	12 ga track x 10" wide		45	.178		10.40	4.57		14.97	19.20
2560	12" wide		40	.200		11.75	5.15		16.90	21.50
05 42 13.25 Framing, Band Joist										
0010	**FRAMING, BAND JOIST** (track) fastened to bearing wall									
0220	18 ga track x 6" deep	2 Carp	1000	.016	L.F.	1.26	.41		1.67	2.09
0230	8" deep		920	.017		1.60	.45		2.05	2.52
0240	10" deep		860	.019		1.97	.48		2.45	2.98
0320	16 ga track x 6" deep		900	.018		1.58	.46		2.04	2.51
0330	8" deep		840	.019		2.02	.49		2.51	3.05
0340	10" deep		780	.021		2.48	.53		3.01	3.62
0350	12" deep		740	.022		2.86	.56		3.42	4.08
0430	14 ga track x 8" deep		750	.021		2.54	.55		3.09	3.73
0440	10" deep		720	.022		3.12	.57		3.69	4.40
0450	12" deep		700	.023		3.60	.59		4.19	4.96
0540	12 ga track x 10" deep		670	.024		4.58	.61		5.19	6.10
0550	12" deep		650	.025		5.20	.63		5.83	6.75
05 42 13.30 Framing, Boxed Headers/Beams										
0010	**FRAMING, BOXED HEADERS/BEAMS**									
0200	Double, 18 ga x 6" deep	2 Carp	220	.073	L.F.	5.35	1.87		7.22	9.05
0210	8" deep		210	.076		6	1.96		7.96	9.90

05 42 Cold-Formed Metal Joist Framing

05 42 13 – Cold-Formed Metal Floor Joist Framing

05 42 13.30 Framing, Boxed Headers/Beams		Crew	Daily Output	Labor-Hours	Unit	Material	2007 Bare Costs Labor	Equipment	Total	Total Incl O&P
0220	10" deep	2 Carp	200	.080	L.F.	7.25	2.06		9.31	11.45
0230	12" deep		190	.084		7.95	2.16		10.11	12.40
0300	16 ga x 8" deep		180	.089		6.85	2.28		9.13	11.45
0310	10" deep		170	.094		8.25	2.42		10.67	13.15
0320	12" deep		160	.100		8.95	2.57		11.52	14.20
0400	14 ga x 10" deep		140	.114		9.55	2.94		12.49	15.50
0410	12" deep		130	.123		10.50	3.16		13.66	16.90
0500	12 ga x 10" deep		110	.145		12.65	3.74		16.39	20.50
0510	12" deep		100	.160		14.05	4.11		18.16	22.50
1210	Triple, 18 ga x 8" deep		170	.094		8.65	2.42		11.07	13.65
1220	10" deep		165	.097		10.40	2.49		12.89	15.65
1230	12" deep		160	.100		11.45	2.57		14.02	16.95
1300	16 ga x 8" deep		145	.110		10	2.84		12.84	15.80
1310	10" deep		140	.114		11.90	2.94		14.84	18.05
1320	12" deep		135	.119		13	3.05		16.05	19.45
1400	14 ga x 10" deep		115	.139		13.90	3.58		17.48	21.50
1410	12" deep		110	.145		15.30	3.74		19.04	23
1500	12 ga x 10" deep		90	.178		18.55	4.57		23.12	28.50
1510	12" deep		85	.188		20.50	4.84		25.34	30.50

05 42 13.40 Framing, Joists		Crew	Daily Output	Labor-Hours	Unit	Material	Labor	Equipment	Total	Total Incl O&P
0010	**FRAMING, JOISTS**, no band joists (track), web stiffeners, headers,									
0020	Beams, bridging or bracing									
0030	Joists (2" flange) and fasteners, materials only									
0220	18 ga x 6" deep				L.F.	1.66			1.66	1.82
0230	8" deep					1.98			1.98	2.18
0240	10" deep					2.32			2.32	2.55
0320	16 ga x 6" deep					2.04			2.04	2.24
0330	8" deep					2.45			2.45	2.69
0340	10" deep					2.85			2.85	3.13
0350	12" deep					3.23			3.23	3.56
0430	14 ga x 8" deep					3.09			3.09	3.40
0440	10" deep					3.55			3.55	3.90
0450	12" deep					4.04			4.04	4.45
0540	12 ga x 10" deep					5.20			5.20	5.70
0550	12" deep					5.90			5.90	6.50
1010	Installation of joists to band joists, beams & headers, labor only									
1220	18 ga x 6" deep	2 Carp	110	.145	Ea.		3.74		3.74	6.35
1230	8" deep		90	.178			4.57		4.57	7.75
1240	10" deep		80	.200			5.15		5.15	8.70
1320	16 ga x 6" deep		95	.168			4.33		4.33	7.35
1330	8" deep		70	.229			5.85		5.85	9.95
1340	10" deep		60	.267			6.85		6.85	11.65
1350	12" deep		55	.291			7.50		7.50	12.70
1430	14 ga x 8" deep		65	.246			6.35		6.35	10.75
1440	10" deep		45	.356			9.15		9.15	15.50
1450	12" deep		35	.457			11.75		11.75	19.95
1540	12 ga x 10" deep		40	.400			10.30		10.30	17.45
1550	12" deep		30	.533			13.70		13.70	23.50

05 42 13.45 Framing, Web Stiffeners		Crew	Daily Output	Labor-Hours	Unit	Material	Labor	Equipment	Total	Total Incl O&P
0010	**FRAMING, WEB STIFFENERS** at joist bearing, fabricated from									
0020	Stud piece (1-5/8" flange) to stiffen joist (2" flange)									
2120	For 6" deep joist, with 18 ga x 2-1/2" stud	1 Carp	120	.067	Ea.	1.98	1.71		3.69	5.10

05 42 Cold-Formed Metal Joist Framing

05 42 13 – Cold-Formed Metal Floor Joist Framing

	05 42 13.45 Framing, Web Stiffeners	Crew	Daily Output	Labor-Hours	Unit	Material	2007 Bare Costs Labor	Equipment	Total	Total Incl O&P
2130	3-5/8" stud	1 Carp	110	.073	Ea.	2.21	1.87		4.08	5.60
2140	4" stud		105	.076		2.15	1.96		4.11	5.70
2150	6" stud		100	.080		2.34	2.06		4.40	6.05
2160	8" stud		95	.084		2.41	2.16		4.57	6.30
2220	8" deep joist, with 2-1/2" stud		120	.067		2.17	1.71		3.88	5.30
2230	3-5/8" stud		110	.073		2.39	1.87		4.26	5.80
2240	4" stud		105	.076		2.35	1.96		4.31	5.90
2250	6" stud		100	.080		2.57	2.06		4.63	6.30
2260	8" stud		95	.084		2.77	2.16		4.93	6.70
2320	10" deep joist, with 2-1/2" stud		110	.073		3.06	1.87		4.93	6.55
2330	3-5/8" stud		100	.080		3.41	2.06		5.47	7.25
2340	4" stud		95	.084		3.38	2.16		5.54	7.40
2350	6" stud		90	.089		3.65	2.28		5.93	7.90
2360	8" stud		85	.094		3.71	2.42		6.13	8.20
2420	12" deep joist, with 2-1/2" stud		110	.073		3.24	1.87		5.11	6.75
2430	3-5/8" stud		100	.080		3.56	2.06		5.62	7.40
2440	4" stud		95	.084		3.50	2.16		5.66	7.50
2450	6" stud		90	.089		3.83	2.28		6.11	8.10
2460	8" stud		85	.094		4.13	2.42		6.55	8.65
3130	For 6" deep joist, with 16 ga x 3-5/8" stud		100	.080		2.31	2.06		4.37	6.05
3140	4" stud		95	.084		2.29	2.16		4.45	6.20
3150	6" stud		90	.089		2.52	2.28		4.80	6.65
3160	8" stud		85	.094		2.66	2.42		5.08	7.05
3230	8" deep joist, with 3-5/8" stud		100	.080		2.56	2.06		4.62	6.30
3240	4" stud		95	.084		2.51	2.16		4.67	6.45
3250	6" stud		90	.089		2.80	2.28		5.08	6.95
3260	8" stud		85	.094		3	2.42		5.42	7.40
3330	10" deep joist, with 3-5/8" stud		85	.094		3.51	2.42		5.93	7.95
3340	4" stud		80	.100		3.58	2.57		6.15	8.30
3350	6" stud		75	.107		3.90	2.74		6.64	8.95
3360	8" stud		70	.114		4.07	2.94		7.01	9.45
3430	12" deep joist, with 3-5/8" stud		85	.094		3.83	2.42		6.25	8.30
3440	4" stud		80	.100		3.75	2.57		6.32	8.50
3450	6" stud		75	.107		4.18	2.74		6.92	9.25
3460	8" stud		70	.114		4.47	2.94		7.41	9.90
4230	For 8" deep joist, with 14 ga x 3-5/8" stud		90	.089		3.34	2.28		5.62	7.55
4240	4" stud		85	.094		3.40	2.42		5.82	7.85
4250	6" stud		80	.100		3.69	2.57		6.26	8.40
4260	8" stud		75	.107		3.95	2.74		6.69	9
4330	10" deep joist, with 3-5/8" stud		75	.107		4.68	2.74		7.42	9.80
4340	4" stud		70	.114		4.65	2.94		7.59	10.10
4350	6" stud		65	.123		5.10	3.16		8.26	10.95
4360	8" stud		60	.133		5.35	3.43		8.78	11.65
4430	12" deep joist, with 3-5/8" stud		75	.107		4.98	2.74		7.72	10.15
4440	4" stud		70	.114		5.10	2.94		8.04	10.60
4450	6" stud		65	.123		5.50	3.16		8.66	11.40
4460	8" stud		60	.133		5.90	3.43		9.33	12.30
5330	For 10" deep joist, with 12 ga x 3-5/8" stud		65	.123		4.94	3.16		8.10	10.80
5340	4" stud		60	.133		5.10	3.43		8.53	11.40
5350	6" stud		55	.145		5.60	3.74		9.34	12.50
5360	8" stud		50	.160		6.15	4.11		10.26	13.80
5430	12" deep joist, with 3-5/8" stud		65	.123		5.45	3.16		8.61	11.35
5440	4" stud		60	.133		5.35	3.43		8.78	11.70

05 42 Cold-Formed Metal Joist Framing

05 42 13 – Cold-Formed Metal Floor Joist Framing

05 42 13.45 Framing, Web Stiffeners		Crew	Daily Output	Labor-Hours	Unit	Material	2007 Bare Costs Labor	Equipment	Total	Total Incl O&P
5450	6" stud	1 Carp	55	.145	Ea.	6.10	3.74		9.84	13.10
5460	8" stud		50	.160		7.05	4.11		11.16	14.75

05 42 23 – Cold-Formed Metal Roof Joist Framing

05 42 23.05 Framing, Bracing

		Crew	Daily Output	Labor-Hours	Unit	Material	Labor	Equipment	Total	Total Incl O&P
0010	**FRAMING, BRACING**									
0020	Continuous bracing, per row									
0100	16 ga x 1-1/2" channel thru rafters/trusses @ 16" O.C.	1 Carp	4.50	1.778	C.L.F.	51.50	45.50		97	135
0120	24" O.C.		6	1.333		51.50	34.50		86	115
0300	2" x 2" angle x 18 ga, rafters/trusses @ 16" O.C.		6	1.333		76	34.50		110.50	142
0320	24" O.C.		8	1		76	25.50		101.50	127
0400	16 ga, rafters/trusses @ 16" O.C.		4.50	1.778		97	45.50		142.50	184
0420	24" O.C.		6.50	1.231		97	31.50		128.50	160

05 42 23.10 Framing, Bridging

		Crew	Daily Output	Labor-Hours	Unit	Material	Labor	Equipment	Total	Total Incl O&P
0010	**FRAMING, BRIDGING**									
0020	Solid, between rafters w/ 1-1/4" leg track, per rafter bay									
1200	Rafters 16" O.C., 18 ga x 4" deep	1 Carp	60	.133	Ea.	1.52	3.43		4.95	7.45
1210	6" deep		57	.140		1.98	3.61		5.59	8.30
1220	8" deep		55	.145		2.51	3.74		6.25	9.10
1230	10" deep		52	.154		3.10	3.95		7.05	10.10
1240	12" deep		50	.160		3.60	4.11		7.71	10.95
2200	24" O.C., 18 ga x 4" deep		60	.133		2.20	3.43		5.63	8.20
2210	6" deep		57	.140		2.86	3.61		6.47	9.25
2220	8" deep		55	.145		3.63	3.74		7.37	10.35
2230	10" deep		52	.154		4.49	3.95		8.44	11.65
2240	12" deep		50	.160		5.20	4.11		9.31	12.70

05 42 23.50 Framing, Parapets

		Crew	Daily Output	Labor-Hours	Unit	Material	Labor	Equipment	Total	Total Incl O&P
0010	**FRAMING, PARAPETS**									
0100	3' high installed on 1st story, 18 ga x 4" wide studs, 12" O.C.	2 Carp	100	.160	L.F.	5.65	4.11		9.76	13.20
0110	16" O.C.		150	.107		4.80	2.74		7.54	9.95
0120	24" O.C.		200	.080		3.95	2.06		6.01	7.85
0200	6" wide studs, 12" O.C.		100	.160		7.20	4.11		11.31	14.90
0210	16" O.C.		150	.107		6.15	2.74		8.89	11.40
0220	24" O.C.		200	.080		5.05	2.06		7.11	9.10
1100	Installed on 2nd story, 18 ga x 4" wide studs, 12" O.C.		95	.168		5.65	4.33		9.98	13.55
1110	16" O.C.		145	.110		4.80	2.84		7.64	10.10
1120	24" O.C.		190	.084		3.95	2.16		6.11	8
1200	6" wide studs, 12" O.C.		95	.168		7.20	4.33		11.53	15.25
1210	16" O.C.		145	.110		6.15	2.84		8.99	11.55
1220	24" O.C.		190	.084		5.05	2.16		7.21	9.25
2100	Installed on gable, 18 ga x 4" wide studs, 12" O.C.		85	.188		5.65	4.84		10.49	14.40
2110	16" O.C.		130	.123		4.80	3.16		7.96	10.65
2120	24" O.C.		170	.094		3.95	2.42		6.37	8.45
2200	6" wide studs, 12" O.C.		85	.188		7.20	4.84		12.04	16.10
2210	16" O.C.		130	.123		6.15	3.16		9.31	12.10
2220	24" O.C.		170	.094		5.05	2.42		7.47	9.70

05 42 23.60 Framing, Roof Rafters

		Crew	Daily Output	Labor-Hours	Unit	Material	Labor	Equipment	Total	Total Incl O&P
0010	**FRAMING, ROOF RAFTERS**									
0100	Boxed ridge beam, double, 18 ga x 6" deep	2 Carp	160	.100	L.F.	5.35	2.57		7.92	10.25
0110	8" deep		150	.107		6	2.74		8.74	11.25
0120	10" deep		140	.114		7.25	2.94		10.19	12.95
0130	12" deep		130	.123		7.95	3.16		11.11	14.10

05 42 Cold-Formed Metal Joist Framing

05 42 23 – Cold-Formed Metal Roof Joist Framing

05 42 23.60 Framing, Roof Rafters

		Crew	Daily Output	Labor-Hours	Unit	Material	2007 Bare Costs Labor	2007 Bare Costs Equipment	Total	Total Incl O&P
0200	16 ga x 6" deep	2 Carp	150	.107	L.F.	6.10	2.74		8.84	11.35
0210	8" deep		140	.114		6.85	2.94		9.79	12.55
0220	10" deep		130	.123		8.25	3.16		11.41	14.40
0230	12" deep		120	.133		8.95	3.43		12.38	15.65
1100	Rafters, 2" flange, material only, 18 ga x 6" deep					1.66			1.66	1.82
1110	8" deep					1.98			1.98	2.18
1120	10" deep					2.32			2.32	2.55
1130	12" deep					2.70			2.70	2.97
1200	16 ga x 6" deep					2.04			2.04	2.24
1210	8" deep					2.45			2.45	2.69
1220	10" deep					2.85			2.85	3.13
1230	12" deep					3.23			3.23	3.56
2100	Installation only, ordinary rafter to 4:12 pitch, 18 ga x 6" deep	2 Carp	35	.457	Ea.		11.75		11.75	19.95
2110	8" deep		30	.533			13.70		13.70	23.50
2120	10" deep		25	.640			16.45		16.45	28
2130	12" deep		20	.800			20.50		20.50	35
2200	16 ga x 6" deep		30	.533			13.70		13.70	23.50
2210	8" deep		25	.640			16.45		16.45	28
2220	10" deep		20	.800			20.50		20.50	35
2230	12" deep		15	1.067			27.50		27.50	46.50
8100	Add to labor, ordinary rafters on steep roofs						25%			
8110	Dormers & complex roofs						50%			
8200	Hip & valley rafters to 4:12 pitch						25%			
8210	Steep roofs						50%			
8220	Dormers & complex roofs						75%			
8300	Hip & valley jack rafters to 4:12 pitch						50%			
8310	Steep roofs						75%			
8320	Dormers & complex roofs						100%			

05 42 23.70 Framing, Soffits and Canopies

		Crew	Daily Output	Labor-Hours	Unit	Material	2007 Bare Costs Labor	2007 Bare Costs Equipment	Total	Total Incl O&P
0010	**FRAMING, SOFFITS & CANOPIES**									
0130	Continuous ledger track @ wall, studs @ 16" O.C., 18 ga x 4" wide	2 Carp	535	.030	L.F.	1.01	.77		1.78	2.41
0140	6" wide		500	.032		1.32	.82		2.14	2.85
0150	8" wide		465	.034		1.67	.88		2.55	3.34
0160	10" wide		430	.037		2.07	.96		3.03	3.89
0230	Studs @ 24" O.C., 18 ga x 4" wide		800	.020		.97	.51		1.48	1.93
0240	6" wide		750	.021		1.26	.55		1.81	2.32
0250	8" wide		700	.023		1.60	.59		2.19	2.76
0260	10" wide		650	.025		1.97	.63		2.60	3.24
1000	Horizontal soffit and canopy members, material only									
1030	1-5/8" flange studs, 18 ga x 4" deep				L.F.	1.36			1.36	1.49
1040	6" deep					1.70			1.70	1.87
1050	8" deep					2.06			2.06	2.27
1140	2" flange joists, 18 ga x 6" deep					1.90			1.90	2.09
1150	8" deep					2.27			2.27	2.49
1160	10" deep					2.65			2.65	2.92
4030	Installation only, 18 ga, 1-5/8" flange x 4" deep	2 Carp	130	.123	Ea.		3.16		3.16	5.35
4040	6" deep		110	.145			3.74		3.74	6.35
4050	8" deep		90	.178			4.57		4.57	7.75
4140	2" flange, 18 ga x 6" deep		110	.145			3.74		3.74	6.35
4150	8" deep		90	.178			4.57		4.57	7.75
4160	10" deep		80	.200			5.15		5.15	8.70
6010	Clips to attach facia to rafter tails, 2" x 2" x 18 ga angle	1 Carp	120	.067		.90	1.71		2.61	3.90

05 42 Cold-Formed Metal Joist Framing

05 42 23 – Cold-Formed Metal Roof Joist Framing

05 42 23.70 Framing, Soffits and Canopies	Crew	Daily Output	Labor-Hours	Unit	Material	2007 Bare Costs Labor	Equipment	Total	Total Incl O&P
6020 16 ga angle	1 Carp	100	.080	Ea.	1.14	2.06		3.20	4.75

05 44 Cold-Formed Metal Trusses

05 44 13 – Cold-Formed Metal Roof Trusses

05 44 13.60 Framing, Roof Trusses

		Crew	Daily Output	Labor-Hours	Unit	Material	Labor	Equipment	Total	Total Incl O&P
0010	**FRAMING, ROOF TRUSSES**									
0020	Fabrication of trusses on ground, Fink (W) or King Post, to 4:12 pitch									
0120	18 ga x 4" chords, 16' span	2 Carp	12	1.333	Ea.	63.50	34.50		98	128
0130	20' span		11	1.455		79	37.50		116.50	151
0140	24' span		11	1.455		95	37.50		132.50	168
0150	28' span		10	1.600		111	41		152	192
0160	32' span		10	1.600		127	41		168	209
0250	6" chords, 28' span		9	1.778		139	45.50		184.50	231
0260	32' span		9	1.778		159	45.50		204.50	253
0270	36' span		8	2		179	51.50		230.50	284
0280	40' span		8	2		199	51.50		250.50	305
1120	5:12 to 8:12 pitch, 18 ga x 4" chords, 16' span		10	1.600		72.50	41		113.50	150
1130	20' span		9	1.778		90.50	45.50		136	177
1140	24' span		9	1.778		108	45.50		153.50	197
1150	28' span		8	2		127	51.50		178.50	226
1160	32' span		8	2		145	51.50		196.50	246
1250	6" chords, 28' span		7	2.286		159	58.50		217.50	275
1260	32' span		7	2.286		182	58.50		240.50	300
1270	36' span		6	2.667		204	68.50		272.50	340
1280	40' span		6	2.667		227	68.50		295.50	365
2120	9:12 to 12:12 pitch, 18 ga x 4" chords, 16' span		8	2		90.50	51.50		142	187
2130	20' span		7	2.286		113	58.50		171.50	224
2140	24' span		7	2.286		136	58.50		194.50	249
2150	28' span		6	2.667		158	68.50		226.50	290
2160	32' span		6	2.667		181	68.50		249.50	315
2250	6" chords, 28' span		5	3.200		199	82		281	360
2260	32' span		5	3.200		227	82		309	390
2270	36' span		4	4		256	103		359	455
2280	40' span		4	4		284	103		387	485
5120	Erection only of roof trusses, to 4:12 pitch, 16' span	F-6	48	.833			19.35	15.10	34.45	49
5130	20' span		46	.870			20	15.75	35.75	51.50
5140	24' span		44	.909			21	16.45	37.45	53.50
5150	28' span		42	.952			22	17.25	39.25	56.50
5160	32' span		40	1			23.50	18.10	41.60	59
5170	36' span		38	1.053			24.50	19.05	43.55	62
5180	40' span		36	1.111			26	20	46	65.50
5220	5:12 to 8:12 pitch, 16' span		42	.952			22	17.25	39.25	56.50
5230	20' span		40	1			23.50	18.10	41.60	59
5240	24' span		38	1.053			24.50	19.05	43.55	62
5250	28' span		36	1.111			26	20	46	65.50
5260	32' span		34	1.176			27.50	21.50	49	69.50
5270	36' span		32	1.250			29	22.50	51.50	74
5280	40' span		30	1.333			31	24	55	78.50
5320	9:12 to 12:12 pitch, 16' span		36	1.111			26	20	46	65.50
5330	20' span		34	1.176			27.50	21.50	49	69.50
5340	24' span		32	1.250			29	22.50	51.50	74

05 44 Cold-Formed Metal Trusses

05 44 13 – Cold-Formed Metal Roof Trusses

05 44 13.60 Framing, Roof Trusses

		Crew	Daily Output	Labor-Hours	Unit	Material	2007 Bare Costs Labor	2007 Bare Costs Equipment	Total	Total Incl O&P
5350	28' span	F-6	30	1.333	Ea.		31	24	55	78.50
5360	32' span		28	1.429			33	26	59	84.50
5370	36' span		26	1.538			36	28	64	91
5380	40' span		24	1.667			39	30	69	98.50

05 51 Metal Stairs

05 51 13 – Metal Pan Stairs

05 51 13.50 Pan Stairs

		Crew	Daily Output	Labor-Hours	Unit	Material	Labor	Equipment	Total	Total Incl O&P
0010	**PAN STAIRS**, shop fabricated, steel stringers									
1700	Pre-erected, steel pan tread, 3'-6" wide, 2 line pipe rail	E-2	87	.552	Riser	430	15.45	17.80	463.25	525
1800	With flat bar picket rail	"	87	.552	"	485	15.45	17.80	518.25	580

05 51 23 – Metal Fire Escapes

05 51 23.50 Fire Escape Stairs

		Crew	Daily Output	Labor-Hours	Unit	Material	Labor	Equipment	Total	Total Incl O&P
0010	**FIRE ESCAPE STAIRS**, shop fabricated									
0020	One story, disappearing, stainless steel	2 Sswk	20	.800	V.L.F.	193	22		215	255
0100	Portable ladder				Ea.	58.50			58.50	64.50

05 52 Metal Railings

05 52 13 – Pipe and Tube Railings

05 52 13.50 Railings, Pipe

		Crew	Daily Output	Labor-Hours	Unit	Material	Labor	Equipment	Total	Total Incl O&P
0010	**RAILINGS, PIPE**, shop fabricated									
0020	Aluminum, 2 rail, satin finish, 1-1/4" diameter	E-4	160	.200	L.F.	23.50	5.65	.72	29.87	38
0030	Clear anodized		160	.200		29.50	5.65	.72	35.87	44
0040	Dark anodized		160	.200		33	5.65	.72	39.37	48.50
0080	1-1/2" diameter, satin finish		160	.200		28.50	5.65	.72	34.87	43
0090	Clear anodized		160	.200		31.50	5.65	.72	37.87	47
0100	Dark anodized		160	.200		35	5.65	.72	41.37	50.50
0140	Aluminum, 3 rail, 1-1/4" diam., satin finish		137	.234		36.50	6.60	.84	43.94	54
0150	Clear anodized		137	.234		45.50	6.60	.84	52.94	64
0160	Dark anodized		137	.234		50.50	6.60	.84	57.94	69.50
0200	1-1/2" diameter, satin finish		137	.234		43.50	6.60	.84	50.94	62
0210	Clear anodized		137	.234		49.50	6.60	.84	56.94	68
0220	Dark anodized		137	.234		54	6.60	.84	61.44	73.50
0500	Steel, 2 rail, on stairs, primed, 1-1/4" diameter		160	.200		19.45	5.65	.72	25.82	33.50
0520	1-1/2" diameter		160	.200		21.50	5.65	.72	27.87	35.50
0540	Galvanized, 1-1/4" diameter		160	.200		27	5.65	.72	33.37	41.50
0560	1-1/2" diameter		160	.200		30	5.65	.72	36.37	45
0580	Steel, 3 rail, primed, 1-1/4" diameter		137	.234		29	6.60	.84	36.44	46
0600	1-1/2" diameter		137	.234		31	6.60	.84	38.44	48
0620	Galvanized, 1-1/4" diameter		137	.234		40.50	6.60	.84	47.94	58.50
0640	1-1/2" diameter		137	.234		48	6.60	.84	55.44	67
0700	Stainless steel, 2 rail, 1-1/4" diam. #4 finish		137	.234		70	6.60	.84	77.44	91
0720	High polish		137	.234		113	6.60	.84	120.44	138
0740	Mirror polish		137	.234		141	6.60	.84	148.44	169
0760	Stainless steel, 3 rail, 1-1/2" diam., #4 finish		120	.267		106	7.50	.96	114.46	132
0770	High polish		120	.267		175	7.50	.96	183.46	208
0780	Mirror finish		120	.267		213	7.50	.96	221.46	250
0900	Wall rail, alum. pipe, 1-1/4" diam., satin finish		213	.150		13.55	4.24	.54	18.33	23.50
0905	Clear anodized		213	.150		16.50	4.24	.54	21.28	27

05 52 Metal Railings

05 52 13 – Pipe and Tube Railings

05 52 13.50 Railings, Pipe

		Crew	Daily Output	Labor-Hours	Unit	Material	2007 Bare Costs Labor	Equipment	Total	Total Incl O&P
0910	Dark anodized	E-4	213	.150	L.F.	20	4.24	.54	24.78	31
0915	1-1/2" diameter, satin finish		213	.150		15	4.24	.54	19.78	25.50
0920	Clear anodized		213	.150		18.85	4.24	.54	23.63	29.50
0925	Dark anodized		213	.150		23.50	4.24	.54	28.28	34.50
0930	Steel pipe, 1-1/4" diameter, primed		213	.150		11.80	4.24	.54	16.58	22
0935	Galvanized		213	.150		17.10	4.24	.54	21.88	27.50
0940	1-1/2" diameter		176	.182		12.15	5.15	.65	17.95	24
0945	Galvanized		213	.150		17.15	4.24	.54	21.93	27.50
0955	Stainless steel pipe, 1-1/2" diam., #4 finish		107	.299		56	8.45	1.08	65.53	79
0960	High polish		107	.299		114	8.45	1.08	123.53	143
0965	Mirror polish		107	.299		134	8.45	1.08	143.53	166
2000	Aluminum pipe & picket railing, double top rail, pickets @ 4-1/2" OC									
2010	36" high, straight & level	2 Sswk	80	.200	L.F.	73.50	5.55		79.05	91.50
2020	Curved & level		60	.267		103	7.40		110.40	128
2030	Straight & sloped		40	.400		84	11.10		95.10	114

05 58 Formed Metal Fabrications

05 58 25 – Formed Lamp Posts

05 58 25.40 Lamp Posts

		Crew	Daily Output	Labor-Hours	Unit	Material	Labor	Equipment	Total	Total Incl O&P
0010	**LAMP POSTS**									
0020	Aluminum, 7' high, stock units, post only	1 Carp	16	.500	Ea.	28	12.85		40.85	53
0100	Mild steel, plain	"	16	.500	"	23	12.85		35.85	47.50

05 71 Decorative Metal Stairs

05 71 13 – Fabricated Metal Spiral Stairs

05 71 13.50 Spiral Stairs

		Crew	Daily Output	Labor-Hours	Unit	Material	Labor	Equipment	Total	Total Incl O&P
0010	**SPIRAL STAIRS**, shop fabricated									
1810	Spiral aluminum, 5'-0" diameter, stock units	E-4	45	.711	Riser	475	20	2.56	497.56	560
1820	Custom units		45	.711		895	20	2.56	917.56	1,025
1900	Spiral, cast iron, 4'-0" diameter, ornamental, minimum		45	.711		425	20	2.56	447.56	505
1920	Maximum		25	1.280		575	36	4.61	615.61	710

Division 6
Wood, Plastics, & Composites

06 05 Common Work Results for Wood, Plastics and Composites

06 05 05 – Selective Wood and Plastics Demolition

06 05 05.10 Selective Demolition Wood Framing		Crew	Daily Output	Labor-Hours	Unit	Material	2007 Bare Costs Labor	Equipment	Total	Total Incl O&P
0010	**SELECTIVE DEMOLITION WOOD FRAMING** R024119-10									
0100	Timber connector, nailed, small	1 Clab	96	.083	Ea.		1.56		1.56	2.65
0110	Medium		60	.133			2.49		2.49	4.23
0120	Large		48	.167			3.12		3.12	5.30
0130	Bolted, small		48	.167			3.12		3.12	5.30
0140	Medium		32	.250			4.68		4.68	7.95
0150	Large		24	.333			6.25		6.25	10.60
2958	Beams, 2" x 6"	2 Clab	1100	.015	L.F.		.27		.27	.46
2960	2" x 8"		825	.019			.36		.36	.62
2965	2" x 10"		665	.024			.45		.45	.76
2970	2" x 12"		550	.029			.54		.54	.92
2972	2" x 14"		470	.034			.64		.64	1.08
2975	4" x 8"	B-1	413	.058			1.13		1.13	1.91
2980	4" x 10"		330	.073			1.41		1.41	2.39
2985	4" x 12"		275	.087			1.69		1.69	2.87
3000	6" x 8"		275	.087			1.69		1.69	2.87
3040	6" x 10"		220	.109			2.11		2.11	3.59
3080	6" x 12"		185	.130			2.51		2.51	4.27
3120	8" x 12"		140	.171			3.32		3.32	5.65
3160	10" x 12"		110	.218			4.23		4.23	7.15
3162	Alternate pricing method		1.10	21.818	M.B.F.		425		425	715
3170	Blocking, in 16" OC wall framing, 2" x 4"	1 Clab	600	.013	L.F.		.25		.25	.42
3172	2" x 6"		400	.020			.37		.37	.64
3174	In 24" OC wall framing, 2" x 4"		600	.013			.25		.25	.42
3176	2" x 6"		400	.020			.37		.37	.64
3178	Alt method, wood blocking removal from wood framing		.40	20	M.B.F.		375		375	635
3179	Wood blocking removal from steel framing		.36	22.222	"		415		415	705
3180	Bracing, let in, 1" x 3", studs 16" OC		1050	.008	L.F.		.14		.14	.24
3181	Studs 24" OC		1080	.007			.14		.14	.24
3182	1" x 4", studs 16" OC		1050	.008			.14		.14	.24
3183	Studs 24" OC		1080	.007			.14		.14	.24
3184	1" x 6", studs 16" OC		1050	.008			.14		.14	.24
3185	Studs 24" OC		1080	.007			.14		.14	.24
3186	2" x 3", studs 16" OC		800	.010			.19		.19	.32
3187	Studs 24" OC		830	.010			.18		.18	.31
3188	2" x 4", studs 16" OC		800	.010			.19		.19	.32
3189	Studs 24" OC		830	.010			.18		.18	.31
3190	2" x 6", studs 16" OC		800	.010			.19		.19	.32
3191	Studs 24" OC		830	.010			.18		.18	.31
3192	2" x 8", studs 16" OC		800	.010			.19		.19	.32
3193	Studs 24" OC		830	.010			.18		.18	.31
3194	"T" shaped metal bracing, studs at 16" OC		1060	.008			.14		.14	.24
3195	Studs at 24" OC		1200	.007			.12		.12	.21
3196	Metal straps, studs at 16" OC		1200	.007			.12		.12	.21
3197	Studs at 24" OC		1240	.006			.12		.12	.20
3200	Columns, round, 8' to 14' tall		40	.200	Ea.		3.74		3.74	6.35
3202	Dimensional lumber sizes	2 Clab	1.10	14.545	M.B.F.		272		272	460
3250	Blocking, between joists	1 Clab	320	.025	Ea.		.47		.47	.79
3252	Bridging, metal strap, between joists		320	.025	Pr.		.47		.47	.79
3254	Wood, between joists		320	.025	"		.47		.47	.79
3260	Door buck, studs, header & access, 8' high 2" x 4" wall, 3' wide		32	.250	Ea.		4.68		4.68	7.95
3261	4' wide		32	.250			4.68		4.68	7.95
3262	5' wide		32	.250			4.68		4.68	7.95

06 05 Common Work Results for Wood, Plastics and Composites

06 05 05 – Selective Wood and Plastics Demolition

06 05 05.10 Selective Demolition Wood Framing		Crew	Daily Output	Labor-Hours	Unit	Material	2007 Bare Costs Labor	2007 Bare Costs Equipment	Total	Total Incl O&P
3263	6' wide	1 Clab	32	.250	Ea.		4.68		4.68	7.95
3264	8' wide		30	.267			4.99		4.99	8.45
3265	10' wide		30	.267			4.99		4.99	8.45
3266	12' wide		30	.267			4.99		4.99	8.45
3267	2" x 6" wall, 3' wide		32	.250			4.68		4.68	7.95
3268	4' wide		32	.250			4.68		4.68	7.95
3269	5' wide		32	.250			4.68		4.68	7.95
3270	6' wide		32	.250			4.68		4.68	7.95
3271	8' wide		30	.267			4.99		4.99	8.45
3272	10' wide		30	.267			4.99		4.99	8.45
3273	12' wide		30	.267			4.99		4.99	8.45
3274	Window buck, studs, header & access, 8' high 2" x 4" wall, 2' wide		24	.333			6.25		6.25	10.60
3275	3' wide		24	.333			6.25		6.25	10.60
3276	4' wide		24	.333			6.25		6.25	10.60
3277	5' wide		24	.333			6.25		6.25	10.60
3278	6' wide		24	.333			6.25		6.25	10.60
3279	7' wide		24	.333			6.25		6.25	10.60
3280	8' wide		22	.364			6.80		6.80	11.55
3281	10' wide		22	.364			6.80		6.80	11.55
3282	12' wide		22	.364			6.80		6.80	11.55
3283	2" x 6" wall, 2' wide		24	.333			6.25		6.25	10.60
3284	3' wide		24	.333			6.25		6.25	10.60
3285	4' wide		24	.333			6.25		6.25	10.60
3286	5' wide		24	.333			6.25		6.25	10.60
3287	6' wide		24	.333			6.25		6.25	10.60
3288	7' wide		24	.333			6.25		6.25	10.60
3289	8' wide		22	.364			6.80		6.80	11.55
3290	10' wide		22	.364			6.80		6.80	11.55
3291	12' wide		22	.364			6.80		6.80	11.55
3400	Fascia boards, 1" x 6"		500	.016	L.F.		.30		.30	.51
3440	1" x 8"		450	.018			.33		.33	.56
3480	1" x 10"		400	.020			.37		.37	.64
3490	2" x 6"		450	.018			.33		.33	.56
3500	2" x 8"		400	.020			.37		.37	.64
3510	2" x 10"		350	.023			.43		.43	.73
3610	Furring, on wood walls or ceiling		4000	.002	S.F.		.04		.04	.06
3620	On masonry or concrete walls or ceiling		1200	.007	"		.12		.12	.21
3800	Headers over openings, 2 @ 2" x 6"		110	.073	L.F.		1.36		1.36	2.31
3840	2 @ 2" x 8"		100	.080			1.50		1.50	2.54
3880	2 @ 2" x 10"		90	.089			1.66		1.66	2.82
3885	Alternate pricing method		.26	30.651	M.B.F.		575		575	975
3920	Joists, 1" x 4"		1250	.006	L.F.		.12		.12	.20
3930	1" x 6"		1135	.007			.13		.13	.22
3950	1" x 10"		895	.009			.17		.17	.28
3960	1" x 12"		765	.010			.20		.20	.33
4200	2" x 4"	2 Clab	1000	.016			.30		.30	.51
4230	2" x 6"		970	.016			.31		.31	.52
4240	2" x 8"		940	.017			.32		.32	.54
4250	2" x 10"		910	.018			.33		.33	.56
4280	2" x 12"		880	.018			.34		.34	.58
4281	2" x 14"		850	.019			.35		.35	.60
4282	Composite joists, 9-1/2"		960	.017			.31		.31	.53
4283	11-7/8"		930	.017			.32		.32	.55

06 05 Common Work Results for Wood, Plastics and Composites

06 05 05 – Selective Wood and Plastics Demolition

06 05 05.10 Selective Demolition Wood Framing		Crew	Daily Output	Labor-Hours	Unit	Material	2007 Bare Costs Labor	Equipment	Total	Total Incl O&P
4284	14"	2 Clab	897	.018	L.F.		.33		.33	.57
4285	16"		865	.019			.35		.35	.59
4290	Wood joists, alternate pricing method		1.50	10.667	M.B.F.		199		199	340
4500	Open web joist, 12" deep		500	.032	L.F.		.60		.60	1.02
4505	14" deep		475	.034			.63		.63	1.07
4510	16" deep		450	.036			.67		.67	1.13
4520	18" deep		425	.038			.70		.70	1.20
4530	24" deep		400	.040			.75		.75	1.27
4550	Ledger strips, 1" x 2"	1 Clab	1200	.007			.12		.12	.21
4560	1" x 3"		1200	.007			.12		.12	.21
4570	1" x 4"		1200	.007			.12		.12	.21
4580	2" x 2"		1100	.007			.14		.14	.23
4590	2" x 4"		1000	.008			.15		.15	.25
4600	2" x 6"		1000	.008			.15		.15	.25
4601	2" x 8" or 2" x 10"		800	.010			.19		.19	.32
4602	4" x 6"		600	.013			.25		.25	.42
4604	4" x 8"		450	.018			.33		.33	.56
5400	Posts, 4" x 4"	2 Clab	800	.020			.37		.37	.64
5405	4" x 6"		550	.029			.54		.54	.92
5410	4" x 8"		440	.036			.68		.68	1.15
5425	4" x 10"		390	.041			.77		.77	1.30
5430	4" x 12"		350	.046			.85		.85	1.45
5440	6" x 6"		400	.040			.75		.75	1.27
5445	6" x 8"		350	.046			.85		.85	1.45
5450	6" x 10"		320	.050			.94		.94	1.59
5455	6" x 12"		290	.055			1.03		1.03	1.75
5480	8" x 8"		300	.053			1		1	1.69
5500	10" x 10"		240	.067			1.25		1.25	2.12
5660	Tongue and groove floor planks		2	8	M.B.F.		150		150	254
5750	Rafters, ordinary, 16" OC, 2" x 4"		880	.018	S.F.		.34		.34	.58
5755	2" x 6"		840	.019			.36		.36	.60
5760	2" x 8"		820	.020			.36		.36	.62
5770	2" x 10"		820	.020			.36		.36	.62
5780	2" x 12"		810	.020			.37		.37	.63
5785	24" OC, 2" x 4"		1170	.014			.26		.26	.43
5786	2" x 6"		1117	.014			.27		.27	.45
5787	2" x 8"		1091	.015			.27		.27	.47
5788	2" x 10"		1091	.015			.27		.27	.47
5789	2" x 12"		1077	.015			.28		.28	.47
5795	Rafters, ordinary, 2" x 4" (alternate method)		862	.019	L.F.		.35		.35	.59
5800	2" x 6" (alternate method)		850	.019			.35		.35	.60
5840	2" x 8" (alternate method)		837	.019			.36		.36	.61
5855	2" x 10" (alternate method)		825	.019			.36		.36	.62
5865	2" x 12" (alternate method)		812	.020			.37		.37	.63
5870	Sill plate, 2" x 4"	1 Clab	1170	.007			.13		.13	.22
5871	2" x 6"		780	.010			.19		.19	.33
5872	2" x 8"		586	.014			.26		.26	.43
5873	Alternate pricing method		.78	10.256	M.B.F.		192		192	325
5885	Ridge board, 1" x 4"	2 Clab	900	.018	L.F.		.33		.33	.56
5886	1" x 6"		875	.018			.34		.34	.58
5887	1" x 8"		850	.019			.35		.35	.60
5888	1" x 10"		825	.019			.36		.36	.62
5889	1" x 12"		800	.020			.37		.37	.64

06 05 Common Work Results for Wood, Plastics and Composites

06 05 05 – Selective Wood and Plastics Demolition

06 05 05.10 Selective Demolition Wood Framing		Crew	Daily Output	Labor-Hours	Unit	Material	2007 Bare Costs Labor	2007 Bare Costs Equipment	Total	Total Incl O&P
5890	2" x 4"	2 Clab	900	.018	L.F.		.33		.33	.56
5892	2" x 6"		875	.018			.34		.34	.58
5894	2" x 8"		850	.019			.35		.35	.60
5896	2" x 10"		825	.019			.36		.36	.62
5898	2" x 12"		800	.020			.37		.37	.64
6050	Rafter tie, 1" x 4"		1250	.013			.24		.24	.41
6052	1" x 6"		1135	.014			.26		.26	.45
6054	2" x 4"		1000	.016			.30		.30	.51
6056	2" x 6"		970	.016			.31		.31	.52
6070	Sleepers, on concrete, 1" x 2"	1 Clab	4700	.002			.03		.03	.05
6075	1" x 3"		4000	.002			.04		.04	.06
6080	2" x 4"		3000	.003			.05		.05	.08
6085	2" x 6"		2600	.003			.06		.06	.10
6086	Sheathing from roof, 5/16"	2 Clab	1600	.010	S.F.		.19		.19	.32
6088	3/8"		1525	.010			.20		.20	.33
6090	1/2"		1400	.011			.21		.21	.36
6092	5/8"		1300	.012			.23		.23	.39
6094	3/4"		1200	.013			.25		.25	.42
6096	Board sheathing from roof		1400	.011			.21		.21	.36
6100	Sheathing, from walls, 1/4"		1200	.013			.25		.25	.42
6110	5/16"		1175	.014			.25		.25	.43
6120	3/8"		1150	.014			.26		.26	.44
6130	1/2"		1125	.014			.27		.27	.45
6140	5/8"		1100	.015			.27		.27	.46
6150	3/4"		1075	.015			.28		.28	.47
6152	Board sheathing from walls		1500	.011			.20		.20	.34
6158	Subfloor, with boards		1050	.015			.29		.29	.48
6160	Plywood, 1/2" thick		768	.021			.39		.39	.66
6162	5/8" thick		760	.021			.39		.39	.67
6164	3/4" thick		750	.021			.40		.40	.68
6165	1-1/8" thick		720	.022			.42		.42	.71
6166	Underlayment, particle board, 3/8" thick	1 Clab	780	.010			.19		.19	.33
6168	1/2" thick		768	.010			.19		.19	.33
6170	5/8" thick		760	.011			.20		.20	.33
6172	3/4" thick		750	.011			.20		.20	.34
6200	Stairs and stringers, minimum	2 Clab	40	.400	Riser		7.50		7.50	12.70
6240	Maximum	"	26	.615	"		11.50		11.50	19.55
6300	Components, tread	1 Clab	110	.073	Ea.		1.36		1.36	2.31
6320	Riser		80	.100	"		1.87		1.87	3.18
6390	Stringer, 2" x 10"		260	.031	L.F.		.58		.58	.98
6400	2" x 12"		260	.031			.58		.58	.98
6410	3" x 10"		250	.032			.60		.60	1.02
6420	3" x 12"		250	.032			.60		.60	1.02
6590	Wood studs, 2" x 3"	2 Clab	3076	.005			.10		.10	.17
6600	2" x 4"		2000	.008			.15		.15	.25
6640	2" x 6"		1600	.010			.19		.19	.32
6720	Wall framing, including studs plates and blocking, 2" x 4"	1 Clab	600	.013	S.F.		.25		.25	.42
6740	2" x 6"		480	.017	"		.31		.31	.53
6750	Headers, 2" x 4"		1125	.007	L.F.		.13		.13	.23
6755	2" x 6"		1125	.007			.13		.13	.23
6760	2" x 8"		1050	.008			.14		.14	.24
6765	2" x 10"		1050	.008			.14		.14	.24
6770	2" x 12"		1000	.008			.15		.15	.25

06 05 Common Work Results for Wood, Plastics and Composites

06 05 05 – Selective Wood and Plastics Demolition

06 05 05.10 Selective Demolition Wood Framing

		Crew	Daily Output	Labor-Hours	Unit	Material	2007 Bare Costs Labor	Equipment	Total	Total Incl O&P
6780	4" x 10"	1 Clab	525	.015	L.F.		.29		.29	.48
6785	4" x 12"		500	.016			.30		.30	.51
6790	6" x 8"		560	.014			.27		.27	.45
6795	6" x 10"		525	.015			.29		.29	.48
6797	6" x 12"	↓	500	.016	↓		.30		.30	.51
7000	Trusses									
7050	12' span	2 Clab	74	.216	Ea.		4.04		4.04	6.85
7150	24' span	F-3	66	.606			14.10	10.95	25.05	36
7200	26' span		64	.625			14.55	11.30	25.85	37
7250	28' span		62	.645			15.05	11.70	26.75	38.50
7300	30' span		58	.690			16.05	12.50	28.55	41
7350	32' span		56	.714			16.65	12.95	29.60	42
7400	34' span		54	.741			17.25	13.40	30.65	44
7450	36' span	↓	52	.769	↓		17.90	13.90	31.80	45.50
8000	Soffit, T & G wood	1 Clab	520	.015	S.F.		.29		.29	.49
8010	Hardboard, vinyl or aluminum	"	640	.013			.23		.23	.40
8030	Plywood	2 Carp	315	.051	↓		1.31		1.31	2.21

06 05 05.20 Selective Demolition Millwork and Trim

		Crew	Daily Output	Labor-Hours	Unit	Material	2007 Bare Costs Labor	Equipment	Total	Total Incl O&P
0010	**SELECTIVE DEMOLITION MILLWORK AND TRIM** R024119-10									
1000	Cabinets, wood, base cabinets, per L.F.	2 Clab	80	.200	L.F.		3.74		3.74	6.35
1020	Wall cabinets, per L.F.		80	.200			3.74		3.74	6.35
1100	Steel, painted, base cabinets		60	.267			4.99		4.99	8.45
1500	Counter top, minimum		200	.080			1.50		1.50	2.54
1510	Maximum		120	.133	↓		2.49		2.49	4.23
2000	Paneling, 4' x 8' sheets		2000	.008	S.F.		.15		.15	.25
2100	Boards, 1" x 4"		700	.023			.43		.43	.73
2120	1" x 6"		750	.021			.40		.40	.68
2140	1" x 8"		800	.020	↓		.37		.37	.64
3000	Trim, baseboard, to 6" wide		1200	.013	L.F.		.25		.25	.42
3040	Greater than 6" and up to 12" wide		1000	.016			.30		.30	.51
3100	Ceiling trim		1000	.016			.30		.30	.51
3120	Chair rail		1200	.013			.25		.25	.42
3140	Railings with balusters		240	.067			1.25		1.25	2.12
3160	Wainscoting	↓	700	.023	S.F.		.43		.43	.73
4000	Curtain rod	1 Clab	80	.100	L.F.		1.87		1.87	3.18
9000	Minimum labor/equipment charge	"	4	2	Job		37.50		37.50	63.50

06 05 23 – Wood, Plastic, and Composite Fastenings

06 05 23.10 Nails

		Crew	Daily Output	Labor-Hours	Unit	Material	2007 Bare Costs Labor	Equipment	Total	Total Incl O&P
0010	**NAILS**									
0020	Copper nails, plain				Lb.	6.60			6.60	7.25
0400	Stainless steel, plain					5.95			5.95	6.55
0500	Box, 3d to 20d, bright					.76			.76	.84
0520	Galvanized					.88			.88	.97
0600	Common, 3d to 60d, plain					.82			.82	.90
0700	Galvanized					1.06			1.06	1.17
0800	Aluminum					4.45			4.45	4.90
1000	Annular or spiral thread, 4d to 60d, plain					1.88			1.88	2.07
1200	Galvanized					1.95			1.95	2.15
1400	Drywall nails, plain					.79			.79	.87
1600	Galvanized					1.59			1.59	1.75
1800	Finish nails, 4d to 10d, plain				↓	.91			.91	1
2000	Galvanized					1.51			1.51	1.66

06 05 Common Work Results for Wood, Plastics and Composites

06 05 23 – Wood, Plastic, and Composite Fastenings

06 05 23.10 Nails		Crew	Daily Output	Labor-Hours	Unit	Material	2007 Bare Costs Labor	Equipment	Total	Total Incl O&P
2100	Aluminum				Lb.	4.10			4.10	4.51
2300	Flooring nails, hardened steel, 2d to 10d, plain					1.61			1.61	1.77
2400	Galvanized					2.25			2.25	2.48
2500	Gypsum lath nails, 1-1/8", 13 ga. flathead, blued					1.54			1.54	1.69
2600	Masonry nails, hardened steel, 3/4" to 3" long, plain					1.61			1.61	1.77
2700	Galvanized					2			2	2.20
2900	Roofing nails, threaded, galvanized					1.05			1.05	1.16
3100	Aluminum					4.80			4.80	5.30
3300	Compressed lead head, threaded, galvanized					1.50			1.50	1.65
3600	Siding nails, plain shank, galvanized					1.20			1.20	1.32
3800	Aluminum					4.11			4.11	4.52
5000	Add to prices above for cement coating					.10			.10	.11
5200	Zinc or tin plating					.13			.13	.14
5500	Vinyl coated sinkers, 8d to 16d					.57			.57	.63

06 05 23.20 Pneumatic Nails										
0010	**PNEUMATIC NAILS**									
0020	Framing, per carton of 5000, 2"				Ea.	37			37	41
0100	2-3/8"					42.50			42.50	46.50
0200	Per carton of 4000, 3"					37.50			37.50	41.50
0300	3-1/4"					40			40	44
0400	Per carton of 5000, 2-3/8", galv.					57.50			57.50	63.50
0500	Per carton of 4000, 3", galv.					65			65	71.50
0600	3-1/4", galv.					80.50			80.50	88.50
0700	Roofing, per carton of 7200, 1"					34			34	37
0800	1-1/4"					31.50			31.50	34.50
0900	1-1/2"					36.50			36.50	40
1000	1-3/4"					45.50			45.50	50

06 05 23.40 Sheet Metal Screws										
0010	**SHEET METAL SCREWS**									
0020	Steel, standard, #8 x 3/4", plain				C	3.29			3.29	3.62
0100	Galvanized					3.44			3.44	3.78
0300	#10 x 1", plain					4.49			4.49	4.94
0400	Galvanized					4.60			4.60	5.05
1500	Self-drilling, with washers, (pinch point) #8 x 3/4", plain					7.10			7.10	7.80
1600	Galvanized					7.20			7.20	7.90
1800	#10 x 3/4", plain					9.30			9.30	10.20
1900	Galvanized					9.40			9.40	10.35
3000	Stainless steel w/aluminum or neoprene washers, #14 x 1", plain					18.35			18.35	20
3100	#14 x 2", plain					25			25	27.50

06 05 23.50 Wood Screws										
0010	**WOOD SCREWS**									
0020	Steel, #8 x 1" long				C	3.29			3.29	3.62
0100	Brass					11.50			11.50	12.65
0200	#8, 2" long, steel					5.60			5.60	6.15
0300	Brass					19.50			19.50	21.50
0400	#10, 1" long, steel					5.90			5.90	6.50
0500	Brass					14			14	15.40
0600	#10, 2" long, steel					6.30			6.30	6.90
0700	Brass					23			23	25.50
0800	#10, 3" long, steel					10.50			10.50	11.55
1000	#12, 2" long, steel					7.50			7.50	8.25
1100	Brass					30			30	33

06 05 Common Work Results for Wood, Plastics and Composites

06 05 23 – Wood, Plastic, and Composite Fastenings

06 05 23.50 Wood Screws

		Crew	Daily Output	Labor-Hours	Unit	Material	2007 Bare Costs Labor	Equipment	Total	Total Incl O&P
1500	#12, 3" long, steel				C	11.50			11.50	12.65
2000	#12, 4" long, steel				↓	43			43	47.50

06 05 23.60 Timber Connectors

		Crew	Daily Output	Labor-Hours	Unit	Material	Labor	Equipment	Total	Total Incl O&P
0010	**TIMBER CONNECTORS**									
0020	Add up cost of each part for total cost of connection									
0100	Connector plates, steel, with bolts, straight	2 Carp	75	.213	Ea.	22.50	5.50		28	34.50
0110	Tee		50	.320		33.50	8.20		41.70	51
0120	T-Strap, 14 gauge, 12" x 8" x 2"		50	.320		33.50	8.20		41.70	51
0150	Anchor plate, 7 gauge, 9" x 7"	↓	75	.213		22.50	5.50		28	34.50
0200	Bolts, machine, sq. hd. with nut & washer, 1/2" diameter, 4" long	1 Carp	140	.057		.41	1.47		1.88	2.94
0300	7-1/2" long		130	.062		.72	1.58		2.30	3.47
0500	3/4" diameter, 7-1/2" long		130	.062		1.92	1.58		3.50	4.79
0610	Machine bolts, w/ nut, washer, 3/4" dia, 15" L, HD's & beam hangers		95	.084	↓	3.52	2.16		5.68	7.55
0720	Machine bolts, sq. hd. w/nut & wash		150	.053	Lb.	2.60	1.37		3.97	5.20
0800	Drilling bolt holes in timber, 1/2" diameter		450	.018	Inch		.46		.46	.78
0900	1" diameter		350	.023	"		.59		.59	1
1100	Framing anchors, 2 or 3 dimensional, 10 gauge, no nails incl.		175	.046	Ea.	.57	1.17		1.74	2.62
1150	Framing anchors, 18 gauge, 4 1/2" x 2 3/4"		175	.046		.57	1.17		1.74	2.62
1160	Framing anchors, 18 gauge, 4 1/2" x 3"		175	.046		.57	1.17		1.74	2.62
1170	Clip anchors plates, 18 gauge, 12" x 1 1/8"		175	.046		.57	1.17		1.74	2.62
1250	Holdowns, 3 gauge base, 10 gauge body		8	1		18.95	25.50		44.45	64.50
1260	Holdowns, 7 gauge 11 1/16" x 3 1/4"		8	1		18.95	25.50		44.45	64.50
1270	Holdowns, 7 gauge 14 3/8" x 3 1/8"		8	1		18.95	25.50		44.45	64.50
1275	Holdowns, 12 gauge 8" x 2 1/2"		8	1		18.95	25.50		44.45	64.50
1300	Joist and beam hangers, 18 ga. galv., for 2" x 4" joist		175	.046		.71	1.17		1.88	2.77
1400	2" x 6" to 2" x 10" joist		165	.048		.81	1.25		2.06	3
1600	16 ga. galv., 3" x 6" to 3" x 10" joist		160	.050		3.19	1.29		4.48	5.70
1700	3" x 10" to 3" x 14" joist		160	.050		3.71	1.29		5	6.25
1800	4" x 6" to 4" x 10" joist		155	.052		2.81	1.33		4.14	5.35
1900	4" x 10" to 4" x 14" joist		155	.052		3.81	1.33		5.14	6.45
2000	Two-2" x 6" to two-2" x 10" joists		150	.053		3.07	1.37		4.44	5.70
2100	Two-2" x 10" to two-2" x 14" joists		150	.053		3.07	1.37		4.44	5.70
2300	3/16" thick, 6" x 8" joist		145	.055		6.70	1.42		8.12	9.75
2400	6" x 10" joist		140	.057		7.85	1.47		9.32	11.15
2500	6" x 12" joist		135	.059		9.45	1.52		10.97	13
2700	1/4" thick, 6" x 14" joist	↓	130	.062		11.75	1.58		13.33	15.65
2900	Plywood clips, extruded aluminum H clip, for 3/4" panels					.18			.18	.20
3000	Galvanized 18 ga. back-up clip					.17			.17	.19
3200	Post framing, 16 ga. galv. for 4" x 4" base, 2 piece	1 Carp	130	.062		6.60	1.58		8.18	9.95
3300	Cap		130	.062		3.24	1.58		4.82	6.25
3500	Rafter anchors, 18 ga. galv., 1-1/2" wide, 5-1/4" long		145	.055		.56	1.42		1.98	3.03
3600	10-3/4" long		145	.055		1.10	1.42		2.52	3.62
3800	Shear plates, 2-5/8" diameter		120	.067		1.98	1.71		3.69	5.10
3900	4" diameter		115	.070		4.55	1.79		6.34	8.05
4000	Sill anchors, embedded in concrete or block, 18-5/8" long		115	.070		1.33	1.79		3.12	4.49
4100	Spike grids, 4" x 4", flat or curved		120	.067		.49	1.71		2.20	3.45
4400	Split rings, 2-1/2" diameter		120	.067		1.60	1.71		3.31	4.67
4500	4" diameter		110	.073		2.47	1.87		4.34	5.90
4550	Tie plate, 20 gauge, 7" x 3 1/8"		110	.073		2.47	1.87		4.34	5.90
4560	Tie plate, 20 gauge, 5" x 4 1/8"		110	.073		2.47	1.87		4.34	5.90
4575	Twist straps, 18 gauge, 12" x 1 1/4"		110	.073		2.47	1.87		4.34	5.90
4580	Twist straps, 18 gauge, 16" x 1 1/4"	↓	110	.073	↓	2.47	1.87		4.34	5.90

06 05 Common Work Results for Wood, Plastics and Composites

06 05 23 – Wood, Plastic, and Composite Fastenings

06 05 23.60 Timber Connectors		Crew	Daily Output	Labor-Hours	Unit	Material	2007 Bare Costs Labor	Equipment	Total	Total Incl O&P
4600	Strap ties, 20 ga., 2-1/16" wide, 12 13/16" long	1 Carp	180	.044	Ea.	1.44	1.14		2.58	3.52
4700	Strap ties, 16 ga., 1-3/8" wide, 12" long		180	.044		1.44	1.14		2.58	3.52
4800	24" long		160	.050		1.99	1.29		3.28	4.37
5000	Toothed rings, 2-5/8" or 4" diameter		90	.089		1.36	2.28		3.64	5.40
5200	Truss plates, nailed, 20 gauge, up to 32' span		17	.471	Truss Ea.	9.85	12.10		21.95	31.50
5400	Washers, 2" x 2" x 1/8"				"	.31			.31	.34
5500	3" x 3" x 3/16"					.81			.81	.89
6101	Beam hangers, polymer painted									
6102	Bolted, 3 ga., (W x H x L)									
6104	3-1/4" x 9" x 12" top flange	1 Carp	1	8	C	7,750	206		7,956	8,850
6106	5-1/4" x 9" x 12" top flange		1	8		8,050	206		8,256	9,200
6108	5-1/4" x 11" x 11-3/4" top flange		1	8		18,100	206		18,306	20,300
6110	6-7/8" x 9" x 12" top flange		1	8		8,325	206		8,531	9,500
6112	6-7/8" x 11" x 13-1/2" top flange		1	8		19,000	206		19,206	21,300
6114	8-7/8" x 11" x 15-1/2" top flange		1	8		20,400	206		20,606	22,800
6116	Nailed, 3 ga., (W x H x L)									
6118	3-1/4" x 10-1/2" x 10" top flange	1 Carp	1.80	4.444	C	5,350	114		5,464	6,075
6120	3-1/4" x 10-1/2" x 12" top flange		1.80	4.444		6,200	114		6,314	7,000
6122	5-1/4" x 9-1/2" x 10" top flange		1.80	4.444		5,750	114		5,864	6,550
6124	5-1/4" x 9-1/2" x 12" top flange		1.80	4.444		6,525	114		6,639	7,400
6126	5-1/2" x 9-1/2" x 12" top flange		1.80	4.444		5,750	114		5,864	6,550
6128	6-7/8" x 8-1/2" x 12" top flange		1.80	4.444		5,950	114		6,064	6,750
6130	7-1/2" x 8-1/2" x 12" top flange		1.80	4.444		6,150	114		6,264	6,975
6132	8-7/8" x 7-1/2" x 14" top flange		1.80	4.444		6,800	114		6,914	7,675
6201	Beam and purlin hangers, galvanized, 12 ga.									
6202	Purlin or joist size, 3" x 8"	1 Carp	1.70	4.706	C	975	121		1,096	1,275
6204	3" x 10"		1.70	4.706		1,100	121		1,221	1,400
6206	3" x 12"		1.65	4.848		1,300	125		1,425	1,625
6208	3" x 14"		1.65	4.848		1,525	125		1,650	1,875
6210	3" x 16"		1.65	4.848		1,750	125		1,875	2,125
6212	4" x 8"		1.65	4.848		975	125		1,100	1,275
6214	4" x 10"		1.65	4.848		1,150	125		1,275	1,450
6216	4" x 12"		1.60	5		1,225	129		1,354	1,575
6218	4" x 14"		1.60	5		1,325	129		1,454	1,700
6220	4" x 16"		1.60	5		1,525	129		1,654	1,900
6224	6" x 10"		1.55	5.161		1,675	133		1,808	2,050
6226	6" x 12"		1.55	5.161		1,800	133		1,933	2,200
6228	6" x 14"		1.50	5.333		1,925	137		2,062	2,350
6230	6" x 16"		1.50	5.333		2,200	137		2,337	2,650
6300	Column bases									
6302	4 x 4, 16 ga.	1 Carp	1.80	4.444	C	1,350	114		1,464	1,700
6306	7 ga.		1.80	4.444		2,525	114		2,639	2,975
6314	6 x 6, 16 ga.		1.75	4.571		1,725	117		1,842	2,100
6318	7 ga.		1.75	4.571		3,550	117		3,667	4,100
6326	8 x 8, 7 ga.		1.65	4.848		5,900	125		6,025	6,700
6330	8 x 10, 7 ga.		1.65	4.848		6,300	125		6,425	7,150
6590	Joist hangers, heavy duty 12 ga., galvanized									
6592	2" x 4"	1 Carp	1.75	4.571	C	1,100	117		1,217	1,425
6594	2" x 6"		1.65	4.848		1,175	125		1,300	1,500
6595	2" x 6", 16 gauge		1.65	4.848		1,175	125		1,300	1,500
6596	2" x 8"		1.65	4.848		1,250	125		1,375	1,600
6597	2" x 8", 16 gauge		1.65	4.848		1,250	125		1,375	1,600
6598	2" x 10"		1.65	4.848		1,350	125		1,475	1,700

06 05 Common Work Results for Wood, Plastics and Composites

06 05 23 – Wood, Plastic, and Composite Fastenings

06 05 23.60 Timber Connectors

		Crew	Daily Output	Labor-Hours	Unit	Material	2007 Bare Costs Labor	Equipment	Total	Total Incl O&P
6600	2" x 12"	1 Carp	1.65	4.848	C	1,525	125		1,650	1,875
6622	(2) 2" x 6"		1.60	5		1,475	129		1,604	1,850
6624	(2) 2" x 8"		1.60	5		1,600	129		1,729	2,000
6626	(2) 2" x 10"		1.55	5.161		1,800	133		1,933	2,200
6628	(2) 2" x 12"		1.55	5.161		2,175	133		2,308	2,625
6890	Purlin hangers, painted									
6892	12 ga., 2" x 6"	1 Carp	1.80	4.444	C	1,350	114		1,464	1,675
6894	2" x 8"		1.80	4.444		1,425	114		1,539	1,775
6896	2" x 10"		1.80	4.444		1,475	114		1,589	1,800
6898	2" x 12"		1.75	4.571		1,575	117		1,692	1,950
6934	(2) 2" x 6"		1.70	4.706		1,375	121		1,496	1,700
6936	(2) 2" x 8"		1.70	4.706		1,500	121		1,621	1,850
6938	(2) 2" x 10"		1.70	4.706		1,650	121		1,771	2,025
6940	(2) 2" x 12"		1.65	4.848		1,775	125		1,900	2,150

06 05 23.70 Rough Hardware

0010	**ROUGH HARDWARE**									
0020	Minimum					.50%				
0200	Maximum					1.50%				
0210	In seismic or hurricane areas, up to					10%				

06 05 23.80 Metal Bracing

0010	**METAL BRACING**									
0302	Let-in, "T" shaped, 22 ga. galv. steel, studs at 16" O.C.	1 Carp	580	.014	L.F.	.53	.35		.88	1.18
0402	Studs at 24" O.C.		600	.013		.53	.34		.87	1.16
0502	16 ga. galv. steel straps, studs at 16" O.C.		600	.013		.80	.34		1.14	1.46
0602	Studs at 24" O.C.		620	.013		.80	.33		1.13	1.44

06 05 73 – Wood Treatment

06 05 73.10 Lumber Treatment

0011	**LUMBER TREATMENT**									
0400	Fire retardant, wet				M.B.F.	330			330	365
0500	KDAT					315			315	345
0700	Salt treated, water borne, .40 lb. retention					150			150	166
0800	Oil borne, 8 lb. retention					176			176	194
1000	Kiln dried lumber, 1" & 2" thick, softwoods					101			101	111
1100	Hardwoods					107			107	118
1500	For small size 1" stock, add					13.60			13.60	14.95
1700	For full size rough lumber, add					20%				

06 05 73.20 Plywood Treatment

0010	**PLYWOOD TREATMENT**									
0020	Fire retardant, 1/4" thick				M.S.F.	252			252	277
0030	3/8" thick					276			276	305
0050	1/2" thick					296			296	325
0070	5/8" thick					315			315	345
0100	3/4" thick					345			345	380
0200	For KDAT, add					75			75	83
0500	Salt treated water borne, .25 lb., wet, 1/4" thick					138			138	152
0530	3/8" thick					144			144	159
0550	1/2" thick					150			150	166
0570	5/8" thick					164			164	180
0600	3/4" thick					170			170	187
0800	For KDAT add					75			75	83
0900	For .40 lb., per C.F. retention, add					63			63	69

06 05 Common Work Results for Wood, Plastics and Composites

06 05 73 – Wood Treatment

06 05 73.20 Plywood Treatment	Crew	Daily Output	Labor-Hours	Unit	Material	2007 Bare Costs Labor	2007 Bare Costs Equipment	Total	Total Incl O&P
1000 For certification stamp, add				M.S.F.	37			37	40.50

06 11 Wood Framing

06 11 10 – Wood Framing

06 11 10.02 Blocking

		Crew	Daily Output	Labor-Hours	Unit	Material	Labor	Equipment	Total	Total Incl O&P
0010	**BLOCKING**									
1950	Miscellaneous, to wood construction									
2000	2" x 4"	1 Carp	250	.032	L.F.	.37	.82		1.19	1.81
2005	Pneumatic nailed		305	.026		.37	.67		1.04	1.55
2050	2" x 6"		222	.036		.58	.93		1.51	2.20
2055	Pneumatic nailed		271	.030		.58	.76		1.34	1.92
2100	2" x 8"		200	.040		.84	1.03		1.87	2.66
2105	Pneumatic nailed		244	.033		.84	.84		1.68	2.35
2150	2" x 10"		178	.045		1.19	1.16		2.35	3.27
2155	Pneumatic nailed		217	.037		1.19	.95		2.14	2.92
2200	2" x 12"		151	.053		1.43	1.36		2.79	3.88
2205	Pneumatic nailed		185	.043		1.43	1.11		2.54	3.46
2300	To steel construction									
2320	2" x 4"	1 Carp	208	.038	L.F.	.37	.99		1.36	2.09
2340	2" x 6"		180	.044		.58	1.14		1.72	2.57
2360	2" x 8"		158	.051		.84	1.30		2.14	3.13
2380	2" x 10"		136	.059		1.19	1.51		2.70	3.87
2400	2" x 12"		109	.073		1.43	1.89		3.32	4.77

06 11 10.04 Bracing Let-In, With 1" X 6" Boards, Studs 16" O.c.

		Crew	Daily Output	Labor-Hours	Unit	Material	Labor	Equipment	Total	Total Incl O&P
0012	**BRACING** Let-in, with 1" x 6" boards, studs @ 16" O.C.		150	.053		.65	1.37		2.02	3.05
0202	Studs @ 24" O.C.		230	.035		.65	.89		1.54	2.24

06 11 10.06 Bridging

		Crew	Daily Output	Labor-Hours	Unit	Material	Labor	Equipment	Total	Total Incl O&P
0012	**BRIDGING** Wood, for joists 16" O.C., 1" x 3"		130	.062		.34	1.58		1.92	3.05
0017	Pneumatic nailed		170	.047		.34	1.21		1.55	2.42
0102	2" x 3" bridging		130	.062		.45	1.58		2.03	3.17
0107	Pneumatic nailed		170	.047		.45	1.21		1.66	2.54
0302	Steel, galvanized, 18 ga., for 2" x 10" joists at 12" O.C.		130	.062		1.28	1.58		2.86	4.09
0402	24" O.C.		140	.057		1.55	1.47		3.02	4.20
0602	For 2" x 14" joists at 16" O.C.		130	.062		1.52	1.58		3.10	4.35
0902	Compression type, 16" O.C., 2" x 8" joists		200	.040		1.50	1.03		2.53	3.39
1002	2" x 12" joists		200	.040		1.50	1.03		2.53	3.39

06 11 10.10 Beam and Girder Framing

		Crew	Daily Output	Labor-Hours	Unit	Material	Labor	Equipment	Total	Total Incl O&P
0010	**BEAM AND GIRDER FRAMING** R061110-30									
1002	Single, 2" x 6"	2 Carp	700	.023	L.F.	.58	.59		1.17	1.63
1007	Pneumatic nailed		812	.020		.58	.51		1.09	1.49
1022	2" x 8"		650	.025		.84	.63		1.47	1.99
1027	Pneumatic nailed		754	.021		.84	.55		1.39	1.85
1042	2" x 10"		600	.027		1.19	.69		1.88	2.47
1047	Pneumatic nailed		696	.023		1.19	.59		1.78	2.31
1062	2" x 12"		550	.029		1.43	.75		2.18	2.84
1067	Pneumatic nailed		638	.025		1.43	.64		2.07	2.66
1082	2" x 14"		500	.032		2.18	.82		3	3.80
1087	Pneumatic nailed		580	.028		2.18	.71		2.89	3.60
1102	3" x 8"		550	.029		2.59	.75		3.34	4.12
1122	3" x 10"		500	.032		3.26	.82		4.08	4.99

06 11 Wood Framing

06 11 10 – Wood Framing

06 11 10.10 Beam and Girder Framing		Crew	Daily Output	Labor-Hours	Unit	Material	2007 Bare Costs Labor	Equipment	Total	Total Incl O&P
1142	3" x 12"	2 Carp	450	.036	L.F.	3.92	.91		4.83	5.85
1162	3" x 14"	↓	400	.040		5.70	1.03		6.73	8
1170	4" x 6"	F-3	1100	.036		2.78	.85	.66	4.29	5.20
1182	4" x 8"		1000	.040		3.74	.93	.72	5.39	6.50
1202	4" x 10"		950	.042		5.50	.98	.76	7.24	8.55
1222	4" x 12"		900	.044		5.10	1.04	.80	6.94	8.25
1242	4" x 14"	↓	850	.047		4.95	1.10	.85	6.90	8.25
2002	Double, 2" x 6"	2 Carp	625	.026		1.15	.66		1.81	2.39
2007	Pneumatic nailed		725	.022		1.15	.57		1.72	2.23
2022	2" x 8"		575	.028		1.68	.72		2.40	3.06
2027	Pneumatic nailed		667	.024		1.68	.62		2.30	2.90
2042	2" x 10"		550	.029		2.38	.75		3.13	3.89
2047	Pneumatic nailed		638	.025		2.38	.64		3.02	3.71
2062	2" x 12"		525	.030		2.86	.78		3.64	4.48
2067	Pneumatic nailed		610	.026		2.86	.67		3.53	4.29
2082	2" x 14"		475	.034		4.36	.87		5.23	6.25
2087	Pneumatic nailed		551	.029		4.36	.75		5.11	6.05
3002	Triple, 2" x 6"		550	.029		1.73	.75		2.48	3.17
3007	Pneumatic nailed		638	.025		1.73	.64		2.37	2.99
3022	2" x 8"		525	.030		2.52	.78		3.30	4.10
3027	Pneumatic nailed		609	.026		2.52	.68		3.20	3.92
3042	2" x 10"		500	.032		3.57	.82		4.39	5.35
3047	Pneumatic nailed		580	.028		3.57	.71		4.28	5.15
3062	2" x 12"		475	.034		4.29	.87		5.16	6.20
3067	Pneumatic nailed		551	.029		4.29	.75		5.04	6
3082	2" x 14"		450	.036		6.55	.91		7.46	8.75
3087	Pneumatic nailed	↓	522	.031	↓	6.55	.79		7.34	8.55
06 11 10.12 Ceiling Framing										
0010	**CEILING FRAMING**									
6002	Suspended, 2" x 3"	2 Carp	1000	.016	L.F.	.30	.41		.71	1.03
6052	2" x 4"		900	.018		.37	.46		.83	1.19
6102	2" x 6"		800	.020		.58	.51		1.09	1.50
6152	2" x 8"	↓	650	.025		.84	.63		1.47	1.99
06 11 10.14 Column Framing										
0010	**COLUMN FRAMING**									
0101	4" x 4"	2 Carp	390	.041	L.F.	1.67	1.05		2.72	3.63
0151	4" x 6"		275	.058		2.78	1.50		4.28	5.60
0201	4" x 8"		220	.073		3.74	1.87		5.61	7.30
0251	6" x 6"		215	.074		4.83	1.91		6.74	8.55
0301	6" x 8"		175	.091		7.60	2.35		9.95	12.35
0351	6" x 10"	↓	150	.107	↓	10.90	2.74		13.64	16.65
06 11 10.18 Joist Framing										
0010	**JOIST FRAMING**									
2002	Joists, 2" x 4"	2 Carp	1250	.013	L.F.	.37	.33		.70	.97
2007	Pneumatic nailed		1438	.011		.37	.29		.66	.90
2100	2" x 6"		1250	.013		.58	.33		.91	1.19
2105	Pneumatic nailed		1438	.011		.58	.29		.87	1.12
2152	2" x 8"		1100	.015		.84	.37		1.21	1.55
2157	Pneumatic nailed		1265	.013		.84	.33		1.17	1.47
2202	2" x 10"		900	.018		1.19	.46		1.65	2.09
2207	Pneumatic nailed		1035	.015		1.19	.40		1.59	1.98
2252	2" x 12"	↓	875	.018	↓	1.43	.47		1.90	2.37

06 11 Wood Framing

06 11 10 – Wood Framing

06 11 10.18 Joist Framing

		Crew	Daily Output	Labor-Hours	Unit	Material	2007 Bare Costs Labor	Equipment	Total	Total Incl O&P
2257	Pneumatic nailed	2 Carp	1006	.016	L.F.	1.43	.41		1.84	2.26
2302	2" x 14"		770	.021		2.18	.53		2.71	3.31
2307	Pneumatic nailed		886	.018		2.18	.46		2.64	3.19
2352	3" x 6"		925	.017		1.92	.44		2.36	2.86
2402	3" x 10"		780	.021		3.26	.53		3.79	4.48
2452	3" x 12"		600	.027		3.92	.69		4.61	5.45
2502	4" x 6"		800	.020		2.78	.51		3.29	3.93
2552	4" x 10"		600	.027		5.50	.69		6.19	7.20
2602	4" x 12"		450	.036		5.10	.91		6.01	7.20
2607	Sister joist, 2" x 6"		800	.020		.58	.51		1.09	1.50
2608	Pneumatic nailed		960	.017		.58	.43		1.01	1.36
3000	Composite wood joist 9-1/2" deep		.90	17.778	M.L.F.	1,850	455		2,305	2,825
3010	11-1/2" deep		.88	18.182		1,975	465		2,440	2,975
3020	14" deep		.82	19.512		2,325	500		2,825	3,400
3030	16" deep		.78	20.513		2,600	525		3,125	3,750
4000	Open web joist 12" deep		.88	18.182		1,825	465		2,290	2,800
4010	14" deep		.82	19.512		2,125	500		2,625	3,175
4020	16" deep		.78	20.513		2,200	525		2,725	3,325
4030	18" deep		.74	21.622		2,250	555		2,805	3,425
6000	Composite rim joist, 1-1/4" x 9-1/2"		90	.178		2,475	4.57		2,479.57	2,700
6010	1-1/4" x 11-1/2"		.88	18.182		2,725	465		3,190	3,800
6020	1-1/4" x 14-1/2"		.82	19.512		3,175	500		3,675	4,350
6030	1-1/4" x 16-1/2"		.78	20.513		3,200	525		3,725	4,400

06 11 10.24 Miscellaneous Framing

		Crew	Daily Output	Labor-Hours	Unit	Material	Labor	Equipment	Total	Total Incl O&P
0010	**MISCELLANEOUS FRAMING**									
2002	Firestops, 2" x 4"	2 Carp	780	.021	L.F.	.37	.53		.90	1.30
2007	Pneumatic nailed		952	.017		.37	.43		.80	1.14
2102	2" x 6"		600	.027		.58	.69		1.27	1.79
2107	Pneumatic nailed		732	.022		.58	.56		1.14	1.58
5002	Nailers, treated, wood construction, 2" x 4"		800	.020		.52	.51		1.03	1.45
5007	Pneumatic nailed		960	.017		.52	.43		.95	1.31
5102	2" x 6"		750	.021		.83	.55		1.38	1.84
5107	Pneumatic nailed		900	.018		.83	.46		1.29	1.69
5122	2" x 8"		700	.023		1.15	.59		1.74	2.26
5127	Pneumatic nailed		840	.019		1.15	.49		1.64	2.09
5202	Steel construction, 2" x 4"		750	.021		.52	.55		1.07	1.51
5222	2" x 6"		700	.023		.83	.59		1.42	1.91
5242	2" x 8"		650	.025		1.15	.63		1.78	2.33
7002	Rough bucks, treated, for doors or windows, 2" x 6"		400	.040		.83	1.03		1.86	2.65
7007	Pneumatic nailed		480	.033		.83	.86		1.69	2.36
7102	2" x 8"		380	.042		1.15	1.08		2.23	3.10
7107	Pneumatic nailed		456	.035		1.15	.90		2.05	2.79
8001	Stair stringers, 2" x 10"		130	.123		1.19	3.16		4.35	6.65
8101	2" x 12"		130	.123		1.43	3.16		4.59	6.90
8151	3" x 10"		125	.128		3.26	3.29		6.55	9.20
8201	3" x 12"		125	.128		3.92	3.29		7.21	9.90
8870	Composite LSL, 1-1/4" x 11-1/2"		130	.123		2.73	3.16		5.89	8.35
8880	1-1/4" x 14-1/2"		130	.123		3.17	3.16		6.33	8.85

06 11 10.26 Partitions

		Crew	Daily Output	Labor-Hours	Unit	Material	Labor	Equipment	Total	Total Incl O&P
0010	**PARTITIONS**									
0020	Single bottom and double top plate, no waste, std. & better lumber									
0182	2" x 4" studs, 8' high, studs 12" O.C.	2 Carp	80	.200	L.F.	4.53	5.15		9.68	13.70

06 11 Wood Framing

06 11 10 - Wood Framing

06 11 10.26 Partitions

		Crew	Daily Output	Labor-Hours	Unit	Material	2007 Bare Costs Labor	Equipment	Total	Total Incl O&P
0187	12" O.C., pneumatic nailed	2 Carp	96	.167	L.F.	4.53	4.28		8.81	12.25
0202	16" O.C.		100	.160		3.71	4.11		7.82	11.10
0207	16" O.C., pneumatic nailed		120	.133		3.71	3.43		7.14	9.90
0302	24" O.C.		125	.128		2.88	3.29		6.17	8.75
0307	24" O.C., pneumatic nailed		150	.107		2.88	2.74		5.62	7.80
0382	10' high, studs 12" O.C.		80	.200		5.35	5.15		10.50	14.60
0387	12" O.C., pneumatic nailed		96	.167		5.35	4.28		9.63	13.15
0402	16" O.C.		100	.160		4.32	4.11		8.43	11.75
0407	16" O.C., pneumatic nailed		120	.133		4.32	3.43		7.75	10.55
0502	24" O.C.		125	.128		3.29	3.29		6.58	9.20
0507	24" O.C., pneumatic nailed		150	.107		3.29	2.74		6.03	8.25
0582	12' high, studs 12" O.C.		65	.246		6.20	6.35		12.55	17.55
0587	12" O.C., pneumatic nailed		78	.205		6.20	5.25		11.45	15.75
0602	16" O.C.		80	.200		4.94	5.15		10.09	14.15
0607	16" O.C., pneumatic nailed		96	.167		4.94	4.28		9.22	12.70
0700	24" O.C.		100	.160		3.37	4.11		7.48	10.70
0705	24" O.C., pneumatic nailed		120	.133		3.37	3.43		6.80	9.50
0782	2" x 6" studs, 8' high, studs 12" O.C.		70	.229		6.95	5.85		12.80	17.60
0787	12" O.C., pneumatic nailed		84	.190		6.95	4.90		11.85	15.95
0802	16" O.C.		90	.178		5.70	4.57		10.27	14
0807	16" O.C., pneumatic nailed		108	.148		5.70	3.81		9.51	12.70
0902	24" O.C.		115	.139		4.43	3.58		8.01	10.90
0907	24" O.C., pneumatic nailed		138	.116		4.43	2.98		7.41	9.90
0982	10' high, studs 12" O.C.		70	.229		8.20	5.85		14.05	19
0987	12" O.C., pneumatic nailed		84	.190		8.20	4.90		13.10	17.35
1002	16" O.C.		90	.178		6.65	4.57		11.22	15.05
1007	16" O.C., pneumatic nailed		108	.148		6.65	3.81		10.46	13.75
1102	24" O.C.		115	.139		5.05	3.58		8.63	11.60
1107	24" O.C., pneumatic nailed		138	.116		5.05	2.98		8.03	10.60
1182	12' high, studs 12" O.C.		55	.291		9.50	7.50		17	23
1187	12" O.C., pneumatic nailed		66	.242		9.50	6.25		15.75	21
1202	16" O.C.		70	.229		7.60	5.85		13.45	18.30
1207	16" O.C., pneumatic nailed		84	.190		7.60	4.90		12.50	16.65
1302	24" O.C.		90	.178		5.70	4.57		10.27	14
1307	24" O.C., pneumatic nailed		108	.148		5.70	3.81		9.51	12.70
1402	For horizontal blocking, 2" x 4", add		600	.027		.41	.69		1.10	1.61
1502	2" x 6", add		600	.027		.63	.69		1.32	1.86
1600	For openings, add		250	.064			1.64		1.64	2.79
1702	Headers for above openings, material only, add				B.F.	.69			.69	.76

06 11 10.28 Porch or Deck Framing

		Crew	Daily Output	Labor-Hours	Unit	Material	Labor	Equipment	Total	Total Incl O&P
0010	**PORCH OR DECK FRAMING**									
0100	Treated lumber, posts or columns, 4" x 4"	2 Carp	390	.041	L.F.	1.41	1.05		2.46	3.34
0110	4" x 6"		275	.058		3.02	1.50		4.52	5.85
0120	4" x 8"		220	.073		3.33	1.87		5.20	6.85
0130	Girder, single, 4" x 4"		675	.024		1.41	.61		2.02	2.58
0140	4" x 6"		600	.027		3.02	.69		3.71	4.48
0150	4" x 8"		525	.030		3.33	.78		4.11	4.99
0160	Double, 2" x 4"		625	.026		1.07	.66		1.73	2.30
0170	2" x 6"		600	.027		1.69	.69		2.38	3.02
0180	2" x 8"		575	.028		2.34	.72		3.06	3.79
0190	2" x 10"		550	.029		3.17	.75		3.92	4.76
0200	2" x 12"		525	.030		3.75	.78		4.53	5.45

06 11 Wood Framing

06 11 10 − Wood Framing

06 11 10.28 Porch or Deck Framing		Crew	Daily Output	Labor-Hours	Unit	Material	2007 Bare Costs Labor	Equipment	Total	Total Incl O&P
0210	Triple, 2" x 4"	2 Carp	575	.028	L.F.	1.60	.72		2.32	2.97
0220	2" x 6"		550	.029		2.54	.75		3.29	4.06
0230	2" x 8"		525	.030		3.51	.78		4.29	5.20
0240	2" x 10"		500	.032		4.76	.82		5.58	6.65
0250	2" x 12"		475	.034		5.65	.87		6.52	7.65
0260	Ledger, bolted 4' O.C., 2" x 4"		400	.040		.63	1.03		1.66	2.44
0270	2" x 6"		395	.041		.94	1.04		1.98	2.80
0280	2" x 8"		390	.041		1.26	1.05		2.31	3.18
0300	2" x 12"		380	.042		1.96	1.08		3.04	3.99
0310	Joists, 2" x 4"		1250	.013		.53	.33		.86	1.15
0320	2" x 6"		1250	.013		.84	.33		1.17	1.49
0330	2" x 8"		1100	.015		1.17	.37		1.54	1.91
0340	2" x 10"		900	.018		1.58	.46		2.04	2.52
0350	2" x 12"		875	.018		1.90	.47		2.37	2.89
0360	Railings and trim, 1" x 4"	1 Carp	300	.027		.61	.69		1.30	1.83
0370	2" x 2"		300	.027		.30	.69		.99	1.49
0380	2" x 4"		300	.027		.52	.69		1.21	1.74
0390	2" x 6"		300	.027		.83	.69		1.52	2.07
0400	Decking, 1" x 4"		275	.029	S.F.	1.91	.75		2.66	3.37
0410	2" x 4"		300	.027		1.79	.69		2.48	3.13
0420	2" x 6"		320	.025		1.81	.64		2.45	3.08
0430	5/4" x 6"		320	.025		1.29	.64		1.93	2.51
0440	Balusters, square, 2" x 2"	2 Carp	660	.024	L.F.	.30	.62		.92	1.39
0450	Turned, 2" x 2"		420	.038		.40	.98		1.38	2.10
0460	Stair stringer, 2" x 10"		130	.123		1.58	3.16		4.74	7.10
0470	2" x 12"		130	.123		1.90	3.16		5.06	7.45
0480	Stair treads, 1" x 4"		140	.114		.64	2.94		3.58	5.70
0490	2" x 4"		140	.114		.53	2.94		3.47	5.55
0500	2" x 6"		160	.100		.80	2.57		3.37	5.25
0510	5/4" x 6"		160	.100		.60	2.57		3.17	5
0520	Turned handrail post, 4" x 4"		64	.250	Ea.	13.80	6.45		20.25	26
0530	Lattice panel, 4' x 8'		1600	.010	S.F.	.74	.26		1	1.25
0540	Cedar, posts or columns, 4" x 4"		390	.041	L.F.	3.02	1.05		4.07	5.10
0550	4" x 6"		275	.058		4.59	1.50		6.09	7.60
0560	4" x 8"		220	.073		6.90	1.87		8.77	10.70
0800	Decking, 1" x 4"		550	.029		1.43	.75		2.18	2.85
0810	2" x 4"		600	.027		2.82	.69		3.51	4.26
0820	2" x 6"		640	.025		4.20	.64		4.84	5.70
0830	5/4" x 6"		640	.025		3.29	.64		3.93	4.71
0840	Railings and trim, 1" x 4"		600	.027		1.43	.69		2.12	2.74
0860	2" x 4"		600	.027		2.82	.69		3.51	4.26
0870	2" x 6"		600	.027		4.20	.69		4.89	5.80
0920	Stair treads, 1" x 4"		140	.114		1.43	2.94		4.37	6.55
0930	2" x 4"		140	.114		2.82	2.94		5.76	8.10
0940	2" x 6"		160	.100		4.20	2.57		6.77	9
0950	5/4" x 6"		160	.100		3.29	2.57		5.86	8
0980	Redwood, posts or columns, 4" x 4"		390	.041		6.50	1.05		7.55	8.95
0990	4" x 6"		275	.058		10.85	1.50		12.35	14.50
1000	4" x 8"		220	.073		19	1.87		20.87	24
1240	Redwood decking, 1" x 4"	1 Carp	275	.029	S.F.	5.10	.75		5.85	6.85
1260	2" x 6"		340	.024		12.70	.60		13.30	15
1270	5/4" x 6"		320	.025		8.20	.64		8.84	10.10
1280	Railings and trim, 1" x 4"	2 Carp	600	.027	L.F.	1.50	.69		2.19	2.81

06 11 Wood Framing

06 11 10 – Wood Framing

06 11 10.28 Porch or Deck Framing

		Crew	Daily Output	Labor-Hours	Unit	Material	2007 Bare Costs Labor	Equipment	Total	Total Incl O&P
1310	2" x 6"	2 Carp	600	.027	L.F.	5.85	.69		6.54	7.55
1420	Alternative decking, wood / plastic composite, 5/4" x 6"		640	.025		2.25	.64		2.89	3.56
1430	Vinyl, 1-1/2" x 5-1/2"		640	.025		3.73	.64		4.37	5.20
1440	1" x 4" square edge fir		550	.029		1.37	.75		2.12	2.77
1450	1" x 4" tongue and groove fir		450	.036		1.31	.91		2.22	2.99
1460	1" x 4" mahogany		550	.029		1.31	.75		2.06	2.71
1470	Accessories, joist hangers, 2" x 4"	1 Carp	160	.050	Ea.	.71	1.29		2	2.96
1480	2" x 6" through 2" x 12"	"	150	.053		.81	1.37		2.18	3.22
1530	Post footing, incl excav, backfill, tube form & concrete, 4' deep, 8" dia	F-7	12	2.667		11.45	59		70.45	113
1540	10" diameter		11	2.909		16.50	64.50		81	128
1550	12" diameter		10	3.200		22	71		93	145

06 11 10.30 Roof Framing

		Crew	Daily Output	Labor-Hours	Unit	Material	2007 Bare Costs Labor	Equipment	Total	Total Incl O&P
0010	**ROOF FRAMING**									
2001	Fascia boards, 2" x 8"	2 Carp	225	.071	L.F.	.84	1.83		2.67	4.02
2101	2" x 10"		180	.089		1.19	2.28		3.47	5.20
5002	Rafters, to 4 in 12 pitch, 2" x 6", ordinary		1000	.016		.58	.41		.99	1.33
5021	On steep roofs		800	.020		.58	.51		1.09	1.50
5041	On dormers or complex roofs		590	.027		.58	.70		1.28	1.81
5062	2" x 8", ordinary		950	.017		.84	.43		1.27	1.65
5081	On steep roofs		750	.021		.84	.55		1.39	1.85
5101	On dormers or complex roofs		540	.030		.84	.76		1.60	2.21
5122	2" x 10", ordinary		630	.025		1.19	.65		1.84	2.42
5141	On steep roofs		495	.032		1.19	.83		2.02	2.72
5161	On dormers or complex roofs		425	.038		1.19	.97		2.16	2.95
5182	2" x 12", ordinary		575	.028		1.43	.72		2.15	2.78
5201	On steep roofs		455	.035		1.43	.90		2.33	3.10
5221	On dormers or complex roofs		395	.041		1.43	1.04		2.47	3.34
5250	Composite rafter, 9-1/2" deep		575	.028		1.86	.72		2.58	3.26
5260	11-1/2" deep		575	.028		1.98	.72		2.70	3.38
5301	Hip and valley rafters, 2" x 6", ordinary		760	.021		.58	.54		1.12	1.55
5321	On steep roofs		585	.027		.58	.70		1.28	1.82
5341	On dormers or complex roofs		510	.031		.58	.81		1.39	2
5361	2" x 8", ordinary		720	.022		.84	.57		1.41	1.89
5381	On steep roofs		545	.029		.84	.75		1.59	2.20
5401	On dormers or complex roofs		470	.034		.84	.87		1.71	2.40
5421	2" x 10", ordinary		570	.028		1.19	.72		1.91	2.53
5441	On steep roofs		440	.036		1.19	.93		2.12	2.90
5461	On dormers or complex roofs		380	.042		1.19	1.08		2.27	3.15
5470										
5481	Hip and valley rafters, 2" x 12", ordinary	2 Carp	525	.030	L.F.	1.43	.78		2.21	2.90
5501	On steep roofs		410	.039		1.43	1		2.43	3.27
5521	On dormers or complex roofs		355	.045		1.43	1.16		2.59	3.54
5541	Hip and valley jacks, 2" x 6", ordinary		600	.027		.58	.69		1.27	1.79
5561	On steep roofs		475	.034		.58	.87		1.45	2.10
5581	On dormers or complex roofs		410	.039		.58	1		1.58	2.33
5601	2" x 8", ordinary		490	.033		.84	.84		1.68	2.34
5621	On steep roofs		385	.042		.84	1.07		1.91	2.73
5641	On dormers or complex roofs		335	.048		.84	1.23		2.07	3
5661	2" x 10", ordinary		450	.036		1.19	.91		2.10	2.86
5681	On steep roofs		350	.046		1.19	1.17		2.36	3.30
5701	On dormers or complex roofs		305	.052		1.19	1.35		2.54	3.60
5721	2" x 12", ordinary		375	.043		1.43	1.10		2.53	3.43

06 11 Wood Framing

06 11 10 – Wood Framing

06 11 10.30 Roof Framing

		Crew	Daily Output	Labor-Hours	Unit	Material	2007 Bare Costs Labor	Equipment	Total	Total Incl O&P
5741	On steep roofs	2 Carp	295	.054	L.F.	1.43	1.39		2.82	3.93
5762	On dormers or complex roofs		255	.063		1.43	1.61		3.04	4.31
5781	Rafter tie, 1" x 4", #3		800	.020		.39	.51		.90	1.30
5791	2" x 4", #3		800	.020		.37	.51		.88	1.28
5801	Ridge board, #2 or better, 1" x 6"		600	.027		.70	.69		1.39	1.93
5821	1" x 8"		550	.029		.89	.75		1.64	2.24
5841	1" x 10"		500	.032		1.30	.82		2.12	2.83
5861	2" x 6"		500	.032		.58	.82		1.40	2.03
5881	2" x 8"		450	.036		.84	.91		1.75	2.47
5901	2" x 10"		400	.040		1.19	1.03		2.22	3.05
5921	Roof cants, split, 4" x 4"		650	.025		1.67	.63		2.30	2.91
5941	6" x 6"		600	.027		4.83	.69		5.52	6.45
5961	Roof curbs, untreated, 2" x 6"		520	.031		.58	.79		1.37	1.97
5981	2" x 12"		400	.040		1.43	1.03		2.46	3.31
6001	Sister rafters, 2" x 6"		800	.020		.58	.51		1.09	1.50
6021	2" x 8"		640	.025		.84	.64		1.48	2.01
6041	2" x 10"		535	.030		1.19	.77		1.96	2.61
6061	2" x 12"		455	.035		1.43	.90		2.33	3.10

06 11 10.32 Sill and Ledger Framing

		Crew	Daily Output	Labor-Hours	Unit	Material	Labor	Equipment	Total	Total Incl O&P
0010	**SILL AND LEDGER FRAMING**									
2002	Ledgers, nailed, 2" x 4"	2 Carp	755	.021	L.F.	.37	.54		.91	1.33
2052	2" x 6"		600	.027		.58	.69		1.27	1.79
2102	Bolted, not including bolts, 3" x 6"		325	.049		1.92	1.27		3.19	4.26
2152	3" x 12"		233	.069		3.92	1.76		5.68	7.30
2602	Mud sills, redwood, construction grade, 2" x 4"		895	.018		2.27	.46		2.73	3.27
2622	2" x 6"		780	.021		3.40	.53		3.93	4.63
4002	Sills, 2" x 4"		600	.027		.37	.69		1.06	1.57
4052	2" x 6"		550	.029		.58	.75		1.33	1.90
4082	2" x 8"		500	.032		.84	.82		1.66	2.32
4101	2" x 10"		450	.036		1.19	.91		2.10	2.86
4121	2" x 12"		400	.040		1.43	1.03		2.46	3.31
4202	Treated, 2" x 4"		550	.029		.52	.75		1.27	1.85
4222	2" x 6"		500	.032		.83	.82		1.65	2.31
4242	2" x 8"		450	.036		1.15	.91		2.06	2.81
4261	2" x 10"		400	.040		1.56	1.03		2.59	3.46
4281	2" x 12"		350	.046		1.85	1.17		3.02	4.02
4402	4" x 4"		450	.036		1.39	.91		2.30	3.08
4422	4" x 6"		350	.046		2.99	1.17		4.16	5.30
4462	4" x 8"		300	.053		3.29	1.37		4.66	5.95
4480	4" x 10"		260	.062		4.10	1.58		5.68	7.20

06 11 10.34 Sleepers

		Crew	Daily Output	Labor-Hours	Unit	Material	Labor	Equipment	Total	Total Incl O&P
0010	**SLEEPERS**									
0100	On concrete, treated, 1" x 2"	2 Carp	2350	.007	L.F.	.20	.18		.38	.52
0150	1" x 3"		2000	.008		.31	.21		.52	.69
0200	2" x 4"		1500	.011		.52	.27		.79	1.05
0250	2" x 6"		1300	.012		.83	.32		1.15	1.45

06 11 10.36 Soffit and Canopy Framing

		Crew	Daily Output	Labor-Hours	Unit	Material	Labor	Equipment	Total	Total Incl O&P
0010	**SOFFIT AND CANOPY FRAMING**									
1002	Canopy or soffit framing, 1" x 4"	2 Carp	900	.018	L.F.	.46	.46		.92	1.29
1021	1" x 6"		850	.019		.70	.48		1.18	1.59
1042	1" x 8"		750	.021		.89	.55		1.44	1.90
1102	2" x 4"		620	.026		.37	.66		1.03	1.54

06 11 Wood Framing

06 11 10 – Wood Framing

06 11 10.36 Soffit and Canopy Framing

		Crew	Daily Output	Labor-Hours	Unit	Material	2007 Bare Costs Labor	2007 Bare Costs Equipment	Total	Total Incl O&P
1121	2" x 6"	2 Carp	560	.029	L.F.	.58	.73		1.31	1.88
1142	2" x 8"		500	.032		.84	.82		1.66	2.32
1202	3" x 4"		500	.032		1.09	.82		1.91	2.60
1221	3" x 6"		400	.040		1.92	1.03		2.95	3.85
1242	3" x 10"		300	.053		3.26	1.37		4.63	5.90
1250										

06 11 10.38 Treated Lumber Framing Material

		Crew	Daily Output	Labor-Hours	Unit	Material	Labor	Equipment	Total	Total Incl O&P
0010	**TREATED LUMBER FRAMING MATERIAL**									
0100	2" x 4"				M.B.F.	785			785	865
0110	2" x 6"					830			830	915
0120	2" x 8"					860			860	950
0130	2" x 10"					935			935	1,025
0140	2" x 12"					925			925	1,025
0200	4" x 4"					1,050			1,050	1,150
0210	4" x 6"					1,500			1,500	1,650
0220	4" x 8"					1,225			1,225	1,350

06 11 10.40 Wall Framing

		Crew	Daily Output	Labor-Hours	Unit	Material	Labor	Equipment	Total	Total Incl O&P
0010	**WALL FRAMING** R061110-30									
0100	Door buck, studs, header, access, 8' H, 2"x4" wall, 3' W	1 Carp	32	.250	Ea.	15.70	6.45		22.15	28
0110	4' wide		32	.250		16.85	6.45		23.30	29.50
0120	5' wide		32	.250		21	6.45		27.45	34
0130	6' wide		32	.250		22.50	6.45		28.95	35.50
0140	8' wide		30	.267		31.50	6.85		38.35	46.50
0150	10' wide		30	.267		41.50	6.85		48.35	57
0160	12' wide		30	.267		65.50	6.85		72.35	83.50
0170	2" x 6" wall, 3' wide		32	.250		22	6.45		28.45	35.50
0180	4' wide		32	.250		23.50	6.45		29.95	36.50
0190	5' wide		32	.250		27	6.45		33.45	41
0200	6' wide		32	.250		29	6.45		35.45	43
0210	8' wide		30	.267		38	6.85		44.85	53.50
0220	10' wide		30	.267		47.50	6.85		54.35	64
0230	12' wide		30	.267		72	6.85		78.85	90.50
0240	Window buck, studs, header & access, 8' high 2" x 4" wall, 2' wide		24	.333		16.65	8.55		25.20	33
0250	3' wide		24	.333		19.45	8.55		28	36
0260	4' wide		24	.333		21.50	8.55		30.05	38
0270	5' wide		24	.333		25.50	8.55		34.05	42.50
0280	6' wide		24	.333		28	8.55		36.55	45.50
0300	8' wide		22	.364		39.50	9.35		48.85	59.50
0310	10' wide		22	.364		50.50	9.35		59.85	71.50
0320	12' wide		22	.364		76.50	9.35		85.85	100
0330	2" x 6" wall, 2' wide		24	.333		24.50	8.55		33.05	41.50
0340	3' wide		24	.333		28	8.55		36.55	45
0350	4' wide		24	.333		30	8.55		38.55	47.50
0360	5' wide		24	.333		34	8.55		42.55	52
0370	6' wide		24	.333		37.50	8.55		46.05	56
0380	7' wide		24	.333		46	8.55		54.55	65
0390	8' wide		22	.364		50	9.35		59.35	71
0400	10' wide		22	.364		61.50	9.35		70.85	83.50
0410	12' wide		22	.364		89	9.35		98.35	114
2002	Headers over openings, 2" x 6"	2 Carp	360	.044	L.F.	.58	1.14		1.72	2.57
2007	2" x 6", pneumatic nailed		432	.037		.58	.95		1.53	2.24
2052	2" x 8"		340	.047		.84	1.21		2.05	2.97

06 11 Wood Framing

06 11 10 – Wood Framing

06 11 10.40 Wall Framing

		Crew	Daily Output	Labor-Hours	Unit	Material	2007 Bare Costs Labor	Equipment	Total	Total Incl O&P
2057	2" x 8", pneumatic nailed	2 Carp	408	.039	L.F.	.84	1.01		1.85	2.63
2101	2" x 10"		320	.050		1.19	1.29		2.48	3.49
2106	2" x 10", pneumatic nailed		384	.042		1.19	1.07		2.26	3.13
2152	2" x 12"		300	.053		1.43	1.37		2.80	3.90
2157	2" x 12", pneumatic nailed		360	.044		1.43	1.14		2.57	3.51
2191	4" x 10"		240	.067		5.50	1.71		7.21	8.95
2196	4" x 10", pneumatic nailed		288	.056		5.50	1.43		6.93	8.45
2202	4" x 12"		190	.084		5.10	2.16		7.26	9.30
2207	4" x 12", pneumatic nailed		228	.070		5.10	1.80		6.90	8.70
2241	6" x 10"		165	.097		10.90	2.49		13.39	16.25
2246	6" x 10", pneumatic nailed		198	.081		10.90	2.08		12.98	15.50
2251	6" x 12"		140	.114		9.15	2.94		12.09	15.10
2256	6" x 12", pneumatic nailed		168	.095		9.15	2.45		11.60	14.25
3000	Radius, 2" x 6"		270	.059		.90	1.52		2.42	3.57
3010	2" x 6", pneumatic nailed		324	.049		.90	1.27		2.17	3.14
3020	2" x 8"		255	.063		1.31	1.61		2.92	4.18
3030	2" x 8", pneumatic nailed		296	.054		1.31	1.39		2.70	3.80
3040	2" x 10"		240	.067		1.86	1.71		3.57	4.96
3050	2" x 10", pneumatic nailed		285	.056		1.86	1.44		3.30	4.50
3060	2" x 12"		225	.071		2.23	1.83		4.06	5.55
3070	2" x 12", pneumatic nailed		270	.059		2.23	1.52		3.75	5.05
5002	Plates, untreated, 2" x 3"		850	.019		.30	.48		.78	1.15
5007	2" x 3", pneumatic nailed		1020	.016		.30	.40		.70	1.01
5022	2" x 4"		800	.020		.37	.51		.88	1.28
5027	2" x 4", pneumatic nailed		960	.017		.37	.43		.80	1.14
5040	2" x 6"		750	.021		.58	.55		1.13	1.56
5045	2" x 6", pneumatic nailed		900	.018		.58	.46		1.04	1.41
5061	Treated, 2" x 3"		850	.019		.42	.48		.90	1.28
5066	2" x 3", treated, pneumatic nailed		1020	.016		.42	.40		.82	1.14
5081	2" x 4"		800	.020		.52	.51		1.03	1.45
5086	2" x 4", treated, pneumatic nailed		960	.017		.52	.43		.95	1.31
5101	2" x 6"		750	.021		.83	.55		1.38	1.84
5106	2" x 6", treated, pneumatic nailed		900	.018		.83	.46		1.29	1.69
5122	Studs, 8' high wall, 2" x 3"		1200	.013		.30	.34		.64	.91
5127	2" x 3", pneumatic nailed		1440	.011		.30	.29		.59	.81
5142	2" x 4"		1100	.015		.37	.37		.74	1.04
5147	2" x 4", pneumatic nailed		1320	.012		.37	.31		.68	.94
5162	2" x 6"		1000	.016		.58	.41		.99	1.33
5167	2" x 6", pneumatic nailed		1200	.013		.58	.34		.92	1.21
5182	3" x 4"		800	.020		1.09	.51		1.60	2.07
5187	3" x 4", pneumatic nailed		960	.017		1.09	.43		1.52	1.93
5201	Installed on second story, 2" x 3"		1170	.014		.30	.35		.65	.93
5206	2" x 3", pneumatic nailed		1200	.013		.30	.34		.64	.91
5221	2" x 4"		1015	.016		.37	.41		.78	1.10
5226	2" x 4", pneumatic nailed		1080	.015		.37	.38		.75	1.06
5241	2" x 6"		890	.018		.58	.46		1.04	1.41
5246	2" x 6", pneumatic nailed		1020	.016		.58	.40		.98	1.31
5261	3" x 4"		800	.020		1.09	.51		1.60	2.07
5266	3" x 4", pneumatic nailed		960	.017		1.09	.43		1.52	1.93
5281	Installed on dormer or gable, 2" x 3"		1045	.015		.30	.39		.69	1
5286	2" x 3", pneumatic nailed		1254	.013		.30	.33		.63	.89
5301	2" x 4"		905	.018		.37	.45		.82	1.18
5306	2" x 4", pneumatic nailed		1086	.015		.37	.38		.75	1.05

06 11 Wood Framing

06 11 10 – Wood Framing

06 11 10.40 Wall Framing		Crew	Daily Output	Labor-Hours	Unit	Material	2007 Bare Costs Labor	Equipment	Total	Total Incl O&P
5321	2" x 6"	2 Carp	800	.020	L.F.	.58	.51		1.09	1.50
5326	2" x 6", pneumatic nailed		960	.017		.58	.43		1.01	1.36
5341	3" x 4"		700	.023		1.09	.59		1.68	2.20
5346	3" x 4", pneumatic nailed		840	.019		1.09	.49		1.58	2.03
5361	6' high wall, 2" x 3"		970	.016		.30	.42		.72	1.05
5366	2" x 3", pneumatic nailed		1164	.014		.30	.35		.65	.93
5381	2" x 4"		850	.019		.37	.48		.85	1.23
5386	2" x 4", pneumatic nailed		1020	.016		.37	.40		.77	1.09
5401	2" x 6"		740	.022		.58	.56		1.14	1.57
5406	2" x 6", pneumatic nailed		888	.018		.58	.46		1.04	1.42
5421	3" x 4"		600	.027		1.09	.69		1.78	2.36
5426	3" x 4", pneumatic nailed		720	.022		1.09	.57		1.66	2.17
5441	Installed on second story, 2" x 3"		950	.017		.30	.43		.73	1.06
5446	2" x 3", pneumatic nailed		1140	.014		.30	.36		.66	.94
5461	2" x 4"		810	.020		.37	.51		.88	1.27
5466	2" x 4", pneumatic nailed		972	.016		.37	.42		.79	1.13
5481	2" x 6"		700	.023		.58	.59		1.17	1.63
5486	2" x 6", pneumatic nailed		840	.019		.58	.49		1.07	1.46
5501	3" x 4"		550	.029		1.09	.75		1.84	2.47
5506	3" x 4", pneumatic nailed		660	.024		1.09	.62		1.71	2.26
5521	Installed on dormer or gable, 2" x 3"		850	.019		.30	.48		.78	1.15
5526	2" x 3", pneumatic nailed		1020	.016		.30	.40		.70	1.01
5541	2" x 4"		720	.022		.37	.57		.94	1.38
5546	2" x 4", pneumatic nailed		864	.019		.37	.48		.85	1.22
5561	2" x 6"		620	.026		.58	.66		1.24	1.76
5566	2" x 6", pneumatic nailed		744	.022		.58	.55		1.13	1.57
5581	3" x 4"		480	.033		1.09	.86		1.95	2.65
5586	3" x 4", pneumatic nailed		576	.028		1.09	.71		1.80	2.41
5601	3' high wall, 2" x 3"		740	.022		.30	.56		.86	1.27
5606	2" x 3", pneumatic nailed		888	.018		.30	.46		.76	1.12
5621	2" x 4"		640	.025		.37	.64		1.01	1.50
5626	2" x 4", pneumatic nailed		768	.021		.37	.54		.91	1.32
5641	2" x 6"		550	.029		.58	.75		1.33	1.90
5646	2" x 6", pneumatic nailed		660	.024		.58	.62		1.20	1.69
5661	3" x 4"		440	.036		1.09	.93		2.02	2.79
5666	3" x 4", pneumatic nailed		528	.030		1.09	.78		1.87	2.52
5681	Installed on second story, 2" x 3"		700	.023		.30	.59		.89	1.33
5686	2" x 3", pneumatic nailed		840	.019		.30	.49		.79	1.16
5701	2" x 4"		610	.026		.37	.67		1.04	1.55
5706	2" x 4", pneumatic nailed		732	.022		.37	.56		.93	1.36
5721	2" x 6"		520	.031		.58	.79		1.37	1.97
5726	2" x 6", pneumatic nailed		624	.026		.58	.66		1.24	1.75
5741	3" x 4"		430	.037		1.09	.96		2.05	2.82
5746	3" x 4", pneumatic nailed		516	.031		1.09	.80		1.89	2.55
5761	Installed on dormer or gable, 2" x 3"		625	.026		.30	.66		.96	1.45
5766	2" x 3", pneumatic nailed		750	.021		.30	.55		.85	1.26
5781	2" x 4"		545	.029		.37	.75		1.12	1.69
5786	2" x 4", pneumatic nailed		654	.024		.37	.63		1	1.48
5801	2" x 6"		465	.034		.58	.88		1.46	2.13
5806	2" x 6", pneumatic nailed		558	.029		.58	.74		1.32	1.88
5821	3" x 4"		380	.042		1.09	1.08		2.17	3.04
5826	3" x 4", pneumatic nailed		456	.035		1.09	.90		1.99	2.73
8250	For second story & above, add						5%			

06 11 Wood Framing

06 11 10 – Wood Framing

06 11 10.40 Wall Framing

		Crew	Daily Output	Labor-Hours	Unit	Material	2007 Bare Costs Labor	Equipment	Total	Total Incl O&P
8300	For dormer & gable, add						15%			

06 11 10.42 Furring

		Crew	Daily Output	Labor-Hours	Unit	Material	Labor	Equipment	Total	Total Incl O&P
0010	**FURRING**									
0012	Wood strips, 1" x 2", on walls, on wood	1 Carp	550	.015	L.F.	.12	.37		.49	.76
0017	Pneumatic nailed		710	.011		.12	.29		.41	.62
0302	On masonry		495	.016		.12	.42		.54	.83
0402	On concrete		260	.031		.12	.79		.91	1.47
0602	1" x 3", on walls, on wood		550	.015		.22	.37		.59	.88
0607	Pneumatic nailed		710	.011		.22	.29		.51	.74
0702	On masonry		495	.016		.22	.42		.64	.95
0802	On concrete		260	.031		.22	.79		1.01	1.59
0852	On ceilings, on wood		350	.023		.22	.59		.81	1.25
0857	Pneumatic nailed		450	.018		.22	.46		.68	1.03
0902	On masonry		320	.025		.22	.64		.86	1.34
0952	On concrete		210	.038		.22	.98		1.20	1.91

06 11 10.44 Grounds

		Crew	Daily Output	Labor-Hours	Unit	Material	Labor	Equipment	Total	Total Incl O&P
0010	**GROUNDS**									
0020	For casework, 1" x 2" wood strips, on wood	1 Carp	330	.024	L.F.	.12	.62		.74	1.19
0102	On masonry		285	.028		.12	.72		.84	1.35
0202	On concrete		250	.032		.12	.82		.94	1.53
0402	For plaster, 3/4" deep, on wood		450	.018		.12	.46		.58	.91
0502	On masonry		225	.036		.12	.91		1.03	1.68
0602	On concrete		175	.046		.12	1.17		1.29	2.12
0702	On metal lath		200	.040		.12	1.03		1.15	1.87

06 12 Structural Panels

06 12 10 – Structural Insulated Panels

06 12 10.10 Structural Insulated Panels

		Crew	Daily Output	Labor-Hours	Unit	Material	Labor	Equipment	Total	Total Incl O&P
0010	**STRUCTURAL INSULATED PANELS**									
0100	Structural insul. panels, 7/16" OSB both faces, EPS insul, 3-5/8" T	F-3	2075	.019	S.F.	3.71	.45	.35	4.51	5.20
0110	5-5/8" thick		1725	.023		4.25	.54	.42	5.21	6.05
0120	7-3/8" thick		1425	.028		4.75	.65	.51	5.91	6.90
0130	9-3/8" thick		1125	.036		5.30	.83	.64	6.77	7.90
0140	7/16" OSB one face, EPS insul, 3-5/8" thick		2175	.018		2.09	.43	.33	2.85	3.39
0150	5-5/8" thick		1825	.022		2.50	.51	.40	3.41	4.05
0160	7-3/8" thick		1525	.026		2.84	.61	.47	3.92	4.67
0170	9-3/8" thick		1225	.033		3.25	.76	.59	4.60	5.50
0180	11-3/8" thick		925	.043		3.48	1.01	.78	5.27	6.40
0190	7/16" OSB - 1/2" GWB faces, EPS insul, 3-5/8" T		2075	.019		3.25	.45	.35	4.05	4.72
0200	5-5/8" thick		1725	.023		3.65	.54	.42	4.61	5.40
0210	7-3/8" thick		1425	.028		4	.65	.51	5.16	6.05
0220	9-3/8" thick		1125	.036		4.40	.83	.64	5.87	6.95
0230	11-3/8" thick		825	.048		4.68	1.13	.88	6.69	8
0240	7/16" OSB - 1/2" MRGWB faces, EPS insul, 3-5/8" T		2075	.019		3.25	.45	.35	4.05	4.72
0250	5-5/8" thick		1725	.023		3.65	.54	.42	4.61	5.40
0260	7-3/8" thick		1425	.028		4	.65	.51	5.16	6.05
0270	9-3/8" thick		1125	.036		4.40	.83	.64	5.87	6.95
0280	11-3/8" thick		825	.048		4.68	1.13	.88	6.69	8
0300	For 1/2" GWB added to OSB skin, add					.60			.60	.66
0310	For 1/2" MRGWB added to OSB skin, add					.75			.75	.83

06 13 Heavy Timber

06 13 13 – Log Construction

06 13 13.10 Log Structures

		Crew	Daily Output	Labor-Hours	Unit	Material	2007 Bare Costs Labor	2007 Bare Costs Equipment	Total	Total Incl O&P
0010	**LOG STRUCTURES**									
0020	Exterior walls, pine, D logs, with double T & G, 6" x 6"	2 Carp	500	.032	L.F.	4.06	.82		4.88	5.85
0030	6" x 8"		375	.043		4.86	1.10		5.96	7.20
0040	8" x 6"		375	.043		4.06	1.10		5.16	6.30
0050	8" x 7"		322	.050		4.06	1.28		5.34	6.65
0060	Square/rectangular logs, with double T & G, 6" x 6"		500	.032		4.06	.82		4.88	5.85
0160	Upper floor framing, pine, posts/columns, 4" x 6"		750	.021		2.78	.55		3.33	3.99
0180	4" x 8"		562	.028		3.74	.73		4.47	5.35
0190	6" x 6"		500	.032		4.83	.82		5.65	6.70
0200	6" x 8"		375	.043		7.60	1.10		8.70	10.20
0210	8" x 8"		281	.057		7.80	1.46		9.26	11.10
0220	8" x 10"		225	.071		9.75	1.83		11.58	13.85
0230	Beams, 4" x 8"		562	.028		3.74	.73		4.47	5.35
0240	4" x 10"		449	.036		5.50	.92		6.42	7.60
0250	4" x 12"		375	.043		5.10	1.10		6.20	7.50
0260	6" x 8"		375	.043		7.60	1.10		8.70	10.20
0270	6" x 10"		300	.053		10.90	1.37		12.27	14.35
0280	6" x 12"		250	.064		9.15	1.64		10.79	12.90
0290	8" x 10"		225	.071		9.75	1.83		11.58	13.85
0300	8" x 12"		188	.085		9.05	2.19		11.24	13.65
0310	Joists, 4" x 8"		562	.028		3.74	.73		4.47	5.35
0320	4" x 10"		449	.036		5.50	.92		6.42	7.60
0330	4" x 12"		375	.043		5.10	1.10		6.20	7.50
0340	6" x 8"		375	.043		7.60	1.10		8.70	10.20
0350	6" x 10"		300	.053		10.90	1.37		12.27	14.35
0390	Decking, 1" x 6" T & G		964	.017	S.F.	1.53	.43		1.96	2.40
0400	1" x 8" T & G		700	.023	"	1.39	.59		1.98	2.53
0420	Gable end roof framing, rafters, 4" x 8"		562	.028	L.F.	3.74	.73		4.47	5.35

06 13 23 – Heavy Timber Construction

06 13 23.10 Heavy Framing

		Crew	Daily Output	Labor-Hours	Unit	Material	2007 Bare Costs Labor	2007 Bare Costs Equipment	Total	Total Incl O&P
0010	**HEAVY FRAMING**									
0020	Beams, single 6" x 10"	2 Carp	1.10	14.545	M.B.F.	2,150	375		2,525	3,000
0100	Single 8" x 16"		1.20	13.333	"	2,700	345		3,045	3,550
0202	Built from 2" lumber, multiple 2" x 14"		900	.018	B.F.	.93	.46		1.39	1.81
0212	Built from 3" lumber, multiple 3" x 6"		700	.023		1.28	.59		1.87	2.41
0222	Multiple 3" x 8"		800	.020		1.30	.51		1.81	2.29
0232	Multiple 3" x 10"		900	.018		1.30	.46		1.76	2.21
0242	Multiple 3" x 12"		1000	.016		1.31	.41		1.72	2.14
0252	Built from 4" lumber, multiple 4" x 6"		800	.020		1.39	.51		1.90	2.40
0262	Multiple 4" x 8"		900	.018		1.40	.46		1.86	2.32
0272	Multiple 4" x 10"		1000	.016		1.65	.41		2.06	2.52
0282	Multiple 4" x 12"		1100	.015		1.28	.37		1.65	2.04
0292	Columns, structural grade, 1500f, 4" x 4"		450	.036	L.F.	2.25	.91		3.16	4.02
0302	6" x 6"		225	.071		4.35	1.83		6.18	7.90
0402	8" x 8"		240	.067		12.15	1.71		13.86	16.30

06 13 Heavy Timber

06 13 23 – Heavy Timber Construction

06 13 23.10 Heavy Framing

		Crew	Daily Output	Labor-Hours	Unit	Material	2007 Bare Costs Labor	Equipment	Total	Total Incl O&P
0502	10" x 10"	2 Carp	90	.178	L.F.	17.95	4.57		22.52	27.50
0602	12" x 12"		70	.229	↓	26	5.85		31.85	38.50
0802	Floor planks, 2" thick, T & G, 2" x 6"		1050	.015	B.F.	1.88	.39		2.27	2.73
0902	2" x 10"		1100	.015		1.88	.37		2.25	2.70
1102	3" thick, 3" x 6"		1050	.015		1.35	.39		1.74	2.15
1202	3" x 10"		1100	.015		1.35	.37		1.72	2.12
1402	Girders, structural grade, 12" x 12"		800	.020		2.03	.51		2.54	3.11
1502	10" x 16"	↓	1000	.016	↓	1.98	.41		2.39	2.88
2050	Roof planks, see division 06 15 16.10									
2302	Roof purlins, 4" thick, structural grade	2 Carp	1050	.015	B.F.	1.39	.39		1.78	2.19
2502	Roof trusses, add timber connectors, division 06 05 23.60	"	450	.036	"	1.36	.91		2.27	3.04

06 15 Wood Decking

06 15 16 – Wood Roof Decking

06 15 16.10 Wood Roof Decking

		Crew	Daily Output	Labor-Hours	Unit	Material	2007 Bare Costs Labor	Equipment	Total	Total Incl O&P
0010	**WOOD ROOF DECKING**									
0400	Cedar planks, 3" thick	2 Carp	320	.050	S.F.	6.40	1.29		7.69	9.25
0500	4" thick		250	.064		8.65	1.64		10.29	12.30
0702	Douglas fir, 3" thick		320	.050		2.40	1.29		3.69	4.82
0802	4" thick		250	.064		3.21	1.64		4.85	6.30
1002	Hemlock, 3" thick		320	.050		2.40	1.29		3.69	4.82
1102	4" thick		250	.064		3.20	1.64		4.84	6.30
1302	Western white spruce, 3" thick		320	.050		2.31	1.29		3.60	4.72
1402	4" thick	↓	250	.064	↓	3.08	1.64		4.72	6.20

06 15 23 – Laminated Wood Decking

06 15 23.10 Laminated Roof Deck

		Crew	Daily Output	Labor-Hours	Unit	Material	2007 Bare Costs Labor	Equipment	Total	Total Incl O&P
0010	**LAMINATED ROOF DECK**									
0020	Pine or hemlock, 3" thick	2 Carp	425	.038	S.F.	3.20	.97		4.17	5.15
0100	4" thick		325	.049		4.26	1.27		5.53	6.85
0300	Cedar, 3" thick		425	.038		3.92	.97		4.89	5.95
0400	4" thick		325	.049		4.99	1.27		6.26	7.65
0600	Fir, 3" thick		425	.038		3.26	.97		4.23	5.25
0700	4" thick	↓	325	.049	↓	4.08	1.27		5.35	6.65

06 16 Sheathing

06 16 23 – Subflooring

06 16 23.10 Subfloor

		Crew	Daily Output	Labor-Hours	Unit	Material	2007 Bare Costs Labor	Equipment	Total	Total Incl O&P
0010	**SUBFLOOR**									
0011	Plywood, CDX, 1/2" thick	2 Carp	1500	.011	SF Flr.	.47	.27		.74	.99
0015	Pneumatic nailed		1860	.009		.47	.22		.69	.90
0017	Pneumatic nailed		1860	.009		.47	.22		.69	.90
0102	5/8" thick		1350	.012		.59	.30		.89	1.17
0107	Pneumatic nailed		1674	.010		.59	.25		.84	1.07
0202	3/4" thick		1250	.013		.71	.33		1.04	1.34
0207	Pneumatic nailed		1550	.010		.71	.27		.98	1.23
0302	1-1/8" thick, 2-4-1 including underlayment		1050	.015		1.54	.39		1.93	2.35
0452	1" x 8" S4S, laid regular		1000	.016		1.28	.41		1.69	2.11
0462	Laid diagonal		850	.019		1.28	.48		1.76	2.23
0502	1" x 10" S4S, laid regular	↓	1100	.015	↓	1.07	.37		1.44	1.80

06 16 Sheathing

06 16 23 – Subflooring

06 16 23.10 Subfloor

		Crew	Daily Output	Labor-Hours	Unit	Material	2007 Bare Costs Labor	2007 Bare Costs Equipment	Total	Total Incl O&P
0602	Laid diagonal	2 Carp	900	.018	SF Flr.	1.07	.46		1.53	1.95
1500	Wafer board, 5/8" thick	↓	1330	.012	S.F.	.41	.31		.72	.97
1600	3/4" thick		1230	.013	"	.64	.33		.97	1.27

06 16 26 – Underlayment

06 16 26.10 Underlayment

		Crew	Daily Output	Labor-Hours	Unit	Material	Labor	Equipment	Total	Total Incl O&P
0010	**UNDERLAYMENT**									
0030	Plywood, underlayment grade, 3/8" thick	2 Carp	1500	.011	SF Flr.	.78	.27		1.05	1.33
0080	Pneumatic nailed		1860	.009		.78	.22		1	1.24
0102	1/2" thick		1450	.011		.90	.28		1.18	1.47
0107	Pneumatic nailed		1798	.009		.90	.23		1.13	1.38
0202	5/8" thick		1400	.011		1.09	.29		1.38	1.70
0207	Pneumatic nailed		1736	.009		1.09	.24		1.33	1.60
0302	3/4" thick		1300	.012		1.11	.32		1.43	1.76
0306	Pneumatic nailed		1612	.010		1.11	.26		1.37	1.65
0502	Particle board, 3/8" thick		1500	.011		.33	.27		.60	.83
0507	Pneumatic nailed		1860	.009		.33	.22		.55	.74
0602	1/2" thick		1450	.011		.37	.28		.65	.89
0607	Pneumatic nailed		1798	.009		.37	.23		.60	.80
0802	5/8" thick		1400	.011		.41	.29		.70	.95
0807	Pneumatic nailed		1736	.009		.41	.24		.65	.85
0902	3/4" thick		1300	.012		.64	.32		.96	1.24
0907	Pneumatic nailed		1612	.010		.64	.26		.90	1.13
1102	Hardboard, underlayment grade, 4' x 4', .215" thick	↓	1500	.011	↓	.39	.27		.66	.90

06 16 36 – Wood Panel Product Sheathing

06 16 36.10 Sheathing

		Crew	Daily Output	Labor-Hours	Unit	Material	Labor	Equipment	Total	Total Incl O&P
0010	**SHEATHING**, plywood on roofs R061636-20									
0012	Plywood on roofs, CDX									
0032	5/16" thick	2 Carp	1600	.010	S.F.	.62	.26		.88	1.12
0037	Pneumatic nailed		1952	.008		.62	.21		.83	1.04
0052	3/8" thick		1525	.010		.44	.27		.71	.94
0057	Pneumatic nailed		1860	.009		.44	.22		.66	.86
0102	1/2" thick		1400	.011		.47	.29		.76	1.02
0103	Pneumatic nailed		1708	.009		.47	.24		.71	.93
0202	5/8" thick		1300	.012		.59	.32		.91	1.19
0207	Pneumatic nailed		1586	.010		.59	.26		.85	1.09
0302	3/4" thick		1200	.013		.71	.34		1.05	1.36
0307	Pneumatic nailed		1464	.011		.71	.28		.99	1.26
0502	Plywood on walls with exterior CDX, 3/8" thick		1200	.013		.44	.34		.78	1.06
0507	Pneumatic nailed		1488	.011		.44	.28		.72	.95
0603	1/2" thick		1125	.014		.47	.37		.84	1.14
0608	Pneumatic nailed		1395	.011		.47	.29		.76	1.02
0702	5/8" thick		1050	.015		.59	.39		.98	1.31
0707	Pneumatic nailed		1302	.012		.59	.32		.91	1.19
0803	3/4" thick		975	.016		.71	.42		1.13	1.50
0808	Pneumatic nailed	↓	1209	.013	↓	.71	.34		1.05	1.36
1000	For shear wall construction, add						20%			
1200	For structural 1 exterior plywood, add				S.F.	10%				
1402	With boards, on roof 1" x 6" boards, laid horizontal	2 Carp	725	.022		1.42	.57		1.99	2.52
1502	Laid diagonal		650	.025		1.42	.63		2.05	2.63
1702	1" x 8" boards, laid horizontal		875	.018		1.28	.47		1.75	2.21
1802	Laid diagonal		725	.022		1.28	.57		1.85	2.37
2000	For steep roofs, add				↓		40%			

06 16 Sheathing

06 16 36 – Wood Panel Product Sheathing

06 16 36.10 Sheathing		Crew	Daily Output	Labor-Hours	Unit	Material	2007 Bare Costs Labor	Equipment	Total	Total Incl O&P
2200	For dormers, hips and valleys, add				S.F.	5%	50%			
2402	Boards on walls, 1" x 6" boards, laid regular	2 Carp	650	.025		1.42	.63		2.05	2.63
2502	Laid diagonal		585	.027		1.42	.70		2.12	2.75
2702	1" x 8" boards, laid regular		765	.021		1.28	.54		1.82	2.32
2802	Laid diagonal		650	.025		1.28	.63		1.91	2.48
2852	Gypsum, weatherproof, 1/2" thick		1050	.015		.62	.39		1.01	1.34
2900	With embedded glass mats		1100	.015		.52	.37		.89	1.20
3000	Wood fiber, regular, no vapor barrier, 1/2" thick		1200	.013		.55	.34		.89	1.19
3100	5/8" thick		1200	.013		.72	.34		1.06	1.37
3300	No vapor barrier, in colors, 1/2" thick		1200	.013		.78	.34		1.12	1.44
3400	5/8" thick		1200	.013		.96	.34		1.30	1.64
3600	With vapor barrier one side, white, 1/2" thick		1200	.013		.55	.34		.89	1.19
3700	Vapor barrier 2 sides, 1/2" thick		1200	.013		.77	.34		1.11	1.43
3800	Asphalt impregnated, 25/32" thick		1200	.013		.28	.34		.62	.89
3850	Intermediate, 1/2" thick		1200	.013		.18	.34		.52	.78
4000	Wafer board on roof, 1/2" thick		1455	.011		.70	.28		.98	1.25
4100	5/8" thick		1330	.012		.80	.31		1.11	1.40

06 17 Shop-Fabricated Structural Wood

06 17 33 – Wood I-Joists

06 17 33.10 Wood I-Joists

		Crew	Daily Output	Labor-Hours	Unit	Material	Labor	Equipment	Total	Total Incl O&P
0010	**WOOD I-JOISTS**									
0100	Plywood webs, incl. bridging & blocking, panels 24" O.C.									
1200	15' to 24' span, 50 psf live load	F-5	2400	.013	SF Flr.	2.12	.30		2.42	2.83
1300	55 psf live load		2250	.014		2.25	.32		2.57	3.02
1400	24' to 30' span, 45 psf live load		2600	.012		2.65	.27		2.92	3.38
1500	55 psf live load		2400	.013		2.96	.30		3.26	3.75
1600	Tubular steel open webs, 45 psf, 24" O.C., 40' span	F-3	6250	.006		2.13	.15	.12	2.40	2.72
1700	55' span		7750	.005		2.07	.12	.09	2.28	2.58
1800	70' span		9250	.004		2.69	.10	.08	2.87	3.22
1900	85 psf live load, 26' span		2300	.017		2.50	.41	.31	3.22	3.78

06 17 53 – Shop-Fabricated Wood Trusses

06 17 53.10 Roof Trusses

		Crew	Daily Output	Labor-Hours	Unit	Material	Labor	Equipment	Total	Total Incl O&P
0010	**ROOF TRUSSES**									
0020	For timber connectors, see div. 06 05 23.60									
5000	Common wood, 2" x 4" metal plate connected, 24" O.C., 4/12 slope									
5010	1' overhang, 12' span	F-5	55	.582	Ea.	42	12.95		54.95	68
5050	20' span	F-6	62	.645		64.50	15	11.70	91.20	109
5100	24' span		60	.667		75	15.50	12.05	102.55	122
5150	26' span		57	.702		77.50	16.30	12.70	106.50	127
5200	28' span		53	.755		69	17.55	13.65	100.20	121
5240	30' span		51	.784		96.50	18.25	14.20	128.95	152
5250	32' span		50	.800		100	18.60	14.50	133.10	157
5280	34' span		48	.833		122	19.35	15.10	156.45	183
5350	8/12 pitch, 1' overhang, 20' span		57	.702		68	16.30	12.70	97	116
5400	24' span		55	.727		86.50	16.90	13.15	116.55	138
5450	26' span		52	.769		90	17.90	13.90	121.80	144
5500	28' span		49	.816		99.50	19	14.80	133.30	157
5550	32' span		45	.889		119	20.50	16.10	155.60	184
5600	36' span		41	.976		149	22.50	17.65	189.15	221

06 17 Shop-Fabricated Structural Wood

06 17 53 – Shop-Fabricated Wood Trusses

06 17 53.10 Roof Trusses

		Crew	Daily Output	Labor-Hours	Unit	Material	2007 Bare Costs Labor	2007 Bare Costs Equipment	Total	Total Incl O&P
5650	38' span	F-6	40	1	Ea.	162	23.50	18.10	203.60	237
5700	40' span	↓	40	1	↓	167	23.50	18.10	208.60	243

06 18 Glued-Laminated Construction

06 18 13 – Glued-Laminated Beams

06 18 13.10 Laminated Beams

		Crew	Daily Output	Labor-Hours	Unit	Material	Labor	Equipment	Total	Total Incl O&P
0010	**LAMINATED BEAMS**									
0050	3-1/2" x 18"	F-3	480	.083	L.F.	23	1.94	1.51	26.45	30.50
0100	5-1/4" x 11-7/8"		450	.089		22.50	2.07	1.61	26.18	30.50
0150	5-1/4" x 16"		360	.111		30.50	2.59	2.01	35.10	40
0200	5-1/4" x 18"		290	.138		34.50	3.21	2.50	40.21	46
0250	5-1/4" x 24"		220	.182		41.50	4.23	3.29	49.02	56.50
0300	7" x 11-7/8"		320	.125		36.50	2.91	2.26	41.67	47.50
0350	7" x 16"		260	.154		49.50	3.58	2.78	55.86	63.50
0400	7" x 18"	↓	210	.190		57	4.44	3.45	64.89	73.50
0500	For premium appearance, add to S.F. prices					5%				
0550	For industrial type, deduct					15%				
0600	For stain and varnish, add					5%				
0650	For 3/4" laminations, add				↓	25%				

06 18 13.20 Laminated Framing

		Crew	Daily Output	Labor-Hours	Unit	Material	Labor	Equipment	Total	Total Incl O&P
0010	**LAMINATED FRAMING**									
0020	30 lb., short term live load, 15 lb. dead load									
0200	Straight roof beams, 20' clear span, beams 8' O.C.	F-3	2560	.016	SF Flr.	1.89	.36	.28	2.53	3
0300	Beams 16' O.C.		3200	.013		1.36	.29	.23	1.88	2.24
0500	40' clear span, beams 8' O.C.		3200	.013		3.60	.29	.23	4.12	4.70
0600	Beams 16' O.C.	↓	3840	.010		2.94	.24	.19	3.37	3.85
0800	60' clear span, beams 8' O.C.	F-4	2880	.014		6.20	.32	.37	6.89	7.75
0900	Beams 16' O.C.	"	3840	.010		4.61	.24	.28	5.13	5.75
1100	Tudor arches, 30' to 40' clear span, frames 8' O.C.	F-3	1680	.024		8.05	.55	.43	9.03	10.30
1200	Frames 16' O.C.	"	2240	.018		6.30	.42	.32	7.04	8
1400	50' to 60' clear span, frames 8' O.C.	F-4	2200	.018		8.70	.42	.48	9.60	10.85
1500	Frames 16' O.C.		2640	.015		7.40	.35	.40	8.15	9.20
1700	Radial arches, 60' clear span, frames 8' O.C.		1920	.021		8.15	.49	.55	9.19	10.40
1800	Frames 16' O.C.		2880	.014		6.25	.32	.37	6.94	7.85
2000	100' clear span, frames 8' O.C.		1600	.025		8.45	.58	.66	9.69	10.95
2100	Frames 16' O.C.		2400	.017		7.40	.39	.44	8.23	9.30
2300	120' clear span, frames 8' O.C.		1440	.028		11.20	.65	.74	12.59	14.20
2400	Frames 16' O.C.	↓	1920	.021		10.25	.49	.55	11.29	12.70
2600	Bowstring trusses, 20' O.C., 40' clear span	F-3	2400	.017		5.05	.39	.30	5.74	6.55
2700	60' clear span	F-4	3600	.011		4.53	.26	.29	5.08	5.75
2800	100' clear span		4000	.010		6.40	.23	.27	6.90	7.75
2900	120' clear span	↓	3600	.011		6.90	.26	.29	7.45	8.35
3100	For premium appearance, add to S.F. prices					5%				
3300	For industrial type, deduct					15%				
3500	For stain and varnish, add					5%				
3900	For 3/4" laminations, add to straight					25%				
4100	Add to curved				↓	15%				
4300	Alternate pricing method: (use nominal footage of									
4310	components). Straight beams, camber less than 6"	F-3	3.50	11.429	M.B.F.	2,800	266	207	3,273	3,750
4400	Columns, including hardware		2	20		3,000	465	360	3,825	4,475
4600	Curved members, radius over 32'	↓	2.50	16	↓	3,075	375	290	3,740	4,325

06 18 Glued-Laminated Construction

06 18 13 – Glued-Laminated Beams

06 18 13.20 Laminated Framing

		Crew	Daily Output	Labor-Hours	Unit	Material	2007 Bare Costs Labor	2007 Bare Costs Equipment	Total	Total Incl O&P
4700	Radius 10' to 32'	F-3	3	13.333	M.B.F.	3,050	310	241	3,601	4,150
4900	For complicated shapes, add maximum					100%				
5100	For pressure treating, add to straight					35%				
5200	Add to curved					45%				
6000	Laminated veneer members, southern pine or western species									
6050	1-3/4" wide x 5-1/2" deep	2 Carp	480	.033	L.F.	3.20	.86		4.06	4.97
6100	9-1/2" deep		480	.033		4.56	.86		5.42	6.45
6150	14" deep		450	.036		7.20	.91		8.11	9.45
6200	18" deep		450	.036		9.70	.91		10.61	12.20
6300	Parallel strand members, southern pine or western species									
6350	1-3/4" wide x 9-1/4" deep	2 Carp	480	.033	L.F.	4.25	.86		5.11	6.15
6400	11-1/4" deep		450	.036		5.20	.91		6.11	7.25
6450	14" deep		400	.040		6.25	1.03		7.28	8.65
6500	3-1/2" wide x 9-1/4" deep		480	.033		10	.86		10.86	12.45
6550	11-1/4" deep		450	.036		12.50	.91		13.41	15.30
6600	14" deep		400	.040		15.20	1.03		16.23	18.45
6650	7" wide x 9-1/4" deep		450	.036		21	.91		21.91	24.50
6700	11-1/4" deep		420	.038		25	.98		25.98	29
6750	14" deep		400	.040		27	1.03		28.03	31

06 22 Millwork

06 22 13 – Standard Pattern Wood Trim

06 22 13.10 Millwork

		Crew	Daily Output	Labor-Hours	Unit	Material	Labor	Equipment	Total	Total Incl O&P
0010	**MILLWORK**									
0020	equals three times cost of lumber									
1000	Typical finish hardwood milled material									
1020	1" x 12", custom birch				L.F.	3.05			3.05	3.36
1040	Cedar					2.73			2.73	3
1060	Oak					5.65			5.65	6.20
1080	Redwood					6			6	6.60
1100	Southern yellow pine					2.64			2.64	2.90
1120	Sugar pine					2.51			2.51	2.76
1140	Teak					14.05			14.05	15.45
1160	Walnut					4.96			4.96	5.45
1180	White pine					3.07			3.07	3.38

06 22 13.15 Moldings, Base

		Crew	Daily Output	Labor-Hours	Unit	Material	Labor	Equipment	Total	Total Incl O&P
0010	**MOLDINGS, BASE**									
0500	Base, stock pine, 9/16" x 3-1/2"	1 Carp	240	.033	L.F.	2.21	.86		3.07	3.88
0501	Oak or birch, 9/16" x 3-1/2"		240	.033		3.38	.86		4.24	5.15
0550	9/16" x 4-1/2"		200	.040		2.40	1.03		3.43	4.38
0561	Base shoe, oak, 3/4" x 1"		240	.033		1.13	.86		1.99	2.69
0570	Base, prefinished, 2-1/2" x 9/16"		242	.033		1.05	.85		1.90	2.60
0580	Shoe, prefinished, 3/8" x 5/8"		266	.030		.45	.77		1.22	1.81
0585	Flooring cant strip, 3/4" x 1/2"		500	.016		.46	.41		.87	1.21

06 22 13.30 Moldings, Casings

		Crew	Daily Output	Labor-Hours	Unit	Material	Labor	Equipment	Total	Total Incl O&P
0010	**MOLDINGS, CASINGS**									
0090	Apron, stock pine, 5/8" x 2"	1 Carp	250	.032	L.F.	1.18	.82		2	2.70
0110	5/8" x 3-1/2"		220	.036		1.83	.93		2.76	3.60
0300	Band, stock pine, 11/16" x 1-1/8"		270	.030		.79	.76		1.55	2.16
0350	11/16" x 1-3/4"		250	.032		1.12	.82		1.94	2.63

06 22 Millwork

06 22 13 – Standard Pattern Wood Trim

06 22 13.30 Moldings, Casings

		Crew	Daily Output	Labor-Hours	Unit	Material	2007 Bare Costs Labor	Equipment	Total	Total Incl O&P
0700	Casing, stock pine, 11/16" x 2-1/2"	1 Carp	240	.033	L.F.	1.35	.86		2.21	2.94
0701	Oak or birch		240	.033		2.58	.86		3.44	4.29
0750	11/16" x 3-1/2"		215	.037		1.83	.96		2.79	3.63
0760	Door & window casing, exterior, 1-1/4" x 2"		200	.040		2.13	1.03		3.16	4.08
0770	Finger jointed, 1-1/4" x 2"		200	.040		1.04	1.03		2.07	2.88
4600	Mullion casing, stock pine, 5/16" x 2"		200	.040		.78	1.03		1.81	2.60
4601	Oak or birch, 9/16" x 2-1/2"		200	.040		2.79	1.03		3.82	4.81
4700	Teak, custom, nominal 1" x 1"		215	.037		1.16	.96		2.12	2.90
4800	Nominal 1" x 3"		200	.040		3.50	1.03		4.53	5.60

06 22 13.35 Moldings, Ceilings

		Crew	Daily Output	Labor-Hours	Unit	Material	Labor	Equipment	Total	Total Incl O&P
0010	**MOLDINGS, CEILINGS**									
0600	Bed, stock pine, 9/16" x 1-3/4"	1 Carp	270	.030	L.F.	.99	.76		1.75	2.38
0650	9/16" x 2"		240	.033		1.19	.86		2.05	2.76
1200	Cornice molding, stock pine, 9/16" x 1-3/4"		330	.024		.98	.62		1.60	2.14
1300	9/16" x 2-1/4"		300	.027		1.21	.69		1.90	2.49
2400	Cove scotia, stock pine, 9/16" x 1-3/4"		270	.030		.98	.76		1.74	2.37
2401	Oak or birch, 9/16" x 1-3/4"		270	.030		.80	.76		1.56	2.17
2500	11/16" x 2-3/4"		255	.031		1.30	.81		2.11	2.80
2600	Crown, stock pine, 9/16" x 3-5/8"		250	.032		1.86	.82		2.68	3.45
2700	11/16" x 4-5/8"		220	.036		3.11	.93		4.04	5

06 22 13.40 Moldings, Exterior

		Crew	Daily Output	Labor-Hours	Unit	Material	Labor	Equipment	Total	Total Incl O&P
0010	**MOLDINGS, EXTERIOR**									
1500	Cornice, boards, pine, 1" x 2"	1 Carp	330	.024	L.F.	.32	.62		.94	1.41
1600	1" x 4"		250	.032		.62	.82		1.44	2.08
1700	1" x 6"		250	.032		1.07	.82		1.89	2.58
1800	1" x 8"		200	.040		1.39	1.03		2.42	3.27
1900	1" x 10"		180	.044		1.72	1.14		2.86	3.83
2000	1" x 12"		180	.044		1.89	1.14		3.03	4.02
2200	Three piece, built-up, pine, minimum		80	.100		2.01	2.57		4.58	6.55
2300	Maximum		65	.123		5	3.16		8.16	10.85
3000	Corner board, sterling pine, 1" x 4"		200	.040		.73	1.03		1.76	2.54
3100	1" x 6"		200	.040		1.59	1.03		2.62	3.49
3200	2" x 6"		165	.048		2.25	1.25		3.50	4.59
3300	2" x 8"		165	.048		3.04	1.25		4.29	5.45
3350	Fascia, sterling pine, 1" x 6"		250	.032		1.59	.82		2.41	3.15
3370	1" x 8"		225	.036		2.01	.91		2.92	3.76
3372	2" x 6"		225	.036		2.45	.91		3.36	4.25
3374	2" x 8"		200	.040		3.32	1.03		4.35	5.40
3376	2" x 10"		180	.044		4.03	1.14		5.17	6.35
3395	Grounds, 1" x 1" redwood		300	.027		.12	.69		.81	1.29
3400	Trim, exterior, sterling pine, back band		250	.032		.78	.82		1.60	2.26
3500	Casing		250	.032		2.08	.82		2.90	3.69
3600	Crown		250	.032		2	.82		2.82	3.60
3700	Porch rail with balusters		22	.364		16.30	9.35		25.65	34
3800	Screen		395	.020		1.26	.52		1.78	2.27
4100	Verge board, sterling pine, 1" x 4"		200	.040		.83	1.03		1.86	2.65
4200	1" x 6"		200	.040		1.14	1.03		2.17	2.99
4300	2" x 6"		165	.048		1.86	1.25		3.11	4.16
4400	2" x 8"		165	.048		2.46	1.25		3.71	4.82
4700	For redwood trim, add					200%				
5000	Casing/fascia, rough-sawn cedar									
5100	1" x 2"	1 Carp	275	.029	L.F.	.30	.75		1.05	1.60

06 22 Millwork

06 22 13 – Standard Pattern Wood Trim

06 22 13.40 Moldings, Exterior		Crew	Daily Output	Labor-Hours	Unit	Material	2007 Bare Costs Labor	Equipment	Total	Total Incl O&P
5200	1" x 6"	1 Carp	250	.032	L.F.	.99	.82		1.81	2.49
5300	1" x 8"		230	.035		1.29	.89		2.18	2.94
5400	2" x 4"		220	.036		1.09	.93		2.02	2.79
5500	2" x 6"		220	.036		1.85	.93		2.78	3.63
5600	2" x 8"		200	.040		2.80	1.03		3.83	4.82
5700	2" x 10"		180	.044		3.07	1.14		4.21	5.30
5800	2" x 12"		170	.047		3.70	1.21		4.91	6.10

06 22 13.45 Moldings, Trim		Crew	Daily Output	Labor-Hours	Unit	Material	Labor	Equipment	Total	Total Incl O&P
0010	**MOLDINGS, TRIM**									
0200	Astragal, stock pine, 11/16" x 1-3/4"	1 Carp	255	.031	L.F.	1.16	.81		1.97	2.65
0250	1-5/16" x 2-3/16"		240	.033		2.71	.86		3.57	4.43
0800	Chair rail, stock pine, 5/8" x 2-1/2"		270	.030		1.44	.76		2.20	2.87
0900	5/8" x 3-1/2"		240	.033		2.20	.86		3.06	3.87
1000	Closet pole, stock pine, 1-1/8" diameter		200	.040		.87	1.03		1.90	2.70
1100	Fir, 1-5/8" diameter		200	.040		1.60	1.03		2.63	3.50
1150	Corner, inside, 5/16" x 1"		225	.036		.33	.91		1.24	1.91
1160	Outside, 1-1/16" x 1-1/16"		240	.033		1.20	.86		2.06	2.77
1161	1-5/16" x 1-5/16"		240	.033		1.65	.86		2.51	3.27
3300	Half round, stock pine, 1/4" x 1/2"		270	.030		.23	.76		.99	1.54
3350	1/2" x 1"		255	.031		.60	.81		1.41	2.03
3400	Handrail, fir, single piece, stock, hardware not included									
3450	1-1/2" x 1-3/4"	1 Carp	80	.100	L.F.	1.77	2.57		4.34	6.30
3470	Pine, 1-1/2" x 1-3/4"		80	.100		1.77	2.57		4.34	6.30
3500	1-1/2" x 2-1/2"		76	.105		1.63	2.71		4.34	6.40
3600	Lattice, stock pine, 1/4" x 1-1/8"		270	.030		.41	.76		1.17	1.74
3700	1/4" x 1-3/4"		250	.032		.65	.82		1.47	2.12
3800	Miscellaneous, custom, pine, 1" x 1"		270	.030		.34	.76		1.10	1.66
3900	1" x 3"		240	.033		.69	.86		1.55	2.21
4100	Birch or oak, nominal 1" x 1"		240	.033		.53	.86		1.39	2.03
4200	Nominal 1" x 3"		215	.037		1.81	.96		2.77	3.61
4400	Walnut, nominal 1" x 1"		215	.037		.87	.96		1.83	2.58
4500	Nominal 1" x 3"		200	.040		2.61	1.03		3.64	4.61
4700	Teak, nominal 1" x 1"		215	.037		1.23	.96		2.19	2.97
4800	Nominal 1" x 3"		200	.040		3.52	1.03		4.55	5.60
4900	Quarter round, stock pine, 1/4" x 1/4"		275	.029		.22	.75		.97	1.51
4950	3/4" x 3/4"		255	.031		.64	.81		1.45	2.07
5600	Wainscot moldings, 1-1/8" x 9/16", 2' high, minimum		76	.105	S.F.	10.75	2.71		13.46	16.40
5700	Maximum		65	.123	"	19.75	3.16		22.91	27

06 22 13.50 Moldings, Window and Door		Crew	Daily Output	Labor-Hours	Unit	Material	Labor	Equipment	Total	Total Incl O&P
0010	**MOLDINGS, WINDOW AND DOOR**									
2800	Door moldings, stock, decorative, 1-1/8" wide, plain	1 Carp	17	.471	Set	46	12.10		58.10	71
2900	Detailed		17	.471	"	87.50	12.10		99.60	117
2960	Clear pine door jamb, no stops, 11/16" x 4-9/16"		240	.033	L.F.	3	.86		3.86	4.75
3150	Door trim set, 1 head and 2 sides, pine, 2-1/2 wide		5.90	1.356	Opng.	23	35		58	84.50
3170	3-1/2" wide		5.30	1.509	"	31	39		70	100
3250	Glass beads, stock pine, 3/8" x 1/2"		275	.029	L.F.	.31	.75		1.06	1.61
3270	3/8" x 7/8"		270	.030		.41	.76		1.17	1.74
4850	Parting bead, stock pine, 3/8" x 3/4"		275	.029		.35	.75		1.10	1.66
4870	1/2" x 3/4"		255	.031		.44	.81		1.25	1.85
5000	Stool caps, stock pine, 11/16" x 3-1/2"		200	.040		2.20	1.03		3.23	4.16
5100	1-1/16" x 3-1/4"		150	.053		3.48	1.37		4.85	6.15
5300	Threshold, oak, 3' long, inside, 5/8" x 3-5/8"		32	.250	Ea.	9	6.45		15.45	21

06 22 Millwork

06 22 13 – Standard Pattern Wood Trim

06 22 13.50 Moldings, Window and Door

		Crew	Daily Output	Labor-Hours	Unit	Material	2007 Bare Costs Labor	2007 Bare Costs Equipment	Total	Total Incl O&P
5400	Outside, 1-1/2" x 7-5/8"	1 Carp	16	.500	Ea.	37	12.85		49.85	62.50
5900	Window trim sets, including casings, header, stops,									
5910	stool and apron, 2-1/2" wide, minimum	1 Carp	13	.615	Opng.	30.50	15.80		46.30	61
5950	Average	↓	10	.800		36	20.50		56.50	74.50
6000	Maximum		6	1.333	↓	62.50	34.50		97	127

06 22 13.60 Moldings, Soffits

		Crew	Daily Output	Labor-Hours	Unit	Material	Labor	Equipment	Total	Total Incl O&P
0010	**MOLDINGS, SOFFITS**									
0200	Soffits, pine, 1" x 4"	2 Carp	420	.038	L.F.	.39	.98		1.37	2.09
0210	1" x 6"		420	.038		.65	.98		1.63	2.38
0220	1" x 8"		420	.038		.80	.98		1.78	2.54
0230	1" x 10"		400	.040		.85	1.03		1.88	2.67
0240	1" x 12"		400	.040		1.10	1.03		2.13	2.94
0250	STK cedar, 1" x 4"		420	.038		.55	.98		1.53	2.27
0260	1" x 6"		420	.038		.99	.98		1.97	2.75
0270	1" x 8"		420	.038		1.29	.98		2.27	3.08
0280	1" x 10"		400	.040		1.64	1.03		2.67	3.54
0290	1" x 12"		400	.040	↓	1.98	1.03		3.01	3.92
1000	Exterior AC plywood, 1/4" thick		420	.038	S.F.	.70	.98		1.68	2.43
1050	3/8" thick		420	.038		.78	.98		1.76	2.52
1100	1/2" thick	↓	420	.038		.90	.98		1.88	2.65
1150	Polyvinyl chloride, white, solid	1 Carp	230	.035		.87	.89		1.76	2.48
1160	Perforated	"	230	.035	↓	.87	.89		1.76	2.48
1170	Accessories, "J" channel 5/8"	2 Carp	700	.023	L.F.	.36	.59		.95	1.40

06 25 Prefinished Paneling

06 25 13 – Prefinished Hardboard Paneling

06 25 13.10 Paneling, Hardboard

		Crew	Daily Output	Labor-Hours	Unit	Material	Labor	Equipment	Total	Total Incl O&P
0010	**PANELING, HARDBOARD**									
0050	Not incl. furring or trim, hardboard, tempered, 1/8" thick	2 Carp	500	.032	S.F.	.35	.82		1.17	1.79
0100	1/4" thick		500	.032		.48	.82		1.30	1.93
0300	Tempered pegboard, 1/8" thick		500	.032		.43	.82		1.25	1.87
0400	1/4" thick		500	.032		.60	.82		1.42	2.06
0600	Untempered hardboard, natural finish, 1/8" thick		500	.032		.36	.82		1.18	1.80
0700	1/4" thick		500	.032		.38	.82		1.20	1.82
0900	Untempered pegboard, 1/8" thick		500	.032		.35	.82		1.17	1.79
1000	1/4" thick		500	.032		.39	.82		1.21	1.83
1200	Plastic faced hardboard, 1/8" thick		500	.032		.59	.82		1.41	2.05
1300	1/4" thick		500	.032		.78	.82		1.60	2.26
1500	Plastic faced pegboard, 1/8" thick		500	.032		.56	.82		1.38	2.02
1600	1/4" thick		500	.032		.69	.82		1.51	2.16
1800	Wood grained, plain or grooved, 1/4" thick, minimum		500	.032		.52	.82		1.34	1.97
1900	Maximum		425	.038	↓	1.09	.97		2.06	2.84
2100	Moldings for hardboard, wood or aluminum, minimum		500	.032	L.F.	.35	.82		1.17	1.79
2200	Maximum	↓	425	.038	"	.99	.97		1.96	2.73

06 25 16 – Prefinished Plywood Paneling

06 25 16.10 Paneling, Plywood

		Crew	Daily Output	Labor-Hours	Unit	Material	Labor	Equipment	Total	Total Incl O&P
0010	**PANELING, PLYWOOD**									
2400	Plywood, prefinished, 1/4" thick, 4' x 8' sheets									
2410	with vertical grooves. Birch faced, minimum	2 Carp	500	.032	S.F.	.85	.82		1.67	2.34
2420	Average	↓	420	.038	↓	1.29	.98		2.27	3.08

06 25 Prefinished Paneling

06 25 16 – Prefinished Plywood Paneling

06 25 16.10 Paneling, Plywood		Crew	Daily Output	Labor-Hours	Unit	Material	2007 Bare Costs Labor	Equipment	Total	Total Incl O&P
2430	Maximum	2 Carp	350	.046	S.F.	1.89	1.17		3.06	4.07
2600	Mahogany, African		400	.040		2.41	1.03		3.44	4.39
2700	Philippine (Lauan)		500	.032		1.04	.82		1.86	2.54
2900	Oak or Cherry, minimum		500	.032		2.02	.82		2.84	3.62
3000	Maximum		400	.040		3.10	1.03		4.13	5.15
3200	Rosewood		320	.050		4.40	1.29		5.69	7
3400	Teak		400	.040		3.10	1.03		4.13	5.15
3600	Chestnut		375	.043		4.59	1.10		5.69	6.90
3800	Pecan		400	.040		1.98	1.03		3.01	3.92
3900	Walnut, minimum		500	.032		2.65	.82		3.47	4.32
3950	Maximum		400	.040		5	1.03		6.03	7.25
4000	Plywood, prefinished, 3/4" thick, stock grades, minimum		320	.050		1.20	1.29		2.49	3.50
4100	Maximum		224	.071		5.15	1.84		6.99	8.80
4300	Architectural grade, minimum		224	.071		3.82	1.84		5.66	7.30
4400	Maximum		160	.100		5.85	2.57		8.42	10.75
4600	Plywood, "A" face, birch, V.C., 1/2" thick, natural		450	.036		1.81	.91		2.72	3.54
4700	Select		450	.036		1.98	.91		2.89	3.73
4900	Veneer core, 3/4" thick, natural		320	.050		1.91	1.29		3.20	4.28
5000	Select		320	.050		2.15	1.29		3.44	4.55
5200	Lumber core, 3/4" thick, natural		320	.050		2.87	1.29		4.16	5.35
5500	Plywood, knotty pine, 1/4" thick, A2 grade		450	.036		1.57	.91		2.48	3.28
5600	A3 grade		450	.036		1.98	.91		2.89	3.73
5800	3/4" thick, veneer core, A2 grade		320	.050		2.03	1.29		3.32	4.41
5900	A3 grade		320	.050		2.28	1.29		3.57	4.69
6100	Aromatic cedar, 1/4" thick, plywood		400	.040		2	1.03		3.03	3.94
6200	1/4" thick, particle board		400	.040		.97	1.03		2	2.81

06 25 26 – Panel System

06 25 26.10 Panel Systems

		Crew	Daily Output	Labor-Hours	Unit	Material	Labor	Equipment	Total	Total Incl O&P
0010	**PANEL SYSTEMS**									
0100	Raised panel, eng. wood core w/ wood veneer, std., paint grade	2 Carp	300	.053	S.F.	11.20	1.37		12.57	14.70
0110	Oak veneer		300	.053		18.60	1.37		19.97	23
0120	Maple veneer		300	.053		24	1.37		25.37	28.50
0130	Cherry veneer		300	.053		30	1.37		31.37	35.50
0300	Class I fire rated, paint grade		300	.053		13.40	1.37		14.77	17.10
0310	Oak veneer		300	.053		22.50	1.37		23.87	27
0320	Maple veneer		300	.053		28.50	1.37		29.87	33.50
0330	Cherry veneer		300	.053		36	1.37		37.37	42
0510	Beadboard, 5/8" MDF, standard, primed		300	.053		7.90	1.37		9.27	11.05
0520	Oak veneer, unfinished		300	.053		13.75	1.37		15.12	17.50
0530	Maple veneer, unfinished		300	.053		16.30	1.37		17.67	20
0610	Rustic paneling, 5/8" MDF, standard, maple veneer, unfinished		300	.053		21	1.37		22.37	25.50
5000	For prefinished paneling, see division 06 25 16.10 & 06 25 13.10									

06 26 Board Paneling

06 26 13 – Profile Board Paneling

06 26 13.10 Paneling, Boards

		Crew	Daily Output	Labor-Hours	Unit	Material	2007 Bare Costs Labor	Equipment	Total	Total Incl O&P
0010	**PANELING, BOARDS**									
6400	Wood board paneling, 3/4" thick, knotty pine	2 Carp	300	.053	S.F.	1.41	1.37		2.78	3.88
6500	Rough sawn cedar		300	.053		1.81	1.37		3.18	4.32
6700	Redwood, clear, 1" x 4" boards		300	.053		4.22	1.37		5.59	6.95
6900	Aromatic cedar, closet lining, boards	↓	275	.058	↓	3.25	1.50		4.75	6.10

06 43 Wood Stairs and Railings

06 43 13 – Wood Stairs

06 43 13.20 Prefabricated Wood Stairs

		Crew	Daily Output	Labor-Hours	Unit	Material	Labor	Equipment	Total	Total Incl O&P
0010	**PREFABRICATED WOOD STAIRS**									
0100	Box stairs, prefabricated, 3'-0" wide									
0110	Oak treads, up to 14 risers	2 Carp	39	.410	Riser	75.50	10.55		86.05	101
0600	With pine treads for carpet, up to 14 risers	"	39	.410	"	49	10.55		59.55	72
1100	For 4' wide stairs, add				Flight	25%				
1550	Stairs, prefabricated stair handrail with balusters	1 Carp	30	.267	L.F.	65	6.85		71.85	83
1700	Basement stairs, prefabricated, pine treads									
1710	Pine risers, 3' wide, up to 14 risers	2 Carp	52	.308	Riser	49	7.90		56.90	67.50
4000	Residential, wood, oak treads, prefabricated		1.50	10.667	Flight	985	274		1,259	1,550
4200	Built in place	↓	.44	36.364	"	1,475	935		2,410	3,200
4400	Spiral, oak, 4'-6" diameter, unfinished, prefabricated,									
4500	incl. railing, 9' high	2 Carp	1.50	10.667	Flight	4,675	274		4,949	5,600

06 43 13.30 Wood Stair Components

		Crew	Daily Output	Labor-Hours	Unit	Material	Labor	Equipment	Total	Total Incl O&P
0010	**WOOD STAIR COMPONENTS**									
0020	Balusters, turned, 3" high, pine, minimum	1 Carp	28	.286	Ea.	3.77	7.35		11.12	16.60
0100	Maximum		26	.308		19.40	7.90		27.30	35
0300	30" high birch balusters, minimum		28	.286		6.85	7.35		14.20	19.95
0400	Maximum		26	.308		27.50	7.90		35.40	43.50
0600	42" high, pine balusters, minimum		27	.296		5	7.60		12.60	18.40
0700	Maximum		25	.320		28	8.20		36.20	44.50
0900	42" high birch balusters, minimum		27	.296		10.90	7.60		18.50	25
1000	Maximum		25	.320	↓	39.50	8.20		47.70	57
1050	Baluster, stock pine, 1-1/4" x 1-1/4"		240	.033	L.F.	3.18	.86		4.04	4.95
1100	1-3/4" x 1-3/4"		220	.036	"	9.10	.93		10.03	11.60
1200	Newels, 3-1/4" wide, starting, minimum		7	1.143	Ea.	38	29.50		67.50	92
1300	Maximum		6	1.333		325	34.50		359.50	420
1500	Landing, minimum		5	1.600		106	41		147	187
1600	Maximum		4	2	↓	355	51.50		406.50	480
1800	Railings, oak, built-up, minimum		60	.133	L.F.	32.50	3.43		35.93	41.50
1900	Maximum		55	.145		47	3.74		50.74	58
2100	Add for sub rail		110	.073		5.40	1.87		7.27	9.10
2300	Risers, beech, 3/4" x 7-1/2" high		64	.125		5.90	3.21		9.11	11.95
2400	Fir, 3/4" x 7-1/2" high		64	.125		1.63	3.21		4.84	7.25
2600	Oak, 3/4" x 7-1/2" high		64	.125		6.20	3.21		9.41	12.30
2800	Pine, 3/4" x 7-1/2" high		66	.121		3.11	3.12		6.23	8.70
2850	Skirt board, pine, 1" x 10"		55	.145		2.96	3.74		6.70	9.60
2900	1" x 12"		52	.154	↓	3.52	3.95		7.47	10.55
3000	Treads, oak, 1-1/4" x 10" wide, 3' long		18	.444	Ea.	59	11.40		70.40	84.50
3100	4' long, oak		17	.471		79	12.10		91.10	107
3300	1-1/4" x 11-1/2" wide, 3' long, oak		18	.444		59	11.40		70.40	84.50
3400	6' long, oak	↓	14	.571		118	14.70		132.70	155
3600	Beech treads, add				↓	40%				

06 43 Wood Stairs and Railings

06 43 13 – Wood Stairs

06 43 13.30 Wood Stair Components	Crew	Daily Output	Labor-Hours	Unit	Material	2007 Bare Costs Labor	Equipment	Total	Total Incl O&P
3800 For mitered return nosings, add				L.F.	3.98			3.98	4.38

06 43 16 – Wood Railings

06 43 16.10 Wood Railings

0010 **WOOD RAILINGS**									
0020 Custom design, architectural grade, hardwood, minimum	1 Carp	38	.211	L.F.	5.60	5.40		11	15.35
0100 Maximum		30	.267		46.50	6.85		53.35	63
0300 Stock interior railing with spindles 6" O.C., 4' long		40	.200		42.50	5.15		47.65	55
0400 8' long	↓	48	.167	↓	21	4.28		25.28	31

06 44 Ornamental Woodwork

06 44 19 – Wood Grilles

06 44 19.10 Grilles

0010 **GRILLES**									
0020 2' x 4' to 4' x 8', custom designs, unfinished, minimum	1 Carp	38	.211	S.F.	12.85	5.40		18.25	23.50
0050 Average		30	.267		28	6.85		34.85	42
0100 Maximum		19	.421		43	10.80		53.80	65.50
0300 As above, but prefinished, minimum		38	.211		12.85	5.40		18.25	23.50
0400 Maximum	↓	19	.421		48	10.80		58.80	71.50

06 44 33 – Wood Mantels

06 44 33.10 Fireplace Mantels

0010 **FIREPLACE MANTELS**									
0015 6" molding, 6' x 3'-6" opening, minimum	1 Carp	5	1.600	Opng.	147	41		188	232
0100 Maximum		5	1.600		183	41		224	272
0300 Prefabricated pine, colonial type, stock, deluxe		2	4		2,300	103		2,403	2,725
0400 Economy	↓	3	2.667	↓	395	68.50		463.50	550

06 44 33.20 Fireplace Mantel Beam

0010 **FIREPLACE MANTEL BEAM**									
0020 Rough texture wood, 4" x 8"	1 Carp	36	.222	L.F.	5.45	5.70		11.15	15.70
0100 4" x 10"		35	.229	"	6.55	5.85		12.40	17.15
0300 Laminated hardwood, 2-1/4" x 10-1/2" wide, 6' long		5	1.600	Ea.	104	41		145	185
0400 8' long		5	1.600	"	145	41		186	229
0600 Brackets for above, rough sawn		12	.667	Pr.	9.60	17.15		26.75	39.50
0700 Laminated	↓	12	.667	"	14.50	17.15		31.65	45

06 44 39 – Wood Posts and Columns

06 44 39.10 Decorative Beams

0010 **DECORATIVE BEAMS**									
0020 Rough sawn cedar, non-load bearing, 4" x 4"	2 Carp	180	.089	L.F.	1.47	2.28		3.75	5.50
0100 4" x 6"		170	.094		2.21	2.42		4.63	6.55
0200 4" x 8"		160	.100		2.95	2.57		5.52	7.60
0300 4" x 10"		150	.107		3.68	2.74		6.42	8.70
0400 4" x 12"		140	.114		4.42	2.94		7.36	9.85
0500 8" x 8"		130	.123		7.50	3.16		10.66	13.60
0600 Plastic beam, "hewn finish", 6" x 2"		240	.067		2.89	1.71		4.60	6.10
0601 6" x 4"	↓	220	.073	↓	3.37	1.87		5.24	6.90
1100 Beam connector plates see div. 06 05 23.60									

06 44 39.20 Columns

0010 **COLUMNS**									
0050 Aluminum, round colonial, 6" diameter	2 Carp	80	.200	V.L.F.	16	5.15		21.15	26.50
0100 8" diameter	↓	62.25	.257	↓	17	6.60		23.60	30

06 44 Ornamental Woodwork

06 44 39 – Wood Posts and Columns

06 44 39.20 Columns		Crew	Daily Output	Labor-Hours	Unit	Material	2007 Bare Costs Labor	Equipment	Total	Total Incl O&P
0200	10" diameter	2 Carp	55	.291	V.L.F.	19.10	7.50		26.60	33.50
0250	Fir, stock units, hollow round, 6" diameter		80	.200		17.20	5.15		22.35	27.50
0300	8" diameter		80	.200		20.50	5.15		25.65	31
0350	10" diameter		70	.229		25.50	5.85		31.35	38.50
0400	Solid turned, to 8' high, 3-1/2" diameter		80	.200		8.40	5.15		13.55	17.95
0500	4-1/2" diameter		75	.213		12.80	5.50		18.30	23.50
0600	5-1/2" diameter		70	.229		16.40	5.85		22.25	28
0800	Square columns, built-up, 5" x 5"		65	.246		10.20	6.35		16.55	22
0900	Solid, 3-1/2" x 3-1/2"		130	.123		6.70	3.16		9.86	12.70
1600	Hemlock, tapered, T & G, 12" diam, 10' high		100	.160		31.50	4.11		35.61	41.50
1700	16' high		65	.246		46.50	6.35		52.85	62
1900	10' high, 14" diameter		100	.160		78.50	4.11		82.61	93.50
2000	18' high		65	.246		63.50	6.35		69.85	81
2200	18" diameter, 12' high		65	.246		114	6.35		120.35	137
2300	20' high		50	.320		80	8.20		88.20	102
2500	20" diameter, 14' high		40	.400		129	10.30		139.30	159
2600	20' high		35	.457		120	11.75		131.75	152
2800	For flat pilasters, deduct					33%				
3000	For splitting into halves, add				Ea.	116			116	128
4000	Rough sawn cedar posts, 4" x 4"	2 Carp	250	.064	V.L.F.	3.70	1.64		5.34	6.85
4100	4" x 6"		235	.068		6.70	1.75		8.45	10.30
4200	6" x 6"		220	.073		12.50	1.87		14.37	16.90
4300	8" x 8"		200	.080		28.50	2.06		30.56	35

06 48 Wood Frames

06 48 13 – Exterior Wood Door Frames

06 48 13.10 Exterior Wood Door Frames		Crew	Daily Output	Labor-Hours	Unit	Material	2007 Bare Costs Labor	Equipment	Total	Total Incl O&P
0010	**EXTERIOR WOOD DOOR FRAMES**									
0400	Exterior frame, incl. ext. trim, pine, 5/4 x 4-9/16" deep	2 Carp	375	.043	L.F.	5.35	1.10		6.45	7.70
0420	5-3/16" deep		375	.043		10.90	1.10		12	13.85
0440	6-9/16" deep		375	.043		8.25	1.10		9.35	10.95
0600	Oak, 5/4 x 4-9/16" deep		350	.046		10.50	1.17		11.67	13.55
0620	5-3/16" deep		350	.046		12.60	1.17		13.77	15.85
0640	6-9/16" deep		350	.046		14.40	1.17		15.57	17.85
0800	Walnut, 5/4 x 4-9/16" deep		350	.046		12	1.17		13.17	15.20
0820	5-3/16" deep		350	.046		16.50	1.17		17.67	20
0840	6-9/16" deep		350	.046		19.20	1.17		20.37	23
1000	Sills, 8/4 x 8" deep, oak, no horns		100	.160		18	4.11		22.11	27
1020	2" horns		100	.160		18.25	4.11		22.36	27
1040	3" horns		100	.160		18.25	4.11		22.36	27
1100	8/4 x 10" deep, oak, no horns		90	.178		23	4.57		27.57	33.50
1120	2" horns		90	.178		21	4.57		25.57	31
1140	3" horns		90	.178		21	4.57		25.57	31
2000	Exterior, colonial, frame & trim, 3' opng., in-swing, minimum		22	.727	Ea.	310	18.70		328.70	370
2010	Average		21	.762		460	19.60		479.60	545
2020	Maximum		20	.800		1,050	20.50		1,070.50	1,175
2100	5'-4" opening, in-swing, minimum		17	.941		355	24		379	430
2120	Maximum		15	1.067		1,050	27.50		1,077.50	1,200
2140	Out-swing, minimum		17	.941		360	24		384	435
2160	Maximum		15	1.067		1,100	27.50		1,127.50	1,250
2400	6'-0" opening, in-swing, minimum		16	1		340	25.50		365.50	415

06 48 Wood Frames

06 48 13 – Exterior Wood Door Frames

06 48 13.10 Exterior Wood Door Frames		Crew	Daily Output	Labor-Hours	Unit	Material	2007 Bare Costs Labor	Equipment	Total	Total Incl O&P
2420	Maximum	2 Carp	10	1.600	Ea.	1,100	41		1,141	1,275
2460	Out-swing, minimum		16	1		365	25.50		390.50	445
2480	Maximum		10	1.600	↓	1,275	41		1,316	1,500
2600	For two sidelights, add, minimum		30	.533	Opng.	345	13.70		358.70	405
2620	Maximum		20	.800	"	1,125	20.50		1,145.50	1,250
2700	Custom birch frame, 3'-0" opening		16	1	Ea.	199	25.50		224.50	263
2750	6'-0" opening		16	1		300	25.50		325.50	375
2900	Exterior, modern, plain trim, 3' opng., in-swing, minimum		26	.615		33.50	15.80		49.30	64
2920	Average		24	.667		40	17.15		57.15	73
2940	Maximum	↓	22	.727	↓	48.50	18.70		67.20	85

06 48 16 – Interior Wood Door Frames

06 48 16.10 Interior Wood Door Frames

		Crew	Daily Output	Labor-Hours	Unit	Material	Labor	Equipment	Total	Total Incl O&P
0010	**INTERIOR WOOD DOOR FRAMES**									
3000	Interior frame, pine, 11/16" x 3-5/8" deep	2 Carp	375	.043	L.F.	5.10	1.10		6.20	7.45
3020	4-9/16" deep		375	.043		6.25	1.10		7.35	8.75
3040	5-3/16" deep		375	.043		3.92	1.10		5.02	6.15
3200	Oak, 11/16" x 3-5/8" deep		350	.046		4	1.17		5.17	6.40
3220	4-9/16" deep		350	.046		4.31	1.17		5.48	6.75
3240	5-3/16" deep		350	.046		4.47	1.17		5.64	6.90
3400	Walnut, 11/16" x 3-5/8" deep		350	.046		6.75	1.17		7.92	9.45
3420	4-9/16" deep		350	.046		7.10	1.17		8.27	9.80
3440	5-3/16" deep		350	.046		7.40	1.17		8.57	10.15
3800	Threshold, oak, 5/8" x 3-5/8" deep		200	.080		3.75	2.06		5.81	7.60
3820	4-5/8" deep		190	.084		4.50	2.16		6.66	8.60
3840	5-5/8" deep	↓	180	.089	↓	7.40	2.28		9.68	12.05
4000	For casing see division 06 22 13.30 & 06 22 13.50									

06 49 Wood Screens and Exterior Wood Shutters

06 49 19 – Exterior Wood Shutters

06 49 19.10 Shutters, Exterior

		Crew	Daily Output	Labor-Hours	Unit	Material	Labor	Equipment	Total	Total Incl O&P
0010	**SHUTTERS, EXTERIOR**									
0012	Aluminum, louvered, 1'-4" wide, 3'-0" long	1 Carp	10	.800	Pr.	48	20.50		68.50	87.50
0200	4'-0" long		10	.800		57	20.50		77.50	98
0300	5'-4" long		10	.800		75.50	20.50		96	118
0400	6'-8" long		9	.889		96.50	23		119.50	145
1000	Pine, louvered, primed, each 1'-2" wide, 3'-3" long		10	.800		91	20.50		111.50	135
1100	4'-7" long		10	.800		123	20.50		143.50	170
1250	Each 1'-4" wide, 3'-0" long		10	.800		93	20.50		113.50	137
1350	5'-3" long		10	.800		139	20.50		159.50	188
1500	Each 1'-6" wide, 3'-3" long		10	.800		96.50	20.50		117	141
1600	4'-7" long		10	.800		136	20.50		156.50	184
1620	Hemlock, louvered, 1'-2" wide, 5'-7" long		10	.800		149	20.50		169.50	199
1630	Each 1'-4" wide, 2'-2" long		10	.800		93	20.50		113.50	137
1640	3'-0" long		10	.800		93	20.50		113.50	137
1650	3'-3" long		10	.800		100	20.50		120.50	145
1660	3'-11" long		10	.800		114	20.50		134.50	160
1670	4'-3" long		10	.800		111	20.50		131.50	157
1680	5'-3" long		10	.800		138	20.50		158.50	187
1690	5'-11" long		10	.800		157	20.50		177.50	207
1700	Door blinds, 6'-9" long, each 1'-3" wide	↓	9	.889		158	23		181	212

06 49 Wood Screens and Exterior Wood Shutters

06 49 19 – Exterior Wood Shutters

06 49 19.10 Shutters, Exterior		Crew	Daily Output	Labor-Hours	Unit	Material	2007 Bare Costs Labor	Equipment	Total	Total Incl O&P
1710	1'-6" wide	1 Carp	9	.889	Pr.	169	23		192	225
1720	Hemlock, solid raised panel, each 1'-4" wide, 3'-3" long		10	.800		150	20.50		170.50	200
1730	3'-11" long		10	.800		176	20.50		196.50	228
1740	4'-3" long		10	.800		190	20.50		210.50	244
1750	4'-7" long		10	.800		205	20.50		225.50	260
1760	4'-11" long		10	.800		223	20.50		243.50	280
1770	5'-11" long		10	.800		254	20.50		274.50	315
1800	Door blinds, 6'-9" long, each 1'-3" wide		9	.889		286	23		309	355
1900	1'-6" wide		9	.889		310	23		333	385
2500	Polystyrene, solid raised panel, each 1'-4" wide, 3'-3" long		10	.800		40	20.50		60.50	79
2600	3'-11" long		10	.800		46	20.50		66.50	86
2700	4'-7" long		10	.800		50.50	20.50		71	90.50
2800	5'-3" long		10	.800		54.50	20.50		75	95
2900	6'-8" long		9	.889		65	23		88	111
4500	Polystyrene, louvered, each 1'-2" wide, 3'-3" long		10	.800		27.50	20.50		48	65
4600	4'-7" long		10	.800		34	20.50		54.50	72
4750	5'-3" long		10	.800		39	20.50		59.50	77.50
4850	6'-8" long		9	.889		46	23		69	89.50
6000	Vinyl, louvered, each 1'-2" x 4'-7" long		10	.800		33.50	20.50		54	72
6200	Each 1'-4" x 6'-8" long	↓	9	.889	↓	47.50	23		70.50	91
8000	PVC exterior rolling shutters									
8100	including crank control	1 Carp	8	1	Ea.	435	25.50		460.50	525
8500	Insulative - 6' x 6'8" stock unit	"	8	1	"	625	25.50		650.50	730

06 65 Plastic Simulated Wood Trim

06 65 10 – Plastic Simulated Wood Trim

06 65 10.10 PVC Trim, Exterior

		Crew	Daily Output	Labor-Hours	Unit	Material	Labor	Equipment	Total	Total Incl O&P
0010	**PVC TRIM, EXTERIOR**									
0100	Cornerboards, 5/4" x 6" x 6"	1 Carp	240	.033	L.F.	6.60	.86		7.46	8.70
0110	Door / window casing, 1" x 4"		200	.040		1.40	1.03		2.43	3.28
0120	1" x 6"		200	.040		2.18	1.03		3.21	4.14
0130	1" x 8"		195	.041		2.87	1.05		3.92	4.95
0140	1" x 10"		195	.041		3.65	1.05		4.70	5.80
0150	1" x 12"		190	.042		4.45	1.08		5.53	6.75
0160	5/4" x 4"		195	.041		1.78	1.05		2.83	3.75
0170	5/4" x 6"		195	.041		2.79	1.05		3.84	4.86
0180	5/4" x 8"		190	.042		3.69	1.08		4.77	5.90
0190	5/4" x 10"		190	.042		4.70	1.08		5.78	7
0200	5/4" x 12"		185	.043		5.70	1.11		6.81	8.20
0210	Fascia, 1" x 4"		250	.032		1.40	.82		2.22	2.94
0220	1" x 6"		250	.032		2.18	.82		3	3.80
0230	1" x 8"		225	.036		2.87	.91		3.78	4.71
0240	1" x 10"		225	.036		3.65	.91		4.56	5.55
0250	1" x 12"		200	.040		4.45	1.03		5.48	6.65
0260	5/4" x 4"		240	.033		1.78	.86		2.64	3.41
0270	5/4" x 6"		240	.033		2.79	.86		3.65	4.52
0280	5/4" x 8"		215	.037		3.69	.96		4.65	5.70
0290	5/4" x 10"		215	.037		4.70	.96		5.66	6.75
0300	5/4" x 12"		190	.042		5.70	1.08		6.78	8.15
0310	Fascia, 1" x 4"		250	.032		1.40	.82		2.22	2.94
0320	1" x 6"	↓	250	.032	↓	2.18	.82		3	3.80

06 65 Plastic Simulated Wood Trim

06 65 10 – Plastic Simulated Wood Trim

06 65 10.10 PVC Trim, Exterior		Crew	Daily Output	Labor-Hours	Unit	Material	2007 Bare Costs Labor	Equipment	Total	Total Incl O&P
0330	1" x 8"	1 Carp	225	.036	L.F.	2.87	.91		3.78	4.71
0340	1" x 10"		225	.036		3.65	.91		4.56	5.55
0350	1" x 12"		200	.040		4.45	1.03		5.48	6.65
0360	5/4" x 4"		240	.033		1.78	.86		2.64	3.41
0370	5/4" x 6"		240	.033		2.79	.86		3.65	4.52
0380	5/4" x 8"		215	.037		3.69	.96		4.65	5.70
0390	5/4" x 10"		215	.037		4.70	.96		5.66	6.75
0400	5/4" x 12"		190	.042		5.70	1.08		6.78	8.15
0410	Rake, 1" x 4"		200	.040		1.40	1.03		2.43	3.28
0420	1" x 6"		200	.040		2.18	1.03		3.21	4.14
0430	1" x 8"		190	.042		2.87	1.08		3.95	5
0440	1" x 10"		190	.042		3.65	1.08		4.73	5.85
0450	1" x 12"		180	.044		4.45	1.14		5.59	6.85
0460	5/4" x 4"		195	.041		1.78	1.05		2.83	3.75
0470	5/4" x 6"		195	.041		2.79	1.05		3.84	4.86
0480	5/4" x 8"		185	.043		3.69	1.11		4.80	5.95
0490	5/4" x 10"		185	.043		4.70	1.11		5.81	7.05
0500	5/4" x 12"		175	.046		5.70	1.17		6.87	8.30
0510	Rake trim, 1" x 4"		225	.036		1.40	.91		2.31	3.09
0520	1" x 6"		225	.036		2.18	.91		3.09	3.95
0560	5/4" x 4"		220	.036		1.78	.93		2.71	3.55
0570	5/4" x 6"		220	.036		2.79	.93		3.72	4.66
0610	Soffit, 1" x 4"	2 Carp	420	.038		1.40	.98		2.38	3.20
0620	1" x 6"		420	.038		2.18	.98		3.16	4.06
0630	1" x 8"		420	.038		2.87	.98		3.85	4.82
0640	1" x 10"		400	.040		3.65	1.03		4.68	5.75
0650	1" x 12"		400	.040		4.45	1.03		5.48	6.65
0660	5/4" x 4"		410	.039		1.78	1		2.78	3.66
0670	5/4" x 6"		410	.039		2.79	1		3.79	4.77
0680	5/4" x 8"		410	.039		3.69	1		4.69	5.75
0690	5/4" x 10"		390	.041		4.70	1.05		5.75	6.95
0700	5/4" x 12"		390	.041		5.70	1.05		6.75	8.10

Division 7
Thermal and Moisture Protection

07 01 Operation and Maint. of Thermal and Moisture Protection

07 01 50 – Maintenance of Membrane Roofing

07 01 50.10 Roof Coatings		Crew	Daily Output	Labor-Hours	Unit	Material	2007 Bare Costs Labor	Equipment	Total	Total Incl O&P
0010	**ROOF COATINGS**									
0012	Asphalt, brush grade, material only				Gal.	7.45			7.45	8.15
0800	Glass fibered roof & patching cement, 5 gallon					4.96			4.96	5.45
1100	Roof patch & flashing cement, 5 gallon					11.25			11.25	12.35

07 05 Common Work Results for Thermal and Moisture Protection

07 05 05 – Selective Demolition

07 05 05.10 Selective Demolition, Roofing and Siding		Crew	Daily Output	Labor-Hours	Unit	Material	2007 Bare Costs Labor	Equipment	Total	Total Incl O&P
0010	**SELECTIVE DEMOLITION, ROOFING AND SIDING** R024119-10									
0200	Waterproofing demo., scrape off vertical waterproofing, to 1/2" thick	2 Clab	2000	.008	S.F.		.15		.15	.25
0210	Over 1/2" thick		1750	.009	"		.17		.17	.29
0250	Protection / drain board		3900	.004	B.F.		.08		.08	.13
1200	Wood, boards, tongue and groove, 2" x 6"		960	.017	S.F.		.31		.31	.53
1220	2" x 10"		1040	.015			.29		.29	.49
1280	Standard planks, 1" x 6"		1080	.015			.28		.28	.47
1320	1" x 8"		1160	.014			.26		.26	.44
1340	1" x 12"		1200	.013			.25		.25	.42
1350	Plywood, to 1" thick		2000	.008			.15		.15	.25
1360	Flashing, aluminum	1 Clab	290	.028			.52		.52	.88
2000	Gutters, aluminum or wood, edge hung	"	240	.033	L.F.		.62		.62	1.06
2010	Remove and reset, aluminum	1 Shee	125	.064			1.84		1.84	3.09
2020	Remove and reset, vinyl	1 Carp	125	.064			1.64		1.64	2.79
2100	Built-in	1 Clab	100	.080			1.50		1.50	2.54
2200	Insulation removal, loose fitting		3000	.003	C.F.		.05		.05	.08
2250	Air barrier		3500	.002	S.F.		.04		.04	.07
2300	Batts or blankets		1400	.006	C.F.		.11		.11	.18
2350	Rigid board		3450	.002	B.F.		.04		.04	.07
2500	Roof accessories, plumbing vent flashing		14	.571	Ea.		10.70		10.70	18.15
2600	Adjustable metal chimney flashing		9	.889	"		16.60		16.60	28
2650	Coping, sheet metal, up to 12" wide		240	.033	L.F.		.62		.62	1.06
2660	Concrete, up to 12" wide	2 Clab	160	.100	"		1.87		1.87	3.18
3000	Roofing, built-up, 5 ply roof, no gravel	B-2	1600	.025	S.F.		.48		.48	.81
3100	Gravel removal, minimum		5000	.008			.15		.15	.26
3120	Maximum		2000	.020			.38		.38	.65
3400	Roof insulation board, up to 2" thick		3900	.010			.20		.20	.33
3405	Over 2" thick		7800	.005	B.F.		.10		.10	.17
3450	Roll roofing, cold adhesive	1 Clab	12	.667	Sq.		12.45		12.45	21
4000	Shingles, asphalt strip, 1 layer	B-2	3500	.011	S.F.		.22		.22	.37
4100	Slate		2500	.016			.31		.31	.52
4300	Wood		2200	.018			.35		.35	.59
4500	Skylight to 10 S.F.	1 Clab	8	1	Ea.		18.70		18.70	32
5000	Siding, metal, horizontal		444	.018	S.F.		.34		.34	.57
5020	Vertical		400	.020			.37		.37	.64
5200	Wood, boards, vertical		400	.020			.37		.37	.64
5220	Clapboards, horizontal		380	.021			.39		.39	.67
5240	Shingles		350	.023			.43		.43	.73
5260	Textured plywood		725	.011			.21		.21	.35

07 11 Dampproofing

07 11 13 – Bituminous Damppoofing

07 11 13.10 Bituminous Asphalt Coating		Crew	Daily Output	Labor-Hours	Unit	Material	2007 Bare Costs Labor	Equipment	Total	Total Incl O&P
0010	**BITUMINOUS ASPHALT COATING**									
0030	Brushed on, below grade, 1 coat	1 Rofc	665	.012	S.F.	.13	.26		.39	.62
0100	2 coat		500	.016		.25	.35		.60	.92
0300	Sprayed on, below grade, 1 coat, 25.6 S.F./gal.		830	.010		.13	.21		.34	.53
0400	2 coat, 20.5 S.F./gal.		500	.016		.25	.35		.60	.91
0600	Troweled on, asphalt with fibers, 1/16" thick		500	.016		.24	.35		.59	.90
0700	1/8" thick		400	.020		.42	.44		.86	1.27
1000	1/2" thick		350	.023		1.36	.50		1.86	2.41

07 11 16 – Cementitious Dampproofing

07 11 16.20 Cementitious Parging										
0010	**CEMENTITIOUS PARGING**									
0020	Portland cement, 2 coats, 1/2" thick	D-1	250	.064	S.F.	.20	1.49		1.69	2.70
0100	Waterproofed Portland cement	"	250	.064	"	2.03	1.49		3.52	4.72

07 19 Water Repellents

07 19 19 – Silicone Water Repellents

07 19 19.10 Silicone Water Repellents										
0010	**SILICONE WATER REPELLENTS**									
0020	Water base liquid, roller applied	2 Rofc	7000	.002	S.F.	.58	.05		.63	.73
0200	Silicone or stearate, sprayed on CMU, 1 coat	1 Rofc	4000	.002		.33	.04		.37	.44
0300	2 coats	"	3000	.003		.66	.06		.72	.83

07 21 Thermal Insulation

07 21 13 – Board Insulation

07 21 13.10 Rigid Insulation										
0010	**RIGID INSULATION**									
0040	Fiberglass, 1.5#/CF, unfaced, 1" thick, R4.1	1 Carp	1000	.008	S.F.	.44	.21		.65	.83
0060	1-1/2" thick, R6.2		1000	.008		.79	.21		1	1.22
0080	2" thick, R8.3		1000	.008		.68	.21		.89	1.10
0120	3" thick, R12.4		800	.010		.79	.26		1.05	1.31
0370	3#/CF, unfaced, 1" thick, R4.3		1000	.008		.49	.21		.70	.89
0390	1-1/2" thick, R6.5		1000	.008		.94	.21		1.15	1.38
0400	2" thick, R8.7		890	.009		1.14	.23		1.37	1.64
0420	2-1/2" thick, R10.9		800	.010		1	.26		1.26	1.54
0440	3" thick, R13		800	.010		1	.26		1.26	1.54
0520	Foil faced, 1" thick, R4.3		1000	.008		.91	.21		1.12	1.35
0540	1-1/2" thick, R6.5		1000	.008		1.35	.21		1.56	1.84
0560	2" thick, R8.7		890	.009		1.69	.23		1.92	2.25
0580	2-1/2" thick, R10.9		800	.010		2	.26		2.26	2.64
0600	3" thick, R13		800	.010		2.17	.26		2.43	2.83
0670	6#/CF, unfaced, 1" thick, R4.3		1000	.008		.97	.21		1.18	1.42
0690	1-1/2" thick, R6.5		890	.009		1.49	.23		1.72	2.03
0700	2" thick, R8.7		800	.010		2.10	.26		2.36	2.75
0721	2-1/2" thick, R10.9		800	.010		2.30	.26		2.56	2.97
0741	3" thick, R13		730	.011		2.75	.28		3.03	3.51
0821	Foil faced, 1" thick, R4.3		1000	.008		1.37	.21		1.58	1.86
0840	1-1/2" thick, R6.5		890	.009		1.96	.23		2.19	2.55
0850	2" thick, R8.7		800	.010		2.56	.26		2.82	3.26
0880	2-1/2" thick, R10.9		800	.010		3.08	.26		3.34	3.83

07 21 Thermal Insulation

07 21 13 – Board Insulation

07 21 13.10 Rigid Insulation

		Crew	Daily Output	Labor-Hours	Unit	Material	2007 Bare Costs Labor	Equipment	Total	Total Incl O&P
0900	3" thick, R13	1 Carp	730	.011	S.F.	3.68	.28		3.96	4.53
1500	Foamglass, 1-1/2" thick, R4.5		800	.010		1.31	.26		1.57	1.88
1550	3" thick, R9	↓	730	.011	↓	3.16	.28		3.44	3.96
1600	Isocyanurate, 4' x 8' sheet, foil faced, both sides									
1610	1/2" thick, R3.9	1 Carp	800	.010	S.F.	.30	.26		.56	.77
1620	5/8" thick, R4.5		800	.010		.50	.26		.76	.99
1630	3/4" thick, R5.4		800	.010		.35	.26		.61	.83
1640	1" thick, R7.2		800	.010		.55	.26		.81	1.05
1650	1-1/2" thick, R10.8		730	.011		.64	.28		.92	1.18
1660	2" thick, R14.4		730	.011		.87	.28		1.15	1.44
1670	3" thick, R21.6		730	.011		1.90	.28		2.18	2.57
1680	4" thick, R28.8		730	.011		2.14	.28		2.42	2.83
1700	Perlite, 1" thick, R2.77		800	.010		.30	.26		.56	.77
1750	2" thick, R5.55		730	.011		.59	.28		.87	1.13
1900	Extruded polystyrene, 25 PSI compressive strength, 1" thick, R5		800	.010		.47	.26		.73	.96
1940	2" thick R10		730	.011		.99	.28		1.27	1.57
1960	3" thick, R15		730	.011		1.34	.28		1.62	1.95
2100	Expanded polystyrene, 1" thick, R3.85		800	.010		.22	.26		.48	.68
2120	2" thick, R7.69		730	.011		.59	.28		.87	1.13
2140	3" thick, R11.49	↓	730	.011	↓	.75	.28		1.03	1.31

07 21 13.13 Foam Board Insulation

		Crew	Daily Output	Labor-Hours	Unit	Material	Labor	Equipment	Total	Total Incl O&P
0010	**FOAM BOARD INSULATION**									
0600	Polystyrene, expanded, 1" thick, R4	1 Carp	680	.012	S.F.	.29	.30		.59	.83
0700	2" thick, R8	"	675	.012	"	.59	.30		.89	1.17

07 21 16 – Blanket Insulation

07 21 16.10 Blanket Insulation for Floors

		Crew	Daily Output	Labor-Hours	Unit	Material	Labor	Equipment	Total	Total Incl O&P
0010	**BLANKET INSULATION FOR FLOORS**									
0020	Including spring type wire fasteners									
2000	Fiberglass, blankets or batts, paper or foil backing									
2100	1 side, 3-1/2" thick, R11	1 Carp	700	.011	S.F.	.45	.29		.74	1
2150	6" thick, R19		600	.013		.51	.34		.85	1.14
2200	8-1/2" thick, R30	↓	550	.015	↓	1.05	.37		1.42	1.79

07 21 16.20 Blanket Insulation for Walls

		Crew	Daily Output	Labor-Hours	Unit	Material	Labor	Equipment	Total	Total Incl O&P
0010	**BLANKET INSULATION FOR WALLS**									
0040	Fiberglass, kraft faced, batts or blankets									
0061	3-1/2" thick, R11, 11" wide	1 Carp	1600	.005	S.F.	.37	.13		.50	.63
0080	15" wide		1600	.005		.37	.13		.50	.63
0141	6" thick, R19, 11" wide		1350	.006		.43	.15		.58	.73
0201	9" thick, R30, 15" wide		1350	.006		.97	.15		1.12	1.33
0241	12" thick, R38, 15" wide	↓	1350	.006	↓	.90	.15		1.05	1.25
0400	Fiberglass, foil faced, batts or blankets									
0420	3-1/2" thick, R11, 15" wide	1 Carp	1600	.005	S.F.	.54	.13		.67	.81
0461	6" thick, R19, 15" wide		1600	.005		.57	.13		.70	.85
0501	9" thick, R30, 15" wide	↓	1350	.006	↓	.83	.15		.98	1.17
0800	Fiberglass, unfaced, batts or blankets									
0821	3-1/2" thick, R11, 15" wide	1 Carp	1600	.005	S.F.	.37	.13		.50	.63
0861	6" thick, R19, 15" wide		1350	.006		.46	.15		.61	.77
0901	9" thick, R30, 15" wide		1150	.007		.83	.18		1.01	1.21
0941	12" thick, R38, 15" wide	↓	1150	.007	↓	.89	.18		1.07	1.28
1300	Mineral fiber batts, kraft faced									
1320	3-1/2" thick, R12	1 Carp	1600	.005	S.F.	.38	.13		.51	.64
1340	6" thick, R19	↓	1600	.005	↓	.50	.13		.63	.77

07 21 Thermal Insulation

07 21 16 – Blanket Insulation

07 21 16.20 Blanket Insulation for Walls

		Crew	Daily Output	Labor-Hours	Unit	Material	2007 Bare Costs Labor	Equipment	Total	Total Incl O&P
1380	10" thick, R30	1 Carp	1350	.006	S.F.	.74	.15		.89	1.07
1850	Friction fit wire insulation supports, 16" O.C.	↓	960	.008	Ea.	.08	.21		.29	.45
1900	For foil backing, add				S.F.	.06			.06	.07

07 21 23 – Loose-Fill Insulation

07 21 23.10 Loose-Fill Insulation

		Crew	Daily Output	Labor-Hours	Unit	Material	Labor	Equipment	Total	Total Incl O&P
0010	**LOOSE-FILL INSULATION**									
0020	R3.8 per inch	1 Carp	200	.040	C.F.	.63	1.03		1.66	2.43
0080	Fiberglass wool, R4 per inch		200	.040		.50	1.03		1.53	2.29
0100	Mineral wool, R3 per inch		200	.040		.39	1.03		1.42	2.17
0300	Polystyrene, R4 per inch		200	.040		3.06	1.03		4.09	5.10
0400	Vermiculite or perlite, R2.7 per inch	↓	200	.040	↓	1.70	1.03		2.73	3.61

07 21 23.20 Masonry Loose-Fill Insulation

		Crew	Daily Output	Labor-Hours	Unit	Material	Labor	Equipment	Total	Total Incl O&P
0010	**MASONRY LOOSE-FILL INSULATION**									
0100	In cores of concrete block, 4" thick wall, .115 CF/SF	D-1	4800	.003	S.F.	.20	.08		.28	.35
0700	Foamed in place, urethane in 2-5/8" cavity	G-2	1035	.023		.41	.48	.12	1.01	1.38
0800	For each 1" added thickness, add	"	2372	.010	↓	.12	.21	.05	.38	.54

07 21 26 – Blown Insulation

07 21 26.10 Blown Insulation

		Crew	Daily Output	Labor-Hours	Unit	Material	Labor	Equipment	Total	Total Incl O&P
0010	**BLOWN INSULATION**									
0020	Cellulose, 3-1/2" thick, R13	G-4	5000	.005	S.F.	.21	.09	.05	.35	.45
0030	5-3/16" thick, R19		3800	.006		.31	.12	.07	.50	.63
0050	6-1/2" thick, R22		3000	.008		.40	.16	.09	.65	.80
1000	Fiberglass, 5" thick, R11		3800	.006		.21	.12	.07	.40	.52
1050	6" thick, R13		3000	.008		.25	.16	.09	.50	.64
1100	8-1/2" thick, R19		2200	.011		.36	.21	.12	.69	.88
1300	12" thick, R26		1500	.016		.50	.31	.18	.99	1.27
2000	Mineral wool, 4" thick, R12		3500	.007		.23	.13	.08	.44	.56
2050	6" thick, R17		2500	.010		.26	.19	.11	.56	.73
2100	9" thick, R23	↓	1750	.014	↓	.34	.27	.15	.76	.99
2500	Wall installation, incl. drilling & patching from outside, two 1"									
2510	diam. holes @ 16" O.C., top & mid-point of wall, add to above									
2700	For masonry	G-4	415	.058	S.F.	.06	1.12	.64	1.82	2.67
2800	For wood siding		840	.029		.06	.55	.31	.92	1.36
2900	For stucco/plaster	↓	665	.036	↓	.06	.70	.40	1.16	1.70

07 21 27 – Reflective Insulation

07 21 27.10 Reflective Insulation, Aluminum Foil On Reinforced Scrim

		Crew	Daily Output	Labor-Hours	Unit	Material	Labor	Equipment	Total	Total Incl O&P
0011	**REFLECTIVE INSULATION**, aluminum foil on reinforced scrim		1900	.004		.14	.11		.25	.33
0101	Reinforced with woven polyolefin		1900	.004		.17	.11		.28	.37
0501	With single bubble air space, R8.8		1500	.005		.25	.14		.39	.51
0601	With double bubble air space, R9.8	↓	1500	.005	↓	.26	.14		.40	.52

07 22 Roof and Deck Insulation

07 22 16 – Roof Board Insulation

07 22 16.10 Roof Deck Insulation		Crew	Daily Output	Labor-Hours	Unit	Material	2007 Bare Costs Labor	Equipment	Total	Total Incl O&P
0010	**ROOF DECK INSULATION**									
0020	Fiberboard low density, 1/2" thick R1.39	1 Rofc	1000	.008	S.F.	.24	.18		.42	.58
0030	1" thick R2.78		800	.010		.42	.22		.64	.86
0080	1 1/2" thick R4.17		800	.010		.63	.22		.85	1.09
0100	2" thick R5.56		800	.010		.84	.22		1.06	1.32
0110	Fiberboard high density, 1/2" thick R1.3		1000	.008		.22	.18		.40	.56
0120	1" thick R2.5		800	.010		.44	.22		.66	.88
0130	1-1/2" thick R3.8		800	.010		.65	.22		.87	1.12
0200	Fiberglass, 3/4" thick R2.78		1000	.008		.55	.18		.73	.93
0400	15/16" thick R3.70		1000	.008		.73	.18		.91	1.12
0460	1-1/16" thick R4.17		1000	.008		.92	.18		1.10	1.33
0600	1-5/16" thick R5.26		1000	.008		1.25	.18		1.43	1.70
0650	2-1/16" thick R8.33		800	.010		1.34	.22		1.56	1.87
0700	2-7/16" thick R10		800	.010		1.53	.22		1.75	2.08
1650	Perlite, 1/2" thick R1.32		1050	.008		.33	.17		.50	.67
1655	3/4" thick R2.08		800	.010		.36	.22		.58	.80
1660	1" thick R2.78		800	.010		.45	.22		.67	.90
1670	1-1/2" thick R4.17		800	.010		.47	.22		.69	.92
1680	2" thick R5.56		700	.011		.78	.25		1.03	1.32
1685	2-1/2" thick R6.67		700	.011		.92	.25		1.17	1.47
1700	Polyisocyanurate, 2#/CF density, 3/4" thick, R5.1		1500	.005		.44	.12		.56	.69
1705	1" thick R7.14		1400	.006		.50	.13		.63	.78
1715	1-1/2" thick R10.87		1250	.006		.63	.14		.77	.95
1725	2" thick R14.29		1100	.007		.83	.16		.99	1.20
1735	2-1/2" thick R16.67		1050	.008		1.01	.17		1.18	1.42
1745	3" thick R21.74		1000	.008		1.26	.18		1.44	1.71
1755	3-1/2" thick R25		1000	.008		1.93	.18		2.11	2.44
1765	Tapered for drainage		1400	.006	B.F.	1.93	.13		2.06	2.35
1900	Extruded Polystyrene									
1910	15 PSI compressive strength, 1" thick, R5	1 Rofc	1500	.005	S.F.	.45	.12		.57	.71
1920	2" thick, R10		1250	.006		.59	.14		.73	.91
1930	3" thick R15		1000	.008		1.21	.18		1.39	1.65
1932	4" thick R20		1000	.008		1.56	.18		1.74	2.04
1934	Tapered for drainage		1500	.005	B.F.	.50	.12		.62	.76
1940	25 PSI compressive strength, 1" thick R5		1500	.005	S.F.	.63	.12		.75	.90
1942	2" thick R10		1250	.006		1.21	.14		1.35	1.59
1944	3" thick R15		1000	.008		1.84	.18		2.02	2.34
1946	4" thick R20		1000	.008		2.59	.18		2.77	3.17
1948	Tapered for drainage		1500	.005	B.F.	.53	.12		.65	.79
1950	40 psi compressive strength, 1" thick R5		1500	.005	S.F.	.47	.12		.59	.73
1952	2" thick R10		1250	.006		.91	.14		1.05	1.26
1954	3" thick R15		1000	.008		1.34	.18		1.52	1.79
1956	4" thick R20		1000	.008		1.79	.18		1.97	2.29
1958	Tapered for drainage		1400	.006	B.F.	.67	.13		.80	.97
1960	60 PSI compressive strength, 1" thick R5		1450	.006	S.F.	.56	.12		.68	.84
1962	2" thick R10		1200	.007		1	.15		1.15	1.37
1964	3" thick R15		975	.008		1.48	.18		1.66	1.96
1966	4" thick R20		950	.008		2.07	.18		2.25	2.62
1968	Tapered for drainage		1400	.006	B.F.	.81	.13		.94	1.12
2010	Expanded polystyrene, 1#/CF density, 3/4" thick R2.89		1500	.005	S.F.	.29	.12		.41	.53
2020	1" thick R3.85		1500	.005		.29	.12		.41	.53
2100	2" thick R7.69		1250	.006		.59	.14		.73	.91
2110	3" thick R11.49		1250	.006		.87	.14		1.01	1.22

07 22 Roof and Deck Insulation

07 22 16 – Roof Board Insulation

07 22 16.10 Roof Deck Insulation

		Crew	Daily Output	Labor-Hours	Unit	Material	2007 Bare Costs Labor	Equipment	Total	Total Incl O&P
2120	4" thick R15.38	1 Rofc	1200	.007	S.F.	.80	.15		.95	1.15
2130	5" thick R19.23		1150	.007		1.01	.15		1.16	1.39
2140	6" thick R23.26		1150	.007		1.18	.15		1.33	1.58
2150	Tapered for drainage		1500	.005	B.F.	.47	.12		.59	.73
2400	Composites with 2" EPS									
2410	1" fiberboard	1 Rofc	950	.008	S.F.	1.04	.18		1.22	1.48
2420	7/16" oriented strand board		800	.010		1.23	.22		1.45	1.75
2430	1/2" plywood		800	.010		1.33	.22		1.55	1.86
2440	1" perlite		800	.010		1.09	.22		1.31	1.60
2450	Composites with 1-1/2" polyisocyanurate									
2460	1" fiberboard	1 Rofc	800	.010	S.F.	1.42	.22		1.64	1.96
2470	1" perlite		850	.009		1.49	.21		1.70	2.02
2480	7/16" oriented strand board		800	.010		1.72	.22		1.94	2.29

07 24 Exterior Insulation and Finish Systems

07 24 13 – Polymer Based Exterior Insulation and Finish Systems

07 24 13.10 Polymer Based Exterior Insulation and Finish Systems

		Crew	Daily Output	Labor-Hours	Unit	Material	2007 Bare Costs Labor	Equipment	Total	Total Incl O&P
0010	**POLYMER BASED EXTERIOR INSULATION AND FINISH SYSTEMS**									
0095	Field applied, 1" EPS insulation	J-1	295	.136	S.F.	2.15	3	.39	5.54	7.75
0100	With 1/2" cement board sheathing		220	.182		2.87	4.03	.52	7.42	10.40
0105	2" EPS insulation		295	.136		2.52	3	.39	5.91	8.15
0110	With 1/2" cement board sheathing		220	.182		3.24	4.03	.52	7.79	10.80
0115	3" EPS insulation		295	.136		2.68	3	.39	6.07	8.35
0120	With 1/2" cement board sheathing		220	.182		3.40	4.03	.52	7.95	10.95
0125	4" EPS insulation		295	.136		3.11	3	.39	6.50	8.80
0130	With 1/2" cement board sheathing		220	.182		4.55	4.03	.52	9.10	12.20
0140	Premium finish add		1265	.032		.31	.70	.09	1.10	1.60
0150	Heavy duty reinforcement add		914	.044		1.07	.97	.12	2.16	2.92
0160	2.5#/S.Y. metal lath substrate add	1 Lath	75	.107	S.Y.	2.35	2.52		4.87	6.70
0170	3.4#/S.Y. metal lath substrate add	"	75	.107	"	2.55	2.52		5.07	6.90
0180	Color or texture change,	J-1	1265	.032	S.F.	.82	.70	.09	1.61	2.16
0190	With substrate leveling base coat	1 Plas	530	.015		.82	.35		1.17	1.49
0210	With substrate sealing base coat	1 Pord	1224	.007		.08	.15		.23	.34
0370	V groove shape in panel face				L.F.	.59			.59	.65
0380	U groove shape in panel face				"	.78			.78	.86

07 26 Vapor Retarders

07 26 10 – Vapor Retarders

07 26 10.10 Building Paper Alum. & Kraft Lamin., Foil 1 Side

		Crew	Daily Output	Labor-Hours	Unit	Material	2007 Bare Costs Labor	Equipment	Total	Total Incl O&P
0011	**BUILDING PAPER** Aluminum and kraft laminated, foil 1 side		3700	.002		.05	.06		.11	.15
0101	Foil 2 sides		3700	.002		.08	.06		.14	.18
0301	Asphalt, two ply, 30#, for subfloors		1900	.004		.14	.11		.25	.34
0401	Asphalt felt sheathing paper, 15#		3700	.002		.04	.06		.10	.14
0450	Housewrap, exterior, spun bonded polypropylene									
0470	Small roll	1 Carp	3800	.002	S.F.	.24	.05		.29	.35
0480	Large roll	"	4000	.002	"	.13	.05		.18	.23
0500	Material only, 3' x 111.1' roll				Ea.	80			80	88
0520	9' x 111.1' roll				"	130			130	143
0601	Polyethylene vapor barrier, standard, .002" thick	1 Carp	3700	.002	S.F.	.01	.06		.07	.10

07 26 Vapor Retarders

07 26 10 – Vapor Retarders

07 26 10.10 Building Paper Alum. & Kraft Lamin., Foil 1 Side		Crew	Daily Output	Labor-Hours	Unit	Material	2007 Bare Costs Labor	Equipment	Total	Total Incl O&P
0701	.004" thick	1 Carp	3700	.002	S.F.	.03	.06		.09	.12
0901	.006" thick		3700	.002		.05	.06		.11	.14
1201	.010" thick		3700	.002		.05	.06		.11	.15
1501	Red rosin paper, 5 sq rolls, 4 lb per square		3700	.002		.02	.06		.08	.11
1601	5 lbs. per square		3700	.002		.03	.06		.09	.12
1801	Reinf. waterproof, .002" polyethylene backing, 1 side		3700	.002		.05	.06		.11	.15
1901	2 sides		3700	.002		.07	.06		.13	.16
3000	Building wrap, spunbonded polyethylene	2 Carp	8000	.002		.14	.05		.19	.24

07 31 Shingles and Shakes

07 31 13 – Asphalt Shingles

07 31 13.10 Asphalt Shingles

		Crew	Daily Output	Labor-Hours	Unit	Material	Labor	Equipment	Total	Total Incl O&P
0010	**ASPHALT SHINGLES**									
0100	Standard strip shingles									
0150	Inorganic, class A, 210-235 lb/sq	1 Rofc	5.50	1.455	Sq.	41.50	32		73.50	105
0155	Pneumatic nailed		7	1.143		41.50	25		66.50	92
0200	Organic, class C, 235-240 lb/sq		5	1.600		44.50	35		79.50	114
0205	Pneumatic nailed		6.25	1.280		44.50	28		72.50	101
0250	Standard, laminated multi-layered shingles									
0300	Class A, 240-260 lb/sq	1 Rofc	4.50	1.778	Sq.	52	39		91	129
0305	Pneumatic nailed		5.63	1.422		52	31		83	115
0350	Class C, 260-300 lb/square, 4 bundles/square		4	2		51.50	44		95.50	138
0355	Pneumatic nailed		5	1.600		51.50	35		86.50	122
0400	Premium, laminated multi-layered shingles									
0450	Class A, 260-300 lb, 4 bundles/sq	1 Rofc	3.50	2.286	Sq.	67.50	50		117.50	166
0455	Pneumatic nailed		4.37	1.831		67.50	40		107.50	148
0500	Class C, 300-385 lb/square, 5 bundles/square		3	2.667		75.50	58.50		134	191
0505	Pneumatic nailed		3.75	2.133		75.50	47		122.50	170
0800	#15 felt underlayment		64	.125		4.23	2.74		6.97	9.70
0825	#30 felt underlayment		58	.138		7.05	3.03		10.08	13.30
0850	Self adhering polyethylene and rubberized asphalt underlayment		22	.364		49.50	8		57.50	69
0900	Ridge shingles		330	.024	L.F.	1.44	.53		1.97	2.56
0905	Pneumatic nailed		412.50	.019	"	1.44	.43		1.87	2.36
1000	For steep roofs (7 to 12 pitch or greater), add						50%			

07 31 19 – Mineral-Fiber Cement Shingles

07 31 19.10 Fiber Cement Shingles

		Crew	Daily Output	Labor-Hours	Unit	Material	Labor	Equipment	Total	Total Incl O&P
0010	**FIBER CEMENT SHINGLES**									
0012	Field shingles, 16" x 9.35", 500 lb per square	1 Rofc	2.20	3.636	Sq.	315	80		395	490
0200	Shakes, 16" x 9.35", 550 lb per square		2.20	3.636	"	283	80		363	455
0301	Hip & ridge, 4.75 x 14"		100	.080	L.F.	7.70	1.76		9.46	11.70
0400	Hexagonal, 16" x 16"		3	2.667	Sq.	212	58.50		270.50	340
0500	Square, 16" x 16"		3	2.667		190	58.50		248.50	315
2000	For steep roofs (7/12 pitch or greater), add						50%			

07 31 26 – Slate Shingles

07 31 26.10 Slate Shingles

			Crew	Daily Output	Labor-Hours	Unit	Material	Labor	Equipment	Total	Total Incl O&P
0010	**SLATE SHINGLES**	R073126-20									
0100	Buckingham Virginia black, 3/16" - 1/4" thick		1 Rots	1.75	4.571	Sq.	395	101		496	620
0200	1/4" thick			1.75	4.571		410	101		511	635
0900	Pennsylvania black, Bangor, #1 clear			1.75	4.571		475	101		576	710
1200	Vermont, unfading, green, mottled green			1.75	4.571		435	101		536	665

07 31 Shingles and Shakes

07 31 26 – Slate Shingles

07 31 26.10 Slate Shingles		Crew	Daily Output	Labor-Hours	Unit	Material	2007 Bare Costs Labor	Equipment	Total	Total Incl O&P
1300	Semi-weathering green & gray	1 Rots	1.75	4.571	Sq.	345	101		446	565
1400	Purple		1.75	4.571		390	101		491	615
1500	Black or gray		1.75	4.571		445	101		546	675
2700	Ridge shingles, slate		200	.040	L.F.	9	.88		9.88	11.50

07 31 29 – Wood Shingles and Shakes

07 31 29.13 Wood Shingles

		Crew	Daily Output	Labor-Hours	Unit	Material	Labor	Equipment	Total	Total Incl O&P
0010	**WOOD SHINGLES**									
0012	16" No. 1 red cedar shingles, 5" exposure, on roof	1 Carp	2.50	3.200	Sq.	189	82		271	350
0015	Pneumatic nailed		3.25	2.462		189	63.50		252.50	315
0200	7-1/2" exposure, on walls		2.05	3.902		126	100		226	310
0205	Pneumatic nailed		2.67	2.996		126	77		203	270
0300	18" No. 1 red cedar perfections, 5-1/2" exposure, on roof		2.75	2.909		167	75		242	310
0305	Pneumatic nailed		3.57	2.241		167	57.50		224.50	282
0500	7-1/2" exposure, on walls		2.25	3.556		123	91.50		214.50	290
0505	Pneumatic nailed		2.92	2.740		123	70.50		193.50	254
0600	Resquared, and rebutted, 5-1/2" exposure, on roof		3	2.667		240	68.50		308.50	380
0605	Pneumatic nailed		3.90	2.051		240	52.50		292.50	355
0900	7-1/2" exposure, on walls		2.45	3.265		176	84		260	335
0905	Pneumatic nailed		3.18	2.516		176	64.50		240.50	305
1000	Add to above for fire retardant shingles, 16" long					42			42	46
1050	18" long					42			42	46
1060	Preformed ridge shingles	1 Carp	400	.020	L.F.	1.75	.51		2.26	2.80
1100	Hand-split red cedar shakes, 1/2" thick x 24" long, 10" exp. on roof		2.50	3.200	Sq.	165	82		247	320
1105	Pneumatic nailed		3.25	2.462		165	63.50		228.50	288
1110	3/4" thick x 24" long, 10" exp. on roof		2.25	3.556		165	91.50		256.50	335
1115	Pneumatic nailed		2.92	2.740		165	70.50		235.50	300
1200	1/2" thick, 18" long, 8-1/2" exp. on roof		2	4		191	103		294	385
1205	Pneumatic nailed		2.60	3.077		191	79		270	345
1210	3/4" thick x 18" long, 8 1/2" exp. on roof		1.80	4.444		191	114		305	405
1215	Pneumatic nailed		2.34	3.419		191	88		279	360
1255	10" exp. on walls		2	4		216	103		319	410
1260	10" exposure on walls, pneumatic nailed		2.60	3.077		216	79		295	370
1700	Add to above for fire retardant shakes, 24" long					42			42	46
1800	18" long					42			42	46
1810	Ridge shakes	1 Carp	350	.023	L.F.	3	.59		3.59	4.30
2000	White cedar shingles, 16" long, extras, 5" exposure, on roof		2.40	3.333	Sq.	184	85.50		269.50	350
2005	Pneumatic nailed		3.12	2.564		184	66		250	315
2050	5" exposure on walls		2	4		184	103		287	375
2055	Pneumatic nailed		2.60	3.077		184	79		263	335
2100	7-1/2" exposure, on walls		2	4		132	103		235	320
2105	Pneumatic nailed		2.60	3.077		132	79		211	279
2150	"B" grade, 5" exposure on walls		2	4		170	103		273	360
2155	Pneumatic nailed		2.60	3.077		170	79		249	320
2300	For 15# organic felt underlayment on roof, 1 layer, add		64	.125		4.23	3.21		7.44	10.10
2400	2 layers, add		32	.250		8.45	6.45		14.90	20
2600	For steep roofs (7/12 pitch or greater), add to above						50%			
3000	Ridge shakes or shingle wood	1 Carp	280	.029	L.F.	3	.73		3.73	4.55

07 32 Roof Tiles

07 32 13 – Clay Roof Tiles

07 32 13.10 Clay Tiles

		Crew	Daily Output	Labor-Hours	Unit	Material	2007 Bare Costs Labor	Equipment	Total	Total Incl O&P
0010	**CLAY TILES**									
0200	Lanai tile or Classic tile, 158 pc per sq	1 Rots	1.65	4.848	Sq.	470	107		577	715
0300	Americana, 158 pc per sq, most colors		1.65	4.848		605	107		712	865
0350	Green, gray or brown		1.65	4.848		590	107		697	840
0400	Blue		1.65	4.848		590	107		697	840
0600	Spanish tile, 171 pc per sq, red		1.80	4.444		298	98		396	505
0800	Blend		1.80	4.444		565	98		663	800
0900	Glazed white		1.80	4.444		625	98		723	870
1100	Mission tile, 192 pc per sq, machine scored finish, red		1.15	6.957		700	153		853	1,050
1700	French tile, 133 pc per sq, smooth finish, red		1.35	5.926		635	131		766	940
1750	Blue or green		1.35	5.926		820	131		951	1,150
1800	Norman black 317 pc per sq		1	8		930	176		1,106	1,350
2200	Williamsburg tile, 158 pc per sq, aged cedar		1.35	5.926		560	131		691	855
2250	Gray or green		1.35	5.926		560	131		691	855
2350	Ridge shingles, clay tile		200	.040	L.F.	10.10	.88		10.98	12.70
2510	One piece mission tile, natural red, 75 pc per square		1.65	4.848	Sq.	218	107		325	435
2530	Mission Tile, 134 pc per square		1.15	6.957		248	153		401	555
3000	For steep roofs (7/12 pitch or greater), add to above						50%			

07 32 16 – Concrete Roof Tiles

07 32 16.10 Concrete Tiles

		Crew	Daily Output	Labor-Hours	Unit	Material	Labor	Equipment	Total	Total Incl O&P
0010	**CONCRETE TILES**									
0020	Corrugated, 13" x 16-1/2", 90 per sq, 950 lb per sq									
0050	Earthtone colors, nailed to wood deck	1 Rots	1.35	5.926	Sq.	93.50	131		224.50	345
0150	Blues		1.35	5.926		94.50	131		225.50	345
0200	Greens		1.35	5.926		94.50	131		225.50	345
0250	Premium colors		1.35	5.926		208	131		339	470
0500	Shakes, 13" x 16-1/2", 90 per sq, 950 lb per sq									
0600	All colors, nailed to wood deck	1 Rots	1.50	5.333	Sq.	245	118		363	485
1500	Accessory pieces, ridge & hip, 10" x 16-1/2", 8 lbs. each				Ea.	3.30			3.30	3.63
1700	Rake, 6-1/2" x 16-3/4", 9 lbs. each					3.30			3.30	3.63
1800	Mansard hip, 10" x 16-1/2", 9.2 lbs. each					3.30			3.30	3.63
1900	Hip starter, 10" x 16-1/2", 10.5 lbs. each					10.40			10.40	11.45
2000	3 or 4 way apex, 10" each side, 11.5 lbs. each					12			12	13.20

07 41 Roof Panels

07 41 13 – Metal Roof Panels

07 41 13.10 Aluminum Roof Panels

		Crew	Daily Output	Labor-Hours	Unit	Material	Labor	Equipment	Total	Total Incl O&P
0010	**ALUMINUM ROOF PANELS**									
0020	Corrugated or ribbed, .0155" thick, natural	G-3	1200	.027	S.F.	.72	.63		1.35	1.86
0300	Painted	"	1200	.027	"	1.05	.63		1.68	2.23

07 41 33 – Plastic Roof Panels

07 41 33.10 Corrugated Fiberglass Panels

		Crew	Daily Output	Labor-Hours	Unit	Material	Labor	Equipment	Total	Total Incl O&P
0010	**CORRUGATED FIBERGLASS PANELS**									
0012	Corrugated, 8 oz per SF	G-3	1000	.032	S.F.	1.53	.76		2.29	2.96
0100	12 oz per SF		1000	.032		3.33	.76		4.09	4.94
0300	Corrugated siding, 6 oz per SF		880	.036		1.53	.86		2.39	3.14
0400	8 oz per SF		880	.036		2.30	.86		3.16	3.99
0600	12 oz. siding, textured		880	.036		3.22	.86		4.08	5
0900	Flat panels, 6 oz per SF, clear or colors		880	.036		1.78	.86		2.64	3.42
1300	8 oz per SF, clear or colors		880	.036		2.31	.86		3.17	4

07 42 Wall Panels

07 42 13 – Metal Wall Panels

07 42 13.20 Aluminum Siding		Crew	Daily Output	Labor-Hours	Unit	Material	2007 Bare Costs Labor	Equipment	Total	Total Incl O&P
0011	**ALUMINUM SIDING**									
6040	.024 thick smooth white single 8" wide	2 Carp	515	.031	S.F.	1.64	.80		2.44	3.15
6060	Double 4" pattern		515	.031		1.17	.80		1.97	2.64
6080	Double 5" pattern		550	.029		1.20	.75		1.95	2.59
6120	Embossed white, 8" wide		515	.031		1.64	.80		2.44	3.15
6140	Double 4" pattern		515	.031		1.73	.80		2.53	3.25
6160	Double 5" pattern		550	.029		1.73	.75		2.48	3.17
6170	Vertical, embossed white, 12" wide		590	.027		1.73	.70		2.43	3.08
6320	.019 thick, insulated, smooth white, 8" wide		515	.031		1.53	.80		2.33	3.03
6340	Double 4" pattern		515	.031		1.51	.80		2.31	3.01
6360	Double 5" pattern		550	.029		1.51	.75		2.26	2.93
6400	Embossed white, 8" wide		515	.031		1.77	.80		2.57	3.30
6420	Double 4" pattern		515	.031		1.80	.80		2.60	3.33
6440	Double 5" pattern		550	.029		1.80	.75		2.55	3.25
6500	Shake finish 10" wide white		550	.029		1.90	.75		2.65	3.36
6600	Vertical pattern, 12" wide, white		590	.027		1.80	.70		2.50	3.16
6640	For colors add					.10			.10	.11
6700	Accessories, white									
6720	Starter strip 2-1/8"	2 Carp	610	.026	L.F.	.29	.67		.96	1.46
6740	Sill trim		450	.036		.43	.91		1.34	2.02
6760	Inside corner		610	.026		1.29	.67		1.96	2.56
6780	Outside corner post		610	.026		2.18	.67		2.85	3.54
6800	Door & window trim		440	.036		.43	.93		1.36	2.06
6820	For colors add					.09			.09	.10
6900	Soffit & fascia 1' overhang solid	2 Carp	110	.145		2.83	3.74		6.57	9.45
6920	Vented		110	.145		2.83	3.74		6.57	9.45
6940	2' overhang solid		100	.160		4.29	4.11		8.40	11.70
6960	Vented		100	.160		4.29	4.11		8.40	11.70

07 42 13.30 Steel Siding		Crew	Daily Output	Labor-Hours	Unit	Material	Labor	Equipment	Total	Total Incl O&P
0010	**STEEL SIDING**									
0020	Beveled, vinyl coated, 8" wide, including fasteners	1 Carp	265	.030	S.F.	1.62	.78		2.40	3.10
0050	10" wide	"	275	.029		1.73	.75		2.48	3.17
0081	Galv., corrugated or ribbed, on steel frame, 30 gauge	G-3	775	.041		1.08	.98		2.06	2.84
0101	28 gauge		775	.041		1.13	.98		2.11	2.89
0301	26 gauge		775	.041		1.59	.98		2.57	3.40
0401	24 gauge		775	.041		1.60	.98		2.58	3.41
0601	22 gauge		775	.041		1.84	.98		2.82	3.67
0701	Colored, corrugated/ribbed, on steel frame, 10 yr fnsh, 28 ga.		775	.041		1.68	.98		2.66	3.50
0901	26 gauge		775	.041		1.76	.98		2.74	3.59
1001	24 gauge		775	.041		2.04	.98		3.02	3.89

07 46 Siding

07 46 23 – Wood Siding

07 46 23.10 Wood Board Siding

		Crew	Daily Output	Labor-Hours	Unit	Material	2007 Bare Costs Labor	Equipment	Total	Total Incl O&P
0010	**WOOD BOARD SIDING**									
2000	Board & batten, cedar, "B" grade, 1" x 10"	1 Carp	400	.020	S.F.	1.98	.51		2.49	3.05
2200	Redwood, clear, vertical grain, 1" x 10"		400	.020		3.65	.51		4.16	4.89
2400	White pine, #2 & better, 1" x 10"		400	.020		.79	.51		1.30	1.74
2410	Board & batten siding, white pine #2, 1" x 12"		450	.018		.79	.46		1.25	1.65
3200	Wood, cedar bevel, A grade, 1/2" x 6"		250	.032		3.05	.82		3.87	4.76
3300	1/2" x 8"		275	.029		3.37	.75		4.12	4.98
3500	3/4" x 10", clear grade		300	.027		4.76	.69		5.45	6.40
3600	"B" grade		300	.027		3.35	.69		4.04	4.85
3800	Cedar, rough sawn, 1" x 4", A grade, natural		240	.033		1.74	.86		2.60	3.36
3900	Stained		240	.033		2.04	.86		2.90	3.69
4100	1" x 12", board & batten, #3 & Btr., natural		260	.031		1.07	.79		1.86	2.52
4200	Stained		260	.031		1.15	.79		1.94	2.61
4400	1" x 8" channel siding, #3 & Btr., natural		250	.032		2.14	.82		2.96	3.75
4500	Stained		250	.032		2.43	.82		3.25	4.07
4700	Redwood, clear, beveled, vertical grain, 1/2" x 4"		200	.040		3.46	1.03		4.49	5.55
4750	1/2" x 6"		225	.036		2.86	.91		3.77	4.70
4800	1/2" x 8"		250	.032		2.31	.82		3.13	3.94
5000	3/4" x 10"		300	.027		3.80	.69		4.49	5.35
5200	Channel siding, 1" x 10", B grade		285	.028		2.50	.72		3.22	3.97
5250	Redwood, T&G boards, B grade, 1" x 4"	2 Carp	300	.053		2.90	1.37		4.27	5.50
5270	1" x 8"	"	375	.043		2.60	1.10		3.70	4.72
5400	White pine, rough sawn, 1" x 8", natural	1 Carp	275	.029		1.07	.75		1.82	2.45
5500	Stained	"	275	.029		1.15	.75		1.90	2.54
5600	Tongue and groove, 1" x 8", horizontal	2 Carp	375	.043		1.07	1.10		2.17	3.04

07 46 26 – Hardboard Siding

07 46 26.10 Hardboard Siding

		Crew	Daily Output	Labor-Hours	Unit	Material	Labor	Equipment	Total	Total Incl O&P
0010	**HARDBOARD SIDING**									
0030	Lap siding, hardboard, 7/16" x 8", primed									
0050	Wood grain texture finish	2 Carp	650	.025	S.F.	1.13	.63		1.76	2.31
0100	Panels, 7/16" thick, smooth, textured or grooved, primed		700	.023		.92	.59		1.51	2.01
0200	Stained		700	.023		1	.59		1.59	2.10
0700	Particle board, overlaid, 3/8" thick		750	.021		.70	.55		1.25	1.70

07 46 29 – Plywood Siding

07 46 29.10 Plywood Siding

		Crew	Daily Output	Labor-Hours	Unit	Material	Labor	Equipment	Total	Total Incl O&P
0010	**PLYWOOD SIDING**									
0900	Plywood, medium density overlaid, 3/8" thick	2 Carp	750	.021	S.F.	.85	.55		1.40	1.87
1000	1/2" thick		700	.023		1.07	.59		1.66	2.18
1100	3/4" thick		650	.025		1.29	.63		1.92	2.49
1600	Texture 1-11, cedar, 5/8" thick, natural		675	.024		2.41	.61		3.02	3.68
1700	Factory stained		675	.024		1.92	.61		2.53	3.14
1900	Texture 1-11, fir, 5/8" thick, natural		675	.024		1.11	.61		1.72	2.25
2000	Factory stained		675	.024		1.70	.61		2.31	2.90
2050	Texture 1-11, S.Y.P., 5/8" thick, natural		675	.024		1.19	.61		1.80	2.34
2100	Factory stained		675	.024		1.23	.61		1.84	2.38
2200	Rough sawn cedar, 3/8" thick, natural		675	.024		1.19	.61		1.80	2.34
2300	Factory stained		675	.024		.95	.61		1.56	2.08
2500	Rough sawn fir, 3/8" thick, natural		675	.024		.75	.61		1.36	1.86
2600	Factory stained		675	.024		.95	.61		1.56	2.08
2800	Redwood, textured siding, 5/8" thick		675	.024		1.90	.61		2.51	3.12

07 46 Siding

07 46 33 – Plastic Siding

07 46 33.10 Vinyl Siding

		Crew	Daily Output	Labor-Hours	Unit	Material	2007 Bare Costs Labor	Equipment	Total	Total Incl O&P
0010	**VINYL SIDING**									
0020	Clapboard profile, woodgrain texture, .048 thick, double 4	1 Carp	255	.031	S.F.	.78	.81		1.59	2.23
2000	Smooth, white, single, 8" wide	2 Carp	495	.032		.89	.83		1.72	2.39
2020	Dutch lap, 10" wide		550	.029		.84	.75		1.59	2.19
2100	Double 4" pattern, 8" wide		495	.032		.84	.83		1.67	2.33
2120	Double 5" pattern, 10" wide		550	.029		.73	.75		1.48	2.07
2200	Embossed, white, single, 8" wide		495	.032		.88	.83		1.71	2.38
2220	10" wide		550	.029		.71	.75		1.46	2.05
2300	Double 4" pattern, 8" wide		495	.032		.69	.83		1.52	2.17
2320	5" pattern, 10" wide		550	.029		.69	.75		1.44	2.03
2400	Shake finish, 10" wide, white		550	.029		2.01	.75		2.76	3.48
2600	Vertical pattern, double 5", 10" wide, white		550	.029		1.70	.75		2.45	3.14
2620										
2700	For colors, add				S.F.	.09			.09	.10
2720	1/4" extruded polystyrene fan folded insulation	2 Carp	2000	.008	"	.21	.21		.42	.58
3000	Accessories, starter strip		700	.023	L.F.	.34	.59		.93	1.37
3100	"J" channel, 1/2"		700	.023		.35	.59		.94	1.39
3120	5/8"		700	.023		.36	.59		.95	1.40
3140	3/4"		695	.023		.37	.59		.96	1.41
3160	1"		690	.023		.38	.60		.98	1.43
3180	1-1/8"		685	.023		.40	.60		1	1.46
3190	1-1/4"		680	.024		.48	.60		1.08	1.56
3200	Under sill trim		500	.032		.40	.82		1.22	1.84
3300	Outside corner post, 3" face, pocket 5/8"		700	.023		1.52	.59		2.11	2.67
3320	7/8"		690	.023		1.58	.60		2.18	2.75
3340	1-1/4"		680	.024		1.39	.60		1.99	2.56
3400	Inside corner post, pocket 5/8"		700	.023		.80	.59		1.39	1.88
3420	7/8"		690	.023		.90	.60		1.50	2
3440	1-1/4"		680	.024		.78	.60		1.38	1.89
3500	Door & window trim, 2-1/2" face, pocket 5/8"		510	.031		.78	.81		1.59	2.23
3520	7/8"		500	.032		.85	.82		1.67	2.34
3540	1-1/4"		490	.033		.93	.84		1.77	2.44
3600	Soffit & fascia, 1' overhang, solid		120	.133		1.48	3.43		4.91	7.45
3620	Vented		120	.133		1.65	3.43		5.08	7.60
3700	2' overhang, solid		110	.145		2.33	3.74		6.07	8.90
3720	Vented		110	.145		2.33	3.74		6.07	8.90

07 46 46 – Mineral-Fiber Cement Siding

07 46 46.10 Fiber Cement Siding

		Crew	Daily Output	Labor-Hours	Unit	Material	2007 Bare Costs Labor	Equipment	Total	Total Incl O&P
0010	**FIBER CEMENT SIDING**									
0020	Lap siding, 5/16" thick, 6" wide, smooth texture	2 Carp	415	.039	S.F.	1.34	.99		2.33	3.15
0025	Woodgrain texture		415	.039		1.34	.99		2.33	3.15
0030	7-1/2" wide, smooth texture		425	.038		.94	.97		1.91	2.67
0035	Woodgrain texture		425	.038		.94	.97		1.91	2.67
0040	8" wide, smooth texture		425	.038		1.10	.97		2.07	2.85
0045	Roughsawn texture		425	.038		1.10	.97		2.07	2.85
0050	9-1/2" wide, smooth texture		440	.036		1.06	.93		1.99	2.76
0055	Woodgrain texture		440	.036		1.06	.93		1.99	2.76
0060	12" wide, smooth texture		455	.035		1.04	.90		1.94	2.67
0065	Woodgrain texture		455	.035		1.04	.90		1.94	2.67
0070	Panel siding, 5/16" thick, smooth texture		750	.021		.87	.55		1.42	1.89
0075	Stucco texture		750	.021		.87	.55		1.42	1.89
0080	Grooved woodgrain texture		750	.021		.87	.55		1.42	1.89

07 46 Siding

07 46 46 – Mineral-Fiber Cement Siding

07 46 46.10 Fiber Cement Siding

		Crew	Daily Output	Labor-Hours	Unit	Material	2007 Bare Costs Labor	Equipment	Total	Total Incl O&P
0085	V - grooved woodgrain texture	2 Carp	750	.021	S.F.	.87	.55		1.42	1.89
0090	Wood starter strip	↓	400	.040	L.F.	.31	1.03		1.34	2.08

07 46 73 – Soffit

07 46 73.10 Soffit

		Crew	Daily Output	Labor-Hours	Unit	Material	Labor	Equipment	Total	Total Incl O&P
0010	**SOFFIT**									
0012	Aluminum, residential, .020" thick	1 Carp	210	.038	S.F.	1.40	.98		2.38	3.20
0100	Baked enamel on steel, 16 or 18 gauge		105	.076		4.97	1.96		6.93	8.75
0300	Polyvinyl chloride, white, solid		230	.035		.87	.89		1.76	2.48
0400	Perforated	↓	230	.035		.87	.89		1.76	2.48
0500	For colors, add					.10			.10	.11

07 51 Built-Up Bituminous Roofing

07 51 13 – Built-Up Asphalt Roofing

07 51 13.10 Built-Up Roofing Components

		Crew	Daily Output	Labor-Hours	Unit	Material	Labor	Equipment	Total	Total Incl O&P
0010	**BUILT-UP ROOFING COMPONENTS**									
0012	Asphalt saturated felt, #30, 2 square per roll	1 Rofc	58	.138	Sq.	7.05	3.03		10.08	13.30
0200	#15, 4 sq per roll, plain or perforated, not mopped		58	.138		4.23	3.03		7.26	10.20
0250	Perforated		58	.138		4.23	3.03		7.26	10.20
0300	Roll roofing, smooth, #65		15	.533		7.35	11.70		19.05	29.50
0500	#90		15	.533		23	11.70		34.70	46.50
0520	Mineralized		15	.533		18.05	11.70		29.75	41.50
0540	D.C. (Double coverage), 19" selvage edge	↓	10	.800		39	17.55		56.55	75
0580	Adhesive (lap cement)				Gal.	4.37			4.37	4.81

07 51 13.20 Built-Up Roofing Systems

		Crew	Daily Output	Labor-Hours	Unit	Material	Labor	Equipment	Total	Total Incl O&P
0010	**BUILT-UP ROOFING SYSTEMS**									
0120	Asphalt flood coat with gravel/slag surfacing, not including									
0140	Insulation, flashing or wood nailers									
0200	Asphalt base sheet, 3 plies #15 asphalt felt, mopped	G-1	22	2.545	Sq.	62.50	52.50	15.10	130.10	182
0350	On nailable decks		21	2.667		67.50	55	15.80	138.30	192
0500	4 plies #15 asphalt felt, mopped		20	2.800		88	57.50	16.60	162.10	221
0550	On nailable decks	↓	19	2.947	↓	79	60.50	17.50	157	217
2000	Asphalt flood coat, smooth surface									
2200	Asphalt base sheet & 3 plies #15 asphalt felt, mopped	G-1	24	2.333	Sq.	66.50	48	13.85	128.35	177
2400	On nailable decks		23	2.435		62.50	50	14.45	126.95	176
2600	4 plies #15 asphalt felt, mopped		24	2.333		78.50	48	13.85	140.35	189
2700	On nailable decks	↓	23	2.435		74	50	14.45	138.45	189
4500	Coal tar pitch with gravel/slag surfacing									
4600	4 plies #15 tarred felt, mopped	G-1	21	2.667	Sq.	131	55	15.80	201.80	263
4800	3 plies glass fiber felt (type IV), mopped	"	19	2.947	"	108	60.50	17.50	186	248

07 51 13.30 Cants

		Crew	Daily Output	Labor-Hours	Unit	Material	Labor	Equipment	Total	Total Incl O&P
0010	**CANTS**									
0012	Lumber, treated, 4" x 4" cut diagonally	1 Rofc	325	.025	L.F.	1.38	.54		1.92	2.51
0100	Foamglass		325	.025		2.17	.54		2.71	3.38
0300	Mineral or fiber, trapezoidal, 1"x 4" x 48"		325	.025		.19	.54		.73	1.20
0400	1-1/2" x 5-5/8" x 48"	↓	325	.025		.31	.54		.85	1.33

07 52 Modified Bituminous Membrane Roofing

07 52 13 – Atactic-Polypropylene-Modified Bituminous Membrane Roofing

07 52 13.10 APP Modified Bituminous Membrane

		Crew	Daily Output	Labor-Hours	Unit	Material	2007 Bare Costs Labor	Equipment	Total	Total Incl O&P
0010	APP MODIFIED BITUMINOUS MEMBRANE R075213-30									
0020	Base sheet, #15 glass fiber felt, nailed to deck	1 Rofc	58	.138	Sq.	5.75	3.03		8.78	11.90
0030	Spot mopped to deck	G-1	295	.190		8.30	3.90	1.13	13.33	17.55
0040	Fully mopped to deck	"	192	.292		11.45	6	1.73	19.18	25.50
0050	#15 organic felt, nailed to deck	1 Rofc	58	.138		4.82	3.03		7.85	10.85
0060	Spot mopped to deck	G-1	295	.190		7.40	3.90	1.13	12.43	16.50
0070	Fully mopped to deck	"	192	.292		10.50	6	1.73	18.23	24.50
2100	APP mod., smooth surf. cap sheet, poly. reinf., torched, 160 mils	G-5	2100	.019	S.F.	.46	.38	.07	.91	1.29
2150	170 mils		2100	.019		.53	.38	.07	.98	1.36
2200	Granule surface cap sheet, poly. reinf., torched, 180 mils		2000	.020		.57	.40	.08	1.05	1.44
2250	Smooth surface flashing, torched, 160 mils		1260	.032		.46	.64	.12	1.22	1.81
2300	170 mils		1260	.032		.53	.64	.12	1.29	1.88
2350	Granule surface flashing, torched, 180 mils		1260	.032		.57	.64	.12	1.33	1.93
2400	Fibrated aluminum coating	1 Rofc	3800	.002		.09	.05		.14	.19

07 52 16 – Styrene-Butadiene-Styrene Modified Bituminous Membrane Roofing

07 52 16.10 SBS Modified Bituminous Membrane

		Crew	Daily Output	Labor-Hours	Unit	Material	Labor	Equipment	Total	Total Incl O&P
0010	SBS MODIFIED BITUMINOUS MEMBRANE									
0080	SBS modified, granule surf cap sheet, polyester rein., mopped									
1500	Glass fiber reinforced, mopped, 160 mils	G-1	2000	.028	S.F.	.47	.58	.17	1.22	1.76
1600	Smooth surface cap sheet, mopped, 145 mils		2100	.027		.47	.55	.16	1.18	1.70
1700	Smooth surface flashing, 145 mils		1260	.044		.47	.91	.26	1.64	2.49
1800	150 mils		1260	.044		.46	.91	.26	1.63	2.48
1900	Granular surface flashing, 150 mils		1260	.044		.51	.91	.26	1.68	2.53
2000	160 mils		1260	.044		.75	.91	.26	1.92	2.80

07 58 Roll Roofing

07 58 10 – Roll Roofing

07 58 10.10 Roll Roofing

		Crew	Daily Output	Labor-Hours	Unit	Material	Labor	Equipment	Total	Total Incl O&P
0010	ROLL ROOFING									
0100	Asphalt, mineral surface									
0200	1 ply #15 organic felt, 1 ply mineral surfaced									
0300	Selvage roofing, lap 19", nailed & mopped	G-1	27	2.074	Sq.	52	42.50	12.30	106.80	149
0400	3 plies glass fiber felt (type IV), 1 ply mineral surfaced									
0500	Selvage roofing, lapped 19", mopped	G-1	25	2.240	Sq.	80	46	13.30	139.30	187
0600	Coated glass fiber base sheet, 2 plies of glass fiber									
0700	Felt (type IV), 1 ply mineral surfaced selvage									
0800	Roofing, lapped 19", mopped	G-1	25	2.240	Sq.	86.50	46	13.30	145.80	195
0900	On nailable decks	"	24	2.333	"	80.50	48	13.85	142.35	192
1000	3 plies glass fiber felt (type III), 1 ply mineral surfaced									
1100	Selvage roofing, lapped 19", mopped	G-1	25	2.240	Sq.	80	46	13.30	139.30	187

07 61 Sheet Metal Roofing

07 61 13 – Standing Seam Sheet Metal Roofing

07 61 13.10 Standing Seam Sheet Metal Roofing

		Crew	Daily Output	Labor-Hours	Unit	Material	2007 Bare Costs Labor	Equipment	Total	Total Incl O&P
0010	STANDING SEAM SHEET METAL ROOFING									
0400	Copper standing seam roofing, over 10 squares, 16 oz, 125 lb per sq	1 Shee	1.30	6.154	Sq.	845	177		1,022	1,225
0600	18 oz, 140 lb per sq	"	1.20	6.667		950	192		1,142	1,375
1200	For abnormal conditions or small areas, add					25%	100%			
1300	For lead-coated copper, add					25%				

07 61 16 – Batten Seam Sheet Metal Roofing

07 61 16.10 Batten Seam Sheet Metal Roofing

0009	BATTEN SEAM SHEET METAL ROOFING									
0012	Copper batten seam roofing, 16 oz, 130 lb per sq	1 Shee	1.10	7.273	Sq.	880	209		1,089	1,325
0100	Zinc / copper alloy batten seam roofing, .020 thick		1.20	6.667		1,100	192		1,292	1,525
0200	18 oz, 145 lb per sq		1	8		980	230		1,210	1,450
0800	.027" thick		1.15	6.957		1,300	200		1,500	1,750
0900	.032" thick		1.10	7.273		1,500	209		1,709	2,000
1000	.040" thick		1.05	7.619		1,800	219		2,019	2,350

07 61 19 – Flat Seam Sheet Metal Roofing

07 61 19.10 Flat Seam Sheet Metal Roofing

0010	FLAT SEAM SHEET METAL ROOFING									
0900	Copper flat seam roofing, over 10 squares, 16 oz, 115 lb per sq	1 Shee	1.20	6.667	Sq.	780	192		972	1,175

07 62 Sheet Metal Flashing and Trim

07 62 10 – Sheet Metal Flashing and Trim

07 62 10.10 Sheet Metal Cladding

0010	SHEET METAL CLADDING									
0100	Aluminum, up to 6 bends, .032" thick, window casing	1 Carp	180	.044	S.F.	.76	1.14		1.90	2.78
0200	Window sill		72	.111	L.F.	.76	2.86		3.62	5.70
0300	Door casing		180	.044	S.F.	.76	1.14		1.90	2.78
0400	Fascia		250	.032		.76	.82		1.58	2.24
0500	Rake trim		225	.036		.76	.91		1.67	2.39
0700	.024" thick, window casing		180	.044		1.40	1.14		2.54	3.48
0800	Window sill		72	.111	L.F.	1.40	2.86		4.26	6.40
0900	Door casing		180	.044	S.F.	1.40	1.14		2.54	3.48
1000	Fascia		250	.032		1.40	.82		2.22	2.94
1100	Rake trim		225	.036		1.40	.91		2.31	3.09
1200	Vinyl coated aluminum, up to 6 bends, window casing		180	.044		.88	1.14		2.02	2.91
1300	Window sill		72	.111	L.F.	.88	2.86		3.74	5.80
1400	Door casing		180	.044	S.F.	.88	1.14		2.02	2.91
1500	Fascia		250	.032		.88	.82		1.70	2.37
1600	Rake trim		225	.036		.88	.91		1.79	2.52

07 65 Flexible Flashing

07 65 10 – Sheet Metal Flashing

07 65 10.10 Sheet Metal Flashing		Crew	Daily Output	Labor-Hours	Unit	Material	2007 Bare Costs Labor	Equipment	Total	Total incl O&P
0010	**SHEET METAL FLASHING**									
0011	Including up to 4 bends									
0020	Aluminum, mill finish, .013" thick	1 Rofc	145	.055	S.F.	.43	1.21		1.64	2.69
0030	.016" thick		145	.055		.63	1.21		1.84	2.91
0060	.019" thick		145	.055		.82	1.21		2.03	3.12
0100	.032" thick		145	.055		1.18	1.21		2.39	3.52
0200	.040" thick		145	.055		1.77	1.21		2.98	4.17
0300	.050" thick		145	.055		2.01	1.21		3.22	4.43
0325	Mill finish 5" x 7" step flashing, .016" thick		1920	.004	Ea.	.13	.09		.22	.31
0350	Mill finish 12" x 12" step flashing, .016" thick		1600	.005	"	.49	.11		.60	.74
0400	Painted finish, add				S.F.	.29			.29	.32
1600	Copper, 16 oz, sheets, under 1000 lbs.	1 Rofc	115	.070		5.05	1.53		6.58	8.35
1900	20 oz sheets, under 1000 lbs.		110	.073		6.50	1.60		8.10	10.10
2200	24 oz sheets, under 1000 lbs.		105	.076		7.95	1.67		9.62	11.80
2500	32 oz sheets, under 1000 lbs.		100	.080		10.60	1.76		12.36	14.85
2700	W shape for valleys, 16 oz, 24" wide		100	.080	L.F.	12.10	1.76		13.86	16.50
5800	Lead, 2.5 lb. per SF, up to 12" wide		135	.059	S.F.	3.03	1.30		4.33	5.70
5900	Over 12" wide		135	.059		3.63	1.30		4.93	6.40
8650	Copper, 16 oz		100	.080		4.33	1.76		6.09	8
8900	Stainless steel sheets, 32 ga, .010" thick		155	.052		3.13	1.13		4.26	5.50
9000	28 ga, .015" thick		155	.052		3.89	1.13		5.02	6.35
9100	26 ga, .018" thick		155	.052		4.71	1.13		5.84	7.30
9200	24 ga, .025" thick		155	.052		6.10	1.13		7.23	8.85
9290	For mechanically keyed flashing, add					40%				
9320	Steel sheets, galvanized, 20 gauge	1 Rofc	130	.062	S.F.	.98	1.35		2.33	3.56
9340	30 gauge		160	.050		.42	1.10		1.52	2.48
9400	Terne coated stainless steel, .015" thick, 28 ga		155	.052		6.05	1.13		7.18	8.80
9500	.018" thick, 26 ga		155	.052		6.85	1.13		7.98	9.65
9600	Zinc and copper alloy (brass), .020" thick		155	.052		4.26	1.13		5.39	6.75
9700	.027" thick		155	.052		5.70	1.13		6.83	8.40
9800	.032" thick		155	.052		6.65	1.13		7.78	9.45
9900	.040" thick		155	.052		8.15	1.13		9.28	11.05

07 65 13 – Laminated Sheet Flashing

07 65 13.10 Laminated Sheet Flashing		Crew	Daily Output	Labor-Hours	Unit	Material	2007 Bare Costs Labor	Equipment	Total	Total incl O&P
0010	**LAMINATED SHEET FLASHING**, Including up to 4 bends									
2800	Copper, paperbacked 1 side, 2 oz	1 Rofc	330	.024	S.F.	1.17	.53		1.70	2.27
2900	3 oz		330	.024		1.53	.53		2.06	2.66
3100	Paperbacked 2 sides, 2 oz		330	.024		1.18	.53		1.71	2.28
3150	3 oz		330	.024		1.52	.53		2.05	2.65
3200	5 oz		330	.024		2.28	.53		2.81	3.49
6100	Lead-coated copper, fabric-backed, 2 oz		330	.024		2	.53		2.53	3.18
6200	5 oz		330	.024		2.30	.53		2.83	3.51
6400	Mastic-backed 2 sides, 2 oz		330	.024		1.56	.53		2.09	2.70
6500	5 oz		330	.024		1.94	.53		2.47	3.11
6700	Paperbacked 1 side, 2 oz		330	.024		1.35	.53		1.88	2.47
6800	3 oz		330	.024		1.59	.53		2.12	2.73
7000	Paperbacked 2 sides, 2 oz		330	.024		1.39	.53		1.92	2.51
7100	5 oz		330	.024		2.27	.53		2.80	3.48
8550	3 ply copper and fabric, 3 oz		155	.052		2.26	1.13		3.39	4.57
8600	7 oz		155	.052		4.69	1.13		5.82	7.25
8700	Lead on copper and fabric, 5 oz		155	.052		2.30	1.13		3.43	4.61
8800	7 oz		155	.052		4.08	1.13		5.21	6.55

07 65 Flexible Flashing

07 65 13 – Laminated Sheet Flashing

07 65 13.10 Laminated Sheet Flashing		Crew	Daily Output	Labor-Hours	Unit	Material	2007 Bare Costs Labor	Equipment	Total	Total Incl O&P
9300	Stainless steel, paperbacked 2 sides, .005" thick	1 Rofc	330	.024	S.F.	2.76	.53		3.29	4.02

07 65 26 – Self-Adhering Sheet Flashing

07 65 26.10 Self-Adhering Sheet Flashing

0010	**SELF-ADHERING SHEET FLASHING**									
8500	Shower pan, bituminous membrane, 7 oz	1 Rofc	155	.052	S.F.	1.51	1.13		2.64	3.74

07 71 Roof Specialties

07 71 19 – Manufactured Gravel Stops and Fascias

07 71 19.10 Gravel Stop

0010	**GRAVEL STOP**									
0020	Aluminum, .050" thick, 4" face height, mill finish	1 Shee	145	.055	L.F.	4.66	1.59		6.25	7.80
0080	Duranodic finish		145	.055		4.50	1.59		6.09	7.60
0100	Painted		145	.055		5.20	1.59		6.79	8.35
1350	Galv steel, 24 ga., 4" leg, plain, with continuous cleat, 4" face		145	.055		1.83	1.59		3.42	4.68
1500	Polyvinyl chloride, 6" face height		135	.059		4	1.70		5.70	7.25
1800	Stainless steel, 24 ga., 6" face height		135	.059		8.75	1.70		10.45	12.45

07 71 19.30 Fascia

0010	**FASCIA**									
0100	Aluminum, reverse board and batten, .032" thick, colored, no furring incl	1 Shee	145	.055	S.F.	4.71	1.59		6.30	7.85
0200	Residential type, aluminum	1 Carp	200	.040	L.F.	1.24	1.03		2.27	3.10
0300	Steel, galv and enameled, stock, no furring, long panels	1 Shee	145	.055	S.F.	3.24	1.59		4.83	6.25
0600	Short panels	"	115	.070	"	4.32	2		6.32	8.10

07 71 23 – Manufactured Gutters and Downspouts

07 71 23.10 Downspouts

0010	**DOWNSPOUTS**									
0020	Aluminum 2" x 3", .020" thick, embossed	1 Shee	190	.042	L.F.	.91	1.21		2.12	3.04
0100	Enameled		190	.042		1.51	1.21		2.72	3.70
0300	Enameled, .024" thick, 2" x 3"		180	.044		1.51	1.28		2.79	3.81
0400	3" x 4"		140	.057		3.49	1.64		5.13	6.60
0600	Round, corrugated aluminum, 3" diameter, .020" thick		190	.042		1.42	1.21		2.63	3.60
0700	4" diameter, .025" thick		140	.057		1.92	1.64		3.56	4.87
0900	Wire strainer, round, 2" diameter		155	.052	Ea.	1.89	1.48		3.37	4.58
1000	4" diameter		155	.052		2.04	1.48		3.52	4.74
1200	Rectangular, perforated, 2" x 3"		145	.055		2.30	1.59		3.89	5.20
1300	3" x 4"		145	.055		3.32	1.59		4.91	6.30
1500	Copper, round, 16 oz., stock, 2" diameter		190	.042	L.F.	9	1.21		10.21	11.95
1600	3" diameter		190	.042		7.35	1.21		8.56	10.15
1800	4" diameter		145	.055		7.85	1.59		9.44	11.30
1900	5" diameter		130	.062		13.85	1.77		15.62	18.25
2100	Rectangular, corrugated copper, stock, 2" x 3"		190	.042		6.50	1.21		7.71	9.20
2200	3" x 4"		145	.055		8.60	1.59		10.19	12.10
2400	Rectangular, plain copper, stock, 2" x 3"		190	.042		6.75	1.21		7.96	9.45
2500	3" x 4"		145	.055		8.40	1.59		9.99	11.90
2700	Wire strainers, rectangular, 2" x 3"		145	.055	Ea.	5	1.59		6.59	8.15
2800	3" x 4"		145	.055		6	1.59		7.59	9.25
3000	Round, 2" diameter		145	.055		4	1.59		5.59	7.05
3100	3" diameter		145	.055		6.10	1.59		7.69	9.40
3300	4" diameter		145	.055		11.30	1.59		12.89	15.05
3400	5" diameter		115	.070		16.60	2		18.60	21.50
3600	Lead-coated copper, round, stock, 2" diameter		190	.042	L.F.	6.90	1.21		8.11	9.65

07 71 Roof Specialties

07 71 23 – Manufactured Gutters and Downspouts

07 71 23.10 Downspouts

		Crew	Daily Output	Labor-Hours	Unit	Material	Labor	Equipment	Total	Total Incl O&P
3700	3" diameter	1 Shee	190	.042	L.F.	6.90	1.21		8.11	9.65
3900	4" diameter		145	.055		9.60	1.59		11.19	13.20
4300	Rectangular, corrugated, stock, 2" x 3"		190	.042		9.75	1.21		10.96	12.80
4500	Plain, stock, 2" x 3"		190	.042		9.75	1.21		10.96	12.80
4600	3" x 4"		145	.055		10	1.59		11.59	13.65
4800	Steel, galvanized, round, corrugated, 2" or 3" diam, 28 ga		190	.042		1.43	1.21		2.64	3.61
4900	4" diameter, 28 gauge		145	.055		2.15	1.59		3.74	5.05
5700	Rectangular, corrugated, 28 gauge, 2" x 3"		190	.042		1.43	1.21		2.64	3.61
5800	3" x 4"		145	.055		1.43	1.59		3.02	4.24
6000	Rectangular, plain, 28 gauge, galvanized, 2" x 3"		190	.042		1.55	1.21		2.76	3.75
6100	3" x 4"		145	.055		1.80	1.59		3.39	4.65
6300	Epoxy painted, 24 gauge, corrugated, 2" x 3"		190	.042		1.70	1.21		2.91	3.91
6400	3" x 4"		145	.055		1.89	1.59		3.48	4.75
6600	Wire strainers, rectangular, 2" x 3"		145	.055	Ea.	1.70	1.59		3.29	4.54
6700	3" x 4"		145	.055		3.38	1.59		4.97	6.40
6900	Round strainers, 2" or 3" diameter		145	.055		1.70	1.59		3.29	4.54
7000	4" diameter		145	.055		3.38	1.59		4.97	6.40
7200	5" diameter		145	.055		2.38	1.59		3.97	5.30
7300	6" diameter		115	.070		2.84	2		4.84	6.50
8200	Vinyl, rectangular, 2" x 3"		210	.038	L.F.	.75	1.10		1.85	2.67
8300	Round, 2-1/2"		220	.036	"	.75	1.05		1.80	2.59

07 71 23.20 Elbows

		Crew	Daily Output	Labor-Hours	Unit	Material	Labor	Equipment	Total	Total Incl O&P
0010	**ELBOWS**									
0020	Aluminum, 2" x 3", embossed	1 Shee	100	.080	Ea.	3.96	2.30		6.26	8.25
0100	Enameled		100	.080		2.95	2.30		5.25	7.10
0200	3" x 4", .025" thick, embossed		100	.080		3.96	2.30		6.26	8.25
0300	Enameled		100	.080		3.84	2.30		6.14	8.10
0400	Round corrugated, 3", embossed, .020" thick		100	.080		2.64	2.30		4.94	6.75
0500	4", .025" thick		100	.080		5.50	2.30		7.80	9.90
0600	Copper, 16 oz. round, 2" diameter		100	.080		13.40	2.30		15.70	18.60
0700	3" diameter		100	.080		11.20	2.30		13.50	16.15
0800	4" diameter		100	.080		17.50	2.30		19.80	23
1000	2" x 3" corrugated		100	.080		9.30	2.30		11.60	14.05
1100	3" x 4" corrugated		100	.080		15.90	2.30		18.20	21.50
1300	Vinyl, 2-1/2" diameter, 45° or 75°		100	.080		2.88	2.30		5.18	7.05
1400	Tee Y junction		75	.107		10.80	3.07		13.87	17.05

07 71 23.30 Gutters

		Crew	Daily Output	Labor-Hours	Unit	Material	Labor	Equipment	Total	Total Incl O&P
0010	**GUTTERS**									
0012	Aluminum, stock units, 5" K type, .027" thick, plain	1 Shee	120	.067	L.F.	1.76	1.92		3.68	5.15
0020	Inside corner		25	.320	Ea.	6.70	9.20		15.90	23
0030	Outside corner		25	.320	"	6.70	9.20		15.90	23
0100	Enameled		120	.067	L.F.	1.85	1.92		3.77	5.25
0110	Inside corner		25	.320	Ea.	6.70	9.20		15.90	23
0120	Outside corner		25	.320	"	6.70	9.20		15.90	23
0300	5" K type type, .032" thick, plain		120	.067	L.F.	2.81	1.92		4.73	6.30
0310	Inside corner		25	.320	Ea.	6.70	9.20		15.90	23
0320	Outside corner		25	.320	"	2.21	9.20		11.41	17.90
0400	Enameled		120	.067	L.F.	2.36	1.92		4.28	5.80
0410	Inside corner		25	.320	Ea.	6.70	9.20		15.90	23
0420	Outside corner		25	.320	"	6.70	9.20		15.90	23
0700	Copper, half round, 16 oz, stock units, 4" wide		120	.067	L.F.	5.75	1.92		7.67	9.50
0900	5" wide		120	.067		7.30	1.92		9.22	11.25

07 71 Roof Specialties

07 71 23 – Manufactured Gutters and Downspouts

07 71 23.30 Gutters

		Crew	Daily Output	Labor-Hours	Unit	Material	2007 Bare Costs Labor	Equipment	Total	Total Incl O&P
1000	6" wide	1 Shee	115	.070	L.F.	7.95	2		9.95	12.10
1200	K type, 16 oz, stock, 4" wide		120	.067		7.05	1.92		8.97	10.95
1300	5" wide		120	.067		7.70	1.92		9.62	11.65
1500	Lead coated copper, half round, stock, 4" wide		120	.067		10.50	1.92		12.42	14.75
1600	6" wide		115	.070		15.95	2		17.95	21
1800	K type, stock, 4" wide		120	.067		11.35	1.92		13.27	15.70
1900	5" wide		120	.067		12.45	1.92		14.37	16.85
2100	Stainless steel, half round or box, stock, 4" wide		120	.067		5.25	1.92		7.17	9
2200	5" wide		120	.067		5.65	1.92		7.57	9.40
2400	Steel, galv, half round or box, 28 ga, 5" wide, plain		120	.067		1.30	1.92		3.22	4.65
2500	Enameled		120	.067		1.35	1.92		3.27	4.71
2700	26 ga, stock, 5" wide		120	.067		1.42	1.92		3.34	4.78
2800	6" wide		120	.067		2.01	1.92		3.93	5.45
3000	Vinyl, O.G., 4" wide	1 Carp	110	.073		.92	1.87		2.79	4.18
3100	5" wide		110	.073		1.08	1.87		2.95	4.36
3200	4" half round, stock units		110	.073		.73	1.87		2.60	3.97
3250	Joint connectors				Ea.	1.47			1.47	1.62
3300	Wood, clear treated cedar, fir or hemlock, 3" x 4"	1 Carp	100	.080	L.F.	6.40	2.06		8.46	10.55
3400	4" x 5"	"	100	.080	"	7.40	2.06		9.46	11.65

07 71 23.35 Gutter Guard

0010	**GUTTER GUARD**									
0020	6" wide strip, aluminum mesh	1 Carp	500	.016	L.F.	.39	.41		.80	1.13
0100	Vinyl mesh	"	500	.016	"	.40	.41		.81	1.14

07 71 43 – Drip Edge

07 71 43.10 Drip Edge

0010	**DRIP EDGE**									
0020	Aluminum, .016" thick, 5" wide, mill finish	1 Carp	400	.020	L.F.	.42	.51		.93	1.33
0100	White finish		400	.020		.39	.51		.90	1.30
0200	8" wide, mill finish		400	.020		.62	.51		1.13	1.55
0300	Ice belt, 28" wide, mill finish		100	.080		3.80	2.06		5.86	7.65
0310	Vented, mill finish		400	.020		1.39	.51		1.90	2.40
0320	Painted finish		400	.020		1.51	.51		2.02	2.53
0400	Galvanized, 5" wide		400	.020		.43	.51		.94	1.34
0500	8" wide, mill finish		400	.020		.52	.51		1.03	1.44
0510	Rake edge, aluminum, 1-1/2" x 1-1/2"		400	.020		.39	.51		.90	1.30
0520	3-1/2" x 1-1/2"		400	.020		.44	.51		.95	1.35

07 72 Roof Accessories

07 72 23 – Relief Vents

07 72 23.20 Vents

0010	**VENTS**									
0100	Soffit or eave, aluminum, mill finish, strips, 2-1/2" wide	1 Carp	200	.040	L.F.	.36	1.03		1.39	2.14
0200	3" wide		200	.040		.38	1.03		1.41	2.16
0300	Enamel finish, 3" wide		200	.040		.50	1.03		1.53	2.29
0400	Mill finish, rectangular, 4" x 16"		72	.111	Ea.	1.52	2.86		4.38	6.50
0500	8" x 16"		72	.111	"	1.83	2.86		4.69	6.85

07 72 26 – Ridge Vents

07 72 26.10 Ridge Vents

0010	**RIDGE VENTS**									
0100	Aluminum strips, mill finish	1 Rofc	160	.050	L.F.	1.64	1.10		2.74	3.82

07 72 Roof Accessories

07 72 26 – Ridge Vents

07 72 26.10 Ridge Vents

		Crew	Daily Output	Labor-Hours	Unit	Material	2007 Bare Costs Labor	Equipment	Total	Total Incl O&P
0150	Painted finish	1 Rofc	160	.050	L.F.	2.08	1.10		3.18	4.31
0200	Connectors		48	.167	Ea.	4.26	3.66		7.92	11.40
0300	End caps		48	.167	"	1.64	3.66		5.30	8.50
0400	Galvanized strips		160	.050	L.F.	2.14	1.10		3.24	4.37
0430	Molded polyethylene, shingles not included		160	.050	"	2.85	1.10		3.95	5.15
0440	End plugs		48	.167	Ea.	1.64	3.66		5.30	8.50
0450	Flexible roll, shingles not included		160	.050	L.F.	2.31	1.10		3.41	4.56
2300	Ridge vent strip, mill finish	1 Shee	155	.052	"	2.68	1.48		4.16	5.45

07 72 53 – Snow Guards

07 72 53.10 Snow Guards

		Crew	Daily Output	Labor-Hours	Unit	Material	Labor	Equipment	Total	Total Incl O&P
0010	**SNOW GUARDS**									
0100	Slate & asphalt shingle roofs	1 Rofc	160	.050	Ea.	9	1.10		10.10	11.90
0200	Standing seam metal roofs		48	.167		17	3.66		20.66	25.50
0300	Surface mount for metal roofs		48	.167		7.25	3.66		10.91	14.70
0400	Double rail pipe type, including pipe		130	.062	L.F.	18.05	1.35		19.40	22.50

07 72 80 – Vents

07 72 80.30 Vents

		Crew	Daily Output	Labor-Hours	Unit	Material	Labor	Equipment	Total	Total Incl O&P
0010	**VENTS**									
0800	Polystyrene baffles, 12" wide for 16" O.C. rafter spacing	1 Carp	90	.089	Ea.	.33	2.28		2.61	4.24

07 92 Joint Sealants

07 92 10 – Caulking and Sealants

07 92 10.10 Caulking and Sealants

		Crew	Daily Output	Labor-Hours	Unit	Material	Labor	Equipment	Total	Total Incl O&P
0010	**CAULKING AND SEALANTS**									
0020	Acoustical sealant, elastomeric, cartridges				Ea.	4.60			4.60	5.05
0100	Acrylic latex caulk, white									
0200	11 fl. oz cartridge				Ea.	2.40			2.40	2.64
0500	1/4" x 1/2"	1 Bric	248	.032	L.F.	.20	.86		1.06	1.65
0600	1/2" x 1/2"		250	.032		.39	.85		1.24	1.85
0800	3/4" x 3/4"		230	.035		.88	.93		1.81	2.51
0900	3/4" x 1"		200	.040		1.18	1.07		2.25	3.06
1000	1" x 1"		180	.044		1.47	1.18		2.65	3.59
1400	Butyl based, bulk				Gal.	24			24	26.50
1500	Cartridges				"	29			29	32
1700	Bulk, in place 1/4" x 1/2", 154 L.F./gal.	1 Bric	230	.035	L.F.	.16	.93		1.09	1.71
1800	1/2" x 1/2", 77 L.F./gal.	"	180	.044	"	.31	1.18		1.49	2.31
2000	Latex acrylic based, bulk				Gal.	25			25	27.50
2100	Cartridges				"	31			31	34
2200	Bulk in place, 1/4" x 1/2", 154 L.F./gal.	1 Bric	230	.035	L.F.	.16	.93		1.09	1.72
2300	Polysulfide compounds, 1 component, bulk				Gal.	47.50			47.50	52
2400	Cartridges				"	50.50			50.50	55.50
2600	1 or 2 component, in place, 1/4" x 1/4", 308 L.F./gal.	1 Bric	145	.055	L.F.	.15	1.47		1.62	2.61
2700	1/2" x 1/4", 154 L.F./gal.		135	.059		.31	1.58		1.89	2.97
2900	3/4" x 3/8", 68 L.F./gal.		130	.062		.70	1.64		2.34	3.50
3000	1" x 1/2", 38 L.F./gal.		130	.062		1.25	1.64		2.89	4.10
3200	Polyurethane, 1 or 2 component				Gal.	52.50			52.50	57.50
3300	Cartridges				"	49			49	54
3500	Bulk, in place, 1/4" x 1/4"	1 Bric	150	.053	L.F.	.17	1.42		1.59	2.55
3600	1/2" x 1/4"		145	.055		.34	1.47		1.81	2.81
3800	3/4" x 3/8", 68 L.F./gal.		130	.062		.77	1.64		2.41	3.58

07 92 Joint Sealants

07 92 10 – Caulking and Sealants

07 92 10.10 Caulking and Sealants		Crew	Daily Output	Labor-Hours	Unit	Material	2007 Bare Costs Labor	Equipment	Total	Total Incl O&P
3900	1" x 1/2"	1 Bric	110	.073	L.F.	1.36	1.94		3.30	4.72
4100	Silicone rubber, bulk				Gal.	40			40	44
4200	Cartridges				"	40			40	44

Division 8
Openings

08 01 Operation and Maintenance of Openings

08 01 53 – Operation and Maintenance of Plastic Windows

08 01 53.81 Solid Vinyl Replacement Windows

		Crew	Daily Output	Labor-Hours	Unit	Material	2007 Bare Costs Labor	Equipment	Total	Total Incl O&P
0010	**SOLID VINYL REPLACEMENT WINDOWS** R085313-20									
0020	Double hung, insulated glass, up to 83 united inches	2 Carp	8	2	Ea.	315	51.50		366.50	430
0040	84 to 93		8	2		340	51.50		391.50	460
0060	94 to 101		6	2.667		340	68.50		408.50	490
0080	102 to 111		6	2.667		360	68.50		428.50	515
0100	112 to 120		6	2.667		395	68.50		463.50	550
0120	For each united inch over 120, add		800	.020	Inch	4.65	.51		5.16	5.95
0140	Casement windows, one operating sash, 42 to 60 united inches		8	2	Ea.	191	51.50		242.50	297
0160	61 to 70		8	2		217	51.50		268.50	325
0180	71 to 80		8	2		236	51.50		287.50	345
0200	81 to 96		8	2		250	51.50		301.50	360
0220	Two operating sash, 58 to 78 united inches		8	2		380	51.50		431.50	505
0240	79 to 88		8	2		405	51.50		456.50	535
0260	89 to 98		8	2		445	51.50		496.50	575
0280	99 to 108		6	2.667		465	68.50		533.50	625
0300	109 to 121		6	2.667		500	68.50		568.50	665
0320	Three operating sash, 73 to 108 united inches		8	2		600	51.50		651.50	745
0340	109 to 118		8	2		635	51.50		686.50	780
0360	119 to 128		6	2.667		650	68.50		718.50	830
0380	129 to 138		6	2.667		695	68.50		763.50	880
0400	139 to 156		6	2.667		735	68.50		803.50	925
0420	Four operating sash, 98 to 118 united inches		8	2		865	51.50		916.50	1,025
0440	119 to 128		8	2		925	51.50		976.50	1,100
0460	129 to 138		6	2.667		980	68.50		1,048.50	1,200
0480	139 to 148		6	2.667		1,025	68.50		1,093.50	1,250
0500	149 to 168		6	2.667		1,100	68.50		1,168.50	1,325
0520	169 to 178		6	2.667		1,200	68.50		1,268.50	1,425
0540	For venting unit to fixed unit, deduct					17.40			17.40	19.15
0560	Fixed picture window, up to 63 united inches	2 Carp	8	2		138	51.50		189.50	239
0580	64 to 83		8	2		163	51.50		214.50	267
0600	84 to 101		8	2		210	51.50		261.50	320
0620	For each united inch over 101, add		900	.018	Inch	2.53	.46		2.99	3.56
0640	Picture window opt., low E glazing, up to 101 united inches				Ea.	19.65			19.65	21.50
0660	102 to 124					25.50			25.50	28
0680	124 and over					38			38	42
0700	Options, low E glazing, up to 101 united inches					9.80			9.80	10.80
0720	102 to 124					12.70			12.70	13.95
0740	124 and over					19.50			19.50	21.50
0760	Muntins, between glazing, square, per lite					2.10			2.10	2.31
0780	Diamond shape, per full or partial diamond					3.10			3.10	3.41
0800	Celluose fiber insulation, poured into sash balance cavity	1 Carp	36	.222	C.F.	.63	5.70		6.33	10.40
0820	Silicone caulking at perimeter	"	800	.010	L.F.	.13	.26		.39	.58

08 05 Common Work Results for Openings

08 05 05 – Selective Windows and Doors Demolition

08 05 05.10 Selective Demolition Doors

		Crew	Daily Output	Labor-Hours	Unit	Material	2007 Bare Costs Labor	2007 Bare Costs Equipment	Total	Total Incl O&P
0010	**SELECTIVE DEMOLITION DOORS** R024119-10									
0200	Doors, exterior, 1-3/4" thick, single, 3' x 7' high	1 Clab	16	.500	Ea.		9.35		9.35	15.90
0220	Double, 6' x 7' high		12	.667			12.45		12.45	21
0500	Interior, 1-3/8" thick, single, 3' x 7' high		20	.400			7.50		7.50	12.70
0520	Double, 6' x 7' high		16	.500			9.35		9.35	15.90
0700	Bi-folding, 3' x 6'-8" high		20	.400			7.50		7.50	12.70
0720	6' x 6'-8" high		18	.444			8.30		8.30	14.10
0900	Bi-passing, 3' x 6'-8" high		16	.500			9.35		9.35	15.90
0940	6' x 6'-8" high		14	.571			10.70		10.70	18.15
1500	Remove and reset, minimum	1 Carp	8	1			25.50		25.50	43.50
1520	Maximum		6	1.333			34.50		34.50	58
2000	Frames, including trim, metal		8	1			25.50		25.50	43.50
2200	Wood	2 Carp	32	.500			12.85		12.85	22
2201	Alternate pricing method	1 Carp	200	.040	L.F.		1.03		1.03	1.74
3000	Special doors, counter doors	2 Carp	6	2.667	Ea.		68.50		68.50	116
3300	Glass, sliding, including frames		12	1.333			34.50		34.50	58
3400	Overhead, commercial, 12' x 12' high		4	4			103		103	174
3500	Residential, 9' x 7' high		8	2			51.50		51.50	87
3540	16' x 7' high		7	2.286			58.50		58.50	99.50
3600	Remove and reset, minimum		4	4			103		103	174
3620	Maximum		2.50	6.400			164		164	279
3660	Remove and reset elec. garage door opener	1 Carp	8	1			25.50		25.50	43.50
4000	Residential lockset, exterior		30	.267			6.85		6.85	11.65
4200	Deadbolt lock		32	.250			6.45		6.45	10.90
9000	Minimum labor/equipment charge		4	2	Job		51.50		51.50	87

08 05 05.20 Selective Demolition of Windows

		Crew	Daily Output	Labor-Hours	Unit	Material	2007 Bare Costs Labor	2007 Bare Costs Equipment	Total	Total Incl O&P
0010	**SELECTIVE DEMOLITION OF WINDOWS** R024119-10									
0200	Aluminum, including trim, to 12 S.F.	1 Clab	16	.500	Ea.		9.35		9.35	15.90
0240	To 25 S.F.		11	.727			13.60		13.60	23
0280	To 50 S.F.		5	1.600			30		30	51
0320	Storm windows/scree, to 12 S.F.		27	.296			5.55		5.55	9.40
0360	To 25 S.F.		21	.381			7.10		7.10	12.10
0400	To 50 S.F.		16	.500			9.35		9.35	15.90
0500	Screens, incl. aluminum frame, small		20	.400			7.50		7.50	12.70
0510	Large		16	.500			9.35		9.35	15.90
0600	Glass, minimum		200	.040	S.F.		.75		.75	1.27
0620	Maximum		150	.053	"		1		1	1.69
2000	Wood, including trim, to 12 S.F.		22	.364	Ea.		6.80		6.80	11.55
2020	To 25 S.F.		18	.444			8.30		8.30	14.10
2060	To 50 S.F.		13	.615			11.50		11.50	19.55
2065	To 180 S.F.		8	1			18.70		18.70	32
4300	Remove bay/bow window	2 Carp	6	2.667			68.50		68.50	116
4410	Remove skylight, plstc domes, flush/curb mtd	G-3	395	.081	S.F.		1.92		1.92	3.24
5020	Remove and reset window, minimum	1 Carp	6	1.333	Ea.		34.50		34.50	58
5040	Average		4	2			51.50		51.50	87
5080	Maximum		2	4			103		103	174
9100	Window awning, residential	1 Clab	80	.100	L.F.		1.87		1.87	3.18

08 11 Metal Doors and Frames

08 11 63 – Metal Screen and Storm Doors and Frames

08 11 63.23 Aluminum Screen and Storm Doors and Frames

		Crew	Daily Output	Labor-Hours	Unit	Material	2007 Bare Costs Labor	2007 Bare Costs Equipment	Total	Total Incl O&P
0010	**ALUMINUM SCREEN AND STORM DOORS AND FRAMES**									
0020	Combination storm and screen									
0400	Clear anodic coating, 6'-8" x 2'-6" wide	2 Carp	15	1.067	Ea.	162	27.50		189.50	226
0420	2'-8" wide		14	1.143		191	29.50		220.50	260
0440	3'-0" wide		14	1.143		191	29.50		220.50	260
0500	For 7' door height, add					5%				
1000	Mill finish, 6'-8" x 2'-6" wide	2 Carp	15	1.067	Ea.	216	27.50		243.50	284
1020	2'-8" wide		14	1.143		216	29.50		245.50	287
1040	3'-0" wide		14	1.143		234	29.50		263.50	305
1100	For 7'-0" door, add					5%				
1500	White painted, 6'-8" x 2'-6" wide	2 Carp	15	1.067		262	27.50		289.50	335
1520	2'-8" wide		14	1.143		223	29.50		252.50	295
1540	3'-0" wide		14	1.143		258	29.50		287.50	335
1541	Storm door, painted, alum., insul., 6'-8" x 2'-6" wide		14	1.143		247	29.50		276.50	320
1545	2'-8" wide		14	1.143		247	29.50		276.50	320
1600	For 7'-0" door, add					5%				
1800	Aluminum screen door, minimum, 6'-8" x 2'-8" wide	2 Carp	14	1.143		98	29.50		127.50	158
1810	3'-0" wide		14	1.143		228	29.50		257.50	300
1820	Average, 6'-8" x 2'-8" wide		14	1.143		163	29.50		192.50	229
1830	3'-0" wide		14	1.143		228	29.50		257.50	300
1840	Maximum, 6'-8" x 2'-8" wide		14	1.143		325	29.50		354.50	410
1850	3'-0" wide		14	1.143		247	29.50		276.50	320
2000	Wood door & screen, see division 08 14 33.20									

08 12 Metal Frames

08 12 13 – Hollow Metal Frames

08 12 13.13 Standard Hollow Metal Frames

		Crew	Daily Output	Labor-Hours	Unit	Material	2007 Bare Costs Labor	2007 Bare Costs Equipment	Total	Total Incl O&P
0010	**STANDARD HOLLOW METAL FRAMES**									
0020	16 ga., up to 5-3/4" jamb depth									
0025	6'-8" high, 3'-0" wide, single	2 Carp	16	1	Ea.	128	25.50		153.50	185
0028	3'-6" wide, single		16	1		109	25.50		134.50	164
0030	4'-0" wide, single		16	1		109	25.50		134.50	164
0040	6'-0" wide, double		14	1.143		143	29.50		172.50	208
0045	8'-0" wide, double		14	1.143		149	29.50		178.50	214
0100	7'-0" high, 3'-0" wide, single		16	1		132	25.50		157.50	189
0110	3'-6" wide, single		16	1		112	25.50		137.50	167
0112	4'-0" wide, single		16	1		143	25.50		168.50	201
0140	6'-0" wide, double		14	1.143		150	29.50		179.50	215
0145	8'-0" wide, double		14	1.143		140	29.50		169.50	204
1000	16 ga., up to 4-7/8" deep, 7'-0" H, 3'-0" W, single		16	1		123	25.50		148.50	179
1140	6'-0" wide, double		14	1.143		149	29.50		178.50	214
2800	14 ga., up to 3-7/8" deep, 7'-0" high, 3'-0" wide, single		16	1		127	25.50		152.50	183
2840	6'-0" wide, double		14	1.143		152	29.50		181.50	217
3000	14 ga., up to 5-3/4" deep, 6'-8" high, 3'-0" wide, single		16	1		133	25.50		158.50	191
3002	3'-6" wide, single		16	1		133	25.50		158.50	190
3005	4'-0" wide, single		16	1		133	25.50		158.50	190
3600	up to 5-3/4" jamb depth, 7'-0" high, 4'-0" wide, single		15	1.067		103	27.50		130.50	161
3620	6'-0" wide, double		12	1.333		162	34.50		196.50	236
3640	8'-0" wide, double		12	1.333		153	34.50		187.50	227
3700	8'-0" high, 4'-0" wide, single		15	1.067		136	27.50		163.50	197
3740	8'-0" wide, double		12	1.333		169	34.50		203.50	244

08 12 Metal Frames

08 12 13 – Hollow Metal Frames

08 12 13.13 Standard Hollow Metal Frames

		Crew	Daily Output	Labor-Hours	Unit	Material	2007 Bare Costs Labor	Equipment	Total	Total Incl O&P
4000	6-3/4" deep, 7'-0" high, 4'-0" wide, single	2 Carp	15	1.067	Ea.	162	27.50		189.50	225
4020	6'-0" wide, double		12	1.333		189	34.50		223.50	266
4040	8'-0", wide double		12	1.333		215	34.50		249.50	295
4100	8'-0" high, 4'-0" wide, single		15	1.067		162	27.50		189.50	225
4140	8'-0" wide, double		12	1.333		189	34.50		223.50	266
4400	8-3/4" deep, 7'-0" high, 4'-0" wide, single		15	1.067		152	27.50		179.50	214
4440	8'-0" wide, double		12	1.333		216	34.50		250.50	296
4500	8'-0" high, 4'-0" wide, single		15	1.067		163	27.50		190.50	226
4540	8'-0" wide, double		12	1.333		190	34.50		224.50	267
4900	For welded frames, add					44.50			44.50	49
5400	14 ga., "B" label, up to 5-3/4" deep, 7'-0" high, 4'-0" wide, single	2 Carp	15	1.067		161	27.50		188.50	224
5440	8'-0" wide, double		12	1.333		196	34.50		230.50	274
5800	6-3/4" deep, 7'-0" high, 4'-0" wide, single		15	1.067		140	27.50		167.50	201
5840	8'-0" wide, double		12	1.333		235	34.50		269.50	315
6200	8-3/4" deep, 7'-0" high, 4'-0" wide, single		15	1.067		201	27.50		228.50	268
6240	8'-0" wide, double		12	1.333		236	34.50		270.50	320
6300	For "A" label use same price as "B" label									
6400	For baked enamel finish, add					30%	15%			
6500	For galvanizing, add					15%				
6600	For hospital stop, add				Ea.	263			263	289
7900	Transom lite frames, fixed, add	2 Carp	155	.103	S.F.	42.50	2.65		45.15	51.50
8000	Movable, add	"	130	.123	"	51.50	3.16		54.66	62

08 13 Metal Doors

08 13 13 – Hollow Metal Doors

08 13 13.15 Metal Fire Doors

		Crew	Daily Output	Labor-Hours	Unit	Material	2007 Bare Costs Labor	Equipment	Total	Total Incl O&P
0010	**METAL FIRE DOORS** R081313-20									
0015	Steel, flush, "B" label, 90 minute									
0020	Full panel, 20 ga., 2'-0" x 6'-8"	2 Carp	20	.800	Ea.	315	20.50		335.50	385
0040	2'-8" x 6'-8"		18	.889		320	23		343	395
0060	3'-0" x 6'-8"		17	.941		325	24		349	395
0080	3'-0" x 7'-0"		17	.941		335	24		359	410
0140	18 ga., 3'-0" x 6'-8"		16	1		370	25.50		395.50	450
0160	2'-8" x 7'-0"		17	.941		390	24		414	465
0180	3'-0" x 7'-0"		16	1		370	25.50		395.50	450
0200	4'-0" x 7'-0"		15	1.067		485	27.50		512.50	575
0220	For "A" label, 3 hour, 18 ga., use same price as "B" label									
0240	For vision lite, add				Ea.	105			105	116
0520	Flush, "B" label 90 min., composite, 20 ga., 2'-0" x 6'-8"	2 Carp	18	.889		405	23		428	485
0540	2'-8" x 6'-8"		17	.941		405	24		429	485
0560	3'-0" x 6'-8"		16	1		410	25.50		435.50	495
0580	3'-0" x 7'-0"		16	1		420	25.50		445.50	505
0640	Flush, "A" label 3 hour, composite, 18 ga., 3'-0" x 6'-8"		15	1.067		465	27.50		492.50	555
0660	2'-8" x 7'-0"		16	1		480	25.50		505.50	575
0680	3'-0" x 7'-0"		15	1.067		475	27.50		502.50	565
0700	4'-0" x 7'-0"		14	1.143		570	29.50		599.50	680

08 13 13.20 Residential Steel Doors

		Crew	Daily Output	Labor-Hours	Unit	Material	2007 Bare Costs Labor	Equipment	Total	Total Incl O&P
0010	**RESIDENTIAL STEEL DOORS**									
0020	Prehung, insulated, exterior									
0030	Embossed, full panel, 2'-8" x 6'-8"	2 Carp	17	.941	Ea.	266	24		290	335
0040	3'-0" x 6'-8"		15	1.067		239	27.50		266.50	310

08 13 Metal Doors

08 13 13 – Hollow Metal Doors

08 13 13.20 Residential Steel Doors		Crew	Daily Output	Labor-Hours	Unit	Material	2007 Bare Costs Labor	Equipment	Total	Total Incl O&P
0060	3'-0" x 7'-0"	2 Carp	15	1.067	Ea.	283	27.50		310.50	355
0070	5'-4" x 6'-8", double		8	2		500	51.50		551.50	635
0220	Half glass, 2'-8" x 6'-8"		17	.941		263	24		287	330
0240	3'-0" x 6'-8"		16	1		264	25.50		289.50	335
0260	3'-0" x 7'-0"		16	1		320	25.50		345.50	395
0270	5'-4" x 6'-8", double		8	2		550	51.50		601.50	690
0720	Raised plastic face, full panel, 2'-8" x 6'-8"		16	1		266	25.50		291.50	335
0740	3'-0" x 6'-8"		15	1.067		269	27.50		296.50	345
0760	3'-0" x 7'-0"		15	1.067		271	27.50		298.50	345
0780	5'-4" x 6'-8", double		8	2		500	51.50		551.50	640
0820	Half glass, 2'-8" x 6'-8"		17	.941		298	24		322	370
0840	3'-0" x 6'-8"		16	1		300	25.50		325.50	375
0860	3'-0" x 7'-0"		16	1		330	25.50		355.50	410
0880	5'-4" x 6'-8", double		8	2		645	51.50		696.50	795
1320	Flush face, full panel, 2'-6" x 6'-8"		16	1		219	25.50		244.50	284
1340	3'-0" x 6'-8"		15	1.067		219	27.50		246.50	287
1360	3'-0" x 7'-0"		15	1.067		281	27.50		308.50	355
1380	5'-4" x 6'-8", double		8	2		455	51.50		506.50	585
1420	Half glass, 2'-8" x 6'-8"		17	.941		273	24		297	340
1440	3'-0" x 6'-8"		16	1		276	25.50		301.50	350
1460	3'-0" x 7'-0"		16	1		320	25.50		345.50	395
1480	5'-4" x 6'-8", double		8	2		535	51.50		586.50	675
1500	Sidelight, full lite, 1'-0" x 6'-8" with grille					213			213	235
1510	1'-0" x 6'-8", low e					245			245	269
1520	1'-0" x 6'-8", half lite					191			191	210
1530	1'-0" x 6'-8", half lite, low e					233			233	256
2300	Interior, residential, closet, bi-fold, 6'-8" x 2'-0" wide	2 Carp	16	1		141	25.50		166.50	200
2330	3'-0" wide		16	1		158	25.50		183.50	218
2360	4'-0" wide		15	1.067		241	27.50		268.50	310
2400	5'-0" wide		14	1.143		279	29.50		308.50	355
2420	6'-0" wide		13	1.231		315	31.50		346.50	400
2510	Bi-passing closet, incl. hardware, no frame or trim incl.									
2511	Mirrored, metal frame, 6'-8" x 4'-0" wide	2 Carp	10	1.600	Opng.	176	41		217	264
2512	5'-0" wide		10	1.600		204	41		245	294
2513	6'-0" wide		10	1.600		233	41		274	325
2514	7'-0" wide		9	1.778		255	45.50		300.50	360
2515	8'-0" wide		9	1.778		485	45.50		530.50	615
2611	Mirrored, metal, 8'-0" x 4'-0" wide		10	1.600		282	41		323	380
2612	5'-0" wide		10	1.600		305	41		346	405
2613	6'-0" wide		10	1.600		335	41		376	435
2614	7'-0" wide		9	1.778		370	45.50		415.50	485
2615	8'-0" wide		9	1.778		400	45.50		445.50	515

08 14 Wood Doors

08 14 13 – Carved Wood Doors

08 14 13.10 Carved Wood Doors

		Crew	Daily Output	Labor-Hours	Unit	Material	2007 Bare Costs Labor	Equipment	Total	Total Incl O&P
0010	**CARVED WOOD DOORS**									
3000	Solid wood, 1-3/4" thick stile and rail									
3020	Mahogany, 3'-0" x 7'-0", minimum	2 Carp	14	1.143	Ea.	790	29.50		819.50	920
3030	Maximum		10	1.600		1,325	41		1,366	1,525
3040	3'-6" x 8'-0", minimum		10	1.600		935	41		976	1,100
3050	Maximum		8	2		1,725	51.50		1,776.50	1,975
3100	Pine, 3'-0" x 7'-0", minimum		14	1.143		415	29.50		444.50	505
3110	Maximum		10	1.600		705	41		746	845
3120	3'-6" x 8'-0", minimum		10	1.600		675	41		716	810
3130	Maximum		8	2		1,125	51.50		1,176.50	1,300
3200	Red oak, 3'-0" x 7'-0", minimum		14	1.143		1,550	29.50		1,579.50	1,750
3210	Maximum		10	1.600		1,825	41		1,866	2,100
3220	3'-6" x 8'-0", minimum		10	1.600		1,700	41		1,741	1,925
3230	Maximum		8	2		3,000	51.50		3,051.50	3,375
4000	Hand carved door, mahogany									
4020	3'-0" x 7'-0", minimum	2 Carp	14	1.143	Ea.	1,525	29.50		1,554.50	1,725
4030	Maximum		11	1.455		2,975	37.50		3,012.50	3,350
4040	3'-6" x 8'-0", minimum		10	1.600		1,900	41		1,941	2,175
4050	Maximum		8	2		2,925	51.50		2,976.50	3,300
4200	Red oak, 3'-0" x 7'-0", minimum		14	1.143		4,625	29.50		4,654.50	5,150
4210	Maximum		11	1.455		12,700	37.50		12,737.50	14,000
4220	3'-6" x 8'-0", minimum		10	1.600		5,200	41		5,241	5,800
4280	For 6'-8" high door, deduct from 7'-0" door					33			33	36.50
4400	For custom finish, add					355			355	390
4600	Side light, mahogany, 7'-0" x 1'-6" wide, minimum	2 Carp	18	.889		850	23		873	975
4610	Maximum		14	1.143		2,500	29.50		2,529.50	2,800
4620	8'-0" x 1'-6" wide, minimum		14	1.143		1,600	29.50		1,629.50	1,825
4630	Maximum		10	1.600		1,825	41		1,866	2,075
4640	Side light, oak, 7'-0" x 1'-6" wide, minimum		18	.889		985	23		1,008	1,125
4650	Maximum		14	1.143		1,750	29.50		1,779.50	1,975
4660	8'-0" x 1-6" wide, minimum		14	1.143		925	29.50		954.50	1,075
4670	Maximum		10	1.600		1,750	41		1,791	2,000

08 14 16 – Flush Wood Doors

08 14 16.09 Flush Wood Doors

		Crew	Daily Output	Labor-Hours	Unit	Material	Labor	Equipment	Total	Total Incl O&P
0010	**FLUSH WOOD DOORS**									
4160	Walnut faced, 1-3/4" x 6'-8" x 3'-0" wide	1 Carp	17	.471	Ea.	269	12.10		281.10	315

08 14 33 – Stile and Rail Wood Doors

08 14 33.10 Wood Doors Paneled

		Crew	Daily Output	Labor-Hours	Unit	Material	Labor	Equipment	Total	Total Incl O&P
0010	**WOOD DOORS PANELED**									
0020	Interior, six panel, hollow core, 1-3/8" thick									
0040	Molded hardboard, 2'-0" x 6'-8"	2 Carp	17	.941	Ea.	49	24		73	94.50
0060	2'-6" x 6'-8"		17	.941		52.50	24		76.50	99
0070	2'-8" x 6'-8"		17	.941		55	24		79	102
0080	3'-0" x 6'-8"		17	.941		58	24		82	105
0140	Embossed print, molded hardboard, 2'-0" x 6'-8"		17	.941		52.50	24		76.50	99
0160	2'-6" x 6'-8"		17	.941		52.50	24		76.50	99
0180	3'-0" x 6'-8"		17	.941		58	24		82	105
0540	Six panel, solid, 1-3/8" thick, pine, 2'-0" x 6'-8"		15	1.067		128	27.50		155.50	187
0560	2'-6" x 6'-8"		14	1.143		143	29.50		172.50	208
0580	3'-0" x 6'-8"		13	1.231		165	31.50		196.50	235
1020	Two panel, bored rail, solid, 1-3/8" thick, pine, 1'-6" x 6'-8"		16	1		233	25.50		258.50	300
1040	2'-0" x 6'-8"		15	1.067		305	27.50		332.50	380

08 14 Wood Doors

08 14 33 – Stile and Rail Wood Doors

08 14 33.10 Wood Doors Paneled		Crew	Daily Output	Labor-Hours	Unit	Material	2007 Bare Costs Labor	Equipment	Total	Total Incl O&P
1060	2'-6" x 6'-8"	2 Carp	14	1.143	Ea.	350	29.50		379.50	435
1340	Two panel, solid, 1-3/8" thick, fir, 2'-0" x 6'-8"		15	1.067		128	27.50		155.50	187
1360	2'-6" x 6'-8"		14	1.143		143	29.50		172.50	208
1380	3'-0" x 6'-8"		13	1.231		350	31.50		381.50	440
1740	Five panel, solid, 1-3/8" thick, fir, 2'-0" x 6'-8"		15	1.067		229	27.50		256.50	298
1760	2'-6" x 6'-8"		14	1.143		365	29.50		394.50	455
1780	3'-0" x 6'-8"	↓	13	1.231	↓	365	31.50		396.50	460
08 14 33.20 Wood Doors Residential										
0010	**WOOD DOORS RESIDENTIAL**									
0200	Exterior, combination storm & screen, pine									
0260	2'-8" wide	2 Carp	10	1.600	Ea.	272	41		313	370
0280	3'-0" wide		9	1.778		278	45.50		323.50	385
0300	7'-1" x 3'-0" wide		9	1.778		310	45.50		355.50	420
0400	Full lite, 6'-9" x 2'-6" wide		11	1.455		290	37.50		327.50	385
0420	2'-8" wide		10	1.600		290	41		331	390
0440	3'-0" wide		9	1.778		299	45.50		344.50	410
0500	7'-1" x 3'-0" wide		9	1.778		320	45.50		365.50	435
0700	Dutch door, pine, 1-3/4" x 6'-8" x 2'-8" wide, minimum		12	1.333		675	34.50		709.50	800
0720	Maximum		10	1.600		710	41		751	855
0800	3'-0" wide, minimum		12	1.333		685	34.50		719.50	815
0820	Maximum		10	1.600		750	41		791	895
1000	Entrance door, colonial, 1-3/4" x 6'-8" x 2'-8" wide		16	1		380	25.50		405.50	460
1020	6 panel pine, 3'-0" wide		15	1.067		400	27.50		427.50	485
1100	8 panel pine, 2'-8" wide		16	1		560	25.50		585.50	660
1120	3'-0" wide	↓	15	1.067		540	27.50		567.50	635
1200	For tempered safety glass lites, (min of 2) add					65			65	71.50
1300	Flush, birch, solid core, 1-3/4" x 6'-8" x 2'-8" wide	2 Carp	16	1		97.50	25.50		123	151
1320	3'-0" wide		15	1.067		101	27.50		128.50	158
1350	7'-0" x 2'-8" wide		16	1		105	25.50		130.50	159
1360	3'-0" wide		15	1.067		114	27.50		141.50	173
1740	Mahogany, 2'-8" x 6'-8"		15	1.067		735	27.50		762.50	855
1760	3'-0" x 6'-8"	↓	15	1.067	↓	765	27.50		792.50	885
2700	Interior, closet, bi-fold, w/hardware, no frame or trim incl.									
2720	Flush, birch, 6'-6" or 6'-8" x 2'-6" wide	2 Carp	13	1.231	Ea.	48.50	31.50		80	107
2740	3'-0" wide		13	1.231		52.50	31.50		84	112
2760	4'-0" wide		12	1.333		96	34.50		130.50	163
2780	5'-0" wide		11	1.455		97	37.50		134.50	171
2800	6'-0" wide		10	1.600		104	41		145	184
3000	Raised panel pine, 6'-6" or 6'-8" x 2'-6" wide		13	1.231		164	31.50		195.50	234
3020	3'-0" wide		13	1.231		209	31.50		240.50	283
3040	4'-0" wide		12	1.333		300	34.50		334.50	390
3060	5'-0" wide		11	1.455		360	37.50		397.50	460
3080	6'-0" wide		10	1.600		395	41		436	505
3200	Louvered, pine 6'-6" or 6'-8" x 2'-6" wide		13	1.231		104	31.50		135.50	169
3220	3'-0" wide		13	1.231		161	31.50		192.50	231
3240	4'-0" wide		12	1.333		195	34.50		229.50	273
3260	5'-0" wide		11	1.455		221	37.50		258.50	305
3280	6'-0" wide	↓	10	1.600	↓	245	41		286	340
4400	Bi-passing closet, incl. hardware and frame, no trim incl.									
4420	Flush, lauan, 6'-8" x 4'-0" wide	2 Carp	12	1.333	Opng.	165	34.50		199.50	240
4440	5'-0" wide		11	1.455		181	37.50		218.50	263
4460	6'-0" wide	↓	10	1.600		194	41		235	283

08 14 Wood Doors

08 14 33 – Stile and Rail Wood Doors

08 14 33.20 Wood Doors Residential		Crew	Daily Output	Labor-Hours	Unit	Material	2007 Bare Costs Labor	Equipment	Total	Total Incl O&P
4600	Flush, birch, 6'-8" x 4'-0" wide	2 Carp	12	1.333	Opng.	207	34.50		241.50	285
4620	5'-0" wide		11	1.455		204	37.50		241.50	289
4640	6'-0" wide		10	1.600		248	41		289	345
4800	Louvered, pine, 6'-8" x 4'-0" wide		12	1.333		405	34.50		439.50	510
4820	5'-0" wide		11	1.455		385	37.50		422.50	485
4840	6'-0" wide		10	1.600		500	41		541	620
4900	Mirrored, 6'-8" x 4'-0" wide		12	1.333	Ea.	295	34.50		329.50	385
5000	Paneled, pine, 6'-8" x 4'-0" wide		12	1.333	Opng.	485	34.50		519.50	590
5020	5'-0" wide		11	1.455		400	37.50		437.50	505
5040	6'-0" wide		10	1.600		575	41		616	700
5061	Hardboard, 6'-8" x 4'-0" wide		10	1.600		184	41		225	272
5062	5'-0" wide		10	1.600		189	41		230	278
5063	6'-0" wide		10	1.600		220	41		261	310
6100	Folding accordion, closet, including track and frame									
6120	Vinyl, 2 layer, stock (see also division 10 22 26.13)	2 Carp	400	.040	Ea.	103	1.03		104.03	115
6200	Rigid PVC	"	400	.040	S.F.	4.88	1.03		5.91	7.10
6220	For custom partition, add					25%	10%			
7310	Passage doors, flush, no frame included									
7320	Hardboard, hollow core, 1-3/8" x 6'-8" x 1'-6" wide	2 Carp	18	.889	Ea.	40.50	23		63.50	84
7330	2'-0" wide		18	.889		40.50	23		63.50	83.50
7340	2'-6" wide		18	.889		44.50	23		67.50	88
7350	2'-8" wide		18	.889		47	23		70	91
7360	3'-0" wide		17	.941		49.50	24		73.50	95.50
7420	Lauan, hollow core, 1-3/8" x 6'-8" x 1'-6" wide		18	.889		28.50	23		51.50	70.50
7440	2'-0" wide		18	.889		27	23		50	69
7450	2'-4" wide		18	.889		30.50	23		53.50	72.50
7460	2'-6" wide		18	.889		30.50	23		53.50	72.50
7480	2'-8" wide		18	.889		32.50	23		55.50	74.50
7500	3'-0" wide		17	.941		33.50	24		57.50	78
7700	Birch, hollow core, 1-3/8" x 6'-8" x 1'-6" wide		18	.889		36	23		59	78.50
7720	2'-0" wide		18	.889		40	23		63	83
7740	2'-6" wide		18	.889		44.50	23		67.50	88
7760	2'-8" wide		18	.889		46	23		69	89.50
7780	3'-0" wide		17	.941		50	24		74	96
8000	Pine louvered, 1-3/8" x 6'-8" x 1'-6" wide		19	.842		100	21.50		121.50	147
8020	2'-0" wide		18	.889		125	23		148	177
8040	2'-6" wide		18	.889		137	23		160	189
8060	2'-8" wide		18	.889		144	23		167	197
8080	3'-0" wide		17	.941		154	24		178	211
8300	Pine paneled, 1-3/8" x 6'-8" x 1'-6" wide		19	.842		109	21.50		130.50	157
8320	2'-0" wide		18	.889		125	23		148	177
8330	2'-4" wide		18	.889		137	23		160	190
8340	2'-6" wide		18	.889		140	23		163	193
8360	2'-8" wide		18	.889		152	23		175	206
8380	3'-0" wide		17	.941		161	24		185	218
8450	French door, pine, 15 lites, 1-3/8"x6'-8"x2'-6" wide		18	.889		172	23		195	228
8470	2'-8" wide		18	.889		222	23		245	283
8490	3'-0" wide		17	.941		237	24		261	300
8550	For over 20 doors, deduct					15%				
8804	Pocket door, 6 panel pine, 2'-6" x 6'-8"	2 Carp	10.50	1.524	Ea.	214	39		253	300
8814	2'-8" x 6'-8"		10.50	1.524		226	39		265	315
8824	3'-0" x 6'-8"		10.50	1.524		234	39		273	325

08 14 Wood Doors

08 14 40 – Interior Cafe Doors

08 14 40.10 Interior Cafe Doors	Crew	Daily Output	Labor-Hours	Unit	Material	2007 Bare Costs Labor	Equipment	Total	Total Incl O&P
0010 **INTERIOR CAFE DOORS**									
6520 Interior cafe doors, 2'-6" opening, stock, panel pine	2 Carp	16	1	Ea.	188	25.50		213.50	250
6540 3'-0" opening	"	16	1	"	196	25.50		221.50	259
6550 Louvered pine									
6560 2'-6" opening	2 Carp	16	1	Ea.	164	25.50		189.50	224
8000 3'-0" opening		16	1		175	25.50		200.50	236
8010 2'-6" opening, hardwood		16	1		282	25.50		307.50	355
8020 3'-0" opening		16	1		310	25.50		335.50	390

08 16 Composite Doors

08 16 13 – Fiberglass Doors

08 16 13.10 Fiberglass Doors	Crew	Daily Output	Labor-Hours	Unit	Material	2007 Bare Costs Labor	Equipment	Total	Total Incl O&P
0010 **FIBERGLASS DOORS**									
0020 Exterior, fiberglass, door, 2'-8" wide x 6'-8" high	2 Carp	15	1.067	Ea.	305	27.50		332.50	380
0040 3'-0" wide x 6'-8" high		15	1.067		305	27.50		332.50	380
0060 3'-0" wide x 7'-0" high		15	1.067		375	27.50		402.50	460
0080 3'-0" wide x 6'-8" high, with two lites		15	1.067		370	27.50		397.50	450
0100 3'-0" wide x 7'-0" high, with two lites		15	1.067		440	27.50		467.50	530
0110 Half glass, 3'-0" wide x 6'-8" high		15	1.067		415	27.50		442.50	500
0120 3'-0" wide x 6'-8" high, low e		15	1.067		440	27.50		467.50	525
0130 3'-0" wide x 7'-0" high		15	1.067		485	27.50		512.50	575
0140 3'-0" wide x 7'-0" high, low e		15	1.067		510	27.50		537.50	605
0150 Side lights, 1'-0" wide x 6'-8" high,					281			281	310
0160 1'-0" wide x 6'-8" high, low e					293			293	325
0180 1'-0" wide x 6'-8" high, full glass					320			320	355
0190 1'-0" wide x 6'-8" high, low e					345			345	380

08 17 Integrated Door Opening Assemblies

08 17 23 – Integrated Wood Door Opening Assemblies

08 17 23.10 Pre-Hung Doors	Crew	Daily Output	Labor-Hours	Unit	Material	2007 Bare Costs Labor	Equipment	Total	Total Incl O&P
0010 **PRE-HUNG DOORS**									
0300 Exterior, wood, comb. storm & screen, 6'-9" x 2'-6" wide	2 Carp	15	1.067	Ea.	283	27.50		310.50	355
0320 2'-8" wide		15	1.067		283	27.50		310.50	355
0340 3'-0" wide		15	1.067		291	27.50		318.50	365
0360 For 7'-0" high door, add					26.50			26.50	29
1600 Entrance door, flush, birch, solid core									
1620 4-5/8" solid jamb, 1-3/4" x 6'-8" x 2'-8" wide	2 Carp	16	1	Ea.	285	25.50		310.50	360
1640 3'-0" wide		16	1		355	25.50		380.50	435
1642 5-5/8" jamb		16	1		320	25.50		345.50	395
1680 For 7'-0" high door, add					18.75			18.75	20.50
2000 Entrance door, colonial, 6 panel pine									
2020 4-5/8" solid jamb, 1-3/4" x 6'-8" x 2'-8" wide	2 Carp	16	1	Ea.	520	25.50		545.50	615
2040 3'-0" wide	"	16	1		520	25.50		545.50	615
2060 For 7'-0" high door, add					49			49	53.50
2200 For 5-5/8" solid jamb, add					38.50			38.50	42
2250 French door, 6'-8" x 4'-0" wide, 1/2" insul. glass and grille	2 Carp	7	2.286	Pr.	1,075	58.50		1,133.50	1,275
2260 5'-0" wide	"	7	2.286	"	1,175	58.50		1,233.50	1,400
2500 Exterior, metal face, insulated, incl. jamb, brickmold and									
2520 threshold, flush, 2'-8" x 6'-8"	2 Carp	16	1	Ea.	235	25.50		260.50	300

08 17 Integrated Door Opening Assemblies

08 17 23 – Integrated Wood Door Opening Assemblies

08 17 23.10 Pre-Hung Doors		Crew	Daily Output	Labor-Hours	Unit	Material	2007 Bare Costs Labor	Equipment	Total	Total Incl O&P
2550	3'-0" x 6'-8"	2 Carp	16	1	Ea.	235	25.50		260.50	300
3500	Embossed, 6 panel, 2'-8" x 6'-8"		16	1		189	25.50		214.50	252
3550	3'-0" x 6'-8"		16	1		203	25.50		228.50	267
3600	2 narrow lites, 2'-8" x 6'-8"		16	1		248	25.50		273.50	315
3650	3'-0" x 6'-8"		16	1		253	25.50		278.50	320
3700	Half glass, 2'-8" x 6'-8"		16	1		262	25.50		287.50	330
3750	3'-0" x 6'-8"		16	1		265	25.50		290.50	335
3800	2 top lites, 2'-8" x 6'-8"		16	1		258	25.50		283.50	330
3850	3'-0" x 6'-8"		16	1		258	25.50		283.50	330
4000	Interior, passage door, 4-5/8" solid jamb									
4400	Lauan, flush, solid core, 1-3/8" x 6'-8" x 2'-6" wide	2 Carp	20	.800	Ea.	178	20.50		198.50	230
4420	2'-8" wide		20	.800		178	20.50		198.50	230
4440	3'-0" wide		19	.842		191	21.50		212.50	247
4600	Hollow core, 1-3/8" x 6'-8" x 2'-6" wide		20	.800		120	20.50		140.50	167
4620	2'-8" wide		20	.800		120	20.50		140.50	167
4640	3'-0" wide		19	.842		121	21.50		142.50	170
4700	For 7'-0" high door, add					23			23	25.50
5000	Birch, flush, solid core, 1-3/8" x 6'-8" x 2'-6" wide	2 Carp	20	.800		166	20.50		186.50	217
5020	2'-8" wide		20	.800		190	20.50		210.50	244
5040	3'-0" wide		19	.842		200	21.50		221.50	257
5200	Hollow core, 1-3/8" x 6'-8" x 2'-6" wide		20	.800		137	20.50		157.50	186
5220	2'-8" wide		20	.800		144	20.50		164.50	193
5240	3'-0" wide		19	.842		144	21.50		165.50	195
5280	For 7'-0" high door, add					19.95			19.95	22
5500	Hardboard paneled, 1-3/8" x 6'-8" x 2'-6" wide	2 Carp	20	.800		138	20.50		158.50	187
5520	2'-8" wide		20	.800		145	20.50		165.50	194
5540	3'-0" wide		19	.842		143	21.50		164.50	194
6000	Pine paneled, 1-3/8" x 6'-8" x 2'-6" wide		20	.800		240	20.50		260.50	299
6020	2'-8" wide		20	.800		261	20.50		281.50	320
6040	3'-0" wide		19	.842		266	21.50		287.50	330
6500	For 5-5/8" solid jamb, add					13.95			13.95	15.35
6520	For split jamb, deduct					15.70			15.70	17.25
7600	Oak, 6 panel, 1-3/4" x 6'-8" x 3'-0"	1 Carp	17	.471		630	12.10		642.10	715

08 31 Access Doors and Panels

08 31 13 – Access Doors and Frames

08 31 13.20 Bulkhead/Cellar Doors

		Crew	Daily Output	Labor-Hours	Unit	Material	Labor	Equipment	Total	Total Incl O&P
0010	**BULKHEAD/CELLAR DOORS**									
0020	Steel, not incl. sides, 44" x 62"	1 Carp	5.50	1.455	Ea.	300	37.50		337.50	395
0100	52" x 73"		5.10	1.569		320	40.50		360.50	420
0500	With sides and foundation plates, 57" x 45" x 24"		4.70	1.702		555	43.50		598.50	690
0600	42" x 49" x 51"		4.30	1.860		645	48		693	790

08 32 Sliding Glass Doors

08 32 19 – Sliding Wood-Framed Glass Doors

08 32 19.10 Sliding Wood Doors

		Crew	Daily Output	Labor-Hours	Unit	Material	2007 Bare Costs Labor	2007 Bare Costs Equipment	Total	Total Incl O&P
0010	**SLIDING WOOD DOORS**									
0020	Wood, 5/8" tempered insul. glass, 6' wide, premium	2 Carp	4	4	Ea.	1,150	103		1,253	1,450
0100	Economy		4	4		790	103		893	1,050
0150	8' wide, wood, premium		3	5.333		1,350	137		1,487	1,700
0200	Economy		3	5.333		915	137		1,052	1,225
0250	12' wide, wood, vinyl clad		2.50	6.400		2,900	164		3,064	3,450
0300	Economy		2.50	6.400		2,100	164		2,264	2,575
0350	Aluminum, 5/8" tempered insulated glass, 6' wide									
0400	Premium	2 Carp	4	4	Ea.	1,425	103		1,528	1,725
0450	Economy		4	4		760	103		863	1,000
0500	8' wide, premium		3	5.333		1,525	137		1,662	1,900
0550	Economy		3	5.333		1,300	137		1,437	1,650
0600	12' wide, premium		2.50	6.400		2,475	164		2,639	2,975
0650	Economy		2.50	6.400		1,500	164		1,664	1,925
1000	Replacement doors, wood									
1050	6' wide, vinyl clad	2 Carp	4	4	Ea.	1,075	103		1,178	1,375

08 32 19.15 Sliding Glass Vinyl-Clad Wood Doors

		Crew	Daily Output	Labor-Hours	Unit	Material	Labor	Equipment	Total	Total Incl O&P
0010	**SLIDING GLASS VINYL-CLAD WOOD DOORS**									
0012	Vinyl clad, 1" insul. glass, 6'-0" x 6'-10" high	2 Carp	4	4	Opng.	1,300	103		1,403	1,625
0030	6'-0" x 8'-0" high		4	4	Ea.	1,900	103		2,003	2,275
0100	8'-0" x 6'-10" high		4	4	Opng.	1,950	103		2,053	2,325
0104	8'-0" x 6'-8" double		4	4	Ea.	1,825	103		1,928	2,175
0500	3 leaf, 9'-0" x 6'-10" high		3	5.333	Opng.	2,275	137		2,412	2,750
0600	12'-0" x 6'-10" high		3	5.333	"	3,100	137		3,237	3,625

08 36 Panel Doors

08 36 13 – Sectional Doors

08 36 13.20 Residential Garage Doors

		Crew	Daily Output	Labor-Hours	Unit	Material	Labor	Equipment	Total	Total Incl O&P
0010	**RESIDENTIAL GARAGE DOORS**									
0050	Hinged, wood, custom, double door, 9' x 7'	2 Carp	4	4	Ea.	400	103		503	615
0070	16' x 7'		3	5.333		680	137		817	980
0200	Overhead, sectional, incl. hardware, fiberglass, 9' x 7', standard		5.28	3.030		625	78		703	820
0220	Deluxe		5.28	3.030		800	78		878	1,000
0300	16' x 7', standard		6	2.667		1,150	68.50		1,218.50	1,375
0320	Deluxe		6	2.667		1,425	68.50		1,493.50	1,700
0500	Hardboard, 9' x 7', standard		8	2		420	51.50		471.50	550
0520	Deluxe		8	2		565	51.50		616.50	710
0600	16' x 7', standard		6	2.667		830	68.50		898.50	1,025
0620	Deluxe		6	2.667		965	68.50		1,033.50	1,200
0700	Metal, 9' x 7', standard		5.28	3.030		495	78		573	675
0720	Deluxe		8	2		670	51.50		721.50	820
0800	16' x 7', standard		3	5.333		630	137		767	930
0820	Deluxe		6	2.667		1,025	68.50		1,093.50	1,250
0900	Wood, 9' x 7', standard		8	2		525	51.50		576.50	660
0920	Deluxe		8	2		1,500	51.50		1,551.50	1,725
1000	16' x 7', standard		6	2.667		1,050	68.50		1,118.50	1,275
1020	Deluxe		6	2.667		2,200	68.50		2,268.50	2,550
1800	Door hardware, sectional	1 Carp	4	2		231	51.50		282.50	340
1810	Door tracks only		4	2		108	51.50		159.50	206
1820	One side only		7	1.143		74.50	29.50		104	132
3000	Swing-up, including hardware, fiberglass, 9' x 7', standard	2 Carp	8	2		725	51.50		776.50	880

Since its founding in 1975, Reed Construction Data has developed an online and print portfolio of innovative products and services for the construction, design and manufacturing community. Our products and services are designed specifically to help construction industry professionals advance their businesses with timely, accurate and actionable project, product and cost data. Reed Construction Data is your all-inclusive source of construction information encompassing all phases of the construction process.

Reed Bulletin and Reed Connect™ deliver the most comprehensive, timely and reliable project information to support contractors, distributors and building product manufacturers in identifying, bidding and tracking projects – private and public, general building and civil. Reed Construction Data also offers in-depth construction activity statistics and forecasts covering major project categories, many at the county and metropolitan level.

Reed Bulletin
www.reedbulletin.com

- Project leads targeted by geographic region and formatted by construction stage – available online or in print.
- Locate those hard-to-find jobs that are more profitable to your business.
- Optional automatic e-mail updates sent whenever project details change.
- Download plans and specs online or order print copies.

Reed Research & Analytics
www.buildingteamforecast.com

Reed Construction Forecast

- Delivers timely construction industry activity combining historical data, current year projections and forecasts.
- Modeled at the individual MSA-level to capture changing local market conditions.
- Covers 21 major project categories.

Reed Construction Starts

- Available in a monthly report or as an interactive database.
- Data provided in square footage and dollar value.
- Highly effective and efficient business planning tool.

Market Fundamentals

- Metropolitan area-specific reporting.
- Five-year forecast of industry performance including major projects in development and underway.
- Property types include office, retail, hotel, warehouse and apartment.

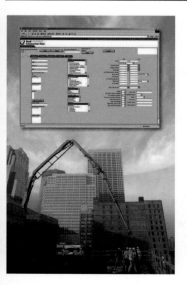

Reed Connect
www.reedconnect.com

- Customized web-based project lead delivery service featuring advanced search capabilities.
- Manage and track actionable leads from planning to quote through winning the job.
- Competitive analysis tool to analyze lost sales opportunities.
- Potential integration with your CRM application.

For more information about Reed Construction Data, please call 877-REED411, visit our website at www.reedconstructiondata.com, or E-mail: marketing@reedbusiness.com

Reed Construction Data®

The design community utilizes the Reed First Source® suite of products to search, select and specify nationally available building products during the formative stages of project design, as well as during other stages of product selection and specification. Reed Design Registry is a detailed database of architecture firms. This tool features sophisticated search and sort capabilities to support the architect selection process and ensure locating the best manufacturers to partner with on your next project.

Reed First Source - The Leading Product Directory to "Find It, Choose It, Use It"

www.reedfirstsource.com

- Comprehensive directory of over 11,000 commercial building product manufacturers classified by MasterFormat™ 2004 categories.

- SPEC-DATA's 10-part format provides performance data along with technical and installation information.

- MANU-SPEC delivers manufacturer guide specifications in the CSI 3-part SectionFormat.

- Search for products, download CAD, research building codes and view catalogs online.

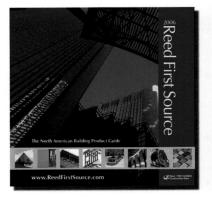

Reed Design Registry – The Premier Source of Architecture Firms

www.reedregistry.com

- Comprehensive website directory contains over 30,000 architect firms.

- Profiles list firm size, firm's specialties and more.

- Official database of AIA member-owned firms.

- Allows architects to share project information with the building community.

- Coming soon – Reed Registry to include engineers and landscape architects.

For more information about Reed Construction Data, please call 877-REED411, visit our website at www.reedconstructiondata.com, or E-mail: marketing@reedbusiness.com

Reed Construction Data
CONNECT • FIRST SOURCE • BULLETIN
RSMEANS • ACP • RESEARCH • PLANSDIRECT

08 36 Panel Doors

08 36 13 – Sectional Doors

08 36 13.20 Residential Garage Doors		Crew	Daily Output	Labor-Hours	Unit	Material	2007 Bare Costs Labor	Equipment	Total	Total Incl O&P
3020	Deluxe	2 Carp	8	2	Ea.	760	51.50		811.50	920
3100	16' x 7', standard		6	2.667		910	68.50		978.50	1,125
3120	Deluxe		6	2.667		940	68.50		1,008.50	1,150
3200	Hardboard, 9' x 7', standard		8	2		330	51.50		381.50	450
3220	Deluxe		8	2		440	51.50		491.50	565
3300	16' x 7', standard		6	2.667		460	68.50		528.50	625
3320	Deluxe		6	2.667		685	68.50		753.50	870
3400	Metal, 9' x 7', standard		8	2		365	51.50		416.50	485
3420	Deluxe		8	2		655	51.50		706.50	810
3500	16' x 7', standard		6	2.667		570	68.50		638.50	740
3520	Deluxe		6	2.667		915	68.50		983.50	1,125
3600	Wood, 9' x 7', standard		8	2		395	51.50		446.50	520
3620	Deluxe		8	2		715	51.50		766.50	870
3700	16' x 7', standard		6	2.667		690	68.50		758.50	875
3720	Deluxe		6	2.667		970	68.50		1,038.50	1,200
3900	Door hardware only, swing up	1 Carp	4	2		117	51.50		168.50	216
3920	One side only		7	1.143		64.50	29.50		94	121
4000	For electric operator, economy, add		8	1		300	25.50		325.50	375
4100	Deluxe, including remote control		8	1		440	25.50		465.50	530
4500	For transmitter/receiver control, add to operator				Total	91			91	100
4600	Transmitters, additional				"	33			33	36.50
6000	Replace section, on sectional door, fiberglass, 9' x 7'	1 Carp	4	2	Ea.	172	51.50		223.50	276
6020	16' x 7'		3.50	2.286		259	58.50		317.50	385
6200	Hardboard, 9' x 7'		4	2		91.50	51.50		143	188
6220	16' x 7'		3.50	2.286		175	58.50		233.50	293
6300	Metal, 9' x 7'		4	2		149	51.50		200.50	250
6320	16' x 7'		3.50	2.286		244	58.50		302.50	370
6500	Wood, 9' x 7'		4	2		90	51.50		141.50	186
6520	16' x 7'		3.50	2.286		175	58.50		233.50	292

08 51 Metal Windows

08 51 13 – Aluminum Windows

08 51 13.20 Aluminum Windows		Crew	Daily Output	Labor-Hours	Unit	Material	2007 Bare Costs Labor	Equipment	Total	Total Incl O&P
0010	**ALUMINUM WINDOWS**, incl. frame and glazing, commercial grade									
1000	Stock units, casement, 3'-1" x 3'-2" opening	2 Sswk	10	1.600	Ea.	315	44.50		359.50	435
1040	Insulating glass	"	10	1.600		405	44.50		449.50	530
1050	Add for storms					102			102	112
1600	Projected, with screen, 3'-1" x 3'-2" opening	2 Sswk	10	1.600		300	44.50		344.50	415
1650	Insulating glass	"	10	1.600		233	44.50		277.50	345
1700	Add for storms					100			100	110
2000	4'-5" x 5'-3" opening	2 Sswk	8	2		340	55.50		395.50	485
2050	Insulating glass	"	8	2		355	55.50		410.50	500
2100	Add for storms					110			110	121
2500	Enamel finish windows, 3'-1" x 3'-2"	2 Sswk	10	1.600		310	44.50		354.50	425
2550	Insulating glass		10	1.600		213	44.50		257.50	320
2600	4'-5" x 5'-3"		8	2		350	55.50		405.50	495
2700	Insulating glass		8	2		385	55.50		440.50	535
3000	Single hung, 2' x 3' opening, enameled, standard glazed		10	1.600		180	44.50		224.50	285
3100	Insulating glass		10	1.600		218	44.50		262.50	325
3300	2'-8" x 6'-8" opening, standard glazed		8	2		315	55.50		370.50	460
3400	Insulating glass		8	2		405	55.50		460.50	560

08 51 Metal Windows

08 51 13 – Aluminum Windows

08 51 13.20 Aluminum Windows

		Crew	Daily Output	Labor-Hours	Unit	Material	2007 Bare Costs Labor	Equipment	Total	Total Incl O&P
3700	3'-4" x 5'-0" opening, standard glazed	2 Sswk	9	1.778	Ea.	260	49		309	380
3800	Insulating glass		9	1.778		287	49		336	410
4000	Sliding aluminum, 3' x 2' opening, standard glazed		10	1.600		190	44.50		234.50	296
4100	Insulating glass		10	1.600		204	44.50		248.50	310
4300	5' x 3' opening, standard glazed		9	1.778		290	49		339	415
4400	Insulating glass		9	1.778		340	49		389	465
4600	8' x 4' opening, standard glazed		6	2.667		305	74		379	480
4700	Insulating glass		6	2.667		495	74		569	685
5000	9' x 5' opening, standard glazed		4	4		465	111		576	725
5100	Insulating glass		4	4		740	111		851	1,025
5500	Sliding, with thermal barrier and screen, 6' x 4', 2 track		8	2		630	55.50		685.50	805
5700	4 track		8	2		800	55.50		855.50	990
6000	For above units with bronze finish, add					12%				
6200	For installation in concrete openings, add					5%				

08 51 23 – Steel Windows

08 51 23.10 Steel Sash

		Crew	Daily Output	Labor-Hours	Unit	Material	Labor	Equipment	Total	Total Incl O&P
0010	**STEEL SASH** Custom units, glazing and trim not included									
1200	1'-11" x 2'-8"	1 Carp	14	.571	Ea.	132	14.70		146.70	170

08 51 23.40 Basement Utility Window

		Crew	Daily Output	Labor-Hours	Unit	Material	Labor	Equipment	Total	Total Incl O&P
0010	**BASEMENT UTILITY WINDOWS**									
0015	1'-3" x 2'-8"	1 Carp	16	.500	Ea.	117	12.85		129.85	151
1100	1'-7" x 2'-8"	"	16	.500	"	130	12.85		142.85	165

08 51 66 – Metal Window Screens

08 51 66.10 Screens

		Crew	Daily Output	Labor-Hours	Unit	Material	Labor	Equipment	Total	Total Incl O&P
0010	**SCREENS**									
0020	For metal sash, aluminum or bronze mesh, flat screen	2 Sswk	1200	.013	S.F.	3.65	.37		4.02	4.74
0500	Wicket screen, inside window	"	1000	.016	"	5.55	.44		5.99	6.95
0600	Residential, aluminum mesh and frame, 2' x 3'	2 Carp	32	.500	Ea.	12.50	12.85		25.35	36
0610	Rescreen		50	.320		10.10	8.20		18.30	25
0620	3' x 5'		32	.500		37	12.85		49.85	63
0630	Rescreen		45	.356		25	9.15		34.15	43
0640	4' x 8'		25	.640		61	16.45		77.45	95
0650	Rescreen		40	.400		40	10.30		50.30	61.50
0660	Patio door		25	.640		146	16.45		162.45	189
0680	Rescreening		1600	.010	S.F.	1.30	.26		1.56	1.87
1000	For solar louvers, add	2 Sswk	160	.100	"	20.50	2.77		23.27	28

08 52 Wood Windows

08 52 10 – Wood Windows

08 52 10.10 Wood Windows

		Crew	Daily Output	Labor-Hours	Unit	Material	Labor	Equipment	Total	Total Incl O&P
0010	**WOOD WINDOWS**, including frame, screens and grills									
0020	Residential, stock units									
0050	Awning type, double insulated glass, 2'-10" x 1'-9" opening	2 Carp	12	1.333	Opng.	195	34.50		229.50	273
0100	2'-10" x 6'-0" opening	1 Carp	8	1		475	25.50		500.50	570
0200	4' x 3'-6" single pane		10	.800		320	20.50		340.50	390
0300	6' x 5' single pane		8	1	Ea.	445	25.50		470.50	535
0301										
1000	Casement, 2'-0" x 3'-4" high	2 Carp	20	.800	Ea.	216	20.50		236.50	273
1020	2'-0" x 4'-0"		18	.889		237	23		260	300
1040	2'-0" x 5'-0"		17	.941		271	24		295	340

08 52 Wood Windows

08 52 10 – Wood Windows

08 52 10.10 Wood Windows		Crew	Daily Output	Labor-Hours	Unit	Material	2007 Bare Costs Labor	2007 Bare Costs Equipment	Total	Total Incl O&P
1060	2'-0" x 6'-0"	2 Carp	16	1	Ea.	268	25.50		293.50	340
1080	4'-0" x 3'-4"		15	1.067		545	27.50		572.50	645
1100	4'-0" x 4'-0"		15	1.067		600	27.50		627.50	705
1120	4'-0" x 5'-0"		14	1.143		680	29.50		709.50	795
1140	4'-0" x 6'-0"		12	1.333		765	34.50		799.50	900
1600	Casement units, 8' x 5', with screens, double insulated glass		2.50	6.400	Opng.	1,250	164		1,414	1,650
1700	Low E glass		2.50	6.400		1,425	164		1,589	1,850
2300	Casements, including screens, 2'-0" x 3'-4", dbl insulated glass		11	1.455		238	37.50		275.50	325
2400	Low E glass		11	1.455		238	37.50		275.50	325
2600	2 lite, 4'-0" x 4'-0", double insulated glass		9	1.778		435	45.50		480.50	560
2700	Low E glass		9	1.778		450	45.50		495.50	575
2900	3 lite, 5'-2" x 5'-0", double insulated glass		7	2.286		665	58.50		723.50	830
3000	Low E glass		7	2.286		705	58.50		763.50	875
3200	4 lite, 7'-0" x 5'-0", double insulated glass		6	2.667		945	68.50		1,013.50	1,175
3300	Low E glass		6	2.667		1,000	68.50		1,068.50	1,225
3500	5 lite, 8'-6" x 5'-0", double insulated glass		5	3.200		1,250	82		1,332	1,525
3600	Low E glass		5	3.200		1,325	82		1,407	1,625
3800	For removable wood grilles, diamond pattern, add				Leaf	32			32	35
3900	Rectangular pattern, add				"	32			32	35
4000	Bow, fixed lites, 8' x 5', double insulated glass	2 Carp	3	5.333	Opng.	1,250	137		1,387	1,600
4100	Low E glass	"	3	5.333	"	1,800	137		1,937	2,225
4150	6'-0" x 5'-0"	1 Carp	8	1	Ea.	1,200	25.50		1,225.50	1,375
4300	Fixed lites, 9'-9" x 5'-0", double insulated glass	2 Carp	2	8	Opng.	865	206		1,071	1,300
4400	Low E glass	"	2	8	"	945	206		1,151	1,400
5000	Bow, casement, 8'-1" x 4'-8" high	3 Carp	8	3	Ea.	1,450	77		1,527	1,725
5020	9'-6" x 4'-8"		8	3		1,600	77		1,677	1,875
5040	8'-1" x 5'-1"		8	3		1,750	77		1,827	2,050
5060	9'-6" x 5'-1"		6	4		1,550	103		1,653	1,875
5080	8'-1" x 6'-0"		6	4		1,750	103		1,853	2,100
5100	9'-6" x 6'-0"		6	4		1,825	103		1,928	2,200
5800	Skylights, hatches, vents, and sky roofs, see division 08 62 13.00									
9000	Minimum labor/equipment charge	2 Carp	2	8	Job		206		206	350

08 52 10.20 Awning Window

		Crew	Daily Output	Labor-Hours	Unit	Material	2007 Bare Costs Labor	2007 Bare Costs Equipment	Total	Total Incl O&P
0010	**AWNING WINDOW**, Including frame, screens and grills									
0100	Average quality, builders model, 34" x 22", double insulated glass	1 Carp	10	.800	Ea.	234	20.50		254.50	292
0200	Low E glass		10	.800		247	20.50		267.50	305
0300	40" x 28", double insulated glass		9	.889		295	23		318	365
0400	Low E Glass		9	.889		315	23		338	385
0500	48" x 36", double insulated glass		8	1		430	25.50		455.50	520
0600	Low E glass		8	1		455	25.50		480.50	545
0800	Vinyl clad, premium, double insulated glass, 24" x 17"		12	.667		198	17.15		215.15	247
0840	24" x 28"		11	.727		238	18.70		256.70	293
0860	36" x 17"		11	.727		239	18.70		257.70	295
0900	36" x 40"		10	.800		370	20.50		390.50	440
1100	40" x 22"		10	.800		271	20.50		291.50	335
1200	36" x 28"		9	.889		289	23		312	360
1300	36" x 36"		9	.889		320	23		343	395
1400	48" x 28"		8	1		345	25.50		370.50	425
1500	60" x 36"		8	1		500	25.50		525.50	595
2000	Bay, casement units, 8' x 5', 30° angle, w/screens, dbl Insul glass	2 Carp	2.50	6.400	Opng.	1,425	164		1,589	1,850
2100	Low E glass	"	2.50	6.400	"	1,500	164		1,664	1,900
2200	36" x 25"	1 Carp	9	.889	Ea.	252	23		275	315

08 52 Wood Windows

08 52 10 – Wood Windows

08 52 10.20 Awning Window		Crew	Daily Output	Labor-Hours	Unit	Material	2007 Bare Costs Labor	Equipment	Total	Total Incl O&P
2300	40" x 30"	1 Carp	9	.889	Ea.	315	23		338	385
2400	48" x 28"		8	1		325	25.50		350.50	400
2500	60" x 36"		8	1		345	25.50		370.50	425

08 52 10.30 Wood Windows

		Crew	Daily Output	Labor-Hours	Unit	Material	Labor	Equipment	Total	Total Incl O&P
0010	**WOOD WINDOWS**, double hung									
0020	Including frame, double insulated glass, screens and grills									
0040	Double hung, 2'-2" x 3'-4" high	2 Carp	15	1.067	Ea.	180	27.50		207.50	245
0060	2'-2" x 4'-4"		14	1.143		202	29.50		231.50	273
0080	2'-6" x 3'-4"		13	1.231		190	31.50		221.50	263
0100	2'-6" x 4'-0"		12	1.333		205	34.50		239.50	283
0120	2'-6" x 4'-8"		12	1.333		225	34.50		259.50	305
0140	2'-10" x 3'-4"		10	1.600		200	41		241	290
0160	2'-10" x 4'-0"		10	1.600		219	41		260	310
0180	3'-7" x 3'-4"		9	1.778		225	45.50		270.50	325
0200	3'-7" x 5'-4"		9	1.778		289	45.50		334.50	400
0220	3'-10" x 5'-4"		8	2		455	51.50		506.50	585
3200	20 S.F. and over	1 Carp	106	.075	S.F.	15.40	1.94		17.34	20
3800	Triple glazing for above, add				"	7.90			7.90	8.65
4010										
4300	Double hung, 2'-4" x 3'-2", double insulated glass	2 Carp	6.80	2.353	Ea.	161	60.50		221.50	280
4400	Low E glass		6.80	2.353		175	60.50		235.50	295
4600	3'-0" x 4'-6" opening, double insulated glass		6	2.667		217	68.50		285.50	355
4700	Low E glass		6	2.667		237	68.50		305.50	375
5000	Picture window, wood, double insulated glass, 4'-6" x 4'-6"		5	3.200		415	82		497	595
5100	5'-8" x 4'-6"		4.50	3.556		470	91.50		561.50	670
5300	Add for snap-in muntins, 4'-6" x 4'-6"					61			61	67
5400	5'-8" x 4'-6"					62.50			62.50	69
6000	Replacement sash, double hung, double glazing, to 12 S.F.	1 Carp	64	.125	S.F.	17.65	3.21		20.86	25
6100	12 S.F. to 20 S.F.		94	.085	"	17.70	2.19		19.89	23
7000	Sash, single lite, 2'-0" x 2'-0" high		20	.400	Ea.	34	10.30		44.30	54.50
7050	2'-6" x 2'-0" high		19	.421		39	10.80		49.80	61.50
7100	2'-6" x 2'-6" high		18	.444		45	11.40		56.40	69
7150	3'-0" x 2'-0" high		17	.471		48	12.10		60.10	73.50

08 52 10.40 Casement Window

		Crew	Daily Output	Labor-Hours	Unit	Material	Labor	Equipment	Total	Total Incl O&P
0010	**CASEMENT WINDOW**, including frame, screen and grills R085216-10									
0100	Avg. quality, bldrs. model, 2'-0" x 3'-0" H, dbl. insulated glass	1 Carp	10	.800	Ea.	263	20.50		283.50	325
0150	Low E glass		10	.800		281	20.50		301.50	345
0200	2'-0" x 4'-6" high, double insulated glass		9	.889		330	23		353	400
0250	Low E glass		9	.889		350	23		373	425
0300	2'-4" x 6'-0" high, double insulated glass		8	1		405	25.50		430.50	490
0350	Low E glass		8	1		425	25.50		450.50	510
0522	Vinyl clad, premium, double insulated glass, 2'-0" x 3'-0"		10	.800		264	20.50		284.50	325
0524	2'-0" x 4'-0"		9	.889		310	23		333	380
0525	2'-0" x 5'-0"		8	1		355	25.50		380.50	435
0528	2'-0" x 6'-0"		8	1		400	25.50		425.50	485
3020	Vinyl clad, premium, double insulated glass, multiple leaf units									
3080	Single unit, 1'-6" x 5'-0"	2 Carp	20	.800	Ea.	305	20.50		325.50	370
3100	2'-0" x 2'-0"		20	.800		203	20.50		223.50	258
3140	2'-0" x 2'-6"		20	.800		264	20.50		284.50	325
3220	2'-0" x 3'-6"		20	.800		262	20.50		282.50	325
3260	2'-0" x 4'-0"		19	.842		310	21.50		331.50	375
3300	2'-0" x 4'-6"		19	.842		305	21.50		326.50	370

08 52 Wood Windows

08 52 10 – Wood Windows

08 52 10.40 Casement Window

		Crew	Daily Output	Labor-Hours	Unit	Material	2007 Bare Costs Labor	Equipment	Total	Total Incl O&P
3340	2'-0" x 5'-0"	2 Carp	18	.889	Ea.	355	23		378	430
3460	2'-4" x 3'-0"		20	.800		264	20.50		284.50	325
3500	2'-4" x 4'-0"		19	.842		330	21.50		351.50	395
3540	2'-4" x 5'-0"		18	.889		390	23		413	470
3700	Double unit, 2'-8" x 5'-0"		18	.889		550	23		573	645
3740	2'-8" x 6'-0"		17	.941		640	24		664	745
3840	3'-0" x 4'-6"		18	.889		495	23		518	585
3860	3'-0" x 5'-0"		17	.941		645	24		669	750
3880	3'-0" x 6'-0"		17	.941		690	24		714	800
3980	3'-4" x 2'-6"		19	.842		420	21.50		441.50	500
4000	3'-4" x 3'-0"		12	1.333		420	34.50		454.50	525
4030	3'-4" x 4'-0"		18	.889		510	23		533	600
4050	3'-4" x 5'-0"		12	1.333		645	34.50		679.50	770
4100	3'-4" x 6'-0"		11	1.455		690	37.50		727.50	825
4200	3'-6" x 3'-0"		18	.889		500	23		523	590
4340	4'-0" x 3'-0"		18	.889		475	23		498	560
4380	4'-0" x 3'-6"		17	.941		510	24		534	600
4420	4'-0" x 4'-0"		16	1		610	25.50		635.50	715
4460	4'-0" x 4'-4"		16	1		600	25.50		625.50	705
4540	4'-0" x 5'-0"		16	1		650	25.50		675.50	760
4580	4'-0" x 6'-0"		15	1.067		740	27.50		767.50	860
4740	4'-8" x 3'-0"		18	.889		535	23		558	625
4780	4'-8" x 3'-6"		17	.941		570	24		594	665
4820	4'-8" x 4'-0"		16	1		675	25.50		700.50	785
4860	4'-8" x 5'-0"		15	1.067		770	27.50		797.50	890
4900	4'-8" x 6'-0"		15	1.067		865	27.50		892.50	1,000
5060	Triple unit, 5'-0" x 5'-0"		15	1.067		995	27.50		1,022.50	1,150
5100	5'-6" x 3'-0"		17	.941		680	24		704	790
5140	5'-6" x 3'-6"		16	1		720	25.50		745.50	840
5180	5'-6" x 4'-6"		15	1.067		815	27.50		842.50	940
5220	5'-6" x 5'-6"		15	1.067		1,100	27.50		1,127.50	1,250
5300	6'-0" x 4'-6"		15	1.067		815	27.50		842.50	940
5850	5'-0" x 3'-0"		12	1.333		720	34.50		754.50	855
5900	5'-0" x 4'-0"		11	1.455		890	37.50		927.50	1,050
6000	5'-0" x 5'-0"		10	1.600		1,025	41		1,066	1,200
6100	5'-0" x 5'-6"		10	1.600		1,050	41		1,091	1,225
6150	5'-0" x 6'-0"		10	1.600		1,150	41		1,191	1,350
6200	6'-0" x 3'-0"		12	1.333		1,125	34.50		1,159.50	1,300
6250	6'-0" x 3'-4"		12	1.333		720	34.50		754.50	855
6300	6'-0" x 4'-0"		11	1.455		775	37.50		812.50	920
6350	6'-0" x 5'-0"		10	1.600		870	41		911	1,025
6400	6'-0" x 6'-0"		10	1.600		1,200	41		1,241	1,400
6500	Quadruple unit, 7'-0" x 4'-0"		9	1.778		1,075	45.50		1,120.50	1,250
6700	8'-0" x 4'-6"		9	1.778		1,325	45.50		1,370.50	1,525
6950	6'-8" x 4'-0"		10	1.600		1,025	41		1,066	1,200
7000	6'-8" x 6'-0"		10	1.600		1,375	41		1,416	1,600
8100	Metal clad, deluxe, dbl. insul. glass, 2'-0" x 3'-0" high	1 Carp	10	.800		206	20.50		226.50	262
8120	2'-0" x 4'-0" high		9	.889		248	23		271	310
8140	2'-0" x 5'-0" high		8	1		282	25.50		307.50	355
8160	2'-0" x 6'-0" high		8	1		325	25.50		350.50	400
8200	For multiple leaf units, deduct for stationary sash									
8220	2' high				Ea.	21.50			21.50	23.50
8240	4'-6" high					24.50			24.50	27

08 52 Wood Windows

08 52 10 – Wood Windows

08 52 10.40 Casement Window		Crew	Daily Output	Labor-Hours	Unit	Material	2007 Bare Costs Labor	Equipment	Total	Total Incl O&P
8260	6' high				Ea.	33			33	36.50
8300	For installation, add per leaf						15%			

08 52 10.50 Double Hung		Crew	Daily Output	Labor-Hours	Unit	Material	Labor	Equipment	Total	Total Incl O&P
0010	**DOUBLE HUNG**, Including frame, screens and grills									
0100	Avg. quality, bldrs. model, 2'-0" x 3'-0" high, dbl insul. glass	1 Carp	10	.800	Ea.	181	20.50		201.50	234
0150	Low E glass		10	.800		190	20.50		210.50	244
0200	3'-0" x 4'-0" high, double insulated glass		9	.889		244	23		267	305
0250	Low E glass		9	.889		256	23		279	320
0300	4'-0" x 4'-6" high, double insulated glass		8	1		310	25.50		335.50	385
0350	Low E glass		8	1		335	25.50		360.50	410
1000	Vinyl clad, premium, double insulated glass, 2'-6" x 3'-0"		10	.800		218	20.50		238.50	275
1100	3'-0" x 3'-6"		10	.800		256	20.50		276.50	315
1200	3'-0" x 4'-0"		9	.889		365	23		388	440
1300	3'-0" x 4'-6"		9	.889		340	23		363	415
1400	3'-0" x 5'-0"		8	1		310	25.50		335.50	385
1500	3'-6" x 6'-0"		8	1		360	25.50		385.50	445
2000	Metal clad, deluxe, dbl. insul. glass, 2'-6" x 3'-0" high		10	.800		238	20.50		258.50	297
2100	3'-0" x 3'-6" high		10	.800		272	20.50		292.50	335
2200	3'-0" x 4'-0" high		9	.889		286	23		309	355
2300	3'-0" x 4'-6" high		9	.889		300	23		323	370
2400	3'-0" x 5'-0" high		8	1		325	25.50		350.50	405
2500	3'-6" x 6'-0" high		8	1		395	25.50		420.50	480

08 52 10.55 Picture Window		Crew	Daily Output	Labor-Hours	Unit	Material	Labor	Equipment	Total	Total Incl O&P
0010	**PICTURE WINDOW**, Including frame and grills									
0100	Average quality, bldrs. model, 3'-6" x 4'-0" high, dbl insulated glass	2 Carp	12	1.333	Ea.	350	34.50		384.50	445
0150	Low E glass		12	1.333		385	34.50		419.50	485
0200	4'-0" x 4'-6" high, double insulated glass		11	1.455		430	37.50		467.50	540
0250	Low E glass		11	1.455		450	37.50		487.50	560
0300	5'-0" x 4'-0" high, double insulated glass		11	1.455		505	37.50		542.50	620
0350	Low E glass		11	1.455		525	37.50		562.50	640
0400	6'-0" x 4'-6" high, double insulated glass		10	1.600		545	41		586	670
0450	Low E glass		10	1.600		565	41		606	690

08 52 10.65 Wood Sash		Crew	Daily Output	Labor-Hours	Unit	Material	Labor	Equipment	Total	Total Incl O&P
0010	**WOOD SASH**, Including glazing but not trim									
0050	Custom, 5'-0" x 4'-0", 1" dbl. glazed, 3/16" thick lites	2 Carp	3.20	5	Ea.	164	129		293	400
0100	1/4" thick lites		5	3.200		169	82		251	325
0200	1" thick, triple glazed		5	3.200		385	82		467	565
0300	7'-0" x 4'-6" high, 1" double glazed, 3/16" thick lites		4.30	3.721		395	95.50		490.50	590
0400	1/4" thick lites		4.30	3.721		445	95.50		540.50	645
0500	1" thick, triple glazed		4.30	3.721		505	95.50		600.50	715
0600	8'-6" x 5'-0" high, 1" double glazed, 3/16" thick lites		3.50	4.571		530	117		647	785
0700	1/4" thick lites		3.50	4.571		580	117		697	840
0800	1" thick, triple glazed		3.50	4.571		585	117		702	845
0900	Window frames only, based on perimeter length				L.F.	3.60			3.60	3.96

08 52 10.70 Sliding Windows		Crew	Daily Output	Labor-Hours	Unit	Material	Labor	Equipment	Total	Total Incl O&P
0010	**SLIDING WINDOWS**									
0100	Average quality, bldrs. model, 3'-0" x 3'-0" high, double insulated	1 Carp	10	.800	Ea.	233	20.50		253.50	291
0120	Low E glass		10	.800		254	20.50		274.50	315
0200	4'-0" x 3'-6" high, double insulated		9	.889		248	23		271	310
0220	Low E glass		9	.889		283	23		306	350
0300	6'-0" x 5'-0" high, double insulated		8	1		400	25.50		425.50	485
0320	Low E glass		8	1		440	25.50		465.50	530

08 52 Wood Windows

08 52 10 - Wood Windows

08 52 10.70 Sliding Windows		Crew	Daily Output	Labor-Hours	Unit	Material	2007 Bare Costs Labor	Equipment	Total	Total Incl O&P
6000	Sliding, insulating glass, including screens,									
6100	3'-0" x 3'-0"	2 Carp	6.50	2.462	Ea.	260	63.50		323.50	395
6200	4'-0" x 3'-6"		6.30	2.540		294	65.50		359.50	435
6300	5'-0" x 4'-0"		6	2.667		360	68.50		428.50	515

08 52 13 - Metal-Clad Wood Windows

08 52 13.10 Metal-Clad Wood Windows

		Crew	Daily Output	Labor-Hours	Unit	Material	Labor	Equipment	Total	Incl O&P
0010	**METAL-CLAD WOOD WINDOWS**									
2000	Metal clad, deluxe, double insulated glass, 34" x 22"	1 Carp	10	.800	Ea.	232	20.50		252.50	290
2100	40" x 22"	"	10	.800	"	273	20.50		293.50	335

08 52 13.35 Metal-Clad Wood Windows

		Crew	Daily Output	Labor-Hours	Unit	Material	Labor	Equipment	Total	Incl O&P
0010	**METAL-CLAD WOOD WINDOWS**									
2000	Metal clad, deluxe, dbl. insul. glass, 4'-0" x 4'-0" high	2 Carp	12	1.333	Ea.	330	34.50		364.50	420
2100	4'-0" x 6'-0" high		11	1.455		485	37.50		522.50	600
2200	5'-0" x 6'-0" high		10	1.600		535	41		576	655
2300	6'-0" x 6'-0" high		10	1.600		615	41		656	745
2400	Metal clad, deluxe, double insulated glass, 3'-0" x 3'-0" high	1 Carp	10	.800		305	20.50		325.50	375
2420	4'-0" x 3'-6" high		9	.889		375	23		398	455
2440	5'-0" x 4'-0" high		9	.889		455	23		478	540
2460	6'-0" x 5'-0" high		8	1		670	25.50		695.50	785

08 52 16 - Plastic-Clad Wood Windows

08 52 16.10 Bow Window

		Crew	Daily Output	Labor-Hours	Unit	Material	Labor	Equipment	Total	Incl O&P
0010	**BOW WINDOW**									
0020	End panels operable									
1000	Bow type, casement, wood, bldrs mdl, 8' x 5' dbl insltd glass, 4 panel	2 Carp	10	1.600	Ea.	1,200	41		1,241	1,400
1050	Low E glass		10	1.600		1,275	41		1,316	1,475
1100	10'-0" x 5'-0", double insulated glass, 6 panels		6	2.667		1,325	68.50		1,393.50	1,600
1200	Low E glass, 6 panels		6	2.667		1,400	68.50		1,468.50	1,675
1300	Vinyl clad, bldrs model, double insulated glass, 6'-0" x 4'-0", 3 panel		10	1.600		1,225	41		1,266	1,425
1340	9'-0" x 4'-0", 4 panel		8	2		1,450	51.50		1,501.50	1,675
1380	10'-0" x 6'-0", 5 panels		7	2.286		2,375	58.50		2,433.50	2,700
1420	12'-0" x 6'-0", 6 panels		6	2.667		3,000	68.50		3,068.50	3,425
1600	Metal clad, casement, bldrs mdl, 6'-0" x 4'-0", dbl insltd gls, 3 panels		10	1.600		855	41		896	1,000
1640	9'-0" x 4'-0", 4 panels		8	2		1,200	51.50		1,251.50	1,400
1680	10'-0" x 5'-0", 5 panels		7	2.286		1,650	58.50		1,708.50	1,925
1720	12'-0" x 6'-0", 6 panels		6	2.667		2,300	68.50		2,368.50	2,650
2000	Bay window, casement, builders model, 8' x 5' dbl insul glass, 4 panels		10	1.600		1,675	41		1,716	1,900
2050	Low E glass,		10	1.600		2,025	41		2,066	2,300
2100	12'-0" x 6'-0", double insulated glass, 6 panels		6	2.667		2,075	68.50		2,143.50	2,425
2200	Low E glass		6	2.667		2,150	68.50		2,218.50	2,475
2280	6'-0" x 4'-0"		11	1.455		1,075	37.50		1,112.50	1,275
2300	Vinyl clad, premium, double insulated glass, 8'-0" x 5'-0"		10	1.600		1,275	41		1,316	1,475
2340	10'-0" x 5'-0"		8	2		1,800	51.50		1,851.50	2,075
2380	10'-0" x 6'-0"		7	2.286		1,900	58.50		1,958.50	2,175
2420	12'-0" x 6'-0"		6	2.667		2,250	68.50		2,318.50	2,600
2430	14'-0" x 3'-0"		7	2.286		1,550	58.50		1,608.50	1,800
2440	14'-0" x 6'-0"		5	3.200		2,475	82		2,557	2,875
2600	Metal clad, deluxe, dbl insul. glass, 8'-0" x 5'-0" high, 4 panels		10	1.600		1,475	41		1,516	1,700
2640	10'-0" x 5'-0" high, 5 panels		8	2		1,575	51.50		1,626.50	1,825
2680	10'-0" x 6'-0" high, 5 panels		7	2.286		1,875	58.50		1,933.50	2,150
2720	12'-0" x 6'-0" high, 6 panels		6	2.667		2,600	68.50		2,668.50	2,975
3000	Double hung, bldrs. model, bay, 8' x 4' high,dbl insulated glass		10	1.600		1,175	41		1,216	1,350

08 52 Wood Windows

08 52 16 – Plastic-Clad Wood Windows

	08 52 16.10 Bow Window	Crew	Daily Output	Labor-Hours	Unit	Material	2007 Bare Costs Labor	Equipment	Total	Total Incl O&P
3050	Low E glass	2 Carp	10	1.600	Ea.	1,250	41		1,291	1,450
3100	9'-0" x 5'-0" high, doublel insulated glass		6	2.667		1,250	68.50		1,318.50	1,525
3200	Low E glass		6	2.667		1,325	68.50		1,393.50	1,600
3300	Vinyl clad, premium, double insulated glass, 7'-0" x 4'-6"		10	1.600		1,225	41		1,266	1,400
3340	8'-0" x 4'-6"		8	2		1,250	51.50		1,301.50	1,450
3380	8'-0" x 5'-0"		7	2.286		1,300	58.50		1,358.50	1,525
3420	9'-0" x 5'-0"		6	2.667		1,325	68.50		1,393.50	1,600
3600	Metal clad, deluxe, dbl insul. glass, 7'-0" x 4'-0" high		10	1.600		1,125	41		1,166	1,300
3640	8'-0" x 4'-0" high		8	2		1,175	51.50		1,226.50	1,350
3680	8'-0" x 5'-0" high		7	2.286		1,200	58.50		1,258.50	1,425
3720	9'-0" x 5'-0" high		6	2.667		1,275	68.50		1,343.50	1,525
08 52 16.20 Half Round, Vinyl Clad										
0010	**HALF ROUND, VINYL CLAD**, double insulated glass, incl. grill									
0800	14" height x 24" base	2 Carp	9	1.778	Ea.	370	45.50		415.50	490
1040	15" height x 25" base		8	2		375	51.50		426.50	495
1060	16" height x 28" base		7	2.286		405	58.50		463.50	545
1080	17" height x 29" base		7	2.286		420	58.50		478.50	560
2000	19" height x 33" base	1 Carp	6	1.333		445	34.50		479.50	550
2100	20" height x 35" base		6	1.333		470	34.50		504.50	575
2200	21" height x 37" base		6	1.333		475	34.50		509.50	580
2250	23" height x 41" base	2 Carp	6	2.667		510	68.50		578.50	680
2300	26" height x 48" base		6	2.667		530	68.50		598.50	700
2350	30" height x 56" base		6	2.667		615	68.50		683.50	790
3000	36" height x 67"base	1 Carp	4	2		1,025	51.50		1,076.50	1,200
3040	38" height x 71" base	2 Carp	5	3.200		955	82		1,037	1,200
3050	40" height x 75" base	"	5	3.200		1,250	82		1,332	1,525
5000	Elliptical, 71" x 16"	1 Carp	11	.727		865	18.70		883.70	980
5100	95" x 21"	"	10	.800		1,175	20.50		1,195.50	1,325
08 52 16.30 Palladian Windows										
0010	**PALLADIAN WINDOWS**									
0020	Vinyl clad, double insulated glass, including frame and grills									
0040	3'-2" x 2'-6" high	2 Carp	11	1.455	Ea.	1,200	37.50		1,237.50	1,400
0060	3'-2" x 4'-10"		11	1.455		1,475	37.50		1,512.50	1,700
0080	3'-2" x 6'-4"		10	1.600		1,675	41		1,716	1,925
0100	4'-0" x 4'-0"		10	1.600		1,450	41		1,491	1,675
0120	4'-0" x 5'-4"	3 Carp	10	2.400		1,750	61.50		1,811.50	2,025
0140	4'-0" x 6'-0"		9	2.667		1,950	68.50		2,018.50	2,275
0160	4'-0" x 7'-4"		9	2.667		1,950	68.50		2,018.50	2,275
0180	5'-5" x 4'-10"		9	2.667		2,050	68.50		2,118.50	2,400
0200	5'-5" x 6'-10"		9	2.667		2,375	68.50		2,443.50	2,725
0220	5'-5" x 7'-9"		9	2.667		2,575	68.50		2,643.50	2,950
0240	6'-0" x 7'-11"		8	3		3,225	77		3,302	3,675
0260	8'-0" x 6'-0"		8	3		2,850	77		2,927	3,275
08 52 16.40 Transom Windows										
0010	**TRANSOM WINDOWS**									
0050	Vinyl clad, premium, double insulated glass, 32" x 8"	1 Carp	16	.500	Ea.	173	12.85		185.85	212
0100	36" x 8"		16	.500		184	12.85		196.85	224
0110	36" x 12"		16	.500		196	12.85		208.85	238
1000	Vinyl clad, premium, dbl. insul. glass, 4'-0" x 4'-0"	2 Carp	12	1.333		470	34.50		504.50	575
1100	4'-0" x 6'-0"		11	1.455		875	37.50		912.50	1,025
1200	5'-0" x 6'-0"		10	1.600		970	41		1,011	1,150
1300	6'-0" x 6'-0"		10	1.600		990	41		1,031	1,150

08 52 Wood Windows

08 52 16 – Plastic-Clad Wood Windows

08 52 16.45 Trapezoid Windows		Crew	Daily Output	Labor-Hours	Unit	Material	2007 Bare Costs Labor	Equipment	Total	Total Incl O&P
0010	**TRAPEZOID WINDOWS**									
0100	Vinyl clad, including frame and exterior trim									
0900	20" base x 44" leg x 53" leg	2 Carp	13	1.231	Ea.	360	31.50		391.50	450
1000	24" base x 90" leg x 102" leg		8	2		600	51.50		651.50	745
3000	36" base x 0" leg x 22" leg		12	1.333		385	34.50		419.50	485
3010	36" base x 4" leg x 25" leg		13	1.231		405	31.50		436.50	500
3050	36" base x 26" leg x 48" leg		9	1.778		420	45.50		465.50	540
3100	36" base x 42" legs, 50" peak		9	1.778		490	45.50		535.50	620
3200	36" base x 60" leg x 81" leg		11	1.455		635	37.50		672.50	765
4320	44" base x 23" leg x 56" leg		11	1.455		505	37.50		542.50	620
4350	44" base x 59" leg x 92" leg		10	1.600		755	41		796	905
4500	46" base x 15" leg x 46" leg		8	2		380	51.50		431.50	505
4550	46" base x 16" leg x 48" leg		8	2		405	51.50		456.50	530
4600	46" base x 50" leg x 80" leg		7	2.286		610	58.50		668.50	770
6600	66" base x 12" leg x 42" leg		8	2		510	51.50		561.50	650
6650	66" base x 12" legs, 28" peak		9	1.778		415	45.50		460.50	540
6700	68" base x 3" legs, 31" peak		8	2		520	51.50		571.50	660

08 52 16.70 Vinyl Clad, Premium, DBL. Insulated Glass		Crew	Daily Output	Labor-Hours	Unit	Material	Labor	Equipment	Total	Total Incl O&P
0010	**VINYL CLAD, PREMIUM, DBL. INSULATED GLASS**									
1000	3'-0" x 3'-0"	1 Carp	10	.800	Ea.	510	20.50		530.50	595
1020	4'-0" x 1'-11"		11	.727		490	18.70		508.70	570
1040	4'-0" x 3'-0"		10	.800		590	20.50		610.50	680
1050	4'-0" x 3'-6"		9	.889		630	23		653	730
1090	4'-0" x 5'-0"		9	.889		775	23		798	890
1100	5'-0" x 4'-0"		9	.889		760	23		783	875
1120	5'-0" x 5'-0"		8	1		855	25.50		880.50	985
1140	6'-0" x 4'-0"		8	1		865	25.50		890.50	995
1150	6'-0" x 5'-0"		8	1		965	25.50		990.50	1,100

08 52 50 – Window Accessories

08 52 50.10 Window Grille or Muntin		Crew	Daily Output	Labor-Hours	Unit	Material	Labor	Equipment	Total	Total Incl O&P
0010	**WINDOW GRILLE OR MUNTIN,** snap in type									
0020	Standard pattern interior grills									
2000	Wood, awning window, glass size 28" x 16" high	1 Carp	30	.267	Ea.	21.50	6.85		28.35	35
2060	44" x 24" high		32	.250		31	6.45		37.45	45.50
2100	Casement, glass size, 20" x 36" high		30	.267		26.50	6.85		33.35	40.50
2180	20" x 56" high		32	.250		38	6.45		44.45	53
2200	Double hung, glass size, 16" x 24" high		24	.333	Set	46	8.55		54.55	65.50
2280	32" x 32" high		34	.235	"	123	6.05		129.05	146
2500	Picture, glass size, 48" x 48" high		30	.267	Ea.	141	6.85		147.85	167
2580	60" x 68" high		28	.286	"	108	7.35		115.35	131
2600	Sliding, glass size, 14" x 36" high		24	.333	Set	25	8.55		33.55	42
2680	36" x 36" high		22	.364	"	38.50	9.35		47.85	58

08 52 66 – Wood Window Screens

08 52 66.10 Wood Screens		Crew	Daily Output	Labor-Hours	Unit	Material	Labor	Equipment	Total	Total Incl O&P
0010	**WOOD SCREENS**									
0020	Over 3 S.F., 3/4" frames	2 Carp	375	.043	S.F.	4.18	1.10		5.28	6.45
0100	1-1/8" frames	"	375	.043	"	6.60	1.10		7.70	9.10

08 52 69 – Wood Storm Windows

08 52 69.10 Storm Windows		Crew	Daily Output	Labor-Hours	Unit	Material	Labor	Equipment	Total	Total Incl O&P
0010	**STORM WINDOWS,** aluminum residential									

08 52 Wood Windows

08 52 69 – Wood Storm Windows

08 52 69.10 Storm Windows		Crew	Daily Output	Labor-Hours	Unit	Material	2007 Bare Costs Labor	Equipment	Total	Total Incl O&P
0300	Basement, mill finish, incl. fiberglass screen									
0320	1'-10" x 1'-0" high	2 Carp	30	.533	Ea.	29.50	13.70		43.20	56
0340	2'-9" x 1'-6" high		30	.533		32	13.70		45.70	59
0360	3'-4" x 2'-0" high		30	.533		39	13.70		52.70	66
1600	Double-hung, combination, storm & screen									
1700	Custom, clear anodic coating, 2'-0" x 3'-5" high	2 Carp	30	.533	Ea.	76.50	13.70		90.20	108
1720	2'-6" x 5'-0" high		28	.571		102	14.70		116.70	138
1740	4'-0" x 6'-0" high		25	.640		217	16.45		233.45	266
1800	White painted, 2'-0" x 3'-5" high		30	.533		91	13.70		104.70	124
1820	2'-6" x 5'-0" high		28	.571		146	14.70		160.70	186
1840	4'-0" x 6'-0" high		25	.640		262	16.45		278.45	315
2000	Average quality, clear anodic coating, 2'-0" x 3'-5" high		30	.533		77.50	13.70		91.20	109
2020	2'-6" x 5'-0" high		28	.571		98	14.70		112.70	133
2040	4'-0" x 6'-0" high		25	.640		115	16.45		131.45	155
2400	White painted, 2'-0" x 3'-5" high		30	.533		76.50	13.70		90.20	108
2420	2'-6" x 5'-0" high		28	.571		84.50	14.70		99.20	118
2440	4'-0" x 6'-0" high		25	.640		92.50	16.45		108.95	130
2600	Mill finish, 2'-0" x 3'-5" high		30	.533		69.50	13.70		83.20	100
2620	2'-6" x 5'-0" high		28	.571		77.50	14.70		92.20	111
2640	4'-0" x 6-8" high		25	.640		87	16.45		103.45	124
4000	Picture window, storm, 1 lite, white or bronze finish									
4020	4'-6" x 4'-6" high	2 Carp	25	.640	Ea.	117	16.45		133.45	156
4040	5'-8" x 4'-6" high		20	.800		132	20.50		152.50	181
4400	Mill finish, 4'-6" x 4'-6" high		25	.640		117	16.45		133.45	156
4420	5'-8" x 4'-6" high		20	.800		132	20.50		152.50	181
4600	3 lite, white or bronze finish									
4620	4'-6" x 4'-6" high	2 Carp	25	.640	Ea.	142	16.45		158.45	184
4640	5'-8" x 4'-6" high		20	.800		158	20.50		178.50	209
4800	Mill finish, 4'-6" x 4'-6" high		25	.640		125	16.45		141.45	165
4820	5'-8" x 4'-6" high		20	.800		142	20.50		162.50	191
5000	Sliding glass door, storm 6' x 6'-8", standard	1 Glaz	2	4		715	102		817	960
5100	Economy	"	2	4		430	102		532	645
6000	Sliding window, storm, 2 lite, white or bronze finish									
6020	3'-4" x 2'-7" high	2 Carp	28	.571	Ea.	105	14.70		119.70	140
6040	4'-4" x 3'-3" high		25	.640		143	16.45		159.45	185
6060	5'-4" x 6'-0" high		20	.800		229	20.50		249.50	287
9000	Magnetic interior storm window									
9100	3/16" plate glass	1 Glaz	107	.075	S.F.	4.50	1.91		6.41	8.10

08 53 Plastic Windows

08 53 13 – Vinyl Windows

08 53 13.10 Solid Vinyl Windows		Crew	Daily Output	Labor-Hours	Unit	Material	2007 Bare Costs Labor	Equipment	Total	Total Incl O&P
0010	**SOLID VINYL WINDOWS**									
0020	Double hung, including frame and screen, 2'-0" x 2'-6"	2 Carp	15	1.067	Ea.	164	27.50		191.50	228
0040	2'-0" x 3'-6"		14	1.143		171	29.50		200.50	238
0060	2'-6" x 4'-6"		13	1.231		206	31.50		237.50	280
0080	3'-0" x 4'-0"		10	1.600		209	41		250	300
0100	3'-0" x 4'-6"		9	1.778		231	45.50		276.50	330
0120	4'-0" x 4'-6"		8	2		255	51.50		306.50	365
0140	4'-0" x 6'-0"		7	2.286		296	58.50		354.50	425
9100	Minimum labor/equipment charge		2	8	Job		206		206	350

08 53 Plastic Windows

08 53 13 – Vinyl Windows

08 53 13.20 Vinyl Single Hung Windows		Crew	Daily Output	Labor-Hours	Unit	Material	2007 Bare Costs Labor	Equipment	Total	Total Incl O&P
0010	**VINYL SINGLE HUNG WINDOWS**									
0100	Grids, low E, J fin, ext. jambs, 21" x 53"	2 Carp	18	.889	Ea.	146	23		169	200
0110	21" x 57"		17	.941		150	24		174	206
0120	21" x 65"		16	1		156	25.50		181.50	215
0130	25" x 41"		20	.800		138	20.50		158.50	187
0140	25" x 49"		18	.889		152	23		175	207
0150	25" x 57"		17	.941		156	24		180	212
0160	25" x 65"		16	1		162	25.50		187.50	222
0170	29" x 41"		18	.889		147	23		170	200
0180	29" x 53"		18	.889		157	23		180	212
0190	29" x 57"		17	.941		161	24		185	218
0200	29" x 65"		16	1		167	25.50		192.50	228
0210	33" x 41"		20	.800		152	20.50		172.50	202
0220	33" x 53"		18	.889		163	23		186	219
0230	33" x 57"		17	.941		167	24		191	225
0240	33" x 65"		16	1		174	25.50		199.50	235
0250	37" x 41"		20	.800		160	20.50		180.50	211
0260	37" x 53"		18	.889		172	23		195	228
0270	37" x 57"		17	.941		175	24		199	234
0280	37" x 65"		16	1		182	25.50		207.50	245
0500	Vinyl clad, premium, double insulated glass, circle, 24" diameter	1 Carp	6	1.333		780	34.50		814.50	915
1000	2'-4" diameter		6	1.333		855	34.50		889.50	1,000
1500	2'-11" diameter		6	1.333		985	34.50		1,019.50	1,125

08 53 13.30 Vinyl Double Hung Windows		Crew	Daily Output	Labor-Hours	Unit	Material	2007 Bare Costs Labor	Equipment	Total	Total Incl O&P
0010	**VINYL DOUBLE HUNG WINDOWS**									
0100	Grids, low E, J fin, ext. jambs, 21" x 53"	2 Carp	18	.889	Ea.	167	23		190	223
0102	21" x 37"		18	.889		149	23		172	203
0104	21" x 41"		18	.889		153	23		176	207
0106	21" x 49"		18	.889		160	23		183	215
0110	21" x 57"		17	.941		171	24		195	229
0120	21" x 65"		16	1		177	25.50		202.50	239
0128	25" x 37"		20	.800		157	20.50		177.50	208
0130	25" x 41"		20	.800		161	20.50		181.50	212
0140	25" x 49"		18	.889		166	23		189	221
0145	25" x 53"		18	.889		172	23		195	229
0150	25" x 57"		17	.941		173	24		197	231
0160	25" x 65"		16	1		184	25.50		209.50	246
0162	25" x 69"		16	1		191	25.50		216.50	254
0164	25" x 77"		16	1		202	25.50		227.50	266
0168	29" x 37"		18	.889		162	23		185	217
0170	29" x 41"		18	.889		166	23		189	221
0172	29" x 49"		18	.889		174	23		197	230
0180	29" x 53"		18	.889		178	23		201	234
0190	29" x 57"		17	.941		181	24		205	240
0200	29" x 65"		16	1		188	25.50		213.50	251
0202	29" x 69"		16	1		195	25.50		220.50	259
0205	29" x 77"		16	1		207	25.50		232.50	271
0208	33" x 37"		20	.800		167	20.50		187.50	218
0210	33" x 41"		20	.800		170	20.50		190.50	222
0215	33" x 49"		20	.800		179	20.50		199.50	232
0220	33" x 53"		18	.889		183	23		206	240
0230	33" x 57"		17	.941		187	24		211	247

08 53 Plastic Windows

08 53 13 – Vinyl Windows

08 53 13.30 Vinyl Double Hung Windows		Crew	Daily Output	Labor-Hours	Unit	Material	2007 Bare Costs Labor	Equipment	Total	Total Incl O&P
0240	33" x 65"	2 Carp	16	1	Ea.	192	25.50		217.50	255
0242	33" x 69"		16	1		204	25.50		229.50	268
0246	33" x 77"		16	1		214	25.50		239.50	279
0250	37" x 41"		20	.800		174	20.50		194.50	227
0255	37" x 49"		20	.800		183	20.50		203.50	237
0260	37" x 53"		18	.889		191	23		214	249
0270	37" x 57"		17	.941		195	24		219	255
0280	37" x 65"		16	1		200	25.50		225.50	264
0282	37" x 69"		16	1		262	25.50		287.50	330
0286	37" x 77"		16	1		274	25.50		299.50	345
0300	Solid vinyl, average quality, double insulated glass, 2'-0" x 3'-0"	1 Carp	10	.800		253	20.50		273.50	315
0310	3'-0" x 4'-0"		9	.889		165	23		188	220
0320	4'-0" x 4'-6"		8	1		264	25.50		289.50	335
0330	Premium, double insulated glass, 2'-6" x 3'-0"		10	.800		180	20.50		200.50	233
0340	3'-0" x 3'-6"		9	.889		210	23		233	270
0350	3'-0" x 4'-0"		9	.889		222	23		245	283
0360	3'-0" x 4'-6"		9	.889		226	23		249	287
0370	3'-0" x 5'-0"		8	1		242	25.50		267.50	310
0380	3'-6" x 6'-0"		8	1		279	25.50		304.50	350
08 53 13.40 Vinyl Casement Windows										
0010	**VINYL CASEMENT WINDOWS**									
0100	Grids, low E, J fin, ext. jambs, 1 lt, 21" x 41"	2 Carp	20	.800	Ea.	216	20.50		236.50	273
0110	21" x 47"		20	.800		236	20.50		256.50	294
0120	21" x 53"		20	.800		255	20.50		275.50	315
0128	24" x 35"		19	.842		208	21.50		229.50	265
0130	24" x 41"		19	.842		226	21.50		247.50	285
0140	24" x 47"		19	.842		244	21.50		265.50	305
0150	24" x 53"		19	.842		263	21.50		284.50	325
0158	28" x 35"		19	.842		221	21.50		242.50	281
0160	28" x 41"		19	.842		239	21.50		260.50	300
0170	28" x 47"		19	.842		258	21.50		279.50	320
0180	28" x 53"		19	.842		284	21.50		305.50	350
0184	28" x 59"		19	.842		290	21.50		311.50	355
0188	Two lites, 33" x 35"		18	.889		355	23		378	430
0190	33" x 41"		18	.889		380	23		403	460
0200	33" x 47"		18	.889		410	23		433	490
0210	33" x 53"		18	.889		440	23		463	520
0212	33" x 59"		18	.889		465	23		488	550
0215	33" x 72"		18	.889		480	23		503	570
0220	41" x 41"		18	.889		415	23		438	500
0230	41" x 47"		18	.889		445	23		468	530
0240	41" x 53"		17	.941		475	24		499	560
0242	41" x 59"		17	.941		500	24		524	590
0246	41" x 72"		17	.941		520	24		544	615
0250	47" x 41"		17	.941		420	24		444	505
0260	47" x 47"		17	.941		450	24		474	535
0270	47" x 53"		17	.941		475	24		499	560
0272	47" x 59"		17	.941		520	24		544	610
0280	56" x 41"		15	1.067		450	27.50		477.50	540
0290	56" x 47"		15	1.067		475	27.50		502.50	565
0300	56" x 53"		15	1.067		520	27.50		547.50	615
0302	56" x 59"		15	1.067		540	27.50		567.50	640

08 53 Plastic Windows

08 53 13 – Vinyl Windows

08 53 13.40 Vinyl Casement Windows

		Crew	Daily Output	Labor-Hours	Unit	Material	2007 Bare Costs Labor	Equipment	Total	Total Incl O&P
0310	56" x 72"	2 Carp	15	1.067	Ea.	590	27.50		617.50	690
0340	Solid vinyl, premium, double insulated glass, 2'-0" x 3'-0" high	1 Carp	10	.800		233	20.50		253.50	291
0360	2'-0" x 4'-0" high		9	.889		267	23		290	335
0380	2'-0" x 5'-0" high		8	1		266	25.50		291.50	335

08 53 13.50 Vinyl Picture Windows

		Crew	Daily Output	Labor-Hours	Unit	Material	Labor	Equipment	Total	Incl O&P
0010	**VINYL PICTURE WINDOWS**									
0100	Grids, low E, J fin, ext. jambs, 33" x 47"	2 Carp	12	1.333	Ea.	217	34.50		251.50	297
0110	35" x 71"		12	1.333		230	34.50		264.50	310
0120	41" x 47"		12	1.333		252	34.50		286.50	335
0130	41" x 71"		12	1.333		273	34.50		307.50	360
0140	47" x 47"		12	1.333		285	34.50		319.50	375
0150	47" x 71"		11	1.455		299	37.50		336.50	395
0160	53" x 47"		11	1.455		280	37.50		317.50	375
0170	53" x 71"		11	1.455		293	37.50		330.50	390
0180	59" x 47"		11	1.455		320	37.50		357.50	415
0190	59" x 71"		11	1.455		340	37.50		377.50	440
0200	71" x 47"		10	1.600		350	41		391	460
0210	71" x 71"		10	1.600		370	41		411	480

08 53 13.60 Vinyl Half Round Windows

		Crew	Daily Output	Labor-Hours	Unit	Material	Labor	Equipment	Total	Incl O&P
0010	**VINYL HALF ROUND WINDOWS**, Including grill, j fin low E, ext. jambs									
0100	10" height x 20" base	2 Carp	8	2	Ea.	247	51.50		298.50	360
0110	15" height x 30" base		8	2		340	51.50		391.50	460
0120	17" height x 34" base		7	2.286		355	58.50		413.50	490
0130	19" height x 38" base		7	2.286		405	58.50		463.50	545
0140	19" height x 33" base		7	2.286		350	58.50		408.50	485
0150	24" height x 48" base	1 Carp	6	1.333		415	34.50		449.50	515
0160	25" height x 50" base	"	6	1.333		460	34.50		494.50	565
0170	30" height x 60" base	2 Carp	6	2.667		560	68.50		628.50	730

08 61 Roof Windows

08 61 13 – Metal Roof Windows

08 61 13.10 Metal Roof Windows

		Crew	Daily Output	Labor-Hours	Unit	Material	Labor	Equipment	Total	Incl O&P
0010	**METAL ROOF WINDOWS**, fixed high perf tmpd glazing									
0020	46" x 21-1/2"	1 Carp	8	1	Ea.	245	25.50		270.50	315
0100	46" x 28"		8	1		278	25.50		303.50	350
0125	57" x 44"		6	1.333		360	34.50		394.50	455
0130	72" x 28"		7	1.143		360	29.50		389.50	445
0150	Venting, high performance tempered glazing, 46" x 21-1/2"		8	1		370	25.50		395.50	455
0175	46" x 28"		8	1		400	25.50		425.50	485
0200	57" x 44"		6	1.333		485	34.50		519.50	595
0500	Flashing set for shingled roof, 46" x 21-1/2"		5	1.600		43.50	41		84.50	118
0525	46" x 28"		5	1.600		43.50	41		84.50	118
0550	57" x 44"		5	1.600		42.50	41		83.50	117
0560	72" x 28"		6	1.333		42.50	34.50		77	105
0575	Flashing set for low pitched roof, 46" x 21-1/2"		5	1.600		179	41		220	267
0600	46" x 28"		5	1.600		184	41		225	272
0625	57" x 44"		5	1.600		212	41		253	305
0650	Flashing set for tile roof 46" x 21-1/2"		5	1.600		105	41		146	186
0675	46" x 28"		5	1.600		109	41		150	190
0700	57" x 44"		5	1.600		125	41		166	207

08 61 Roof Windows

08 61 16 – Wood Roof Windows

08 61 16.16 Wood Roof Windows		Crew	Daily Output	Labor-Hours	Unit	Material	2007 Bare Costs Labor	Equipment	Total	Total Incl O&P
0010	**WOOD ROOF WINDOWS**									
5600	Roof window incl. frame, flashing, double insulated glass & screens,									
5610	complete unit, 22" x 38"	2 Carp	3	5.333	Ea.	218	137		355	475
5650	2'-5" x 3'-8"		3.20	5		280	129		409	530
5700	3'-5" x 4'-9"	↓	3.40	4.706	↓	360	121		481	600

08 62 Unit Skylights

08 62 13 – Domed Unit Skylights

08 62 13.20 Skylights

		Crew	Daily Output	Labor-Hours	Unit	Material	Labor	Equipment	Total	Total Incl O&P
0010	**SKYLIGHTS**, Plastic domes, flush or curb mounted ten or									
0100	more units, curb not included									
0300	Nominal size under 10 S.F., double	G-3	130	.246	S.F.	24.50	5.85		30.35	36.50
0400	Single		160	.200		19.20	4.75		23.95	29
0600	10 S.F. to 20 S.F., double		315	.102		18.55	2.41		20.96	24.50
0700	Single		395	.081		17.25	1.92		19.17	22
0900	20 S.F. to 30 S.F., double		395	.081		16.45	1.92		18.37	21.50
1000	Single		465	.069		15	1.63		16.63	19.25
1200	30 S.F. to 65 S.F., double		465	.069		14.50	1.63		16.13	18.70
1300	Single	↓	610	.052	↓	16	1.24		17.24	19.70
1500	For insulated 4" curbs, double, add					25%				
1600	Single, add					30%				
1800	For integral insulated 9" curbs, double, add					30%				
1900	Single, add					40%				
2120	Ventilating insulated plexiglass dome with									
2130	curb mounting, 36" x 36"	G-3	12	2.667	Ea.	380	63.50		443.50	525
2150	52" x 52"		12	2.667		570	63.50		633.50	730
2160	28" x 52"		10	3.200		445	76		521	620
2170	36" x 52"	↓	10	3.200		480	76		556	660
2180	For electric opening system, add					285			285	315
2210	Operating skylight, with thermopane glass, 24" x 48"	G-3	10	3.200		545	76		621	730
2220	32" x 48"	"	9	3.556	↓	570	84.50		654.50	770
2310	Non venting insulated plexiglass dome skylight with									
2320	Flush mount 22" x 46"	G-3	15.23	2.101	Ea.	310	50		360	425
2330	30" x 30"		16	2		285	47.50		332.50	395
2340	46" x 46"		13.91	2.301		525	54.50		579.50	670
2350	Curb mount 22" x 46"		15.23	2.101		272	50		322	385
2360	30" x 30"		16	2		260	47.50		307.50	365
2370	46" x 46"		13.91	2.301		490	54.50		544.50	625
2381	Non-insulated flush mount 22" x 46"		15.23	2.101		209	50		259	315
2382	30" x 30"		16	2		190	47.50		237.50	289
2383	46" x 46"		13.91	2.301		355	54.50		409.50	480
2384	Curb mount 22" x 46"		15.23	2.101		177	50		227	279
2385	30" x 30"	↓	16	2	↓	171	47.50		218.50	268

08 71 Door Hardware

08 71 20 – Hardware

08 71 20.15 Hardware

		Crew	Daily Output	Labor-Hours	Unit	Material	2007 Bare Costs Labor	Equipment	Total	Total Incl O&P
0009	**HARDWARE**									
0010	Average hardware percentage for hardware, total job cost									
0025	Minimum									.75%
0050	Maximum									3.50%
0500	Total hardware for building, average distribution					85%	15%			
1000	Door hardware, apartment, interior				Door	129			129	142
2100	Pocket door				Ea.	129			129	142
4000	Door knocker, bright brass	1 Carp	32	.250		43	6.45		49.45	58.50
4100	Mail slot, bright brass, 2" x 11"	"	25	.320		57	8.20		65.20	76.50
4200	Peep hole, add to price of door					15.70			15.70	17.30

08 71 20.40 Lockset

		Crew	Daily Output	Labor-Hours	Unit	Material	Labor	Equipment	Total	Total Incl O&P
0010	**LOCKSET**, Standard duty									
0020	Non-keyed, passage	1 Carp	12	.667	Ea.	43	17.15		60.15	76.50
0100	Privacy		12	.667		54	17.15		71.15	88.50
0400	Keyed, single cylinder function		10	.800		75.50	20.50		96	118
0500	Lever handled, keyed, single cylinder function		10	.800		134	20.50		154.50	182
1700	Residential, interior door, minimum		16	.500		15.10	12.85		27.95	38.50
1720	Maximum		8	1		40	25.50		65.50	87.50
1800	Exterior, minimum		14	.571		33.50	14.70		48.20	62
1810	Average		8	1		67	25.50		92.50	117
1820	Maximum		8	1		141	25.50		166.50	200

08 71 20.50 Door Stops

		Crew	Daily Output	Labor-Hours	Unit	Material	Labor	Equipment	Total	Total Incl O&P
0010	**DOOR STOPS**									
0020	Holder & bumper, floor or wall	1 Carp	32	.250	Ea.	31.50	6.45		37.95	46
1300	Wall bumper, 4" diameter, with rubber pad, aluminum		32	.250		9.50	6.45		15.95	21.50
1600	Door bumper, floor type, aluminum		32	.250		4.94	6.45		11.39	16.35
1900	Plunger type, door mounted		32	.250		26	6.45		32.45	40

08 71 20.60 Entrance Locks

		Crew	Daily Output	Labor-Hours	Unit	Material	Labor	Equipment	Total	Total Incl O&P
0010	**ENTRANCE LOCKS**									
0015	Cylinder, grip handle deadlocking latch	1 Carp	9	.889	Ea.	122	23		145	173
0020	Deadbolt		8	1		148	25.50		173.50	207
0100	Push and pull plate, dead bolt		8	1		141	25.50		166.50	199
0900	For handicapped lever, add					154			154	169

08 71 20.65 Thresholds

		Crew	Daily Output	Labor-Hours	Unit	Material	Labor	Equipment	Total	Total Incl O&P
0010	**THRESHOLDS**									
0011	Threshold 3' long saddles aluminum	1 Carp	48	.167	L.F.	3.70	4.28		7.98	11.30
0100	Aluminum, 8" wide, 1/2" thick		12	.667	Ea.	35	17.15		52.15	67.50
0500	Bronze		60	.133	L.F.	32	3.43		35.43	41.50
0600	Bronze, panic threshold, 5" wide, 1/2" thick		12	.667	Ea.	60	17.15		77.15	95
0700	Rubber, 1/2" thick, 5-1/2" wide		20	.400		34	10.30		44.30	55
0800	2-3/4" wide		20	.400		14.75	10.30		25.05	33.50

08 71 20.75 Door Hardware Accessories

		Crew	Daily Output	Labor-Hours	Unit	Material	Labor	Equipment	Total	Total Incl O&P
0010	**DOOR HARDWARE ACCESSORIES**									
1000	Knockers, brass, standard	1 Carp	16	.500	Ea.	39	12.85		51.85	64.50
1100	Deluxe		10	.800		114	20.50		134.50	161
4100	Deluxe		18	.444		40	11.40		51.40	63.50
4500	Rubber door silencers		540	.015		.13	.38		.51	.79

08 71 20.90 Hinges

		Crew	Daily Output	Labor-Hours	Unit	Material	Labor	Equipment	Total	Total Incl O&P
0010	**HINGES** R087120-10									
0012	Full mortise, avg. freq., steel base, USP, 4-1/2" x 4-1/2"				Pr.	21.50			21.50	23.50
0100	5" x 5", USP					36			36	39.50

08 71 Door Hardware

08 71 20 – Hardware

08 71 20.90 Hinges

		Crew	Daily Output	Labor-Hours	Unit	Material	2007 Bare Costs Labor	Equipment	Total	Total Incl O&P
0200	6" x 6", USP				Pr.	76.50			76.50	84.50
0400	Brass base, 4-1/2" x 4-1/2", US10					44.50			44.50	49
0500	5" x 5", US10					65			65	71.50
0600	6" x 6", US10					110			110	121
0800	Stainless steel base, 4-1/2" x 4-1/2", US32				▼	66.50			66.50	73
0900	For non removable pin, add (security item)				Ea.	2.42			2.42	2.66
0910	For floating pin, driven tips, add					2.77			2.77	3.05
0930	For hospital type tip on pin, add					11.95			11.95	13.15
0940	For steeple type tip on pin, add				▼	10.45			10.45	11.50
0950	Full mortise, high frequency, steel base, 3-1/2" x 3-1/2", US26D				Pr.	23			23	25.50
1000	4-1/2" x 4-1/2", USP					52.50			52.50	57.50
1100	5" x 5", USP					49			49	54
1200	6" x 6", USP					119			119	131
1400	Brass base, 3-1/2" x 3-1/2", US4					41			41	45
1430	4-1/2" x 4-1/2", US10					70			70	77.50
1500	5" x 5", US10					105			105	115
1600	6" x 6", US10					152			152	167
1800	Stainless steel base, 4-1/2" x 4-1/2", US32					113			113	124
1810	5" x 4-1/2", US32				▼	157			157	172
1930	For hospital type tip on pin, add				Ea.	7.10			7.10	7.80
1950	Full mortise, low frequency, steel base, 3-1/2" x 3-1/2", US26D				Pr.	9.40			9.40	10.30
2000	4-1/2" x 4-1/2", USP					10.65			10.65	11.75
2100	5" x 5", USP					26.50			26.50	29
2200	6" x 6", USP					53			53	58
2300	4-1/2" x 4-1/2", US3					15.45			15.45	17
2310	5" x 5", US3					38			38	42
2400	Brass bass, 4-1/2" x 4-1/2", US10					37			37	41
2500	5" x 5", US10					56.50			56.50	62
2800	Stainless steel base, 4-1/2" x 4-1/2", US32				▼	64			64	70.50

08 71 20.92 Mortised Hinge

		Crew	Daily Output	Labor-Hours	Unit	Material	Labor	Equipment	Total	Total Incl O&P
0010	**MORTISED HINGES**									
0200	Average frequency, steel plated, ball bearing				Pr.	24			24	26.50
0300	Bronze, ball bearing					27			27	30
0900	High frequency, steel plated, ball bearing					72.50			72.50	80
1100	Bronze, ball bearing					75.50			75.50	83
1300	Average frequency, steel plated, ball bearing					29.50			29.50	32.50
1500	Bronze, ball bearing, to 36" wide					31.50			31.50	34.50
1700	Low frequency, steel, plated, plain bearing					13.25			13.25	14.60
1900	Bronze, plain bearing				▼	15.10			15.10	16.60

08 71 20.95 Kick Plates

		Crew	Daily Output	Labor-Hours	Unit	Material	Labor	Equipment	Total	Total Incl O&P
0010	**KICK PLATES**									
0020	Stainless steel	1 Carp	15	.533	Ea.	28	13.70		41.70	54.50
0500	Bronze	"	15	.533	"	35.50	13.70		49.20	62.50

08 71 21 – Astragals

08 71 21.10 Astragals

		Crew	Daily Output	Labor-Hours	Unit	Material	Labor	Equipment	Total	Total Incl O&P
0010	**ASTRAGALS**									
4170	Astragal for double doors, aluminum	1 Carp	4	2	Opng.	20.50	51.50		72	110
4174	Bronze	"	4	2	"	31	51.50		82.50	121

08 71 25 – Weatherstripping

08 71 25.10 Weatherstripping

0010	**WEATHERSTRIPPING**

08 71 Door Hardware

08 71 25 – Weatherstripping

08 71 25.10 Weatherstripping

		Crew	Daily Output	Labor-Hours	Unit	Material	2007 Bare Costs Labor	Equipment	Total	Total Incl O&P
1000	Doors, wood frame, interlocking, for 3' x 7' door, zinc	1 Carp	3	2.667	Opng.	13.95	68.50		82.45	131
1100	Bronze		3	2.667		22	68.50		90.50	140
1300	6' x 7' opening, zinc		2	4		15.25	103		118.25	191
1400	Bronze		2	4		29	103		132	206
1500	Vinyl V strip		6.40	1.250	Ea.	7.55	32		39.55	63
1700	Wood frame, spring type, bronze									
1800	3' x 7' door	1 Carp	7.60	1.053	Opng.	17.90	27		44.90	65.50
1900	6' x 7' door		7	1.143		21.50	29.50		51	73.50
1920	Felt, 3' x 7' door		14	.571		2.15	14.70		16.85	27.50
1930	6' x 7' door		13	.615		2.32	15.80		18.12	29.50
1950	Rubber, 3' x 7' door		7.60	1.053		4.65	27		31.65	51
1960	6' x 7' door		7	1.143		5.30	29.50		34.80	56
2200	Metal frame, spring type, bronze									
2300	3' x 7' door	1 Carp	3	2.667	Opng.	30	68.50		98.50	149
2400	6' x 7' door	"	2.50	3.200	"	41.50	82		123.50	186
2500	For stainless steel, spring type, add					133%				
2700	Metal frame, extruded sections, 3' x 7' door, aluminum	1 Carp	2	4	Opng.	40.50	103		143.50	219
2800	Bronze		2	4		102	103		205	287
3100	6' x 7' door, aluminum		1.20	6.667		51.50	171		222.50	350
3200	Bronze		1.20	6.667		121	171		292	425
3500	Threshold weatherstripping									
3650	Door sweep, flush mounted, aluminum	1 Carp	25	.320	Ea.	12	8.20		20.20	27
3700	Vinyl		25	.320		14.20	8.20		22.40	29.50
5000	Garage door bottom weatherstrip, 12' aluminum, clear		14	.571		19.20	14.70		33.90	46
5010	Bronze		14	.571		73	14.70		87.70	106
5050	Bottom protection, 12' aluminum, clear		14	.571		22	14.70		36.70	49
5100	Bronze		14	.571		90.50	14.70		105.20	125

08 75 Window Hardware

08 75 10 – Window Hardware

08 75 10.10 Window Hardware

		Crew	Daily Output	Labor-Hours	Unit	Material	2007 Bare Costs Labor	Equipment	Total	Total Incl O&P
0010	**WINDOW HARDWARE**									
1000	Handles, surface mounted, aluminum	1 Carp	24	.333	Ea.	1.95	8.55		10.50	16.70
1020	Brass		24	.333		2.27	8.55		10.82	17.05
1040	Chrome		24	.333		2.08	8.55		10.63	16.85
1500	Recessed, aluminum		12	.667		1.13	17.15		18.28	30
1520	Brass		12	.667		1.25	17.15		18.40	30.50
1540	Chrome		12	.667		1.19	17.15		18.34	30.50
2000	Latches, aluminum		20	.400		1.62	10.30		11.92	19.25
2020	Brass		20	.400		1.95	10.30		12.25	19.60
2040	Chrome		20	.400		1.83	10.30		12.13	19.45

08 75 30 – Weatherstripping

08 75 30.10 Weatherstripping

		Crew	Daily Output	Labor-Hours	Unit	Material	2007 Bare Costs Labor	Equipment	Total	Total Incl O&P
0010	**WEATHERSTRIPPING**, Window, double hung, 3' X 5'									
0020	Zinc	1 Carp	7.20	1.111	Opng.	12.05	28.50		40.55	62
0100	Bronze		7.20	1.111		24	28.50		52.50	75
0200	Vinyl V strip		7	1.143		3.94	29.50		33.44	54.50
0500	As above but heavy duty, zinc		4.60	1.739		15.55	44.50		60.05	93
0600	Bronze		4.60	1.739		27	44.50		71.50	106
9000	Minimum labor/equipment charge	1 Clab	4.60	1.739	Job		32.50		32.50	55

08 79 Hardware Accessories

08 79 20 – Door Accessories

08 79 20.10 Door Accessories

		Crew	Daily Output	Labor-Hours	Unit	Material	2007 Bare Costs Labor	Equipment	Total	Total Incl O&P
0010	**DOOR ACCESSORIES**									
0140	Door bolt, surface, 4"	1 Carp	32	.250	Ea.	8.45	6.45		14.90	20
0160	Door latch	"	12	.667	"	6.50	17.15		23.65	36
0200	Sliding closet door									
0220	Track and hanger, single	1 Carp	10	.800	Ea.	46	20.50		66.50	85.50
0240	Double		8	1		65	25.50		90.50	115
0260	Door guide, single		48	.167		21.50	4.28		25.78	31
0280	Double		48	.167		29	4.28		33.28	39.50
0600	Deadbolt and lock cover plate, brass or stainless steel		30	.267		24	6.85		30.85	38
0620	Hole cover plate, brass or chrome		35	.229		6.20	5.85		12.05	16.80
2240	Mortise lockset, passage, lever handle		9	.889		174	23		197	230
4000	Security chain, standard		18	.444		6.85	11.40		18.25	27

08 81 Glass Glazing

08 81 10 – Float Glass

08 81 10.10 Float Glass

		Crew	Daily Output	Labor-Hours	Unit	Material	2007 Bare Costs Labor	Equipment	Total	Total Incl O&P
0010	**FLOAT GLASS**, 3/16" thick									
0020	3/16" Plain	2 Glaz	130	.123	S.F.	4.87	3.15		8.02	10.55
0200	Tempered, clear		130	.123		5.70	3.15		8.85	11.50
0300	Tinted		130	.123		7.25	3.15		10.40	13.20
0600	1/4" thick, clear, plain		120	.133		5.95	3.41		9.36	12.20
0700	Tinted		120	.133		5.60	3.41		9.01	11.80
0800	Tempered, clear		120	.133		7.10	3.41		10.51	13.45
0900	Tinted		120	.133		10.10	3.41		13.51	16.75
1600	3/8" thick, clear, plain		75	.213		9.60	5.45		15.05	19.60
1700	Tinted		75	.213		11.20	5.45		16.65	21.50
1800	Tempered, clear		75	.213		14.30	5.45		19.75	25
1900	Tinted		75	.213		18.10	5.45		23.55	29
2200	1/2" thick, clear, plain		55	.291		18.95	7.45		26.40	33.50
2300	Tinted		55	.291		19.65	7.45		27.10	34
2400	Tempered, clear		55	.291		21.50	7.45		28.95	36
2500	Tinted		55	.291		27.50	7.45		34.95	42.50
2800	5/8" thick, clear, plain		45	.356		19.65	9.10		28.75	36.50
2900	Tempered, clear		45	.356		22.50	9.10		31.60	39.50
8900	For low emissivity coating for 3/16" & 1/4" only, add to above					15%				

08 81 25 – Glazing Variables

08 81 25.10 Glazing Variables

		Crew	Daily Output	Labor-Hours	Unit	Material	2007 Bare Costs Labor	Equipment	Total	Total Incl O&P
0010	**GLAZING VARIABLES**									
0600	For glass replacement, add				S.F.		100%			
0700	For gasket settings, add				L.F.	4.20			4.20	4.62
0900	For sloped glazing, add				S.F.		25%			
2000	Fabrication, polished edges, 1/4" thick				Inch	.36			.36	.40
2100	1/2" thick					.92			.92	1.01
2500	Mitered edges, 1/4" thick					.92			.92	1.01
2600	1/2" thick					1.48			1.48	1.63

08 81 30 – Insulating Glass

08 81 30.10 Insulating Glass

		Crew	Daily Output	Labor-Hours	Unit	Material	2007 Bare Costs Labor	Equipment	Total	Total Incl O&P
0010	**INSULATING GLASS**, 2 lites 1/8" float, 1/2" thk under 15 S.F.									
0100	Tinted	2 Glaz	95	.168	S.F.	12.70	4.31		17.01	21
0280	Double glazed, 5/8" thk unit, 3/16" float, 15-30 S.F., clear		90	.178		9.80	4.55		14.35	18.35

08 81 Glass Glazing

08 81 30 – Insulating Glass

08 81 30.10 Insulating Glass

		Crew	Daily Output	Labor-Hours	Unit	Material	2007 Bare Costs Labor	Equipment	Total	Total Incl O&P
0400	1" thk, dbl. glazed, 1/4" float, 30-70 S.F., clear	2 Glaz	75	.213	S.F.	14.45	5.45		19.90	25
0500	Tinted		75	.213		16.85	5.45		22.30	27.50
2000	Both lites, light & heat reflective		85	.188		22.50	4.82		27.32	32.50
2500	Heat reflective, film inside, 1" thick unit, clear		85	.188		19.65	4.82		24.47	29.50
2600	Tinted		85	.188		21	4.82		25.82	31.50
3000	Film on weatherside, clear, 1/2" thick unit		95	.168		14.05	4.31		18.36	22.50
3100	5/8" thick unit		90	.178		16.70	4.55		21.25	26
3200	1" thick unit		85	.188		19.35	4.82		24.17	29.50

08 81 40 – Plate Glass

08 81 40.10 Plate Glass

		Crew	Daily Output	Labor-Hours	Unit	Material	2007 Bare Costs Labor	Equipment	Total	Total Incl O&P
0010	**PLATE GLASS** Twin ground, polished,									
0015	3/16" thick, material				S.F.	4.40			4.40	4.84
0020	3/16" thick	2 Glaz	100	.160		4.40	4.10		8.50	11.65
0100	1/4" thick		94	.170		6	4.36		10.36	13.80
0200	3/8" thick		60	.267		10.15	6.85		17	22.50
0300	1/2" thick		40	.400		19.25	10.25		29.50	38

08 81 55 – Window Glass

08 81 55.10 Window Glass

		Crew	Daily Output	Labor-Hours	Unit	Material	2007 Bare Costs Labor	Equipment	Total	Total Incl O&P
0010	**WINDOW GLASS**, clear float, stops, putty bed									
0015	1/8" thick, clear float	2 Glaz	480	.033	S.F.	3.94	.85		4.79	5.75
0500	3/16" thick, clear		480	.033		4.86	.85		5.71	6.75
0600	Tinted		480	.033		5.40	.85		6.25	7.35
0700	Tempered		480	.033		6.55	.85		7.40	8.60

08 83 Mirrors

08 83 13 – Mirrored Glass Glazing

08 83 13.10 Mirrors

		Crew	Daily Output	Labor-Hours	Unit	Material	2007 Bare Costs Labor	Equipment	Total	Total Incl O&P
0010	**MIRRORS**, No frames, wall type, 1/4" plate glass, polished edge									
0100	Up to 5 S.F.	2 Glaz	125	.128	S.F.	6.90	3.28		10.18	12.95
0200	Over 5 S.F.		160	.100		6.65	2.56		9.21	11.60
0500	Door type, 1/4" plate glass, up to 12 S.F.		160	.100		7.30	2.56		9.86	12.25
1000	Float glass, up to 10 S.F., 1/8" thick		160	.100		4.41	2.56		6.97	9.10
1100	3/16" thick		150	.107		5.15	2.73		7.88	10.15
1500	12" x 12" wall tiles, square edge, clear		195	.082		1.66	2.10		3.76	5.30
1600	Veined		195	.082		4.49	2.10		6.59	8.40
2010	Bathroom, unframed, laminated		160	.100		12	2.56		14.56	17.45

08 87 Glazing Surface Films

08 87 53 – Security Films

08 87 53.10 Security Film

		Crew	Daily Output	Labor-Hours	Unit	Material	2007 Bare Costs Labor	Equipment	Total	Total Incl O&P
0010	**SECURITY FILM**, clear, 32000psi tensile strength, adhered to glass									
0100	.002" thick, daylight installation	H-2	950	.025	S.F.	.85	.59		1.44	1.92
0150	.004" thick, daylight installation		800	.030		1.50	.70		2.20	2.81
0200	.006" thick, daylight installation		700	.034		1.60	.80		2.40	3.09
0210	Install for anchorage		600	.040		1.78	.93		2.71	3.51
0400	.007" thick, daylight istallation		600	.040		1.70	.93		2.63	3.42
0410	Install for anchorage		500	.048		1.89	1.12		3.01	3.94
0500	.008" thick, daylight installation		500	.048		2	1.12		3.12	4.06

08 87 Glazing Surface Films

08 87 53 – Security Films

08 87 53.10 Security Film		Crew	Daily Output	Labor-Hours	Unit	Material	2007 Bare Costs Labor	Equipment	Total	Total Incl O&P
0510	Install for anchorage	H-2	500	.048	S.F.	2.22	1.12		3.34	4.30
0600	.015" thick, daylight installation		400	.060		3.10	1.40		4.50	5.75
0610	Install for anchorage		400	.060		2.22	1.40		3.62	4.77
0900	Security Film Anchorage, mechanical attachment and cover plate	H-3	370	.043	L.F.	7.30	.96		8.26	9.65
0950	Security film anchorage, wet glaze structural caulking	1 Glaz	225	.036	"	.80	.91		1.71	2.39
1000	Adhered security film removal	1 Clab	275	.029	S.F.		.54		.54	.92

08 91 Louvers

08 91 19 – Fixed Louvers

08 91 19.10 Louvers		Crew	Daily Output	Labor-Hours	Unit	Material	Labor	Equipment	Total	Total Incl O&P
0010	**LOUVERS**									
0020	Aluminum with screen, residential, 8" x 8"	1 Carp	38	.211	Ea.	10.20	5.40		15.60	20.50
0100	12" x 12"		38	.211		11.50	5.40		16.90	22
0200	12" x 18"		35	.229		13	5.85		18.85	24.50
0250	14" x 24"		30	.267		19.50	6.85		26.35	33
0300	18" x 24"		27	.296		24	7.60		31.60	39.50
0500	24" x 30"		24	.333		33	8.55		41.55	51
0700	Triangle, adjustable, small		20	.400		28.50	10.30		38.80	48.50
0800	Large		15	.533		49	13.70		62.70	77
2100	Midget, aluminum, 3/4" deep, 1" diameter		85	.094		.76	2.42		3.18	4.94
2150	3" diameter		60	.133		1.62	3.43		5.05	7.60
2200	4" diameter		50	.160		2.98	4.11		7.09	10.30
2250	6" diameter		30	.267		3.52	6.85		10.37	15.50

08 95 Vents

08 95 13 – Soffit Vents

08 95 13.10 Wall Louvers		Crew	Daily Output	Labor-Hours	Unit	Material	Labor	Equipment	Total	Total Incl O&P
0010	**WALL LOUVERS**									
2400	Under eaves vent, aluminum, mill finish, 16" x 4"	1 Carp	48	.167	Ea.	1.94	4.28		6.22	9.40
2500	16" x 8"	"	48	.167	"	2.04	4.28		6.32	9.50

08 95 16 – Wall Vents

08 95 16.10 Louvers		Crew	Daily Output	Labor-Hours	Unit	Material	Labor	Equipment	Total	Total Incl O&P
0010	**LOUVERS**									
0020	Redwood, 2'-0" diameter, full circle	1 Carp	16	.500	Ea.	144	12.85		156.85	181
0100	Half circle		16	.500		138	12.85		150.85	174
0200	Octagonal		16	.500		110	12.85		122.85	143
0300	Triangular, 5/12 pitch, 5'-0" at base		16	.500		233	12.85		245.85	279
7000	Vinyl gable vent, 8" x 8"		38	.211		10.90	5.40		16.30	21
7020	12" x 12"		38	.211		22.50	5.40		27.90	34
7080	12" x 18"		35	.229		29	5.85		34.85	42
7200	18" x 24"		30	.267		35	6.85		41.85	50

Division 9
Finishes

09 01 Maintenance of Finishes

09 01 70 – Maintenance of Wall Finishes

09 01 70.10 Gypsum Wallboard Repairs

		Crew	Daily Output	Labor-Hours	Unit	Material	2007 Bare Costs Labor	Equipment	Total	Total Incl O&P
0010	**GYPSUM WALLBOARD REPAIRS**									
0100	Fill and sand, pin / nail holes	1 Carp	960	.008	Ea.		.21		.21	.36
0110	Screw head pops		480	.017			.43		.43	.73
0120	Dents, up to 2" square		48	.167		.01	4.28		4.29	7.25
0130	2" to 4" square		24	.333		.03	8.55		8.58	14.60
0140	Cut square, patch, sand and finish, holes, up to 2" square		12	.667		.04	17.15		17.19	29
0150	2" to 4" square		11	.727		.09	18.70		18.79	31.50
0160	4" to 8" square		10	.800		.24	20.50		20.74	35.50
0170	8" to 12" square		8	1		.46	25.50		25.96	44

09 05 Common Work Results for Finishes

09 05 05 – Selective Finishes Demolition

09 05 05.10 Selective Demolition, Ceilings

			Crew	Daily Output	Labor-Hours	Unit	Material	Labor	Equipment	Total	Total Incl O&P
0010	**SELECTIVE DEMOLITION, CEILINGS**	R024119-10									
0200	Ceiling, drywall, furred and nailed or screwed		2 Clab	800	.020	S.F.		.37		.37	.64
1000	Plaster, lime and horse hair, on wood lath, incl. lath			700	.023			.43		.43	.73
1200	Suspended ceiling, mineral fiber, 2' x 2' or 2' x 4'			1500	.011			.20		.20	.34
1250	On suspension system, incl. system			1200	.013			.25		.25	.42
1500	Tile, wood fiber, 12" x 12", glued			900	.018			.33		.33	.56
1540	Stapled			1500	.011			.20		.20	.34
2000	Wood, tongue and groove, 1" x 4"			1000	.016			.30		.30	.51
2040	1" x 8"			1100	.015			.27		.27	.46
2400	Plywood or wood fiberboard, 4' x 8' sheets			1200	.013			.25		.25	.42

09 05 05.20 Selective Demolition, Flooring

			Crew	Daily Output	Labor-Hours	Unit	Material	Labor	Equipment	Total	Total Incl O&P
0010	**SELECTIVE DEMOLITION, FLOORING**	R024119-10									
0200	Brick with mortar		2 Clab	475	.034	S.F.		.63		.63	1.07
0400	Carpet, bonded, including surface scraping			2000	.008			.15		.15	.25
0480	Tackless			9000	.002			.03		.03	.06
0800	Resilient, sheet goods			1400	.011			.21		.21	.36
0850	Vinyl or rubber cove base		1 Clab	1000	.008	L.F.		.15		.15	.25
0860	Vinyl or rubber cove base, molded corner		"	1000	.008	Ea.		.15		.15	.25
0900	Vinyl composition tile, 12" x 12"		2 Clab	1000	.016	S.F.		.30		.30	.51
2000	Tile, ceramic, thin set			675	.024			.44		.44	.75
2020	Mud set			625	.026			.48		.48	.81
3000	Wood, block, on end		1 Carp	400	.020			.51		.51	.87
3200	Parquet			450	.018			.46		.46	.78
3400	Strip flooring, interior, 2-1/4" x 25/32" thick			325	.025			.63		.63	1.07
3500	Exterior, porch flooring, 1" x 4"			220	.036			.93		.93	1.59
3800	Subfloor, tongue and groove, 1" x 6"			325	.025			.63		.63	1.07
3820	1" x 8"			430	.019			.48		.48	.81
3840	1" x 10"			520	.015			.40		.40	.67
4000	Plywood, nailed			600	.013			.34		.34	.58
4100	Glued and nailed			400	.020			.51		.51	.87

09 05 05.30 Selective Demolition, Walls and Partitions

			Crew	Daily Output	Labor-Hours	Unit	Material	Labor	Equipment	Total	Total Incl O&P
0010	**SELECTIVE DEMOLITION, WALLS AND PARTITIONS**	R024119-10									
1000	Drywall, nailed or screwed		1 Clab	1000	.008	S.F.		.15		.15	.25
1500	Fiberboard, nailed			900	.009			.17		.17	.28
1568	Plenum barrier, sheet lead			300	.027			.50		.50	.85
2200	Metal or wood studs, finish 2 sides, fiberboard		B-1	520	.046			.89		.89	1.52
2250	Lath and plaster			260	.092			1.79		1.79	3.04

09 05 Common Work Results for Finishes

09 05 05 – Selective Finishes Demolition

09 05 05.30 Selective Demolition, Walls and Partitions		Crew	Daily Output	Labor-Hours	Unit	Material	2007 Bare Costs Labor	Equipment	Total	Total Incl O&P
2300	Plasterboard (drywall)	B-1	520	.046	S.F.		.89		.89	1.52
2350	Plywood	↓	450	.053			1.03		1.03	1.75
3000	Plaster, lime and horsehair, on wood lath	1 Clab	400	.020			.37		.37	.64
3020	On metal lath	"	335	.024	↓		.45		.45	.76

09 22 Supports for Plaster and Gypsum Board

09 22 03 – Fastening Methods for Finishes

09 22 03.20 Drilling Plaster/Drywall

		Crew	Daily Output	Labor-Hours	Unit	Material	Labor	Equipment	Total	Total Incl O&P
0010	**DRILLING PLASTER/DRYWALL**									
1100	Drilling & layout for drywall/plaster walls, up to 1" deep, no anchor									
1200	Holes, 1/4" diameter	1 Carp	150	.053	Ea.	.01	1.37		1.38	2.34
1300	3/8" diameter		140	.057		.01	1.47		1.48	2.50
1400	1/2" diameter		130	.062		.01	1.58		1.59	2.69
1500	3/4" diameter		120	.067		.03	1.71		1.74	2.94
1600	1" diameter		110	.073		.04	1.87		1.91	3.21
1700	1-1/4" diameter		100	.080		.05	2.06		2.11	3.54
1800	1-1/2" diameter	↓	90	.089		.08	2.28		2.36	3.96
1900	For ceiling installations, add				↓		40%			

09 22 13 – Metal Furring

09 22 13.13 Metal Channel Furring

		Crew	Daily Output	Labor-Hours	Unit	Material	Labor	Equipment	Total	Total Incl O&P
0010	**METAL CHANNEL FURRING**									
0030	Beams and columns, 7/8" galv. channels, 12" O.C.	1 Lath	155	.052	S.F.	.27	1.22		1.49	2.28
0050	16" O.C.		170	.047		.22	1.11		1.33	2.05
0070	24" O.C.		185	.043		.15	1.02		1.17	1.82
0100	Ceilings, on steel, 7/8" channels, galvanized, 12" O.C.		210	.038		.24	.90		1.14	1.73
0300	16" O.C.		290	.028		.22	.65		.87	1.30
0400	24" O.C.		420	.019		.15	.45		.60	.89
0600	1-5/8" channels, galvanized, 12" O.C.		190	.042		.37	.99		1.36	2.03
0700	16" O.C.		260	.031		.33	.73		1.06	1.55
0900	24" O.C.		390	.021		.22	.48		.70	1.03
0930	7/8" channels with sound isolation clips, 12" O.C.		120	.067		1.27	1.57		2.84	3.96
0940	16" O.C.		100	.080		1.79	1.89		3.68	5.05
0950	24" O.C.		165	.048		1.18	1.14		2.32	3.15
0960	1-5/8" channels, galvanized, 12" O.C.		110	.073		1.40	1.72		3.12	4.34
0970	16" O.C.		100	.080		1.91	1.89		3.80	5.20
0980	24" O.C.		155	.052		1.25	1.22		2.47	3.36
1000	Walls, 7/8" channels, galvanized, 12" O.C.		235	.034		.24	.80		1.04	1.58
1200	16" O.C.		265	.030		.22	.71		.93	1.40
1300	24" O.C.		350	.023		.15	.54		.69	1.04
1500	1-5/8" channels, galvanized, 12" O.C.		210	.038		.37	.90		1.27	1.87
1600	16" O.C.		240	.033		.33	.79		1.12	1.65
1800	24" O.C.		305	.026		.22	.62		.84	1.25
1920	7/8" channels with sound isolation clips, 12" O.C.		125	.064		1.27	1.51		2.78	3.86
1940	16" O.C.		100	.080		1.79	1.89		3.68	5.05
1950	24" O.C.		150	.053		1.18	1.26		2.44	3.34
1960	1-5/8" channels, galvanized, 12" O.C.		115	.070		1.40	1.64		3.04	4.22
1970	16" O.C.		95	.084		1.91	1.99		3.90	5.35
1980	24" O.C.	↓	140	.057	↓	1.25	1.35		2.60	3.58

09 22 Supports for Plaster and Gypsum Board

09 22 16 – Non-Structural Metal Framing

09 22 16.13 Metal Studs and Track		Crew	Daily Output	Labor-Hours	Unit	Material	2007 Bare Costs Labor	Equipment	Total	Total Incl O&P
0010	**METAL STUDS AND TRACK**									
1600	Non-load bearing, galv, 8' high, 25 ga. 1-5/8" wide, 16" O.C.	1 Carp	619	.013	S.F.	.30	.33		.63	.89
1610	24" O.C.		950	.008		.23	.22		.45	.62
1620	2-1/2" wide, 16" O.C.		613	.013		.36	.34		.70	.96
1630	24" O.C.		938	.009		.27	.22		.49	.67
1640	3-5/8" wide, 16" O.C.		600	.013		.41	.34		.75	1.03
1650	24" O.C.		925	.009		.31	.22		.53	.72
1660	4" wide, 16" O.C.		594	.013		.50	.35		.85	1.14
1670	24" O.C.		925	.009		.37	.22		.59	.79
1680	6" wide, 16" O.C.		588	.014		.62	.35		.97	1.27
1690	24" O.C.		906	.009		.46	.23		.69	.90
1700	20 ga. studs, 1-5/8" wide, 16" O.C.		494	.016		.50	.42		.92	1.26
1710	24" O.C.		763	.010		.37	.27		.64	.87
1720	2-1/2" wide, 16" O.C.		488	.016		.58	.42		1	1.35
1730	24" O.C.		750	.011		.44	.27		.71	.95
1740	3-5/8" wide, 16" O.C.		481	.017		.65	.43		1.08	1.45
1750	24" O.C.		738	.011		.49	.28		.77	1.01
1760	4" wide, 16" O.C.		475	.017		.78	.43		1.21	1.59
1770	24" O.C.		738	.011		.58	.28		.86	1.11
1780	6" wide, 16" O.C.		469	.017		.90	.44		1.34	1.73
1790	24" O.C.		725	.011		.68	.28		.96	1.23
2000	Non-load bearing, galv, 10' high, 25 ga. 1-5/8" wide, 16" O.C.		495	.016		.28	.42		.70	1.01
2100	24" O.C.		760	.011		.21	.27		.48	.69
2200	2-1/2" wide, 16" O.C.		490	.016		.34	.42		.76	1.08
2250	24" O.C.		750	.011		.25	.27		.52	.75
2300	3-5/8" wide, 16" O.C.		480	.017		.39	.43		.82	1.16
2350	24" O.C.		740	.011		.29	.28		.57	.79
2400	4" wide, 16" O.C.		475	.017		.47	.43		.90	1.25
2450	24" O.C.		740	.011		.35	.28		.63	.85
2500	6" wide, 16" O.C.		470	.017		.58	.44		1.02	1.38
2550	24" O.C.		725	.011		.43	.28		.71	.95
2600	20 ga. studs, 1-5/8" wide, 16" O.C.		395	.020		.47	.52		.99	1.40
2650	24" O.C.		610	.013		.35	.34		.69	.95
2700	2-1/2" wide, 16" O.C.		390	.021		.55	.53		1.08	1.49
2750	24" O.C.		600	.013		.41	.34		.75	1.03
2800	3-5/8" wide, 16" O.C.		385	.021		.62	.53		1.15	1.59
2850	24" O.C.		590	.014		.46	.35		.81	1.09
2900	4" wide, 16" O.C.		380	.021		.74	.54		1.28	1.73
2950	24" O.C.		590	.014		.54	.35		.89	1.19
3000	6" wide, 16" O.C.		375	.021		.86	.55		1.41	1.87
3050	24" O.C.		580	.014		.63	.35		.98	1.30
3060	Non-load bearing, galv, 12' high, 25 ga. 1-5/8" wide, 16" O.C.		413	.019		.27	.50		.77	1.14
3070	24" O.C.		633	.013		.20	.32		.52	.77
3080	2-1/2" wide, 16" O.C.		408	.020		.32	.50		.82	1.22
3090	24" O.C.		625	.013		.24	.33		.57	.82
3100	3-5/8" wide, 16" O.C.		400	.020		.37	.51		.88	1.28
3110	24" O.C.		617	.013		.27	.33		.60	.87
3120	4" wide, 16" O.C.		396	.020		.45	.52		.97	1.37
3130	24" O.C.		617	.013		.33	.33		.66	.93
3140	6" wide, 16" O.C.		392	.020		.56	.52		1.08	1.50
3150	24" O.C.		604	.013		.41	.34		.75	1.03
3160	20 ga. studs, 1-5/8" wide, 16" O.C.		329	.024		.45	.63		1.08	1.56

09 22 Supports for Plaster and Gypsum Board

09 22 16 – Non-Structural Metal Framing

09 22 16.13 Metal Studs and Track

		Crew	Daily Output	Labor-Hours	Unit	Material	2007 Bare Costs Labor	Equipment	Total	Total Incl O&P
3170	24" O.C.	1 Carp	508	.016	S.F.	.33	.40		.73	1.05
3180	2-1/2" wide, 16" O.C.		325	.025		.52	.63		1.15	1.65
3190	24" O.C.		500	.016		.38	.41		.79	1.12
3200	3-5/8" wide, 16" O.C.		321	.025		.59	.64		1.23	1.74
3210	24" O.C.		492	.016		.43	.42		.85	1.18
3220	4" wide, 16" O.C.		317	.025		.71	.65		1.36	1.88
3230	24" O.C.		492	.016		.51	.42		.93	1.27
3240	6" wide, 16" O.C.		313	.026		.82	.66		1.48	2.01
3250	24" O.C.		483	.017		.60	.43		1.03	1.38
5000	Load bearing studs, see division 05 41 13.30									

09 22 26 – Suspension Systems

09 22 26.13 Ceiling Suspension Systems

		Crew	Daily Output	Labor-Hours	Unit	Material	Labor	Equipment	Total	Total Incl O&P
0010	**CEILING SUSPENSION SYSTEMS** For gypsum board or plaster									
8000	Suspended ceilings, including carriers									
8200	1-1/2" carriers, 24" O.C. with:									
8300	7/8" channels, 16" O.C.	1 Lath	165	.048	S.F.	.44	1.14		1.58	2.34
8320	24" O.C.		200	.040		.37	.94		1.31	1.94
8400	1-5/8" channels, 16" O.C.		155	.052		.55	1.22		1.77	2.59
8420	24" O.C.		190	.042		.44	.99		1.43	2.11
8600	2" carriers, 24" O.C. with:									
8700	7/8" channels, 16" O.C.	1 Lath	155	.052	S.F.	.46	1.22		1.68	2.49
8720	24" O.C.		190	.042		.39	.99		1.38	2.05
8800	1-5/8" channels, 16" O.C.		145	.055		.57	1.30		1.87	2.75
8820	24" O.C.		180	.044		.46	1.05		1.51	2.22

09 22 36 – Lath

09 22 36.13 Gypsum Lath

		Crew	Daily Output	Labor-Hours	Unit	Material	Labor	Equipment	Total	Total Incl O&P
0011	**GYPSUM LATH** Plain or perforated, nailed, 3/8" thick		765	.010		.50	.25		.75	.95
0101	1/2" thick, nailed		720	.011		.47	.26		.73	.95
0301	Clipped to steel studs, 3/8" thick		675	.012		.50	.28		.78	1.01
0401	1/2" thick		630	.013		.47	.30		.77	1.01
0601	Firestop gypsum base, to steel studs, 3/8" thick		630	.013		.39	.30		.69	.92
0701	1/2" thick		585	.014		.51	.32		.83	1.09
0901	Foil back, to steel studs, 3/8" thick		675	.012		.43	.28		.71	.93
1001	1/2" thick		630	.013		.45	.30		.75	.99
1501	For ceiling installations, add		1950	.004			.10		.10	.16
1601	For columns and beams, add		1550	.005			.12		.12	.20

09 22 36.23 Metal Lath

		Crew	Daily Output	Labor-Hours	Unit	Material	Labor	Equipment	Total	Total Incl O&P
0010	**METAL LATH** R092000-50									
3601	2.5 lb. diamond painted, on wood framing, on walls	1 Lath	765	.010	S.F.	.29	.25		.54	.72
3701	On ceilings		675	.012		.29	.28		.57	.78
4201	3.4 lb. diamond painted, wired to steel framing, on walls		675	.012		.48	.28		.76	.99
4301	On ceilings		540	.015		.48	.35		.83	1.10
5101	Rib lath, painted, wired to steel, on walls, 2.75 lb.		675	.012		.36	.28		.64	.85
5201	3.4 lb.		630	.013		.51	.30		.81	1.05
5701	Suspended ceiling system, incl. 3.4 lb. diamond lath, painted		135	.059		1.52	1.40		2.92	3.95
5801	Galvanized		135	.059		1.56	1.40		2.96	4

09 22 36.83 Accessories, Plaster

		Crew	Daily Output	Labor-Hours	Unit	Material	Labor	Equipment	Total	Total Incl O&P
0010	**ACCESSORIES, PLASTER**									
0020	Casing bead, expanded flange, galvanized	1 Lath	2.70	2.963	C.L.F.	40.50	70		110.50	159
0900	Channels, cold rolled, 16 ga., 3/4" deep, galvanized					24.50			24.50	27
1620	Corner bead, expanded bullnose, 3/4" radius, #10, galvanized	1 Lath	2.60	3.077		31.50	72.50		104	153

09 22 Supports for Plaster and Gypsum Board

09 22 36 – Lath

09 22 36.83 Accessories, Plaster

		Crew	Daily Output	Labor-Hours	Unit	Material	2007 Bare Costs Labor	2007 Bare Costs Equipment	Total	Total Incl O&P
1650	#1, galvanized	1 Lath	2.55	3.137	C.L.F.	48.50	74		122.50	175
1670	Expanded wing, 2-3/4" wide, galv. #1		2.65	3.019		32	71.50		103.50	152
1700	Inside corner, (corner rite) 3" x 3", painted		2.60	3.077		24.50	72.50		97	145
1750	Strip-ex, 4" wide, painted		2.55	3.137		25.50	74		99.50	149
1800	Expansion joint, 3/4" grounds, limited expansion, galv., 1 piece		2.70	2.963		96	70		166	220
2100	Extreme expansion, galvanized, 2 piece		2.60	3.077		150	72.50		222.50	283

09 23 Gypsum Plastering

09 23 13 – Acoustical Gypsum Plastering

09 23 13.10 Perlite or Vermiculite Plaster

			Crew	Daily Output	Labor-Hours	Unit	Material	Labor	Equipment	Total	Total Incl O&P
0010	**PERLITE OR VERMICULITE PLASTER**	R092000-50									
0020	In 100 lb. bags, under 200 bags					Bag	13.65			13.65	15.05
0301	2 coats, no lath included, on walls		J-1	830	.048	S.F.	.40	1.07	.14	1.61	2.36
0401	On ceilings			710	.056		.40	1.25	.16	1.81	2.68
0901	3 coats, no lath included, on walls			665	.060		.68	1.33	.17	2.18	3.14
1001	On ceilings			565	.071		.68	1.57	.20	2.45	3.56
1700	For irregular or curved surfaces, add to above					S.Y.		30%			
1800	For columns and beams, add to above							50%			
1900	For soffits, add to ceiling prices							40%			

09 23 20 – Gypsum Plaster

09 23 20.10 Gypsum Plaster

			Crew	Daily Output	Labor-Hours	Unit	Material	Labor	Equipment	Total	Total Incl O&P
0010	**GYPSUM PLASTER**	R092000-50									
0020	80# bag, less than 1 ton					Bag	13.60			13.60	14.95
0302	2 coats, no lath included, on walls		J-1	750	.053	S.F.	.37	1.18	.15	1.70	2.53
0402	On ceilings			660	.061		.37	1.34	.17	1.88	2.82
0903	3 coats, no lath included, on walls			620	.065		.52	1.43	.18	2.13	3.14
1002	On ceilings			560	.071		.85	1.58	.20	2.63	3.77
1600	For irregular or curved surfaces, add							30%			
1800	For columns & beams, add							50%			

09 24 Portland Cement Plastering

09 24 23 – Portland Cement Stucco

09 24 23.40 Stucco

			Crew	Daily Output	Labor-Hours	Unit	Material	Labor	Equipment	Total	Total Incl O&P
0010	**STUCCO**	R092000-50									
0011	3 coats 1" thick, float finish, with mesh, on wood frame		J-2	470	.102	S.F.	.75	2.29	.24	3.28	4.86
0101	On masonry construction		J-1	495	.081		.22	1.79	.23	2.24	3.45
0151	2 coats, 3/4" thick, float finish, no lath incl.		"	980	.041		.23	.90	.12	1.25	1.88
0301	For trowel finish, add		1 Plas	1530	.005			.12		.12	.20
0600	For coloring and special finish, add, minimum		J-1	685	.058	S.Y.	.39	1.29	.17	1.85	2.75
0700	Maximum			200	.200	"	1.35	4.43	.57	6.35	9.45
1001	Exterior stucco, with bonding agent, 3 coats, on walls			1800	.022	S.F.	.35	.49	.06	.90	1.27
1201	Ceilings			1620	.025		.35	.55	.07	.97	1.37
1301	Beams			720	.056		.35	1.23	.16	1.74	2.60
1501	Columns			900	.044		.35	.98	.13	1.46	2.16
1601	Mesh, painted, nailed to wood, 1.8 lb.		1 Lath	540	.015		.53	.35		.88	1.16
1801	3.6 lb.			495	.016		.39	.38		.77	1.05
1901	Wired to steel, painted, 1.8 lb.			477	.017		.53	.40		.93	1.23
2101	3.6 lb.			450	.018		.39	.42		.81	1.11

09 26 Veneer Plastering

09 26 13 – Gypsum Veneer Plastering

09 26 13.20 Blueboard

		Crew	Daily Output	Labor-Hours	Unit	Material	2007 Bare Costs Labor	Equipment	Total	Total Incl O&P
0010	**BLUEBOARD** For use with thin coat									
0100	plaster application (see division 09 26 13.80)									
1000	3/8" thick, on walls or ceilings, standard, no finish included	2 Carp	1900	.008	S.F.	.30	.22		.52	.70
1100	With thin coat plaster finish		875	.018		.38	.47		.85	1.22
1400	On beams, columns, or soffits, standard, no finish included		675	.024		.35	.61		.96	1.41
1450	With thin coat plaster finish		475	.034		.42	.87		1.29	1.93
3000	1/2" thick, on walls or ceilings, standard, no finish included		1900	.008		.32	.22		.54	.72
3100	With thin coat plaster finish		875	.018		.40	.47		.87	1.24
3300	Fire resistant, no finish included		1900	.008		.32	.22		.54	.72
3400	With thin coat plaster finish		875	.018		.40	.47		.87	1.24
3450	On beams, columns, or soffits, standard, no finish included		675	.024		.37	.61		.98	1.43
3500	With thin coat plaster finish		475	.034		.45	.87		1.32	1.96
3700	Fire resistant, no finish included		675	.024		.37	.61		.98	1.43
3800	With thin coat plaster finish		475	.034		.45	.87		1.32	1.96
5000	5/8" thick, on walls or ceilings, fire resistant, no finish included		1900	.008		.45	.22		.67	.87
5100	With thin coat plaster finish		875	.018		.53	.47		1	1.38
5500	On beams, columns, or soffits, no finish included		675	.024		.52	.61		1.13	1.60
5600	With thin coat plaster finish		475	.034		.60	.87		1.47	2.12
6000	For high ceilings, over 8' high, add		3060	.005			.13		.13	.23
6500	For over 3 stories high, add per story		6100	.003			.07		.07	.11

09 26 13.80 Thin Coat Plaster

		Crew	Daily Output	Labor-Hours	Unit	Material	Labor	Equipment	Total	Total Incl O&P
0010	**THIN COAT PLASTER** R092000-50									
0012	1 coat veneer, not incl. lath	J-1	3600	.011	S.F.	.08	.25	.03	.36	.53
1000	In 50 lb. bags				Bag	10.50			10.50	11.55

09 28 Backing Boards and Underlayments

09 28 13 – Cementitious Backing Boards

09 28 13.10 Cementitious Backerboard

		Crew	Daily Output	Labor-Hours	Unit	Material	Labor	Equipment	Total	Total Incl O&P
0010	**CEMENTITIOUS BACKERBOARD**									
0070	Cementitious backerboard, on floor, 3' x 4'x 1/2" sheets	2 Carp	525	.030	S.F.	.96	.78		1.74	2.39
0080	3' x 5' x 1/2" sheets		525	.030		.93	.78		1.71	2.35
0090	3' x 6' x 1/2" sheets		525	.030		.72	.78		1.50	2.13
0100	3' x 4' x 5/8" sheets		525	.030		1.08	.78		1.86	2.52
0110	3' x 5' x 5/8" sheets		525	.030		1.06	.78		1.84	2.50
0120	3' x 6' x 5/8" sheets		525	.030		1.12	.78		1.90	2.56
0150	On wall, 3' x 4'x 1/2" sheets		350	.046		.96	1.17		2.13	3.05
0160	3' x 5' x 1/2" sheets		350	.046		.93	1.17		2.10	3.01
0170	3' x 6' x 1/2" sheets		350	.046		.72	1.17		1.89	2.79
0180	3' x 4' x 5/8" sheets		350	.046		1.08	1.17		2.25	3.18
0190	3' x 5' x 5/8" sheets		350	.046		1.06	1.17		2.23	3.16
0200	3' x 6' x 5/8" sheets		350	.046		1.12	1.17		2.29	3.22
0250	On counter, 3' x 4'x 1/2" sheets		180	.089		.96	2.28		3.24	4.94
0260	3' x 5' x 1/2" sheets		180	.089		.93	2.28		3.21	4.90
0270	3' x 6' x 1/2" sheets		180	.089		.72	2.28		3	4.68
0300	3' x 4' x 5/8" sheets		180	.089		1.08	2.28		3.36	5.05
0310	3' x 5' x 5/8" sheets		180	.089		1.06	2.28		3.34	5.05
0320	3' x 6' x 5/8" sheets		180	.089		1.12	2.28		3.40	5.10

09 29 Gypsum Board

09 29 10 – Gypsum Board

09 29 10.20 Taping and Finishing

		Crew	Daily Output	Labor-Hours	Unit	Material	2007 Bare Costs Labor	2007 Bare Costs Equipment	Total	Total Incl O&P
0010	**TAPING AND FINISHING**									
3600	For taping and finishing joints, add	2 Carp	2000	.008	S.F.	.04	.21		.25	.39
4500	For thin coat plaster instead of taping, add	J-1	3600	.011	"	.08	.25	.03	.36	.53

09 29 10.30 Gypsum Board

		Crew	Daily Output	Labor-Hours	Unit	Material	Labor	Equipment	Total	Total Incl O&P
0010	**GYPSUM BOARD** on walls & ceilings R092910-10									
0100	Nailed or screwed to studs unless otherwise noted									
0150	3/8" thick, on walls, standard, no finish included	2 Carp	2000	.008	S.F.	.33	.21		.54	.71
0200	On ceilings, standard, no finish included		1800	.009		.33	.23		.56	.75
0250	On beams, columns, or soffits, no finish included		675	.024		.33	.61		.94	1.39
0300	1/2" thick, on walls, standard, no finish included		2000	.008		.32	.21		.53	.70
0350	Taped and finished (level 4 finish)		965	.017		.36	.43		.79	1.12
0390	With compound skim coat (level 5 finish)		775	.021		.40	.53		.93	1.34
0400	Fire resistant, no finish included		2000	.008		.38	.21		.59	.77
0450	Taped and finished (level 4 finish)		965	.017		.42	.43		.85	1.18
0490	With compound skim coat (level 5 finish)		775	.021		.46	.53		.99	1.41
0500	Water resistant, no finish included		2000	.008		.39	.21		.60	.78
0550	Taped and finished (level 4 finish)		965	.017		.43	.43		.86	1.19
0590	With compound skim coat (level 5 finish)		775	.021		.47	.53		1	1.42
0600	Prefinished, vinyl, clipped to studs		900	.018		.60	.46		1.06	1.44
0700	Mold resistant, no finish included		2000	.008		.43	.21		.64	.82
0710	Taped and finished (level 4 finish)		965	.017		.47	.43		.90	1.24
0720	With compound skim coat (level 5 finish)		775	.021		.51	.53		1.04	1.46
1000	On ceilings, standard, no finish included		1800	.009		.32	.23		.55	.74
1050	Taped and finished (level 4 finish)		765	.021		.36	.54		.90	1.31
1090	With compound skim coat (level 5 finish)		610	.026		.40	.67		1.07	1.58
1100	Fire resistant, no finish included		1800	.009		.38	.23		.61	.81
1150	Taped and finished (level 4 finish)		765	.021		.42	.54		.96	1.37
1195	With compound skim coat (level 5 finish)		610	.026		.46	.67		1.13	1.65
1200	Water resistant, no finish included		1800	.009		.39	.23		.62	.82
1250	Taped and finished (level 4 finish)		765	.021		.43	.54		.97	1.38
1290	With compound skim coat (level 5 finish)		610	.026		.47	.67		1.14	1.66
1310	Mold resistant, no finish included		1800	.009		.43	.23		.66	.86
1320	Taped and finished (level 4 finish)		765	.021		.47	.54		1.01	1.43
1330	With compound skim coat (level 5 finish)		610	.026		.51	.67		1.18	1.70
1500	On beams, columns, or soffits, standard, no finish included		675	.024		.37	.61		.98	1.43
1550	Taped and finished (level 4 finish)		475	.034		.36	.87		1.23	1.87
1590	With compound skim coat (level 5 finish)		540	.030		.40	.76		1.16	1.73
1600	Fire resistant, no finish included		675	.024		.38	.61		.99	1.45
1650	Taped and finished (level 4 finish)		475	.034		.42	.87		1.29	1.93
1690	With compound skim coat (level 5 finish)		540	.030		.46	.76		1.22	1.80
1700	Water resistant, no finish included		675	.024		.45	.61		1.06	1.52
1750	Taped and finished (level 4 finish)		475	.034		.43	.87		1.30	1.94
1790	With compound skim coat (level 5 finish)		540	.030		.47	.76		1.23	1.81
1800	Mold resistant, no finish included		675	.024		.49	.61		1.10	1.57
1810	Taped and finished (level 4 finish)		475	.034		.47	.87		1.34	1.99
1820	With compound skim coat (level 5 finish)		540	.030		.51	.76		1.27	1.85
2000	5/8" thick, on walls, standard, no finish included		2000	.008		.39	.21		.60	.78
2050	Taped and finished (level 4 finish)		965	.017		.43	.43		.86	1.19
2090	With compound skim coat (level 5 finish)		775	.021		.47	.53		1	1.42
2100	Fire resistant, no finish included		2000	.008		.38	.21		.59	.77
2150	Taped and finished (level 4 finish)		965	.017		.42	.43		.85	1.18
2195	With compound skim coat (level 5 finish)		775	.021		.46	.53		.99	1.41

09 29 Gypsum Board

09 29 10 - Gypsum Board

09 29 10.30 Gypsum Board		Crew	Daily Output	Labor-Hours	Unit	Material	2007 Bare Costs Labor	Equipment	Total	Total Incl O&P
2200	Water resistant, no finish included	2 Carp	2000	.008	S.F.	.39	.21		.60	.78
2250	Taped and finished (level 4 finish)		965	.017		.43	.43		.86	1.19
2290	With compound skim coat (level 5 finish)		775	.021		.47	.53		1	1.42
2300	Prefinished, vinyl, clipped to studs		900	.018		.70	.46		1.16	1.55
2510	Mold resistant, no finish included		2000	.008		.49	.21		.70	.89
2520	Taped and finished (level 4 finish)		965	.017		.53	.43		.96	1.30
2530	With compound skim coat (level 5 finish)		775	.021		.57	.53		1.10	1.53
3000	On ceilings, standard, no finish included		1800	.009		.39	.23		.62	.82
3050	Taped and finished (level 4 finish)		765	.021		.43	.54		.97	1.38
3090	With compound skim coat (level 5 finish)		615	.026		.47	.67		1.14	1.65
3100	Fire resistant, no finish included		1800	.009		.38	.23		.61	.81
3150	Taped and finished (level 4 finish)		765	.021		.42	.54		.96	1.37
3190	With compound skim coat (level 5 finish)		615	.026		.46	.67		1.13	1.64
3200	Water resistant, no finish included		1800	.009		.39	.23		.62	.82
3250	Taped and finished (level 4 finish)		765	.021		.43	.54		.97	1.38
3290	With compound skim coat (level 5 finish)		615	.026		.47	.67		1.14	1.65
3300	Mold resistant, no finish included		1800	.009		.49	.23		.72	.93
3310	Taped and finished (level 4 finish)		765	.021		.53	.54		1.07	1.49
3320	With compound skim coat (level 5 finish)		615	.026		.57	.67		1.24	1.76
3500	On beams, columns, or soffits, no finish included		675	.024		.45	.61		1.06	1.52
3550	Taped and finished (level 4 finish)		475	.034		.49	.87		1.36	2.01
3590	With compound skim coat (level 5 finish)		380	.042		.54	1.08		1.62	2.44
3600	Fire resistant, no finish included		675	.024		.44	.61		1.05	1.51
3650	Taped and finished (level 4 finish)		475	.034		.48	.87		1.35	2
3690	With compound skim coat (level 5 finish)		380	.042		.46	1.08		1.54	2.35
3700	Water resistant, no finish included		675	.024		.45	.61		1.06	1.52
3750	Taped and finished (level 4 finish)		475	.034		.49	.87		1.36	2.01
3790	With compound skim coat (level 5 finish)		380	.042		.47	1.08		1.55	2.36
3800	Mold resistant, no finish included		675	.024		.56	.61		1.17	1.65
3810	Taped and finished (level 4 finish)		475	.034		.61	.87		1.48	2.14
3820	With compound skim coat (level 5 finish)		380	.042		.57	1.08		1.65	2.47
4000	Fireproofing, beams or columns, 2 layers, 1/2" thick, incl finish		330	.048		.80	1.25		2.05	2.99
4010	Mold resistant		330	.048		.90	1.25		2.15	3.10
4050	5/8" thick		300	.053		.84	1.37		2.21	3.25
4060	Mold resistant		300	.053		1.06	1.37		2.43	3.50
4100	3 layers, 1/2" thick		225	.071		1.18	1.83		3.01	4.40
4110	Mold resistant		225	.071		1.33	1.83		3.16	4.56
4150	5/8" thick		210	.076		1.26	1.96		3.22	4.71
4160	Mold resistant		210	.076		1.59	1.96		3.55	5.05
5200	For work over 8' high, add		3060	.005			.13		.13	.23
5270	For textured spray, add	2 Lath	1600	.010		.05	.24		.29	.44
5350	For finishing inner corners, add	2 Carp	950	.017	L.F.	.09	.43		.52	.83
5355	For finishing outer corners, add	"	1250	.013		.20	.33		.53	.78
5500	For acoustical sealant, add per bead	1 Carp	500	.016		.03	.41		.44	.73
5550	Sealant, 1 quart tube				Ea.	5.15			5.15	5.65
5600	Sound deadening board, 1/4" gypsum	2 Carp	1800	.009	S.F.	.31	.23		.54	.73
5650	1/2" wood fiber	"	1800	.009	"	.41	.23		.64	.84

09 29 15 - Gypsum Board Accessories

09 29 15.10 Accessories, Drywall

0011	**ACCESSORIES, DRYWALL** Casing bead, galvanized steel		290	.028		.20	.71		.91	1.42
0101	Vinyl		290	.028		.22	.71		.93	1.44
0401	Corner bead, galvanized steel, 1-1/4" x 1-1/4"		350	.023		.21	.59		.80	1.23

09 29 Gypsum Board

09 29 15 – Gypsum Board Accessories

09 29 15.10 Accessories, Drywall

		Crew	Daily Output	Labor-Hours	Unit	Material	2007 Bare Costs Labor	Equipment	Total	Total Incl O&P
0411	1-1/4" x 1-1/4", 10' long	2 Carp	35	.229	Ea.	2.10	5.85		7.95	12.25
0601	Vinyl corner bead		400	.020	L.F.	.21	.51		.72	1.10
0901	Furring channel, galv. steel, 7/8" deep, standard		260	.031		.26	.79		1.05	1.63
1001	Resilient		260	.031		.27	.79		1.06	1.64
1101	J trim, galvanized steel, 1/2" wide		300	.027			.69		.69	1.16
1121	5/8" wide		300	.027		.22	.69		.91	1.40
1160	Screws #6 x 1" A				M	5.80			5.80	6.40
1170	#6 x 1-5/8" A				"	9.20			9.20	10.10
1501	Z stud, galvanized steel, 1-1/2" wide	1 Carp	260	.031	L.F.	.37	.79		1.16	1.74

09 30 Tiling

09 30 13 – Ceramic Tiling

09 30 13.10 Ceramic Tile

		Crew	Daily Output	Labor-Hours	Unit	Material	2007 Bare Costs Labor	Equipment	Total	Total Incl O&P
0010	**CERAMIC TILE**									
0050	Base, using 1' x 4" high pc. with 1" x 1" tiles, mud set	D-7	82	.195	L.F.	4.48	4.29		8.77	11.85
0100	Thin set	"	128	.125		4.26	2.75		7.01	9.10
0300	For 6" high base, 1" x 1" tile face, add					.70			.70	.77
0400	For 2" x 2" tile face, add to above					.37			.37	.41
0600	Cove base, 4-1/4" x 4-1/4" high, mud set	D-7	91	.176		3.48	3.86		7.34	10.05
0700	Thin set		128	.125		3.50	2.75		6.25	8.25
0900	6" x 4-1/4" high, mud set		100	.160		3.23	3.52		6.75	9.20
1000	Thin set		137	.117		3.23	2.57		5.80	7.70
1200	Sanitary cove base, 6" x 4-1/4" high, mud set		93	.172		3.55	3.78		7.33	10
1300	Thin set		124	.129		4.04	2.84		6.88	9
1500	6" x 6" high, mud set		84	.190		4.47	4.19		8.66	11.65
1600	Thin set		117	.137		4.47	3.01		7.48	9.75
1800	Bathroom accessories, average		82	.195	Ea.	10.35	4.29		14.64	18.30
1900	Bathtub, 5', rec. 4-1/4" x 4-1/4" tile wainscot, adhesive set 6' high		2.90	5.517		152	121		273	360
2100	7' high wainscot		2.50	6.400		174	141		315	415
2200	8' high wainscot		2.20	7.273		184	160		344	460
2400	Bullnose trim, 4-1/4" x 4-1/4", mud set		82	.195	L.F.	3.08	4.29		7.37	10.30
2500	Thin set		128	.125		2.85	2.75		5.60	7.55
2700	6" x 4-1/4" bullnose trim, mud set		84	.190		2.54	4.19		6.73	9.55
2800	Thin set		124	.129		2.54	2.84		5.38	7.35
3000	Floors, natural clay, random or uniform, thin set, color group 1		183	.087	S.F.	3.95	1.92		5.87	7.45
3100	Color group 2		183	.087		4.26	1.92		6.18	7.80
3255	Floors, glazed, thin set, 6" x 6", color group 1		200	.080		3.36	1.76		5.12	6.55
3260	8" x 8" tile		250	.064		3.36	1.41		4.77	5.95
3270	12" x 12" tile		325	.049		4.22	1.08		5.30	6.40
3280	16" x 16" tile		550	.029		5.85	.64		6.49	7.50
3285	Border, 6" x 12" tile		275	.058		11.10	1.28		12.38	14.25
3290	3" x 12" tile		200	.080		32.50	1.76		34.26	39
3300	Porcelain type, 1 color, color group 2, 1" x 1"		183	.087		4.57	1.92		6.49	8.15
3310	2" x 2" or 2" x 1", thin set		190	.084		4.90	1.85		6.75	8.40
3350	For random blend, 2 colors, add					.85			.85	.94
3360	4 colors, add					1.20			1.20	1.32
4300	Specialty tile, 4-1/4" x 4-1/4" x 1/2", decorator finish	D-7	183	.087		9.70	1.92		11.62	13.80
4500	Add for epoxy grout, 1/16" joint, 1" x 1" tile		800	.020		.60	.44		1.04	1.37
4600	2" x 2" tile		820	.020		.54	.43		.97	1.28
4800	Pregrouted sheets, walls, 4-1/4" x 4-1/4", 6" x 4-1/4"									
4810	and 8-1/2" x 4-1/4", 4 S.F. sheets, silicone grout	D-7	240	.067	S.F.	4.59	1.47		6.06	7.40

09 30 Tiling

09 30 13 – Ceramic Tiling

09 30 13.10 Ceramic Tile

		Crew	Daily Output	Labor-Hours	Unit	Material	2007 Bare Costs Labor	Equipment	Total	Total Incl O&P
5100	Floors, unglazed, 2 S.F. sheets,									
5110	urethane adhesive	D-7	180	.089	S.F.	4.57	1.95		6.52	8.20
5400	Walls, interior, thin set, 4-1/4" x 4-1/4" tile		190	.084		2.16	1.85		4.01	5.35
5500	6" x 4-1/4" tile		190	.084		2.51	1.85		4.36	5.75
5700	8-1/2" x 4-1/4" tile		190	.084		3.55	1.85		5.40	6.90
5800	6" x 6" tile		200	.080		2.98	1.76		4.74	6.10
5810	8" x 8" tile		225	.071		3.97	1.56		5.53	6.90
5820	12" x 12" tile		300	.053		3.25	1.17		4.42	5.45
5830	16" x 16" tile		500	.032		3.52	.70		4.22	5
6000	Decorated wall tile, 4-1/4" x 4-1/4", minimum		270	.059		3.95	1.30		5.25	6.45
6100	Maximum		180	.089		41.50	1.95		43.45	49
6600	Crystalline glazed, 4-1/4" x 4-1/4", mud set, plain		100	.160		3.63	3.52		7.15	9.65
6700	4-1/4" x 4-1/4", scored tile		100	.160		4.42	3.52		7.94	10.50
6900	6" x 6" plain		93	.172		4.82	3.78		8.60	11.40
7000	For epoxy grout, 1/16" joints, 4-1/4" tile, add		800	.020		.35	.44		.79	1.10
7200	For tile set in dry mortar, add		1735	.009			.20		.20	.33
7300	For tile set in portland cement mortar, add		290	.055			1.21		1.21	1.95

09 30 16 – Quarry Tiling

09 30 16.10 Quarry Tile

		Crew	Daily Output	Labor-Hours	Unit	Material	2007 Bare Costs Labor	Equipment	Total	Total Incl O&P
0010	**QUARRY TILE** Base, cove or sanitary, 2" or 5" high, mud set									
0100	1/2" thick	D-7	110	.145	L.F.	4.79	3.20		7.99	10.40
0300	Bullnose trim, red, mud set, 6" x 6" x 1/2" thick		120	.133		4.02	2.93		6.95	9.15
0400	4" x 4" x 1/2" thick		110	.145		4.53	3.20		7.73	10.15
0600	4" x 8" x 1/2" thick, using 8" as edge		130	.123		3.99	2.71		6.70	8.75
0700	Floors, mud set, 1,000 S.F. lots, red, 4" x 4" x 1/2" thick		120	.133	S.F.	4.08	2.93		7.01	9.20
0900	6" x 6" x 1/2" thick		140	.114		3.36	2.51		5.87	7.75
1000	4" x 8" x 1/2" thick		130	.123		4.08	2.71		6.79	8.85
1300	For waxed coating, add					.65			.65	.72
1500	For non-standard colors, add					.39			.39	.43
1600	For abrasive surface, add					.46			.46	.51
1800	Brown tile, imported, 6" x 6" x 3/4"	D-7	120	.133		4.85	2.93		7.78	10.05
1900	8" x 8" x 1"		110	.145		5.50	3.20		8.70	11.20
2100	For thin set mortar application, deduct		700	.023			.50		.50	.81
2700	Stair tread, 6" x 6" x 3/4", plain		50	.320		5.10	7.05		12.15	16.90
2800	Abrasive		47	.340		5.15	7.50		12.65	17.75
3000	Wainscot, 6" x 6" x 1/2", thin set, red		105	.152		3.85	3.35		7.20	9.65
3100	Non-standard colors		105	.152		4.29	3.35		7.64	10.10
3300	Window sill, 6" wide, 3/4" thick		90	.178	L.F.	4.94	3.91		8.85	11.75
3400	Corners		80	.200	Ea.	5.40	4.40		9.80	13

09 30 29 – Metal Tiling

09 30 29.10 Metal Tile

		Crew	Daily Output	Labor-Hours	Unit	Material	2007 Bare Costs Labor	Equipment	Total	Total Incl O&P
0010	**METAL TILE** 4' x 4' sheet, 24 ga., tile pattern, nailed									
0200	Stainless steel	2 Carp	512	.031	S.F.	24.50	.80		25.30	28.50
0400	Aluminized steel	"	512	.031	"	13.15	.80		13.95	15.80

09 51 Acoustical Ceilings

09 51 23 – Acoustical Tile Ceilings

09 51 23.10 Suspended Acoustic Ceiling Tiles

		Crew	Daily Output	Labor-Hours	Unit	Material	2007 Bare Costs Labor	Equipment	Total	Total Incl O&P
0010	**SUSPENDED ACOUSTIC CEILING TILES**, Not including									
0100	suspension system									
0300	Fiberglass boards, film faced, 2' x 2' or 2' x 4', 5/8" thick	1 Carp	625	.013	S.F.	.59	.33		.92	1.21
0400	3/4" thick		600	.013		1.36	.34		1.70	2.08
0500	3" thick, thermal, R11		450	.018		1.43	.46		1.89	2.35
0600	Glass cloth faced fiberglass, 3/4" thick		500	.016		1.74	.41		2.15	2.61
0700	1" thick		485	.016		1.93	.42		2.35	2.84
0820	1-1/2" thick, nubby face		475	.017		2.48	.43		2.91	3.46
1110	Mineral fiber tile, lay-in, 2' x 2' or 2' x 4', 5/8" thick, fine texture		625	.013		.47	.33		.80	1.08
1115	Rough textured		625	.013		1.14	.33		1.47	1.81
1125	3/4" thick, fine textured		600	.013		1.32	.34		1.66	2.03
1130	Rough textured		600	.013		1.65	.34		1.99	2.40
1135	Fissured		600	.013		2.09	.34		2.43	2.88
1150	Tegular, 5/8" thick, fine textured		470	.017		1.24	.44		1.68	2.10
1155	Rough textured		470	.017		1.62	.44		2.06	2.52
1165	3/4" thick, fine textured		450	.018		1.77	.46		2.23	2.73
1170	Rough textured		450	.018		2	.46		2.46	2.98
1175	Fissured		450	.018		3.10	.46		3.56	4.19
1180	For aluminum face, add					5.15			5.15	5.70
1185	For plastic film face, add					.82			.82	.90
1190	For fire rating, add					.38			.38	.42
1300	Mirror faced panels, 15/16" thick, 2' x 2'	1 Carp	500	.016		10.70	.41		11.11	12.50
1900	Eggcrate, acrylic, 1/2" x 1/2" x 1/2" cubes		500	.016		1.55	.41		1.96	2.41
2100	Polystyrene eggcrate, 3/8" x 3/8" x 1/2" cubes		510	.016		1.30	.40		1.70	2.11
2200	1/2" x 1/2" x 1/2" cubes		500	.016		1.74	.41		2.15	2.61
2400	Luminous panels, prismatic, acrylic		400	.020		1.89	.51		2.40	2.95
2500	Polystyrene		400	.020		.97	.51		1.48	1.94
2700	Flat white acrylic		400	.020		3.29	.51		3.80	4.49
2800	Polystyrene		400	.020		2.25	.51		2.76	3.35
3000	Drop pan, white, acrylic		400	.020		4.82	.51		5.33	6.15
3100	Polystyrene		400	.020		4.03	.51		4.54	5.30
3600	Perforated aluminum sheets, .024" thick, corrugated, painted		490	.016		1.92	.42		2.34	2.82
3700	Plain		500	.016		3.25	.41		3.66	4.28
3750	Wood fiber in cementitious binder, 2' x 2' or 4', painted, 1" thick		600	.013		1.31	.34		1.65	2.02
3760	2" thick		550	.015		2.16	.37		2.53	3.01
3770	2-1/2" thick		500	.016		2.96	.41		3.37	3.96
3780	3" thick		450	.018		3.32	.46		3.78	4.43

09 51 23.30 Suspended Ceilings, Complete

		Crew	Daily Output	Labor-Hours	Unit	Material	Labor	Equipment	Total	Total Incl O&P
0010	**SUSPENDED CEILINGS, COMPLETE** Including standard									
0100	suspension system but not incl. 1-1/2" carrier channels									
0600	Fiberglass ceiling board, 2' x 4' x 5/8", plain faced,	1 Carp	500	.016	S.F.	1.14	.41		1.55	1.95
0700	Offices, 2' x 4' x 3/4"		380	.021		1.91	.54		2.45	3.02
1800	Tile, Z bar suspension, 5/8" mineral fiber tile		150	.053		1.88	1.37		3.25	4.40
1900	3/4" mineral fiber tile		150	.053		2.01	1.37		3.38	4.54

09 51 53 – Direct-Applied Acoustical Ceilings

09 51 53.10 Ceiling Tile

		Crew	Daily Output	Labor-Hours	Unit	Material	Labor	Equipment	Total	Total Incl O&P
0010	**CEILING TILE**, Stapled or cemented									
0100	12" x 12" or 12" x 24", not including furring									
0600	Mineral fiber, vinyl coated, 5/8" thick	1 Carp	1000	.008	S.F.	1.65	.21		1.86	2.17
0700	3/4" thick		1000	.008		1.52	.21		1.73	2.02
0900	Fire rated, 3/4" thick, plain faced		1000	.008		1.39	.21		1.60	1.88
1000	Plastic coated face		1000	.008		1.18	.21		1.39	1.65

09 51 Acoustical Ceilings

09 51 53 – Direct-Applied Acoustical Ceilings

09 51 53.10 Ceiling Tile

		Crew	Daily Output	Labor-Hours	Unit	Material	2007 Bare Costs Labor	Equipment	Total	Total Incl O&P
1200	Aluminum faced, 5/8" thick, plain	1 Carp	1000	.008	S.F.	1.31	.21		1.52	1.79
3300	For flameproofing, add					.10			.10	.11
3400	For sculptured 3 dimensional, add					.28			.28	.31
3900	For ceiling primer, add					.13			.13	.14
4000	For ceiling cement, add					.36			.36	.40

09 53 Acoustical Ceiling Suspension Assemblies

09 53 23 – Metal Acoustical Ceiling Suspension Assemblies

09 53 23.30 Ceiling Suspension Systems

		Crew	Daily Output	Labor-Hours	Unit	Material	Labor	Equipment	Total	Total Incl O&P
0010	**CEILING SUSPENSION SYSTEMS** for boards and tile									
0050	Class A suspension system, 15/16" T bar, 2' x 4' grid	1 Carp	800	.010	S.F.	.55	.26		.81	1.04
0300	2' x 2' grid	"	650	.012		.68	.32		1	1.29
0350	For 9/16" grid, add					.16			.16	.18
0360	For fire rated grid, add					.09			.09	.10
0370	For colored grid, add					.20			.20	.22
0400	Concealed Z bar suspension system, 12" module	1 Carp	520	.015		.65	.40		1.05	1.39
0600	1-1/2" carrier channels, 4' O.C., add		470	.017		.12	.44		.56	.87
0650	1-1/2" x 3-1/2" channels		470	.017		.31	.44		.75	1.08
0700	Carrier channels for ceilings with									
0900	recessed lighting fixtures, add	1 Carp	460	.017	S.F.	.22	.45		.67	1
5000	Wire hangers, #12 wire	"	300	.027	Ea.	1.16	.69		1.85	2.44

09 63 Masonry Flooring

09 63 13 – Brick Flooring

09 63 13.10 Brick Flooring

		Crew	Daily Output	Labor-Hours	Unit	Material	Labor	Equipment	Total	Total Incl O&P
0010	**BRICK FLOORING**									
0020	Acid proof shales, red, 8" x 3-3/4" x 1-1/4" thick	D-7	.43	37.209	M	780	820		1,600	2,175
0050	2-1/4" thick	D-1	.40	40		865	935		1,800	2,500
0200	Acid proof clay brick, 8" x 3-3/4" x 2-1/4" thick	"	.40	40		805	935		1,740	2,450
0260	Cast ceramic, pressed, 4" x 8" x 1/2", unglazed	D-7	100	.160	S.F.	5.40	3.52		8.92	11.60
0270	Glazed		100	.160		7.20	3.52		10.72	13.55
0280	Hand molded flooring, 4" x 8" x 3/4", unglazed		95	.168		7.15	3.70		10.85	13.80
0290	Glazed		95	.168		8.95	3.70		12.65	15.80
0300	8" hexagonal, 3/4" thick, unglazed		85	.188		7.80	4.14		11.94	15.25
0310	Glazed		85	.188		14.10	4.14		18.24	22
0450	Acid proof joints, 1/4" wide	D-1	65	.246		1.24	5.75		6.99	10.90
0500	Pavers, 8" x 4", 1" to 1-1/4" thick, red	D-7	95	.168		3.15	3.70		6.85	9.40
0510	Ironspot	"	95	.168		4.44	3.70		8.14	10.85
0540	1-3/8" to 1-3/4" thick, red	D-1	95	.168		3.03	3.93		6.96	9.90
0560	Ironspot		95	.168		4.39	3.93		8.32	11.40
0580	2-1/4" thick, red		90	.178		3.09	4.15		7.24	10.30
0590	Ironspot		90	.178		4.79	4.15		8.94	12.15
0870	For epoxy joints, add		600	.027		2.35	.62		2.97	3.62
0880	For Furan underlayment, add		600	.027		1.94	.62		2.56	3.16
0890	For waxed surface, steam cleaned, add	A-1H	1000	.008		.17	.15	.06	.38	.51

09 63 40 – Stone Flooring

09 63 40.10 Marble

		Crew	Daily Output	Labor-Hours	Unit	Material	Labor	Equipment	Total	Total Incl O&P
0010	**MARBLE**									
0020	Thin gauge tile, 12" x 6", 3/8", White Carara	D-7	60	.267	S.F.	9.10	5.85		14.95	19.40

09 63 Masonry Flooring

09 63 40 – Stone Flooring

09 63 40.10 Marble

		Crew	Daily Output	Labor-Hours	Unit	Material	2007 Bare Costs Labor	Equipment	Total	Total Incl O&P
0100	Travertine	D-7	60	.267	S.F.	10.70	5.85		16.55	21
0200	12" x 12" x 3/8", thin set, floors		60	.267		6.45	5.85		12.30	16.50
0300	On walls		52	.308		9.35	6.75		16.10	21
1000	Marble Threshold, 4" Wide x 36" Long x 5/8" Thick, White		60	.267	Ea.	6.45	5.85		12.30	16.50

09 63 40.20 Slate Tile

0010	**SLATE TILE**									
0020	Vermont, 6" x 6" x 1/4" thick, thin set	D-7	180	.089	S.F.	5	1.95		6.95	8.65

09 64 Wood Flooring

09 64 23 – Wood Parquet Flooring

09 64 23.10 Wood Parquet

		Crew	Daily Output	Labor-Hours	Unit	Material	Labor	Equipment	Total	Total Incl O&P
0010	**WOOD PARQUET** flooring									
5200	Parquetry, standard, 5/16" thick, not incl. finish, oak, minimum	1 Carp	160	.050	S.F.	4.16	1.29		5.45	6.75
5300	Maximum		100	.080		4.91	2.06		6.97	8.90
5500	Teak, minimum		160	.050		4.69	1.29		5.98	7.35
5600	Maximum		100	.080		8.20	2.06		10.26	12.50
5650	13/16" thick, select grade oak, minimum		160	.050		9.05	1.29		10.34	12.15
5700	Maximum		100	.080		13.75	2.06		15.81	18.60
5800	Custom parquetry, including finish, minimum		100	.080		15.15	2.06		17.21	20
5900	Maximum		50	.160		20	4.11		24.11	29
6700	Parquetry, prefinished white oak, 5/16" thick, minimum		160	.050		3.61	1.29		4.90	6.15
6800	Maximum		100	.080		7.30	2.06		9.36	11.55
7000	Walnut or teak, parquetry, minimum		160	.050		5.25	1.29		6.54	8
7100	Maximum		100	.080		9.15	2.06		11.21	13.60
7200	Acrylic wood parquet blocks, 12" x 12" x 5/16",									
7210	irradiated, set in epoxy	1 Carp	160	.050	S.F.	7.65	1.29		8.94	10.60

09 64 29 – Wood Strip and Plank Flooring

09 64 29.10 Wood

		Crew	Daily Output	Labor-Hours	Unit	Material	Labor	Equipment	Total	Total Incl O&P
0010	**WOOD**									
0020	Fir, vertical grain, 1" x 4", not incl. finish, B & better	1 Carp	255	.031	S.F.	2.69	.81		3.50	4.33
0100	C grade & better		255	.031		2.53	.81		3.34	4.15
0300	Flat grain, 1" x 4", not incl. finish, B & better		255	.031		3.07	.81		3.88	4.75
0400	C & better		255	.031		2.95	.81		3.76	4.62
4000	Maple, strip, 25/32" x 2-1/4", not incl. finish, select		170	.047		5.15	1.21		6.36	7.70
4100	#2 & better		170	.047		3.22	1.21		4.43	5.60
4300	33/32" x 3-1/4", not incl. finish, #1 grade		170	.047		3.99	1.21		5.20	6.45
4400	#2 & better		170	.047		3.55	1.21		4.76	5.95
4600	Oak, white or red, 25/32" x 2-1/4", not incl. finish									
4700	#1 common	1 Carp	170	.047	S.F.	3.09	1.21		4.30	5.45
4900	Select quartered, 2-1/4" wide		170	.047		2.82	1.21		4.03	5.15
5000	Clear		170	.047		3.89	1.21		5.10	6.35
6100	Prefinished, white oak, prime grade, 2-1/4" wide		170	.047		6.40	1.21		7.61	9.10
6200	3-1/4" wide		185	.043		8.20	1.11		9.31	10.90
6400	Ranch plank		145	.055		7.90	1.42		9.32	11.10
6500	Hardwood blocks, 9" x 9", 25/32" thick		160	.050		5.65	1.29		6.94	8.45
7400	Yellow pine, 3/4" x 3-1/8", T & G, C & better, not incl. finish		200	.040		2.35	1.03		3.38	4.33
7500	Refinish wood floor, sand, 2 cts poly, wax, soft wood, min.	1 Clab	400	.020		.76	.37		1.13	1.48
7600	Hard wood, max		130	.062		1.14	1.15		2.29	3.20
7800	Sanding and finishing, 2 coats polyurethane		295	.027		.76	.51		1.27	1.70
7900	Subfloor and underlayment, see division 06 16 36.00									

09 64 Wood Flooring

09 64 29 – Wood Strip and Plank Flooring

09 64 29.10 Wood

		Crew	Daily Output	Labor-Hours	Unit	Material	2007 Bare Costs Labor	Equipment	Total	Total Incl O&P
8015	Transition molding, 2 1/4" wide, 5' long	1 Carp	19.20	.417	Ea.	15	10.70		25.70	34.50
8300	Floating floor, wood composition strip, complete.	1 Clab	133	.060	S.F.	4.24	1.12		5.36	6.55
8310	Floating floor components, T & G wood composite strips					3.64			3.64	4.01
8320	Film					.15			.15	.16
8330	Foam					.26			.26	.29
8340	Adhesive					.24			.24	.27
8350	Installation kit					.17			.17	.19
8360	Trim, 2" wide x 3' long				L.F.	2.44			2.44	2.68
8370	Reducer moulding				"	4.21			4.21	4.63

09 65 Resilient Flooring

09 65 10 – Resilient Tile Underlayment

09 65 10.10 Resilient Tile Underlayment

		Crew	Daily Output	Labor-Hours	Unit	Material	Labor	Equipment	Total	Total Incl O&P
0010	**RESILIENT TILE UNDERLAYMENT**									
3600	Latex underlayment, 1/8" thk., cementitious for resilient flooring	1 Tilf	160	.050	S.F.	1.56	1.24		2.80	3.72
4000	Latex underlayment, liquid, fortified				Gal.	35			35	38.50

09 65 13 – Resilient Base and Accessories

09 65 13.13 Resilient Base

		Crew	Daily Output	Labor-Hours	Unit	Material	Labor	Equipment	Total	Total Incl O&P
0010	**RESILIENT BASE**									
0800	Base, cove, rubber or vinyl, .080" thick									
1100	Standard colors, 2-1/2" high	1 Tilf	315	.025	L.F.	.65	.63		1.28	1.73
1150	4" high		315	.025		.70	.63		1.33	1.78
1200	6" high		315	.025		1.10	.63		1.73	2.22
1450	1/8" thick, standard colors, 2-1/2" high		315	.025		.75	.63		1.38	1.84
1500	4" high		315	.025		.64	.63		1.27	1.71
1550	6" high		315	.025		1.10	.63		1.73	2.22
1600	Corners, 2-1/2" high		315	.025	Ea.	1.35	.63		1.98	2.50
1630	4" high		315	.025		2.25	.63		2.88	3.49
1660	6" high		315	.025		2.35	.63		2.98	3.60

09 65 16 – Resilient Sheet Flooring

09 65 16.10 Resilient Sheet Flooring

		Crew	Daily Output	Labor-Hours	Unit	Material	Labor	Equipment	Total	Total Incl O&P
0010	**RESILIENT SHEET FLOORING**									
5900	Rubber, sheet goods, 36" wide, 1/8" thick	1 Tilf	120	.067	S.F.	5.50	1.66		7.16	8.70
5950	3/16" thick		100	.080		7.50	1.99		9.49	11.45
6000	1/4" thick		90	.089		9	2.21		11.21	13.45
8000	Vinyl sheet goods, backed, .065" thick, minimum		250	.032		2.75	.80		3.55	4.31
8050	Maximum		200	.040		3.03	.99		4.02	4.93
8100	.080" thick, minimum		230	.035		3	.86		3.86	4.69
8150	Maximum		200	.040		4.25	.99		5.24	6.30
8200	.125" thick, minimum		230	.035		3.35	.86		4.21	5.10
8250	Maximum		200	.040		5	.99		5.99	7.10
8700	Adhesive cement, 1 gallon does 200 to 300 S.F.				Gal.	18.05			18.05	19.85
8800	Asphalt primer, 1 gallon per 300 S.F.					10.80			10.80	11.90
8900	Emulsion, 1 gallon per 140 S.F.					13.75			13.75	15.15

09 65 19 – Resilient Tile Flooring

09 65 19.10 Resilient Tile Flooring

		Crew	Daily Output	Labor-Hours	Unit	Material	Labor	Equipment	Total	Total Incl O&P
0010	**RESILIENT TILE FLOORING**									
2200	Cork tile, standard finish, 1/8" thick	1 Tilf	315	.025	S.F.	4.74	.63		5.37	6.20
2250	3/16" thick		315	.025		5.30	.63		5.93	6.85

09 65 Resilient Flooring

09 65 19 – Resilient Tile Flooring

09 65 19.10 Resilient Tile Flooring

		Crew	Daily Output	Labor-Hours	Unit	Material	2007 Bare Costs Labor	2007 Bare Costs Equipment	Total	Total Incl O&P
2300	5/16" thick	1 Tilf	315	.025	S.F.	6.15	.63		6.78	7.80
2350	1/2" thick		315	.025		7.10	.63		7.73	8.80
2500	Urethane finish, 1/8" thick		315	.025		5.65	.63		6.28	7.25
2550	3/16" thick		315	.025		6.25	.63		6.88	7.85
2600	5/16" thick		315	.025		7.85	.63		8.48	9.60
2650	1/2" thick		315	.025		10.85	.63		11.48	12.95
6050	Tile, marbleized colors, 12" x 12", 1/8" thick		400	.020		5.60	.50		6.10	6.95
6100	3/16" thick		400	.020		8.25	.50		8.75	9.90
6300	Special tile, plain colors, 1/8" thick		400	.020		6.50	.50		7	7.95
6350	3/16" thick		400	.020		8.85	.50		9.35	10.55
7000	Vinyl composition tile, 12" x 12", 1/16" thick		500	.016		.97	.40		1.37	1.71
7050	Embossed		500	.016		1.75	.40		2.15	2.57
7100	Marbleized		500	.016		1.75	.40		2.15	2.57
7150	Solid		500	.016		2.25	.40		2.65	3.12
7200	3/32" thick, embossed		500	.016		1.17	.40		1.57	1.93
7250	Marbleized		500	.016		2	.40		2.40	2.84
7300	Solid		500	.016		2	.40		2.40	2.84
7350	1/8" thick, marbleized		500	.016		1.39	.40		1.79	2.17
7400	Solid		500	.016		2.07	.40		2.47	2.92
7450	Conductive		500	.016		4.82	.40		5.22	5.95
7500	Vinyl tile, 12" x 12", .050" thick, minimum		500	.016		2.20	.40		2.60	3.06
7550	Maximum		500	.016		5	.40		5.40	6.15
7600	1/8" thick, minimum		500	.016		3.60	.40		4	4.60
7650	Solid colors		500	.016		5.35	.40		5.75	6.50
7700	Marbleized or Travertine pattern		500	.016		5.15	.40		5.55	6.30
7750	Florentine pattern		500	.016		5.20	.40		5.60	6.35
7800	Maximum		500	.016		11	.40		11.40	12.75

09 65 33 – Conductive Resilient Flooring

09 65 33.10 Conductive Resilient Flooring

		Crew	Daily Output	Labor-Hours	Unit	Material	Labor	Equipment	Total	Total Incl O&P
0010	**CONDUCTIVE RESILIENT FLOORING**									
1700	Conductive flooring, rubber tile, 1/8" thick	1 Tilf	315	.025	S.F.	4.50	.63		5.13	5.95
1800	Homogeneous vinyl tile, 1/8" thick	"	315	.025	"	5.70	.63		6.33	7.25

09 66 Terrazzo Flooring

09 66 13 – Portland Cement Terrazzo Flooring

09 66 13.10 Portland Cement Terrazzo

		Crew	Daily Output	Labor-Hours	Unit	Material	Labor	Equipment	Total	Total Incl O&P
0010	**PORTLAND CEMENT TERRAZZO**, cast-in-place									
4300	Stone chips, onyx gemstone, per 50 lb. bag				Bag	15.55			15.55	17.10

09 66 13.30 Terrazzo, Precast

		Crew	Daily Output	Labor-Hours	Unit	Material	Labor	Equipment	Total	Total Incl O&P
0010	**TERRAZZO, PRECAST**									
0020	Base, 6" high, straight	1 Mstz	35	.229	L.F.	10.15	5.60		15.75	20.50
0100	Cove		30	.267		11.55	6.55		18.10	23.50
0300	8" high base, straight		30	.267		10.30	6.55		16.85	22
0400	Cove		25	.320		15.20	7.85		23.05	29.50
0600	For white cement, add					.41			.41	.45
0700	For 16 ga. zinc toe strip, add					1.52			1.52	1.67
0900	Curbs, 4" x 4" high	1 Mstz	19	.421		29	10.35		39.35	48.50
1000	8" x 8" high	"	15	.533		33	13.10		46.10	57.50
1200	Floor tiles, non-slip, 1" thick, 12" x 12"	D-1	29	.552	S.F.	16.95	12.90		29.85	40
1300	1-1/4" thick, 12" x 12"		29	.552		19	12.90		31.90	42.50

09 66 Terrazzo Flooring

09 66 13 – Portland Cement Terrazzo Flooring

09 66 13.30 Terrazzo, Precast		Crew	Daily Output	Labor-Hours	Unit	Material	2007 Bare Costs Labor	Equipment	Total	Total Incl O&P
1500	16" x 16"	D-1	23	.696	S.F.	20.50	16.25		36.75	49.50
1600	1-1/2" thick, 16" x 16"	↓	21	.762		18.90	17.80		36.70	50.50
4800	Wainscot, 12" x 12" x 1" tiles	1 Mstz	12	.667		6.20	16.40		22.60	33.50
4900	16" x 16" x 1-1/2" tiles	"	8	1	↓	13.30	24.50		37.80	54

09 68 Carpeting

09 68 05 – Carpet Accessories

09 68 05.11 Flooring Transition Strip

		Crew	Daily Output	Labor-Hours	Unit	Material	Labor	Equipment	Total	Total Incl O&P
0010	**FLOORING TRANSITION STRIP**									
0107	Clamp down brass divider, 12' strip, vinyl to carpet	1 Tilf	31.25	.256	Ea.	25	6.35		31.35	38
0117	Vinyl to hard surface	"	31.25	.256	"	25	6.35		31.35	38

09 68 10 – Carpet Pad

09 68 10.10 Carpet Pad

		Crew	Daily Output	Labor-Hours	Unit	Material	Labor	Equipment	Total	Total Incl O&P
0010	**CARPET PAD**, commercial grade									
9001	Sponge rubber pad, minimum	1 Tilf	1350	.006	S.F.	.43	.15		.58	.72
9101	Maximum		1350	.006		.96	.15		1.11	1.30
9201	Felt pad, minimum		1350	.006		.44	.15		.59	.72
9301	Maximum		1350	.006		.78	.15		.93	1.10
9401	Bonded urethane pad, minimum		1350	.006		.46	.15		.61	.75
9501	Maximum		1350	.006		.79	.15		.94	1.11
9601	Prime urethane pad, minimum		1350	.006		.26	.15		.41	.53
9701	Maximum	↓	1350	.006	↓	.49	.15		.64	.78

09 68 13 – Tile Carpeting

09 68 13.10 Carpet Tile

		Crew	Daily Output	Labor-Hours	Unit	Material	Labor	Equipment	Total	Total Incl O&P
0010	**CARPET TILE**									
0100	Tufted nylon, 18" x 18", hard back, 20 oz.	1 Tilf	150	.053	S.Y.	21.50	1.33		22.83	26
0110	26 oz.		150	.053		37	1.33		38.33	42.50
0200	Cushion back, 20 oz.		150	.053		27	1.33		28.33	32
0210	26 oz.	↓	150	.053	↓	42.50	1.33		43.83	49

09 68 16 – Sheet Carpeting

09 68 16.10 Sheet Carpet

		Crew	Daily Output	Labor-Hours	Unit	Material	Labor	Equipment	Total	Total Incl O&P
0010	**SHEET CARPET**									
0701	Nylon, level loop, 26 oz., light to medium traffic	1 Tilf	445	.018	S.F.	2.50	.45		2.95	3.47
0901	32 oz., medium traffic		445	.018		3.17	.45		3.62	4.20
1101	40 oz., medium to heavy traffic		445	.018		4.67	.45		5.12	5.85
2101	Nylon, plush, 20 oz., light traffic		445	.018		1.50	.45		1.95	2.37
2801	24 oz., light to medium traffic		445	.018		1.58	.45		2.03	2.46
2901	30 oz., medium traffic		445	.018		2.33	.45		2.78	3.29
3001	36 oz., medium traffic		445	.018		3.06	.45		3.51	4.08
3101	42 oz., medium to heavy traffic		370	.022		3.54	.54		4.08	4.75
3201	46 oz., medium to heavy traffic		370	.022		4.06	.54		4.60	5.30
3301	54 oz., heavy traffic		370	.022		4.44	.54		4.98	5.75
3501	Olefin, 15 oz., light traffic		445	.018		.83	.45		1.28	1.64
3651	22 oz., light traffic		445	.018		.99	.45		1.44	1.81
4501	50 oz., medium to heavy traffic, level loop		445	.018		10.85	.45		11.30	12.60
4701	32 oz., medium to heavy traffic, patterned		400	.020		10.65	.50		11.15	12.55
4901	48 oz., heavy traffic, patterned	↓	400	.020	↓	10.90	.50		11.40	12.80
5000	For less than full roll, add					25%				
5100	For small rooms, less than 12' wide, add						25%			

09 68 Carpeting

09 68 16 – Sheet Carpeting

09 68 16.10 Sheet Carpet

		Crew	Daily Output	Labor-Hours	Unit	Material	2007 Bare Costs Labor	2007 Bare Costs Equipment	Total	Total Incl O&P
5200	For large open areas (no cuts), deduct						25%			
5600	For bound carpet baseboard, add	1 Tilf	300	.027	L.F.	1.65	.66		2.31	2.89
5610	For stairs, not incl. price of carpet, add	"	30	.267	Riser		6.65		6.65	10.65
8950	For tackless, stretched installation, add padding to above									
9850	For "branded" fiber, add				S.Y.	25%				

09 68 20 – Athletic Carpet

09 68 20.10 Indoor Athletic Carpet

		Crew	Daily Output	Labor-Hours	Unit	Material	Labor	Equipment	Total	Total Incl O&P
0010	**INDOOR ATHLETIC CARPET**									
3700	Polyethylene, in rolls, no base incl., landscape surfaces	1 Tilf	275	.029	S.F.	2.95	.72		3.67	4.41
3800	Nylon action surface, 1/8" thick		275	.029		2.90	.72		3.62	4.35
3900	1/4" thick		275	.029		4.18	.72		4.90	5.75
4000	3/8" thick	↓	275	.029	↓	5.25	.72		5.97	6.95

09 72 Wall Coverings

09 72 23 – Wallpapering

09 72 23.10 Wallpaper

		Crew	Daily Output	Labor-Hours	Unit	Material	Labor	Equipment	Total	Total Incl O&P
0010	**WALLPAPER** including sizing; add 10-30 percent waste @ takeoff R097223-10									
0050	Aluminum foil	1 Pape	275	.029	S.F.	.91	.68		1.59	2.12
0100	Copper sheets, .025" thick, vinyl backing		240	.033		4.89	.78		5.67	6.70
0300	Phenolic backing		240	.033		6.35	.78		7.13	8.25
0600	Cork tiles, light or dark, 12" x 12" x 3/16"		240	.033		3.93	.78		4.71	5.60
0700	5/16" thick		235	.034		3.35	.79		4.14	5
0900	1/4" basketweave		240	.033		5.15	.78		5.93	6.95
1000	1/2" natural, non-directional pattern		240	.033		6.45	.78		7.23	8.40
1100	3/4" natural, non-directional pattern		240	.033		9.90	.78		10.68	12.15
1200	Granular surface, 12" x 36", 1/2" thick		385	.021		1.11	.49		1.60	2.02
1300	1" thick		370	.022		1.44	.50		1.94	2.41
1500	Polyurethane coated, 12" x 12" x 3/16" thick		240	.033		3.48	.78		4.26	5.10
1600	5/16" thick		235	.034		4.95	.79		5.74	6.75
1800	Cork wallpaper, paperbacked, natural		480	.017		1.86	.39		2.25	2.69
1900	Colors		480	.017		2.45	.39		2.84	3.34
2100	Flexible wood veneer, 1/32" thick, plain woods		100	.080		2.08	1.87		3.95	5.35
2200	Exotic woods	↓	95	.084	↓	3.16	1.97		5.13	6.70
2400	Gypsum-based, fabric-backed, fire									
2500	resistant for masonry walls, minimum, 21 oz./S.Y.	1 Pape	800	.010	S.F.	.72	.23		.95	1.17
2600	Average		720	.011		1.08	.26		1.34	1.62
2700	Maximum, (small quantities)	↓	640	.013		1.20	.29		1.49	1.80
2750	Acrylic, modified, semi-rigid PVC, .028" thick	2 Carp	330	.048		1	1.25		2.25	3.21
2800	.040" thick	"	320	.050		1.31	1.29		2.60	3.62
3000	Vinyl wall covering, fabric-backed, lightweight, (12-15 oz./S.Y.)	1 Pape	640	.013		.60	.29		.89	1.14
3300	Medium weight, type 2, (20-24 oz./S.Y.)		480	.017		.72	.39		1.11	1.43
3400	Heavy weight, type 3, (28 oz./S.Y.)	↓	435	.018	↓	1.46	.43		1.89	2.32
3600	Adhesive, 5 gal. lots, (18SY/Gal.)				Gal.	9.40			9.40	10.30
3700	Wallpaper, average workmanship, solid pattern, low cost paper	1 Pape	640	.013	S.F.	.32	.29		.61	.83
3900	basic patterns (matching required), avg. cost paper		535	.015		.65	.35		1	1.29
4000	Paper at $85 per double roll, quality workmanship	↓	435	.018	↓	2.17	.43		2.60	3.10
4100	Linen wall covering, paper backed									
4150	Flame treatment, minimum				S.F.	.79			.79	.87
4180	Maximum					1.37			1.37	1.51
4200	Grass cloths with lining paper, minimum	1 Pape	400	.020	↓	.68	.47		1.15	1.52

09 72 Wall Coverings

09 72 23 – Wallpapering

09 72 23.10 Wallpaper	Crew	Daily Output	Labor-Hours	Unit	Material	2007 Bare Costs Labor	Equipment	Total	Total Incl O&P
4300　　Maximum	1 Pape	350	.023	S.F.	2.19	.53		2.72	3.29

09 91 Painting

09 91 03 – Paint Restoration

09 91 03.20 Sanding

		Crew	Daily Output	Labor-Hours	Unit	Material	2007 Bare Costs Labor	Equipment	Total	Total Incl O&P
0010	**SANDING** and puttying interior trim, compared to									
0100	Painting 1 coat, on quality work				L.F.		100%			
0300	Medium work						50%			
0400	Industrial grade				↓		25%			
0500	Surface protection, placement and removal									
0510	Basic drop cloths	1 Pord	6400	.001	S.F.		.03		.03	.05
0520	Masking with paper		800	.010		.06	.23		.29	.45
0530	Volume cover up (using plastic sheathing, or building paper)	↓	16000	.001	↓		.01		.01	.02

09 91 03.30 Exterior Surface Preparation

		Crew	Daily Output	Labor-Hours	Unit	Material	2007 Bare Costs Labor	Equipment	Total	Total Incl O&P
0010	**EXTERIOR SURFACE PREPARATION**									
0015	Doors, per side, not incl. frames or trim									
0020	Scrape & sand									
0030	Wood, flush	1 Pord	616	.013	S.F.		.30		.30	.50
0040	Wood, detail		496	.016			.37		.37	.62
0050	Wood, louvered		280	.029			.66		.66	1.09
0060	Wood, overhead		616	.013			.30		.30	.50
0070	Wire brush									
0080	Metal, flush	1 Pord	640	.013	S.F.		.29		.29	.48
0090	Metal, detail		520	.015			.36		.36	.59
0100	Metal, louvered		360	.022			.52		.52	.85
0110	Metal or fibr., overhead		640	.013			.29		.29	.48
0120	Metal, roll up		560	.014			.33		.33	.55
0130	Metal, bulkhead	↓	640	.013	↓		.29		.29	.48
0140	Power wash, based on 2500 lb. operating pressure									
0150	Metal, flush	B-9	2240	.018	S.F.		.34	.08	.42	.67
0160	Metal, detail		2120	.019			.36	.08	.44	.70
0170	Metal, louvered		2000	.020			.38	.09	.47	.75
0180	Metal or fibr., overhead		2400	.017			.32	.08	.40	.62
0190	Metal, roll up		2400	.017			.32	.08	.40	.62
0200	Metal, bulkhead	↓	2200	.018	↓		.35	.08	.43	.68
0400	Windows, per side, not incl. trim									
0410	Scrape & sand									
0420	Wood, 1-2 lite	1 Pord	320	.025	S.F.		.58		.58	.95
0430	Wood, 3-6 lite		280	.029			.66		.66	1.09
0440	Wood, 7-10 lite		240	.033			.77		.77	1.27
0450	Wood, 12 lite		200	.040			.93		.93	1.53
0460	Wood, Bay / Bow	↓	320	.025	↓		.58		.58	.95
0470	Wire brush									
0480	Metal, 1-2 lite	1 Pord	480	.017	S.F.		.39		.39	.64
0490	Metal, 3-6 lite		400	.020			.46		.46	.76
0500	Metal, Bay / Bow	↓	480	.017	↓		.39		.39	.64
0510	Power wash, based on 2500 lb. operating pressure									
0520	1-2 lite	B-9	4400	.009	S.F.		.17	.04	.21	.33
0530	3-6 lite		4320	.009			.18	.04	.22	.35
0540	7-10 lite		4240	.009			.18	.04	.22	.36
0550	12 lite	↓	4160	.010	↓		.18	.04	.22	.36

09 91 Painting

09 91 03 – Paint Restoration

09 91 03.30 Exterior Surface Preparation

		Crew	Daily Output	Labor-Hours	Unit	Material	2007 Bare Costs Labor	Equipment	Total	Total Incl O&P
0560	Bay / Bow	B-9	4400	.009	S.F.		.17	.04	.21	.33
0600	Siding, scrape and sand, light=10-30%, med.=30-70%									
0610	Heavy=70-100%, % of surface to sand									
0650	Texture 1-11, light	1 Pord	480	.017	S.F.		.39		.39	.64
0660	Med.		440	.018			.42		.42	.69
0670	Heavy		360	.022			.52		.52	.85
0680	Wood shingles, shakes, light		440	.018			.42		.42	.69
0690	Med.		360	.022			.52		.52	.85
0700	Heavy		280	.029			.66		.66	1.09
0710	Clapboard, light		520	.015			.36		.36	.59
0720	Med.		480	.017			.39		.39	.64
0730	Heavy	▼	400	.020	▼		.46		.46	.76
0740	Wire brush									
0750	Aluminum, light	1 Pord	600	.013	S.F.		.31		.31	.51
0760	Med.		520	.015			.36		.36	.59
0770	Heavy	▼	440	.018	▼		.42		.42	.69
0780	Pressure wash, based on 2500 lb.. operating pressure									
0790	Stucco	B-9	3080	.013	S.F.		.25	.06	.31	.48
0800	Aluminum or vinyl		3200	.013			.24	.06	.30	.47
0810	Siding, masonry, brick & block	▼	2400	.017	▼		.32	.08	.40	.62
1300	Miscellaneous, wire brush									
1310	Metal, pedestrian gate	1 Pord	100	.080	S.F.		1.86		1.86	3.05

09 91 03.40 Interior Surface Preparation

		Crew	Daily Output	Labor-Hours	Unit	Material	Labor	Equipment	Total	Total Incl O&P
0010	**INTERIOR SURFACE PREPARATION**									
0020	Doors									
0030	Scrape & sand									
0040	Wood, flush	1 Pord	616	.013	S.F.		.30		.30	.50
0050	Wood, detail		496	.016			.37		.37	.62
0060	Wood, louvered	▼	280	.029	▼		.66		.66	1.09
0070	Wire brush									
0080	Metal, flush	1 Pord	640	.013	S.F.		.29		.29	.48
0090	Metal, detail		520	.015			.36		.36	.59
0100	Metal, louvered	▼	360	.022	▼		.52		.52	.85
0110	Hand wash									
0120	Wood, flush	1 Pord	2160	.004	S.F.		.09		.09	.14
0130	Wood, detailed		2000	.004			.09		.09	.15
0140	Wood, louvered		1360	.006			.14		.14	.22
0150	Metal, flush		2160	.004			.09		.09	.14
0160	Metal, detail		2000	.004			.09		.09	.15
0170	Metal, louvered	▼	1360	.006	▼		.14		.14	.22
0400	Windows, per side, not incl. trim									
0410	Scrape & sand									
0420	Wood, 1-2 lite	1 Pord	360	.022	S.F.		.52		.52	.85
0430	Wood, 3-6 lite		320	.025			.58		.58	.95
0440	Wood, 7-10 lite		280	.029			.66		.66	1.09
0450	Wood, 12 lite		240	.033			.77		.77	1.27
0460	Wood, Bay / Bow	▼	360	.022	▼		.52		.52	.85
0470	Wire brush									
0480	Metal, 1-2 lite	1 Pord	520	.015	S.F.		.36		.36	.59
0490	Metal, 3-6 lite		440	.018			.42		.42	.69
0500	Metal, Bay / Bow	▼	520	.015	▼		.36		.36	.59
0600	Walls, sanding, light=10-30%									

09 91 Painting

09 91 03 – Paint Restoration

09 91 03.40 Interior Surface Preparation

		Crew	Daily Output	Labor-Hours	Unit	Material	2007 Bare Costs Labor	Equipment	Total	Total Incl O&P
0610	Med.=30-70%, heavy=70-100%, % of surface to sand									
0650	Walls, sand									
0660	Drywall, gypsum, plaster, light	1 Pord	3077	.003	S.F.		.06		.06	.10
0670	Drywall, gypsum, plaster, med.		2160	.004			.09		.09	.14
0680	Drywall, gypsum, plaster, heavy		923	.009			.20		.20	.33
0690	Wood, T&G, light		2400	.003			.08		.08	.13
0700	Wood, T&G, med.		1600	.005			.12		.12	.19
0710	Wood, T&G, heavy		800	.010			.23		.23	.38
0720	Walls, wash									
0730	Drywall, gypsum, plaster	1 Pord	3200	.003	S.F.		.06		.06	.10
0740	Wood, T&G		3200	.003			.06		.06	.10
0750	Masonry, brick & block, smooth		2800	.003			.07		.07	.11
0760	Masonry, brick & block, coarse		2000	.004			.09		.09	.15
8000	For Chemical Washing, see Division 04 01 30.00									

09 91 03.41 Scrape After Fire Damage

		Crew	Daily Output	Labor-Hours	Unit	Material	Labor	Equipment	Total	Total Incl O&P
0010	**SCRAPE AFTER FIRE DAMAGE**									
0050	Boards, 1" x 4"	1 Pord	336	.024	L.F.		.55		.55	.91
0060	1" x 6"		260	.031			.71		.71	1.17
0070	1" x 8"		207	.039			.90		.90	1.47
0080	1" x 10"		174	.046			1.07		1.07	1.75
0500	Framing, 2" x 4"		265	.030			.70		.70	1.15
0510	2" x 6"		221	.036			.84		.84	1.38
0520	2" x 8"		190	.042			.98		.98	1.61
0530	2" x 10"		165	.048			1.12		1.12	1.85
0540	2" x 12"		144	.056			1.29		1.29	2.12
1000	Heavy framing, 3" x 4"		226	.035			.82		.82	1.35
1010	4" x 4"		210	.038			.88		.88	1.45
1020	4" x 6"		191	.042			.97		.97	1.60
1030	4" x 8"		165	.048			1.12		1.12	1.85
1040	4" x 10"		144	.056			1.29		1.29	2.12
1060	4" x 12"		131	.061			1.42		1.42	2.33
2900	For sealing, minimum		825	.010	S.F.	.14	.23		.37	.52
2920	Maximum		460	.017	"	.27	.40		.67	.96

09 91 13 – Exterior Painting

09 91 13.30 Fences

		Crew	Daily Output	Labor-Hours	Unit	Material	Labor	Equipment	Total	Total Incl O&P
0010	**FENCES** R099100-20									
0100	Chain link or wire metal, one side, water base									
0110	Roll & brush, first coat	1 Pord	960	.008	S.F.	.06	.19		.25	.39
0120	Second coat		1280	.006		.06	.15		.21	.30
0130	Spray, first coat		2275	.004		.06	.08		.14	.20
0140	Second coat		2600	.003		.06	.07		.13	.19
0150	Picket, water base									
0160	Roll & brush, first coat	1 Pord	865	.009	S.F.	.06	.21		.27	.42
0170	Second coat		1050	.008		.06	.18		.24	.36
0180	Spray, first coat		2275	.004		.06	.08		.14	.20
0190	Second coat		2600	.003		.06	.07		.13	.19
0200	Stockade, water base									
0210	Roll & brush, first coat	1 Pord	1040	.008	S.F.	.06	.18		.24	.36
0220	Second coat		1200	.007		.06	.15		.21	.32
0230	Spray, first coat		2275	.004		.06	.08		.14	.20
0240	Second coat		2600	.003		.06	.07		.13	.19

09 91 Painting

09 91 13 – Exterior Painting

09 91 13.42 Miscellaneous, Exterior

		Crew	Daily Output	Labor-Hours	Unit	Material	2007 Bare Costs Labor	2007 Bare Costs Equipment	Total	Total Incl O&P
0010	**MISCELLANEOUS, EXTERIOR** R099100-20									
0100	Railing, ext., decorative wood, incl. cap & baluster									
0110	newels & spindles @ 12" O.C.									
0120	Brushwork, stain, sand, seal & varnish									
0130	First coat	1 Pord	90	.089	L.F.	.50	2.06		2.56	3.94
0140	Second coat	"	120	.067	"	.50	1.55		2.05	3.09
0150	Rough sawn wood, 42" high, 2" x 2" verticals, 6" O.C.									
0160	Brushwork, stain, each coat	1 Pord	90	.089	L.F.	.16	2.06		2.22	3.57
0170	Wrought iron, 1" rail, 1/2" sq. verticals									
0180	Brushwork, zinc chromate, 60" high, bars 6" O.C.									
0190	Primer	1 Pord	130	.062	L.F.	.52	1.43		1.95	2.92
0200	Finish coat		130	.062		.16	1.43		1.59	2.53
0210	Additional coat	↓	190	.042	↓	.19	.98		1.17	1.82
0220	Shutters or blinds, single panel, 2' x 4', paint all sides									
0230	Brushwork, primer	1 Pord	20	.400	Ea.	.55	9.30		9.85	15.85
0240	Finish coat, exterior latex		20	.400		.49	9.30		9.79	15.80
0250	Primer & 1 coat, exterior latex		13	.615		.92	14.30		15.22	24.50
0260	Spray, primer		35	.229		.81	5.30		6.11	9.60
0270	Finish coat, exterior latex		35	.229		1.03	5.30		6.33	9.85
0280	Primer & 1 coat, exterior latex	↓	20	.400	↓	.87	9.30		10.17	16.20
0290	For louvered shutters, add				S.F.	10%				
0300	Stair stringers, exterior, metal									
0310	Roll & brush, zinc chromate, to 14", each coat	1 Pord	320	.025	L.F.	.05	.58		.63	1.01
0320	Rough sawn wood, 4" x 12"									
0330	Roll & brush, exterior latex, each coat	1 Pord	215	.037	L.F.	.07	.86		.93	1.50
0340	Trellis/lattice, 2" x 2" @ 3" O.C. with 2" x 8" supports									
0350	Spray, latex, per side, each coat	1 Pord	475	.017	S.F.	.07	.39		.46	.72
0450	Decking, Ext., sealer, alkyd, brushwork, sealer coat		1140	.007		.06	.16		.22	.34
0460	1st coat		1140	.007		.06	.16		.22	.34
0470	2nd coat		1300	.006		.04	.14		.18	.28
0500	Paint, alkyd, brushwork, primer coat		1140	.007		.07	.16		.23	.34
0510	1st coat		1140	.007		.07	.16		.23	.35
0520	2nd coat		1300	.006		.05	.14		.19	.29
0600	Sand paint, alkyd, brushwork, 1 coat	↓	150	.053	↓	.10	1.24		1.34	2.14

09 91 13.60 Siding Exterior

		Crew	Daily Output	Labor-Hours	Unit	Material	2007 Bare Costs Labor	2007 Bare Costs Equipment	Total	Total Incl O&P
0010	**SIDING EXTERIOR**, Alkyd (oil base)									
0450	Steel siding, oil base, paint 1 coat, brushwork	2 Pord	2015	.008	S.F.	.06	.18		.24	.36
0500	Spray		4550	.004		.09	.08		.17	.23
0800	Paint 2 coats, brushwork		1300	.012		.11	.29		.40	.60
1000	Spray		4550	.004		.15	.08		.23	.30
1200	Stucco, rough, oil base, paint 2 coats, brushwork		1300	.012		.11	.29		.40	.60
1400	Roller		1625	.010		.12	.23		.35	.51
1600	Spray		2925	.005		.13	.13		.26	.35
1800	Texture 1-11 or clapboard, oil base, primer coat, brushwork		1300	.012		.09	.29		.38	.57
2000	Spray		4550	.004		.09	.08		.17	.23
2100	Paint 1 coat, brushwork		1300	.012		.08	.29		.37	.56
2200	Spray		4550	.004		.08	.08		.16	.22
2400	Paint 2 coats, brushwork		810	.020		.17	.46		.63	.93
2600	Spray		2600	.006		.19	.14		.33	.43
3000	Stain 1 coat, brushwork		1520	.011		.05	.24		.29	.46
3200	Spray		5320	.003		.06	.07		.13	.17
3400	Stain 2 coats, brushwork	↓	950	.017	↓	.11	.39		.50	.76

09 91 Painting

09 91 13 – Exterior Painting

09 91 13.60 Siding Exterior

		Crew	Daily Output	Labor-Hours	Unit	Material	2007 Bare Costs Labor	Equipment	Total	Total Incl O&P
4000	Spray	2 Pord	3050	.005	S.F.	.12	.12		.24	.33
4200	Wood shingles, oil base primer coat, brushwork		1300	.012		.08	.29		.37	.56
4400	Spray		3900	.004		.07	.10		.17	.24
4600	Paint 1 coat, brushwork		1300	.012		.07	.29		.36	.55
4800	Spray		3900	.004		.09	.10		.19	.26
5000	Paint 2 coats, brushwork		810	.020		.14	.46		.60	.90
5200	Spray		2275	.007		.13	.16		.29	.42
5800	Stain 1 coat, brushwork		1500	.011		.05	.25		.30	.47
6000	Spray		3900	.004		.05	.10		.15	.22
6500	Stain 2 coats, brushwork		950	.017		.11	.39		.50	.76
7000	Spray		2660	.006		.15	.14		.29	.39
8000	For latex paint, deduct					10%				
8100	For work over 12' H, from pipe scaffolding, add						15%			
8200	For work over 12' H, from extension ladder, add						25%			
8300	For work over 12' H, from swing staging, add						35%			

09 91 13.62 Siding, Misc.

		Crew	Daily Output	Labor-Hours	Unit	Material	2007 Bare Costs Labor	Equipment	Total	Total Incl O&P
0010	**SIDING, MISC.** R099100-20									
0100	Aluminum siding									
0110	Brushwork, primer	2 Pord	2275	.007	S.F.	.05	.16		.21	.33
0120	Finish coat, exterior latex		2275	.007		.05	.16		.21	.32
0130	Primer & 1 coat exterior latex		1300	.012		.11	.29		.40	.59
0140	Primer & 2 coats exterior latex		975	.016		.15	.38		.53	.80
0150	Mineral Fiber shingles									
0160	Brushwork, primer	2 Pord	1495	.011	S.F.	.09	.25		.34	.51
0170	Finish coat, industrial enamel		1495	.011		.10	.25		.35	.52
0180	Primer & 1 coat enamel		810	.020		.19	.46		.65	.96
0190	Primer & 2 coats enamel		540	.030		.29	.69		.98	1.45
0200	Roll, primer		1625	.010		.10	.23		.33	.49
0210	Finish coat, industrial enamel		1625	.010		.11	.23		.34	.50
0220	Primer & 1 coat enamel		975	.016		.21	.38		.59	.86
0230	Primer & 2 coats enamel		650	.025		.32	.57		.89	1.29
0240	Spray, primer		3900	.004		.07	.10		.17	.24
0250	Finish coat, industrial enamel		3900	.004		.09	.10		.19	.26
0260	Primer & 1 coat enamel		2275	.007		.17	.16		.33	.45
0270	Primer & 2 coats enamel		1625	.010		.26	.23		.49	.67
0280	Waterproof sealer, first coat		4485	.004		.07	.08		.15	.22
0290	Second coat		5235	.003		.07	.07		.14	.19
0300	Rough wood incl. shingles, shakes or rough sawn siding									
0310	Brushwork, primer	2 Pord	1280	.013	S.F.	.11	.29		.40	.60
0320	Finish coat, exterior latex		1280	.013		.08	.29		.37	.57
0330	Primer & 1 coat exterior latex		960	.017		.20	.39		.59	.86
0340	Primer & 2 coats exterior latex		700	.023		.28	.53		.81	1.18
0350	Roll, primer		2925	.005		.15	.13		.28	.37
0360	Finish coat, exterior latex		2925	.005		.10	.13		.23	.32
0370	Primer & 1 coat exterior latex		1790	.009		.25	.21		.46	.62
0380	Primer & 2 coats exterior latex		1300	.012		.35	.29		.64	.86
0390	Spray, primer		3900	.004		.13	.10		.23	.30
0400	Finish coat, exterior latex		3900	.004		.08	.10		.18	.25
0410	Primer & 1 coat exterior latex		2600	.006		.20	.14		.34	.45
0420	Primer & 2 coats exterior latex		2080	.008		.28	.18		.46	.60
0430	Waterproof sealer, first coat		4485	.004		.12	.08		.20	.28
0440	Second coat		4485	.004		.07	.08		.15	.22

09 91 Painting

09 91 13 – Exterior Painting

09 91 13.62 Siding, Misc.

		Crew	Daily Output	Labor-Hours	Unit	Material	2007 Bare Costs Labor	Equipment	Total	Total Incl O&P
0450	Smooth wood incl. butt, T&G, beveled, drop or B&B siding									
0460	Brushwork, primer	2 Pord	2325	.007	S.F.	.08	.16		.24	.35
0470	Finish coat, exterior latex		1280	.013		.08	.29		.37	.57
0480	Primer & 1 coat exterior latex		800	.020		.16	.46		.62	.94
0490	Primer & 2 coats exterior latex		630	.025		.25	.59		.84	1.24
0500	Roll, primer		2275	.007		.09	.16		.25	.37
0510	Finish coat, exterior latex		2275	.007		.09	.16		.25	.37
0520	Primer & 1 coat exterior latex		1300	.012		.18	.29		.47	.67
0530	Primer & 2 coats exterior latex		975	.016		.27	.38		.65	.93
0540	Spray, primer		4550	.004		.07	.08		.15	.21
0550	Finish coat, exterior latex		4550	.004		.08	.08		.16	.22
0560	Primer & 1 coat exterior latex		2600	.006		.15	.14		.29	.39
0570	Primer & 2 coats exterior latex		1950	.008		.23	.19		.42	.56
0580	Waterproof sealer, first coat		5230	.003		.07	.07		.14	.20
0590	Second coat	▼	5980	.003	▼	.07	.06		.13	.18
0600	For oil base paint, add					10%				

09 91 13.70 Doors and Windows, Exterior

		Crew	Daily Output	Labor-Hours	Unit	Material	2007 Bare Costs Labor	Equipment	Total	Total Incl O&P
0010	**DOORS AND WINDOWS, EXTERIOR** R099100-20									
0100	Door frames & trim, only									
0110	Brushwork, primer	1 Pord	512	.016	L.F.	.05	.36		.41	.66
0120	Finish coat, exterior latex		512	.016		.06	.36		.42	.67
0130	Primer & 1 coat, exterior latex		300	.027		.11	.62		.73	1.14
0140	Primer & 2 coats, exterior latex	▼	265	.030	▼	.17	.70		.87	1.34
0150	Doors, flush, both sides, incl. frame & trim									
0160	Roll & brush, primer	1 Pord	10	.800	Ea.	3.83	18.55		22.38	34.50
0170	Finish coat, exterior latex		10	.800		4.63	18.55		23.18	35.50
0180	Primer & 1 coat, exterior latex		7	1.143		8.45	26.50		34.95	53
0190	Primer & 2 coats, exterior latex		5	1.600		13.10	37		50.10	75.50
0200	Brushwork, stain, sealer & 2 coats polyurethane	▼	4	2	▼	17.50	46.50		64	96
0210	Doors, French, both sides, 10-15 lite, incl. frame & trim									
0220	Brushwork, primer	1 Pord	6	1.333	Ea.	1.92	31		32.92	53
0230	Finish coat, exterior latex		6	1.333		2.31	31		33.31	53.50
0240	Primer & 1 coat, exterior latex		3	2.667		4.23	62		66.23	107
0250	Primer & 2 coats, exterior latex		2	4		6.40	93		99.40	160
0260	Brushwork, stain, sealer & 2 coats polyurethane	▼	2.50	3.200	▼	6.35	74		80.35	129
0270	Doors, louvered, both sides, incl. frame & trim									
0280	Brushwork, primer	1 Pord	7	1.143	Ea.	3.83	26.50		30.33	47.50
0290	Finish coat, exterior latex		7	1.143		4.63	26.50		31.13	48.50
0300	Primer & 1 coat, exterior latex		4	2		8.45	46.50		54.95	86
0310	Primer & 2 coats, exterior latex		3	2.667		12.80	62		74.80	116
0320	Brushwork, stain, sealer & 2 coats polyurethane	▼	4.50	1.778	▼	17.50	41		58.50	87.50
0330	Doors, panel, both sides, incl. frame & trim									
0340	Roll & brush, primer	1 Pord	6	1.333	Ea.	3.83	31		34.83	55
0350	Finish coat, exterior latex		6	1.333		4.63	31		35.63	56
0360	Primer & 1 coat, exterior latex		3	2.667		8.45	62		70.45	111
0370	Primer & 2 coats, exterior latex		2.50	3.200		12.80	74		86.80	136
0380	Brushwork, stain, sealer & 2 coats polyurethane	▼	3	2.667	▼	17.50	62		79.50	121
0400	Windows, per ext. side, based on 15 SF									
0410	1 to 6 lite									
0420	Brushwork, primer	1 Pord	13	.615	Ea.	.76	14.30		15.06	24.50
0430	Finish coat, exterior latex		13	.615		.91	14.30		15.21	24.50
0440	Primer & 1 coat, exterior latex	▼	8	1	▼	1.67	23		24.67	40

09 91 Painting

09 91 13 – Exterior Painting

09 91 13.70 Doors and Windows, Exterior		Crew	Daily Output	Labor-Hours	Unit	Material	2007 Bare Costs Labor	Equipment	Total	Total Incl O&P
0450	Primer & 2 coats, exterior latex	1 Pord	6	1.333	Ea.	2.53	31		33.53	54
0460	Stain, sealer & 1 coat varnish	↓	7	1.143	↓	2.50	26.50		29	46.50
0470	7 to 10 lite									
0480	Brushwork, primer	1 Pord	11	.727	Ea.	.76	16.85		17.61	29
0490	Finish coat, exterior latex		11	.727		.91	16.85		17.76	29
0500	Primer & 1 coat, exterior latex		7	1.143		1.67	26.50		28.17	45.50
0510	Primer & 2 coats, exterior latex		5	1.600		2.53	37		39.53	64
0520	Stain, sealer & 1 coat varnish	↓	6	1.333		2.50	31		33.50	54
0530	12 lite									
0540	Brushwork, primer	1 Pord	10	.800	Ea.	.76	18.55		19.31	31.50
0550	Finish coat, exterior latex		10	.800		.91	18.55		19.46	31.50
0560	Primer & 1 coat, exterior latex		6	1.333		1.67	31		32.67	53
0570	Primer & 2 coats, exterior latex		5	1.600		2.53	37		39.53	64
0580	Stain, sealer & 1 coat varnish	↓	6	1.333		2.49	31		33.49	53.50
0590	For oil base paint, add				↓	10%				

09 91 13.80 Trim, Exterior		Crew	Daily Output	Labor-Hours	Unit	Material	Labor	Equipment	Total	Total Incl O&P
0010	**TRIM, EXTERIOR**	R099100-20								
0100	Door frames & trim (see Doors, interior or exterior)									
0110	Fascia, latex paint, one coat coverage									
0120	1" x 4", brushwork	1 Pord	640	.013	L.F.	.02	.29		.31	.50
0130	Roll		1280	.006		.02	.15		.17	.26
0140	Spray		2080	.004		.02	.09		.11	.17
0150	1" x 6" to 1" x 10", brushwork		640	.013		.06	.29		.35	.55
0160	Roll		1230	.007		.07	.15		.22	.32
0170	Spray		2100	.004		.05	.09		.14	.21
0180	1" x 12", brushwork		640	.013		.06	.29		.35	.55
0190	Roll		1050	.008		.07	.18		.25	.36
0200	Spray	↓	2200	.004	↓	.05	.08		.13	.20
0210	Gutters & downspouts, metal, zinc chromate paint									
0220	Brushwork, gutters, 5", first coat	1 Pord	640	.013	L.F.	.06	.29		.35	.54
0230	Second coat		960	.008		.05	.19		.24	.38
0240	Third coat		1280	.006		.04	.15		.19	.29
0250	Downspouts, 4", first coat		640	.013		.06	.29		.35	.54
0260	Second coat		960	.008		.05	.19		.24	.38
0270	Third coat	↓	1280	.006	↓	.04	.15		.19	.29
0280	Gutters & downspouts, wood									
0290	Brushwork, gutters, 5", primer	1 Pord	640	.013	L.F.	.05	.29		.34	.54
0300	Finish coat, exterior latex		640	.013		.05	.29		.34	.54
0310	Primer & 1 coat exterior latex		400	.020		.11	.46		.57	.88
0320	Primer & 2 coats exterior latex		325	.025		.17	.57		.74	1.13
0330	Downspouts, 4", primer		640	.013		.05	.29		.34	.54
0340	Finish coat, exterior latex		640	.013		.05	.29		.34	.54
0350	Primer & 1 coat exterior latex		400	.020		.11	.46		.57	.88
0360	Primer & 2 coats exterior latex	↓	325	.025		.09	.57		.66	1.03
0370	Molding, exterior, up to 14" wide									
0380	Brushwork, primer	1 Pord	640	.013	L.F.	.06	.29		.35	.55
0390	Finish coat, exterior latex		640	.013		.07	.29		.36	.55
0400	Primer & 1 coat exterior latex		400	.020		.13	.46		.59	.91
0410	Primer & 2 coats exterior latex		315	.025		.13	.59		.72	1.12
0420	Stain & fill		1050	.008		.06	.18		.24	.36
0430	Shellac		1850	.004		.06	.10		.16	.23
0440	Varnish	↓	1275	.006	↓	.07	.15		.22	.32

09 91 Painting

09 91 13 – Exterior Painting

09 91 13.90 Walls, Masonry (CMU), Exterior

		Crew	Daily Output	Labor-Hours	Unit	Material	2007 Bare Costs Labor	2007 Bare Costs Equipment	Total	Total Incl O&P
0350	**WALLS, MASONRY (CMU), EXTERIOR**									
0360	Concrete masonry units (CMU), smooth surface									
0370	Brushwork, latex, first coat	1 Pord	640	.013	S.F.	.05	.29		.34	.54
0380	Second coat		960	.008		.04	.19		.23	.37
0390	Waterproof sealer, first coat		736	.011		.24	.25		.49	.67
0400	Second coat		1104	.007		.24	.17		.41	.54
0410	Roll, latex, paint, first coat		1465	.005		.06	.13		.19	.28
0420	Second coat		1790	.004		.05	.10		.15	.22
0430	Waterproof sealer, first coat		1680	.005		.24	.11		.35	.44
0440	Second coat		2060	.004		.24	.09		.33	.41
0450	Spray, latex, paint, first coat		1950	.004		.05	.10		.15	.22
0460	Second coat		2600	.003		.04	.07		.11	.16
0470	Waterproof sealer, first coat		2245	.004		.24	.08		.32	.40
0480	Second coat		2990	.003		.24	.06		.30	.36
0490	Concrete masonry unit (CMU), porous									
0500	Brushwork, latex, first coat	1 Pord	640	.013	S.F.	.11	.29		.40	.60
0510	Second coat		960	.008		.05	.19		.24	.38
0520	Waterproof sealer, first coat		736	.011		.24	.25		.49	.67
0530	Second coat		1104	.007		.24	.17		.41	.54
0540	Roll latex, first coat		1465	.005		.08	.13		.21	.30
0550	Second coat		1790	.004		.05	.10		.15	.23
0560	Waterproof sealer, first coat		1680	.005		.24	.11		.35	.44
0570	Second coat		2060	.004		.24	.09		.33	.41
0580	Spray latex, first coat		1950	.004		.06	.10		.16	.22
0590	Second coat		2600	.003		.04	.07		.11	.16
0600	Waterproof sealer, first coat		2245	.004		.24	.08		.32	.40
0610	Second coat		2990	.003		.24	.06		.30	.36

09 91 23 – Interior Painting

09 91 23.20 Cabinets and Casework

		Crew	Daily Output	Labor-Hours	Unit	Material	2007 Bare Costs Labor	2007 Bare Costs Equipment	Total	Total Incl O&P
0010	**CABINETS AND CASEWORK**									
1000	Primer coat, oil base, brushwork	1 Pord	650	.012	S.F.	.05	.29		.34	.53
2000	Paint, oil base, brushwork, 1 coat		650	.012		.07	.29		.36	.54
2500	2 coats		400	.020		.13	.46		.59	.90
3000	Stain, brushwork, wipe off		650	.012		.05	.29		.34	.53
4000	Shellac, 1 coat, brushwork		650	.012		.05	.29		.34	.53
4500	Varnish, 3 coats, brushwork, sand after 1st coat		325	.025		.18	.57		.75	1.13
5000	For latex paint, deduct					10%				

09 91 23.33 Doors and Windows, Interior Alkyd (Oil Base)

		Crew	Daily Output	Labor-Hours	Unit	Material	2007 Bare Costs Labor	2007 Bare Costs Equipment	Total	Total Incl O&P
0010	**DOORS AND WINDOWS, INTERIOR ALKYD (OIL BASE)**									
0500	Flush door & frame, 3' x 7', oil, primer, brushwork	1 Pord	10	.800	Ea.	2.16	18.55		20.71	33
1000	Paint, 1 coat		10	.800		2.38	18.55		20.93	33
1200	2 coats		6	1.333		3.86	31		34.86	55.50
1400	Stain, brushwork, wipe off		18	.444		1.12	10.30		11.42	18.20
1600	Shellac, 1 coat, brushwork		25	.320		1.13	7.40		8.53	13.45
1800	Varnish, 3 coats, brushwork, sand after 1st coat		9	.889		3.69	20.50		24.19	38
2000	Panel door & frame, 3' x 7', oil, primer, brushwork		6	1.333		2.05	31		33.05	53.50
2200	Paint, 1 coat		6	1.333		2.38	31		33.38	53.50
2400	2 coats		3	2.667		6.80	62		68.80	110
2600	Stain, brushwork, panel door, 3' x 7', not incl. frame		16	.500		1.12	11.60		12.72	20.50
2800	Shellac, 1 coat, brushwork		22	.364		1.13	8.45		9.58	15.10
3000	Varnish, 3 coats, brushwork, sand after 1st coat		7.50	1.067		3.69	25		28.69	44.50

09 91 Painting

09 91 23 – Interior Painting

09 91 23.33 Doors and Windows, Interior Alkyd (Oil Base)

		Crew	Daily Output	Labor-Hours	Unit	Material	2007 Bare Costs Labor	Equipment	Total	Total Incl O&P
3020	French door, incl. 3' x 7', 6 lites, frame & trim									
3022	Paint, 1 coat, over existing paint	1 Pord	5	1.600	Ea.	4.77	37		41.77	66.50
3024	2 coats, over existing paint		5	1.600		9.25	37		46.25	71
3026	Primer & 1 coat		3.50	2.286		8.85	53		61.85	97
3028	Primer & 2 coats		3	2.667		13.65	62		75.65	117
3032	Varnish or polyurethane, 1 coat		5	1.600		4.80	37		41.80	66.50
3034	2 coats, sanding between		3	2.667		9.60	62		71.60	113
4400	Windows, including frame and trim, per side									
4600	Colonial type, 6/6 lites, 2' x 3', oil, primer, brushwork	1 Pord	14	.571	Ea.	.32	13.25		13.57	22.50
5800	Paint, 1 coat		14	.571		.38	13.25		13.63	22.50
6000	2 coats		9	.889		.73	20.50		21.23	35
6200	3' x 5' opening, 6/6 lites, primer coat, brushwork		12	.667		.81	15.45		16.26	26.50
6400	Paint, 1 coat		12	.667		.94	15.45		16.39	26.50
6600	2 coats		7	1.143		1.83	26.50		28.33	45.50
6800	4' x 8' opening, 6/6 lites, primer coat, brushwork		8	1		1.73	23		24.73	40
7000	Paint, 1 coat		8	1		2.01	23		25.01	40
7200	2 coats		5	1.600		3.90	37		40.90	65.50
8000	Single lite type, 2' x 3', oil base, primer coat, brushwork		33	.242		.32	5.60		5.92	9.60
8200	Paint, 1 coat		33	.242		.38	5.60		5.98	9.65
8400	2 coats		20	.400		.73	9.30		10.03	16.05
8600	3' x 5' opening, primer coat, brushwork		20	.400		.81	9.30		10.11	16.15
8800	Paint, 1 coat		20	.400		.94	9.30		10.24	16.30
8900	2 coats		13	.615		1.83	14.30		16.13	25.50
9200	4' x 8' opening, primer coat, brushwork		14	.571		1.73	13.25		14.98	24
9400	Paint, 1 coat		14	.571		2.01	13.25		15.26	24
9600	2 coats		8	1		3.90	23		26.90	42.50

09 91 23.35 Doors and Windows, Interior Latex

		Crew	Daily Output	Labor-Hours	Unit	Material	2007 Bare Costs Labor	Equipment	Total	Total Incl O&P
0010	**DOORS & WINDOWS, INTERIOR LATEX** R099100-20									
0100	Doors flush, both sides, incl. frame & trim									
0110	Roll & brush, primer	1 Pord	10	.800	Ea.	3.95	18.55		22.50	35
0120	Finish coat, latex		10	.800		4.43	18.55		22.98	35.50
0130	Primer & 1 coat latex		7	1.143		8.40	26.50		34.90	52.50
0140	Primer & 2 coats latex		5	1.600		12.55	37		49.55	75
0160	Spray, both sides, primer		20	.400		4.16	9.30		13.46	19.80
0170	Finish coat, latex		20	.400		4.65	9.30		13.95	20.50
0180	Primer & 1 coat latex		11	.727		8.85	16.85		25.70	38
0190	Primer & 2 coats latex		8	1		13.30	23		36.30	52.50
0200	Doors, French, both sides, 10-15 lite, incl. frame & trim									
0210	Roll & brush, primer	1 Pord	6	1.333	Ea.	1.98	31		32.98	53
0220	Finish coat, latex		6	1.333		2.22	31		33.22	53.50
0230	Primer & 1 coat latex		3	2.667		4.19	62		66.19	107
0240	Primer & 2 coats latex		2	4		6.30	93		99.30	160
0260	Doors, louvered, both sides, incl. frame & trim									
0270	Roll & brush, primer	1 Pord	7	1.143	Ea.	3.95	26.50		30.45	48
0280	Finish coat, latex		7	1.143		4.43	26.50		30.93	48.50
0290	Primer & 1 coat, latex		4	2		8.15	46.50		54.65	85.50
0300	Primer & 2 coats, latex		3	2.667		12.80	62		74.80	116
0320	Spray, both sides, primer		20	.400		4.16	9.30		13.46	19.80
0330	Finish coat, latex		20	.400		4.65	9.30		13.95	20.50
0340	Primer & 1 coat, latex		11	.727		8.85	16.85		25.70	38
0350	Primer & 2 coats, latex		8	1		13.55	23		36.55	53
0360	Doors, panel, both sides, incl. frame & trim									

09 91 Painting

09 91 23 – Interior Painting

09 91 23.35 Doors and Windows, Interior Latex

		Crew	Daily Output	Labor-Hours	Unit	Material	2007 Bare Costs Labor	2007 Bare Costs Equipment	Total	Total Incl O&P
0370	Roll & brush, primer	1 Pord	6	1.333	Ea.	4.16	31		35.16	55.50
0380	Finish coat, latex		6	1.333		4.43	31		35.43	56
0390	Primer & 1 coat, latex		3	2.667		8.40	62		70.40	111
0400	Primer & 2 coats, latex		2.50	3.200		12.80	74		86.80	136
0420	Spray, both sides, primer		10	.800		4.16	18.55		22.71	35
0430	Finish coat, latex		10	.800		4.65	18.55		23.20	35.50
0440	Primer & 1 coat, latex		5	1.600		8.85	37		45.85	71
0450	Primer & 2 coats, latex		4	2		13.55	46.50		60.05	91.50
0460	Windows, per interior side, based on 15 SF									
0470	1 to 6 lite									
0480	Brushwork, primer	1 Pord	13	.615	Ea.	.78	14.30		15.08	24.50
0490	Finish coat, enamel		13	.615		.87	14.30		15.17	24.50
0500	Primer & 1 coat enamel		8	1		1.65	23		24.65	40
0510	Primer & 2 coats enamel		6	1.333		2.53	31		33.53	54
0530	7 to 10 lite									
0540	Brushwork, primer	1 Pord	11	.727	Ea.	.78	16.85		17.63	29
0550	Finish coat, enamel		11	.727		.87	16.85		17.72	29
0560	Primer & 1 coat enamel		7	1.143		1.65	26.50		28.15	45.50
0570	Primer & 2 coats enamel		5	1.600		2.53	37		39.53	64
0590	12 lite									
0600	Brushwork, primer	1 Pord	10	.800	Ea.	.78	18.55		19.33	31.50
0610	Finish coat, enamel		10	.800		.87	18.55		19.42	31.50
0620	Primer & 1 coat enamel		6	1.333		1.65	31		32.65	53
0630	Primer & 2 coats enamel		5	1.600		2.53	37		39.53	64
0650	For oil base paint, add					10%				

09 91 23.40 Floors, Interior

		Crew	Daily Output	Labor-Hours	Unit	Material	Labor	Equipment	Total	Total Incl O&P
0010	**FLOORS, INTERIOR**									
0100	Concrete paint									
0110	Brushwork, latex, block filler									
0120	1st coat	1 Pord	975	.008	S.F.	.13	.19		.32	.46
0130	2nd coat		1150	.007		.09	.16		.25	.37
0140	3rd coat		1300	.006		.07	.14		.21	.31
0150	Roll, latex, block filler									
0160	1st coat	1 Pord	2600	.003	S.F.	.18	.07		.25	.32
0170	2nd coat		3250	.002		.11	.06		.17	.21
0180	3rd coat		3900	.002		.08	.05		.13	.17
0190	Spray, latex, block filler									
0200	1st coat	1 Pord	2600	.003	S.F.	.15	.07		.22	.29
0210	2nd coat		3250	.002		.08	.06		.14	.18
0220	3rd coat		3900	.002		.07	.05		.12	.15
0300	Acid stain and sealer									
0310	Stain, one coat	1 Pord	650	.012	S.F.	.11	.29		.40	.59
0320	Two coats		570	.014		.22	.33		.55	.78
0330	Acrylic sealer, one coat		2600	.003		.15	.07		.22	.29
0340	Two coats		1400	.006		.30	.13		.43	.55

09 91 23.52 Miscellaneous, Interior

		Crew	Daily Output	Labor-Hours	Unit	Material	Labor	Equipment	Total	Total Incl O&P
0010	**MISCELLANEOUS, INTERIOR**									
2400	Floors, conc./wood, oil base, primer/sealer coat, brushwork	2 Pord	1950	.008	S.F.	.05	.19		.24	.37
2450	Roller		5200	.003		.06	.07		.13	.18
2600	Spray		6000	.003		.06	.06		.12	.16
2650	Paint 1 coat, brushwork		1950	.008		.06	.19		.25	.38
2800	Roller		5200	.003		.06	.07		.13	.19

09 91 Painting

09 91 23 – Interior Painting

09 91 23.52 Miscellaneous, Interior		Crew	Daily Output	Labor-Hours	Unit	Material	2007 Bare Costs Labor	Equipment	Total	Total Incl O&P
2850	Spray	2 Pord	6000	.003	S.F.	.07	.06		.13	.17
3000	Stain, wood floor, brushwork, 1 coat		4550	.004		.05	.08		.13	.19
3200	Roller		5200	.003		.06	.07		.13	.18
3250	Spray		6000	.003		.06	.06		.12	.16
3400	Varnish, wood floor, brushwork		4550	.004		.06	.08		.14	.19
3450	Roller		5200	.003		.06	.07		.13	.19
3600	Spray	↓	6000	.003		.07	.06		.13	.17
3800	Grilles, per side, oil base, primer coat, brushwork	1 Pord	520	.015		.11	.36		.47	.71
3850	Spray		1140	.007		.11	.16		.27	.39
3880	Paint 1 coat, brushwork		520	.015		.13	.36		.49	.73
3900	Spray		1140	.007		.14	.16		.30	.42
3920	Paint 2 coats, brushwork		325	.025		.24	.57		.81	1.21
3940	Spray	↓	650	.012	↓	.28	.29		.57	.78
4250	Paint 1 coat, brushwork	2 Pord	1300	.012	L.F.	.11	.29		.40	.59
4500	Louvers, one side, primer, brushwork	1 Pord	524	.015	S.F.	.05	.35		.40	.64
4520	Paint one coat, brushwork		520	.015		.05	.36		.41	.65
4530	Spray		1140	.007		.06	.16		.22	.33
4540	Paint two coats, brushwork		325	.025		.10	.57		.67	1.05
4550	Spray		650	.012		.11	.29		.40	.60
4560	Paint three coats, brushwork		270	.030		.15	.69		.84	1.30
4570	Spray	↓	500	.016	↓	.17	.37		.54	.80
5000	Pipe, thru 4" diameter, primer or sealer coat, oil base, brushwork	2 Pord	1250	.013	L.F.	.06	.30		.36	.55
5100	Spray		2165	.007		.05	.17		.22	.34
5200	Paint 1 coat, brushwork		1250	.013		.06	.30		.36	.56
5300	Spray		2165	.007		.05	.17		.22	.34
5350	Paint 2 coats, brushwork		775	.021		.11	.48		.59	.91
5400	Spray		1240	.013		.12	.30		.42	.62
5450	Tthru 8" diameter, primer or sealer coat, brushwork		620	.026		.11	.60		.71	1.11
5500	Spray		1085	.015		.19	.34		.53	.77
5550	Paint 1 coat, brushwork		620	.026		.16	.60		.76	1.16
5600	Spray		1085	.015		.18	.34		.52	.76
5650	Paint 2 coats, brushwork		385	.042		.22	.96		1.18	1.83
5700	Spray	↓	620	.026	↓	.24	.60		.84	1.24
6600	Radiators, per side, primer, brushwork	1 Pord	520	.015	S.F.	.05	.36		.41	.65
6620	Paint one coat, brushwork		520	.015		.05	.36		.41	.65
6640	Paint two coats, brushwork		340	.024		.10	.55		.65	1.01
6660	Paint three coats, brushwork	↓	283	.028	↓	.15	.66		.81	1.25
7000	Trim, wood, incl. puttying, under 6" wide									
7200	Primer coat, oil base, brushwork	1 Pord	650	.012	L.F.	.03	.29		.32	.50
7250	Paint, 1 coat, brushwork		650	.012		.03	.29		.32	.50
7400	2 coats		400	.020		.06	.46		.52	.83
7450	3 coats		325	.025		.09	.57		.66	1.04
7500	Over 6" wide, primer coat, brushwork		650	.012		.05	.29		.34	.53
7550	Paint, 1 coat, brushwork		650	.012		.06	.29		.35	.54
7600	2 coats		400	.020		.12	.46		.58	.89
7650	3 coats		325	.025	↓	.18	.57		.75	1.14
8000	Cornice, simple design, primer coat, oil base, brushwork		650	.012	S.F.	.05	.29		.34	.53
8250	Paint, 1 coat		650	.012		.06	.29		.35	.54
8300	2 coats		400	.020		.12	.46		.58	.89
8350	Ornate design, primer coat		350	.023		.05	.53		.58	.93
8400	Paint, 1 coat		350	.023		.06	.53		.59	.94
8450	2 coats		400	.020		.12	.46		.58	.89
8600	Balustrades, primer coat, oil base, brushwork	↓	520	.015	↓	.05	.36		.41	.65

09 91 Painting

09 91 23 – Interior Painting

09 91 23.52 Miscellaneous, Interior

		Crew	Daily Output	Labor-Hours	Unit	Material	2007 Bare Costs Labor	Equipment	Total	Total Incl O&P
8650	Paint, 1 coat	1 Pord	520	.015	S.F.	.06	.36		.42	.66
8700	2 coats		325	.025		.12	.57		.69	1.07
8900	Trusses and wood frames, primer coat, oil base, brushwork		800	.010		.05	.23		.28	.44
8950	Spray		1200	.007		.06	.15		.21	.31
9000	Paint 1 coat, brushwork		750	.011		.06	.25		.31	.48
9200	Spray		1200	.007		.07	.15		.22	.33
9220	Paint 2 coats, brushwork		500	.016		.12	.37		.49	.74
9240	Spray		600	.013		.14	.31		.45	.66
9260	Stain, brushwork, wipe off		600	.013		.05	.31		.36	.57
9280	Varnish, 3 coats, brushwork		275	.029		.18	.67		.85	1.30
9350	For latex paint, deduct					10%				

09 91 23.72 Walls and Ceilings, Interior

		Crew	Daily Output	Labor-Hours	Unit	Material	2007 Bare Costs Labor	Equipment	Total	Total Incl O&P
0010	**WALLS AND CEILINGS, INTERIOR**									
0100	Concrete, dry wall or plaster, oil base, primer or sealer coat									
0200	Smooth finish, brushwork	1 Pord	1150	.007	S.F.	.05	.16		.21	.33
0240	Roller		1350	.006		.05	.14		.19	.29
0280	Spray		2750	.003		.04	.07		.11	.15
0300	Sand finish, brushwork		975	.008		.05	.19		.24	.37
0340	Roller		1150	.007		.06	.16		.22	.33
0380	Spray		2275	.004		.04	.08		.12	.18
0400	Paint 1 coat, smooth finish, brushwork		1200	.007		.05	.15		.20	.31
0440	Roller		1300	.006		.05	.14		.19	.29
0480	Spray		2275	.004		.05	.08		.13	.18
0500	Sand finish, brushwork		1050	.008		.05	.18		.23	.35
0540	Roller		1600	.005		.05	.12		.17	.25
0580	Spray		2100	.004		.05	.09		.14	.20
0800	Paint 2 coats, smooth finish, brushwork		680	.012		.11	.27		.38	.57
0840	Roller		800	.010		.11	.23		.34	.50
0880	Spray		1625	.005		.09	.11		.20	.29
0900	Sand finish, brushwork		605	.013		.10	.31		.41	.61
0940	Roller		1020	.008		.11	.18		.29	.42
0980	Spray		1700	.005		.09	.11		.20	.28
1200	Paint 3 coats, smooth finish, brushwork		510	.016		.16	.36		.52	.77
1240	Roller		650	.012		.16	.29		.45	.65
1280	Spray		1625	.005		.14	.11		.25	.35
1300	Sand finish, brushwork		454	.018		.16	.41		.57	.84
1340	Roller		680	.012		.16	.27		.43	.63
1380	Spray		1133	.007		.14	.16		.30	.43
1600	Glaze coating, 2 coats, spray, clear		1200	.007		.42	.15		.57	.71
1640	Multicolor		1200	.007		.87	.15		1.02	1.20
1700	For latex paint, deduct					10%				
1800	For ceiling installations, add						25%			
2000	Masonry or concrete block, oil base, primer or sealer coat									
2100	Smooth finish, brushwork	1 Pord	1224	.007	S.F.	.07	.15		.22	.33
2180	Spray		2400	.003		.08	.08		.16	.21
2200	Sand finish, brushwork		1089	.007		.08	.17		.25	.37
2280	Spray		2400	.003		.08	.08		.16	.21
2400	Paint 1 coat, smooth finish, brushwork		1100	.007		.08	.17		.25	.37
2480	Spray		2400	.003		.08	.08		.16	.21
2500	Sand finish, brushwork		979	.008		.08	.19		.27	.40
2580	Spray		2400	.003		.08	.08		.16	.21
2800	Paint 2 coats, smooth finish, brushwork		756	.011		.16	.25		.41	.58

09 91 Painting

09 91 23 – Interior Painting

09 91 23.72 Walls and Ceilings, Interior

		Crew	Daily Output	Labor-Hours	Unit	Material	2007 Bare Costs Labor	Equipment	Total	Total Incl O&P
2880	Spray	1 Pord	1360	.006	S.F.	.15	.14		.29	.39
2900	Sand finish, brushwork		672	.012		.16	.28		.44	.63
2980	Spray		1360	.006		.15	.14		.29	.39
3200	Paint 3 coats, smooth finish, brushwork		560	.014		.25	.33		.58	.82
3280	Spray		1088	.007		.23	.17		.40	.53
3300	Sand finish, brushwork		498	.016		.25	.37		.62	.88
3380	Spray		1088	.007		.23	.17		.40	.53
3600	Glaze coating, 3 coats, spray, clear		900	.009		.60	.21		.81	1
3620	Multicolor		900	.009		1	.21		1.21	1.44
4000	Block filler, 1 coat, brushwork		425	.019		.13	.44		.57	.87
4100	Silicone, water repellent, 2 coats, spray	▼	2000	.004		.27	.09		.36	.44
4120	For latex paint, deduct					10%				
8200	For work 8 - 15' H, add						10%			
8300	For work over 15' H, add				▼		20%			

09 91 23.75 Dry Fall Painting

		Crew	Daily Output	Labor-Hours	Unit	Material	2007 Bare Costs Labor	Equipment	Total	Total Incl O&P
0010	**DRY FALL PAINTING**									
0100	Walls									
0200	Wallboard and smooth plaster, one coat, brush	1 Pord	910	.009	S.F.	.04	.20		.24	.39
0210	Roll		1560	.005		.04	.12		.16	.25
0220	Spray		2600	.003		.04	.07		.11	.17
0230	Two coats, brush		520	.015		.09	.36		.45	.69
0240	Roll		877	.009		.09	.21		.30	.45
0250	Spray		1560	.005		.09	.12		.21	.30
0260	Concrete or textured plaster, one coat, brush		747	.011		.04	.25		.29	.46
0270	Roll		1300	.006		.04	.14		.18	.28
0280	Spray		1560	.005		.04	.12		.16	.25
0290	Two coats, brush		422	.019		.09	.44		.53	.82
0300	Roll		747	.011		.09	.25		.34	.51
0310	Spray		1300	.006		.09	.14		.23	.33
0320	Concrete block, one coat, brush		747	.011		.04	.25		.29	.46
0330	Roll		1300	.006		.04	.14		.18	.28
0340	Spray		1560	.005		.04	.12		.16	.25
0350	Two coats, brush		422	.019		.09	.44		.53	.82
0360	Roll		747	.011		.09	.25		.34	.51
0370	Spray		1300	.006		.09	.14		.23	.33
0380	Wood, one coat, brush		747	.011		.04	.25		.29	.46
0390	Roll		1300	.006		.04	.14		.18	.28
0400	Spray		877	.009		.04	.21		.25	.40
0410	Two coats, brush		487	.016		.09	.38		.47	.73
0420	Roll		747	.011		.09	.25		.34	.51
0430	Spray	▼	650	.012	▼	.09	.29		.38	.57
0440	Ceilings									
0450	Wallboard and smooth plaster, one coat, brush	1 Pord	600	.013	S.F.	.04	.31		.35	.56
0460	Roll		1040	.008		.04	.18		.22	.34
0470	Spray		1560	.005		.04	.12		.16	.25
0480	Two coats, brush		341	.023		.09	.54		.63	1
0490	Roll		650	.012		.09	.29		.38	.57
0500	Spray		1300	.006		.09	.14		.23	.33
0510	Concrete or textured plaster, one coat, brush		487	.016		.04	.38		.42	.68
0520	Roll		877	.009		.04	.21		.25	.40
0530	Spray		1560	.005		.04	.12		.16	.25
0540	Two coats, brush	▼	276	.029	▼	.09	.67		.76	1.21

09 91 Painting

09 91 23 – Interior Painting

09 91 23.75 Dry Fall Painting

		Crew	Daily Output	Labor-Hours	Unit	Material	2007 Bare Costs Labor	2007 Bare Costs Equipment	Total	Total Incl O&P
0550	Roll	1 Pord	520	.015	S.F.	.09	.36		.45	.69
0560	Spray		1300	.006		.09	.14		.23	.33
0570	Structural steel, bar joists or metal deck, one coat, spray		1560	.005		.04	.12		.16	.25
0580	Two coats, spray		1040	.008		.09	.18		.27	.39

09 93 Staining and Transparent Finishing

09 93 23 – Interior Staining and Finishing

09 93 23.10 Varnish

		Crew	Daily Output	Labor-Hours	Unit	Material	Labor	Equipment	Total	Total Incl O&P
0010	**VARNISH**									
0012	1 coat + sealer, on wood trim, no sanding included	1 Pord	400	.020	S.F.	.07	.46		.53	.84
0100	Hardwood floors, 2 coats, no sanding included, roller	"	1890	.004	"	.14	.10		.24	.31

09 96 High-Performance Coatings

09 96 56 – Epoxy Coatings

09 96 56.20 Wall Coatings

		Crew	Daily Output	Labor-Hours	Unit	Material	Labor	Equipment	Total	Total Incl O&P
0010	**WALL COATINGS**									
0100	Acrylic glazed coatings, minimum	1 Pord	525	.015	S.F.	.27	.35		.62	.88
0200	Maximum		305	.026		.57	.61		1.18	1.63
0300	Epoxy coatings, minimum		525	.015		.35	.35		.70	.97
0400	Maximum		170	.047		1.07	1.09		2.16	2.98
0600	Exposed aggregate, troweled on, 1/16" to 1/4", minimum		235	.034		.53	.79		1.32	1.88
0700	Maximum (epoxy or polyacrylate)		130	.062		1.14	1.43		2.57	3.60
0900	1/2" to 5/8" aggregate, minimum		130	.062		1.06	1.43		2.49	3.52
1000	Maximum		80	.100		1.81	2.32		4.13	5.80
1200	1" aggregate size, minimum		90	.089		1.84	2.06		3.90	5.40
1300	Maximum		55	.145		2.81	3.37		6.18	8.65
1500	Exposed aggregate, sprayed on, 1/8" aggregate, minimum		295	.027		.49	.63		1.12	1.57
1600	Maximum		145	.055		.91	1.28		2.19	3.10

Division 10
Specialties

10 21 Compartments and Cubicles

10 21 16 – Shower and Dressing Compartments

10 21 16.10 Partitions, Shower		Crew	Daily Output	Labor-Hours	Unit	Material	2007 Bare Costs Labor	Equipment	Total	Total Incl O&P
0010	**PARTITIONS, SHOWER** Floor mounted, no plumbing									
0100	Cabinet, incl. base, no door, painted steel, 1" thick walls	2 Shee	5	3.200	Ea.	825	92		917	1,050
0300	With door, fiberglass		4.50	3.556		665	102		767	905
0600	Galvanized and painted steel, 1" thick walls		5	3.200		865	92		957	1,100
0800	Stall, 1" thick wall, no base, enameled steel		5	3.200		940	92		1,032	1,175
1500	Circular fiberglass, cabinet 36" diameter,		4	4		685	115		800	945
1700	One piece, 36" diameter, less door		4	4		575	115		690	830
1800	With door		3.50	4.571		945	131		1,076	1,275
2400	Glass stalls, with doors, no receptors, chrome on brass		3	5.333		1,350	153		1,503	1,750
2700	Anodized aluminum		4	4		935	115		1,050	1,225
3200	Receptors, precast terrazzo, 32" x 32"	2 Marb	14	1.143		240	29		269	310
3300	48" x 34"		9.50	1.684		390	42.50		432.50	500
3500	Plastic, simulated terrazzo receptor, 32" x 32"		14	1.143		115	29		144	175
3600	32" x 48"		12	1.333		162	34		196	234
3800	Precast concrete, colors, 32" x 32"		14	1.143		180	29		209	246
3900	48" x 48"		8	2		310	50.50		360.50	425
4100	Shower doors, economy plastic, 24" wide	1 Shee	9	.889		110	25.50		135.50	164
4200	Tempered glass door, economy		8	1		160	29		189	225
4400	Folding, tempered glass, aluminum frame		6	1.333		355	38.50		393.50	455
4700	Deluxe, tempered glass, chrome on brass frame, minimum		8	1		204	29		233	274
4800	Maximum		1	8		805	230		1,035	1,275
4850	On anodized aluminum frame, minimum		2	4		132	115		247	340
4900	Maximum		1	8		475	230		705	910
5100	Shower enclosure, tempered glass, anodized alum. frame									
5120	2 panel & door, corner unit, 32" x 32"	1 Shee	2	4	Ea.	465	115		580	705
5140	Neo-angle corner unit, 16" x 24" x 16"	"	2	4		855	115		970	1,125
5200	Shower surround, 3 wall, polypropylene, 32" x 32"	1 Carp	4	2		249	51.50		300.50	360
5220	PVC, 32" x 32"		4	2		292	51.50		343.50	405
5240	Fiberglass		4	2		340	51.50		391.50	455
5250	2 wall, polypropylene, 32" x 32"		4	2		242	51.50		293.50	355
5270	PVC		4	2		300	51.50		351.50	415
5290	Fiberglass		4	2		335	51.50		386.50	455
5300	Tub doors, tempered glass & frame, minimum	1 Shee	8	1		187	29		216	254
5400	Maximum		6	1.333		435	38.50		473.50	545
5600	Chrome plated, brass frame, minimum		8	1		247	29		276	320
5700	Maximum		6	1.333		605	38.50		643.50	730
5900	Tub/shower enclosure, temp. glass, alum. frame, minimum		2	4		335	115		450	565
6200	Maximum		1.50	5.333		695	153		848	1,025
6500	On chrome-plated brass frame, minimum		2	4		460	115		575	705
6600	Maximum		1.50	5.333		985	153		1,138	1,325
6800	Tub surround, 3 wall, polypropylene	1 Carp	4	2		196	51.50		247.50	300
6900	PVC		4	2		298	51.50		349.50	415
7000	Fiberglass, minimum		4	2		335	51.50		386.50	455
7100	Maximum		3	2.667		570	68.50		638.50	745

10 28 Toilet, Bath, and Laundry Accessories

10 28 13 – Toilet Accessories

10 28 13.13 Commercial Toilet Accessories

	10 28 13.13 Commercial Toilet Accessories	Crew	Daily Output	Labor-Hours	Unit	Material	2007 Bare Costs Labor	Equipment	Total	Total Incl O&P
0010	**COMMERCIAL TOILET ACCESSORIES**									
0200	Curtain rod, stainless steel, 5' long, 1" diameter	1 Carp	13	.615	Ea.	33	15.80		48.80	63.50
0300	1-1/4" diameter		13	.615		32	15.80		47.80	62
0800	Grab bar, straight, 1-1/4" diameter, stainless steel, 18" long		24	.333		23.50	8.55		32.05	40
1100	36" long		20	.400		24	10.30		34.30	43.50
1105	42" long		20	.400		27.50	10.30		37.80	47.50
3000	Mirror, with stainless steel 3/4" square frame, 18" x 24"		20	.400		53.50	10.30		63.80	76.50
3100	36" x 24"		15	.533		135	13.70		148.70	172
3300	72" x 24"		6	1.333		300	34.50		334.50	390
4300	Robe hook, single, regular		36	.222		12.60	5.70		18.30	23.50
4400	Heavy duty, concealed mounting		36	.222		13.85	5.70		19.55	25
6400	Towel bar, stainless steel, 18" long		23	.348		39	8.95		47.95	58
6500	30" long		21	.381		87.50	9.80		97.30	113
7400	Tumbler holder, tumbler only		30	.267		35	6.85		41.85	50
7500	Soap, tumbler & toothbrush		30	.267		25	6.85		31.85	39

10 28 16 – Bath Accessories

10 28 16.20 Medicine Cabinets

	10 28 16.20 Medicine Cabinets	Crew	Daily Output	Labor-Hours	Unit	Material	2007 Bare Costs Labor	Equipment	Total	Total Incl O&P
0010	**MEDICINE CABINETS**									
0020	With mirror, st. st. frame, 16" x 22", unlighted	1 Carp	14	.571	Ea.	76.50	14.70		91.20	110
0100	Wood frame		14	.571		106	14.70		120.70	142
0300	Sliding mirror doors, 20" x 16" x 4-3/4", unlighted		7	1.143		95	29.50		124.50	155
0400	24" x 19" x 8-1/2", lighted		5	1.600		150	41		191	235
0600	Triple door, 30" x 32", unlighted, plywood body		7	1.143		225	29.50		254.50	298
0700	Steel body		7	1.143		296	29.50		325.50	375
0900	Oak door, wood body, beveled mirror, single door		7	1.143		145	29.50		174.50	209
1000	Double door		6	1.333		350	34.50		384.50	445

10 28 23 – Laundry Accessories

10 28 23.13 Built-In Ironing Boards

	10 28 23.13 Built-In Ironing Boards	Crew	Daily Output	Labor-Hours	Unit	Material	2007 Bare Costs Labor	Equipment	Total	Total Incl O&P
0010	**BUILT-IN IRONING BOARDS**									
0020	Including cabinet, board & light, minimum	1 Carp	2	4	Ea.	297	103		400	500

10 31 Manufactured Fireplaces

10 31 13 – Manufactured Fireplace Chimneys

10 31 13.10 Fireplace Chimneys

	10 31 13.10 Fireplace Chimneys	Crew	Daily Output	Labor-Hours	Unit	Material	2007 Bare Costs Labor	Equipment	Total	Total Incl O&P
0010	**FIREPLACE CHIMNEYS**									
0500	Chimney dbl. wall, all stainless, over 8'-6", 7" diam., add to fireplace	1 Carp	33	.242	V.L.F.	60.50	6.25		66.75	77
0600	10" diameter, add to fireplace		32	.250		61.50	6.45		67.95	79
0700	12" diameter, add to fireplace		31	.258		84	6.65		90.65	104
0800	14" diameter, add to fireplace		30	.267		111	6.85		117.85	134
1000	Simulated brick chimney top, 4' high, 16" x 16"		10	.800	Ea.	226	20.50		246.50	284
1100	24" x 24"		7	1.143	"	420	29.50		449.50	515

10 31 13.20 Chimney Accessories

	10 31 13.20 Chimney Accessories	Crew	Daily Output	Labor-Hours	Unit	Material	2007 Bare Costs Labor	Equipment	Total	Total Incl O&P
0010	**CHIMNEY ACCESSORIES**									
0020	Chimney screens, galv., 13" x 13" flue	1 Bric	8	1	Ea.	48.50	26.50		75	97.50
0050	24" x 24" flue		5	1.600		112	42.50		154.50	194
0200	Stainless steel, 13" x 13" flue		8	1		296	26.50		322.50	370
0250	20" x 20" flue		5	1.600		400	42.50		442.50	515
2400	Squirrel and bird screens, galvanized, 8" x 8" flue		16	.500		42	13.35		55.35	68
2450	13" x 13" flue		12	.667		45.50	17.75		63.25	79.50

10 31 Manufactured Fireplaces

10 31 16 – Manufactured Fireplace Forms

10 31 16.10 Fireplace Forms

		Crew	Daily Output	Labor-Hours	Unit	Material	2007 Bare Costs Labor	2007 Bare Costs Equipment	Total	Total Incl O&P
0010	**FIREPLACE FORMS**									
1800	Fireplace forms, no accessories, 32" opening	1 Bric	3	2.667	Ea.	570	71		641	750
1900	36" opening		2.50	3.200		725	85.50		810.50	940
2000	40" opening		2	4		965	107		1,072	1,225
2100	78" opening	↓	1.50	5.333	↓	1,400	142		1,542	1,775

10 31 23 – Manufactured Fireplaces

10 31 23.10 Fireplace, Prefabricated

		Crew	Daily Output	Labor-Hours	Unit	Material	Labor	Equipment	Total	Total Incl O&P
0010	**FIREPLACE, PREFABRICATED**, free standing or wall hung									
0100	With hood & screen, minimum	1 Carp	1.30	6.154	Ea.	1,075	158		1,233	1,450
0150	Average		1	8		1,550	206		1,756	2,050
0200	Maximum		.90	8.889		3,825	228		4,053	4,600
1500	Simulated logs, gas fired, 40,000 BTU, 2' long, minimum		7	1.143	Set	560	29.50		589.50	665
1600	Maximum		6	1.333		775	34.50		809.50	915
1700	Electric, 1,500 BTU, 1'-6" long, minimum		7	1.143		154	29.50		183.50	219
1800	11,500 BTU, maximum		6	1.333	↓	345	34.50		379.50	440
2000	Fireplace, built-in, 36" hearth, radiant		1.30	6.154	Ea.	605	158		763	935
2100	Recirculating, small fan		1	8		965	206		1,171	1,400
2150	Large fan		.90	8.889		1,675	228		1,903	2,250
2200	42" hearth, radiant		1.20	6.667		815	171		986	1,200
2300	Recirculating, small fan		.90	8.889		1,025	228		1,253	1,550
2350	Large fan		.80	10		1,675	257		1,932	2,275
2400	48" hearth, radiant		1.10	7.273		1,575	187		1,762	2,050
2500	Recirculating, small fan		.80	10		1,950	257		2,207	2,575
2550	Large fan		.70	11.429		3,025	294		3,319	3,825
3000	See through, including doors		.80	10		2,575	257		2,832	3,250
3200	Corner (2 wall)	↓	1	8	↓	2,575	206		2,781	3,175

10 32 Fireplace Specialties

10 32 13 – Fireplace Dampers

10 32 13.10 Dampers

		Crew	Daily Output	Labor-Hours	Unit	Material	Labor	Equipment	Total	Total Incl O&P
0010	**DAMPERS**									
0800	Damper, rotary control, steel, 30" opening	1 Bric	6	1.333	Ea.	71.50	35.50		107	138
0850	Cast iron, 30" opening		6	1.333		79.50	35.50		115	147
1200	Steel plate, poker control, 60" opening		8	1		252	26.50		278.50	320
1250	84" opening, special opening		5	1.600		460	42.50		502.50	575
1400	"Universal" type, chain operated, 32" x 20" opening		8	1		196	26.50		222.50	261
1450	48" x 24" opening	↓	5	1.600	↓	292	42.50		334.50	390

10 32 23 – Fireplace Doors

10 32 23.10 Doors

		Crew	Daily Output	Labor-Hours	Unit	Material	Labor	Equipment	Total	Total Incl O&P
0010	**DOORS**									
0400	Cleanout doors and frames, cast iron, 8" x 8"	1 Bric	12	.667	Ea.	33.50	17.75		51.25	66.50
0450	12" x 12"		10	.800		41	21.50		62.50	80.50
0500	18" x 24"		8	1		118	26.50		144.50	174
0550	Cast iron frame, steel door, 24" x 30"		5	1.600		254	42.50		296.50	350
1600	Dutch Oven door and frame, cast iron, 12" x 15" opening		13	.615		103	16.40		119.40	141
1650	Copper plated, 12" x 15" opening	↓	13	.615	↓	201	16.40		217.40	249

10 35 Stoves

10 35 13 – Heating Stoves

10 35 13.10 Woodburning Stoves

		Crew	Daily Output	Labor-Hours	Unit	Material	2007 Bare Costs Labor	Equipment	Total	Total Incl O&P
0010	**WOODBURNING STOVES**									
0015	Cast iron, minimum	2 Carp	1.30	12.308	Ea.	1,100	315		1,415	1,725
0020	Average	↓	1	16		1,600	410		2,010	2,450
0030	Maximum	↓	.80	20	↓	2,650	515		3,165	3,800
0050	For gas log lighter, add					45.50			45.50	50

10 44 Fire Protection Specialties

10 44 16 – Fire Extinguishers

10 44 16.13 Portable Fire Extinguishers

		Crew	Daily Output	Labor-Hours	Unit	Material	2007 Bare Costs Labor	Equipment	Total	Total Incl O&P
0010	**PORTABLE FIRE EXTINGUISHERS**									
0120	CO_2, portable with swivel horn, 5 lb.				Ea.	130			130	143
0140	With hose and "H" horn, 10 lb.				"	160			160	176
1000	Dry chemical, pressurized									
1040	Standard type, portable, painted, 2-1/2 lb.				Ea.	27.50			27.50	30.50
1080	10 lb.					80			80	88
1100	20 lb.					110			110	121
1120	30 lb.					209			209	230
2000	ABC all purpose type, portable, 2-1/2 lb.					30			30	33
2080	9-1/2 lb.				↓	68			68	75

10 55 Postal Specialties

10 55 23 – Mail Boxes

10 55 23.10 Mail Boxes

		Crew	Daily Output	Labor-Hours	Unit	Material	2007 Bare Costs Labor	Equipment	Total	Total Incl O&P
0011	**MAIL BOXES**									
1900	Letter slot, residential	1 Carp	20	.400	Ea.	64.50	10.30		74.80	88.50
2400	Residential, galv. steel, small 20" x 7" x 9"	1 Clab	16	.500		109	9.35		118.35	136
2410	With galv. steel post, 54" long		6	1.333		209	25		234	273
2420	Large, 24" x 12" x 15"		16	.500		134	9.35		143.35	163
2430	With galv. steel post, 54" long		6	1.333		241	25		266	310
2440	Decorative, polyethylene, 22" x 10" x 10"		16	.500		43	9.35		52.35	63
2450	With alum. post, decorative, 54" long	↓	6	1.333	↓	75	25		100	125

10 56 Storage Assemblies

10 56 13 – Metal Storage Shelving

10 56 13.10 Shelving

		Crew	Daily Output	Labor-Hours	Unit	Material	2007 Bare Costs Labor	Equipment	Total	Total Incl O&P
0010	**SHELVING**									
0020	Metal, industrial, cross-braced, 3' wide, 12" deep	1 Sswk	175	.046	SF Shlf	9	1.27		10.27	12.35
0100	24" deep		330	.024		6.20	.67		6.87	8.10
2200	Wide span, 1600 lb. capacity per shelf, 6' wide, 24" deep		380	.021		9	.58		9.58	11.05
2400	36" deep	↓	440	.018	↓	7.90	.50		8.40	9.65
3000	Residential, vinyl covered wire, wardrobe, 12" deep	1 Carp	195	.041	L.F.	8.95	1.05		10	11.65
3100	16" deep		195	.041		5.70	1.05		6.75	8.10
3200	Standard, 6" deep		195	.041		3.14	1.05		4.19	5.25
3300	9" deep		195	.041		4.72	1.05		5.77	7
3400	12" deep		195	.041		6.30	1.05		7.35	8.70
3500	16" deep		195	.041		10.20	1.05		11.25	13.05
3600	20" deep	↓	195	.041		12.75	1.05		13.80	15.85

10 56 Storage Assemblies

10 56 13 – Metal Storage Shelving

10 56 13.10 Shelving		Crew	Daily Output	Labor-Hours	Unit	Material	2007 Bare Costs Labor	Equipment	Total	Total Incl O&P
3700	Support bracket	1 Carp	80	.100	Ea.	4	2.57		6.57	8.75

10 57 Wardrobe and Closet Specialties

10 57 23 – Closet and Utility Shelving

10 57 23.19 Wood Closet and Utility Shelving

		Crew	Daily Output	Labor-Hours	Unit	Material	Labor	Equipment	Total	Total Incl O&P
0010	**WOOD CLOSET AND UTILITY SHELVING**									
0020	Pine, clear grade, no edge band, 1" x 8"	1 Carp	115	.070	L.F.	2	1.79		3.79	5.25
0100	1" x 10"		110	.073		2.24	1.87		4.11	5.65
0200	1" x 12"		105	.076		2.67	1.96		4.63	6.25
0450	1" x 18"		95	.084		4.01	2.16		6.17	8.10
0460	1" x 24"		85	.094		5.35	2.42		7.77	9.95
0600	Plywood, 3/4" thick with lumber edge, 12" wide		75	.107		1.45	2.74		4.19	6.25
0700	24" wide		70	.114		2.56	2.94		5.50	7.80
0900	Bookcase, clear grade pine, shelves 12" O.C., 8" deep, /SF shelf		70	.114	S.F.	6.50	2.94		9.44	12.15
1000	12" deep shelves		65	.123	"	8.70	3.16		11.86	14.90
1200	Adjustable closet rod and shelf, 12" wide, 3' long		20	.400	Ea.	8.70	10.30		19	27
1300	8' long		15	.533	"	24	13.70		37.70	50
1500	Prefinished shelves with supports, stock, 8" wide		75	.107	L.F.	3.83	2.74		6.57	8.85
1600	10" wide		70	.114	"	4.26	2.94		7.20	9.65

10 73 Protective Covers

10 73 16 – Canopies

10 73 16.10 Canopies, Residential

		Crew	Daily Output	Labor-Hours	Unit	Material	Labor	Equipment	Total	Total Incl O&P
0010	**CANOPIES, RESIDENTIAL** Prefabricated									
0500	Carport, free standing, baked enamel, alum., .032", 40 psf									
0520	16' x 8', 4 posts	2 Carp	3	5.333	Ea.	3,600	137		3,737	4,175
0600	20' x 10', 6 posts		2	8		3,750	206		3,956	4,475
0605	30' x 10', 8 posts		2	8		5,625	206		5,831	6,525
1000	Door canopies, extruded alum., .032", 42" projection, 4' wide	1 Carp	8	1		305	25.50		330.50	380
1020	6' wide	"	6	1.333		375	34.50		409.50	475
1040	8' wide	2 Carp	9	1.778		420	45.50		465.50	545
1060	10' wide		7	2.286		650	58.50		708.50	815
1080	12' wide		5	3.200		750	82		832	965
1200	54" projection, 4' wide	1 Carp	8	1		375	25.50		400.50	460
1220	6' wide	"	6	1.333		420	34.50		454.50	520
1240	8' wide	2 Carp	9	1.778		475	45.50		520.50	605
1260	10' wide		7	2.286		700	58.50		758.50	870
1280	12' wide		5	3.200		750	82		832	965
1300	Painted, add					20%				
1310	Bronze anodized, add					50%				
3000	Window awnings, aluminum, window 3' high, 4' wide	1 Carp	10	.800		279	20.50		299.50	340
3020	6' wide	"	8	1		288	25.50		313.50	360
3040	9' wide	2 Carp	9	1.778		300	45.50		345.50	410
3060	12' wide	"	5	3.200		350	82		432	525
3100	Window, 4' high, 4' wide	1 Carp	10	.800		157	20.50		177.50	208
3120	6' wide	"	8	1		223	25.50		248.50	289
3140	9' wide	2 Carp	9	1.778		370	45.50		415.50	490
3160	12' wide	"	5	3.200		400	82		482	580
3200	Window, 6' high, 4' wide	1 Carp	10	.800		450	20.50		470.50	530

10 73 Protective Covers

10 73 16 – Canopies

10 73 16.10 Canopies, Residential

		Crew	Daily Output	Labor-Hours	Unit	Material	2007 Bare Costs Labor	2007 Bare Costs Equipment	Total	Total Incl O&P
3220	6' wide	1 Carp	8	1	Ea.	625	25.50		650.50	735
3240	9' wide	2 Carp	9	1.778		850	45.50		895.50	1,025
3260	12' wide	"	5	3.200		1,300	82		1,382	1,575
3400	Roll-up aluminum, 2'-6" wide	1 Carp	14	.571		160	14.70		174.70	201
3420	3' wide		12	.667		167	17.15		184.15	213
3440	4' wide		10	.800		189	20.50		209.50	243
3460	6' wide		8	1		233	25.50		258.50	300
3480	9' wide	2 Carp	9	1.778		335	45.50		380.50	450
3500	12' wide	"	5	3.200		410	82		492	590
3600	Window awnings, canvas, 24" drop, 3' wide	1 Carp	30	.267	L.F.	46.50	6.85		53.35	62.50
3620	4' wide		40	.200		40.50	5.15		45.65	53
3700	30" drop, 3' wide		30	.267		55.50	6.85		62.35	72.50
3720	4' wide		40	.200		47.50	5.15		52.65	60.50
3740	5' wide		45	.178		42	4.57		46.57	54.50
3760	6' wide		48	.167		39	4.28		43.28	50
3780	8' wide		48	.167		35.50	4.28		39.78	46.50
3800	10' wide		50	.160		32	4.11		36.11	42

10 74 Manufactured Exterior Specialties

10 74 23 – Cupolas

10 74 23.10 Cupolas

		Crew	Daily Output	Labor-Hours	Unit	Material	2007 Bare Costs Labor	2007 Bare Costs Equipment	Total	Total Incl O&P
0010	**CUPOLAS**									
0020	Stock units, pine, painted, 18" sq., 28" high, alum. roof	1 Carp	4.10	1.951	Ea.	158	50		208	258
0100	Copper roof		3.80	2.105		184	54		238	294
0300	23" square, 33" high, aluminum roof		3.70	2.162		345	55.50		400.50	475
0400	Copper roof		3.30	2.424		460	62.50		522.50	615
0600	30" square, 37" high, aluminum roof		3.70	2.162		485	55.50		540.50	625
0700	Copper roof		3.30	2.424		520	62.50		582.50	675
0900	Hexagonal, 31" wide, 46" high, copper roof		4	2		650	51.50		701.50	800
1000	36" wide, 50" high, copper roof		3.50	2.286		715	58.50		773.50	885
1200	For deluxe stock units, add to above					25%				
1400	For custom built units, add to above					50%	50%			

10 74 33 – Weathervanes

10 74 33.10 Weathervanes

		Crew	Daily Output	Labor-Hours	Unit	Material	2007 Bare Costs Labor	2007 Bare Costs Equipment	Total	Total Incl O&P
0010	**WEATHERVANES**									
0020	Residential types, minimum	1 Carp	8	1	Ea.	130	25.50		155.50	187
0100	Maximum	"	2	4	"	2,300	103		2,403	2,725

10 75 Flagpoles

10 75 16 – Ground-Set Flagpoles

10 75 16.10 Flagpoles

		Crew	Daily Output	Labor-Hours	Unit	Material	2007 Bare Costs Labor	2007 Bare Costs Equipment	Total	Total Incl O&P
0010	**FLAGPOLES**, Ground set									
0050	Not including base or foundation									
0100	Aluminum, tapered, ground set 20' high	K-1	2	8	Ea.	790	184	92.50	1,066.50	1,275
0200	25' high		1.70	9.412		1,025	216	109	1,350	1,600
0300	30' high		1.50	10.667		1,050	245	124	1,419	1,700
0500	40' high		1.20	13.333		2,075	305	154	2,534	2,975

Division 11
Equipment

11 24 Maintenance Equipment

11 24 19 – Vacuum Cleaning Systems

11 24 19.10 Vacuum Cleaning		Crew	Daily Output	Labor-Hours	Unit	Material	2007 Bare Costs Labor	2007 Bare Costs Equipment	Total	Total Incl O&P
0010	**VACUUM CLEANING**									
0020	Central, 3 inlet, residential	1 Skwk	.90	8.889	Total	680	232		912	1,150
0400	5 inlet system, residential		.50	16		1,025	420		1,445	1,825
0600	7 inlet system, commercial		.40	20		1,150	520		1,670	2,150
0800	9 inlet system, residential	↓	.30	26.667		1,450	695		2,145	2,775
4010	Rule of thumb: First 1200 S.F., installed									1,180
4020	For each additional S.F., add				S.F.					.19

11 26 Unit Kitchens

11 26 13 – Metal Unit Kitchens

11 26 13.10 Unit Kitchens		Crew	Daily Output	Labor-Hours	Unit	Material	2007 Bare Costs Labor	2007 Bare Costs Equipment	Total	Total Incl O&P
0010	**UNIT KITCHENS**									
1500	Combination range, refrigerator and sink, 30" wide, minimum	L-1	2	5	Ea.	840	148		988	1,175
1550	Maximum		1	10		2,675	295		2,970	3,425
1570	60" wide, average		1.40	7.143		2,225	211		2,436	2,775
1590	72" wide, average	↓	1.20	8.333	↓	3,100	246		3,346	3,825

11 31 Residential Appliances

11 31 13 – Residential Kitchen Appliances

11 31 13.13 Cooking Equipment		Crew	Daily Output	Labor-Hours	Unit	Material	2007 Bare Costs Labor	2007 Bare Costs Equipment	Total	Total Incl O&P
0010	**COOKING EQUIPMENT**									
0020	Cooking range, 30" free standing, 1 oven, minimum	2 Clab	10	1.600	Ea.	270	30		300	350
0050	Maximum		4	4		1,550	75		1,625	1,825
0150	2 oven, minimum		10	1.600		1,575	30		1,605	1,800
0200	Maximum	↓	10	1.600		1,725	30		1,755	1,950
0350	Built-in, 30" wide, 1 oven, minimum	1 Elec	6	1.333		470	39		509	580
0400	Maximum	2 Carp	2	8		1,450	206		1,656	1,950
0500	2 oven, conventional, minimum		4	4		1,300	103		1,403	1,600
0550	1 conventional, 1 microwave, maximum	↓	2	8		1,650	206		1,856	2,150
0700	Free-standing, 1 oven, 21" wide range, minimum	2 Clab	10	1.600		310	30		340	390
0750	21" wide, maximum	"	4	4		325	75		400	485
0900	Counter top cook tops, 4 burner, standard, minimum	1 Elec	6	1.333		215	39		254	300
0950	Maximum		3	2.667		585	78.50		663.50	775
1050	As above, but with grille and griddle attachment, minimum		6	1.333		520	39		559	635
1100	Maximum		3	2.667		870	78.50		948.50	1,075
1250	Microwave oven, minimum		4	2		85.50	59		144.50	190
1300	Maximum		2	4		420	118		538	650
5380	Oven, built in, standard		4	2		425	59		484	560
5390	Deluxe	↓	2	4	↓	1,875	118		1,993	2,250

11 31 13.23 Refrigeration Equipment		Crew	Daily Output	Labor-Hours	Unit	Material	2007 Bare Costs Labor	2007 Bare Costs Equipment	Total	Total Incl O&P
0010	**REFRIGERATION EQUIPMENT**									
2000	Deep freeze, 15 to 23 C.F., minimum	2 Clab	10	1.600	Ea.	475	30		505	575
2050	Maximum		5	3.200		600	60		660	760
2200	30 C.F., minimum		8	2		805	37.50		842.50	950
2250	Maximum	↓	3	5.333		915	99.50		1,014.50	1,175
5200	Icemaker, automatic, 20 lb. per day	1 Plum	7	1.143		850	34		884	990
5350	51 lb. per day	"	2	4		1,075	118		1,193	1,375
5450	Refrigerator, no frost, 6 C.F.	2 Clab	15	1.067		249	19.95		268.95	310
5500	Refrigerator, no frost, 10 C.F. to 12 C.F. minimum	↓	10	1.600		410	30		440	500

11 31 Residential Appliances

11 31 13 – Residential Kitchen Appliances

	11 31 13.23 Refrigeration Equipment	Crew	Daily Output	Labor-Hours	Unit	Material	2007 Bare Costs Labor	Equipment	Total	Total Incl O&P
5600	Maximum	2 Clab	6	2.667	Ea.	500	50		550	635
5750	14 C.F. to 16 C.F., minimum		9	1.778		445	33		478	545
5800	Maximum		5	3.200		435	60		495	580
5950	18 C.F. to 20 C.F., minimum		8	2		530	37.50		567.50	645
6000	Maximum		4	4		795	75		870	1,000
6150	21 C.F. to 29 C.F., minimum		7	2.286		705	42.50		747.50	850
6200	Maximum		3	5.333		2,250	99.50		2,349.50	2,650

11 31 13.33 Kitchen Cleaning Equipment

		Crew	Daily Output	Labor-Hours	Unit	Material	Labor	Equipment	Total	Total Incl O&P
0010	**KITCHEN CLEANING EQUIPMENT**									
2750	Dishwasher, built-in, 2 cycles, minimum	L-1	4	2.500	Ea.	268	74		342	415
2800	Maximum		2	5		310	148		458	580
2950	4 or more cycles, minimum		4	2.500		285	74		359	435
2960	Average		4	2.500		380	74		454	535
3000	Maximum		2	5		530	148		678	825

11 31 13.43 Waste Disposal Equipment

		Crew	Daily Output	Labor-Hours	Unit	Material	Labor	Equipment	Total	Total Incl O&P
0010	**WASTE DISPOSAL EQUIPMENT**									
1750	Compactor, residential size, 4 to 1 compaction, minimum	1 Carp	5	1.600	Ea.	440	41		481	555
1800	Maximum	"	3	2.667		520	68.50		588.50	690
3300	Garbage disposal, sink type, minimum	L-1	10	1		50	29.50		79.50	104
3350	Maximum	"	10	1		159	29.50		188.50	224

11 31 13.53 Kitchen Ventilation Equipment

		Crew	Daily Output	Labor-Hours	Unit	Material	Labor	Equipment	Total	Total Incl O&P
0010	**KITCHEN VENTILATION EQUIPMENT**									
4150	Hood for range, 2 speed, vented, 30" wide, minimum	L-3	5	2	Ea.	40.50	53		93.50	134
4200	Maximum		3	3.333		595	88		683	800
4300	42" wide, minimum		5	2		242	53		295	355
4330	Custom		5	2		670	53		723	825
4350	Maximum		3	3.333		820	88		908	1,050
4500	For ventless hood, 2 speed, add					16.15			16.15	17.75
4650	For vented 1 speed, deduct from maximum					42			42	46

11 31 23 – Residential Laundry Appliances

11 31 23.13 Washers

		Crew	Daily Output	Labor-Hours	Unit	Material	Labor	Equipment	Total	Total Incl O&P
0010	**WASHERS**									
6650	Washing machine, automatic, minimum	1 Plum	3	2.667	Ea.	310	79		389	470
6700	Maximum	"	1	8	"	1,100	236		1,336	1,625

11 31 23.23 Dryers

		Crew	Daily Output	Labor-Hours	Unit	Material	Labor	Equipment	Total	Total Incl O&P
0010	**DRYERS**									
3200	Dryer, automatic, minimum	L-2	3	5.333	Ea.	269	119		388	495
3250	Maximum	"	2	8		665	178		843	1,025
7450	Vent kits for dryers	1 Carp	10	.800		14.65	20.50		35.15	51

11 31 33 – Miscellaneous Residential Appliances

11 31 33.13 Sump Pumps

		Crew	Daily Output	Labor-Hours	Unit	Material	Labor	Equipment	Total	Total Incl O&P
0010	**SUMP PUMPS**									
6400	Sump pump cellar drainer, pedestal, 1/3 H.P., molded PVC base	1 Plum	3	2.667	Ea.	96	79		175	236
6450	Solid brass	"	2	4	"	201	118		319	415
6460	Sump pump, see also division 22 14 29.16									

11 31 33.23 Water Heaters

		Crew	Daily Output	Labor-Hours	Unit	Material	Labor	Equipment	Total	Total Incl O&P
0010	**WATER HEATERS**									
6900	Water heater, electric, glass lined, 30 gallon, minimum	L-1	5	2	Ea.	345	59		404	475
6950	Maximum		3	3.333		480	98.50		578.50	690
7100	80 gallon, minimum		2	5		630	148		778	930

11 31 Residential Appliances

11 31 33 – Miscellaneous Residential Appliances

11 31 33.23 Water Heaters

		Crew	Daily Output	Labor-Hours	Unit	Material	2007 Bare Costs Labor	Equipment	Total	Total Incl O&P
7150	Maximum	L-1	1	10	Ea.	875	295		1,170	1,450
7180	Water heater, gas, glass lined, 30 gallon, minimum	2 Plum	5	3.200		565	94.50		659.50	775
7220	Maximum		3	5.333		785	158		943	1,125
7260	50 gallon, minimum		2.50	6.400		595	189		784	965
7300	Maximum		1.50	10.667		830	315		1,145	1,425
7310	Water heater, see also division 22 33 30.13									

11 31 33.43 Air Quality

		Crew	Daily Output	Labor-Hours	Unit	Material	2007 Bare Costs Labor	Equipment	Total	Total Incl O&P
0010	**AIR QUALITY**									
2450	Dehumidifier, portable, automatic, 15 pint				Ea.	159			159	175
2550	40 pint					194			194	213
3550	Heater, electric, built-in, 1250 watt, ceiling type, minimum	1 Elec	4	2		74.50	59		133.50	178
3600	Maximum		3	2.667		122	78.50		200.50	262
3700	Wall type, minimum		4	2		106	59		165	212
3750	Maximum		3	2.667		140	78.50		218.50	282
3900	1500 watt wall type, with blower		4	2		131	59		190	240
3950	3000 watt		3	2.667		267	78.50		345.50	420
4850	Humidifier, portable, 8 gallons per day					160			160	176
5000	15 gallons per day					193			193	212

11 33 Retractable Stairs

11 33 10 – Disappearing Stairs

11 33 10.10 Disappearing Stairway

		Crew	Daily Output	Labor-Hours	Unit	Material	2007 Bare Costs Labor	Equipment	Total	Total Incl O&P
0010	**DISAPPEARING STAIRWAY** No trim included									
0020	One piece, yellow pine, 8'-0" ceiling	2 Carp	4	4	Ea.	1,125	103		1,228	1,425
0030	9'-0" ceiling		4	4		1,175	103		1,278	1,450
0040	10'-0" ceiling		3	5.333		1,225	137		1,362	1,575
0050	11'-0" ceiling		3	5.333		1,375	137		1,512	1,725
0060	12'-0" ceiling		3	5.333		1,475	137		1,612	1,850
0100	Custom grade, pine, 8'-6" ceiling, minimum	1 Carp	4	2		122	51.50		173.50	221
0150	Average		3.50	2.286		123	58.50		181.50	235
0200	Maximum		3	2.667		204	68.50		272.50	340
0500	Heavy duty, pivoted, from 7'-7" to 12'-10" floor to floor		3	2.667		400	68.50		468.50	555
0600	16'-0" ceiling		2	4		1,325	103		1,428	1,650
0800	Economy folding, pine, 8'-6" ceiling		4	2		111	51.50		162.50	209
0900	9'-6" ceiling		4	2		121	51.50		172.50	220
1000	Fire escape, galvanized steel, 8'-0" to 10'-4" ceiling	2 Carp	1	16		1,400	410		1,810	2,250
1010	10'-6" to 13'-6" ceiling		1	16		1,775	410		2,185	2,650
1100	Automatic electric, aluminum, floor to floor height, 8' to 9'		1	16		7,300	410		7,710	8,725

11 41 Food Storage Equipment

11 41 13 – Refrigerated Food Storage Cases

11 41 13.30 Wine Cellar		Crew	Daily Output	Labor-Hours	Unit	Material	2007 Bare Costs Labor	Equipment	Total	Total Incl O&P
0010	**WINE CELLAR**, refrigerated, Redwood interior, carpeted, walk-in type									
0020	6'-8" high, including racks									
0200	80"W x 48"D for 900 bottles	2 Carp	1.50	10.667	Ea.	3,125	274		3,399	3,900
0250	80" W x 72" D for 1300 bottles		1.33	12.030		4,125	310		4,435	5,075
0300	80" W x 94" D for 1900 bottles		1.17	13.675		5,350	350		5,700	6,500

Division 12
Furnishings

12 21 Window Blinds

12 21 13 – Horizontal Louver Blinds

12 21 13.13 Metal Horizontal Louver Blinds		Crew	Daily Output	Labor-Hours	Unit	Material	2007 Bare Costs Labor	Equipment	Total	Total Incl O&P
0010	**METAL HORIZONTAL LOUVER BLINDS**									
0020	Horizontal, 1" aluminum slats, solid color, stock	1 Carp	590	.014	S.F.	3.52	.35		3.87	4.46
0090	Custom, minimum		590	.014		3.10	.35		3.45	4
0100	Maximum		440	.018		7.50	.47		7.97	9.05
0450	Stock, minimum		590	.014		4.72	.35		5.07	5.80
0500	Maximum	↓	440	.018	↓	7.70	.47		8.17	9.25

12 22 Curtains and Drapes

12 22 16 – Drapery Track and Accessories

12 22 16.10 Drapery Hardware

		Crew	Daily Output	Labor-Hours	Unit	Material	Labor	Equipment	Total	Total Incl O&P
0010	**DRAPERY HARDWARE**									
0030	Standard traverse, per foot, minimum	1 Carp	59	.136	L.F.	2.66	3.48		6.14	8.85
0100	Maximum		51	.157	"	10.90	4.03		14.93	18.85
0200	Decorative traverse, 28"-48", minimum		22	.364	Ea.	15.40	9.35		24.75	33
0220	Maximum		21	.381		36	9.80		45.80	56
0300	48"-84", minimum		20	.400		20.50	10.30		30.80	40
0320	Maximum		19	.421		58.50	10.80		69.30	82.50
0400	66"-120", minimum		18	.444		23	11.40		34.40	45
0420	Maximum		17	.471		91.50	12.10		103.60	122
0500	84"-156", minimum		16	.500		25.50	12.85		38.35	50
0520	Maximum		15	.533		96	13.70		109.70	130
0600	130"-240", minimum		14	.571		31	14.70		45.70	59
0620	Maximum	↓	13	.615		122	15.80		137.80	161
0700	Slide rings, each, minimum					.71			.71	.78
0720	Maximum					2.41			2.41	2.65
3000	Ripplefold, snap-a-pleat system, 3' or less, minimum	1 Carp	15	.533		47.50	13.70		61.20	76
3020	Maximum	"	14	.571	↓	69.50	14.70		84.20	102
3200	Each additional foot, add, minimum				L.F.	2.19			2.19	2.41
3220	Maximum				"	6.65			6.65	7.30
4000	Traverse rods, adjustable, 28" to 48"	1 Carp	22	.364	Ea.	20.50	9.35		29.85	38.50
4020	48" to 84"		20	.400		25.50	10.30		35.80	45.50
4040	66" to 120"		18	.444		29.50	11.40		40.90	51.50
4060	84" to 156"		16	.500		32.50	12.85		45.35	57.50
4080	100" to 180"		14	.571		41.50	14.70		56.20	70.50
4100	228" to 312"		13	.615		58.50	15.80		74.30	91
4500	Curtain rod, 28" to 48", single		22	.364		5	9.35		14.35	21.50
4510	Double		22	.364		8.55	9.35		17.90	25.50
4520	48" to 86", single		20	.400		8.60	10.30		18.90	27
4530	Double		20	.400		14.30	10.30		24.60	33
4540	66" to 120", single		18	.444		14.40	11.40		25.80	35.50
4550	Double	↓	18	.444		22.50	11.40		33.90	44
4600	Valance, pinch pleated fabric, 12" deep, up to 54" long, minimum					36			36	39.50
4610	Maximum					90			90	99
4620	Up to 77" long, minimum					55			55	60.50
4630	Maximum					145			145	160
5000	Stationary rods, first 2 feet				↓	9.70			9.70	10.70

12 23 Interior Shutters

12 23 10 − Interior Shutters

12 23 10.10 Interior Shutters

12 23 10.10 Interior Shutters	Crew	Daily Output	Labor-Hours	Unit	Material	2007 Bare Costs Labor	Equipment	Total	Total Incl O&P
0010	**INTERIOR SHUTTERS**, wood, louvered								
0200	Two panel, 27" wide, 36" high	1 Carp	5	1.600	Set	114	41		155	195
0300	33" wide, 36" high		5	1.600		149	41		190	234
0500	47" wide, 36" high		5	1.600		199	41		240	288
1000	Four panel, 27" wide, 36" high		5	1.600		134	41		175	218
1100	33" wide, 36" high		5	1.600		165	41		206	251
1300	47" wide, 36" high		5	1.600		234	41		275	330

12 23 13 − Wood Interior Shutters

12 23 13.13 Wood Panels

		Crew	Daily Output	Labor-Hours	Unit	Material	Labor	Equipment	Total	Total Incl O&P
0010	**WOOD PANELS**									
3000	Wood folding panels with movable louvers, 7" x 20" each	1 Carp	17	.471	Pr.	48	12.10		60.10	73.50
3300	8" x 28" each		17	.471		69.50	12.10		81.60	97
3450	9" x 36" each		17	.471		83	12.10		95.10	112
3600	10" x 40" each		17	.471		93.50	12.10		105.60	124
4000	Fixed louver type, stock units, 8" x 20" each		17	.471		72	12.10		84.10	100
4150	10" x 28" each		17	.471		80	12.10		92.10	109
4300	12" x 36" each		17	.471		96	12.10		108.10	127
4450	18" x 40" each		17	.471		156	12.10		168.10	192
5000	Insert panel type, stock, 7" x 20" each		17	.471		17.80	12.10		29.90	40
5150	8" x 28" each		17	.471		32.50	12.10		44.60	56.50
5300	9" x 36" each		17	.471		41.50	12.10		53.60	66
5450	10" x 40" each		17	.471		44.50	12.10		56.60	69.50
5600	Raised panel type, stock, 10" x 24" each		17	.471		191	12.10		203.10	232
5650	12" x 26" each		17	.471		144	12.10		156.10	180
5700	14" x 30" each		17	.471		163	12.10		175.10	200
5750	16" x 36" each		17	.471		183	12.10		195.10	222
6000	For custom built pine, add					22%				
6500	For custom built hardwood blinds, add					42%				

12 24 Window Shades

12 24 13 − Roller Window Shades

12 24 13.10 Shades

		Crew	Daily Output	Labor-Hours	Unit	Material	Labor	Equipment	Total	Total Incl O&P
0011	**SHADES** Basswood roll-up, stain finish, 3/8" slats		300	.027		10.80	.69		11.49	13.05
5011	Insulative shades		125	.064		9.95	1.64		11.59	13.75
6011	Solar screening, fiberglass		85	.094		4.38	2.42		6.80	8.90
8011	Interior insulative shutter									
8111	Stock unit, 15" x 60"	1 Carp	17	.471	Pr.	9.95	12.10		22.05	31.50

12 32 Manufactured Wood Casework

12 32 13 − Manufactured Wood-Veneer-Faced Casework

12 32 13.10 Manufactured Wood Casework

		Crew	Daily Output	Labor-Hours	Unit	Material	Labor	Equipment	Total	Total Incl O&P
0010	**MANUFACTURED WOOD CASEWORK**									
0700	Kitchen base cabinets, hardwood, not incl. counter tops,									
0710	24" deep, 35" high, prefinished									
0800	One top drawer, one door below, 12" wide	2 Carp	24.80	.645	Ea.	195	16.60		211.60	242
0820	15" wide		24	.667		204	17.15		221.15	254
0840	18" wide		23.30	.687		223	17.65		240.65	275
0860	21" wide		22.70	.705		242	18.10		260.10	298
0880	24" wide		22.30	.717		265	18.45		283.45	325

12 32 Manufactured Wood Casework

12 32 13 – Manufactured Wood-Veneer-Faced Casework

12 32 13.10 Manufactured Wood Casework		Crew	Daily Output	Labor-Hours	Unit	Material	2007 Bare Costs Labor	Equipment	Total	Total Incl O&P
1000	Four drawers, 12" wide	2 Carp	24.80	.645	Ea.	350	16.60		366.60	415
1020	15" wide		24	.667		288	17.15		305.15	345
1040	18" wide		23.30	.687		320	17.65		337.65	380
1060	24" wide		22.30	.717		345	18.45		363.45	410
1200	Two top drawers, two doors below, 27" wide		22	.727		295	18.70		313.70	355
1220	30" wide		21.40	.748		320	19.20		339.20	385
1240	33" wide		20.90	.766		355	19.65		374.65	425
1260	36" wide		20.30	.788		365	20.50		385.50	440
1280	42" wide		19.80	.808		390	21		411	465
1300	48" wide		18.90	.847		410	22		432	485
1500	Range or sink base, two doors below, 30" wide		21.40	.748		289	19.20		308.20	355
1520	33" wide		20.90	.766		305	19.65		324.65	375
1540	36" wide		20.30	.788		320	20.50		340.50	385
1560	42" wide		19.80	.808		340	21		361	410
1580	48" wide	▼	18.90	.847		355	22		377	425
1800	For sink front units, deduct					56			56	61.50
2000	Corner base cabinets, 36" wide, standard	2 Carp	18	.889		420	23		443	500
2100	Lazy Susan with revolving door	"	16.50	.970	▼	415	25		440	500
4000	Kitchen wall cabinets, hardwood, 12" deep with two doors									
4050	12" high, 30" wide	2 Carp	24.80	.645	Ea.	157	16.60		173.60	200
4100	36" wide		24	.667		183	17.15		200.15	230
4400	15" high, 30" wide		24	.667		167	17.15		184.15	213
4420	33" wide		23.30	.687		180	17.65		197.65	228
4440	36" wide		22.70	.705		189	18.10		207.10	239
4450	42" wide		22.70	.705		260	18.10		278.10	315
4700	24" high, 30" wide		23.30	.687		207	17.65		224.65	258
4720	36" wide		22.70	.705		229	18.10		247.10	283
4740	42" wide		22.30	.717		250	18.45		268.45	305
5000	30" high, one door, 12" wide		22	.727		140	18.70		158.70	186
5020	15" wide		21.40	.748		158	19.20		177.20	207
5040	18" wide		20.90	.766		173	19.65		192.65	225
5060	24" wide		20.30	.788		195	20.50		215.50	250
5300	Two doors, 27" wide		19.80	.808		234	21		255	292
5320	30" wide		19.30	.829		234	21.50		255.50	293
5340	36" wide		18.80	.851		267	22		289	330
5360	42" wide		18.50	.865		291	22		313	360
5380	48" wide		18.40	.870		330	22.50		352.50	400
6000	Corner wall, 30" high, 24" wide		18	.889		159	23		182	214
6050	30" wide		17.20	.930		193	24		217	253
6100	36" wide		16.50	.970		209	25		234	273
6500	Revolving Lazy Susan		15.20	1.053		355	27		382	435
7000	Broom cabinet, 84" high, 24" deep, 18" wide		10	1.600		430	41		471	540
7500	Oven cabinets, 84" high, 24" deep, 27" wide		8	2	▼	630	51.50		681.50	780
7750	Valance board trim		396	.040	L.F.	9	1.04		10.04	11.65
9000	For deluxe models of all cabinets, add					40%				
9500	For custom built in place, add					25%	10%			
9550	Rule of thumb, kitchen cabinets not including									
9560	appliances & counter top, minimum	2 Carp	30	.533	L.F.	104	13.70		117.70	139
9600	Maximum	"	25	.640	"	250	16.45		266.45	305

12 32 13.15 Manufactured Wood Casework Frames

0010	**MANUFACTURED WOOD CASEWORK FRAMES**									
0050	Base cabinets, counter storage, 36" high, one bay									

12 32 Manufactured Wood Casework

12 32 13 – Manufactured Wood-Veneer-Faced Casework

12 32 13.15 Manufactured Wood Casework Frames		Crew	Daily Output	Labor-Hours	Unit	Material	2007 Bare Costs Labor	Equipment	Total	Total Incl O&P
0100	18" wide	1 Carp	2.70	2.963	Ea.	115	76		191	255
0400	Two bay, 36" wide		2.20	3.636		175	93.50		268.50	350
1100	Three bay, 54" wide		1.50	5.333		208	137		345	460
2800	Book cases, one bay, 7' high, 18" wide		2.40	3.333		135	85.50		220.50	293
3500	Two bay, 36" wide		1.60	5		195	129		324	435
4100	Three bay, 54" wide		1.20	6.667		325	171		496	645
6100	Wall mounted cabinet, one bay, 24" high, 18" wide		3.60	2.222		74	57		131	179
6800	Two bay, 36" wide		2.20	3.636		108	93.50		201.50	278
7400	Three bay, 54" wide		1.70	4.706		135	121		256	355
8400	30" high, one bay, 18" wide		3.60	2.222		80.50	57		137.50	186
9000	Two bay, 36" wide		2.15	3.721		107	95.50		202.50	280
9400	Three bay, 54" wide		1.60	5		133	129		262	365
9800	Wardrobe, 7' high, single, 24" wide		2.70	2.963		148	76		224	292
9950	Partition, adjustable shelves & drawers, 48" wide		1.40	5.714		283	147		430	560

12 32 13.20 Manufactured Wood Casework Doors		Crew	Daily Output	Labor-Hours	Unit	Material	2007 Bare Costs Labor	Equipment	Total	Total Incl O&P
0010	**MANUFACTURED WOOD CASEWORK DOORS**									
2000	Glass panel, hardwood frame									
2200	12" wide, 18" high	1 Carp	34	.235	Ea.	14.30	6.05		20.35	26
2400	24" high		33	.242		19.10	6.25		25.35	31.50
2600	30" high		32	.250		24	6.45		30.45	37.50
2800	36" high		30	.267		28.50	6.85		35.35	43
3000	48" high		23	.348		38	8.95		46.95	57
3200	60" high		17	.471		48	12.10		60.10	73
3400	72" high		15	.533		57.50	13.70		71.20	86.50
3600	15" wide x 18" high		33	.242		17.90	6.25		24.15	30.50
3800	24" high		32	.250		24	6.45		30.45	37.50
4000	30" high		30	.267		30	6.85		36.85	44.50
4250	36" high		28	.286		36	7.35		43.35	52
4300	48" high		22	.364		48	9.35		57.35	68.50
4350	60" high		16	.500		59.50	12.85		72.35	87.50
4400	72" high		14	.571		71.50	14.70		86.20	104
4450	18" wide, 18" high		32	.250		21.50	6.45		27.95	34.50
4500	24" high		30	.267		28.50	6.85		35.35	43
4550	30" high		29	.276		26.50	7.10		33.60	41
4600	36" high		27	.296		43	7.60		50.60	60.50
4650	48" high		21	.381		57.50	9.80		67.30	79.50
4700	60" high		15	.533		71.50	13.70		85.20	103
4750	72" high		13	.615		86	15.80		101.80	122
5000	Hardwood, raised panel									
5100	12" wide, 18" high	1 Carp	16	.500	Ea.	24	12.85		36.85	48.50
5150	24" high		15.50	.516		32	13.25		45.25	57.50
5200	30" high		15	.533		40	13.70		53.70	67.50
5250	36" high		14	.571		48	14.70		62.70	78
5300	48" high		11	.727		64	18.70		82.70	102
5320	60" high		8	1		80	25.50		105.50	131
5340	72" high		7	1.143		95.50	29.50		125	155
5360	15" wide x 18" high		15.50	.516		30	13.25		43.25	55.50
5380	24" high		15	.533		40	13.70		53.70	67.50
5400	30" high		14.50	.552		50	14.20		64.20	79
5420	36" high		13.50	.593		60	15.25		75.25	92
5440	48" high		10.50	.762		80	19.60		99.60	121
5460	60" high		7.50	1.067		99.50	27.50		127	157

12 32 Manufactured Wood Casework

12 32 13 – Manufactured Wood-Veneer-Faced Casework

12 32 13.20 Manufactured Wood Casework Doors		Crew	Daily Output	Labor-Hours	Unit	Material	2007 Bare Costs Labor	Equipment	Total	Total Incl O&P
5480	72" high	1 Carp	6.50	1.231	Ea.	120	31.50		151.50	186
5500	18" wide, 18" high		15	.533		36	13.70		49.70	63
5550	24" high		14.50	.552		48	14.20		62.20	76.50
5600	30" high		14	.571		60	14.70		74.70	91
5650	36" high		13	.615		72	15.80		87.80	106
5700	48" high		10	.800		95.50	20.50		116	140
5750	60" high		7	1.143		120	29.50		149.50	182
5800	72" high		6	1.333		144	34.50		178.50	216
6000	Plastic laminate on particle board									
6100	12" wide, 18" high	1 Carp	25	.320	Ea.	10.35	8.20		18.55	25.50
6120	24" high		24	.333		13.80	8.55		22.35	30
6140	30" high		23	.348		17.25	8.95		26.20	34
6160	36" high		21	.381		20.50	9.80		30.30	39.50
6200	48" high		16	.500		27.50	12.85		40.35	52.50
6250	60" high		13	.615		34.50	15.80		50.30	65
6300	72" high		12	.667		41.50	17.15		58.65	74.50
6320	15" wide x 18" high		24.50	.327		13.75	8.40		22.15	29.50
6340	24" high		23.50	.340		18.35	8.75		27.10	35
6360	30" high		22.50	.356		23	9.15		32.15	41
6380	36" high		20.50	.390		27.50	10.05		37.55	47.50
6400	48" high		15.50	.516		36.50	13.25		49.75	63
6450	60" high		12.50	.640		46	16.45		62.45	78.50
6480	72" high		11.50	.696		55	17.90		72.90	91
6500	18" wide, 18" high		24	.333		15.50	8.55		24.05	31.50
6550	24" high		23	.348		20.50	8.95		29.45	38
6600	30" high		22	.364		26	9.35		35.35	44.50
6650	36" high		20	.400		31	10.30		41.30	51.50
6700	48" high		15	.533		41.50	13.70		55.20	69
6750	60" high		12	.667		52	17.15		69.15	86
6800	72" high		11	.727		62	18.70		80.70	100
12 32 13.25 Manufactured Wood Casework Drawer Fronts										
0010	**MANUFACTURED WOOD CASEWORK DRAWER FRONTS**									
0100	Solid hardwood front									
1000	4" high, 12" wide	1 Carp	17	.471	Ea.	3.50	12.10		15.60	24.50
1200	18" wide		16	.500		5.05	12.85		17.90	27.50
1400	24" wide		15	.533		6.75	13.70		20.45	31
1600	6" high, 12" wide		16	.500		5.10	12.85		17.95	27.50
1800	18" wide		15	.533		7.70	13.70		21.40	32
2000	24" wide		14	.571		10.25	14.70		24.95	36.50
2200	9" high, 12" wide		15	.533		7.70	13.70		21.40	32
2400	18" wide		14	.571		11.55	14.70		26.25	37.50
2600	24" wide		13	.615		15.35	15.80		31.15	44
2800	Plastic laminate on particle board front									
3000	4" high, 12" wide	1 Carp	17	.471	Ea.	4.39	12.10		16.49	25.50
3200	18" wide		16	.500		6.60	12.85		19.45	29.50
3600	24" wide		15	.533		8.75	13.70		22.45	33
3800	6" high, 12" wide		16	.500		6.65	12.85		19.50	29.50
4000	18" wide		15	.533		10	13.70		23.70	34.50
4500	24" wide		14	.571		13.30	14.70		28	39.50
4800	9" high, 12" wide		15	.533		10	13.70		23.70	34.50
5000	18" wide		14	.571		15	14.70		29.70	41.50
5200	24" wide		13	.615		19.95	15.80		35.75	49

12 32 Manufactured Wood Casework

12 32 13 – Manufactured Wood-Veneer-Faced Casework

12 32 13.30 Manufactured Wood Casework Vanities

		Crew	Daily Output	Labor-Hours	Unit	Material	2007 Bare Costs Labor	Equipment	Total	Total Incl O&P
0010	**MANUFACTURED WOOD CASEWORK VANITIES**									
8000	Vanity bases, 2 doors, 30" high, 21" deep, 24" wide	2 Carp	20	.800	Ea.	197	20.50		217.50	251
8050	30" wide		16	1		226	25.50		251.50	293
8100	36" wide		13.33	1.200		300	31		331	385
8150	48" wide	↓	11.43	1.400		360	36		396	455
9000	For deluxe models of all vanities, add to above					40%				
9500	For custom built in place, add to above				↓	25%	10%			

12 32 13.35 Manufactured Wood Casework Hardware

		Crew	Daily Output	Labor-Hours	Unit	Material	2007 Bare Costs Labor	Equipment	Total	Total Incl O&P
0010	**MANUFACTURED WOOD CASEWORK HARDWARE**									
1000	Catches, minimum	1 Carp	235	.034	Ea.	.84	.87		1.71	2.40
1020	Average		119.40	.067		2.75	1.72		4.47	5.95
1040	Maximum	↓	80	.100	↓	5.20	2.57		7.77	10.10
2000	Door/drawer pulls, handles									
2200	Handles and pulls, projecting, metal, minimum	1 Carp	160	.050	Ea.	3.70	1.29		4.99	6.25
2220	Average		95.24	.084		5.75	2.16		7.91	10
2240	Maximum		68	.118		7.85	3.02		10.87	13.80
2300	Wood, minimum		160	.050		3.90	1.29		5.19	6.45
2320	Average		95.24	.084		5.20	2.16		7.36	9.40
2340	Maximum		68	.118		7.15	3.02		10.17	13.05
2600	Flush, metal, minimum		160	.050		3.90	1.29		5.19	6.45
2620	Average		95.24	.084		5.20	2.16		7.36	9.40
2640	Maximum		68	.118	↓	7.15	3.02		10.17	13.05
3000	Drawer tracks/glides, minimum		48	.167	Pr.	6.60	4.28		10.88	14.55
3020	Average		32	.250		11.20	6.45		17.65	23.50
3040	Maximum		24	.333		19.25	8.55		27.80	35.50
4000	Cabinet hinges, minimum		160	.050		2.25	1.29		3.54	4.66
4020	Average		95.24	.084		5	2.16		7.16	9.15
4040	Maximum	↓	68	.118	↓	8.50	3.02		11.52	14.50

12 34 Manufactured Plastic Casework

12 34 16 – Manufactured Solid-Plastic Casework

12 34 16.10 Outdoor Casework

		Crew	Daily Output	Labor-Hours	Unit	Material	2007 Bare Costs Labor	Equipment	Total	Total Incl O&P
0010	**OUTDOOR CASEWORK**									
0020	Cabinet, base, sink/range, 36"	2 Carp	20.30	.788	Ea.	1,300	20.50		1,320.50	1,475
0100	Base, 36"		20.30	.788		1,875	20.50		1,895.50	2,100
0200	Filler strip, 1" x 30"	↓	158	.101	↓	26	2.60		28.60	33

12 36 Countertops

12 36 23 – Plastic Countertops

12 36 23.10 Countertops

		Crew	Daily Output	Labor-Hours	Unit	Material	2007 Bare Costs Labor	Equipment	Total	Total Incl O&P
0010	**COUNTERTOPS**									
0020	Stock plastic laminate, 24" wide w/ backsplash, minimum	1 Carp	30	.267	L.F.	9.55	6.85		16.40	22
0100	Maximum		25	.320		17.30	8.20		25.50	33
0300	Custom plastic, 7/8" thick, aluminum molding, no splash		30	.267		19.10	6.85		25.95	32.50
0400	Cove splash		30	.267		25	6.85		31.85	39
0600	1-1/4" thick, no splash		28	.286		28.50	7.35		35.85	44
0700	Square splash		28	.286		27.50	7.35		34.85	42.50
0900	Square edge, plastic face, 7/8" thick, no splash	↓	30	.267	↓	24	6.85		30.85	38

12 36 Countertops

12 36 23 – Plastic Countertops

12 36 23.10 Countertops

		Crew	Daily Output	Labor-Hours	Unit	Material	2007 Bare Costs Labor	Equipment	Total	Total Incl O&P
1000	With splash	1 Carp	30	.267	L.F.	30.50	6.85		37.35	45
1200	For stainless channel edge, 7/8" thick, add					2.51			2.51	2.76
1300	1-1/4" thick, add					2.95			2.95	3.25
1500	For solid color suede finish, add				↓	2.45			2.45	2.70
1700	For end splash, add				Ea.	16.35			16.35	18
1901	For cut outs, standard, add, minimum	1 Carp	32	.250			6.45		6.45	10.90
2000	Maximum		8	1		5.45	25.50		30.95	49.50
2010	Cut out in blacksplash for elec. wall outlet		38	.211			5.40		5.40	9.20
2020	Cut out for sink		20	.400			10.30		10.30	17.45
2030	Cut out for stove top		18	.444	↓		11.40		11.40	19.40
2100	Postformed, including backsplash and front edge		30	.267	L.F.	9.80	6.85		16.65	22.50
2110	Mitred, add		12	.667	Ea.		17.15		17.15	29
2200	Built-in place, 25" wide, plastic laminate		25	.320	L.F.	13.10	8.20		21.30	28.50
2300	Ceramic tile mosaic	↓	25	.320		28.50	8.20		36.70	45
2500	Marble, stock, with splash, 1/2" thick, minimum	1 Bric	17	.471		35	12.55		47.55	59.50
2700	3/4" thick, maximum	"	13	.615		87.50	16.40		103.90	124
2900	Maple, solid, laminated, 1-1/2" thick, no splash	1 Carp	28	.286		57	7.35		64.35	75.50
3000	With square splash		28	.286	↓	68	7.35		75.35	87
3200	Stainless steel		24	.333	S.F.	131	8.55		139.55	159
3400	Recessed cutting block with trim, 16" x 20" x 1"		8	1	Ea.	66.50	25.50		92	117
3411	Replace cutting block only	↓	16	.500	"	32.50	12.85		45.35	58

12 36 61 – Simulated Stone Countertops

12 36 61.16 Solid Surface Countertops

		Crew	Daily Output	Labor-Hours	Unit	Material	Labor	Equipment	Total	Total Incl O&P
0010	**SOLID SURFACE COUNTERTOPS**, Acrylic polymer									
2000	Pricing for order of 1 - 50 L.F.									
2100	25" wide, solid colors	2 Carp	20	.800	L.F.	60	20.50		80.50	101
2200	Patterned colors		20	.800		76.50	20.50		97	119
2300	Premium patterned colors		20	.800		95.50	20.50		116	140
2400	With silicone attached 4" backsplash, solid colors		19	.842		66	21.50		87.50	109
2500	Patterned colors		19	.842		83.50	21.50		105	129
2600	Premium patterned colors		19	.842		104	21.50		125.50	152
2700	With hard seam attached 4" backsplash, solid colors		15	1.067		66	27.50		93.50	119
2800	Patterned colors		15	1.067		83.50	27.50		111	139
2900	Premium patterned colors	↓	15	1.067	↓	104	27.50		131.50	162
3800	Sinks, pricing for order of 1 - 50 units									
3900	Single bowl, hard seamed, solid colors, 13" x 17"	1 Carp	2	4	Ea.	405	103		508	620
4000	10" x 15"		4.55	1.758		188	45		233	284
4100	Cutouts for sinks	↓	5.25	1.524	↓		39		39	66.50

12 36 61.17 Solid Surface Vanity Tops

		Crew	Daily Output	Labor-Hours	Unit	Material	Labor	Equipment	Total	Total Incl O&P
0010	**SOLID SURFACE VANITY TOPS**									
0015	Solid surface, center bowl, 17" x 19"	1 Carp	12	.667	Ea.	184	17.15		201.15	231
0020	19" x 25"		12	.667		223	17.15		240.15	274
0030	19" x 31"		12	.667		270	17.15		287.15	325
0040	19" x 37"		12	.667		315	17.15		332.15	375
0050	22" x 25"		10	.800		196	20.50		216.50	251
0060	22" x 31"		10	.800		229	20.50		249.50	287
0070	22" x 37"		10	.800		266	20.50		286.50	330
0080	22" x 43"		10	.800		305	20.50		325.50	370
0090	22" x 49"		10	.800		335	20.50		355.50	405
0110	22" x 55"		8	1		380	25.50		405.50	465
0120	22" x 61"		8	1		435	25.50		460.50	525
0220	Double bowl, 22" x 61"	↓	8	1		495	25.50		520.50	585

12 36 Countertops

12 36 61 – Simulated Stone Countertops

12 36 61.17 Solid Surface Vanity Tops

		Crew	Daily Output	Labor-Hours	Unit	Material	2007 Bare Costs Labor	Equipment	Total	Total Incl O&P
0230	Double bowl, 22" x 73"	1 Carp	8	1	Ea.	685	25.50		710.50	795
0240	For aggregate colors, add					35%				
0250	For faucets and fittings see 22 41 39.10									

12 36 61.19 Engineered Stone Countertops

		Crew	Daily Output	Labor-Hours	Unit	Material	Labor	Equipment	Total	Total Incl O&P
0010	**ENGINEERED STONE COUNTERTOPS**									
0100	25" wide, 4" backsplash, color group A, minimum	2 Carp	15	1.067	L.F.	16.15	27.50		43.65	64.50
0110	Maximum		15	1.067		41.50	27.50		69	92
0120	Color group B, minimum		15	1.067		21	27.50		48.50	69.50
0130	Maximum		15	1.067		48.50	27.50		76	99.50
0140	Color group C, minimum		15	1.067		29	27.50		56.50	78.50
0150	Maximum		15	1.067		59.50	27.50		87	112
0160	Color group D, minimum		15	1.067		36	27.50		63.50	86.50
0170	Maximum		15	1.067		69.50	27.50		97	123

12 93 Site Furnishings

12 93 33 – Manufactured Planters

12 93 33.10 Planters

		Crew	Daily Output	Labor-Hours	Unit	Material	Labor	Equipment	Total	Total Incl O&P
0010	**PLANTERS**									
0012	Concrete, sandblasted, precast, 48" diameter, 24" high	2 Clab	15	1.067	Ea.	615	19.95		634.95	710
0300	Fiberglass, circular, 36" diameter, 24" high		15	1.067		450	19.95		469.95	530
1200	Wood, square, 48" side, 24" high		15	1.067		950	19.95		969.95	1,075
1300	Circular, 48" diameter, 30" high		10	1.600		780	30		810	910
1600	Planter/bench, 72"		5	3.200		3,075	60		3,135	3,475

12 93 43 – Site Seating and Tables

12 93 43.13 Site Seating

		Crew	Daily Output	Labor-Hours	Unit	Material	Labor	Equipment	Total	Total Incl O&P
0010	**SITE SEATING**									
0012	Seating, benches, park, precast conc, w/backs, wood rails, 4' long	2 Clab	5	3.200	Ea.	350	60		410	485
0100	8' long		4	4		735	75		810	935
0500	Steel barstock pedestals w/backs, 2" x 3" wood rails, 4' long		10	1.600		875	30		905	1,000
0510	8' long		7	2.286		1,025	42.50		1,067.50	1,200
0800	Cast iron pedestals, back & arms, wood slats, 4' long		8	2		325	37.50		362.50	420
0820	8' long		5	3.200		940	60		1,000	1,125
1700	Steel frame, fir seat, 10' long		10	1.600		200	30		230	270

Division 13
Special Construction

13 11 Swimming Pools

13 11 13 – Below-Grade Swimming Pools

13 11 13.50 Swimming Pools

		Crew	Daily Output	Labor-Hours	Unit	Material	2007 Bare Costs Labor	2007 Bare Costs Equipment	Total	Total Incl O&P
0010	**SWIMMING POOLS** Residential in-ground, vinyl lined, concrete									
0020	Sides including equipment, sand bottom	B-52	300	.187	SF Surf	12.25	3.95	1.39	17.59	22
0100	Metal or polystyrene sides R131113-20	B-14	410	.117		10.25	2.36	.59	13.20	15.90
0200	Add for vermiculite bottom					.79			.79	.87
0500	Gunite bottom and sides, white plaster finish									
0600	12' x 30' pool	B-52	145	.386	SF Surf	23	8.20	2.88	34.08	42
0720	16' x 32' pool		155	.361		20.50	7.65	2.69	30.84	38.50
0750	20' x 40' pool		250	.224		18.35	4.74	1.67	24.76	30
0810	Concrete bottom and sides, tile finish									
0820	12' x 30' pool	B-52	80	.700	SF Surf	23	14.85	5.20	43.05	56.50
0830	16' x 32' pool		95	.589		19.05	12.50	4.39	35.94	47
0840	20' x 40' pool		130	.431		15.15	9.10	3.21	27.46	35.50
1600	For water heating system, see division 23 52 28.10									
1700	Filtration and deck equipment only, as % of total				Total				20%	20%
1800	Deck equipment, rule of thumb, 20' x 40' pool				SF Pool					1.30
3000	Painting pools, preparation + 3 coats, 20' x 40' pool, epoxy	2 Pord	.33	48.485	Total	700	1,125		1,825	2,625
3100	Rubber base paint, 18 gallons	"	.33	48.485	"	530	1,125		1,655	2,425

13 11 46 – Swimming Pool Accessories

13 11 46.50 Swimming Pool Equipment

		Crew	Daily Output	Labor-Hours	Unit	Material	2007 Bare Costs Labor	2007 Bare Costs Equipment	Total	Total Incl O&P
0010	**SWIMMING POOL EQUIPMENT**									
0020	Diving stand, stainless steel, 3 meter	2 Carp	.40	40	Ea.	6,150	1,025		7,175	8,525
0600	Diving boards, 16' long, aluminum		2.70	5.926		3,050	152		3,202	3,600
0700	Fiberglass		2.70	5.926		2,450	152		2,602	2,950
1200	Ladders, heavy duty, stainless steel, 2 tread		7	2.286		535	58.50		593.50	690
1500	4 tread		6	2.667		620	68.50		688.50	795
2100	Lights, underwater, 12 volt, with transformer, 300 watt	1 Elec	.40	20		175	590		765	1,150
2200	110 volt, 500 watt, standard	"	.40	20		122	590		712	1,100
3000	Pool covers, reinforced vinyl	3 Clab	1800	.013	S.F.	.33	.25		.58	.78
3100	Vinyl water tube		3200	.008		.46	.14		.60	.75
3200	Maximum		3000	.008		.58	.15		.73	.89
3300	Slides, tubular, fiberglass, aluminum handrails & ladder, 5'-0", straight	2 Carp	1.60	10	Ea.	2,325	257		2,582	2,975
3320	8'-0", curved	"	3	5.333	"	8,875	137		9,012	10,000

13 12 Fountains

13 12 13 – Exterior Fountains

13 12 13.10 Yard Fountains

		Crew	Daily Output	Labor-Hours	Unit	Material	2007 Bare Costs Labor	2007 Bare Costs Equipment	Total	Total Incl O&P
0010	**YARD FOUNTAINS**									
0100	Outdoor fountain, 48" high with bowl and figures	2 Clab	2	8	Ea.	189	150		339	460

13 17 Tubs and Pools

13 17 13 – Redwood Hot Tub System

13 17 13.10 Redwood Hot Tub System

		Crew	Daily Output	Labor-Hours	Unit	Material	2007 Bare Costs Labor	Equipment	Total	Total Incl O&P
0010	**REDWOOD HOT TUB SYSTEM**									
7050	4' diameter x 4' deep	Q-1	1	16	Ea.	1,750	425		2,175	2,625
7150	6' diameter x 4' deep		.80	20		2,575	530		3,105	3,700
7200	8' diameter x 4' deep		.80	20		3,950	530		4,480	5,225

13 17 33 – Whirlpool Tubs

13 17 33.10 Whirlpool Bath

		Crew	Daily Output	Labor-Hours	Unit	Material	Labor	Equipment	Total	Total Incl O&P
0010	**WHIRLPOOL BATH**									
6000	Whirlpool, bath with vented overflow, molded fiberglass									
6100	66" x 48" x 24"	Q-1	1	16	Ea.	2,850	425		3,275	3,825
6400	72" x 36" x 24"		1	16		2,825	425		3,250	3,800
6500	60" x 30" x 21"		1	16		2,425	425		2,850	3,375
6600	72" x 42" x 22"		1	16		3,975	425		4,400	5,075
6700	83" x 65"		.30	53.333		4,925	1,425		6,350	7,750

13 24 Special Activity Rooms

13 24 16 – Saunas

13 24 16.50 Saunas

		Crew	Daily Output	Labor-Hours	Unit	Material	Labor	Equipment	Total	Total Incl O&P
0010	**SAUNAS**									
0020	Prefabricated, incl. heater & controls, 7' high, 6' x 4', C/C	L-7	2.20	11.818	Ea.	3,875	257		4,132	4,675
0050	6' x 4', C/P		2	13		3,575	283		3,858	4,425
0400	6' x 5', C/C		2	13		4,325	283		4,608	5,225
0450	6' x 5', C/P		2	13		4,050	283		4,333	4,925
0600	6' x 6', C/C		1.80	14.444		4,625	315		4,940	5,600
0650	6' x 6', C/P		1.80	14.444		4,300	315		4,615	5,250
0800	6' x 9', C/C		1.60	16.250		5,775	355		6,130	6,950
0850	6' x 9', C/P		1.60	16.250		5,475	355		5,830	6,625
1000	8' x 12', C/C		1.10	23.636		8,950	515		9,465	10,700
1050	8' x 12', C/P		1.10	23.636		8,200	515		8,715	9,900
1400	8' x 10', C/C		1.20	21.667		7,575	470		8,045	9,125
1450	8' x 10', C/P		1.20	21.667		7,025	470		7,495	8,525
1600	10' x 12', C/C		1	26		9,475	565		10,040	11,400
1650	10' x 12', C/P		1	26		8,575	565		9,140	10,400
1700	Door only, cedar, 2'x6', with tempered insulated glass window	2 Carp	3.40	4.706		505	121		626	760
1800	Prehung, incl. jambs, pulls & hardware	"	12	1.333		515	34.50		549.50	630
2500	Heaters only (incl. above), wall mounted, to 200 C.F.					500			500	550
2750	To 300 C.F.					615			615	675
3000	Floor standing, to 720 C.F., 10,000 watts, w/controls	1 Elec	3	2.667		1,575	78.50		1,653.50	1,850
3250	To 1,000 C.F., 16,000 watts	"	3	2.667		1,750	78.50		1,828.50	2,050

13 24 26 – Steam Baths

13 24 26.50 Steam Baths

		Crew	Daily Output	Labor-Hours	Unit	Material	Labor	Equipment	Total	Total Incl O&P
0010	**STEAM BATHS**									
0020	Heater, timer & head, single, to 140 C.F.	1 Plum	1.20	6.667	Ea.	1,125	197		1,322	1,575
0500	To 300 C.F.	"	1.10	7.273		1,275	215		1,490	1,750
2700	Conversion unit for residential tub, including door					3,550			3,550	3,900

13 34 Fabricated Engineered Structures

13 34 13 – Glazed Structures

13 34 13.13 Greenhouses

		Crew	Daily Output	Labor-Hours	Unit	Material	2007 Bare Costs Labor	2007 Bare Costs Equipment	Total	Total Incl O&P
0010	**GREENHOUSES**, Shell only, stock units, not incl. 2' stub walls,									
0020	foundation, floors, heat or compartments									
0300	Residential type, free standing, 8'-6" long x 7'-6" wide	2 Carp	59	.271	SF Flr.	46.50	6.95		53.45	63.50
0400	10'-6" wide		85	.188		36	4.84		40.84	47.50
0600	13'-6" wide		108	.148		32	3.81		35.81	41.50
0700	17'-0" wide		160	.100		36	2.57		38.57	44
0900	Lean-to type, 3'-10" wide		34	.471		41.50	12.10		53.60	66
1000	6'-10" wide		58	.276		32	7.10		39.10	47
1100	Wall mounted, to existing window, 3' x 3'	1 Carp	4	2	Ea.	445	51.50		496.50	575
1120	4' x 5'	"	3	2.667	"	665	68.50		733.50	850
1200	Deluxe quality, free standing, 7'-6" wide	2 Carp	55	.291	SF Flr.	92	7.50		99.50	114
1220	10'-6" wide		81	.198		85.50	5.10		90.60	103
1240	13'-6" wide		104	.154		80	3.95		83.95	94.50
1260	17'-0" wide		150	.107		68	2.74		70.74	79.50
1400	Lean-to type, 3'-10" wide		31	.516		107	13.25		120.25	140
1420	6'-10" wide		55	.291		100	7.50		107.50	123
1440	8'-0" wide		97	.165		93.50	4.24		97.74	110

13 34 13.19 Swimming Pool Enclosures

		Crew	Daily Output	Labor-Hours	Unit	Material	2007 Bare Costs Labor	2007 Bare Costs Equipment	Total	Total Incl O&P
0010	**SWIMMING POOL ENCLOSURES** Translucent, free standing									
0020	not including foundations, heat or light									
0200	Economy, minimum	2 Carp	200	.080	SF Hor.	12.85	2.06		14.91	17.65
0300	Maximum		100	.160		36	4.11		40.11	46.50
0400	Deluxe, minimum		100	.160		42.50	4.11		46.61	53.50
0600	Maximum		70	.229		196	5.85		201.85	225

Division 14
Conveying Equipment

14 21 Electric Traction Elevators

14 21 33 – Electric Traction Residential Elevators

14 21 33.20 Electric Traction Residential Elevators	Crew	Daily Output	Labor-Hours	Unit	Material	2007 Bare Costs Labor	Equipment	Total	Total Incl O&P
0010 **ELECTRIC TRACTION RESIDENTIAL ELEVATORS**									
7000 Residential, cab type, 1 floor, 2 stop, minimum	2 Elev	.20	80	Ea.	8,925	2,875		11,800	14,500
7100 Maximum		.10	160		15,100	5,725		20,825	26,000
7200 2 floor, 3 stop, minimum		.12	133		13,200	4,775		17,975	22,400
7300 Maximum	↓	.06	266	↓	21,600	9,550		31,150	39,400

14 42 Wheelchair Lifts

14 42 13 – Inclined Wheelchair Lifts

14 42 13.10 Inclined Wheelchair Lifts

	Crew	Daily Output	Labor-Hours	Unit	Material	Labor	Equipment	Total	Total Incl O&P
0010 **INCLINED WHEELCHAIR LIFTS**									
7700 Stair climber (chair lift), single seat, minimum	2 Elev	1	16	Ea.	4,325	575		4,900	5,675
7800 Maximum	"	.20	80	"	5,925	2,875		8,800	11,200

Division 22
Plumbing

22 05 Common Work Results for Plumbing

22 05 05 – Selective Plumbing Demolition

22 05 05.10 Plumbing Demolition

		Crew	Daily Output	Labor-Hours	Unit	Material	2007 Bare Costs Labor	2007 Bare Costs Equipment	Total	Total Incl O&P
0010	**PLUMBING DEMOLITION**									
1020	Fixtures, including 10' piping									
1100	Bath tubs, cast iron	1 Plum	4	2	Ea.		59		59	97
1120	Fiberglass		6	1.333			39.50		39.50	65
1140	Steel		5	1.600			47.50		47.50	78
1200	Lavatory, wall hung		10	.800			23.50		23.50	39
1220	Counter top		8	1			29.50		29.50	48.50
1300	Sink, single compartment		8	1			29.50		29.50	48.50
1320	Double compartment		7	1.143			34		34	55.50
1400	Water closet, floor mounted		8	1			29.50		29.50	48.50
1420	Wall mounted		7	1.143	↓		34		34	55.50
2000	Piping, metal, up thru 1-1/2" diameter		200	.040	L.F.		1.18		1.18	1.94
2050	2" thru 3-1/2" diameter	↓	150	.053			1.58		1.58	2.59
2100	4" thru 6" diameter	2 Plum	100	.160	↓		4.73		4.73	7.80
2250	Water heater, 40 gal.	1 Plum	6	1.333	Ea.		39.50		39.50	65
3000	Submersible sump pump		24	.333			9.85		9.85	16.20
6000	Remove and reset fixtures, minimum		6	1.333			39.50		39.50	65
6100	Maximum		4	2	↓		59		59	97
9000	Minimum labor/equipment charge	↓	2	4	Job		118		118	194

22 05 23 – General-Duty Valves for Plumbing Piping

22 05 23.20 Valves, Bronze

		Crew	Daily Output	Labor-Hours	Unit	Material	2007 Bare Costs Labor	2007 Bare Costs Equipment	Total	Total Incl O&P
0010	**VALVES, BRONZE**									
1750	Check, swing, class 150, regrinding disc, threaded									
1860	3/4"	1 Plum	20	.400	Ea.	45	11.80		56.80	69
1870	1"	"	19	.421	"	65.50	12.45		77.95	92.50
2850	Gate, N.R.S., soldered, 125 psi									
2940	3/4"	1 Plum	20	.400	Ea.	25	11.80		36.80	47
2950	1"	"	19	.421	"	36	12.45		48.45	60
5600	Relief, pressure & temperature, self-closing, ASME, threaded									
5650	1"	1 Plum	24	.333	Ea.	131	9.85		140.85	160
5660	1-1/4"	"	20	.400	"	262	11.80		273.80	305
6400	Pressure, water, ASME, threaded									
6440	3/4"	1 Plum	28	.286	Ea.	58	8.45		66.45	78
6450	1"	"	24	.333	"	124	9.85		133.85	152
6900	Reducing, water pressure									
6940	1/2"	1 Plum	24	.333	Ea.	165	9.85		174.85	198
6960	1"	"	19	.421	"	256	12.45		268.45	300
8350	Tempering, water, sweat connections									
8400	1/2"	1 Plum	24	.333	Ea.	57	9.85		66.85	79
8440	3/4"	"	20	.400	"	70	11.80		81.80	96.50
8650	Threaded connections									
8700	1/2"	1 Plum	24	.333	Ea.	70	9.85		79.85	93
8740	3/4"	"	20	.400	"	279	11.80		290.80	325

22 05 76 – Facility Drainage Piping Cleanouts

22 05 76.10 Cleanouts

		Crew	Daily Output	Labor-Hours	Unit	Material	2007 Bare Costs Labor	2007 Bare Costs Equipment	Total	Total Incl O&P
0010	**CLEANOUTS**									
0080	Round or square, scoriated nickel bronze top									
0100	2" pipe size	1 Plum	10	.800	Ea.	112	23.50		135.50	162
0140	4" pipe size	"	6	1.333	"	168	39.50		207.50	250

22 05 76.20 Cleanout Tees

0010	**CLEANOUT TEES**									

22 05 Common Work Results for Plumbing

22 05 76 – Facility Drainage Piping Cleanouts

22 05 76.20 Cleanout Tees		Crew	Daily Output	Labor-Hours	Unit	Material	2007 Bare Costs Labor	2007 Bare Costs Equipment	Total	Total Incl O&P
0100	Cast iron, B&S, with countersunk plug									
0220	3" pipe size	1 Plum	3.60	2.222	Ea.	167	65.50		232.50	291
0240	4" pipe size	"	3.30	2.424		207	71.50		278.50	345
0500	For round smooth access cover, same price									
4000	Plastic, tees and adapters. Add plugs									
4010	ABS, DWV									
4020	Cleanout tee, 1-1/2" pipe size	1 Plum	15	.533	Ea.	7.75	15.75		23.50	34.50

22 07 Plumbing Insulation

22 07 16 – Plumbing Equipment Insulation

22 07 16.10 Plumbing Equipment Insulation

		Crew	Daily Output	Labor-Hours	Unit	Material	Labor	Equipment	Total	Total Incl O&P
0010	**PLUMBING EQUIPMENT INSULATION**									
2900	Domestic water heater wrap kit									
2920	1-1/2" with vinyl jacket, 20-60 gal.	1 Plum	8	1	Ea.	16.05	29.50		45.55	66

22 07 19 – Plumbing Piping Insulation

22 07 19.10 Piping Insulation

		Crew	Daily Output	Labor-Hours	Unit	Material	Labor	Equipment	Total	Total Incl O&P
0010	**PIPING INSULATION**									
2930	Insulated protectors, (ADA)									
2935	For exposed piping under sinks or lavatories.									
2940	Vinyl coated foam, velcro tabs									
2945	P Trap, 1-1/4" or 1-1/2"	1 Plum	32	.250	Ea.	17.60	7.40		25	31.50
2960	Valve and supply cover									
2965	1/2", 3/8", and 7/16" pipe size	1 Plum	32	.250	Ea.	17.60	7.40		25	31.50
2970	Extension drain cover									
2975	1-1/4", or 1-1/2" pipe size	1 Plum	32	.250	Ea.	18.40	7.40		25.80	32
2985	1-1/4" pipe size	"	32	.250	"	20.50	7.40		27.90	34.50
4000	Pipe covering (price copper tube one size less than IPS)									
6600	Fiberglass, with all service jacket									
6840	1" wall, 1/2" iron pipe size	Q-14	240	.067	L.F.	.89	1.64		2.53	3.81
6860	3/4" iron pipe size		230	.070		.97	1.71		2.68	4.02
6870	1" iron pipe size		220	.073		1.04	1.79		2.83	4.23
6900	2" iron pipe size		200	.080		1.31	1.97		3.28	4.83
7879	Rubber tubing, flexible closed cell foam									
8100	1/2" wall, 1/4" iron pipe size	1 Asbe	90	.089	L.F.	.36	2.44		2.80	4.59
8130	1/2" iron pipe size		89	.090		.44	2.46		2.90	4.72
8140	3/4" iron pipe size		89	.090		.49	2.46		2.95	4.78
8150	1" iron pipe size		88	.091		.54	2.49		3.03	4.88
8170	1-1/2" iron pipe size		87	.092		.76	2.52		3.28	5.20
8180	2" iron pipe size		86	.093		.97	2.55		3.52	5.45
8300	3/4" wall, 1/4" iron pipe size		90	.089		.56	2.44		3	4.81
8330	1/2" iron pipe size		89	.090		.73	2.46		3.19	5.05
8340	3/4" iron pipe size		89	.090		.89	2.46		3.35	5.20
8350	1" iron pipe size		88	.091		1.01	2.49		3.50	5.40
8380	2" iron pipe size		86	.093		1.81	2.55		4.36	6.40
8444	1" wall, 1/2" iron pipe size		86	.093		1.41	2.55		3.96	5.95
8445	3/4" iron pipe size		84	.095		1.71	2.61		4.32	6.35
8446	1" iron pipe size		84	.095		1.99	2.61		4.60	6.70
8447	1-1/4" iron pipe size		82	.098		2.25	2.67		4.92	7.10
8448	1-1/2" iron pipe size		82	.098		2.61	2.67		5.28	7.45
8449	2" iron pipe size		80	.100		3.49	2.74		6.23	8.55

22 07 Plumbing Insulation

22 07 19 – Plumbing Piping Insulation

22 07 19.10 Piping Insulation		Crew	Daily Output	Labor-Hours	Unit	Material	2007 Bare Costs Labor	Equipment	Total	Total Incl O&P
8450	2-1/2" iron pipe size	1 Asbe	80	.100	L.F.	4.55	2.74		7.29	9.70
8456	Rubber insulation tape, 1/8" x 2" x 30'				Ea.	11.40			11.40	12.50

22 11 Facility Water Distribution

22 11 13 – Facility Water Distribution Piping

22 11 13.23 Pipe, Copper

		Crew	Daily Output	Labor-Hours	Unit	Material	Labor	Equipment	Total	Total Incl O&P
0010	**PIPE, COPPER**, Solder joints									
1000	Type K tubing, couplings & clevis hangers 10' O.C.									
1180	3/4" diameter	1 Plum	74	.108	L.F.	7.40	3.19		10.59	13.40
1200	1" diameter	"	66	.121	"	9.80	3.58		13.38	16.65
2000	Type L tubing, couplings & hangers 10' O.C.									
2140	1/2" diameter	1 Plum	81	.099	L.F.	3.41	2.92		6.33	8.55
2160	5/8" diameter		79	.101		4.89	2.99		7.88	10.30
2180	3/4" diameter		76	.105		5.20	3.11		8.31	10.80
2200	1" diameter		68	.118		7.40	3.48		10.88	13.85
2220	1-1/4" diameter		58	.138		10.45	4.08		14.53	18.15
3000	Type M tubing, couplings & hangers 10' O.C.									
3140	1/2" diameter	1 Plum	84	.095	L.F.	2.56	2.81		5.37	7.45
3180	3/4" diameter		78	.103		3.91	3.03		6.94	9.30
3200	1" diameter		70	.114		5.55	3.38		8.93	11.65
3220	1-1/4" diameter		60	.133		8.40	3.94		12.34	15.70
3240	1-1/2" diameter		54	.148		11.50	4.38		15.88	19.85
3260	2" diameter		44	.182		18.15	5.35		23.50	29
4000	Type DWV tubing, couplings & hangers 10' O.C.									
4100	1-1/4" diameter	1 Plum	60	.133	L.F.	9.20	3.94		13.14	16.60
4120	1-1/2" diameter		54	.148		11.55	4.38		15.93	19.90
4140	2" diameter		44	.182		15.45	5.35		20.80	26
4160	3" diameter	Q-1	58	.276		27	7.35		34.35	41.50
4180	4" diameter	"	40	.400		47	10.65		57.65	69

22 11 13.25 Pipe, Copper, Fittings

		Crew	Daily Output	Labor-Hours	Unit	Material	Labor	Equipment	Total	Total Incl O&P
0010	**PIPE, COPPER, FITTINGS**, Wrought unless otherwise noted.									
0040	Solder joints, copper x copper									
0100	1/2"	1 Plum	20	.400	Ea.	.68	11.80		12.48	20
0120	3/4"		19	.421		1.52	12.45		13.97	22
0250	45° elbow, 1/4"		22	.364		3.64	10.75		14.39	21.50
0280	1/2"		20	.400		1.24	11.80		13.04	21
0290	5/8"		19	.421		6.50	12.45		18.95	27.50
0300	3/4"		19	.421		2.17	12.45		14.62	23
0310	1"		16	.500		5.70	14.80		20.50	31
0320	1-1/4"		15	.533		7.80	15.75		23.55	34.50
0450	Tee, 1/4"		14	.571		4.72	16.90		21.62	33
0480	1/2"		13	.615		1.15	18.20		19.35	31.50
0490	5/8"		12	.667		7.85	19.70		27.55	41
0500	3/4"		12	.667		2.79	19.70		22.49	35.50
0510	1"		10	.800		8.65	23.50		32.15	48.50
0520	1-1/4"		9	.889		12.40	26.50		38.90	56.50
0612	Tee, reducing on the outlet, 1/4"		15	.533		7.60	15.75		23.35	34.50
0613	3/8"		15	.533		7.15	15.75		22.90	34
0614	1/2"		14	.571		6.25	16.90		23.15	35
0615	5/8"		13	.615		12.65	18.20		30.85	44
0616	3/4"		12	.667		3.93	19.70		23.63	37

22 11 Facility Water Distribution

22 11 13 – Facility Water Distribution Piping

22 11 13.25 Pipe, Copper, Fittings

		Crew	Daily Output	Labor-Hours	Unit	Material	2007 Bare Costs Labor	Equipment	Total	Total Incl O&P
0617	1"	1 Plum	11	.727	Ea.	9.10	21.50		30.60	45.50
0618	1-1/4"		10	.800		13.40	23.50		36.90	54
0619	1-1/2"		9	.889		14.25	26.50		40.75	58.50
0620	2"		8	1		23	29.50		52.50	74
0621	2-1/2"	Q-1	9	1.778		70	47.50		117.50	155
0622	3"		8	2		77	53		130	173
0623	4"		6	2.667		142	71		213	273
0624	5"		5	3.200		725	85		810	940
0625	6"	Q-2	7	3.429		945	88		1,033	1,200
0626	8"	"	6	4		3,075	102		3,177	3,550
0630	Tee, reducing on the run, 1/4"	1 Plum	15	.533		9.55	15.75		25.30	36.50
0631	3/8"		15	.533		12.75	15.75		28.50	40
0632	1/2"		14	.571		8.75	16.90		25.65	37.50
0633	5/8"		13	.615		13.45	18.20		31.65	45
0634	3/4"		12	.667		5.85	19.70		25.55	39
0635	1"		11	.727		10.85	21.50		32.35	47.50
0636	1-1/4"		10	.800		17.60	23.50		41.10	58.50
0637	1-1/2"		9	.889		29	26.50		55.50	75
0638	2"		8	1		41.50	29.50		71	94.50
0639	2-1/2"	Q-1	9	1.778		94.50	47.50		142	182
0640	3"		8	2		142	53		195	244
0641	4"		6	2.667		296	71		367	440
0642	5"		5	3.200		690	85		775	900
0643	6"	Q-2	7	3.429		995	88		1,083	1,250
0644	8"	"	6	4		3,825	102		3,927	4,375
0650	Coupling, 1/4"	1 Plum	24	.333		.52	9.85		10.37	16.75
0680	1/2"		22	.364		.52	10.75		11.27	18.20
0690	5/8"		21	.381		1.62	11.25		12.87	20.50
0700	3/4"		21	.381		1.01	11.25		12.26	19.60
0710	1"		18	.444		2.05	13.15		15.20	24
0715	1-1/4"		17	.471		3.84	13.90		17.74	27
2000	DWV, solder joints, copper x copper									
2030	90° Elbow, 1-1/4"	1 Plum	13	.615	Ea.	6.90	18.20		25.10	37.50
2050	1-1/2"		12	.667		10.30	19.70		30	44
2070	2"		10	.800		13.45	23.50		36.95	54
2090	3"	Q-1	10	1.600		33.50	42.50		76	107
2100	4"	"	9	1.778		115	47.50		162.50	204
2250	Tee, Sanitary, 1-1/4"	1 Plum	9	.889		13.60	26.50		40.10	58
2270	1-1/2"		8	1		16.95	29.50		46.45	67
2290	2"		7	1.143		19.75	34		53.75	77.50
2310	3"	Q-1	7	2.286		71.50	61		132.50	179
2330	4"	"	6	2.667		110	71		181	238
2400	Coupling, 1-1/4"	1 Plum	14	.571		3.23	16.90		20.13	31.50
2420	1-1/2"		13	.615		4.02	18.20		22.22	34.50
2440	2"		11	.727		5.55	21.50		27.05	41.50
2460	3"	Q-1	11	1.455		10.80	38.50		49.30	75.50
2480	4"	"	10	1.600		34.50	42.50		77	108

22 11 13.44 Pipe, Steel

		Crew	Daily Output	Labor-Hours	Unit	Material	2007 Bare Costs Labor	Equipment	Total	Total Incl O&P
0010	**PIPE, STEEL**									
0050	Schedule 40, threaded, with couplings, and clevis type									
0060	hangers sized for covering, 10' O.C.									
0540	Black, 1/4" diameter	1 Plum	66	.121	L.F.	1.95	3.58		5.53	8.05

22 11 Facility Water Distribution

22 11 13 – Facility Water Distribution Piping

22 11 13.44 Pipe, Steel

		Crew	Daily Output	Labor-Hours	Unit	Material	2007 Bare Costs Labor	Equipment	Total	Total Incl O&P
0570	3/4" diameter	1 Plum	61	.131	L.F.	2.10	3.88		5.98	8.65
0580	1" diameter	↓	53	.151		3.08	4.46		7.54	10.75
0590	1-1/4" diameter	Q-1	89	.180		4.06	4.78		8.84	12.30
0600	1-1/2" diameter		80	.200		4.78	5.30		10.08	14
0610	2" diameter	↓	64	.250	↓	6.40	6.65		13.05	17.95

22 11 13.45 Pipe, Steel, Fittings, Threaded

		Crew	Daily Output	Labor-Hours	Unit	Material	Labor	Equipment	Total	Total Incl O&P
0010	**PIPE, STEEL, FITTINGS, THREADED**									
5000	Malleable iron, 150 lb.									
5020	Black									
5040	90° elbow, straight									
5090	3/4"	1 Plum	14	.571	Ea.	1.98	16.90		18.88	30
5100	1"	"	13	.615		3.44	18.20		21.64	34
5120	1-1/2"	Q-1	20	.800		7.45	21.50		28.95	43
5130	2"	"	18	.889	↓	12.85	23.50		36.35	53
5450	Tee, straight									
5500	3/4"	1 Plum	9	.889	Ea.	3.15	26.50		29.65	46.50
5510	1"	"	8	1		5.40	29.50		34.90	54.50
5520	1-1/4"	Q-1	14	1.143		8.70	30.50		39.20	59.50
5530	1-1/2"		13	1.231		10.85	32.50		43.35	66
5540	2"	↓	11	1.455	↓	18.45	38.50		56.95	84
5650	Coupling									
5700	3/4"	1 Plum	18	.444	Ea.	2.65	13.15		15.80	24.50
5710	1"	"	15	.533		3.97	15.75		19.72	30.50
5730	1-1/2"	Q-1	24	.667		6.95	17.75		24.70	36.50
5740	2"	"	21	.762		10.30	20.50		30.80	45

22 11 13.74 Pipe, Plastic

		Crew	Daily Output	Labor-Hours	Unit	Material	Labor	Equipment	Total	Total Incl O&P
0010	**PIPE, PLASTIC**									
1800	PVC, couplings 10' O.C., hangers 3 per 10'									
1820	Schedule 40									
1860	1/2" diameter	1 Plum	54	.148	L.F.	1.04	4.38		5.42	8.35
1870	3/4" diameter		51	.157		1.21	4.64		5.85	8.95
1880	1" diameter		46	.174		1.47	5.15		6.62	10.05
1890	1-1/4" diameter		42	.190		1.92	5.65		7.57	11.35
1900	1-1/2" diameter	↓	36	.222		2.07	6.55		8.62	13.10
1910	2" diameter	Q-1	59	.271		2.75	7.20		9.95	14.90
1920	2-1/2" diameter		56	.286		3.88	7.60		11.48	16.75
1930	3" diameter		53	.302		5.15	8.05		13.20	18.85
1940	4" diameter	↓	48	.333		6.85	8.85		15.70	22
4100	DWV type, schedule 40, couplings 10' O.C., hangers 3 per 10'									
4120	ABS									
4140	1-1/4" diameter	1 Plum	42	.190	L.F.	1.52	5.65		7.17	10.90
4150	1-1/2" diameter	"	36	.222		1.53	6.55		8.08	12.50
4160	2" diameter	Q-1	59	.271	↓	1.91	7.20		9.11	13.95
4400	PVC									
4410	1-1/4" diameter	1 Plum	42	.190	L.F.	1.64	5.65		7.29	11.05
4420	1-1/2" diameter	"	36	.222		1.43	6.55		7.98	12.40
4460	2" diameter	Q-1	59	.271		1.75	7.20		8.95	13.75
4470	3" diameter		53	.302		3.42	8.05		11.47	16.95
4480	4" diameter	↓	48	.333	↓	4.58	8.85		13.43	19.65
5360	CPVC, couplings 10' O.C., hangers 3 per 10'									
5380	Schedule 40									
5460	1/2" diameter	1 Plum	54	.148	L.F.	2.34	4.38		6.72	9.75

22 11 Facility Water Distribution

22 11 13 – Facility Water Distribution Piping

22 11 13.74 Pipe, Plastic

		Crew	Daily Output	Labor-Hours	Unit	Material	2007 Bare Costs Labor	2007 Bare Costs Equipment	Total	Total Incl O&P
5470	3/4" diameter	1 Plum	51	.157	L.F.	3.21	4.64		7.85	11.15
5480	1" diameter		46	.174		4	5.15		9.15	12.85
5490	1-1/4" diameter		42	.190		4.73	5.65		10.38	14.45
5500	1-1/2" diameter		36	.222		5.30	6.55		11.85	16.65
5510	2" diameter	Q-1	59	.271		6.60	7.20		13.80	19.15
6500	Residential installation, plastic pipe									
6510	Couplings 10' O.C., strap hangers 3 per 10'									
6520	PVC, Schedule 40									
6530	1/2" diameter	1 Plum	138	.058	L.F.	1.06	1.71		2.77	3.98
6540	3/4" diameter		128	.063		1.19	1.85		3.04	4.35
6550	1" diameter		119	.067		1.44	1.99		3.43	4.85
6560	1-1/4" diameter		111	.072		1.73	2.13		3.86	5.40
6570	1-1/2" diameter		104	.077		2.14	2.27		4.41	6.10
6580	2" diameter	Q-1	197	.081		2.72	2.16		4.88	6.55
6590	2-1/2" diameter		162	.099		4.37	2.63		7	9.15
6600	4" diameter		123	.130		7.25	3.46		10.71	13.65
6700	PVC, DWV, Schedule 40									
6720	1-1/4" diameter	1 Plum	100	.080	L.F.	1.70	2.36		4.06	5.75
6730	1-1/2" diameter	"	94	.085		1.77	2.52		4.29	6.10
6740	2" diameter	Q-1	178	.090		2.06	2.39		4.45	6.20
6760	4" diameter	"	110	.145		5.70	3.87		9.57	12.65

22 11 13.76 Pipe, Plastic, Fittings

		Crew	Daily Output	Labor-Hours	Unit	Material	Labor	Equipment	Total	Total Incl O&P
0010	**PIPE, PLASTIC, FITTINGS**									
2700	PVC (white), schedule 40, socket joints									
2760	90° elbow, 1/2"	1 Plum	33.30	.240	Ea.	.33	7.10		7.43	12.05
2770	3/4"		28.60	.280		.37	8.25		8.62	14
2780	1"		25	.320		.66	9.45		10.11	16.30
2790	1-1/4"		22.20	.360		1.17	10.65		11.82	18.80
2800	1-1/2"		20	.400		1.25	11.80		13.05	21
2810	2"	Q-1	36.40	.440		1.97	11.70		13.67	21.50
2820	2-1/2"		26.70	.599		5.95	15.95		21.90	32.50
2830	3"		22.90	.699		7.15	18.60		25.75	38.50
2840	4"		18.20	.879		12.80	23.50		36.30	52.50
3180	Tee, 1/2"	1 Plum	22.20	.360		.41	10.65		11.06	17.95
3190	3/4"		19	.421		.47	12.45		12.92	21
3200	1"		16.70	.479		.88	14.15		15.03	24.50
3210	1-1/4"		14.80	.541		1.37	15.95		17.32	28
3220	1-1/2"		13.30	.602		1.67	17.75		19.42	31
3230	2"	Q-1	24.20	.661		2.42	17.60		20.02	31.50
3240	2-1/2"		17.80	.899		8	24		32	48.50
3250	3"		15.20	1.053		10.50	28		38.50	57.50
3260	4"		12.10	1.322		19	35		54	79
3380	Coupling, 1/2"	1 Plum	33.30	.240		.22	7.10		7.32	11.95
3390	3/4"		28.60	.280		.30	8.25		8.55	13.95
3400	1"		25	.320		.52	9.45		9.97	16.10
3410	1-1/4"		22.20	.360		.71	10.65		11.36	18.30
3420	1-1/2"		20	.400		.77	11.80		12.57	20.50
3430	2"	Q-1	36.40	.440		1.18	11.70		12.88	20.50
3440	2-1/2"		26.70	.599		2.60	15.95		18.55	29
3450	3"		22.90	.699		4.06	18.60		22.66	35
3460	4"		18.20	.879		5.85	23.50		29.35	45
4500	DWV, ABS, non pressure, socket joints									

22 11 Facility Water Distribution

22 11 13 – Facility Water Distribution Piping

22 11 13.76 Pipe, Plastic, Fittings		Crew	Daily Output	Labor-Hours	Unit	Material	2007 Bare Costs Labor	Equipment	Total	Total Incl O&P
4540	1/4 Bend, 1-1/4"	1 Plum	20.20	.396	Ea.	3.62	11.70		15.32	23
4560	1-1/2"	"	18.20	.440		2.69	13		15.69	24.50
4570	2"	Q-1	33.10	.483		4.19	12.85		17.04	25.50
4800	Tee, sanitary									
4820	1-1/4"	1 Plum	13.50	.593	Ea.	4.33	17.50		21.83	34
4830	1-1/2"	"	12.10	.661		3.69	19.55		23.24	36
4840	2"	Q-1	20	.800		5.85	21.50		27.35	41.50
5000	DWV, PVC, schedule 40, socket joints									
5040	1/4 bend, 1-1/4"	1 Plum	20.20	.396	Ea.	5.85	11.70		17.55	25.50
5060	1-1/2"	"	18.20	.440		1.92	13		14.92	23.50
5070	2"	Q-1	33.10	.483		2.82	12.85		15.67	24
5080	3"		20.80	.769		8.80	20.50		29.30	43
5090	4"		16.50	.970		15.20	26		41.20	59.50
5110	1/4 bend, long sweep, 1-1/2"	1 Plum	18.20	.440		4.19	13		17.19	26
5112	2"	Q-1	33.10	.483		4.51	12.85		17.36	26
5114	3"		20.80	.769		10.40	20.50		30.90	45
5116	4"		16.50	.970		19.80	26		45.80	64.50
5250	Tee, sanitary 1-1/4"	1 Plum	13.50	.593		7.10	17.50		24.60	37
5254	1-1/2"	"	12.10	.661		3.30	19.55		22.85	35.50
5255	2"	Q-1	20	.800		4.63	21.50		26.13	40
5256	3"		13.90	1.151		11.85	30.50		42.35	63.50
5257	4"		11	1.455		20.50	38.50		59	86
5259	6"		6.70	2.388		90.50	63.50		154	204
5261	8"	Q-2	6.20	3.871		270	99		369	460
5264	2" x 1-1/2"	Q-1	22	.727		6.95	19.35		26.30	39.50
5266	3" x 1-1/2"		15.50	1.032		9.75	27.50		37.25	55.50
5268	4" x 3"		12.10	1.322		21.50	35		56.50	81.50
5271	6" x 4"		6.90	2.319		68	61.50		129.50	176
5314	Combination Y & 1/8 bend, 1-1/2"	1 Plum	12.10	.661		6.85	19.55		26.40	39.50
5315	2"	Q-1	20	.800		9.05	21.50		30.55	45
5317	3"		13.90	1.151		22	30.50		52.50	74.50
5318	4"		11	1.455		42	38.50		80.50	110
5324	Combination Y & 1/8 bend, reducing									
5325	2" x 2" x 1-1/2"	Q-1	22	.727	Ea.	9.85	19.35		29.20	43
5327	3" x 3" x 1-1/2"		15.50	1.032		17.05	27.50		44.55	64
5328	3" x 3" x 2"		15.30	1.046		13.05	28		41.05	60.50
5329	4" x 4" x 2"		12.20	1.311		22	35		57	81.50
5331	Wye, 1-1/4"	1 Plum	13.50	.593		6.40	17.50		23.90	36
5332	1-1/2"	"	12.10	.661		4.81	19.55		24.36	37.50
5333	2"	Q-1	20	.800		5.65	21.50		27.15	41
5334	3"		13.90	1.151		16.20	30.50		46.70	68.50
5335	4"		11	1.455		27	38.50		65.50	93
5336	6"		6.70	2.388		70.50	63.50		134	182
5337	8"	Q-2	6.20	3.871		120	99		219	295
5341	2" x 1-1/2"	Q-1	22	.727		6.95	19.35		26.30	39.50
5342	3" x 1-1/2"		15.50	1.032		9.75	27.50		37.25	55.50
5343	4" x 3"		12.10	1.322		27	35		62	87.50
5344	6" x 4"		6.90	2.319		61.50	61.50		123	169
5345	8" x 6"	Q-2	6.40	3.750		91	96		187	258
5347	Double wye, 1-1/2"	1 Plum	9.10	.879		10.90	26		36.90	54.50
5348	2"	Q-1	16.60	.964		11.65	25.50		37.15	55
5349	3"		10.40	1.538		30	41		71	101
5350	4"		8.25	1.939		61.50	51.50		113	153

22 11 Facility Water Distribution

22 11 13 – Facility Water Distribution Piping

22 11 13.76 Pipe, Plastic, Fittings

		Crew	Daily Output	Labor-Hours	Unit	Material	2007 Bare Costs Labor	2007 Bare Costs Equipment	Total	Total Incl O&P
5354	2" x 1-1/2"	Q-1	16.80	.952	Ea.	10.65	25.50		36.15	53
5355	3" x 2"		10.60	1.509		22.50	40		62.50	91
5356	4" x 3"		8.45	1.893		48.50	50.50		99	137
5357	6" x 4"		7.25	2.207		123	58.50		181.50	232
5410	Reducer bushing, 2" x 1-1/4"		36.50	.438		4.63	11.65		16.28	24.50
5412	3" x 1-1/2"		27.30	.586		7.10	15.60		22.70	33.50
5414	4" x 2"		18.20	.879		13.80	23.50		37.30	53.50
5416	6" x 4"		11.10	1.441		36.50	38.50		75	103
5418	8" x 6"	Q-2	10.20	2.353		79.50	60.50		140	187
5500	CPVC, Schedule 80, threaded joints									
5540	90° Elbow, 1/4"	1 Plum	32	.250	Ea.	9.15	7.40		16.55	22.50
5560	1/2"		30.30	.264		5.35	7.80		13.15	18.70
5570	3/4"		26	.308		7.95	9.10		17.05	23.50
5580	1"		22.70	.352		11.20	10.40		21.60	29.50
5590	1-1/4"		20.20	.396		21.50	11.70		33.20	43
5600	1-1/2"		18.20	.440		23	13		36	47
5610	2"	Q-1	33.10	.483		31	12.85		43.85	55
6000	Coupling, 1/4"	1 Plum	32	.250		11.70	7.40		19.10	25
6020	1/2"		30.30	.264		9.60	7.80		17.40	23.50
6030	3/4"		26	.308		15.55	9.10		24.65	32
6040	1"		22.70	.352		17.65	10.40		28.05	36.50
6050	1-1/4"		20.20	.396		18.70	11.70		30.40	40
6060	1-1/2"		18.20	.440		20	13		33	43.50
6070	2"	Q-1	33.10	.483		23.50	12.85		36.35	47

22 11 19 – Domestic Water Piping Specialties

22 11 19.38 Water Supply Meters

		Crew	Daily Output	Labor-Hours	Unit	Material	2007 Bare Costs Labor	2007 Bare Costs Equipment	Total	Total Incl O&P
0010	**WATER SUPPLY METERS**									
2000	Domestic/commercial, bronze									
2020	Threaded									
2060	5/8" diameter, to 20 GPM	1 Plum	16	.500	Ea.	40	14.80		54.80	68.50
2080	3/4" diameter, to 30 GPM		14	.571		67.50	16.90		84.40	103
2100	1" diameter, to 50 GPM		12	.667		94	19.70		113.70	136

22 11 19.42 Backflow Preventers

		Crew	Daily Output	Labor-Hours	Unit	Material	2007 Bare Costs Labor	2007 Bare Costs Equipment	Total	Total Incl O&P
0010	**BACKFLOW PREVENTERS**, Includes valves									
0020	and four test cocks, corrosion resistant, automatic operation									
4100	Threaded, bronze, valves are ball									
4120	3/4" pipe size	1 Plum	16	.500	Ea.	230	14.80		244.80	278

22 11 19.46 Vacuum Breakers

		Crew	Daily Output	Labor-Hours	Unit	Material	2007 Bare Costs Labor	2007 Bare Costs Equipment	Total	Total Incl O&P
0010	**VACUUM BREAKERS**, Hot or cold water									
1030	Anti-siphon, brass									
1060	1/2" size	1 Plum	24	.333	Ea.	31	9.85		40.85	50
1080	3/4" size		20	.400		36.50	11.80		48.30	59.50
1100	1" size		19	.421		57	12.45		69.45	83.50

22 11 19.54 Water Hammer Arresters/Shock Absorbers

		Crew	Daily Output	Labor-Hours	Unit	Material	2007 Bare Costs Labor	2007 Bare Costs Equipment	Total	Total Incl O&P
0010	**WATER HAMMER ARRESTERS/SHOCK ABSORBERS**									
0490	Copper									
0500	3/4" male I.P.S. For 1 to 11 fixtures	1 Plum	12	.667	Ea.	15.55	19.70		35.25	49.50

22 13 Facility Sanitary Sewerage

22 13 16 – Sanitary Waste and Vent Piping

22 13 16.20 Pipe, Cast Iron

		Crew	Daily Output	Labor-Hours	Unit	Material	2007 Bare Costs Labor	2007 Bare Costs Equipment	Total	Total Incl O&P
0010	**PIPE, CAST IRON**, Soil, on hangers 5' O.C. R221113-50									
0020	Single hub, service wt., lead & oakum joints 10' O.C.									
2120	2" diameter	Q-1	63	.254	L.F.	5.20	6.75		11.95	16.80
2140	3" diameter		60	.267		7.25	7.10		14.35	19.65
2160	4" diameter		55	.291		9.35	7.75		17.10	23
4000	No hub, couplings 10' O.C.									
4100	1-1/2" diameter	Q-1	71	.225	L.F.	5.45	6		11.45	15.85
4120	2" diameter		67	.239		5.60	6.35		11.95	16.65
4140	3" diameter		64	.250		7.60	6.65		14.25	19.30
4160	4" diameter		58	.276		9.65	7.35		17	22.50

22 13 16.30 Pipe, Cast Iron Fittings

		Crew	Daily Output	Labor-Hours	Unit	Material	2007 Bare Costs Labor	2007 Bare Costs Equipment	Total	Total Incl O&P
0010	**PIPE, CAST IRON FITTINGS**, Soil									
0040	Hub and spigot, service weight, lead & oakum joints									
0080	1/4 bend, 2"	Q-1	16	1	Ea.	11.20	26.50		37.70	56.50
0120	3"		14	1.143		15	30.50		45.50	66.50
0140	4"		13	1.231		23.50	32.50		56	80
0340	1/8 bend, 2"		16	1		8	26.50		34.50	53
0350	3"		14	1.143		12.55	30.50		43.05	64
0360	4"		13	1.231		18.35	32.50		50.85	74
0500	Sanitary tee, 2"		10	1.600		15.75	42.50		58.25	87.50
0540	3"		9	1.778		25.50	47.50		73	106
0620	4"		8	2		31	53		84	122
5990	No hub									
6000	Cplg. & labor required at joints not incl. in fitting									
6010	price. Add 1 coupling per joint for installed price									
6020	1/4 Bend, 1-1/2"				Ea.	6.30			6.30	6.95
6060	2"					6.95			6.95	7.60
6080	3"					9.50			9.50	10.45
6120	4"					13.75			13.75	15.10
6184	1/4 Bend, long sweep, 1-1/2"					14.85			14.85	16.35
6186	2"					14.85			14.85	16.35
6188	3"					17.60			17.60	19.35
6189	4"					28			28	31
6190	5"					52			52	57
6191	6"					63			63	69.50
6192	8"					153			153	168
6193	10"					275			275	305
6200	1/8 Bend, 1-1/2"					5.30			5.30	5.80
6210	2"					5.85			5.85	6.45
6212	3"					7.95			7.95	8.75
6214	4"					10.10			10.10	11.10
6380	Sanitary Tee, tapped, 1-1/2"					11.60			11.60	12.75
6382	2" x 1-1/2"					11.60			11.60	12.75
6384	2"					11.70			11.70	12.90
6386	3" x 2"					16.30			16.30	17.95
6388	3"					30			30	33
6390	4" x 1-1/2"					14.50			14.50	15.95
6392	4" x 2"					16.40			16.40	18
6394	6" x 1-1/2"					34			34	37
6396	6" x 2"					34			34	37.50
6459	Sanitary Tee, 1-1/2"					8.75			8.75	9.65
6460	2"					9.50			9.50	10.45

22 13 Facility Sanitary Sewerage

22 13 16 – Sanitary Waste and Vent Piping

22 13 16.30 Pipe, Cast Iron Fittings		Crew	Daily Output	Labor-Hours	Unit	Material	2007 Bare Costs Labor	Equipment	Total	Total Incl O&P
6470	3"				Ea.	11.60			11.60	12.75
6472	4"				↓	17.95			17.95	19.75
8000	Coupling, standard (by CISPI Mfrs.)									
8020	1-1/2"	Q-1	48	.333	Ea.	6.35	8.85		15.20	21.50
8040	2"		44	.364		6.35	9.65		16	23
8080	3"		38	.421		7.55	11.20		18.75	26.50
8120	4"	↓	33	.485	↓	8.90	12.90		21.80	31

22 13 16.60 Traps										
0010	**TRAPS**									
0030	Cast iron, service weight									
0050	Running P trap, without vent									
1100	2"	Q-1	16	1	Ea.	31	26.50		57.50	78
1150	4"	"	13	1.231		90	32.50		122.50	153
1160	6"	Q-2	17	1.412		390	36		426	490
3000	P trap, B&S, 2" pipe size	Q-1	16	1		21.50	26.50		48	67.50
3040	3" pipe size	"	14	1.143	↓	32	30.50		62.50	85
4700	Copper, drainage, drum trap									
4840	3" x 6" swivel, 1-1/2" pipe size	1 Plum	16	.500	Ea.	98	14.80		112.80	133
5100	P trap, standard pattern									
5200	1-1/4" pipe size	1 Plum	18	.444	Ea.	45.50	13.15		58.65	71.50
5240	1-1/2" pipe size		17	.471		44	13.90		57.90	71.50
5260	2" pipe size		15	.533		68	15.75		83.75	101
5280	3" pipe size	↓	11	.727	↓	163	21.50		184.50	216
6710	ABS DWV P trap, solvent weld joint									
6720	1-1/2" pipe size	1 Plum	18	.444	Ea.	7.30	13.15		20.45	29.50
6722	2" pipe size		17	.471		9.90	13.90		23.80	34
6724	3" pipe size		15	.533		38	15.75		53.75	68
6726	4" pipe size		14	.571		78	16.90		94.90	114
6860	PVC DWV hub x hub, basin trap, 1-1/4" pipe size		18	.444		9.40	13.15		22.55	32
6870	Sink P trap, 1-1/2" pipe size		18	.444		9.40	13.15		22.55	32
6880	Tubular S trap, 1-1/2" pipe size	↓	17	.471	↓	17.40	13.90		31.30	42
6890	PVC sch. 40 DWV, drum trap									
6900	1-1/2" pipe size	1 Plum	16	.500	Ea.	25	14.80		39.80	52
6910	P trap, 1-1/2" pipe size		18	.444		6.15	13.15		19.30	28.50
6920	2" pipe size		17	.471		8.10	13.90		22	32
6930	3" pipe size		15	.533		28	15.75		43.75	57
6940	4" pipe size		14	.571		64.50	16.90		81.40	99
6950	P trap w/clean out, 1-1/2" pipe size		18	.444		10.20	13.15		23.35	32.50
6960	2" pipe size		17	.471		17.30	13.90		31.20	42

22 13 16.80 Vent Flashing Caps										
0010	**VENT FLASHING CAPS**									
0120	Vent caps									
0140	Cast iron									
0160	1-1/4" - 1-1/2" pipe	1 Plum	23	.348	Ea.	27.50	10.30		37.80	47.50
0170	2" - 2-1/8" pipe		22	.364		32	10.75		42.75	53
0180	2-1/2" - 3-5/8" pipe		21	.381		36.50	11.25		47.75	58.50
0190	4" - 4-1/8" pipe		19	.421		44	12.45		56.45	68.50
0200	5" - 6" pipe	↓	17	.471	↓	65	13.90		78.90	94.50
0300	PVC									
0320	1-1/4" - 1-1/2" pipe	1 Plum	24	.333	Ea.	7.30	9.85		17.15	24
0330	2" - 2-1/8" pipe	"	23	.348	"	7.85	10.30		18.15	25.50

22 13 Facility Sanitary Sewerage

22 13 19 – Sanitary Waste Piping Specialties

22 13 19.13 Sanitary Drains

		Crew	Daily Output	Labor-Hours	Unit	Material	2007 Bare Costs Labor	2007 Bare Costs Equipment	Total	Total Incl O&P
0010	**SANITARY DRAINS**									
2000	Floor, medium duty, C.I., deep flange, 7" dia top									
2040	2" and 3" pipe size	Q-1	12	1.333	Ea.	115	35.50		150.50	185
2080	For galvanized body, add					48.50			48.50	53
2120	For polished bronze top, add					63.50			63.50	70

22 14 Facility Storm Drainage

22 14 26 – Facility Storm Drains

22 14 26.13 Roof Drains

		Crew	Daily Output	Labor-Hours	Unit	Material	Labor	Equipment	Total	Total Incl O&P
0010	**ROOF DRAINS**									
3860	Roof, flat metal deck, C.I. body, 12" C.I. dome									
3890	3" pipe size	Q-1	14	1.143	Ea.	225	30.50		255.50	297

22 14 29 – Sump Pumps

22 14 29.16 Submersible Sump Pumps

		Crew	Daily Output	Labor-Hours	Unit	Material	Labor	Equipment	Total	Total Incl O&P
0010	**SUBMERSIBLE SUMP PUMPS**									
7000	Sump pump, automatic									
7100	Plastic, 1-1/4" discharge, 1/4 HP	1 Plum	6	1.333	Ea.	122	39.50		161.50	199
7500	Cast iron, 1-1/4" discharge, 1/4 HP	"	6	1.333	"	140	39.50		179.50	219

22 31 Domestic Water Softeners

22 31 13 – Domestic Water Softeners

22 31 13.10 Residential Water Softeners

		Crew	Daily Output	Labor-Hours	Unit	Material	Labor	Equipment	Total	Total Incl O&P
0010	**RESIDENTIAL WATER SOFTENERS**									
7350	Water softener, automatic, to 30 grains per gallon	2 Plum	5	3.200	Ea.	465	94.50		559.50	665
7400	To 100 grains per gallon	"	4	4	"	645	118		763	905

22 33 Electric Domestic Water Heaters

22 33 30 – Residential, Electric Domestic Water Heaters

22 33 30.13 Residential, Small-Capacity Elec. Water Heaters

		Crew	Daily Output	Labor-Hours	Unit	Material	Labor	Equipment	Total	Total Incl O&P
0010	**RESIDENTIAL, SMALL-CAPACITY ELECTRIC DOMESTIC WATER HEATERS**									
1000	Residential, electric, glass lined tank, 5 yr, 10 gal., single element	1 Plum	2.30	3.478	Ea.	273	103		376	470
1060	30 gallon, double element		2.20	3.636		385	107		492	600
1080	40 gallon, double element		2	4		410	118		528	645
1100	52 gallon, double element		2	4		460	118		578	700
1120	66 gallon, double element		1.80	4.444		625	131		756	905
1140	80 gallon, double element		1.60	5		700	148		848	1,025

22 34 Fuel-Fired Domestic Water Heaters

22 34 30 – Residential Gas Domestic Water Heaters

22 34 30.13 Residential, Atmos, Gas Domestic Wtr Heaters	Crew	Daily Output	Labor-Hours	Unit	Material	2007 Bare Costs Labor	Equipment	Total	Total Incl O&P
0010 **RESIDENTIAL, ATMOSPHERIC, GAS DOMESTIC WATER HEATERS**									
2000 Gas fired, foam lined tank, 10 yr, vent not incl.,									
2040 30 gallon	1 Plum	2	4	Ea.	625	118		743	885
2100 75 gallon	"	1.50	5.333	"	880	158		1,038	1,225

22 34 46 – Oil-Fired Domestic Water Heaters

22 34 46.10 Residential Oil-Fired Water Heaters

	Crew	Daily Output	Labor-Hours	Unit	Material	Labor	Equipment	Total	Total Incl O&P
0010 **RESIDENTIAL OIL-FIRED WATER HEATERS**									
3000 Oil fired, glass lined tank, 5 yr, vent not included, 30 gallon	1 Plum	2	4	Ea.	665	118		783	925
3040 50 gallon	"	1.80	4.444	"	1,375	131		1,506	1,725

22 41 Residential Plumbing Fixtures

22 41 13 – Residential Water Closets, Urinals, and Bidets

22 41 13.40 Water Closets

	Crew	Daily Output	Labor-Hours	Unit	Material	Labor	Equipment	Total	Total Incl O&P
0010 **WATER CLOSETS**									
0150 Tank type, vitreous china, incl. seat, supply pipe w/stop									
0200 Wall hung									
0400 Two piece, close coupled	Q-1	5.30	3.019	Ea.	550	80.50		630.50	735
0960 For rough-in, supply, waste, vent and carrier		2.73	5.861		490	156		646	795
1000 Floor mounted, one piece		5.30	3.019		530	80.50		610.50	715
1020 One piece, low profile		5.30	3.019		400	80.50		480.50	570
1100 Two piece, close coupled		5.30	3.019		183	80.50		263.50	335
1960 For color, add					30%				
1980 For rough-in, supply, waste and vent	Q-1	3.05	5.246	Ea.	218	140		358	470

22 41 16 – Residential Lavatories and Sinks

22 41 16.10 Lavatories

	Crew	Daily Output	Labor-Hours	Unit	Material	Labor	Equipment	Total	Total Incl O&P
0010 **LAVATORIES**, With trim, white unless noted otherwise									
0500 Vanity top, porcelain enamel on cast iron									
0600 20" x 18"	Q-1	6.40	2.500	Ea.	220	66.50		286.50	350
0640 33" x 19" oval		6.40	2.500		460	66.50		526.50	615
0720 19" round		6.40	2.500		207	66.50		273.50	335
0860 For color, add					25%				
1000 Cultured marble, 19" x 17", single bowl	Q-1	6.40	2.500	Ea.	169	66.50		235.50	295
1120 25" x 22", single bowl		6.40	2.500		181	66.50		247.50	310
1160 37" x 22", single bowl		6.40	2.500		215	66.50		281.50	345
1560									
1900 Stainless steel, self-rimming, 25" x 22", single bowl, ledge	Q-1	6.40	2.500	Ea.	253	66.50		319.50	385
1960 17" x 22", single bowl		6.40	2.500		246	66.50		312.50	380
2600 Steel, enameled, 20" x 17", single bowl		5.80	2.759		145	73.50		218.50	281
2900 Vitreous china, 20" x 16", single bowl		5.40	2.963		265	79		344	420
3200 22" x 13", single bowl		5.40	2.963		247	79		326	400
3580 Rough-in, supply, waste and vent for all above lavatories		2.30	6.957		169	185		354	490
4000 Wall hung									
4040 Porcelain enamel on cast iron, 16" x 14", single bowl	Q-1	8	2	Ea.	360	53		413	485
4180 20" x 18", single bowl	"	8	2	"	279	53		332	395
4580 For color, add					30%				
6000 Vitreous china, 18" x 15", single bowl with backsplash	Q-1	7	2.286	Ea.	213	61		274	335
6060 19" x 17", single bowl		7	2.286		179	61		240	297
6960 Rough-in, supply, waste and vent for above lavatories		1.66	9.639		365	256		621	820

22 41 16.30 Sinks

	Crew	Daily Output	Labor-Hours	Unit	Material	Labor	Equipment	Total	Total Incl O&P
0010 **SINKS**, With faucets and drain.									

22 41 Residential Plumbing Fixtures

22 41 16 – Residential Lavatories and Sinks

22 41 16.30 Sinks

		Crew	Daily Output	Labor-Hours	Unit	Material	2007 Bare Costs Labor	2007 Bare Costs Equipment	Total	Total Incl O&P
2000	Kitchen, counter top style, P.E. on C.I., 24" x 21" single bowl R224000-40	Q-1	5.60	2.857	Ea.	251	76		327	400
2100	31" x 22" single bowl		5.60	2.857		237	76		313	385
2200	32" x 21" double bowl		4.80	3.333		325	88.50		413.50	505
3000	Stainless steel, self rimming, 19" x 18" single bowl		5.60	2.857		380	76		456	545
3100	25" x 22" single bowl		5.60	2.857		425	76		501	590
3200	33" x 22" double bowl		4.80	3.333		610	88.50		698.50	820
3300	43" x 22" double bowl		4.80	3.333		710	88.50		798.50	925
4000	Steel, enameled, with ledge, 24" x 21" single bowl		5.60	2.857		145	76		221	284
4100	32" x 21" double bowl		4.80	3.333		166	88.50		254.50	330
4960	For color sinks except stainless steel, add					10%				
4980	For rough-in, supply, waste and vent, counter top sinks	Q-1	2.14	7.477		203	199		402	550
5000	Kitchen, raised deck, P.E. on C.I.									
5100	32" x 21", dual level, double bowl	Q-1	2.60	6.154	Ea.	268	164		432	565
5790	For rough-in, supply, waste & vent, sinks	"	1.85	8.649	"	203	230		433	605

22 41 19 – Residential Bathtubs

22 41 19.10 Baths

		Crew	Daily Output	Labor-Hours	Unit	Material	Labor	Equipment	Total	Total Incl O&P
0010	**BATHS** R224000-40									
0100	Tubs, recessed porcelain enamel on cast iron, with trim									
0180	48" x 42"	Q-1	4	4	Ea.	1,675	106		1,781	2,000
0220	72" x 36"	"	3	5.333	"	1,725	142		1,867	2,125
0300	Mat bottom									
0380	5' long	Q-1	4.40	3.636	Ea.	755	96.50		851.50	990
0480	Above floor drain, 5' long		4	4		700	106		806	945
0560	Corner 48" x 44"		4.40	3.636		1,675	96.50		1,771.50	1,975
2000	Enameled formed steel, 4'-6" long		5.80	2.759		345	73.50		418.50	500
4600	Module tub & showerwall surround, molded fiberglass									
4610	5' long x 34" wide x 76" high	Q-1	4	4	Ea.	545	106		651	770
9600	Rough-in, supply, waste and vent, for all above tubs, add	"	2.07	7.729	"	219	206		425	580

22 41 23 – Residential Shower Receptors and Basins

22 41 23.20 Showers

		Crew	Daily Output	Labor-Hours	Unit	Material	Labor	Equipment	Total	Total Incl O&P
0010	**SHOWERS**									
1500	Stall, with drain only. Add for valve and door/curtain									
1520	32" square	Q-1	5	3.200	Ea.	345	85		430	520
1530	36" square		4.80	3.333		450	88.50		538.50	640
1540	Terrazzo receptor, 32" square		5	3.200		740	85		825	955
1560	36" square		4.80	3.333		875	88.50		963.50	1,100
1580	36" corner angle		4.80	3.333		770	88.50		858.50	990
3000	Fiberglass, one piece, with 3 walls, 32" x 32" square		5.50	2.909		455	77.50		532.50	625
3100	36" x 36" square		5.50	2.909		525	77.50		602.50	700
4200	Rough-in, supply, waste and vent for above showers		2.05	7.805		310	208		518	685

22 41 36 – Residential Laundry Trays

22 41 36.10 Laundry Sinks

		Crew	Daily Output	Labor-Hours	Unit	Material	Labor	Equipment	Total	Total Incl O&P
0010	**LAUNDRY SINKS**, With trim									
0020	Porcelain enamel on cast iron, black iron frame									
0050	24" x 21", single compartment	Q-1	6	2.667	Ea.	365	71		436	515
0100	26" x 21", single compartment	"	6	2.667	"	375	71		446	525
3000	Plastic, on wall hanger or legs									
3020	18" x 23", single compartment	Q-1	6.50	2.462	Ea.	97	65.50		162.50	215
3100	20" x 24", single compartment		6.50	2.462		129	65.50		194.50	249
3200	36" x 23", double compartment		5.50	2.909		154	77.50		231.50	297
3300	40" x 24", double compartment		5.50	2.909		227	77.50		304.50	375

22 41 Residential Plumbing Fixtures

22 41 36 – Residential Laundry Trays

22 41 36.10 Laundry Sinks		Crew	Daily Output	Labor-Hours	Unit	Material	2007 Bare Costs Labor	Equipment	Total	Total Incl O&P
5000	Stainless steel, counter top, 22" x 17" single compartment	Q-1	6	2.667	Ea.	57.50	71		128.50	180
5200	33" x 22", double compartment		5	3.200		69.50	85		154.50	217
9600	Rough-in, supply, waste and vent, for all laundry sinks	↓	2.14	7.477	↓	203	199		402	550

22 41 39 – Residential Faucets, Supplies, and Trim

22 41 39.10 Faucets and Fittings

		Crew	Daily Output	Labor-Hours	Unit	Material	Labor	Equipment	Total	Total Incl O&P
0010	**FAUCETS AND FITTINGS**									
0150	Bath, faucets, diverter spout combination, sweat	1 Plum	8	1	Ea.	104	29.50		133.50	163
0200	For integral stops, IPS unions, add					109			109	120
0420	Bath, press-bal mix valve w/diverter, spout, shower hd, arm/flange	1 Plum	8	1		119	29.50		148.50	180
0500	Drain, central lift, 1-1/2" IPS male		20	.400		40.50	11.80		52.30	64
0600	Trip lever, 1-1/2" IPS male		20	.400		41	11.80		52.80	64.50
1000	Kitchen sink faucets, top mount, cast spout		10	.800		55	23.50		78.50	99.50
1100	For spray, add	↓	24	.333		16.15	9.85		26	34
1300	Single control lever handle									
1310	With pull out spray									
1320	Polished chrome	1 Plum	10	.800	Ea.	167	23.50		190.50	222
1330	Polished brass		10	.800		200	23.50		223.50	259
1340	White	↓	10	.800	↓	184	23.50		207.50	241
1348	With spray thru escutcheon									
1350	Polished chrome	1 Plum	10	.800	Ea.	117	23.50		140.50	167
1360	Polished brass		10	.800		140	23.50		163.50	193
1370	White		10	.800		128	23.50		151.50	180
2000	Laundry faucets, shelf type, IPS or copper unions	↓	12	.667	↓	45	19.70		64.70	82
2020										
2100	Lavatory faucet, centerset, without drain	1 Plum	10	.800	Ea.	40	23.50		63.50	83
2120	With pop-up drain	"	6.66	1.201	"	55.50	35.50		91	120
2210	Porcelain cross handles and pop-up drain									
2220	Polished chrome	1 Plum	6.66	1.201	Ea.	138	35.50		173.50	211
2230	Polished brass	"	6.66	1.201	"	208	35.50		243.50	288
2260	Single lever handle and pop-up drain									
2270	Black nickel	1 Plum	6.66	1.201	Ea.	220	35.50		255.50	300
2280	Polished brass		6.66	1.201		220	35.50		255.50	300
2290	Polished chrome		6.66	1.201		161	35.50		196.50	236
2800	Self-closing, center set		10	.800		191	23.50		214.50	249
3000	Service sink faucet, cast spout, pail hook, hose end		14	.571		76.50	16.90		93.40	113
4000	Shower by-pass valve with union		18	.444		54	13.15		67.15	81
4200	Shower thermostatic mixing valve, concealed	↓	8	1	↓	246	29.50		275.50	320
4204	For inlet strainer, check, and stops, add					34.50			34.50	38
4220	Shower pressure balancing mixing valve,									
4230	With shower head, arm, flange and diverter tub spout									
4240	Chrome	1 Plum	6.14	1.303	Ea.	141	38.50		179.50	219
4250	Polished brass		6.14	1.303		198	38.50		236.50	281
4260	Satin		6.14	1.303		198	38.50		236.50	281
4270	Polished chrome/brass		6.14	1.303		162	38.50		200.50	242
5000	Sillcock, compact, brass, IPS or copper to hose	↓	24	.333		6.65	9.85		16.50	23.50

22 42 Commercial Plumbing Fixtures

22 42 13 – Commercial Water Closets, Urinals, and Bidets

22 42 13.40 Water Closets

		Crew	Daily Output	Labor-Hours	Unit	Material	2007 Bare Costs Labor	2007 Bare Costs Equipment	Total	Total Incl O&P
0010	**WATER CLOSETS**									
3000	Bowl only, with flush valve, seat									
3100	Wall hung	Q-1	5.80	2.759	Ea.	315	73.50		388.50	465
3200	For rough-in, supply, waste and vent, single WC		2.56	6.250		520	166		686	845
3300	Floor mounted		5.80	2.759		375	73.50		448.50	535
3400	For rough-in, supply, waste and vent, single WC		2.84	5.634		248	150		398	520

22 42 16 – Commercial Lavatories and Sinks

22 42 16.40 Service Sinks

		Crew	Daily Output	Labor-Hours	Unit	Material	Labor	Equipment	Total	Total Incl O&P
0010	**SERVICE SINKS**									
6650	Service, floor, corner, P.E. on C.I., 28" x 28"	Q-1	4.40	3.636	Ea.	605	96.50		701.50	825
6750	Vinyl coated rim guard, add					67.50			67.50	74
6760	Mop sink, molded stone, 24" x 36"	1 Plum	3.33	2.402		176	71		247	310
6770	Mop sink, molded stone, 24" x 36", w/rim 3 sides	"	3.33	2.402		195	71		266	330
6790	For rough-in, supply, waste & vent, floor service sinks	Q-1	1.64	9.756		500	260		760	975

22 42 39 – Commercial Faucets, Supplies, and Trim

22 42 39.30 Carriers and Supports

		Crew	Daily Output	Labor-Hours	Unit	Material	Labor	Equipment	Total	Total Incl O&P
0010	**CARRIERS AND SUPPORTS**, For plumbing fixtures.									
0600	Plate type with studs, top back plate	1 Plum	7	1.143	Ea.	51	34		85	112
3000	Lavatory, concealed arm									
3050	Floor mounted, single									
3100	High back fixture	1 Plum	6	1.333	Ea.	285	39.50		324.50	380
3200	Flat slab fixture	"	6	1.333	"	246	39.50		285.50	335
8200	Water closet, residential									
8220	Vertical centerline, floor mount									
8240	Single, 3" caulk, 2" or 3" vent	1 Plum	6	1.333	Ea.	330	39.50		369.50	425
8260	4" caulk, 2" or 4" vent	"	6	1.333	"	425	39.50		464.50	530

22 51 Swimming Pool Plumbing Systems

22 51 19 – Swimming Pool Water Treatment Equipment

22 51 19.50 Swimming Pool Filtration Equipment

		Crew	Daily Output	Labor-Hours	Unit	Material	Labor	Equipment	Total	Total Incl O&P
0010	**SWIMMING POOL FILTRATION EQUIPMENT**									
0900	Filter system, sand or diatomite type, incl. pump, 6,000 gal./hr.	2 Plum	1.80	8.889	Total	1,225	263		1,488	1,775
1020	Add for chlorination system, 800 S.F. pool	"	3	5.333	Ea.	211	158		369	490

Division 23
Heating, Ventilating, and Air Conditioning

23 05 Common Work Results for HVAC

23 05 05 – Selective HVAC Demolition

23 05 05.10 HVAC Demolition

		Crew	Daily Output	Labor-Hours	Unit	Material	2007 Bare Costs Labor	2007 Bare Costs Equipment	Total	Total Incl O&P
0010	**HVAC DEMOLITION**									
0100	Air conditioner, split unit, 3 ton	Q-5	2	8	Ea.		215		215	355
0150	Package unit, 3 ton	Q-6	3	8	"		207		207	340
0260	Baseboard, hydronic fin tube, 1/2"	Q-5	117	.137	L.F.		3.68		3.68	6.05
0298	Boilers									
0300	Electric, up thru 148 kW	Q-19	2	12	Ea.		335		335	545
0310	150 thru 518 kW	"	1	24			665		665	1,100
0320	550 thru 2000 kW	Q-21	.40	80			2,250		2,250	3,700
0330	2070 kW and up	"	.30	106			3,025		3,025	4,950
0340	Gas and/or oil, up thru 150 MBH	Q-7	2.20	14.545			390		390	645
0350	160 thru 2000 MBH		.80	40			1,075		1,075	1,775
0360	2100 thru 4500 MBH		.50	64			1,725		1,725	2,825
0370	4600 thru 7000 MBH		.30	106			2,875		2,875	4,725
0390	12,200 thru 25,000 MBH	↓	.12	266	↓		7,175		7,175	11,800
1000	Ductwork, 4" high, 8" wide	1 Clab	200	.040	L.F.		.75		.75	1.27
1100	6" high, 8" wide		165	.048			.91		.91	1.54
1200	10" high, 12" wide		125	.064			1.20		1.20	2.03
1300	12"-14" high, 16"-18" wide		85	.094			1.76		1.76	2.99
1500	30" high, 36" wide	↓	56	.143	↓		2.67		2.67	4.54
2200	Furnace, electric	Q-20	2	10	Ea.		266		266	445
2300	Gas or oil, under 120 MBH	Q-9	4	4			104		104	174
2340	Over 120 MBH	"	3	5.333			138		138	232
2800	Heat pump, package unit, 3 ton	Q-5	2.40	6.667			179		179	295
2840	Split unit, 3 ton		2	8			215		215	355
2950	Tank, steel, oil, 275 gal., above ground		10	1.600			43		43	70.50
2960	Remove and reset	↓	3	5.333	↓		143		143	236
9000	Minimum labor/equipment charge	Q-6	3	8	Job		207		207	340

23 07 HVAC Insulation

23 07 13 – Duct Insulation

23 07 13.10 Duct Insulation

		Crew	Daily Output	Labor-Hours	Unit	Material	2007 Bare Costs Labor	2007 Bare Costs Equipment	Total	Total Incl O&P
0010	**DUCT INSULATION**									
3000	Ductwork									
3020	Blanket type, fiberglass, flexible									
3030	Fire resistant liner, black coating one side									
3050	1/2" thick, 2 lb. density	Q-14	380	.042	S.F.	.47	1.04		1.51	2.31
3060	1" thick, 1-1/2 lb. density	"	350	.046	"	.62	1.13		1.75	2.62
3140	FRK vapor barrier wrap, .75 lb. density									
3160	1" thick	Q-14	350	.046	S.F.	.30	1.13		1.43	2.27
3170	1-1/2" thick	"	320	.050	"	.32	1.23		1.55	2.47
9600	Minimum labor/equipment charge	1 Stpi	4	2	Job		59.50		59.50	98

23 07 16 – HVAC Equipment Insulation

23 07 16.10 HVAC Equipment Insulation

0010	**HVAC EQUIPMENT INSULATION**

23 09 Instrumentation and Control for HVAC

23 09 53 – Pneumatic and Electric Control System for HVAC

23 09 53.10 Control Components	Crew	Daily Output	Labor-Hours	Unit	Material	2007 Bare Costs Labor	Equipment	Total	Total Incl O&P
0010 **CONTROL COMPONENTS**									
5000 Thermostats									
5030 Manual	1 Shee	8	1	Ea.	27	29		56	78
5040 1 set back, electric, timed	↓	8	1		95	29		124	153
5050 2 set back, electric, timed		8	1	↓	209	29		238	278

23 13 Facility Fuel-Storage Tanks

23 13 13 – Facility Underground Fuel-Oil, Storage Tanks

23 13 13.09 Single-Wall Steel Fuel-Oil Tanks

		Crew	Daily Output	Labor-Hours	Unit	Material	Labor	Equipment	Total	Total Incl O&P
0010	**SINGLE-WALL STEEL FUEL-OIL TANKS**									
5000	Steel underground, sti-P3, set in place, not incl. hold-down bars.									
5500	Excavation, pad, pumps and piping not included									
5510	Single wall, 500 gallon capacity, 7 gauge shell	Q-5	2.70	5.926	Ea.	1,000	159		1,159	1,350
5520	1,000 gallon capacity, 7 gauge shell	"	2.50	6.400		2,300	172		2,472	2,800
5530	2,000 gallon capacity, 1/4" thick shell	Q-7	4.60	6.957		3,700	187		3,887	4,375
5535	2,500 gallon capacity, 7 gauge shell	Q-5	3	5.333		4,025	143		4,168	4,650
5610	25,000 gallon capacity, 3/8" thick shell	Q-7	1.30	24.615		22,500	660		23,160	25,900
5630	40,000 gallon capacity, 3/8" thick shell		.90	35.556		37,400	955		38,355	42,700
5640	50,000 gallon capacity, 3/8" thick shell	↓	.80	40	↓	46,800	1,075		47,875	53,500

23 13 13.23 Glass-Fiber-Reinfcd-Plastic, Fuel-Oil, Storage

		Crew	Daily Output	Labor-Hours	Unit	Material	Labor	Equipment	Total	Total Incl O&P
0010	**GLASS-FIBER-REINFCD-PLASTIC, UNDERGRND FUEL-OIL, STORAGE**									
0210	Fiberglass, underground, single wall, U.L. listed, not including									
0220	manway or hold-down strap									
0230	1,000 gallon capacity	Q-5	2.46	6.504	Ea.	3,075	175		3,250	3,675
0240	2,000 gallon capacity	Q-7	4.57	7.002		4,925	188		5,113	5,725
0500	For manway, fittings and hold-downs, add				↓	20%	15%			
2210	Fiberglass, underground, single wall, U.L. listed, including									
2220	hold-down straps, no manways									
2230	1,000 gallon capacity	Q-5	1.88	8.511	Ea.	3,400	229		3,629	4,100
2240	2,000 gallon capacity	Q-7	3.55	9.014	"	5,250	242		5,492	6,175

23 13 23 – Facility Aboveground Fuel-Oil, Storage Tanks

23 13 23.16 Steel

		Crew	Daily Output	Labor-Hours	Unit	Material	Labor	Equipment	Total	Total Incl O&P
3001	**STEEL**, storage, above ground, including supports, coating									
3020	fittings, not including fdn, pumps or piping									
3040	Single wall, interior, 275 gallon	Q-5	5	3.200	Ea.	330	86		416	505
3060	550 gallon	"	2.70	5.926		1,575	159		1,734	1,975
3080	1,000 gallon	Q-7	5	6.400		2,525	172		2,697	3,075
3320	Double wall, 500 gallon capacity	Q-5	2.40	6.667		2,575	179		2,754	3,150
3330	2000 gallon capacity	Q-7	4.15	7.711		6,450	207		6,657	7,450
3340	4000 gallon capacity		3.60	8.889		11,500	239		11,739	13,000
3350	6000 gallon capacity		2.40	13.333		13,600	360		13,960	15,500
3360	8000 gallon capacity		2	16		17,400	430		17,830	19,800
3370	10000 gallon capacity		1.80	17.778		19,200	480		19,680	21,900
3380	15000 gallon capacity		1.50	21.333		29,100	575		29,675	33,000
3390	20000 gallon capacity		1.30	24.615		33,200	660		33,860	37,600
3400	25000 gallon capacity		1.15	27.826		40,300	750		41,050	45,600
3410	30000 gallon capacity	↓	1	32	↓	44,200	860		45,060	50,000

23 21 Hydronic Piping and Pumps

23 21 20 – Hydronic HVAC Piping Specialties

23 21 20.46 Expansion Tanks

		Crew	Daily Output	Labor-Hours	Unit	Material	2007 Bare Costs Labor	Equipment	Total	Total Incl O&P
0010	**EXPANSION TANKS**									
1507	Fiberglass and steel single / double wall storage, see Div 23 13 23.19									
2000	Steel, liquid expansion, ASME, painted, 15 gallon capacity	Q-5	17	.941	Ea.	410	25.50		435.50	495
2040	30 gallon capacity		12	1.333		460	36		496	565
3000	Steel ASME expansion, rubber diaphragm, 19 gal. cap. accept.		12	1.333		1,950	36		1,986	2,175
3020	31 gallon capacity		8	2		2,150	54		2,204	2,475

23 21 23 – Hydronic Pumps

23 21 23.13 In-Line Centrifugal Hydronic Pumps

		Crew	Daily Output	Labor-Hours	Unit	Material	2007 Bare Costs Labor	Equipment	Total	Total Incl O&P
0010	**IN-LINE CENTRIFUGAL HYDRONIC PUMPS**									
0600	Bronze, sweat connections, 1/40 HP, in line									
0640	3/4" size	Q-1	16	1	Ea.	151	26.50		177.50	210
1000	Flange connection, 3/4" to 1-1/2" size									
1040	1/12 HP	Q-1	6	2.667	Ea.	410	71		481	570
1060	1/8 HP	"	6	2.667	"	710	71		781	895

23 31 HVAC Ducts and Casings

23 31 13 – Metal Ducts

23 31 13.13 Rectangular Metal Ducts

		Crew	Daily Output	Labor-Hours	Unit	Material	2007 Bare Costs Labor	Equipment	Total	Total Incl O&P
0010	**RECTANGULAR METAL DUCTS**									
0020	Fabricated rectangular, includes fittings, joints, supports,									
0030	allowance for flexible connections, no insulation									
0031	NOTE: Fabrication and installation are combined									
0040	as LABOR cost. Approx. 25% fittings assumed.									
0100	Aluminum, alloy 3003-H14, under 100 lb.	Q-10	75	.320	Lb.	3.37	8.60		11.97	18.15
0110	100 to 500 lb.		80	.300		2.25	8.05		10.30	16.05
0120	500 to 1,000 lb.		95	.253		2.13	6.80		8.93	13.75
0140	1,000 to 2,000 lb.		120	.200		2.05	5.35		7.40	11.30
0500	Galvanized steel, under 200 lb.		235	.102		.95	2.74		3.69	5.65
0520	200 to 500 lb.		245	.098		.81	2.63		3.44	5.30
0540	500 to 1,000 lb.		255	.094		.78	2.53		3.31	5.10

23 31 13.19 Metal Duct Fittings

		Crew	Daily Output	Labor-Hours	Unit	Material	2007 Bare Costs Labor	Equipment	Total	Total Incl O&P
0010	**METAL DUCT FITTINGS**									
0050	Air extractors, 12" x 4"	1 Shee	24	.333	Ea.	16.50	9.60		26.10	34.50
0100	8" x 6"	"	22	.364	"	16.50	10.45		26.95	36

23 33 Air Duct Accessories

23 33 13 – Dampers

23 33 13.13 Volume-Control Dampers

		Crew	Daily Output	Labor-Hours	Unit	Material	2007 Bare Costs Labor	Equipment	Total	Total Incl O&P
0010	**VOLUME-CONTROL DAMPERS**									
6000	12" x 12"	1 Shee	21	.381	Ea.	27	10.95		37.95	48.50
8000	Multi-blade dampers, parallel blade									
8100	8" x 8"	1 Shee	24	.333	Ea.	59	9.60		68.60	81

23 33 13.16 Fire Dampers

		Crew	Daily Output	Labor-Hours	Unit	Material	2007 Bare Costs Labor	Equipment	Total	Total Incl O&P
0010	**FIRE DAMPERS**									
3000	Fire damper, curtain type, 1-1/2 hr rated, vertical, 6" x 6"	1 Shee	24	.333	Ea.	14.75	9.60		24.35	32.50
3020	8" x 6"	"	22	.364	"	15.10	10.45		25.55	34

23 33 Air Duct Accessories

23 33 46 – Flexible Ducts

23 33 46.10 Flexible Ducts

		Crew	Daily Output	Labor-Hours	Unit	Material	2007 Bare Costs Labor	Equipment	Total	Total Incl O&P
0010	**FLEXIBLE DUCTS**									
1300	Flexible, coated fiberglass fabric on corr. resist. metal helix									
1400	pressure to 12" (WG) UL-181									
1500	Non-insulated, 3" diameter	Q-9	400	.040	L.F.	1.08	1.04		2.12	2.93
1540	5" diameter		320	.050		1.35	1.29		2.64	3.67
1560	6" diameter		280	.057		1.60	1.48		3.08	4.25
1580	7" diameter		240	.067		1.90	1.73		3.63	4.99
1900	Insulated, 1" thick, PE jacket, 3" diameter		380	.042		2.22	1.09		3.31	4.27
1910	4" diameter		340	.047		2.22	1.22		3.44	4.49
1920	5" diameter		300	.053		2.39	1.38		3.77	4.95
1940	6" diameter		260	.062		2.54	1.59		4.13	5.45
1960	7" diameter		220	.073		2.95	1.88		4.83	6.40
1980	8" diameter		180	.089		3.15	2.30		5.45	7.35
2040	12" diameter		100	.160		4.62	4.14		8.76	12.05

23 33 53 – Duct Liners

23 33 53.10 Duct Liners

		Crew	Daily Output	Labor-Hours	Unit	Material	2007 Bare Costs Labor	Equipment	Total	Total Incl O&P
0010	**DUCT LINERS**									
3490	Board type, fiberglass liner, 3 lb. density									
3500	Fire resistant, black pigmented, 1 side									
3520	1" thick	Q-14	150	.107	S.F.	1.77	2.63		4.40	6.50
3540	1-1/2" thick	"	130	.123	"	2.17	3.03		5.20	7.60

23 34 HVAC Fans

23 34 23 – HVAC Power Ventilators

23 34 23.10 HVAC Power Ventilators

		Crew	Daily Output	Labor-Hours	Unit	Material	2007 Bare Costs Labor	Equipment	Total	Total Incl O&P
0010	**HVAC POWER VENTILATORS**									
8020	Attic, roof type									
8030	Aluminum dome, damper & curb									
8040	6" diameter, 300 CFM	1 Elec	16	.500	Ea.	288	14.70		302.70	340
8050	7" diameter, 450 CFM		15	.533		315	15.70		330.70	370
8060	9" diameter, 900 CFM		14	.571		505	16.80		521.80	585
8080	12" diameter, 1000 CFM (gravity)		10	.800		355	23.50		378.50	430
8090	16" diameter, 1500 CFM (gravity)		9	.889		430	26		456	520
8100	20" diameter, 2500 CFM (gravity)		8	1		525	29.50		554.50	630
8160	Plastic, ABS dome									
8180	1050 CFM	1 Elec	14	.571	Ea.	105	16.80		121.80	143
8200	1600 CFM	"	12	.667	"	157	19.60		176.60	205
8240	Attic, wall type, with shutter, one speed									
8250	12" diameter, 1000 CFM	1 Elec	14	.571	Ea.	227	16.80		243.80	277
8260	14" diameter, 1500 CFM		12	.667		245	19.60		264.60	300
8270	16" diameter, 2000 CFM		9	.889		278	26		304	350
8290	Whole house, wall type, with shutter, one speed									
8300	30" diameter, 4800 CFM	1 Elec	7	1.143	Ea.	595	33.50		628.50	710
8310	36" diameter, 7000 CFM		6	1.333		645	39		684	775
8320	42" diameter, 10,000 CFM		5	1.600		725	47		772	870
8330	48" diameter, 16,000 CFM		4	2		900	59		959	1,075
8340	For two speed, add					54			54	59.50
8350	Whole house, lay-down type, with shutter, one speed									
8360	30" diameter, 4500 CFM	1 Elec	8	1	Ea.	635	29.50		664.50	745
8370	36" diameter, 6500 CFM		7	1.143		680	33.50		713.50	805

23 34 HVAC Fans

23 34 23 – HVAC Power Ventilators

	23 34 23.10 HVAC Power Ventilators	Crew	Daily Output	Labor-Hours	Unit	Material	2007 Bare Costs Labor	2007 Bare Costs Equipment	Total	Total Incl O&P
8380	42" diameter, 9000 CFM	1 Elec	6	1.333	Ea.	750	39		789	885
8390	48" diameter, 12,000 CFM	↓	5	1.600		850	47		897	1,000
8440	For two speed, add					40.50			40.50	45
8450	For 12 hour timer switch, add	1 Elec	32	.250	↓	40.50	7.35		47.85	57

23 37 Air Outlets and Inlets

23 37 13 – Diffusers, Registers, and Grilles

23 37 13.10 Diffusers

		Crew	Daily Output	Labor-Hours	Unit	Material	2007 Bare Costs Labor	2007 Bare Costs Equipment	Total	Total Incl O&P
0010	**DIFFUSERS**, Aluminum, opposed blade damper unless noted									
0100	Ceiling, linear, also for sidewall									
0120	2" wide	1 Shee	32	.250	L.F.	32.50	7.20		39.70	47.50
0160	4" wide		26	.308	"	42	8.85		50.85	61.50
0500	Perforated, 24" x 24" lay-in panel size, 6" x 6"		16	.500	Ea.	75	14.40		89.40	107
0520	8" x 8"		15	.533		77	15.35		92.35	111
0530	9" x 9"		14	.571		80.50	16.45		96.95	116
0590	16" x 16"		11	.727		116	21		137	163
1000	Rectangular, 1 to 4 way blow, 6" x 6"		16	.500		45.50	14.40		59.90	74
1010	8" x 8"		15	.533		54	15.35		69.35	85.50
1014	9" x 9"		15	.533		58	15.35		73.35	90
1016	10" x 10"		15	.533		68.50	15.35		83.85	101
1020	12" x 6"		15	.533		77.50	15.35		92.85	111
1040	12" x 9"		14	.571		84.50	16.45		100.95	121
1060	12" x 12"		12	.667		80	19.15		99.15	120
1070	14" x 6"		13	.615		84	17.70		101.70	122
1074	14" x 14"		12	.667		105	19.15		124.15	147
1150	18" x 18"		9	.889		147	25.50		172.50	205
1170	24" x 12"		10	.800		139	23		162	191
1180	24" x 24"		7	1.143		300	33		333	385
1500	Round, butterfly damper, steel, 6" diameter		18	.444		18.70	12.80		31.50	42
1520	8" diameter		16	.500		20	14.40		34.40	46
2000	T bar mounting, 24" x 24" lay-in frame, 6" x 6"		16	.500		94.50	14.40		108.90	128
2020	9" x 9"		14	.571		105	16.45		121.45	143
2040	12" x 12"		12	.667		135	19.15		154.15	180
2060	15" x 15"		11	.727		173	21		194	225
2080	18" x 18"	↓	10	.800	↓	190	23		213	249
6000	For steel diffusers instead of aluminum, deduct					10%				

23 37 13.30 Grilles

		Crew	Daily Output	Labor-Hours	Unit	Material	2007 Bare Costs Labor	2007 Bare Costs Equipment	Total	Total Incl O&P
0010	**GRILLES**									
0020	Aluminum									
1000	Air return, 6" x 6"	1 Shee	26	.308	Ea.	14.30	8.85		23.15	30.50
1020	10" x 6"		24	.333		16.90	9.60		26.50	34.50
1080	16" x 8"		22	.364		25.50	10.45		35.95	45.50
1100	12" x 12"		22	.364		25.50	10.45		35.95	45.50
1120	24" x 12"		18	.444		45	12.80		57.80	71
1180	16" x 16"	↓	22	.364	↓	36.50	10.45		46.95	57.50

23 37 13.60 Registers

		Crew	Daily Output	Labor-Hours	Unit	Material	2007 Bare Costs Labor	2007 Bare Costs Equipment	Total	Total Incl O&P
0010	**REGISTERS**									
0980	Air supply									
3000	Baseboard, hand adj. damper, enameled steel									
3012	8" x 6"	1 Shee	26	.308	Ea.	14.10	8.85		22.95	30.50

23 37 Air Outlets and Inlets

23 37 13 – Diffusers, Registers, and Grilles

23 37 13.60 Registers		Crew	Daily Output	Labor-Hours	Unit	Material	2007 Bare Costs Labor	2007 Bare Costs Equipment	Total	Total Incl O&P
3020	10" x 6"	1 Shee	24	.333	Ea.	15.90	9.60		25.50	33.50
3040	12" x 5"	↓	23	.348		18.15	10		28.15	37
3060	12" x 6"	↓	23	.348	↓	16.75	10		26.75	35
4000	Floor, toe operated damper, enameled steel									
4020	4" x 8"	1 Shee	32	.250	Ea.	20	7.20		27.20	34
4040	4" x 12"	"	26	.308	"	23.50	8.85		32.35	41
4300	Spiral pipe supply register									
4310	Aluminum, double deflection, w/damper extractor									
4320	4" x 12", for 6" thru 10" diameter duct	1 Shee	25	.320	Ea.	62	9.20		71.20	83.50
4330	4" x 18", for 6" thru 10" diameter duct		18	.444		76.50	12.80		89.30	106
4340	6" x 12", for 8" thru 12" diameter duct		19	.421		67.50	12.10		79.60	95
4350	6" x 18", for 8" thru 12" diameter duct		18	.444		88	12.80		100.80	118
4360	6" x 24", for 8" thru 12" diameter duct		16	.500		109	14.40		123.40	143
4370	6" x 30", for 8" thru 12" diameter duct		15	.533		138	15.35		153.35	178
4380	8" x 18", for 10" thru 14" diameter duct		18	.444		93.50	12.80		106.30	125
4390	8" x 24", for 10" thru 14" diameter duct		15	.533		117	15.35		132.35	155
4400	8" x 30", for 10" thru 14" diameter duct		14	.571		155	16.45		171.45	199
4410	10" x 24", for 12" thru 18" diameter duct		13	.615		129	17.70		146.70	172
4420	10" x 30", for 12" thru 18" diameter duct		12	.667		170	19.15		189.15	219
4430	10" x 36", for 12" thru 18" diameter duct	↓	11	.727	↓	211	21		232	267

23 41 Particulate Air Filtration

23 41 13 – Panel Air Filters

23 41 13.10 Panel Air Filters

		Crew	Daily Output	Labor-Hours	Unit	Material	Labor	Equipment	Total	Total Incl O&P
0010	**PANEL AIR FILTERS**									
2950	Mechanical media filtration units									
3000	High efficiency type, with frame, non-supported				MCFM	45			45	49.50
3100	Supported type				"	60			60	66
5500	Throwaway glass or paper media type				Ea.	7.20			7.20	7.90

23 41 16 – Renewable-Media Air Filters

23 41 16.10 Renewable-Media Air Filters

		Crew	Daily Output	Labor-Hours	Unit	Material	Labor	Equipment	Total	Total Incl O&P
0010	**RENEWABLE-MEDIA AIR FILTERS**									
5000	Renewable disposable roll				MCFM	200			200	220

23 41 19 – Washable Air Filters

23 41 19.10 Washable Air Filters

		Crew	Daily Output	Labor-Hours	Unit	Material	Labor	Equipment	Total	Total Incl O&P
0010	**WASHABLE AIR FILTERS**									
4500	Permanent washable				MCFM	20			20	22

23 41 23 – Extended Surface Filters

23 41 23.10 Extended Surface Filters

		Crew	Daily Output	Labor-Hours	Unit	Material	Labor	Equipment	Total	Total Incl O&P
0010	**EXTENDED SURFACE FILTERS**									
4000	Medium efficiency, extended surface				MCFM	5.50			5.50	6.05

23 42 Gas-Phase Air Filtration

23 42 13 – Activated-Carbon Air Filtration

23 42 13.10 Activated-Carbon Air Filtration

		Crew	Daily Output	Labor-Hours	Unit	Material	2007 Bare Costs Labor	Equipment	Total	Total Incl O&P
0010	**ACTIVATED-CARBON AIR FILTRATION**									
0050	Activated charcoal type, full flow				MCFM	600			600	660
0060	Activated charcoal type, full flow, impregnated media 12" deep					200			200	220
0070	Activated charcoal type, HEPA filter & frame for field erection					200			200	220
0080	Activated charcoal type, HEPA filter-diffuser, ceiling install.				↓	275			275	305

23 43 Electronic Air Cleaners

23 43 13 – Washable Electronic Air Cleaners

23 43 13.10 Washable Electronic Air Cleaners

		Crew	Daily Output	Labor-Hours	Unit	Material	Labor	Equipment	Total	Total Incl O&P
0010	**WASHABLE ELECTRONIC AIR CLEANERS**									
2000	Electronic air cleaner, duct mounted									
2150	400 - 1000 CFM	1 Shee	2.30	3.478	Ea.	695	100		795	935
2200	1000 - 1400 CFM		2.20	3.636		725	105		830	970
2250	1400 - 2000 CFM	↓	2.10	3.810	↓	800	110		910	1,075

23 51 Breechings, Chimneys, and Stacks

23 51 23 – Gas Vents

23 51 23.10 Gas Vents

		Crew	Daily Output	Labor-Hours	Unit	Material	Labor	Equipment	Total	Total Incl O&P
0010	**GAS VENTS**, Prefab metal, U.L. listed									
0020	Gas, double wall, galvanized steel									
0080	3" diameter	Q-9	72	.222	V.L.F.	3.96	5.75		9.71	14
0100	4" diameter	"	68	.235	"	4.95	6.10		11.05	15.70

23 52 Heating Boilers

23 52 13 – Electric Boilers

23 52 13.10 Electric Boilers, ASME

		Crew	Daily Output	Labor-Hours	Unit	Material	Labor	Equipment	Total	Total Incl O&P
0010	**ELECTRIC BOILERS, ASME**, Standard controls and trim.									
1000	Steam, 6 KW, 20.5 MBH	Q-19	1.20	20	Ea.	3,175	555		3,730	4,375
1160	60 KW, 205 MBH		1	24		5,150	665		5,815	6,775
2000	Hot water, 7.5 KW, 25.6 MBH		1.30	18.462		3,225	510		3,735	4,375
2040	30 KW, 102 MBH		1.20	20		3,425	555		3,980	4,650
2060	45 KW, 164 MBH	↓	1.20	20	↓	3,850	555		4,405	5,125

23 52 23 – Cast-Iron Boilers

23 52 23.20 Gas-Fired Boilers

		Crew	Daily Output	Labor-Hours	Unit	Material	Labor	Equipment	Total	Total Incl O&P
0010	**GAS-FIRED BOILERS**, Natural or propane, standard controls.									
1000	Cast iron, with insulated jacket									
3000	Hot water, gross output, 80 MBH	Q-7	1.46	21.918	Ea.	1,575	590		2,165	2,700
3020	100 MBH	"	1.35	23.704		1,800	635		2,435	3,025
7000	For tankless water heater, add					10%				
7050	For additional zone valves up to 312 MBH add				↓	130			130	143

23 52 23.30 Gas/Oil Fired Boilers

		Crew	Daily Output	Labor-Hours	Unit	Material	Labor	Equipment	Total	Total Incl O&P
0010	**GAS/OIL FIRED BOILERS**, Combination with burners and controls.									
1000	Cast iron with insulated jacket									
2000	Steam, gross output, 720 MBH	Q-7	.43	74.074	Ea.	7,975	2,000		9,975	12,100
2900	Hot water, gross output									
2910	200 MBH	Q-6	.62	39.024	Ea.	6,525	1,000		7,525	8,825
2920	300 MBH	↓	.49	49.080	↓	6,525	1,275		7,800	9,275

23 52 Heating Boilers

23 52 23 – Cast-Iron Boilers

23 52 23.30 Gas/Oil Fired Boilers

		Crew	Daily Output	Labor-Hours	Unit	Material	2007 Bare Costs Labor	2007 Bare Costs Equipment	Total	Total Incl O&P
2930	400 MBH	Q-6	.41	57.971	Ea.	7,625	1,500		9,125	10,900
2940	500 MBH	↓	.36	67.039	↓	8,225	1,725		9,950	11,900
3000	584 MBH	Q-7	.44	72.072		10,700	1,925		12,625	15,000

23 52 23.40 Oil-Fired Boilers

		Crew	Daily Output	Labor-Hours	Unit	Material	Labor	Equipment	Total	Total Incl O&P
0010	**OIL-FIRED BOILERS,** Standard controls, flame retention burner.									
1000	Cast iron, with insulated flush jacket									
2000	Steam, gross output, 109 MBH	Q-7	1.20	26.667	Ea.	2,350	715		3,065	3,775
2060	207 MBH	"	.90	35.556	"	3,225	955		4,180	5,100
3000	Hot water, same price as steam									

23 52 26 – Steel Boilers

23 52 26.40 Oil-Fired Boilers

		Crew	Daily Output	Labor-Hours	Unit	Material	Labor	Equipment	Total	Total Incl O&P
0010	**OIL-FIRED BOILERS,** Standard controls, flame retention burner									
5000	Steel, with insulated flush jacket									
7000	Hot water, gross output, 103 MBH	Q-6	1.60	15	Ea.	1,450	390		1,840	2,250
7020	122 MBH		1.45	16.506		2,750	425		3,175	3,725
7060	168 MBH		1.30	18.405		3,450	475		3,925	4,575
7080	225 MBH	↓	1.22	19.704	↓	3,550	510		4,060	4,775

23 52 28 – Swimming Pool Boilers

23 52 28.10 Swimming Pool Heaters

		Crew	Daily Output	Labor-Hours	Unit	Material	Labor	Equipment	Total	Total Incl O&P
0010	**SWIMMING POOL HEATERS,** Not including wiring, external									
0020	piping, base or pad,									
0160	Gas fired, input, 155 MBH	Q-6	1.50	16	Ea.	1,500	415		1,915	2,325
0200	199 MBH		1	24		1,575	620		2,195	2,775
0280	500 MBH	↓	.40	60		6,400	1,550		7,950	9,600
2000	Electric, 12 KW, 4,800 gallon pool	Q-19	3	8		2,050	222		2,272	2,625
2020	15 KW, 7,200 gallon pool		2.80	8.571		2,075	238		2,313	2,675
2040	24 KW, 9,600 gallon pool		2.40	10		2,775	277		3,052	3,500
2100	55 KW, 24,000 gallon pool	↓	1.20	20	↓	3,950	555		4,505	5,250

23 52 88 – Burners

23 52 88.10 Burners

		Crew	Daily Output	Labor-Hours	Unit	Material	Labor	Equipment	Total	Total Incl O&P
0010	**BURNERS**									
0990	Residential, conversion, gas fired, LP or natural									
1000	Gun type, atmospheric input 72 to 200 MBH	Q-1	2.50	6.400	Ea.	685	170		855	1,025
1020	120 to 360 MBH		2	8		765	213		978	1,200
1040	280 to 800 MBH	↓	1.70	9.412	↓	1,475	250		1,725	2,025

23 54 Furnaces

23 54 13 – Electric-Resistance Furnaces

23 54 13.10 Electric-Resistance Furnaces

		Crew	Daily Output	Labor-Hours	Unit	Material	Labor	Equipment	Total	Total Incl O&P
0010	**ELECTRIC-RESISTANCE FURNACES,** Hot air, blowers, std. controls.									
0011	not including gas, oil or flue piping									
1000	Electric, UL listed									
1020	10.2 MBH	Q-20	5	4	Ea.	325	106		431	540
1100	34.1 MBH	"	4.40	4.545	"	425	121		546	665

23 54 16 – Fuel-Fired Furnaces

23 54 16.13 Gas-Fired Furnaces

		Crew	Daily Output	Labor-Hours	Unit	Material	Labor	Equipment	Total	Total Incl O&P
0010	**GAS-FIRED FURNACES**									
3000	Gas, AGA certified, upflow, direct drive models									

23 54 Furnaces

23 54 16 – Fuel-Fired Furnaces

23 54 16.13 Gas-Fired Furnaces

		Crew	Daily Output	Labor-Hours	Unit	Material	2007 Bare Costs Labor	2007 Bare Costs Equipment	Total	Total Incl O&P
3020	45 MBH input	Q-9	4	4	Ea.	455	104		559	675
3040	60 MBH input		3.80	4.211		605	109		714	850
3060	75 MBH input		3.60	4.444		640	115		755	900
3100	100 MBH input		3.20	5		680	129		809	965
3120	125 MBH input		3	5.333		785	138		923	1,100
3130	150 MBH input		2.80	5.714		910	148		1,058	1,250
3140	200 MBH input		2.60	6.154		2,175	159		2,334	2,675
4000	For starter plenum, add	↓	16	1	↓	70	26		96	121

23 54 16.16 Oil-Fired Furnaces

		Crew	Daily Output	Labor-Hours	Unit	Material	Labor	Equipment	Total	Total Incl O&P
0010	**OIL-FIRED FURNACES**									
6000	Oil, UL listed, atomizing gun type burner									
6020	56 MBH output	Q-9	3.60	4.444	Ea.	1,550	115		1,665	1,900
6030	84 MBH output		3.50	4.571		1,575	118		1,693	1,925
6040	95 MBH output		3.40	4.706		1,575	122		1,697	1,950
6060	134 MBH output		3.20	5		1,625	129		1,754	2,025
6080	151 MBH output		3	5.333		1,900	138		2,038	2,325
6100	200 MBH input	↓	2.60	6.154	↓	2,300	159		2,459	2,825

23 54 24 – Furnace Components and Combinations

23 54 24.10 Furnace Components and Combinations

		Crew	Daily Output	Labor-Hours	Unit	Material	Labor	Equipment	Total	Total Incl O&P
0010	**FURNACE COMPONENTS AND COMBINATIONS**									
0080	Coils, A/C evaporator, for gas or oil furnaces									
0090	Add-on, with holding charge									
0100	Upflow									
0120	1-1/2 ton cooling	Q-5	4	4	Ea.	114	108		222	300
0130	2 ton cooling		3.70	4.324		136	116		252	340
0140	3 ton cooling		3.30	4.848		172	130		302	405
0150	4 ton cooling		3	5.333		237	143		380	495
0160	5 ton cooling	↓	2.70	5.926	↓	280	159		439	570
0300	Downflow									
0330	2-1/2 ton cooling	Q-5	3	5.333	Ea.	167	143		310	420
0340	3-1/2 ton cooling		2.60	6.154		199	165		364	490
0350	5 ton cooling	↓	2.20	7.273	↓	280	195		475	630
0600	Horizontal									
0630	2 ton cooling	Q-5	3.90	4.103	Ea.	151	110		261	345
0640	3 ton cooling		3.50	4.571		172	123		295	390
0650	4 ton cooling		3.20	5		221	134		355	465
0660	5 ton cooling	↓	2.90	5.517	↓	244	148		392	510
2000	Cased evaporator coils for air handlers									
2100	1-1/2 ton cooling	Q-5	4.40	3.636	Ea.	222	98		320	405
2110	2 ton cooling		4.10	3.902		225	105		330	420
2120	2-1/2 ton cooling		3.90	4.103		235	110		345	440
2130	3 ton cooling		3.70	4.324		260	116		376	475
2140	3-1/2 ton cooling		3.50	4.571		272	123		395	500
2150	4 ton cooling		3.20	5		325	134		459	580
2160	5 ton cooling	↓	2.90	5.517	↓	375	148		523	660
3010	Air handler, modular									
3100	With cased evaporator cooling coil									
3120	1-1/2 ton cooling	Q-5	3.80	4.211	Ea.	560	113		673	805
3130	2 ton cooling		3.50	4.571		590	123		713	850
3140	2-1/2 ton cooling		3.30	4.848		645	130		775	925
3150	3 ton cooling		3.10	5.161		710	139		849	1,000
3160	3-1/2 ton cooling	↓	2.90	5.517	↓	865	148		1,013	1,200

23 54 Furnaces

23 54 24 – Furnace Components and Combinations

23 54 24.10 Furnace Components and Combinations		Crew	Daily Output	Labor-Hours	Unit	Material	2007 Bare Costs Labor	Equipment	Total	Total Incl O&P
3170	4 ton cooling	Q-5	2.50	6.400	Ea.	980	172		1,152	1,350
3180	5 ton cooling	↓	2.10	7.619	↓	1,050	205		1,255	1,475
3500	With no cooling coil									
3520	1-1/2 ton coil size	Q-5	12	1.333	Ea.	395	36		431	495
3530	2 ton coil size		10	1.600		430	43		473	540
3540	2-1/2 ton coil size		10	1.600		470	43		513	585
3554	3 ton coil size		9	1.778		525	48		573	655
3560	3-1/2 ton coil size		9	1.778		580	48		628	720
3570	4 ton coil size		8.50	1.882		745	50.50		795.50	905
3580	5 ton coil size	↓	8	2	↓	805	54		859	975
4000	With heater									
4120	5 kW, 17.1 MBH	Q-5	16	1	Ea.	435	27		462	525
4130	7.5 kW, 25.6 MBH		15.60	1.026		475	27.50		502.50	570
4140	10 kW, 34.2 MBH		15.20	1.053		580	28.50		608.50	685
4150	12.5 KW, 42.7 MBH		14.80	1.081		705	29		734	825
4160	15 KW, 51.2 MBH		14.40	1.111		870	30		900	1,000
4170	25 KW, 85.4 MBH		14	1.143		980	30.50		1,010.50	1,125
4180	30 KW, 102 MBH	↓	13	1.231	↓	1,150	33		1,183	1,325

23 62 Packaged Compressor and Condenser Units

23 62 13 – Packaged Air-Cooled Refrigerant Compressor and Condenser Units

23 62 13.10 Packaged Air-Cooled Refrigerant Condensing Units

		Crew	Daily Output	Labor-Hours	Unit	Material	Labor	Equipment	Total	Total Incl O&P
0010	**PACKAGED AIR-COOLED REFRIGERANT CONDENSING UNITS**									
0030	Air cooled, compressor, standard controls									
0050	1.5 ton	Q-5	2.50	6.400	Ea.	805	172		977	1,175
0100	2 ton		2.10	7.619		830	205		1,035	1,250
0200	2.5 ton		1.70	9.412		865	253		1,118	1,375
0300	3 ton		1.30	12.308		990	330		1,320	1,625
0350	3.5 ton		1.10	14.545		1,150	390		1,540	1,900
0400	4 ton		.90	17.778		1,300	480		1,780	2,200
0500	5 ton	↓	.60	26.667	↓	1,575	715		2,290	2,900

23 74 Packaged Outdoor HVAC Equipment

23 74 33 – Packaged, Outdoor, Heating and Cooling Makeup Air-Conditioners

23 74 33.10 Roof Top Air Conditioners

		Crew	Daily Output	Labor-Hours	Unit	Material	Labor	Equipment	Total	Total Incl O&P
0010	**ROOF TOP AIR CONDITIONERS**, Standard controls, curb, economizer.									
1000	Single zone, electric cool, gas heat									
1140	5 ton cooling, 112 MBH heating	Q-5	.56	28.521	Ea.	4,900	765		5,665	6,650
1160	10 ton cooling, 200 MBH heating	Q-6	.67	35.982	"	9,175	930		10,105	11,600

23 81 Decentralized Unitary HVAC Equipment

23 81 13 – Packaged Terminal Air-Conditioners

23 81 13.10 Packaged Terminal Air-Conditioners		Crew	Daily Output	Labor-Hours	Unit	Material	2007 Bare Costs Labor	Equipment	Total	Total Incl O&P
0010	PACKAGED TERMINAL AIR-CONDITIONERS, Cabinet, wall sleeve,									
0100	louver, electric heat, thermostat, manual changeover, 208 V									
0200	6,000 BTUH cooling, 8800 BTU heat	Q-5	6	2.667	Ea.	1,000	71.50		1,071.50	1,225
0220	9,000 BTUH cooling, 13,900 BTU heat		5	3.200		1,050	86		1,136	1,325
0240	12,000 BTUH cooling, 13,900 BTU heat		4	4		1,150	108		1,258	1,450
0260	15,000 BTUH cooling, 13,900 BTU heat	↓	3	5.333	↓	1,375	143		1,518	1,750

23 81 19 – Self-Contained Air-Conditioners

23 81 19.10 Window Unit Air Conditioners

		Crew	Daily Output	Labor-Hours	Unit	Material	Labor	Equipment	Total	Total Incl O&P
0010	WINDOW UNIT AIR CONDITIONERS									
4000	Portable/window, 15 amp 125V grounded receptacle required									
4060	5000 BTUH	1 Carp	8	1	Ea.	139	25.50		164.50	197
4340	6000 BTUH		8	1		189	25.50		214.50	252
4480	8000 BTUH		6	1.333		299	34.50		333.50	390
4500	10,000 BTUH	↓	6	1.333		350	34.50		384.50	445
4520	12,000 BTUH	L-2	8	2	↓	540	44.50		584.50	670
4600	Window/thru-the-wall, 15 amp 230V grounded receptacle required									
4780	17,000 BTUH	L-2	6	2.667	Ea.	615	59.50		674.50	775
4940	25,000 BTUH		4	4		1,175	89		1,264	1,425
4960	29,000 BTUH	↓	4	4	↓	1,450	89		1,539	1,750

23 81 19.20 Self-Contained Single Package

		Crew	Daily Output	Labor-Hours	Unit	Material	Labor	Equipment	Total	Total Incl O&P
0010	SELF-CONTAINED SINGLE PACKAGE									
0100	Air cooled, for free blow or duct, not incl. remote condenser									
0200	3 ton cooling	Q-5	1	16	Ea.	3,125	430		3,555	4,125
0210	4 ton cooling	"	.80	20	"	3,400	540		3,940	4,600
1000	Water cooled for free blow or duct, not including tower									
1100	3 ton cooling	Q-6	1	24	Ea.	2,975	620		3,595	4,300

23 81 26 – Split-System Air-Conditioners

23 81 26.10 Split Ductless Systems

		Crew	Daily Output	Labor-Hours	Unit	Material	Labor	Equipment	Total	Total Incl O&P
0010	SPLIT DUCTLESS SYSTEMS									
0100	Cooling only, single zone									
0110	Wall mount									
0120	3/4 ton cooling	Q-5	2	8	Ea.	1,000	215		1,215	1,450
0130	1 ton cooling		1.80	8.889		1,150	239		1,389	1,650
0140	1-1/2 ton cooling		1.60	10		1,400	269		1,669	2,000
0150	2 ton cooling	↓	1.40	11.429	↓	1,800	305		2,105	2,475
1000	Ceiling mount									
1020	2 ton cooling	Q-5	1.40	11.429	Ea.	1,350	305		1,655	2,000
1030	3 ton cooling	"	1.20	13.333	"	3,300	360		3,660	4,225
2000	T-Bar mount									
2010	2 ton cooling	Q-5	1.40	11.429	Ea.	2,450	305		2,755	3,200
2020	3 ton cooling		1.20	13.333		2,950	360		3,310	3,850
2030	3-1/2 ton cooling	↓	1.10	14.545	↓	3,550	390		3,940	4,550
3000	Multizone									
3010	Wall mount									
3020	2 @ 3/4 ton cooling	Q-5	1.80	8.889	Ea.	1,325	239		1,564	1,875
5000	Cooling / Heating									
5110	1 ton cooling	Q-5	1.70	9.412	Ea.	980	253		1,233	1,500
5120	1-1/2 ton cooling	"	1.50	10.667	"	1,575	287		1,862	2,200
5300	Ceiling mount									
5310	3 ton cooling	Q-5	1	16	Ea.	3,825	430		4,255	4,900
7000	Accessories for all split ductless systems									

23 81 Decentralized Unitary HVAC Equipment

23 81 26 – Split-System Air-Conditioners

23 81 26.10 Split Ductless Systems		Crew	Daily Output	Labor-Hours	Unit	Material	2007 Bare Costs Labor	Equipment	Total	Total Incl O&P
7010	Add for ambient frost control	Q-5	8	2	Ea.	120	54		174	221
7020	Add for tube / wiring kit									
7030	15' kit	Q-5	32	.500	Ea.	35	13.45		48.45	60.50
7040	35' kit	"	24	.667	"	112	17.90		129.90	153

23 81 43 – Air-Source Unitary Heat Pumps

23 81 43.10 Air-Source Heat Pumps

0010	**AIR-SOURCE HEAT PUMPS**, Not including interconnecting tubing.									
1000	Air to air, split system, not including curbs, pads, or ductwork									
1020	2 ton cooling, 8.5 MBH heat @ 0° F	Q-5	1.20	13.333	Ea.	1,600	360		1,960	2,350
1054	4 ton cooling, 24 MBH heat @ 0° F	"	.60	26.667	"	2,300	715		3,015	3,700
1500	Single package, not including curbs, pads, or plenums									
1520	2 ton cooling, 6.5 MBH heat @ 0° F	Q-5	1.50	10.667	Ea.	2,650	287		2,937	3,400
1580	4 ton cooling, 13 MBH heat @ 0° F	"	.96	16.667	"	3,725	450		4,175	4,825

23 81 46 – Water-Source Unitary Heat Pumps

23 81 46.10 Water Source Heat Pumps

0010	**WATER SOURCE HEAT PUMPS**, Not incl. conn. tubing or wtr. source									
2000	Water source to air, single package									
2100	1 ton cooling, 13 MBH heat @ 75° F	Q-5	2	8	Ea.	1,200	215		1,415	1,675
2200	4 ton cooling, 31 MBH heat @ 75° F	"	1.20	13.333	"	1,800	360		2,160	2,575

23 82 Convection Heating and Cooling Units

23 82 29 – Radiators

23 82 29.10 Hydronic Heating

0010	**HYDRONIC HEATING**, Terminal units, not incl. main supply pipe									
1000	Radiation									
3000	Radiators, cast iron									
3100	Free standing or wall hung, 6 tube, 25" high	Q-5	96	.167	Section	24	4.48		28.48	34

23 82 36 – Finned-Tube Radiation Heaters

23 82 36.10 Finned Tube Radiation

0010	**FINNED TUBE RADIATION**, Terminal units, not incl. main supply pipe									
1310	Baseboard, pkgd, 1/2" copper tube, alum. fin, 7" high	Q-5	60	.267	L.F.	6.65	7.15		13.80	19.15
1320	3/4" copper tube, alum. fin, 7" high		58	.276		7.05	7.40		14.45	20
1340	1" copper tube, alum. fin, 8-7/8" high		56	.286		14.80	7.70		22.50	29
1360	1-1/4" copper tube, alum. fin, 8-7/8" high		54	.296		22	7.95		29.95	37

23 83 Radiant Heating Units

23 83 33 – Electric Radiant Heaters

23 83 33.10 Electric Heating

0010	**ELECTRIC HEATING**, Not incl. conduit or feed wiring.									
1100	Rule of thumb: Baseboard units, including control	1 Elec	4.40	1.818	kW	84	53.50		137.50	180
1300	Baseboard heaters, 2' long, 375 watt		8	1	Ea.	35	29.50		64.50	86.50
1400	3' long, 500 watt		8	1		39	29.50		68.50	91
1600	4' long, 750 watt		6.70	1.194		49	35		84	111
1800	5' long, 935 watt		5.70	1.404		58	41.50		99.50	131
2000	6' long, 1125 watt		5	1.600		64.50	47		111.50	148
2400	8' long, 1500 watt		4	2		81.50	59		140.50	186
2800	10' long, 1875 watt		3.30	2.424		101	71.50		172.50	227
2950	Wall heaters with fan, 120 to 277 volt									

23 83 Radiant Heating Units

23 83 33 – Electric Radiant Heaters

23 83 33.10 Electric Heating		Crew	Daily Output	Labor-Hours	Unit	Material	2007 Bare Costs Labor	Equipment	Total	Total Incl O&P
3600	Thermostats, integral	1 Elec	16	.500	Ea.	19.50	14.70		34.20	45.50
3800	Line voltage, 1 pole		8	1		23.50	29.50		53	74
5000	Radiant heating ceiling panels, 2' x 4', 500 watt		16	.500		203	14.70		217.70	247
5050	750 watt		16	.500		223	14.70		237.70	270
5300	Infrared quartz heaters, 120 volts, 1000 watts		6.70	1.194		135	35		170	206
5350	1500 watt		5	1.600		135	47		182	226
5400	240 volts, 1500 watt		5	1.600		135	47		182	226
5450	2000 watt		4	2		135	59		194	245
5500	3000 watt		3	2.667		155	78.50		233.50	299

Division 26
Electrical

26 05 Common Work Results for Electrical

26 05 05 – Selective Electrical Demolition

26 05 05.10 Electrical Demolition

		Crew	Daily Output	Labor-Hours	Unit	Material	2007 Bare Costs Labor	2007 Bare Costs Equipment	Total	Total Incl O&P
0010	**ELECTRICAL DEMOLITION**									
0020	Conduit to 15' high, including fittings & hangers									
0100	Rigid galvanized steel, 1/2" to 1" diameter	1 Elec	242	.033	L.F.		.97		.97	1.58
0120	1-1/4" to 2"	"	200	.040	"		1.18		1.18	1.92
0270	Armored cable, (BX) avg. 50' runs									
0290	#14, 3 wire	1 Elec	571	.014	L.F.		.41		.41	.67
0300	#12, 2 wire		605	.013			.39		.39	.63
0310	#12, 3 wire		514	.016			.46		.46	.75
0320	#10, 2 wire		514	.016			.46		.46	.75
0330	#10, 3 wire		425	.019			.55		.55	.90
0340	#8, 3 wire		342	.023			.69		.69	1.12
0350	Non metallic sheathed cable (Romex)									
0360	#14, 2 wire	1 Elec	720	.011	L.F.		.33		.33	.53
0370	#14, 3 wire		657	.012			.36		.36	.58
0380	#12, 2 wire		629	.013			.37		.37	.61
0390	#10, 3 wire		450	.018			.52		.52	.85
0400	Wiremold raceway, including fittings & hangers									
0420	No. 3000	1 Elec	250	.032	L.F.		.94		.94	1.53
0440	No. 4000		217	.037			1.08		1.08	1.77
0460	No. 6000		166	.048			1.42		1.42	2.31
0465	Telephone/power pole		12	.667	Ea.		19.60		19.60	32
0470	Non-metallic, straight section		480	.017	L.F.		.49		.49	.80
0500	Channels, steel, including fittings & hangers									
0520	3/4" x 1-1/2"	1 Elec	308	.026	L.F.		.76		.76	1.24
0540	1-1/2" x 1-1/2"		269	.030			.87		.87	1.42
0560	1-1/2" x 1-7/8"		229	.035			1.03		1.03	1.67
1180	400 amp	2 Elec	6.80	2.353	Ea.		69		69	113
1210	Panel boards, incl. removal of all breakers,									
1220	conduit terminations & wire connections									
1230	3 wire, 120/240V, 100A, to 20 circuits	1 Elec	2.60	3.077	Ea.		90.50		90.50	147
1240	200 amps, to 42 circuits	2 Elec	2.60	6.154			181		181	295
1260	4 wire, 120/208V, 125A, to 20 circuits	1 Elec	2.40	3.333			98		98	160
1270	200 amps, to 42 circuits	2 Elec	2.40	6.667			196		196	320
1720	Junction boxes, 4" sq. & oct.	1 Elec	80	.100			2.94		2.94	4.79
1760	Switch box		107	.075			2.20		2.20	3.58
1780	Receptacle & switch plates		257	.031			.92		.92	1.49
1800	Wire, THW-THWN-THHN, removed from									
1810	in place conduit, to 15' high									
1830	#14	1 Elec	65	.123	C.L.F.		3.62		3.62	5.90
1840	#12		55	.145			4.28		4.28	6.95
1850	#10		45.50	.176			5.15		5.15	8.40
2000	Interior fluorescent fixtures, incl. supports									
2010	& whips, to 15' high									
2100	Recessed drop-in 2' x 2', 2 lamp	2 Elec	35	.457	Ea.		13.45		13.45	22
2140	2' x 4', 4 lamp	"	30	.533	"		15.70		15.70	25.50
2180	Surface mount, acrylic lens & hinged frame									
2220	2' x 2', 2 lamp	2 Elec	44	.364	Ea.		10.70		10.70	17.40
2260	2' x 4', 4 lamp	"	33	.485	"		14.25		14.25	23
2300	Strip fixtures, surface mount									
2320	4' long, 1 lamp	2 Elec	53	.302	Ea.		8.90		8.90	14.45
2380	8' long, 2 lamp	"	40	.400	"		11.75		11.75	19.15
2460	Interior incandescent, surface, ceiling									
2470	or wall mount, to 12' high									

26 05 Common Work Results for Electrical

26 05 05 – Selective Electrical Demolition

26 05 05.10 Electrical Demolition		Crew	Daily Output	Labor-Hours	Unit	Material	2007 Bare Costs Labor	Equipment	Total	Total Incl O&P
2480	Metal cylinder type, 75 Watt	2 Elec	62	.258	Ea.		7.60		7.60	12.35
2600	Exterior fixtures, incandescent, wall mount									
2620	100 Watt	2 Elec	50	.320	Ea.		9.40		9.40	15.35
3000	Ceiling fan, tear out and remove	1 Elec	18	.444	"		13.05		13.05	21.50
9000	Minimum labor/equipment charge	"	4	2	Job		59		59	96

26 05 19 – Low-Voltage Electrical Power Conductors and Cables

26 05 19.20 Armored Cable

		Crew	Daily Output	Labor-Hours	Unit	Material	Labor	Equipment	Total	Total Incl O&P
0010	**ARMORED CABLE**									
0051	600 volt, copper (BX), #14, 2 conductor, solid	1 Elec	240	.033	L.F.	.75	.98		1.73	2.42
0101	3 conductor, solid		200	.040		1.18	1.18		2.36	3.22
0151	#12, 2 conductor, solid		210	.038		.76	1.12		1.88	2.66
0201	3 conductor, solid		180	.044		1.21	1.31		2.52	3.46
0251	#10, 2 conductor, solid		180	.044		1.38	1.31		2.69	3.64
0301	3 conductor, solid		150	.053		1.90	1.57		3.47	4.64
0351	#8, 3 conductor, solid		120	.067		3.56	1.96		5.52	7.10

26 05 19.55 Non-Metallic Sheathed Cable

		Crew	Daily Output	Labor-Hours	Unit	Material	Labor	Equipment	Total	Total Incl O&P
0010	**NON-METALLIC SHEATHED CABLE** 600 volt									
0100	Copper with ground wire, (Romex)									
0151	#14, 2 wire	1 Elec	250	.032	L.F.	.38	.94		1.32	1.95
0201	3 wire		230	.035		.53	1.02		1.55	2.26
0251	#12, 2 wire		220	.036		.58	1.07		1.65	2.38
0301	3 wire		200	.040		.88	1.18		2.06	2.88
0351	#10, 2 wire		200	.040		.97	1.18		2.15	2.98
0401	3 wire		140	.057		1.38	1.68		3.06	4.26
0451	#8, 3 conductor		130	.062		2.24	1.81		4.05	5.40
0501	#6, 3 wire		120	.067		3.57	1.96		5.53	7.10
0550	SE type SER aluminum cable, 3 RHW and									
0601	1 bare neutral, 3 #8 & 1 #8	1 Elec	150	.053	L.F.	1.33	1.57		2.90	4.01
0651	3 #6 & 1 #6	"	130	.062		1.50	1.81		3.31	4.60
0701	3 #4 & 1 #6	2 Elec	220	.073		1.68	2.14		3.82	5.35
0751	3 #2 & 1 #4		200	.080		2.48	2.35		4.83	6.55
0801	3 #1/0 & 1 #2		180	.089		3.75	2.61		6.36	8.40
0851	3 #2/0 & 1 #1		160	.100		4.42	2.94		7.36	9.65
0901	3 #4/0 & 1 #2/0		140	.114		6.30	3.36		9.66	12.35
2401	SEU service entrance cable, copper 2 conductors, #8 + #8 neut.	1 Elec	150	.053		2	1.57		3.57	4.75
2601	#6 + #8 neutral		130	.062		2.76	1.81		4.57	6
2801	#6 + #6 neutral		130	.062		3.07	1.81		4.88	6.35
3001	#4 + #6 neutral	2 Elec	220	.073		4.07	2.14		6.21	7.95
3201	#4 + #4 neutral		220	.073		4.57	2.14		6.71	8.55
3401	#3 + #5 neutral		210	.076		5.35	2.24		7.59	9.55
6500	Service entrance cap for copper SEU									
6600	100 amp	1 Elec	12	.667	Ea.	12.15	19.60		31.75	45.50
6700	150 amp		10	.800		22	23.50		45.50	63
6800	200 amp		8	1		33.50	29.50		63	85

26 05 19.90 Wire

		Crew	Daily Output	Labor-Hours	Unit	Material	Labor	Equipment	Total	Total Incl O&P
0010	**WIRE**									
0021	600 volt type THW, copper solid, #14	1 Elec	1300	.006	L.F.	.10	.18		.28	.40
0031	#12		1100	.007		.15	.21		.36	.52
0041	#10		1000	.008		.23	.24		.47	.64
0161	#6		650	.012		.75	.36		1.11	1.42
0181	#4	2 Elec	1060	.015		1.18	.44		1.62	2.02
0201	#3		1000	.016		1.48	.47		1.95	2.39

26 05 Common Work Results for Electrical

26 05 19 — Low-Voltage Electrical Power Conductors and Cables

26 05 19.90 Wire

		Crew	Daily Output	Labor-Hours	Unit	Material	2007 Bare Costs Labor	Equipment	Total	Total Incl O&P
0221	#2	2 Elec	900	.018	L.F.	1.86	.52		2.38	2.89
0241	#1		800	.020		2.35	.59		2.94	3.54
0261	1/0		660	.024		2.66	.71		3.37	4.09
0281	2/0		580	.028		3.33	.81		4.14	4.98
0301	3/0		500	.032		4.17	.94		5.11	6.10
0351	4/0		440	.036		5.20	1.07		6.27	7.50

26 05 26 — Grounding and Bonding for Electrical Systems

26 05 26.80 Grounding

		Crew	Daily Output	Labor-Hours	Unit	Material	Labor	Equipment	Total	Total Incl O&P
0010	**GROUNDING**									
0030	Rod, copper clad, 8' long, 1/2" diameter	1 Elec	5.50	1.455	Ea.	14.05	43		57.05	85
0050	3/4" diameter		5.30	1.509		29.50	44.50		74	105
0080	10' long, 1/2" diameter		4.80	1.667		18.10	49		67.10	100
0100	3/4" diameter		4.40	1.818		32.50	53.50		86	123
0261	Wire, ground bare armored, #8-1 conductor		200	.040	L.F.	1.09	1.18		2.27	3.12
0271	#6-1 conductor		180	.044	"	1.31	1.31		2.62	3.57
0390	Bare copper wire, stranded, #8		11	.727	C.L.F.	40	21.50		61.50	79
0401	Bare copper, #6 wire		1000	.008	L.F.	.73	.24		.97	1.18
0601	#2 stranded	2 Elec	1000	.016	"	1.67	.47		2.14	2.61
1800	Water pipe ground clamps, heavy duty									
2000	Bronze, 1/2" to 1" diameter	1 Elec	8	1	Ea.	21	29.50		50.50	71

26 05 33 — Raceway and Boxes for Electrical Systems

26 05 33.05 Conduit

		Crew	Daily Output	Labor-Hours	Unit	Material	Labor	Equipment	Total	Total Incl O&P
0010	**CONDUIT** To 15' high, includes 2 terminations, 2 elbows,									
0020	11 beam clamps, and 11 couplings per 100 L.F.									
1750	Rigid galvanized steel, 1/2" diameter	1 Elec	90	.089	L.F.	2.48	2.61		5.09	7
1770	3/4" diameter		80	.100		2.83	2.94		5.77	7.90
1800	1" diameter		65	.123		3.89	3.62		7.51	10.20
1830	1-1/4" diameter		60	.133		5.40	3.92		9.32	12.35
1850	1-1/2" diameter		55	.145		6.25	4.28		10.53	13.80
1870	2" diameter		45	.178		8	5.25		13.25	17.30
5000	Electric metallic tubing (EMT), 1/2" diameter		170	.047		.60	1.38		1.98	2.91
5020	3/4" diameter		130	.062		.99	1.81		2.80	4.04
5040	1" diameter		115	.070		1.74	2.05		3.79	5.25
5060	1-1/4" diameter		100	.080		2.76	2.35		5.11	6.85
5080	1-1/2" diameter		90	.089		3.50	2.61		6.11	8.10
9100	PVC, #40, 1/2" diameter		190	.042		1.02	1.24		2.26	3.14
9110	3/4" diameter		145	.055		1.23	1.62		2.85	3.99
9120	1" diameter		125	.064		2.01	1.88		3.89	5.30
9130	1-1/4" diameter		110	.073		2.61	2.14		4.75	6.35
9140	1-1/2" diameter		100	.080		2.99	2.35		5.34	7.10
9150	2" diameter		90	.089		3.79	2.61		6.40	8.45

26 05 33.25 Conduit Fittings for Rigid Galvanized Steel

		Crew	Daily Output	Labor-Hours	Unit	Material	Labor	Equipment	Total	Total Incl O&P
0010	**CONDUIT FITTINGS FOR RIGID GALVANIZED STEEL**									
2280	LB, LR or LL fittings & covers, 1/2" diameter	1 Elec	16	.500	Ea.	10.25	14.70		24.95	35.50
2290	3/4" diameter		13	.615		12.40	18.10		30.50	43
2300	1" diameter		11	.727		18.30	21.50		39.80	55
2330	1-1/4" diameter		8	1		27.50	29.50		57	78.50
2350	1-1/2" diameter		6	1.333		34	39		73	102
2370	2" diameter		5	1.600		57	47		104	139
5280	Service entrance cap, 1/2" diameter		16	.500		7.30	14.70		22	32
5300	3/4" diameter		13	.615		8.50	18.10		26.60	39

26 05 Common Work Results for Electrical

26 05 33 – Raceway and Boxes for Electrical Systems

26 05 33.25 Conduit Fittings for Rigid Galvanized Steel		Crew	Daily Output	Labor-Hours	Unit	Material	2007 Bare Costs Labor	Equipment	Total	Total Incl O&P
5320	1" diameter	1 Elec	10	.800	Ea.	10.90	23.50		34.40	50.50
5340	1-1/4" diameter		8	1		14.80	29.50		44.30	64.50
5360	1-1/2" diameter		6.50	1.231		31.50	36		67.50	93.50
5380	2" diameter		5.50	1.455		40.50	43		83.50	114

26 05 33.35 Flexible Metallic Conduit

0010	**FLEXIBLE METALLIC CONDUIT**									
0050	Steel, 3/8" diameter	1 Elec	200	.040	L.F.	.34	1.18		1.52	2.29
0100	1/2" diameter		200	.040		.37	1.18		1.55	2.33
0200	3/4" diameter		160	.050		.51	1.47		1.98	2.96
0250	1" diameter		100	.080		.94	2.35		3.29	4.86
0300	1-1/4" diameter		70	.114		1.23	3.36		4.59	6.80
0350	1-1/2" diameter		50	.160		2.16	4.70		6.86	10.05
0370	2" diameter		40	.200		2.65	5.90		8.55	12.50

26 05 33.50 Outlet Boxes

0010	**OUTLET BOXES**									
0021	Pressed steel, octagon, 4"	1 Elec	18	.444	Ea.	2.73	13.05		15.78	24.50
0060	Covers, blank		64	.125		1.14	3.68		4.82	7.25
0100	Extension rings		40	.200		4.53	5.90		10.43	14.60
0151	Square 4"		18	.444		2.35	13.05		15.40	24
0200	Extension rings		40	.200		4.58	5.90		10.48	14.65
0250	Covers, blank		64	.125		1.29	3.68		4.97	7.40
0300	Plaster rings		64	.125		2.51	3.68		6.19	8.75
0651	Switchbox		24	.333		4.38	9.80		14.18	21
1101	Concrete, floor, 1 gang		4.80	1.667		73	49		122	160

26 05 33.60 Outlet Boxes, Plastic

0010	**OUTLET BOXES, PLASTIC**									
0051	4" diameter, round, with 2 mounting nails	1 Elec	23	.348	Ea.	2.64	10.25		12.89	19.55
0101	Bar hanger mounted		23	.348		4.79	10.25		15.04	22
0201	Square with 2 mounting nails		23	.348		3.94	10.25		14.19	21
0300	Plaster ring		64	.125		1.65	3.68		5.33	7.80
0401	Switch box with 2 mounting nails, 1 gang		27	.296		1.79	8.70		10.49	16.15
0501	2 gang		23	.348		3.66	10.25		13.91	20.50
0601	3 gang		18	.444		5.75	13.05		18.80	28

26 05 33.65 Pull Boxes

0010	**PULL BOXES**									
0100	Sheet metal, pull box, NEMA 1, type SC, 6" W x 6" H x 4" D	1 Elec	8	1	Ea.	11.65	29.50		41.15	61
0200	8" W x 8" H x 4" D		8	1		16	29.50		45.50	65.50
0300	10" W x 12" H x 6" D		5.30	1.509		28	44.50		72.50	104

26 05 39 – Underfloor Raceways for Electrical Systems

26 05 39.30 Conduit In Concrete Slab

0010	**CONDUIT IN CONCRETE SLAB** Including terminations,									
0020	fittings and supports									
3230	PVC, schedule 40, 1/2" diameter	1 Elec	270	.030	L.F.	.73	.87		1.60	2.23
3250	3/4" diameter		230	.035		.88	1.02		1.90	2.64
3270	1" diameter		200	.040		1.20	1.18		2.38	3.24
3300	1-1/4" diameter		170	.047		1.68	1.38		3.06	4.10
3330	1-1/2" diameter		140	.057		2.02	1.68		3.70	4.96
3350	2" diameter		120	.067		2.60	1.96		4.56	6.05
4350	Rigid galvanized steel, 1/2" diameter		200	.040		2.26	1.18		3.44	4.41
4400	3/4" diameter		170	.047		2.61	1.38		3.99	5.10
4450	1" diameter		130	.062		3.67	1.81		5.48	7

26 05 Common Work Results for Electrical

26 05 39 – Underfloor Raceways for Electrical Systems

26 05 39.30 Conduit In Concrete Slab

		Crew	Daily Output	Labor-Hours	Unit	Material	2007 Bare Costs Labor	Equipment	Total	Total Incl O&P
4500	1-1/4" diameter	1 Elec	110	.073	L.F.	4.96	2.14		7.10	8.95
4600	1-1/2" diameter		100	.080		5.80	2.35		8.15	10.20
4800	2" diameter		90	.089		7.35	2.61		9.96	12.35

26 05 39.40 Conduit In Trench

		Crew	Daily Output	Labor-Hours	Unit	Material	Labor	Equipment	Total	Total Incl O&P
0010	**CONDUIT IN TRENCH** Includes terminations and fittings									
0200	Rigid galvanized steel, 2" diameter	1 Elec	150	.053	L.F.	7.05	1.57		8.62	10.30
0400	2-1/2" diameter	"	100	.080		13.75	2.35		16.10	19
0600	3" diameter	2 Elec	160	.100		17.05	2.94		19.99	23.50
0800	3-1/2" diameter	"	140	.114		22	3.36		25.36	29.50

26 05 80 – Wiring Connections

26 05 80.10 Motor Connections

		Crew	Daily Output	Labor-Hours	Unit	Material	Labor	Equipment	Total	Total Incl O&P
0010	**MOTOR CONNECTIONS**									
0020	Flexible conduit and fittings, 115 volt, 1 phase, up to 1 HP motor	1 Elec	8	1	Ea.	8.75	29.50		38.25	57.50

26 05 90 – Residential Wiring

26 05 90.10 Residential Wiring

		Crew	Daily Output	Labor-Hours	Unit	Material	Labor	Equipment	Total	Total Incl O&P
0010	**RESIDENTIAL WIRING**									
0020	20' avg. runs and #14/2 wiring incl. unless otherwise noted									
1000	Service & panel, includes 24' SE-AL cable, service eye, meter,									
1010	Socket, panel board, main bkr., ground rod, 15 or 20 amp									
1020	1-pole circuit breakers, and misc. hardware									
1100	100 amp, with 10 branch breakers	1 Elec	1.19	6.723	Ea.	515	198		713	885
1110	With PVC conduit and wire		.92	8.696		575	256		831	1,050
1120	With RGS conduit and wire		.73	10.959		735	320		1,055	1,325
1150	150 amp, with 14 branch breakers		1.03	7.767		800	228		1,028	1,250
1170	With PVC conduit and wire		.82	9.756		925	287		1,212	1,500
1180	With RGS conduit and wire		.67	11.940		1,250	350		1,600	1,925
1200	200 amp, with 18 branch breakers	2 Elec	1.80	8.889		1,050	261		1,311	1,575
1220	With PVC conduit and wire		1.46	10.959		1,175	320		1,495	1,800
1230	With RGS conduit and wire		1.24	12.903		1,600	380		1,980	2,375
1800	Lightning surge suppressor for above services, add	1 Elec	32	.250		46	7.35		53.35	62.50
2000	Switch devices									
2100	Single pole, 15 amp, Ivory, with a 1-gang box, cover plate,									
2110	Type NM (Romex) cable	1 Elec	17.10	.468	Ea.	12.65	13.75		26.40	36.50
2120	Type MC (BX) cable		14.30	.559		26.50	16.45		42.95	56
2130	EMT & wire		5.71	1.401		32	41		73	102
2150	3-way, #14/3, type NM cable		14.55	.550		17.35	16.15		33.50	45.50
2170	Type MC cable		12.31	.650		37	19.10		56.10	71.50
2180	EMT & wire		5	1.600		35.50	47		82.50	116
2200	4-way, #14/3, type NM cable		14.55	.550		31.50	16.15		47.65	61
2220	Type MC cable		12.31	.650		51	19.10		70.10	87
2230	EMT & wire		5	1.600		49.50	47		96.50	131
2250	S.P., 20 amp, #12/2, type NM cable		13.33	.600		21	17.65		38.65	52
2270	Type MC cable		11.43	.700		31	20.50		51.50	68
2280	EMT & wire		4.85	1.649		41.50	48.50		90	125
2290	S.P. rotary dimmer, 600W, no wiring		17	.471		17.45	13.85		31.30	41.50
2300	S.P. rotary dimmer, 600W, type NM cable		14.55	.550		25	16.15		41.15	54
2320	Type MC cable		12.31	.650		39	19.10		58.10	74
2330	EMT & wire		5	1.600		46.50	47		93.50	128
2350	3-way rotary dimmer, type NM cable		13.33	.600		23	17.65		40.65	54.50
2370	Type MC cable		11.43	.700		37	20.50		57.50	74
2380	EMT & wire		4.85	1.649		44.50	48.50		93	128

26 05 Common Work Results for Electrical

26 05 90 – Residential Wiring

26 05 90.10 Residential Wiring		Crew	Daily Output	Labor-Hours	Unit	Material	2007 Bare Costs Labor	2007 Bare Costs Equipment	Total	Total Incl O&P
2400	Interval timer wall switch, 20 amp, 1-30 min., #12/2									
2410	Type NM cable	1 Elec	14.55	.550	Ea.	46	16.15		62.15	77
2420	Type MC cable		12.31	.650		51	19.10		70.10	87
2430	EMT & wire		5	1.600		66.50	47		113.50	150
2500	Decorator style									
2510	S.P., 15 amp, type NM cable	1 Elec	17.10	.468	Ea.	16.75	13.75		30.50	41
2520	Type MC cable		14.30	.559		30.50	16.45		46.95	60.50
2530	EMT & wire		5.71	1.401		36	41		77	107
2550	3-way, #14/3, type NM cable		14.55	.550		21.50	16.15		37.65	50
2570	Type MC cable		12.31	.650		41	19.10		60.10	76
2580	EMT & wire		5	1.600		39.50	47		86.50	120
2600	4-way, #14/3, type NM cable		14.55	.550		35.50	16.15		51.65	65.50
2620	Type MC cable		12.31	.650		55	19.10		74.10	91.50
2630	EMT & wire		5	1.600		53.50	47		100.50	136
2650	S.P., 20 amp, #12/2, type NM cable		13.33	.600		25	17.65		42.65	56.50
2670	Type MC cable		11.43	.700		35.50	20.50		56	72.50
2680	EMT & wire		4.85	1.649		45.50	48.50		94	129
2700	S.P., slide dimmer, type NM cable		17.10	.468		31	13.75		44.75	57
2720	Type MC cable		14.30	.559		45	16.45		61.45	76.50
2730	EMT & wire		5.71	1.401		52.50	41		93.50	125
2750	S.P., touch dimmer, type NM cable		17.10	.468		28	13.75		41.75	53.50
2770	Type MC cable		14.30	.559		42	16.45		58.45	73
2780	EMT & wire		5.71	1.401		49.50	41		90.50	122
2800	3-way touch dimmer, type NM cable		13.33	.600		48	17.65		65.65	82
2820	Type MC cable		11.43	.700		62	20.50		82.50	102
2830	EMT & wire		4.85	1.649		69.50	48.50		118	156
3000	Combination devices									
3100	S.P. switch/15 amp recpt., Ivory, 1-gang box, plate									
3110	Type NM cable	1 Elec	11.43	.700	Ea.	23	20.50		43.50	59
3120	Type MC cable		10	.800		37	23.50		60.50	79
3130	EMT & wire		4.40	1.818		44	53.50		97.50	136
3150	S.P. switch/pilot light, type NM cable		11.43	.700		23.50	20.50		44	59.50
3170	Type MC cable		10	.800		37.50	23.50		61	80
3180	EMT & wire		4.43	1.806		45	53		98	136
3190	2-S.P. switches, 2-#14/2, no wiring		14	.571		7.35	16.80		24.15	35.50
3200	2-S.P. switches, 2-#14/2, type NM cables		10	.800		29.50	23.50		53	71
3220	Type MC cable		8.89	.900		51	26.50		77.50	99
3230	EMT & wire		4.10	1.951		49.50	57.50		107	148
3250	3-way switch/15 amp recpt., #14/3, type NM cable		10	.800		31.50	23.50		55	73
3270	Type MC cable		8.89	.900		51	26.50		77.50	99
3280	EMT & wire		4.10	1.951		49.50	57.50		107	148
3300	2-3 way switches, 2-#14/3, type NM cables		8.89	.900		44.50	26.50		71	92
3320	Type MC cable		8	1		77	29.50		106.50	133
3330	EMT & wire		4	2		58	59		117	160
3350	S.P. switch/20 amp recpt., #12/2, type NM cable		10	.800		35.50	23.50		59	77.50
3370	Type MC cable		8.89	.900		40	26.50		66.50	87
3380	EMT & wire		4.10	1.951		55.50	57.50		113	155
3400	Decorator style									
3410	S.P. switch/15 amp recpt., type NM cable	1 Elec	11.43	.700	Ea.	27	20.50		47.50	63
3420	Type MC cable		10	.800		41	23.50		64.50	83.50
3430	EMT & wire		4.40	1.818		48.50	53.50		102	140
3450	S.P. switch/pilot light, type NM cable		11.43	.700		28	20.50		48.50	64
3470	Type MC cable		10	.800		41.50	23.50		65	84.50

26 05 Common Work Results for Electrical

26 05 90 – Residential Wiring

26 05 90.10 Residential Wiring

		Crew	Daily Output	Labor-Hours	Unit	Material	2007 Bare Costs Labor	Equipment	Total	Total Incl O&P
3480	EMT & wire	1 Elec	4.40	1.818	Ea.	49	53.50		102.50	141
3500	2-S.P. switches, 2-#14/2, type NM cables		10	.800		34	23.50		57.50	75.50
3520	Type MC cable		8.89	.900		55	26.50		81.50	104
3530	EMT & wire		4.10	1.951		53.50	57.50		111	153
3550	3-way/15 amp recpt., #14/3, type NM cable		10	.800		35.50	23.50		59	77.50
3570	Type MC cable		8.89	.900		55	26.50		81.50	104
3580	EMT & wire		4.10	1.951		53.50	57.50		111	153
3650	2-3 way switches, 2-#14/3, type NM cables		8.89	.900		48.50	26.50		75	96.50
3670	Type MC cable		8	1		81	29.50		110.50	137
3680	EMT & wire		4	2		62.50	59		121.50	165
3700	S.P. switch/20 amp recpt., #12/2, type NM cable		10	.800		39.50	23.50		63	82
3720	Type MC cable		8.89	.900		44	26.50		70.50	91.50
3730	EMT & wire		4.10	1.951		60	57.50		117.50	160
4000	Receptacle devices									
4010	Duplex outlet, 15 amp recpt., Ivory, 1-gang box, plate									
4015	Type NM cable	1 Elec	14.55	.550	Ea.	11	16.15		27.15	38.50
4020	Type MC cable		12.31	.650		25	19.10		44.10	58.50
4030	EMT & wire		5.33	1.501		30.50	44		74.50	106
4050	With #12/2, type NM cable		12.31	.650		14.95	19.10		34.05	47.50
4070	Type MC cable		10.67	.750		25	22		47	63.50
4080	EMT & wire		4.71	1.699		35.50	50		85.50	121
4100	20 amp recpt., #12/2, type NM cable		12.31	.650		23	19.10		42.10	56.50
4120	Type MC cable		10.67	.750		33	22		55	72.50
4130	EMT & wire		4.71	1.699		43.50	50		93.50	130
4140	For GFI see line 4300 below									
4150	Decorator style, 15 amp recpt., type NM cable	1 Elec	14.55	.550	Ea.	15.10	16.15		31.25	43
4170	Type MC cable		12.31	.650		29	19.10		48.10	63
4180	EMT & wire		5.33	1.501		34.50	44		78.50	110
4200	With #12/2, type NM cable		12.31	.650		19.05	19.10		38.15	52
4220	Type MC cable		10.67	.750		29	22		51	68
4230	EMT & wire		4.71	1.699		39.50	50		89.50	125
4250	20 amp recpt. #12/2, type NM cable		12.31	.650		27	19.10		46.10	61
4270	Type MC cable		10.67	.750		37.50	22		59.50	77
4280	EMT & wire		4.71	1.699		47.50	50		97.50	134
4300	GFI, 15 amp recpt., type NM cable		12.31	.650		39.50	19.10		58.60	74.50
4320	Type MC cable		10.67	.750		53.50	22		75.50	94.50
4330	EMT & wire		4.71	1.699		58.50	50		108.50	146
4350	GFI with #12/2, type NM cable		10.67	.750		43.50	22		65.50	83.50
4370	Type MC cable		9.20	.870		53.50	25.50		79	101
4380	EMT & wire		4.21	1.900		64	56		120	161
4400	20 amp recpt., #12/2 type NM cable		10.67	.750		45	22		67	85.50
4420	Type MC cable		9.20	.870		55	25.50		80.50	102
4430	EMT & wire		4.21	1.900		65.50	56		121.50	163
4500	Weather-proof cover for above receptacles, add		32	.250		4.55	7.35		11.90	17
4550	Air conditioner outlet, 20 amp-240 volt recpt.									
4560	30' of #12/2, 2 pole circuit breaker									
4570	Type NM cable	1 Elec	10	.800	Ea.	59.50	23.50		83	104
4580	Type MC cable		9	.889		71	26		97	121
4590	EMT & wire		4	2		78.50	59		137.50	183
4600	Decorator style, type NM cable		10	.800		63.50	23.50		87	109
4620	Type MC cable		9	.889		75.50	26		101.50	126
4630	EMT & wire		4	2		83	59		142	187
4650	Dryer outlet, 30 amp-240 volt recpt., 20' of #10/3									

26 05 Common Work Results for Electrical

26 05 90 – Residential Wiring

26 05 90.10 Residential Wiring		Crew	Daily Output	Labor-Hours	Unit	Material	2007 Bare Costs Labor	Equipment	Total	Total Incl O&P
4660	2 pole circuit breaker									
4670	Type NM cable	1 Elec	6.41	1.248	Ea.	78.50	36.50		115	147
4680	Type MC cable		5.71	1.401		76	41		117	151
4690	EMT & wire		3.48	2.299		81.50	67.50		149	200
4700	Range outlet, 50 amp-240 volt recpt., 30' of #8/3									
4710	Type NM cable	1 Elec	4.21	1.900	Ea.	116	56		172	218
4720	Type MC cable		4	2		160	59		219	272
4730	EMT & wire		2.96	2.703		114	79.50		193.50	254
4750	Central vacuum outlet, Type NM cable		6.40	1.250		72	37		109	140
4770	Type MC cable		5.71	1.401		86.50	41		127.50	163
4780	EMT & wire		3.48	2.299		88.50	67.50		156	208
4800	30 amp-110 volt locking recpt., #10/2 circ. bkr.									
4810	Type NM cable	1 Elec	6.20	1.290	Ea.	83	38		121	153
4820	Type MC cable		5.40	1.481		101	43.50		144.50	182
4830	EMT & wire		3.20	2.500		101	73.50		174.50	231
4900	Low voltage outlets									
4910	Telephone recpt., 20' of 4/C phone wire	1 Elec	26	.308	Ea.	9.35	9.05		18.40	25
4920	TV recpt., 20' of RG59U coax wire, F type connector	"	16	.500	"	16.25	14.70		30.95	42
4950	Door bell chime, transformer, 2 buttons, 60' of bellwire									
4970	Economy model	1 Elec	11.50	.696	Ea.	60	20.50		80.50	99.50
4980	Custom model		11.50	.696		93.50	20.50		114	137
4990	Luxury model, 3 buttons		9.50	.842		243	25		268	310
6000	Lighting outlets									
6050	Wire only (for fixture), type NM cable	1 Elec	32	.250	Ea.	9.65	7.35		17	22.50
6070	Type MC cable		24	.333		17.85	9.80		27.65	35.50
6080	EMT & wire		10	.800		22	23.50		45.50	62.50
6100	Box (4"), and wire (for fixture), type NM cable		25	.320		17.60	9.40		27	34.50
6120	Type MC cable		20	.400		26	11.75		37.75	47.50
6130	EMT & wire		11	.727		29.50	21.50		51	67.50
6200	Fixtures (use with lines 6050 or 6100 above)									
6210	Canopy style, economy grade	1 Elec	40	.200	Ea.	27	5.90		32.90	39
6220	Custom grade		40	.200		50	5.90		55.90	64.50
6250	Dining room chandelier, economy grade		19	.421		81	12.40		93.40	109
6260	Custom grade		19	.421		239	12.40		251.40	283
6270	Luxury grade		15	.533		530	15.70		545.70	610
6310	Kitchen fixture (fluorescent), economy grade		30	.267		57	7.85		64.85	75.50
6320	Custom grade		25	.320		165	9.40		174.40	197
6350	Outdoor, wall mounted, economy grade		30	.267		28	7.85		35.85	44
6360	Custom grade		30	.267		106	7.85		113.85	130
6370	Luxury grade		25	.320		240	9.40		249.40	279
6410	Outdoor PAR floodlights, 1 lamp, 150 watt		20	.400		27	11.75		38.75	48.50
6420	2 lamp, 150 watt each		20	.400		44	11.75		55.75	67.50
6430	For infrared security sensor, add		32	.250		92	7.35		99.35	113
6450	Outdoor, quartz-halogen, 300 watt flood		20	.400		40	11.75		51.75	63
6600	Recessed downlight, round, pre-wired, 50 or 75 watt trim		30	.267		37	7.85		44.85	53.50
6610	With shower light trim		30	.267		46	7.85		53.85	63.50
6620	With wall washer trim		28	.286		55	8.40		63.40	74
6630	With eye-ball trim		28	.286		55	8.40		63.40	74
6640	For direct contact with insulation, add					1.85			1.85	2.04
6700	Porcelain lamp holder	1 Elec	40	.200		3.80	5.90		9.70	13.80
6710	With pull switch		40	.200		4.13	5.90		10.03	14.15
6750	Fluorescent strip, 1-20 watt tube, wrap around diffuser, 24"		24	.333		53	9.80		62.80	74.50
6760	1-40 watt tube, 48"		24	.333		65	9.80		74.80	87.50

26 05 Common Work Results for Electrical

26 05 90 – Residential Wiring

26 05 90.10 Residential Wiring		Crew	Daily Output	Labor-Hours	Unit	Material	2007 Bare Costs Labor	Equipment	Total	Total Incl O&P
6770	2-40 watt tubes, 48"	1 Elec	20	.400	Ea.	79	11.75		90.75	106
6780	With residential ballast		20	.400		89.50	11.75		101.25	118
6800	Bathroom heat lamp, 1-250 watt		28	.286		41	8.40		49.40	58.50
6810	2-250 watt lamps		28	.286		67	8.40		75.40	87
6820	For timer switch, see line 2400									
6900	Outdoor post lamp, incl. post, fixture, 35' of #14/2									
6910	Type NMC cable	1 Elec	3.50	2.286	Ea.	195	67		262	325
6920	Photo-eye, add		27	.296		32	8.70		40.70	49
6950	Clock dial time switch, 24 hr., w/enclosure, type NM cable		11.43	.700		60	20.50		80.50	99.50
6970	Type MC cable		11	.727		74	21.50		95.50	117
6980	EMT & wire		4.85	1.649		79.50	48.50		128	166
7000	Alarm systems									
7050	Smoke detectors, box, #14/3, type NM cable	1 Elec	14.55	.550	Ea.	35.50	16.15		51.65	65.50
7070	Type MC cable		12.31	.650		49.50	19.10		68.60	85.50
7080	EMT & wire		5	1.600		48.50	47		95.50	130
7090	For relay output to security system, add					12.90			12.90	14.20
8000	Residential equipment									
8050	Disposal hook-up, incl. switch, outlet box, 3' of flex									
8060	20 amp-1 pole circ. bkr., and 25' of #12/2									
8070	Type NM cable	1 Elec	10	.800	Ea.	33.50	23.50		57	75
8080	Type MC cable		8	1		44.50	29.50		74	97
8090	EMT & wire		5	1.600		56	47		103	138
8100	Trash compactor or dishwasher hook-up, incl. outlet box,									
8110	3' of flex, 15 amp-1 pole circ. bkr., and 25' of #14/2									
8130	Type MC cable	1 Elec	8	1	Ea.	39.50	29.50		69	91.50
8140	EMT & wire	"	5	1.600	"	47.50	47		94.50	129
8150	Hot water sink dispensor hook-up, use line 8100									
8200	Vent/exhaust fan hook-up, type NM cable	1 Elec	32	.250	Ea.	9.65	7.35		17	22.50
8220	Type MC cable		24	.333		17.85	9.80		27.65	35.50
8230	EMT & wire		10	.800		22	23.50		45.50	62.50
8250	Bathroom vent fan, 50 CFM (use with above hook-up)									
8260	Economy model	1 Elec	15	.533	Ea.	22.50	15.70		38.20	50.50
8270	Low noise model		15	.533		32	15.70		47.70	60.50
8280	Custom model		12	.667		117	19.60		136.60	161
8300	Bathroom or kitchen vent fan, 110 CFM									
8310	Economy model	1 Elec	15	.533	Ea.	60.50	15.70		76.20	92
8320	Low noise model	"	15	.533	"	79	15.70		94.70	113
8350	Paddle fan, variable speed (w/o lights)									
8360	Economy model (AC motor)	1 Elec	10	.800	Ea.	105	23.50		128.50	155
8370	Custom model (AC motor)		10	.800		182	23.50		205.50	239
8380	Luxury model (DC motor)		8	1		360	29.50		389.50	445
8390	Remote speed switch for above, add		12	.667		26	19.60		45.60	60.50
8500	Whole house exhaust fan, ceiling mount, 36", variable speed									
8510	Remote switch, incl. shutters, 20 amp-1 pole circ. bkr.									
8520	30' of #12/2, type NM cable	1 Elec	4	2	Ea.	750	59		809	920
8530	Type MC cable		3.50	2.286		765	67		832	950
8540	EMT & wire		3	2.667		780	78.50		858.50	985
8600	Whirlpool tub hook-up, incl. timer switch, outlet box									
8610	3' of flex, 20 amp-1 pole GFI circ. bkr.									
8620	30' of #12/2, type NM cable	1 Elec	5	1.600	Ea.	116	47		163	205
8630	Type MC cable		4.20	1.905		121	56		177	224
8640	EMT & wire		3.40	2.353		131	69		200	257
8650	Hot water heater hook-up, incl. 1-2 pole circ. bkr., box;									

26 05 Common Work Results for Electrical

26 05 90 – Residential Wiring

26 05 90.10 Residential Wiring

		Crew	Daily Output	Labor-Hours	Unit	Material	2007 Bare Costs Labor	Equipment	Total	Total Incl O&P
8660	3' of flex, 20' of #10/2, type NM cable	1 Elec	5	1.600	Ea.	37	47		84	118
8670	Type MC cable		4.20	1.905		52	56		108	148
8680	EMT & wire		3.40	2.353		50	69		119	168
9000	Heating/air conditioning									
9050	Furnace/boiler hook-up, incl. firestat, local on-off switch									
9060	Emergency switch, and 40' of type NM cable	1 Elec	4	2	Ea.	55.50	59		114.50	158
9070	Type MC cable		3.50	2.286		77	67		144	194
9080	EMT & wire		1.50	5.333		87	157		244	350
9100	Air conditioner hook-up, incl. local 60 amp disc. switch									
9110	3' sealtite, 40 amp, 2 pole circuit breaker									
9130	40' of #8/2, type NM cable	1 Elec	3.50	2.286	Ea.	210	67		277	340
9140	Type MC cable		3	2.667		291	78.50		369.50	450
9150	EMT & wire		1.30	6.154		236	181		417	555
9200	Heat pump hook-up, 1-40 & 1-100 amp 2 pole circ. bkr.									
9210	Local disconnect switch, 3' sealtite									
9220	40' of #8/2 & 30' of #3/2									
9230	Type NM cable	1 Elec	1.30	6.154	Ea.	610	181		791	970
9240	Type MC cable		1.08	7.407		720	218		938	1,150
9250	EMT & wire		.94	8.511		645	250		895	1,125
9500	Thermostat hook-up, using low voltage wire									
9520	Heating only	1 Elec	24	.333	Ea.	7.10	9.80		16.90	24
9530	Heating/cooling	"	20	.400	"	8.60	11.75		20.35	28.50

26 24 Switchboards and Panelboards

26 24 16 – Panelboards

26 24 16.10 Load Centers

		Crew	Daily Output	Labor-Hours	Unit	Material	2007 Bare Costs Labor	Equipment	Total	Total Incl O&P
0010	**LOAD CENTERS** (residential type)									
0100	3 wire, 120/240V, 1 phase, including 1 pole plug-in breakers									
0200	100 amp main lugs, indoor, 8 circuits	1 Elec	1.40	5.714	Ea.	162	168		330	450
0300	12 circuits		1.20	6.667		226	196		422	570
0400	Rainproof, 8 circuits		1.40	5.714		194	168		362	490
0500	12 circuits		1.20	6.667		274	196		470	620
0600	200 amp main lugs, indoor, 16 circuits	R-1A	1.80	8.889		315	214		529	700
0700	20 circuits		1.50	10.667		395	257		652	860
0800	24 circuits		1.30	12.308		515	297		812	1,050
1200	Rainproof, 16 circuits		1.80	8.889		375	214		589	765
1300	20 circuits		1.50	10.667		450	257		707	920
1400	24 circuits		1.30	12.308		670	297		967	1,225

26 27 Low-Voltage Distribution Equipment

26 27 13 – Electricity Metering

26 27 13.10 Meter Centers and Sockets

		Crew	Daily Output	Labor-Hours	Unit	Material	2007 Bare Costs Labor	Equipment	Total	Total Incl O&P
0010	**METER CENTERS AND SOCKETS**									
0100	Sockets, single position, 4 terminal, 100 amp	1 Elec	3.20	2.500	Ea.	37.50	73.50		111	161
0200	150 amp		2.30	3.478		43.50	102		145.50	215
0300	200 amp		1.90	4.211		56	124		180	264
0500	Double position, 4 terminal, 100 amp		2.80	2.857		153	84		237	305
0600	150 amp		2.10	3.810		180	112		292	380
0700	200 amp		1.70	4.706		385	138		523	650

26 27 Low-Voltage Distribution Equipment

26 27 13 – Electricity Metering

26 27 13.10 Meter Centers and Sockets

		Crew	Daily Output	Labor-Hours	Unit	Material	2007 Bare Costs Labor	Equipment	Total	Total Incl O&P
2590	Basic meter device									
2600	1P 3W 120/240V 4 jaw 125A sockets, 3 meter	2 Elec	1	16	Ea.	605	470		1,075	1,425
2620	5 meter		.80	20		910	590		1,500	1,950
2640	7 meter		.56	28.571		1,325	840		2,165	2,850
2660	10 meter		.48	33.333		1,825	980		2,805	3,600
2680	Rainproof 1P 3W 120/240V 4 jaw 125A sockets									
2690	3 meter	2 Elec	1	16	Ea.	605	470		1,075	1,425
2710	6 meter		.60	26.667		1,050	785		1,835	2,425
2730	8 meter		.52	30.769		1,450	905		2,355	3,075
2750	1P 3W 120/240V 4 jaw sockets									
2760	with 125A circuit breaker, 3 meter	2 Elec	1	16	Ea.	1,125	470		1,595	2,025
2780	5 meter		.80	20		1,800	590		2,390	2,925
2800	7 meter		.56	28.571		2,575	840		3,415	4,200
2820	10 meter		.48	33.333		3,575	980		4,555	5,525
2830	Rainproof 1P 3W 120/240V 4 jaw sockets									
2840	with 125A circuit breaker, 3 meter	2 Elec	1	16	Ea.	1,125	470		1,595	2,025
2870	6 meter		.60	26.667		2,100	785		2,885	3,575
2890	8 meter		.52	30.769		2,850	905		3,755	4,625
3250	1P 3W 120/240V 4 jaw sockets									
3260	with 200A circuit breaker, 3 meter	2 Elec	1	16	Ea.	1,700	470		2,170	2,650
3290	6 meter		.60	26.667		3,400	785		4,185	5,025
3310	8 meter		.56	28.571		4,600	840		5,440	6,425
3330	Rainproof 1P 3W 120/240V 4 jaw sockets									
3350	with 200A circuit breaker, 3 meter	2 Elec	1	16	Ea.	1,700	470		2,170	2,650
3380	6 meter		.60	26.667		3,400	785		4,185	5,025
3400	8 meter		.52	30.769		4,600	905		5,505	6,525

26 27 23 – Indoor Service Poles

26 27 23.40 Surface Raceway

		Crew	Daily Output	Labor-Hours	Unit	Material	2007 Bare Costs Labor	Equipment	Total	Total Incl O&P
0010	**SURFACE RACEWAY**									
0100	No. 500	1 Elec	100	.080	L.F.	.95	2.35		3.30	4.88
0110	No. 700		100	.080		1.07	2.35		3.42	5
0200	No. 1000		90	.089		1.81	2.61		4.42	6.25
0400	No. 1500, small pancake		90	.089		1.93	2.61		4.54	6.40
0600	No. 2000, base & cover, blank		90	.089		1.88	2.61		4.49	6.35
0800	No. 3000, base & cover, blank		75	.107		3.80	3.14		6.94	9.30
2400	Fittings, elbows, No. 500		40	.200	Ea.	1.72	5.90		7.62	11.50
2800	Elbow cover, No. 2000		40	.200		3.28	5.90		9.18	13.20
2880	Tee, No. 500		42	.190		3.30	5.60		8.90	12.75
2900	No. 2000		27	.296		10.30	8.70		19	25.50
3000	Switch box, No. 500		16	.500		11.20	14.70		25.90	36.50
3400	Telephone outlet, No. 1500		16	.500		12.40	14.70		27.10	37.50
3600	Junction box, No. 1500		16	.500		8.70	14.70		23.40	33.50
3800	Plugmold wired sections, No. 2000									
4000	1 circuit, 6 outlets, 3 ft. long	1 Elec	8	1	Ea.	31.50	29.50		61	83
4100	2 circuits, 8 outlets, 6 ft. long	"	5.30	1.509	"	52.50	44.50		97	131

26 27 26 – Wiring Devices

26 27 26.10 Low Voltage Switching

		Crew	Daily Output	Labor-Hours	Unit	Material	2007 Bare Costs Labor	Equipment	Total	Total Incl O&P
0010	**LOW VOLTAGE SWITCHING**									
3600	Relays, 120 V or 277 V standard	1 Elec	12	.667	Ea.	28	19.60		47.60	63
3800	Flush switch, standard		40	.200		9.75	5.90		15.65	20.50
4000	Interchangeable		40	.200		12.75	5.90		18.65	23.50
4100	Surface switch, standard		40	.200		7.15	5.90		13.05	17.45

26 27 Low-Voltage Distribution Equipment

26 27 26 – Wiring Devices

26 27 26.10 Low Voltage Switching

		Crew	Daily Output	Labor-Hours	Unit	Material	2007 Bare Costs Labor	Equipment	Total	Total Incl O&P
4200	Transformer 115 V to 25 V	1 Elec	12	.667	Ea.	100	19.60		119.60	142
4400	Master control, 12 circuit, manual		4	2		102	59		161	208
4500	25 circuit, motorized		4	2		110	59		169	217
4600	Rectifier, silicon		12	.667		33	19.60		52.60	68
4800	Switchplates, 1 gang, 1, 2 or 3 switch, plastic		80	.100		3.24	2.94		6.18	8.35
5000	Stainless steel		80	.100		8.75	2.94		11.69	14.45
5400	2 gang, 3 switch, stainless steel		53	.151		16.90	4.44		21.34	26
5500	4 switch, plastic		53	.151		7.25	4.44		11.69	15.20
5600	2 gang, 4 switch, stainless steel		53	.151		17.65	4.44		22.09	26.50
5700	6 switch, stainless steel		53	.151		39	4.44		43.44	50.50
5800	3 gang, 9 switch, stainless steel		32	.250		54	7.35		61.35	71.50

26 27 26.20 Wiring Devices

		Crew	Daily Output	Labor-Hours	Unit	Material	Labor	Equipment	Total	Total Incl O&P
0010	**WIRING DEVICES**									
0200	Toggle switch, quiet type, single pole, 15 amp	1 Elec	40	.200	Ea.	5.10	5.90		11	15.25
0600	3 way, 15 amp		23	.348		7.65	10.25		17.90	25
0900	4 way, 15 amp		15	.533		21.50	15.70		37.20	49.50
1650	Dimmer switch, 120 volt, incandescent, 600 watt, 1 pole		16	.500		11.25	14.70		25.95	36.50
2460	Receptacle, duplex, 120 volt, grounded, 15 amp		40	.200		1.23	5.90		7.13	10.95
2470	20 amp		27	.296		9.40	8.70		18.10	24.50
2490	Dryer, 30 amp		15	.533		5.85	15.70		21.55	32
2500	Range, 50 amp		11	.727		11.95	21.50		33.45	48
2600	Wall plates, stainless steel, 1 gang		80	.100		2.06	2.94		5	7.05
2800	2 gang		53	.151		3.80	4.44		8.24	11.45
3200	Lampholder, keyless		26	.308		11.20	9.05		20.25	27
3400	Pullchain with receptacle		22	.364		12.20	10.70		22.90	31

26 27 73 – Door Chimes

26 27 73.10 Doorbell System

		Crew	Daily Output	Labor-Hours	Unit	Material	Labor	Equipment	Total	Total Incl O&P
0010	**DOORBELL SYSTEM** Incl. transformer, button & signal									
1000	Door chimes, 2 notes, minimum	1 Elec	16	.500	Ea.	23	14.70		37.70	49.50
1020	Maximum		12	.667		122	19.60		141.60	166
1100	Tube type, 3 tube system		12	.667		173	19.60		192.60	223
1180	4 tube system		10	.800		277	23.50		300.50	345
1900	For transformer & button, minimum add		5	1.600		13.10	47		60.10	91
1960	Maximum, add		4.50	1.778		39.50	52.50		92	128
3000	For push button only, minimum		24	.333		2.65	9.80		12.45	18.85
3100	Maximum		20	.400		20.50	11.75		32.25	42

26 28 Low-Voltage Circuit Protective Devices

26 28 16 – Enclosed Switches and Circuit Breakers

26 28 16.10 Circuit Breakers

		Crew	Daily Output	Labor-Hours	Unit	Material	Labor	Equipment	Total	Total Incl O&P
0010	**CIRCUIT BREAKERS** (in enclosure)									
0100	Enclosed (NEMA 1), 600 volt, 3 pole, 30 amp	1 Elec	3.20	2.500	Ea.	530	73.50		603.50	705
0200	60 amp		2.80	2.857		530	84		614	720
0400	100 amp		2.30	3.478		605	102		707	835

26 28 16.20 Safety Switches

		Crew	Daily Output	Labor-Hours	Unit	Material	Labor	Equipment	Total	Total Incl O&P
0010	**SAFETY SWITCHES**									
0100	General duty 240 volt, 3 pole NEMA 1, fusible, 30 amp	1 Elec	3.20	2.500	Ea.	92	73.50		165.50	221
0200	60 amp		2.30	3.478		155	102		257	340
0300	100 amp		1.90	4.211		267	124		391	495
0400	200 amp		1.30	6.154		575	181		756	930

26 28 Low-Voltage Circuit Protective Devices

26 28 16 – Enclosed Switches and Circuit Breakers

26 28 16.20 Safety Switches		Crew	Daily Output	Labor-Hours	Unit	Material	2007 Bare Costs Labor	Equipment	Total	Total Incl O&P
0500	400 amp	2 Elec	1.80	8.889	Ea.	1,450	261		1,711	2,025
9010	Disc. switch, 600V 3 pole fusible, 30 amp, to 10 HP motor	1 Elec	3.20	2.500		268	73.50		341.50	415
9050	60 amp, to 30 HP motor		2.30	3.478		615	102		717	845
9070	100 amp, to 60 HP motor		1.90	4.211		615	124		739	880
26 28 16.40 Time Switches										
0010	**TIME SWITCHES**									
0100	Single pole, single throw, 24 hour dial	1 Elec	4	2	Ea.	87	59		146	192
0200	24 hour dial with reserve power		3.60	2.222		370	65.50		435.50	510
0300	Astronomic dial		3.60	2.222		149	65.50		214.50	270
0400	Astronomic dial with reserve power		3.30	2.424		480	71.50		551.50	645
0500	7 day calendar dial		3.30	2.424		134	71.50		205.50	263
0600	7 day calendar dial with reserve power		3.20	2.500		520	73.50		593.50	690
0700	Photo cell 2000 watt		8	1		16.05	29.50		45.55	65.50

26 32 Packaged Generator Assemblies

26 32 13 – Engine Generators

26 32 13.16 Gas-Engine-Driven Generator Sets

		Crew	Daily Output	Labor-Hours	Unit	Material	Labor	Equipment	Total	Total Incl O&P
0010	**GAS-ENGINE-DRIVEN GENERATOR SETS**									
0020	Gas or gasoline operated, includes battery,									
0050	charger, muffler & transfer switch									
0200	3 phase 4 wire, 277/480 volt, 7.5 kW	R-3	.83	24.096	Ea.	6,350	705	197	7,252	8,375
0300	11.5 kW		.71	28.169		9,000	825	230	10,055	11,500
0400	20 kW		.63	31.746		10,600	925	259	11,784	13,500

26 36 Transfer Switches

26 36 23 – Automatic Transfer Switches

26 36 23.10 Automatic Transfer Switches

		Crew	Daily Output	Labor-Hours	Unit	Material	Labor	Equipment	Total	Total Incl O&P
0010	**AUTOMATIC TRANSFER SWITCHES**									
0100	Switches, enclosed 480 volt, 3 pole, 30 amp	1 Elec	2.30	3.478	Ea.	3,275	102		3,377	3,775
0200	60 amp	"	1.90	4.211	"	3,275	124		3,399	3,800

26 41 Facility Lightning Protection

26 41 13 – Lightning Protection for Structures

26 41 13.13 Lightning Protection for Buildings

		Crew	Daily Output	Labor-Hours	Unit	Material	Labor	Equipment	Total	Total Incl O&P
0010	**LIGHTNING PROTECTION FOR BUILDINGS**									
0200	Air terminals & base, copper									
0400	3/8" diameter x 10" (to 75' high)	1 Elec	8	1	Ea.	17.70	29.50		47.20	67.50
1000	Aluminum, 1/2" diameter x 12" (to 75' high)		8	1	"	3.05	29.50		32.55	51.50
2000	Cable, copper, 220 lb. per thousand ft. (to 75' high)		320	.025	L.F.	1.43	.74		2.17	2.77
2500	Aluminum, 101 lb. per thousand ft. (to 75' high)		280	.029	"	.72	.84		1.56	2.16
3000	Arrester, 175 volt AC to ground		8	1	Ea.	62.50	29.50		92	117

26 51 Interior Lighting

26 51 13 – Interior Lighting Fixtures, Lamps, and Ballasts

26 51 13.50 Interior Lighting Fixtures

		Crew	Daily Output	Labor-Hours	Unit	Material	2007 Bare Costs Labor	Equipment	Total	Total Incl O&P
0010	**INTERIOR LIGHTING FIXTURES** Including lamps, mounting									
0030	hardware and connections									
0100	Fluorescent, C.W. lamps, troffer, recess mounted in grid, RS									
0130	grid ceiling mount									
0200	Acrylic lens, 1'W x 4'L, two 40 watt	1 Elec	5.70	1.404	Ea.	50.50	41.50		92	123
0300	2'W x 2'L, two U40 watt		5.70	1.404		54	41.50		95.50	126
0600	2'W x 4'L, four 40 watt		4.70	1.702		61	50		111	149
1000	Surface mounted, RS									
1030	Acrylic lens with hinged & latched door frame									
1100	1'W x 4'L, two 40 watt	1 Elec	7	1.143	Ea.	73.50	33.50		107	136
1200	2'W x 2'L, two U40 watt		7	1.143		79	33.50		112.50	141
1500	2'W x 4'L, four 40 watt		5.30	1.509		93.50	44.50		138	176
2100	Strip fixture									
2200	4' long, one 40 watt RS	1 Elec	8.50	.941	Ea.	29.50	27.50		57	77.50
2300	4' long, two 40 watt RS	"	8	1		31.50	29.50		61	83
2600	8' long, one 75 watt, SL	2 Elec	13.40	1.194		44	35		79	106
2700	8' long, two 75 watt, SL	"	12.40	1.290		53	38		91	121
4450	Incandescent, high hat can, round alzak reflector, prewired									
4470	100 watt	1 Elec	8	1	Ea.	62.50	29.50		92	117
4480	150 watt	"	8	1	"	93	29.50		122.50	150
5200	Ceiling, surface mounted, opal glass drum									
5300	8", one 60 watt lamp	1 Elec	10	.800	Ea.	40	23.50		63.50	82.50
5400	10", two 60 watt lamps		8	1		44.50	29.50		74	97
5500	12", four 60 watt lamps		6.70	1.194		60	35		95	123
6900	Mirror light, fluorescent, RS, acrylic enclosure, two 40 watt		8	1		87.50	29.50		117	144
6910	One 40 watt		8	1		69.50	29.50		99	124
6920	One 20 watt		12	.667		55	19.60		74.60	92.50

26 51 13.70 Residential Fixtures

		Crew	Daily Output	Labor-Hours	Unit	Material	Labor	Equipment	Total	Total Incl O&P
0010	**RESIDENTIAL FIXTURES**									
0400	Fluorescent, interior, surface, circline, 32 watt & 40 watt	1 Elec	20	.400	Ea.	77.50	11.75		89.25	105
0500	2' x 2', two U 40 watt		8	1		103	29.50		132.50	162
0700	Shallow under cabinet, two 20 watt		16	.500		44	14.70		58.70	72.50
0900	Wall mounted, 4'L, one 40 watt, with baffle		10	.800		119	23.50		142.50	170
2000	Incandescent, exterior lantern, wall mounted, 60 watt		16	.500		31.50	14.70		46.20	59
2100	Post light, 150W, with 7' post		4	2		110	59		169	217
2500	Lamp holder, weatherproof with 150W PAR		16	.500		19.50	14.70		34.20	45.50
2550	With reflector and guard		12	.667		54.50	19.60		74.10	91.50
2600	Interior pendent, globe with shade, 150 watt		20	.400		128	11.75		139.75	160

26 55 Special Purpose Lighting

26 55 59 – Display Lighting

26 55 59.10 Track Lighting

		Crew	Daily Output	Labor-Hours	Unit	Material	Labor	Equipment	Total	Total Incl O&P
0010	**TRACK LIGHTING**									
0100	8' section	1 Elec	5.30	1.509	Ea.	68.50	44.50		113	148
0300	3 circuits, 4' section		6.70	1.194		54.50	35		89.50	117
0400	8' section		5.30	1.509		84.50	44.50		129	166
0500	12' section		4.40	1.818		163	53.50		216.50	266
1000	Feed kit, surface mounting		16	.500		10.05	14.70		24.75	35
1100	End cover		24	.333		3.50	9.80		13.30	19.80
1200	Feed kit, stem mounting, 1 circuit		16	.500		29.50	14.70		44.20	56
1300	3 circuit		16	.500		29.50	14.70		44.20	56

26 55 Special Purpose Lighting

26 55 59 – Display Lighting

26 55 59.10 Track Lighting

		Crew	Daily Output	Labor-Hours	Unit	Material	2007 Bare Costs Labor	Equipment	Total	Total Incl O&P
2000	Electrical joiner, for continuous runs, 1 circuit	1 Elec	32	.250	Ea.	14.50	7.35		21.85	28
2100	3 circuit		32	.250		34	7.35		41.35	49.50
2200	Fixtures, spotlight, 75W PAR halogen		16	.500		87	14.70		101.70	120
2210	50W MR16 halogen		16	.500		106	14.70		120.70	141
3000	Wall washer, 250 watt tungsten halogen		16	.500		101	14.70		115.70	135
3100	Low voltage, 25/50 watt, 1 circuit		16	.500		102	14.70		116.70	136
3120	3 circuit		16	.500		105	14.70		119.70	140

26 56 Exterior Lighting

26 56 13 – Lighting Poles and Standards

26 56 13.10 Lighting Poles

		Crew	Daily Output	Labor-Hours	Unit	Material	Labor	Equipment	Total	Total Incl O&P
0010	**LIGHTING POLES**									
6420	Wood pole, 4-1/2" x 5-1/8", 8' high	1 Elec	6	1.333	Ea.	269	39		308	360
6440	12' high		5.70	1.404		385	41.50		426.50	490
6460	20' high		4	2		545	59		604	695

26 56 23 – Area Lighting

26 56 23.10 Exterior Fixtures

		Crew	Daily Output	Labor-Hours	Unit	Material	Labor	Equipment	Total	Total Incl O&P
0010	**EXTERIOR FIXTURES** With lamps									
0400	Quartz, 500 watt	1 Elec	5.30	1.509	Ea.	54	44.50		98.50	132
1100	Wall pack, low pressure sodium, 35 watt		4	2		220	59		279	340
1150	55 watt		4	2		263	59		322	385

26 56 26 – Landscape Lighting

26 56 26.20 Landscape Fixtures

		Crew	Daily Output	Labor-Hours	Unit	Material	Labor	Equipment	Total	Total Incl O&P
0010	**LANDSCAPE FIXTURES**									
7380	Landscape recessed uplight, incl. housing, ballast, transformer									
7390	& reflector, not incl. conduit, wire, trench									
7420	Incandescent, 250 watt	1 Elec	5	1.600	Ea.	490	47		537	615
7440	Quartz, 250 watt	"	5	1.600	"	465	47		512	590

26 56 33 – Walkway Lighting

26 56 33.10 Walkway Luminaire

		Crew	Daily Output	Labor-Hours	Unit	Material	Labor	Equipment	Total	Total Incl O&P
0010	**WALKWAY LUMINAIRE**									
6500	Bollard light, lamp & ballast, 42" high with polycarbonate lens									
7200	Incandescent, 150 watt	1 Elec	3	2.667	Ea.	485	78.50		563.50	665

26 61 Lighting Systems and Accessories

26 61 23 – Lamps

26 61 23.10 Lamps

		Crew	Daily Output	Labor-Hours	Unit	Material	Labor	Equipment	Total	Total Incl O&P
0010	**LAMPS**									
0081	Fluorescent, rapid start, cool white, 2' long, 20 watt	1 Elec	100	.080	Ea.	3.10	2.35		5.45	7.25
0101	4' long, 40 watt		90	.089		3.47	2.61		6.08	8.05
1351	High pressure sodium, 70 watt		30	.267		41	7.85		48.85	58
1371	150 watt		30	.267		44	7.85		51.85	61.50

Division 27
Communications

27 41 Audio-Video Systems

27 41 19 – Portable Audio-Video Equipment

27 41 19.10 T.V. Systems		Crew	Daily Output	Labor-Hours	Unit	Material	2007 Bare Costs Labor	Equipment	Total	Total Incl O&P
0010	**T.V. SYSTEMS** not including rough-in wires, cables & conduits									
0100	Master TV antenna system									
0200	VHF reception & distribution, 12 outlets	1 Elec	6	1.333	Outlet	183	39		222	265
0800	VHF & UHF reception & distribution, 12 outlets		6	1.333	"	182	39		221	264
5000	T.V. Antenna only, minimum		6	1.333	Ea.	40	39		79	108
5100	Maximum	↓	4	2	"	169	59		228	282

Division 28
Electronic Safety and Security

28 16 Intrusion Detection

28 16 16 – Intrusion Detection Systems Infrastructure

28 16 16.50 Intrusion Detection		Crew	Daily Output	Labor-Hours	Unit	Material	2007 Bare Costs Labor	Equipment	Total	Total Incl O&P
0010	**INTRUSION DETECTION**, not including wires & conduits									
0100	Burglar alarm, battery operated, mechanical trigger	1 Elec	4	2	Ea.	254	59		313	375
0200	Electrical trigger		4	2		305	59		364	430
0400	For outside key control, add		8	1		72	29.50		101.50	127
0600	For remote signaling circuitry, add		8	1		114	29.50		143.50	174
0800	Card reader, flush type, standard		2.70	2.963		850	87		937	1,075
1000	Multi-code		2.70	2.963		1,100	87		1,187	1,350
1200	Door switches, hinge switch		5.30	1.509		53.50	44.50		98	132
1400	Magnetic switch		5.30	1.509		63	44.50		107.50	142
2800	Ultrasonic motion detector, 12 volt		2.30	3.478		210	102		312	400
3000	Infrared photoelectric detector		2.30	3.478		173	102		275	360
3200	Passive infrared detector		2.30	3.478		259	102		361	450
3420	Switchmats, 30" x 5'		5.30	1.509		77.50	44.50		122	158
3440	30" x 25'		4	2		186	59		245	300
3460	Police connect panel		4	2		223	59		282	340
3480	Telephone dialer		5.30	1.509		350	44.50		394.50	460
3500	Alarm bell		4	2		71	59		130	174
3520	Siren		4	2		134	59		193	243

28 31 Fire Detection and Alarm

28 31 23 – Fire Detection and Alarm Annunciation Panels and Fire Stations

28 31 23.50 Alarm Panels and Devices		Crew	Daily Output	Labor-Hours	Unit	Material	2007 Bare Costs Labor	Equipment	Total	Total Incl O&P
0010	**ALARM PANELS AND DEVICES**									
5600	Strobe and horn	1 Elec	5.30	1.509	Ea.	95	44.50		139.50	178
5800	Fire alarm horn		6.70	1.194		36.50	35		71.50	97
6600	Drill switch		8	1		86.50	29.50		116	143
6800	Master box		2.70	2.963		3,100	87		3,187	3,550
7800	Remote annunciator, 8 zone lamp		1.80	4.444		201	131		332	435
8000	12 zone lamp	2 Elec	2.60	6.154		345	181		526	675
8200	16 zone lamp	"	2.20	7.273		345	214		559	730
8400	Standpipe or sprinkler alarm, alarm device	1 Elec	8	1		125	29.50		154.50	186
8600	Actuating device	"	8	1		290	29.50		319.50	370

28 31 46 – Smoke Detection Sensors

28 31 46.50 Smoke Detectors		Crew	Daily Output	Labor-Hours	Unit	Material	2007 Bare Costs Labor	Equipment	Total	Total Incl O&P
0010	**SMOKE DETECTORS**									
5200	Smoke detector, ceiling type	1 Elec	6.20	1.290	Ea.	79	38		117	149

Division 31
Earthwork

31 05 Common Work Results for Earthwork

31 05 13 – Soils for Earthwork

31 05 13.10 Borrow

		Crew	Daily Output	Labor-Hours	Unit	Material	2007 Bare Costs Labor	Equipment	Total	Total Incl O&P
0010	**BORROW**									
0020	Spread, 200 H.P. dozer, no compaction, 2 mi. RT haul									
0200	Common borrow	B-15	600	.047	C.Y.	6	1.04	3.41	10.45	12.10

31 05 16 – Aggregates for Earthwork

31 05 16.10 Borrow

		Crew	Daily Output	Labor-Hours	Unit	Material	2007 Bare Costs Labor	Equipment	Total	Total Incl O&P
0010	**BORROW**									
0020	Spread, with 200 H.P. dozer, no compaction, 2 mi RT haul									
0100	Bank run gravel	B-15	600	.047	C.Y.	21.50	1.04	3.41	25.95	29
0300	Crushed stone (1.40 tons per CY), 1-1/2"		600	.047		32	1.04	3.41	36.45	40.50
0320	3/4"		600	.047		32	1.04	3.41	36.45	40.50
0340	1/2"		600	.047		36	1.04	3.41	40.45	45
0360	3/8"		600	.047		27	1.04	3.41	31.45	35
0400	Sand, washed, concrete		600	.047		29.50	1.04	3.41	33.95	38
0500	Dead or bank sand		600	.047		6.25	1.04	3.41	10.70	12.40

31 11 Clearing and Grubbing

31 11 10 – Clearing and Grubbing

31 11 10.10 Clear and Grub

		Crew	Daily Output	Labor-Hours	Unit	Material	2007 Bare Costs Labor	Equipment	Total	Total Incl O&P
0010	**CLEAR AND GRUB**									
0020	Cut & chip light trees to 6" diam.	B-7	1	48	Acre		975	1,075	2,050	2,850
0150	Grub stumps and remove	B-30	2	12			274	920	1,194	1,450
0200	Cut & chip medium, trees to 12" diam.	B-7	.70	68.571			1,400	1,550	2,950	4,050
0250	Grub stumps and remove	B-30	1	24			550	1,825	2,375	2,950
0300	Cut & chip heavy, trees to 24" diam.	B-7	.30	160			3,250	3,625	6,875	9,475
0350	Grub stumps and remove	B-30	.50	48			1,100	3,675	4,775	5,875
0400	If burning is allowed, reduce cut & chip									40%

31 13 Selective Tree and Shrub Removal and Trimming

31 13 13 – Selective Tree and Shrub Removal

31 13 13.10 Selective Clearing

		Crew	Daily Output	Labor-Hours	Unit	Material	2007 Bare Costs Labor	Equipment	Total	Total Incl O&P
0010	**SELECTIVE CLEARING**									
0020	Clearing brush with brush saw	A-1C	.25	32	Acre		600	92	692	1,125
0100	By hand	1 Clab	.12	66.667			1,250		1,250	2,125
0300	With dozer, ball and chain, light clearing	B-11A	2	8			181	495	676	845
0400	Medium clearing	"	1.50	10.667			241	660	901	1,125

31 14 Earth Stripping and Stockpiling

31 14 13 – Soil Stripping and Stockpiling

31 14 13.23 Topsoil Stripping and Stockpiling

		Crew	Daily Output	Labor-Hours	Unit	Material	2007 Bare Costs Labor	Equipment	Total	Total Incl O&P
0010	**TOPSOIL STRIPPING AND STOCKPILING**									
1400	Loam or topsoil, remove and stockpile on site									
1420	6" deep, 200' haul	B-10B	865	.009	C.Y.		.25	1.14	1.39	1.66
1430	300' haul		520	.015			.41	1.90	2.31	2.76
1440	500' haul		225	.036			.94	4.39	5.33	6.40
1450	Alternate method: 6" deep, 200' haul		5090	.002	S.Y.		.04	.19	.23	.28
1460	500' haul		1325	.006	"		.16	.75	.91	1.08

31 22 Grading

31 22 16 – Fine Grading

31 22 16.10 Finish Grading

		Crew	Daily Output	Labor-Hours	Unit	Material	2007 Bare Costs Labor	2007 Bare Costs Equipment	Total	Total Incl O&P
0010	**FINISH GRADING**									
0012	Finish grading area to be paved with grader, small area	B-11L	400	.040	S.Y.		.90	1.26	2.16	2.90
0100	Large area		2000	.008			.18	.25	.43	.58
0200	Grade subgrade for base course, roadways	↓	3500	.005			.10	.14	.24	.33
1020	For large parking lots	B-32C	5000	.010			.22	.34	.56	.75
1050	For small irregular areas	"	2000	.024			.55	.85	1.40	1.86
1100	Fine grade for slab on grade, machine	B-11L	1040	.015			.35	.49	.84	1.11
1150	Hand grading	B-18	700	.034			.66	.05	.71	1.19
1200	Fine grade granular base for sidewalks and bikeways	B-62	1200	.020	↓		.42	.10	.52	.81
2550	Hand grade select gravel	2 Clab	60	.267	C.S.F.		4.99		4.99	8.45
3000	Hand grade select gravel, including compaction, 4" deep	B-18	555	.043	S.Y.		.84	.07	.91	1.49
3100	6" deep		400	.060			1.16	.09	1.25	2.07
3120	8" deep	↓	300	.080			1.55	.12	1.67	2.76
3300	Finishing grading slopes, gentle	B-11L	8900	.002			.04	.06	.10	.13
3310	Steep slopes	"	7100	.002	↓		.05	.07	.12	.16

31 23 Excavation and Fill

31 23 16 – Excavation

31 23 16.13 Excavating, Trench

		Crew	Daily Output	Labor-Hours	Unit	Material	2007 Bare Costs Labor	2007 Bare Costs Equipment	Total	Total Incl O&P
0010	**EXCAVATING, TRENCH** or continuous footing									
0050	1' to 4' deep, 3/8 C.Y. excavator	B-11C	150	.107	B.C.Y.		2.41	1.62	4.03	5.80
0060	1/2 C.Y. excavator	B-11M	200	.080			1.81	1.43	3.24	4.59
0090	4' to 6' deep, 1/2 C.Y. excavator	"	200	.080			1.81	1.43	3.24	4.59
0100	5/8 C.Y. excavator	B-12Q	250	.064			1.48	1.85	3.33	4.49
0300	1/2 C.Y. excavator, truck mounted	B-12J	200	.080			1.85	4.53	6.38	8.05
1352	4' to 6' deep, 1/2 C.Y. excavator w/ trench box	B-13H	188	.085			1.96	5.25	7.21	9.05
1354	5/8 C.Y. excavator	"	235	.068			1.57	4.19	5.76	7.25
1400	By hand with pick and shovel 2' to 6' deep, light soil	1 Clab	8	1			18.70		18.70	32
1500	Heavy soil	"	4	2	↓		37.50		37.50	63.50
5020	Loam & sandy clay with no sheeting or dewatering included									
5050	1' to 4' deep, 3/8 C.Y. excavator	B-11C	162	.099	B.C.Y.		2.23	1.50	3.73	5.35
5060	1/2 C.Y. excavator	B-11M	216	.074			1.67	1.32	2.99	4.24
5080	4' to 6' deep, 1/2 C.Y. excavator	"	216	.074			1.67	1.32	2.99	4.24
5090	5/8 C.Y. excavator	B-12Q	276	.058			1.34	1.67	3.01	4.07
5130	1/2 C.Y. excavator, truck mounted	B-12J	216	.074			1.71	4.19	5.90	7.45
5352	4' to 6' deep, 1/2 C.Y. excavator w/trench box	B-13H	205	.078			1.80	4.81	6.61	8.30
5354	5/8 C.Y. excavator	"	257	.062	↓		1.44	3.83	5.27	6.60
6020	Sand & gravel with no sheeting or dewatering included									
6050	1' to 4' deep, 3/8 C.Y. excavator	B-11C	165	.097	B.C.Y.		2.19	1.47	3.66	5.30
6060	1/2 C.Y. excavator	B-11M	220	.073			1.64	1.30	2.94	4.17
6080	4' to 6' deep, 1/2 C.Y. excavator	"	220	.073			1.64	1.30	2.94	4.17
6090	5/8 C.Y. excavator	B-12Q	275	.058			1.34	1.68	3.02	4.09
6130	1/2 C.Y. excavator, truck mounted	B-12J	220	.073			1.68	4.12	5.80	7.35
6352	4' to 6' deep, 1/2 C.Y. excavator w/ trench box	B-13H	209	.077			1.77	4.71	6.48	8.15
6354	5/8 C.Y. excavator	"	261	.061	↓		1.41	3.77	5.18	6.50
7020	Dense hard clay with no sheeting or dewatering included									
7050	1' to 4' deep, 3/8 C.Y. excavator	B-11C	132	.121	B.C.Y.		2.74	1.84	4.58	6.60
7060	1/2 C.Y. excavator	B-11M	176	.091			2.05	1.62	3.67	5.20
7080	4' to 6' deep, 1/2 C.Y. excavator	"	176	.091			2.05	1.62	3.67	5.20
7090	5/8 C.Y. excavator	B-12Q	220	.073			1.68	2.10	3.78	5.10
7130	1/2 C.Y. excavator, truck mounted	B-12J	176	.091	↓		2.10	5.15	7.25	9.15

31 23 Excavation and Fill

31 23 16 – Excavation

31 23 16.14 Excavating, Utility Trench

		Crew	Daily Output	Labor-Hours	Unit	Material	2007 Bare Costs Labor	Equipment	Total	Total Incl O&P
0010	**EXCAVATING, UTILITY TRENCH** Common earth									
0050	Trenching with chain trencher, 12 H.P., operator walking									
0100	4" wide trench, 12" deep	B-53	800	.010	L.F.		.19	.06	.25	.39
1000	Backfill by hand including compaction, add									
1050	4" wide trench, 12" deep	A-1G	800	.010	L.F.		.19	.05	.24	.37

31 23 16.16 Structural Excavation for Minor Structures

		Crew	Daily Output	Labor-Hours	Unit	Material	Labor	Equipment	Total	Total Incl O&P
0010	**STRUCTURAL EXCAVATION FOR MINOR STRUCTURES**									
0015	Hand, pits to 6' deep, sandy soil	1 Clab	8	1	B.C.Y.		18.70		18.70	32
0100	Heavy soil or clay		4	2			37.50		37.50	63.50
1100	Hand loading trucks from stock pile, sandy soil		12	.667			12.45		12.45	21
1300	Heavy soil or clay	↓	8	1	↓		18.70		18.70	32
1500	For wet or muck hand excavation, add to above				%				50%	50%

31 23 16.42 Excavating, Bulk Bank Measure

		Crew	Daily Output	Labor-Hours	Unit	Material	Labor	Equipment	Total	Total Incl O&P
0010	**EXCAVATING, BULK BANK MEASURE** Common earth piled									
0020	For loading onto trucks, add								15%	15%
0200	Backhoe, hydraulic, crawler mtd., 1 C.Y. cap. = 75 C.Y./hr.	B-12A	600	.027	B.C.Y.		.62	1	1.62	2.13
0310	Wheel mounted, 1/2 C.Y. cap. = 30 C.Y./hr.	B-12E	240	.067			1.54	1.49	3.03	4.21
1200	Front end loader, track mtd., 1-1/2 C.Y. cap. = 70 C.Y./hr.	B-10N	560	.014			.38	.60	.98	1.28
1500	Wheel mounted, 3/4 C.Y. cap. = 45 C.Y./hr.	B-10R	360	.022	↓		.59	.58	1.17	1.60
8000	For hauling excavated material, see div. 31 23 23.18									

31 23 23 – Fill

31 23 23.13 Backfill

		Crew	Daily Output	Labor-Hours	Unit	Material	Labor	Equipment	Total	Total Incl O&P
0010	**BACKFILL**									
0015	By hand, no compaction, light soil	1 Clab	14	.571	L.C.Y.		10.70		10.70	18.15
0100	Heavy soil		11	.727	"		13.60		13.60	23
0300	Compaction in 6" layers, hand tamp, add to above	↓	20.60	.388	E.C.Y.		7.25		7.25	12.35
0500	Air tamp, add	B-9D	190	.211			4.02	1.13	5.15	8.10
0600	Vibrating plate, add	A-1D	60	.133			2.49	.49	2.98	4.77
0800	Compaction in 12" layers, hand tamp, add to above	1 Clab	34	.235	↓		4.40		4.40	7.45
1300	Dozer backfilling, bulk, up to 300' haul, no compaction	B-10B	1200	.007	L.C.Y.		.18	.82	1	1.20
1400	Air tamped, add	B-11B	80	.200	E.C.Y.		4.42	2.92	7.34	10.55

31 23 23.16 Fill By Borrow and Utility Bedding

		Crew	Daily Output	Labor-Hours	Unit	Material	Labor	Equipment	Total	Total Incl O&P
0010	**FILL BY BORROW AND UTILITY BEDDING**									
0049	Utility bedding, for pipe & conduit, not incl. compaction									
0050	Crushed or screened bank run gravel	B-6	150	.160	L.C.Y.	24	3.35	1.62	28.97	34
0100	Crushed stone 3/4" to 1/2"		150	.160		32	3.35	1.62	36.97	42.50
0200	Sand, dead or bank	↓	150	.160	↓	6.25	3.35	1.62	11.22	14.30
0500	Compacting bedding in trench	A-1D	90	.089	E.C.Y.		1.66	.32	1.98	3.18
0600	If material source exceeds 2 miles, add for extra mileage.									
0610	See 31 23 23.18 for hauling kilometer add.									

31 23 23.17 Fill

		Crew	Daily Output	Labor-Hours	Unit	Material	Labor	Equipment	Total	Total Incl O&P
0010	**FILL**, spread dumped material, no compaction									
0020	By dozer, no compaction	B-10B	1000	.008	L.C.Y.		.21	.99	1.20	1.44
0100	By hand	1 Clab	12	.667	"		12.45		12.45	21
0500	Gravel fill, compacted, under floor slabs, 4" deep	B-37	10000	.005	S.F.	.19	.10	.01	.30	.38
0600	6" deep		8600	.006		.28	.11	.01	.40	.52
0700	9" deep		7200	.007		.47	.13	.02	.62	.77
0800	12" deep		6000	.008	↓	.66	.16	.02	.84	1.02
1000	Alternate pricing method, 4" deep		120	.400	E.C.Y.	14.20	8.05	1.02	23.27	30.50
1100	6" deep		160	.300	↓	14.20	6.05	.77	21.02	26.50

31 23 Excavation and Fill

31 23 23 – Fill

31 23 23.17 Fill		Crew	Daily Output	Labor-Hours	Unit	Material	2007 Bare Costs Labor	Equipment	Total	Total Incl O&P
1200	9" deep	B-37	200	.240	E.C.Y.	14.20	4.84	.61	19.65	24.50
1300	12" deep	↓	220	.218	↓	14.20	4.40	.56	19.16	23.50

31 23 23.18 Hauling

		Crew	Daily Output	Labor-Hours	Unit	Material	Labor	Equipment	Total	Total Incl O&P
0010	**HAULING**, excavated or borrow, loose cubic yards									
0012	no loading included, highway haulers									
0020	6 C.Y. dump truck, 1/4 mile round trip, 5.0 loads/hr.	B-34A	195	.041	L.C.Y.		.86	1.89	2.75	3.53
0200	4 mile round trip, 1.8 loads/hr.	"	70	.114			2.40	5.25	7.65	9.85
0310	12 C.Y. dump truck, 1/4 mile round trip 3.7 loads/hr.	B-34B	288	.028			.58	1.84	2.42	3
0500	4 mile round trip, 1.6 loads/hr.	"	125	.064	↓		1.34	4.24	5.58	6.90

31 23 23.24 Compaction, Structural

		Crew	Daily Output	Labor-Hours	Unit	Material	Labor	Equipment	Total	Total Incl O&P
0010	**COMPACTION, STRUCTURAL**									
0020	Steel wheel tandem roller, 5 tons	B-10E	8	1	Hr.		26.50	15.30	41.80	60.50
0050	Air tamp, 6" to 8" lifts, common fill	B-9	250	.160	E.C.Y.		3.06	.72	3.78	6
0060	Select fill	"	300	.133			2.55	.60	3.15	4.98
0600	Vibratory plate, 8" lifts, common fill	A-1D	200	.040			.75	.15	.90	1.43
0700	Select fill	"	216	.037	↓		.69	.14	.83	1.33

31 25 Erosion and Sedimentation Controls

31 25 13 – Erosion Controls

31 25 13.10 Synthetic Erosion Control

		Crew	Daily Output	Labor-Hours	Unit	Material	Labor	Equipment	Total	Total Incl O&P
0010	**SYNTHETIC EROSION CONTROL**									
0020	Jute mesh, 100 SY per roll, 4' wide, stapled	B-80A	2400	.010	S.Y.	.94	.19	.08	1.21	1.43
0100	Plastic netting, stapled, 2" x 1" mesh, 20 mil	B-1	2500	.010		.71	.19		.90	1.10
0200	Polypropylene mesh, stapled, 6.5 oz./S.Y.		2500	.010		1.47	.19		1.66	1.94
0300	Tobacco netting, or jute mesh #2, stapled	↓	2500	.010	↓	.08	.19		.27	.41
1000	Silt fence, polypropylene, 3' high, ideal conditions	2 Clab	1600	.010	L.F.	.34	.19		.53	.69
1100	Adverse conditions	"	950	.017	"	.34	.31		.65	.90

31 31 Soil Treatment

31 31 16 – Termite Control

31 31 16.13 Chemical Termite Control

		Crew	Daily Output	Labor-Hours	Unit	Material	Labor	Equipment	Total	Total Incl O&P
0010	**CHEMICAL TERMITE CONTROL**									
0020	Slab and walls, residential	1 Skwk	1200	.007	SF Flr.	.29	.17		.46	.62
0400	Insecticides for termite control, minimum		14.20	.563	Gal.	11.85	14.70		26.55	38
0500	Maximum	↓	11	.727	"	20.50	19		39.50	54.50

Division 32
Exterior Improvements

32 01 Operation and Maintenance of Exterior Improvements

32 01 13 – Flexible Paving Surface Treatment

32 01 13.66 Fog Seal		Crew	Daily Output	Labor-Hours	Unit	Material	2007 Bare Costs Labor	Equipment	Total	Total Incl O&P
0010	**FOG SEAL**									
0012	Sealcoating, 2 coat coal tar pitch emulsion over 10,000 SY	B-45	5000	.003	S.Y.	.66	.06	.05	.77	.90
0030	1000 to 10,000 S.Y.	"	3000	.005		.66	.11	.09	.86	1.01
0100	Under 1000 S.Y.	B-1	1050	.023		.66	.44		1.10	1.48
0300	Petroleum resistant, over 10,000 S.Y.	B-45	5000	.003		.83	.06	.05	.94	1.08
0320	1000 to 10,000 S.Y.	"	3000	.005		.83	.11	.09	1.03	1.19
0400	Under 1000 S.Y.	B-1	1050	.023		.83	.44		1.27	1.66

32 06 Schedules for Exterior Improvements

32 06 10 – Schedules for Bases, Ballasts, and Paving

32 06 10.10 Sidewalks, Driveways and Patios

		Crew	Daily Output	Labor-Hours	Unit	Material	Labor	Equipment	Total	Total Incl O&P
0010	**SIDEWALKS, DRIVEWAYS AND PATIOS** No base									
0021	Asphaltic concrete, 2" thick	B-37	6480	.007	S.F.	.56	.15	.02	.73	.89
0101	2-1/2" thick	"	5950	.008	"	.72	.16	.02	.90	1.08
0300	Concrete, 3000 psi, CIP, 6 x 6 - W1.4 x W1.4 mesh,									
0310	broomed finish, no base, 4" thick	B-24	600	.040	S.F.	1.73	.92		2.65	3.45
0350	5" thick		545	.044		2.31	1.02		3.33	4.23
0400	6" thick		510	.047		2.70	1.09		3.79	4.78
0450	For bank run gravel base, 4" thick, add	B-18	2500	.010		.50	.19	.01	.70	.89
0520	8" thick, add	"	1600	.015		1	.29	.02	1.31	1.61
1000	Crushed stone, 1" thick, white marble	2 Clab	1700	.009		.21	.18		.39	.53
1050	Bluestone	"	1700	.009		.23	.18		.41	.55
1700	Redwood, prefabricated, 4' x 4' sections	2 Carp	316	.051		8.20	1.30		9.50	11.20
1750	Redwood planks, 1" thick, on sleepers	"	240	.067		5.75	1.71		7.46	9.20
2250	Stone dust, 4" thick	B-62	900	.027	S.Y.	3.08	.56	.14	3.78	4.48

32 06 10.20 Steps

		Crew	Daily Output	Labor-Hours	Unit	Material	Labor	Equipment	Total	Total Incl O&P
0010	**STEPS** Incl. excav., borrow & concrete base, where applicable									
0100	Brick steps	B-24	35	.686	LF Riser	11.15	15.85		27	39
0200	Railroad ties	2 Clab	25	.640		3.19	11.95		15.14	24
0300	Bluestone treads, 12" x 2" or 12" x 1-1/2"	B-24	30	.800		25.50	18.50		44	59
0600	Precast concrete, see Division 03 41 23.50									
4025	Steel edge strips, incl. stakes, 1/4" x 5"	B-1	390	.062	L.F.	3.65	1.19		4.84	6.05
4050	Edging, landscape timber or railroad ties, 6" x 8"	2 Carp	170	.094	"	2.72	2.42		5.14	7.10

32 11 Base Courses

32 11 23 – Aggregate Base Courses

32 11 23.23 Base Course Drainage Layers

		Crew	Daily Output	Labor-Hours	Unit	Material	Labor	Equipment	Total	Total Incl O&P
0010	**BASE COURSE DRAINAGE LAYERS**									
0051	3/4" stone compacted to 3" deep	B-36	36000	.001	S.F.	.34	.02	.03	.39	.46
0101	6" deep		35100	.001		.69	.03	.04	.76	.84
0201	9" deep		25875	.002		1	.03	.05	1.08	1.21
0305	12" deep		21150	.002		1.60	.04	.06	1.70	1.89
0306	Crushed 1-1/2" stone base, compacted to 4" deep		47000	.001		.06	.02	.03	.11	.12
0307	6" deep		35100	.001		.80	.03	.04	.87	.96
0308	8" deep		27000	.001		1.07	.03	.05	1.15	1.29
0309	12" deep		16200	.002		1.60	.05	.08	1.73	1.93
0350	Bank run gravel, spread and compacted									
0371	6" deep	B-32	54000	.001	S.F.	.46	.01	.03	.50	.56
0391	9" deep		39600	.001		.67	.02	.04	.73	.82

32 11 Base Courses

32 11 23 – Aggregate Base Courses

32 11 23.23 Base Course Drainage Layers

		Crew	Daily Output	Labor-Hours	Unit	Material	2007 Bare Costs Labor	Equipment	Total	Total Incl O&P
0401	12" deep	B-32	32400	.001	S.F.	.92	.02	.05	.99	1.11
6900	For small and irregular areas, add						50%	50%		

32 11 26 – Asphaltic Base Courses

32 11 26.19 Bituminous-Stabilized Base Courses

		Crew	Daily Output	Labor-Hours	Unit	Material	Labor	Equipment	Total	Total Incl O&P
0010	**BITUMINOUS-STABILIZED BASE COURSES**									
0020	and large paved areas									
0700	Liquid application to gravel base, asphalt emulsion	B-45	6000	.003	Gal.	3.90	.05	.05	4	4.43
0800	Prime and seal, cut back asphalt		6000	.003	"	4.60	.05	.05	4.70	5.20
1000	Macadam penetration crushed stone, 2 gal. per S.Y., 4" thick		6000	.003	S.Y.	7.80	.05	.05	7.90	8.75
1100	6" thick, 3 gal. per S.Y.		4000	.004		11.70	.08	.07	11.85	13.05
1200	8" thick, 4 gal. per S.Y.		3000	.005		15.60	.11	.09	15.80	17.45
8900	For small and irregular areas, add						50%	50%		

32 12 Flexible Paving

32 12 16 – Asphalt Paving

32 12 16.14 Paving

		Crew	Daily Output	Labor-Hours	Unit	Material	Labor	Equipment	Total	Total Incl O&P
0010	**PAVING** Asphaltic concrete									
0020	6" stone base, 2" binder course, 1" topping	B-25C	9000	.005	S.F.	1.73	.12	.22	2.07	2.33
0300	Binder course, 1-1/2" thick		35000	.001		.44	.03	.06	.53	.59
0400	2" thick		25000	.002		.56	.04	.08	.68	.78
0500	3" thick		15000	.003		.87	.07	.13	1.07	1.22
0600	4" thick		10800	.004		1.14	.10	.18	1.42	1.62
0800	Sand finish course, 3/4" thick		41000	.001		.25	.03	.05	.33	.37
0900	1" thick		34000	.001		.31	.03	.06	.40	.45
1000	Fill pot holes, hot mix, 2" thick	B-16	4200	.008		.59	.15	.13	.87	1.05
1100	4" thick		3500	.009		.87	.18	.15	1.20	1.43
1120	6" thick		3100	.010		1.16	.20	.17	1.53	1.82
1140	Cold patch, 2" thick	B-51	3000	.016		.66	.31	.05	1.02	1.30
1160	4" thick		2700	.018		1.26	.34	.05	1.65	2.03
1180	6" thick		1900	.025		1.96	.49	.07	2.52	3.07

32 13 Rigid Paving

32 13 13 – Concrete Paving

32 13 13.23 Concrete Paving Surface Treatment

		Crew	Daily Output	Labor-Hours	Unit	Material	Labor	Equipment	Total	Total Incl O&P
0010	**CONCRETE PAVING SURFACE TREATMENT**									
0015	Including joints, finishing and curing									
0021	Fixed form, 12' pass, unreinforced, 6" thick	B-26	18000	.005	S.F.	3.19	.11	.16	3.46	3.86
0101	8" thick	"	13500	.007		4.29	.14	.21	4.64	5.20
0701	Finishing, broom finish small areas	2 Cefi	1215	.013			.33		.33	.53

32 14 Unit Paving

32 14 16 – Brick Unit Paving

32 14 16.10 Brick Paving

		Crew	Daily Output	Labor-Hours	Unit	Material	2007 Bare Costs Labor	Equipment	Total	Total Incl O&P
0010	**BRICK PAVING**									
0012	4" x 8" x 1-1/2", without joints (4.5 brick/S.F.)	D-1	110	.145	S.F.	2.73	3.40		6.13	8.65
0100	Grouted, 3/8" joint (3.9 brick/S.F.)		90	.178		3.19	4.15		7.34	10.40
0200	4" x 8" x 2-1/4", without joints (4.5 bricks/S.F.)		110	.145		3.53	3.40		6.93	9.55
0300	Grouted, 3/8" joint (3.9 brick/S.F.)		90	.178		3.26	4.15		7.41	10.50
0500	Bedding, asphalt, 3/4" thick	B-25	5130	.017		.48	.36	.44	1.28	1.62
0540	Course washed sand bed, 1" thick	B-18	5000	.005		.20	.09	.01	.30	.39
0580	Mortar, 1" thick	D-1	300	.053		.59	1.25		1.84	2.72
0620	2" thick		200	.080		1.18	1.87		3.05	4.40
1500	Brick on 1" thick sand bed laid flat, 4.5 per S.F.		100	.160		2.67	3.74		6.41	9.15
2000	Brick pavers, laid on edge, 7.2 per S.F.		70	.229		2.64	5.35		7.99	11.75

32 14 23 – Asphalt Unit Paving

32 14 23.10 Asphalt Blocks

		Crew	Daily Output	Labor-Hours	Unit	Material	Labor	Equipment	Total	Total Incl O&P
0010	**ASPHALT BLOCKS**									
0020	Rectangular, 6" x 12" x 1-1/4", w/bed & neopr. adhesive	D-1	135	.119	S.F.	5.80	2.77		8.57	11
0100	3" thick		130	.123		8.10	2.87		10.97	13.75
0300	Hexagonal tile, 8" wide, 1-1/4" thick		135	.119		5.80	2.77		8.57	11
0400	2" thick		130	.123		8.10	2.87		10.97	13.75
0500	Square, 8" x 8", 1-1/4" thick		135	.119		5.80	2.77		8.57	11
0600	2" thick		130	.123		8.10	2.87		10.97	13.75

32 14 40 – Stone Paving

32 14 40.10 Stone Pavers

		Crew	Daily Output	Labor-Hours	Unit	Material	Labor	Equipment	Total	Total Incl O&P
0010	**STONE PAVERS**									
1100	Flagging, bluestone, irregular, 1" thick,	D-1	81	.198	S.F.	5.65	4.61		10.26	13.90
1150	Snapped random rectangular, 1" thick		92	.174		8.55	4.06		12.61	16.20
1200	1-1/2" thick		85	.188		10.30	4.40		14.70	18.65
1250	2" thick		83	.193		12	4.50		16.50	20.50
1300	Slate, natural cleft, irregular, 3/4" thick		92	.174		6.30	4.06		10.36	13.70
1310	1" thick		85	.188		7.35	4.40		11.75	15.40
1351	Random rectangular, gauged, 1/2" thick		105	.152		13.65	3.56		17.21	21
1400	Random rectangular, butt joint, gauged, 1/4" thick		150	.107		14.70	2.49		17.19	20.50
1450	For sand rubbed finish, add					6.85			6.85	7.55
1500	For interior setting, add								25%	25%
1550	Granite blocks, 3-1/2" x 3-1/2" x 3-1/2"	D-1	92	.174	S.F.	7.65	4.06		11.71	15.20

32 16 Curbs and Gutters

32 16 13 – Concrete Curbs and Gutters

32 16 13.13 Cast-in-Place Concrete Curbs and Gutters

		Crew	Daily Output	Labor-Hours	Unit	Material	Labor	Equipment	Total	Total Incl O&P
0010	**CAST-IN-PLACE CONCRETE CURBS AND GUTTERS**									
0300	Concrete, wood forms, 6" x 18", straight	C-2A	500	.096	L.F.	2.71	2.37		5.08	6.95
0400	6" x 18", radius	"	200	.240	"	2.82	5.95		8.77	13.10

32 16 13.26 Precast Concrete Curbs

		Crew	Daily Output	Labor-Hours	Unit	Material	Labor	Equipment	Total	Total Incl O&P
0010	**PRECAST CONCRETE CURBS**									
0550	Precast, 6" x 18", straight	B-29	700	.069	L.F.	9.05	1.41	1.29	11.75	13.75
0600	6" x 18", radius	"	325	.148	"	10.35	3.03	2.79	16.17	19.55

32 16 19 – Asphalt Curbs

32 16 19.10 Bituminous Concrete Curbs

		Crew	Daily Output	Labor-Hours	Unit	Material	Labor	Equipment	Total	Total Incl O&P
0010	**BITUMINOUS CONCRETE CURBS**									
0012	Curbs, asphaltic, machine formed, 8" wide, 6" high, 40 L.F./ton	B-27	1000	.032	L.F.	1.06	.61	.25	1.92	2.49

32 16 Curbs and Gutters

32 16 19 – Asphalt Curbs

32 16 19.10 Bituminous Concrete Curbs	Crew	Daily Output	Labor-Hours	Unit	Material	2007 Bare Costs Labor	Equipment	Total	Total Incl O&P
0100 8" wide, 8" high, 30 L.F. per ton	B-27	900	.036	L.F.	1.22	.68	.28	2.18	2.81
0150 Asphaltic berm, 12" W, 3"-6" H, 35 L.F./ton, before pavement	↓	700	.046		1.10	.88	.36	2.34	3.09
0200 12" W, 1-1/2" to 4" H, 60 L.F. per ton, laid with pavement	B-2	1050	.038	↓	.67	.73		1.40	1.98

32 16 40 – Stone Curbs

32 16 40.13 Manufactured Stone Curbs

	Crew	Daily Output	Labor-Hours	Unit	Material	2007 Bare Costs Labor	Equipment	Total	Total Incl O&P
0010 **MANUFACTURED STONE CURBS**									
1000 Granite, split face, straight, 5" x 16"	D-13	500	.096	L.F.	10.65	2.39	1.15	14.19	16.95
1100 6" x 18"	"	450	.107		13.95	2.66	1.28	17.89	21
1300 Radius curbing, 6" x 18", over 10' radius	B-29	260	.185	↓	17.10	3.78	3.48	24.36	29
1400 Corners, 2' radius		80	.600	Ea.	57.50	12.30	11.30	81.10	96
1600 Edging, 4-1/2" x 12", straight		300	.160	L.F.	5.30	3.28	3.02	11.60	14.70
1800 Curb inlets, (guttermouth) straight	↓	41	1.171	Ea.	128	24	22	174	205
2000 Indian granite (belgian block)									
2100 Jumbo, 10-1/2" x 7-1/2" x 4", grey	D-1	150	.107	L.F.	1.95	2.49		4.44	6.30
2150 Pink		150	.107		2.54	2.49		5.03	6.95
2200 Regular, 9" x 4-1/2" x 4-1/2", grey		160	.100		1.77	2.34		4.11	5.80
2250 Pink		160	.100		2.43	2.34		4.77	6.55
2300 Cubes, 4" x 4" x 4", grey		175	.091		1.68	2.13		3.81	5.40
2350 Pink		175	.091		1.76	2.13		3.89	5.50
2400 6" x 6" x 6", pink	↓	155	.103	↓	4.37	2.41		6.78	8.80
2500 Alternate pricing method for indian granite									
2550 Jumbo, 10-1/2" x 7-1/2" x 4" (30 lb), grey				Ton	111			111	122
2600 Pink					147			147	162
2650 Regular, 9" x 4-1/2" x 4-1/2" (20 lb), grey					125			125	137
2700 Pink					170			170	187
2750 Cubes, 4" x 4" x 4" (5 lb), grey					204			204	225
2800 Pink					227			227	249
2850 6" x 6" x 6" (25 lb), pink					170			170	187
2900 For pallets, add				↓	16.50			16.50	18.15

32 31 Fences and Gates

32 31 13 – Chain Link Fences and Gates

32 31 13.15 Chain Link Fence

	Crew	Daily Output	Labor-Hours	Unit	Material	2007 Bare Costs Labor	Equipment	Total	Total Incl O&P
0010 **CHAIN LINK FENCE** 11 ga. wire									
0020 1-5/8" post 10'O.C.,1-3/8" top rail,2" corner post galv. stl., 3' high	B-1	185	.130	L.F.	6.80	2.51		9.31	11.70
0050 4' high		170	.141		10.10	2.73		12.83	15.75
0100 6' high		115	.209	↓	11.45	4.04		15.49	19.40
0150 Add for gate 3' wide, 1-3/8" frame 3' high		12	2	Ea.	61	38.50		99.50	133
0170 4' high		10	2.400		75	46.50		121.50	162
0190 6' high		10	2.400		136	46.50		182.50	228
0200 Add for gate 4' wide, 1-3/8" frame 3' high		9	2.667		71	51.50		122.50	166
0220 4' high		9	2.667		93	51.50		144.50	190
0240 6' high		8	3	↓	172	58		230	288
0350 Aluminized steel, 9 ga. wire, 3' high		185	.130	L.F.	8.30	2.51		10.81	13.40
0380 4' high		170	.141		9.45	2.73		12.18	15.05
0400 6' high		115	.209	↓	12.15	4.04		16.19	20
0450 Add for gate 3' wide, 1-3/8" frame 3' high		12	2	Ea.	81	38.50		119.50	155
0470 4' high		10	2.400		111	46.50		157.50	201
0490 6' high		10	2.400		166	46.50		212.50	261
0500 Add for gate 4' wide, 1-3/8" frame 3' high	↓	10	2.400	↓	111	46.50		157.50	201

32 31 Fences and Gates

32 31 13 – Chain Link Fences and Gates

32 31 13.15 Chain Link Fence

		Crew	Daily Output	Labor-Hours	Unit	Material	2007 Bare Costs Labor	2007 Bare Costs Equipment	Total	Total Incl O&P
0520	4' high	B-1	9	2.667	Ea.	147	51.50		198.50	250
0540	6' high		8	3	↓	230	58		288	350
0620	Vinyl covered 9 ga. wire, 3' high		185	.130	L.F.	7.35	2.51		9.86	12.35
0640	4' high		170	.141		12.10	2.73		14.83	17.95
0660	6' high		115	.209	↓	13.80	4.04		17.84	22
0720	Add for gate 3' wide, 1-3/8" frame 3' high		12	2	Ea.	90.50	38.50		129	166
0740	4' high		10	2.400		118	46.50		164.50	208
0760	6' high		10	2.400		181	46.50		227.50	278
0780	Add for gate 4' wide, 1-3/8" frame 3' high		10	2.400		123	46.50		169.50	214
0800	4' high		9	2.667		163	51.50		214.50	267
0820	6' high		8	3		235	58		293	360
0860	Tennis courts, 11 ga. wire, 2 1/2" post 10' O.C., 1-5/8" top rail									
0900	2-1/2" corner post, 10' high	B-1	95	.253	L.F.	18.05	4.89		22.94	28
0920	12' high		80	.300	"	21.50	5.80		27.30	34
1000	Add for gate 3' wide, 1-5/8" frame 10' high		10	2.400	Ea.	226	46.50		272.50	330
1040	Aluminized, 11 ga. wire 10' high		95	.253	L.F.	26	4.89		30.89	37
1100	12' high		80	.300	"	26.50	5.80		32.30	39.50
1140	Add for gate 3' wide, 1-5/8" frame, 10' high		10	2.400	Ea.	295	46.50		341.50	405
1250	Vinyl covered 11 ga. wire, 10' high		95	.253	L.F.	21.50	4.89		26.39	32.50
1300	12' high		80	.300	"	25.50	5.80		31.30	38
1400	Add for gate 3' wide, 1-3/8" frame, 10' high	↓	10	2.400	Ea.	325	46.50		371.50	440

32 31 23 – Plastic Fences and Gates

32 31 23.10 Fence, Vinyl

		Crew	Daily Output	Labor-Hours	Unit	Material	Labor	Equipment	Total	Total Incl O&P
0010	**FENCE, VINYL**, white, steel reinforced, stainless steel fasteners									
0020	Picket, 4" x 4" posts @ 6'-0" OC, 3' high	B-1	140	.171	L.F.	18	3.32		21.32	25.50
0030	4' high		130	.185		21	3.58		24.58	29
0040	5' high		120	.200		24	3.87		27.87	32.50
0100	Board (semi-privacy), 5" x 5" posts @ 7'-6" OC, 5' high		130	.185		24.50	3.58		28.08	33
0120	6' high		125	.192		28	3.72		31.72	37.50
0200	Basketweave, 5" x 5" posts @ 7'-6" OC, 5' high		160	.150		22	2.91		24.91	29
0220	6' high		150	.160		26	3.10		29.10	34
0300	Privacy, 5" x 5" posts @ 7'-6" OC, 5' high		130	.185		26.50	3.58		30.08	35.50
0320	6' high		150	.160	↓	30.50	3.10		33.60	39
0350	Gate, 5' high		9	2.667	Ea.	375	51.50		426.50	505
0360	6' high		9	2.667		385	51.50		436.50	515
0400	For posts set in concrete, add	↓	25	.960	↓	8.65	18.60		27.25	41

32 31 26 – Wire Fences and Gates

32 31 26.10 Fences, Misc. Metal

		Crew	Daily Output	Labor-Hours	Unit	Material	Labor	Equipment	Total	Total Incl O&P
0010	**FENCES, MISC. METAL**									
0012	Chicken wire, posts @ 4', 1" mesh, 4' high	B-80C	410	.059	L.F.	1.90	1.13	.36	3.39	4.39
0100	2" mesh, 6' high		350	.069		1.72	1.32	.42	3.46	4.58
0200	Galv. steel, 12 ga., 2" x 4" mesh, posts 5' O.C., 3' high		300	.080		2.58	1.54	.49	4.61	6
0300	5' high		300	.080		3.45	1.54	.49	5.48	6.95
0400	14 ga., 1" x 2" mesh, 3' high		300	.080		2.74	1.54	.49	4.77	6.15
0500	5' high	↓	300	.080	↓	3.79	1.54	.49	5.82	7.30
1000	Kennel fencing, 1-1/2" mesh, 6' long, 3'-6" wide, 6'-2" high	2 Clab	4	4	Ea.	430	75		505	600
1050	12' long		4	4		515	75		590	695
1200	Top covers, 1-1/2" mesh, 6' long		15	1.067		87.50	19.95		107.45	130
1250	12' long	↓	12	1.333	↓	140	25		165	197

Increase your profits & guarantee accurate estimates!

See how **Means CostWorks** can take your residential estimating to a new level!

Order Today!

Call **1-800-334-3509** or visit **www.rsmeans.com**

Recently added Residential Repair & Remodeling content from one of our most popular Contractor's Pricing Guides organizes information in a different way than other RSMeans cost data titles:

Costs are grouped the way you build – from frame to finish – covering every step from demolition and installation through painting and cleaning. All information is formatted by category and cost element breakouts/tasks.

You'll find unit prices for all aspects of residential repair & remodeling that contain simplified estimating methods with mark-ups. Includes crew tables, location factors, and other supporting text.

From the benchmark *RSMeans Residential Cost Data* title:

Over 9,000 current unit costs and 100 updated assemblies appear in the standard cost data format that you have relied upon for years for simplified system selection and accurate design-stage estimating.

And, like all RSMeans cost data, costs are localized to over 930 locations nationwide – so you can easily determine best-cost solutions.

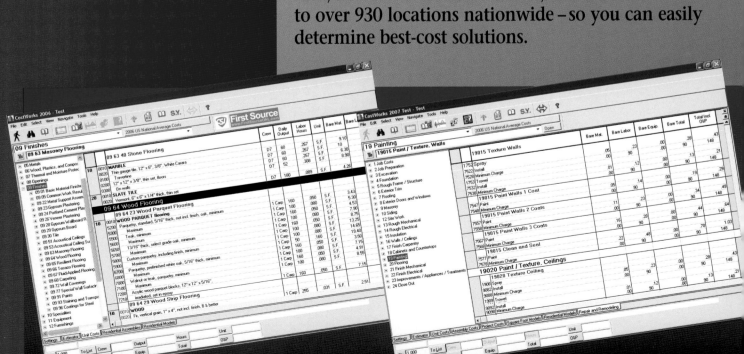

32 31 Fences and Gates

32 31 29 – Wood Fences and Gates

32 31 29.10 Fence, Wood

		Crew	Daily Output	Labor-Hours	Unit	Material	2007 Bare Costs Labor	Equipment	Total	Total Incl O&P
0010	**FENCE, WOOD** Basket weave, 3/8" x 4" boards, 2" x 4"									
0020	stringers on spreaders, 4" x 4" posts									
0050	No. 1 cedar, 6' high	B-80C	160	.150	L.F.	8.85	2.88	.92	12.65	15.65
0070	Treated pine, 6' high		150	.160		10.75	3.08	.98	14.81	18.15
0090	Vertical weave 6' high	↓	145	.166	↓	13	3.18	1.02	17.20	21
0200	Board fence, 1" x 4" boards, 2" x 4" rails, 4" x 4" post									
0220	Preservative treated, 2 rail, 3' high	B-80C	145	.166	L.F.	6.60	3.18	1.02	10.80	13.75
0240	4' high		135	.178		7.25	3.42	1.09	11.76	14.95
0260	3 rail, 5' high		130	.185		8.15	3.55	1.13	12.83	16.20
0300	6' high		125	.192		9.35	3.69	1.18	14.22	17.80
0320	No. 2 grade western cedar, 2 rail, 3' high		145	.166		7.20	3.18	1.02	11.40	14.40
0340	4' high		135	.178		8.50	3.42	1.09	13.01	16.35
0360	3 rail, 5' high		130	.185		9.80	3.55	1.13	14.48	18.05
0400	6' high		125	.192		10.75	3.69	1.18	15.62	19.40
0420	No. 1 grade cedar, 2 rail, 3' high		145	.166		10.80	3.18	1.02	15	18.35
0440	4' high		135	.178		12.25	3.42	1.09	16.76	20.50
0460	3 rail, 5' high		130	.185		14.20	3.55	1.13	18.88	23
0500	6' high	↓	125	.192	↓	15.80	3.69	1.18	20.67	25
0540	Shadow box, 1" x 6" board, 2" x 4" rail, 4" x 4"post									
0560	Pine, pressure treated, 3 rail, 6' high	B-80C	150	.160	L.F.	12.10	3.08	.98	16.16	19.60
0600	Gate, 3'-6" wide		8	3	Ea.	69.50	57.50	18.40	145.40	195
0620	No. 1 cedar, 3 rail, 4' high		130	.185	L.F.	14.85	3.55	1.13	19.53	23.50
0640	6' high		125	.192		18.35	3.69	1.18	23.22	27.50
0860	Open rail fence, split rails, 2 rail 3' high, no. 1 cedar		160	.150		5.95	2.88	.92	9.75	12.45
0870	No. 2 cedar		160	.150		4.64	2.88	.92	8.44	11
0880	3 rail, 4' high, no. 1 cedar		150	.160		8.05	3.08	.98	12.11	15.15
0890	No. 2 cedar		150	.160		5.30	3.08	.98	9.36	12.10
0920	Rustic rails, 2 rail 3' high, no. 1 cedar		160	.150		3.72	2.88	.92	7.52	10
0930	No. 2 cedar		160	.150		3.57	2.88	.92	7.37	9.80
0940	3 rail, 4' high		150	.160		4.99	3.08	.98	9.05	11.80
0950	No. 2 cedar	↓	150	.160	↓	3.77	3.08	.98	7.83	10.45
0960	Picket fence, gothic, pressure treated pine									
1000	2 rail, 3' high	B-80C	140	.171	L.F.	4.59	3.29	1.05	8.93	11.80
1020	3 rail, 4' high		130	.185	"	5.40	3.55	1.13	10.08	13.20
1040	Gate, 3'-6" wide		9	2.667	Ea.	48.50	51.50	16.35	116.35	158
1060	No. 2 cedar, 2 rail, 3' high		140	.171	L.F.	5.70	3.29	1.05	10.04	13.05
1100	3 rail, 4' high		130	.185	"	5.85	3.55	1.13	10.53	13.70
1120	Gate, 3'-6" wide		9	2.667	Ea.	57	51.50	16.35	124.85	168
1140	No. 1 cedar, 2 rail 3' high		140	.171	L.F.	11.45	3.29	1.05	15.79	19.35
1160	3 rail, 4' high		130	.185		13.35	3.55	1.13	18.03	22
1200	Rustic picket, molded pine, 2 rail, 3' high		140	.171		5.25	3.29	1.05	9.59	12.55
1220	No. 1 cedar, 2 rail, 3' high		140	.171		7.15	3.29	1.05	11.49	14.65
1240	Stockade fence, no. 1 cedar, 3-1/4" rails, 6' high		160	.150		10.80	2.88	.92	14.60	17.80
1260	8' high		155	.155		14	2.98	.95	17.93	21.50
1300	No. 2 cedar, treated wood rails, 6' high		160	.150	↓	10.80	2.88	.92	14.60	17.80
1320	Gate, 3'-6" wide		8	3	Ea.	63.50	57.50	18.40	139.40	188
1360	Treated pine, treated rails, 6' high		160	.150	L.F.	10.55	2.88	.92	14.35	17.55
1400	8' high	↓	150	.160	"	15.90	3.08	.98	19.96	24

32 31 29.20 Fence, Wood Rail

		Crew	Daily Output	Labor-Hours	Unit	Material	2007 Bare Costs Labor	Equipment	Total	Total Incl O&P
0010	**FENCE, WOOD RAIL**									
0012	Picket, No. 2 cedar, Gothic, 2 rail, 3' high	B-1	160	.150	L.F.	5.70	2.91		8.61	11.20
0050	Gate, 3'-6" wide	B-80C	9	2.667	Ea.	49	51.50	16.35	116.85	159

32 31 Fences and Gates

32 31 29 – Wood Fences and Gates

32 31 29.20 Fence, Wood Rail

		Crew	Daily Output	Labor-Hours	Unit	Material	2007 Bare Costs Labor	2007 Bare Costs Equipment	Total	Total Incl O&P
0400	3 rail, 4' high	B-80C	150	.160	L.F.	6.55	3.08	.98	10.61	13.50
0500	Gate, 3'-6" wide		9	2.667	Ea.	59	51.50	16.35	126.85	170
1200	Stockade, No. 2 cedar, treated wood rails, 6' high		160	.150	L.F.	7.10	2.88	.92	10.90	13.70
1250	Gate, 3' wide		9	2.667	Ea.	58	51.50	16.35	125.85	169
1300	No. 1 cedar, 3-1/4" cedar rails, 6' high		160	.150	L.F.	17.80	2.88	.92	21.60	25.50
1500	Gate, 3' wide		9	2.667	Ea.	148	51.50	16.35	215.85	268
2700	Prefabricated redwood or cedar, 4' high		160	.150	L.F.	13.45	2.88	.92	17.25	20.50
2800	6' high		150	.160		17.85	3.08	.98	21.91	26
3300	Board, shadow box, 1" x 6", treated pine, 6' high		160	.150		10.50	2.88	.92	14.30	17.45
3400	No. 1 cedar, 6' high		150	.160		20.50	3.08	.98	24.56	29.50
3900	Basket weave, No. 1 cedar, 6' high	▼	160	.150	▼	20.50	2.88	.92	24.30	28.50
4200	Gate, 3'-6" wide	B-1	9	2.667	Ea.	65.50	51.50		117	160
5000	Fence rail, redwood, 2" x 4", merch grade 8'	"	2400	.010	L.F.	1.11	.19		1.30	1.55

32 32 Retaining Walls

32 32 13 – Cast-in-place Concrete Retaining Walls

32 32 13.10 Cast-in-Place Retaining Walls

		Crew	Daily Output	Labor-Hours	Unit	Material	Labor	Equipment	Total	Total Incl O&P
0010	**CAST-IN-PLACE RETAINING WALLS**									
1800	Concrete gravity wall with vertical face including excavation & backfill									
1850	No reinforcing									
1900	6' high, level embankment	C-17C	36	2.306	L.F.	76	61	12.65	149.65	201
2000	33° slope embankment	"	32	2.594	"	69.50	69	14.20	152.70	209
2800	Reinforced concrete cantilever, incl. excavation, backfill & reinf.									
2900	6' high, 33° slope embankment	C-17C	35	2.371	L.F.	69.50	63	13	145.50	198

32 32 23 – Segmental Retaining Walls

32 32 23.13 Segmental Conc. Unit Masonry Retaining Walls

		Crew	Daily Output	Labor-Hours	Unit	Material	Labor	Equipment	Total	Total Incl O&P
0010	**SEGMENTAL CONC. UNIT MASONRY RETAINING WALLS**									
7100	Segmental Retaining Wall system, incl pins, and void fill									
7120	base not included									
7140	Large unit, 8" high x 18" wide x 20" deep, 3 plane split	B-62	300	.080	S.F.	9.35	1.68	.41	11.44	13.55
7150	Straight split		300	.080		9.35	1.68	.41	11.44	13.55
7160	Medium, ltwt, 8" high x 18" wide x 12" deep, 3 plane split		400	.060		10.45	1.26	.30	12.01	13.95
7170	Straight split		400	.060		7.85	1.26	.30	9.41	11.10
7180	Small unit, 4" x 18" x 10" deep, 3 plane split		400	.060		11.15	1.26	.30	12.71	14.70
7190	Straight split		400	.060		11.75	1.26	.30	13.31	15.40
7200	Cap unit, 3 plane split		300	.080		13.20	1.68	.41	15.29	17.75
7210	Cap unit, st split	▼	300	.080		13.20	1.68	.41	15.29	17.75
7260	For reinforcing, add								2.20	2.75
8000	For higher walls, add components as necessary									

32 32 26 – Metal Crib Retaining Walls

32 32 26.10 Metal Bin Retaining Walls

		Crew	Daily Output	Labor-Hours	Unit	Material	Labor	Equipment	Total	Total Incl O&P
0010	**METAL BIN RETAINING WALLS**, Aluminized steel bin, excavation									
0020	and backfill not included, 10' wide									
0100	4' high, 5.5' deep	B-13	650	.074	S.F.	19.40	1.51	1.14	22.05	25.50
0200	8' high, 5.5' deep		615	.078		22.50	1.60	1.20	25.30	28.50
0300	10' high, 7.7' deep		580	.083		23.50	1.70	1.28	26.48	30.50
0400	12' high, 7.7' deep		530	.091		25.50	1.86	1.40	28.76	32.50
0500	16' high, 7.7' deep	▼	515	.093	▼	26.50	1.91	1.44	29.85	34.50

32 32 Retaining Walls
32 32 60 – Stone Retaining Walls

32 32 60.10 Stone Retaining Walls	Crew	Daily Output	Labor-Hours	Unit	Material	2007 Bare Costs Labor	Equipment	Total	Total Incl O&P
0010 **STONE RETAINING WALLS**									
0015 Including excavation, concrete footing and									
0020 stone 3' below grade. Price is exposed face area.									
0200 Decorative random stone, to 6' high, 1'-6" thick, dry set	D-1	35	.457	S.F.	43	10.65		53.65	65
0300 Mortar set		40	.400		43	9.35		52.35	62.50
0500 Cut stone, to 6' high, 1'-6" thick, dry set		35	.457		43	10.65		53.65	65
0600 Mortar set		40	.400		43	9.35		52.35	62.50
0800 Random stone, 6' to 10' high, 2' thick, dry set		45	.356		43	8.30		51.30	61
0900 Mortar set		50	.320		43	7.45		50.45	59.50
1100 Cut stone, 6' to 10' high, 2' thick, dry set		45	.356		43	8.30		51.30	61
1200 Mortar set	↓	50	.320	↓	43	7.45		50.45	59.50

32 84 Planting Irrigation
32 84 23 – Underground Sprinklers
32 84 23.10 Sprinkler Irrigation System

	Crew	Daily Output	Labor-Hours	Unit	Material	2007 Bare Costs Labor	Equipment	Total	Total Incl O&P
0010 **SPRINKLER IRRIGATION SYSTEM** For lawns									
0800 Residential system, custom, 1" supply	B-20	2000	.012	S.F.	.29	.23		.52	.71
0900 1-1/2" supply	"	1800	.013	"	.35	.26		.61	.83

32 91 Planting Preparation
32 91 13 – Soil Preparation
32 91 13.16 Mulching

	Crew	Daily Output	Labor-Hours	Unit	Material	2007 Bare Costs Labor	Equipment	Total	Total Incl O&P
0010 **MULCHING**									
0100 Aged barks, 3" deep, hand spread	1 Clab	100	.080	S.Y.	2.46	1.50		3.96	5.25
0150 Skid steer loader	B-63	13.50	2.963	M.S.F.	273	55.50	9.05	337.55	405
0200 Hay, 1" deep, hand spread	1 Clab	475	.017	S.Y.	.40	.31		.71	.97
0250 Power mulcher, small	B-64	180	.089	M.S.F.	44.50	1.73	1.45	47.68	53.50
0350 Large	B-65	530	.030	"	44.50	.59	.65	45.74	50.50
0400 Humus peat, 1" deep, hand spread	1 Clab	700	.011	S.Y.	4.12	.21		4.33	4.89
0450 Push spreader	"	2500	.003	"	4.12	.06		4.18	4.63
0550 Tractor spreader	B-66	700	.011	M.S.F.	460	.29	.28	460.57	505
0600 Oat straw, 1" deep, hand spread	1 Clab	475	.017	S.Y.	.43	.31		.74	1
0650 Power mulcher, small	B-64	180	.089	M.S.F.	48	1.73	1.45	51.18	57
0700 Large	B-65	530	.030	"	48	.59	.65	49.24	54
0750 Add for asphaltic emulsion	B-45	1770	.009	Gal.	2.21	.18	.15	2.54	2.90
0800 Peat moss, 1" deep, hand spread	1 Clab	900	.009	S.Y.	1.80	.17		1.97	2.26
0850 Push spreader	"	2500	.003	"	1.80	.06		1.86	2.08
0950 Tractor spreader	B-66	700	.011	M.S.F.	200	.29	.28	200.57	221
1000 Polyethylene film, 6 mil.	2 Clab	2000	.008	S.Y.	.19	.15		.34	.46
1100 Redwood nuggets, 3" deep, hand spread	1 Clab	150	.053	"	4.61	1		5.61	6.75
1150 Skid steer loader	B-63	13.50	2.963	M.S.F.	510	55.50	9.05	574.55	670
1200 Stone mulch, hand spread, ceramic chips, economy	1 Clab	125	.064	S.Y.	6.85	1.20		8.05	9.60
1250 Deluxe	"	95	.084	"	10.55	1.57		12.12	14.30
1300 Granite chips	B-1	10	2.400	C.Y.	34.50	46.50		81	117
1400 Marble chips		10	2.400		129	46.50		175.50	221
1600 Pea gravel		28	.857		58	16.60		74.60	91.50
1700 Quartz	↓	10	2.400	↓	166	46.50		212.50	261
1800 Tar paper, 15 Lb. felt	1 Clab	800	.010	S.Y.	.38	.19		.57	.74
1900 Wood chips, 2" deep, hand spread	"	220	.036	"	1.86	.68		2.54	3.20

32 91 Planting Preparation

32 91 13 – Soil Preparation

32 91 13.16 Mulching

		Crew	Daily Output	Labor-Hours	Unit	Material	2007 Bare Costs Labor	2007 Bare Costs Equipment	Total	Total Incl O&P
1950	Skid steer loader	B-63	20.30	1.970	M.S.F.	207	37	6	250	296

32 91 13.26 Planting Beds

		Crew	Daily Output	Labor-Hours	Unit	Material	Labor	Equipment	Total	Total Incl O&P
0010	**PLANTING BEDS**									
0100	Backfill planting pit, by hand, on site topsoil	2 Clab	18	.889	C.Y.		16.60		16.60	28
0200	Prepared planting mix	"	24	.667			12.45		12.45	21
0300	Skid steer loader, on site topsoil	B-62	340	.071			1.48	.36	1.84	2.87
0400	Prepared planting mix	"	410	.059			1.23	.30	1.53	2.39
1000	Excavate planting pit, by hand, sandy soil	2 Clab	16	1			18.70		18.70	32
1100	Heavy soil or clay	"	8	2			37.50		37.50	63.50
1200	1/2 C.Y. backhoe, sandy soil	B-11C	150	.107			2.41	1.62	4.03	5.80
1300	Heavy soil or clay	"	115	.139			3.14	2.12	5.26	7.60
2000	Mix planting soil, incl. loam, manure, peat, by hand	2 Clab	60	.267		38.50	4.99		43.49	51
2100	Skid steer loader	B-62	150	.160		38.50	3.35	.81	42.66	49
3000	Pile sod, skid steer loader	"	2800	.009	S.Y.		.18	.04	.22	.35
3100	By hand	2 Clab	400	.040			.75		.75	1.27
4000	Remove sod, F.E. loader	B-10S	2000	.004			.11	.14	.25	.32
4100	Sod cutter	B-12K	3200	.005			.12	.32	.44	.54
4200	By hand	2 Clab	240	.067			1.25		1.25	2.12

32 91 19 – Landscape Grading

32 91 19.13 Topsoil Placement and Grading

		Crew	Daily Output	Labor-Hours	Unit	Material	Labor	Equipment	Total	Total Incl O&P
0010	**TOPSOIL PLACEMENT AND GRADING**									
0300	Fine grade, base course for paving, see div. 32 11 23.23									
0701	Furnish and place, truck dumped, unscreened, 4" deep	B-10S	12000	.001	S.F.	.37	.02	.02	.41	.46
0801	6" deep	"	7400	.001	"	.37	.03	.04	.44	.50
0900	Fine grading and seeding, incl. lime, fertilizer & seed,									
1001	With equipment	B-14	9000	.005	S.F.	.06	.11	.03	.20	.27

32 92 Turf and Grasses

32 92 19 – Seeding

32 92 19.13 Mechanical Seeding

		Crew	Daily Output	Labor-Hours	Unit	Material	Labor	Equipment	Total	Total Incl O&P
0010	**MECHANICAL SEEDING** R329219-50									
0020	Mechanical seeding, 215 lb./acre	B-66	1.50	5.333	Acre	550	136	131	817	970
0101	$2.00/lb., 44 lb./M.S.Y.	1 Clab	13950	.001	S.F.	.02	.01		.03	.04
0300	Fine grading and seeding incl. lime, fertilizer & seed,									
0310	with equipment	B-14	1000	.048	S.Y.	.17	.97	.24	1.38	2.09
0600	Limestone hand push spreader, 50 lbs. per M.S.F.	1 Clab	180	.044	M.S.F.	3.60	.83		4.43	5.35
0800	Grass seed hand push spreader, 4.5 lbs. per M.S.F.	"	180	.044	"	17.80	.83		18.63	21

32 92 23 – Sodding

32 92 23.10 Sodding

		Crew	Daily Output	Labor-Hours	Unit	Material	Labor	Equipment	Total	Total Incl O&P
0010	**SODDING**									
0020	Sodding, 1" deep, bluegrass sod, on level ground, over 8 MSF	B-63	22	1.818	M.S.F.	234	34	5.55	273.55	320
0200	4 M.S.F.		17	2.353		261	44	7.20	312.20	370
0300	1000 S.F.		13.50	2.963		285	55.50	9.05	349.55	420
0500	Sloped ground, over 8 M.S.F.		6	6.667		234	125	20.50	379.50	490
0600	4 M.S.F.		5	8		261	150	24.50	435.50	570
0700	1000 S.F.		4	10		285	187	30.50	502.50	670
1000	Bent grass sod, on level ground, over 6 M.S.F.		20	2		525	37.50	6.10	568.60	645
1100	3 M.S.F.		18	2.222		580	41.50	6.80	628.30	720
1200	Sodding 1000 S.F. or less		14	2.857		665	53.50	8.70	727.20	830
1500	Sloped ground, over 6 M.S.F.		15	2.667		525	50	8.15	583.15	670

32 92 Turf and Grasses

32 92 23 – Sodding

32 92 23.10 Sodding		Crew	Daily Output	Labor-Hours	Unit	Material	2007 Bare Costs Labor	Equipment	Total	Total Incl O&P
1600	3 M.S.F.	B-63	13.50	2.963	M.S.F.	580	55.50	9.05	644.55	745
1700	1000 S.F.	↓	12	3.333	↓	665	62.50	10.15	737.65	845

32 93 Plants

32 93 13 – Ground Covers

32 93 13.10 Ground Cover

		Crew	Daily Output	Labor-Hours	Unit	Material	Labor	Equipment	Total	Total Incl O&P
0010	**GROUND COVER**									
0012	Plants, pachysandra, in prepared beds	B-1	15	1.600	C	27.50	31		58.50	82.50
0200	Vinca minor, 1 yr, bare root		12	2	"	28.50	38.50		67	97
0600	Stone chips, in 50 lb. bags, Georgia marble		520	.046	Bag	2.56	.89		3.45	4.34
0700	Onyx gemstone		260	.092		18.55	1.79		20.34	23.50
0800	Quartz		260	.092	↓	6.95	1.79		8.74	10.65
0900	Pea gravel, truckload lots	↓	28	.857	Ton	25	16.60		41.60	55.50

32 93 33 – Shrubs

32 93 33.10 Shrubs and Trees

		Crew	Daily Output	Labor-Hours	Unit	Material	Labor	Equipment	Total	Total Incl O&P
0010	**SHRUBS AND TREES** Evergreen, in prepared beds, B & B									
0100	Arborvitae pyramidal, 4'-5'	B-17	30	1.067	Ea.	45	22.50	20.50	88	110
0150	Globe, 12"-15"	B-1	96	.250		11.60	4.84		16.44	21
0300	Cedar, blue, 8'-10'	B-17	18	1.778		192	37.50	34	263.50	310
0500	Hemlock, canadian, 2-1/2'-3'	B-1	36	.667		24	12.90		36.90	48.50
0550	Holly, Savannah, 8' - 10' H		9.68	2.479		545	48		593	680
0600	Juniper, andorra, 18"-24"		80	.300		16.70	5.80		22.50	28.50
0620	Wiltoni, 15"-18"	↓	80	.300		16.40	5.80		22.20	28
0640	Skyrocket, 4-1/2'-5'	B-17	55	.582		50	12.20	11.10	73.30	88
0660	Blue pfitzer, 2'-2-1/2'	B-1	44	.545		26.50	10.55		37.05	47
0680	Ketleerie, 2-1/2'-3'		50	.480		33.50	9.30		42.80	52.50
0700	Pine, black, 2-1/2'-3'		50	.480		41	9.30		50.30	61
0720	Mugo, 18"-24"	↓	60	.400		36	7.75		43.75	52.50
0740	White, 4'-5'	B-17	75	.427		53	8.95	8.15	70.10	82.50
0800	Spruce, blue, 18"-24"	B-1	60	.400		40	7.75		47.75	57
0840	Norway, 4'-5'	B-17	75	.427		90	8.95	8.15	107.10	123
0900	Yew, denisforma, 12"-15"	B-1	60	.400		24	7.75		31.75	39.50
1000	Capitata, 18"-24"		30	.800		20.50	15.50		36	49
1100	Hicksi, 2'-2-1/2'	↓	30	.800	↓	30	15.50		45.50	59.50

32 93 33.20 Shrubs

		Crew	Daily Output	Labor-Hours	Unit	Material	Labor	Equipment	Total	Total Incl O&P
0010	**SHRUBS** Broadleaf evergreen, planted in prepared beds									
0100	Andromeda, 15"-18", container	B-1	96	.250	Ea.	25.50	4.84		30.34	36
0200	Azalea, 15" - 18", container		96	.250		28	4.84		32.84	39
0300	Barberry, 9"-12", container		130	.185		10.85	3.58		14.43	18
0400	Boxwood, 15"-18", B & B		96	.250		30	4.84		34.84	41
0500	Euonymus, emerald gaiety, 12" to 15", container		115	.209		18.15	4.04		22.19	27
0600	Holly, 15"-18", B & B		96	.250		16.70	4.84		21.54	26.50
0900	Mount laurel, 18" - 24", B & B		80	.300		56	5.80		61.80	71.50
1000	Paxistema, 9 - 12" high		130	.185		18.15	3.58		21.73	26
1100	Rhododendron, 18"-24", container		48	.500		31	9.70		40.70	51
1200	Rosemary, 1 gal container		600	.040		61	.77		61.77	68.50
2000	Deciduous, amelanchier, 2'-3', B & B		57	.421		90	8.15		98.15	113
2100	Azalea, 15"-18", B & B		96	.250		24	4.84		28.84	34.50
2300	Bayberry, 2'-3', B & B		57	.421		27.50	8.15		35.65	44.50
2600	Cotoneaster, 15"-18", B & B	↓	80	.300		16.45	5.80		22.25	28

32 93 Plants

32 93 33 – Shrubs

32 93 33.20 Shrubs

		Crew	Daily Output	Labor-Hours	Unit	Material	2007 Bare Costs Labor	Equipment	Total	Total Incl O&P
2800	Dogwood, 3'-4', B & B	B-17	40	.800	Ea.	26	16.75	15.30	58.05	73.50
2900	Euonymus, alatus compacta, 15" to 18", container	B-1	80	.300		21.50	5.80		27.30	34
3200	Forsythia, 2'-3', container	"	60	.400		19	7.75		26.75	34
3300	Hibiscus, 3'-4', B & B	B-17	75	.427		14.85	8.95	8.15	31.95	40.50
3400	Honeysuckle, 3'-4', B & B	B-1	60	.400		21	7.75		28.75	36.50
3500	Hydrangea, 2'-3', B & B	"	57	.421		24.50	8.15		32.65	41
3600	Lilac, 3'-4', B & B	B-17	40	.800		25.50	16.75	15.30	57.55	73
3900	Privet, bare root, 18"-24"	B-1	80	.300		13.15	5.80		18.95	24.50
4100	Quince, 2'-3', B & B	"	57	.421		22	8.15		30.15	38
4200	Russian olive, 3'-4', B & B	B-17	75	.427		23.50	8.95	8.15	40.60	50
4400	Spirea, 3'-4', B & B	B-1	70	.343		27	6.65		33.65	41.50
4500	Viburnum, 3'-4', B & B	B-17	40	.800		28	16.75	15.30	60.05	75.50

32 93 43 – Trees

32 93 43.20 Trees

		Crew	Daily Output	Labor-Hours	Unit	Material	Labor	Equipment	Total	Total Incl O&P
0010	**TREES** Deciduous, in prep. beds, balled & burlapped (B&B)									
0100	Ash, 2" caliper	B-17	8	4	Ea.	113	84	76.50	273.50	350
0200	Beech, 5'-6'		50	.640		222	13.40	12.25	247.65	280
0300	Birch, 6'-8', 3 stems		20	1.600		122	33.50	30.50	186	224
0500	Crabapple, 6'-8'		20	1.600		159	33.50	30.50	223	265
0600	Dogwood, 4'-5'		40	.800		67	16.75	15.30	99.05	118
0700	Eastern redbud 4'-5'		40	.800		134	16.75	15.30	166.05	193
0800	Elm, 8'-10'		20	1.600		112	33.50	30.50	176	213
0900	Ginkgo, 6'-7'		24	1.333		164	28	25.50	217.50	256
1000	Hawthorn, 8'-10', 1" caliper		20	1.600		125	33.50	30.50	189	228
1100	Honeylocust, 10'-12', 1-1/2" caliper		10	3.200		149	67	61	277	345
1300	Larch, 8'		32	1		96.50	21	19.10	136.60	162
1400	Linden, 8'-10', 1" caliper		20	1.600		104	33.50	30.50	168	204
1500	Magnolia, 4'-5'		20	1.600		70	33.50	30.50	134	167
1600	Maple, red, 8'-10', 1-1/2" caliper		10	3.200		160	67	61	288	355
1700	Mountain ash, 8'-10', 1" caliper		16	2		164	42	38	244	293
1800	Oak, 2-1/2"-3" caliper		6	5.333		251	112	102	465	575
2100	Planetree, 9'-11', 1-1/4" caliper		10	3.200		102	67	61	230	294
2200	Plum, 6'-8', 1" caliper		20	1.600		90.50	33.50	30.50	154.50	190
2300	Poplar, 9'-11', 1-1/4" caliper		10	3.200		47.50	67	61	175.50	233
2500	Sumac, 2'-3'		75	.427		22.50	8.95	8.15	39.60	49
2700	Tulip, 5'-6'		40	.800		48.50	16.75	15.30	80.55	98.50
2800	Willow, 6'-8', 1" caliper		20	1.600		59.50	33.50	30.50	123.50	156

32 94 Planting Accessories

32 94 13 – Landscape Edging

32 94 13.20 Edging

		Crew	Daily Output	Labor-Hours	Unit	Material	Labor	Equipment	Total	Total Incl O&P
0010	**EDGING**									
0050	Aluminum alloy, including stakes, 1/8" x 4", mill finish	B-1	390	.062	L.F.	2.50	1.19		3.69	4.77
0051	Black paint		390	.062		2.90	1.19		4.09	5.20
0052	Black anodized		390	.062		3.35	1.19		4.54	5.70
0100	Brick, set horizontally, 1-1/2 bricks per L.F.	D-1	370	.043		1.09	1.01		2.10	2.88
0150	Set vertically, 3 bricks per L.F.	"	135	.119		3	2.77		5.77	7.90
0200	Corrugated aluminum, roll, 4" wide	1 Carp	650	.012		.40	.32		.72	.98
0250	6" wide	"	550	.015		.50	.37		.87	1.18
0600	Railroad ties, 6" x 8"	2 Carp	170	.094		2.72	2.42		5.14	7.10

32 94 Planting Accessories

32 94 13 – Landscape Edging

32 94 13.20 Edging

		Crew	Daily Output	Labor-Hours	Unit	Material	2007 Bare Costs Labor	Equipment	Total	Total Incl O&P
0650	7" x 9"	2 Carp	136	.118	L.F.	3.02	3.02		6.04	8.45
0700	Redwood									
0750	2" x 4"	2 Carp	330	.048	L.F.	2.27	1.25		3.52	4.60
0800	Steel edge strips, incl. stakes, 1/4" x 5"	B-1	390	.062		3.65	1.19		4.84	6.05
0850	3/16" x 4"	"	390	.062		2.88	1.19		4.07	5.20

32 94 50 – Tree Guying

32 94 50.10 Tree Guying

		Crew	Daily Output	Labor-Hours	Unit	Material	2007 Bare Costs Labor	Equipment	Total	Total Incl O&P
0010	**TREE GUYING**									
0015	Tree guying Including stakes, guy wire and wrap									
0100	Less than 3" caliper, 2 stakes	2 Clab	35	.457	Ea.	16.25	8.55		24.80	32.50
0200	3" to 4" caliper, 3 stakes	"	21	.762	"	19.25	14.25		33.50	45
1000	Including arrowhead anchor, cable, turnbuckles and wrap									
1100	Less than 3" caliper, 3" anchors	2 Clab	20	.800	Ea.	51	14.95		65.95	81.50
1200	3" to 6" caliper, 4" anchors		15	1.067		73	19.95		92.95	115
1300	6" caliper, 6" anchors		12	1.333		90.50	25		115.50	142
1400	8" caliper, 8" anchors		9	1.778		104	33		137	171

32 96 Transplanting

32 96 23 – Plant and Bulb Transplanting

32 96 23.23 Planting

		Crew	Daily Output	Labor-Hours	Unit	Material	2007 Bare Costs Labor	Equipment	Total	Total Incl O&P
0010	**PLANTING**									
0012	Moving shrubs on site, 12" ball	B-62	28	.857	Ea.	17.95	4.35		22.30	35
0100	24" ball	"	22	1.091	"	23	5.55		28.55	44.50

32 96 23.43 Moving Trees

		Crew	Daily Output	Labor-Hours	Unit	Material	2007 Bare Costs Labor	Equipment	Total	Total Incl O&P
0290	**MOVING TREES**, On site									
0300	Moving trees on site, 36" ball	B-6	3.75	6.400	Ea.		134	65	199	297
0400	60" ball	"	1	24	"		505	243	748	1,125

Division 33
Utilities

33 05 Common Work Results for Utilities

33 05 23 – Trenchless Utility Installation

33 05 23.19 Microtunneling

		Crew	Daily Output	Labor-Hours	Unit	Material	2007 Bare Costs Labor	Equipment	Total	Total Incl O&P
0010	**MICROTUNNELING** Not including excavation, backfill, shoring,									
0020	or dewatering, average 50'/day, slurry method									
0100	24" to 48" outside diameter, minimum				L.F.					640
0110	Adverse conditions, add				%					50%
1000	Rent microtunneling machine, average monthly lease				Month					85,500
1010	Operating technician				Day					640
1100	Mobilization and demobilization, minimum				Job					42,800
1110	Maximum				"					430,000

33 11 Water Utility Distribution Piping

33 11 13 – Public Water Utility Distribution Piping

33 11 13.15 Water Supply, Ductile Iron Pipe

		Crew	Daily Output	Labor-Hours	Unit	Material	Labor	Equipment	Total	Total Incl O&P
0010	**WATER SUPPLY, DUCTILE IRON PIPE** cement lined									
0020	Not including excavation or backfill									
2000	Pipe, class 50 water piping, 18' lengths									
2020	Mechanical joint, 4" diameter	B-21A	200	.200	L.F.	14.15	4.80	2.88	21.83	27
2040	6" diameter		160	.250		16.45	6	3.60	26.05	32
3000	Tyton, push-on joint, 4" diameter		400	.100		8.35	2.40	1.44	12.19	14.75
3020	6" diameter		333.33	.120		9.80	2.88	1.73	14.41	17.45
8000	Fittings, mechanical joint									
8006	90° bend, 4" diameter	B-20A	16	2	Ea.	182	46.50		228.50	277
8020	6" diameter		12.80	2.500		220	58		278	340
8200	Wye or tee, 4" diameter		10.67	2.999		281	69.50		350.50	425
8220	6" diameter		8.53	3.751		375	87		462	560
8398	45° bends, 4" diameter		16	2		165	46.50		211.50	259
8400	6" diameter		12.80	2.500		224	58		282	345
8450	Decreaser, 6" x 4" diameter		14.22	2.250		201	52		253	310
8460	8" x 6" diameter		11.64	2.749		293	63.50		356.50	425
8550	Butterfly valves with boxes, cast iron									
8560	4" diameter	B-20	6	4	Ea.	625	77.50		702.50	815
9600	Steel sleeve and tap, 4" diameter		3	8		555	155		710	875
9620	6" diameter		2	12		595	232		827	1,050

33 11 13.25 Water Supply, Polyvinyl Chloride Pipe

		Crew	Daily Output	Labor-Hours	Unit	Material	Labor	Equipment	Total	Total Incl O&P
0010	**WATER SUPPLY, POLYVINYL CHLORIDE PIPE**									
2100	AWWA Class 160, S.D.R. 26, 1-1/2" diameter	Q-1A	750	.013	L.F.	.49	.40		.89	1.20
2120	2" diameter		686	.015		1.17	.44		1.61	2.01
2140	2-1/2" diameter		500	.020		1.73	.60		2.33	2.89
2160	3" diameter	B-20	430	.056		2.48	1.08		3.56	4.57
2180	4" diameter	"	375	.064		4.05	1.24		5.29	6.55

33 12 Water Utility Distribution Equipment

33 12 13 – Water Service Connections

33 12 13.15 Tapping, Crosses and Sleeves	Crew	Daily Output	Labor-Hours	Unit	Material	2007 Bare Costs Labor	Equipment	Total	Total Incl O&P
0010 **TAPPING, CROSSES AND SLEEVES**									
4000 Drill and tap pressurized main (labor only)									
4100 6" main, 1" to 2" service	Q-1	3	5.333	Ea.		142		142	233
4150 8" main, 1" to 2" service	"	2.75	5.818	"		155		155	255
4500 Tap and insert gate valve									
4600 8" main, 4" branch	B-21	3.20	8.750	Ea.		180	51	231	360
4650 6" branch		2.70	10.370			213	60.50	273.50	425
4700 10" Main, 4" branch		2.70	10.370			213	60.50	273.50	425
4750 6" branch		2.35	11.915			244	69.50	313.50	490
4800 12" main, 6" branch		2.35	11.915			244	69.50	313.50	490

33 12 19 – Water Utility Distribution Fire Hydrants

33 12 19.40 Utility Boxes	Crew	Daily Output	Labor-Hours	Unit	Material	2007 Bare Costs Labor	Equipment	Total	Total Incl O&P
0010 **UTILITY BOXES** Precast concrete, 6" thick									
0050 5' x 10' x 6' high, I.D.	B-13	2	24	Ea.	1,725	490	370	2,585	3,125
0350 Hand hole, precast concrete, 1-1/2" thick									
0400 1'-0" x 2'-0" x 1'-9", I.D., light duty	B-1	4	6	Ea.	292	116		408	515
0450 4'-6" x 3'-2" x 2'-0", O.D., heavy duty	B-6	3	8	"	895	168	81	1,144	1,350

33 21 Water Supply Wells

33 21 13 – Public Water Supply Wells

33 21 13.10 Wells and Accessories	Crew	Daily Output	Labor-Hours	Unit	Material	2007 Bare Costs Labor	Equipment	Total	Total Incl O&P
0010 **WELLS & ACCESSORIES**, domestic									
0100 Drilled, 4" to 6" diameter	B-23	120	.333	L.F.		6.35	30.50	36.85	45
1500 Pumps, installed in wells to 100' deep, 4" submersible									
1520 3/4 H.P.	Q-1	2.66	6.015	Ea.	435	160		595	740
1600 1 H.P.	"	2.29	6.987	"	455	186		641	810

33 31 Sanitary Utility Sewerage Piping

33 31 13 – Public Sanitary Utility Sewerage Piping

33 31 13.15 Sewage Collection, Concrete Pipe	Crew	Daily Output	Labor-Hours	Unit	Material	2007 Bare Costs Labor	Equipment	Total	Total Incl O&P
0010 **SEWAGE COLLECTION, CONCRETE PIPE**									
0020 See 33 41 13.60 for sewage/drainage collection, concrete pipe									

33 31 13.25 Sewage Collection, Polyvinyl Chloride Pipe	Crew	Daily Output	Labor-Hours	Unit	Material	2007 Bare Costs Labor	Equipment	Total	Total Incl O&P
0010 **SEWAGE COLLECTION, POLYVINYL CHLORIDE PIPE**									
0020 Not including excavation or backfill									
2000 10' lengths, S.D.R. 35, B&S, 4" diameter	B-20	375	.064	L.F.	2.33	1.24		3.57	4.66
2040 6" diameter		350	.069		4.35	1.33		5.68	7.05
2080 8" diameter		335	.072		9.05	1.39		10.44	12.30
2120 10" diameter	B-21	330	.085		14.30	1.74	.50	16.54	19.25
4000 Piping, DWV PVC, no exc/bkfill, 10' L, Sch 40, 4" dia	B-20	375	.064		3.08	1.24		4.32	5.50
4010 6" dia		350	.069		6.30	1.33		7.63	9.20
4020 8" dia		335	.072		12.55	1.39		13.94	16.15

33 36 Utility Septic Tanks

33 36 13 – Utility Septic Tank and Effluent Wet Wells

33 36 13.10 Septic Tanks

		Crew	Daily Output	Labor-Hours	Unit	Material	2007 Bare Costs Labor	2007 Bare Costs Equipment	Total	Total Incl O&P
0010	**SEPTIC TANKS**									
0015	Septic tanks, not incl exc or piping, precast, 1,000 gal	B-21	8	3.500	Ea.	635	72	20.50	727.50	840
0100	2,000 gallon		5	5.600		1,975	115	32.50	2,122.50	2,400
0600	High density polyethylene, 1,000 gallon		6	4.667		1,050	96	27.50	1,173.50	1,350
0700	1,500 gallon		4	7		1,350	144	41	1,535	1,775
1000	Distribution boxes, concrete, 7 outlets	2 Clab	16	1		117	18.70		135.70	161
1100	9 outlets	"	8	2		435	37.50		472.50	540
1150	Leaching field chambers, 13' x 3'-7" x 1'-4", standard	B-13	16	3		645	61.50	46	752.50	865
1420	Leaching pit, 6', dia, 3' deep complete					970			970	1,075
2200	Excavation for septic tank, 3/4 C.Y. backhoe	B-12F	145	.110	C.Y.		2.55	3.56	6.11	8.15
2400	4' trench for disposal field, 3/4 C.Y. backhoe	"	335	.048	L.F.		1.10	1.54	2.64	3.54
2600	Gravel fill, run of bank	B-6	150	.160	C.Y.	19.35	3.35	1.62	24.32	29
2800	Crushed stone, 3/4"	"	150	.160	"	23	3.35	1.62	27.97	33

33 41 Storm Utility Drainage Piping

33 41 13 – Public Storm Utility Drainage Piping

33 41 13.60 Sewage/Drainage Collection, Concrete Pipe

		Crew	Daily Output	Labor-Hours	Unit	Material	Labor	Equipment	Total	Total Incl O&P
0010	**SEWAGE/DRAINAGE COLLECTION, CONCRETE PIPE**									
0020	Not including excavation or backfill									
1020	8" diameter	B-14	224	.214	L.F.	5.30	4.32	1.09	10.71	14.35
1030	10" diameter	"	216	.222	"	5.90	4.48	1.13	11.51	15.25
3780	Concrete slotted pipe, class 4 mortar joint									
3800	12" diameter	B-21	168	.167	L.F.	14.90	3.42	.97	19.29	23
3840	18" diameter	"	152	.184	"	23	3.78	1.08	27.86	33
3900	Class 4 O-ring									
3940	12" diameter	B-21	168	.167	L.F.	15.60	3.42	.97	19.99	24
3960	18" diameter	"	152	.184	"	21	3.78	1.08	25.86	30.50

33 44 Storm Utility Water Drains

33 44 13 – Utility Area Drains

33 44 13.13 Catch Basin Grates and Frames

		Crew	Daily Output	Labor-Hours	Unit	Material	Labor	Equipment	Total	Total Incl O&P
0010	**CATCH BASIN GRATES AND FRAMES** not including footing, excavation									
1600	Frames & covers, C.I., 24" square, 500 lb.	B-6	7.80	3.077	Ea.	310	64.50	31	405.50	485

33 46 Subdrainage

33 46 16 – Subdrainage Piping

33 46 16.20 Piping, Subdrainage, Concrete

		Crew	Daily Output	Labor-Hours	Unit	Material	Labor	Equipment	Total	Total Incl O&P
0010	**PIPING, SUBDRAINAGE, CONCRETE**									
0021	Not including excavation and backfill									
3000	Porous wall concrete underdrain, std. strength, 4" diameter	B-20	335	.072	L.F.	2.87	1.39		4.26	5.50
3020	6" diameter	"	315	.076		3.73	1.48		5.21	6.60
3040	8" diameter	B-21	310	.090		4.61	1.85	.53	6.99	8.75

33 46 16.25 Piping, Subdrainage, Corrugated Metal

		Crew	Daily Output	Labor-Hours	Unit	Material	Labor	Equipment	Total	Total Incl O&P
0010	**PIPING, SUBDRAINAGE, CORRUGATED METAL**									
0021	Not including excavation and backfill									
2010	Aluminum, perforated									
2020	6" diameter, 18 ga.	B-14	380	.126	L.F.	4.25	2.55	.64	7.44	9.70

33 46 Subdrainage

33 46 16 – Subdrainage Piping

33 46 16.25 Piping, Subdrainage, Corrugated Metal

		Crew	Daily Output	Labor-Hours	Unit	Material	2007 Bare Costs Labor	Equipment	Total	Total Incl O&P
2200	8" diameter, 16 ga.	B-14	370	.130	L.F.	6.10	2.62	.66	9.38	11.85
2220	10" diameter, 16 ga.	↓	360	.133	↓	7.60	2.69	.68	10.97	13.60
3000	Uncoated galvanized, perforated									
3020	6" diameter, 18 ga.	B-20	380	.063	L.F.	6.75	1.22		7.97	9.50
3200	8" diameter, 16 ga.	"	370	.065		9.25	1.26		10.51	12.35
3220	10" diameter, 16 ga.	B-21	360	.078		13.90	1.60	.45	15.95	18.45
3240	12" diameter, 16 ga.	"	285	.098	↓	14.50	2.02	.57	17.09	20
4000	Steel, perforated, asphalt coated									
4020	6" diameter 18 ga.	B-20	380	.063	L.F.	5.40	1.22		6.62	8
4030	8" diameter 18 ga	"	370	.065		8.45	1.26		9.71	11.40
4040	10" diameter 16 ga	B-21	360	.078		9.65	1.60	.45	11.70	13.85
4050	12" diameter 16 ga	↓	285	.098		11.10	2.02	.57	13.69	16.25
4060	18" diameter 16 ga	↓	205	.137	↓	15.15	2.80	.80	18.75	22.50

33 49 Storm Drainage Structures

33 49 13 – Storm Drainage Manholes, Frames, and Covers

33 49 13.10 Storm Drainage Manholes, Frames and Covers

		Crew	Daily Output	Labor-Hours	Unit	Material	Labor	Equipment	Total	Total Incl O&P
0010	**STORM DRAINAGE MANHOLES, FRAMES & COVERS** not including									
0020	footing, excavation, backfill (See line items for frame & cover)									
0050	Brick, 4' inside diameter, 4' deep	D-1	1	16	Ea.	390	375		765	1,050
1110	Precast, 4' I.D., 4' deep	B-22	4.10	7.317	"	850	154	60	1,064	1,250

33 51 Natural-Gas Distribution

33 51 13 – Natural-Gas Piping

33 51 13.10 Piping, Gas Service and Distribution, Polyethylene

		Crew	Daily Output	Labor-Hours	Unit	Material	Labor	Equipment	Total	Total Incl O&P
0010	**PIPING, GAS SERVICE AND DISTRIBUTION, POLYETHYLENE**									
0020	not including excavation or backfill									
1000	60 psi coils, comp cplg @ 100', 1/2" diameter, SDR 9.3	B-20A	608	.053	L.F.	1.02	1.22		2.24	3.15
1040	1-1/4" diameter, SDR 11		544	.059		1.88	1.36		3.24	4.33
1100	2" diameter, SDR 11		488	.066		2.33	1.52		3.85	5.10
1160	3" diameter, SDR 11	↓	408	.078		4.88	1.82		6.70	8.40
1500	60 PSI 40' joints with coupling, 3" diameter, SDR 11	B-21A	408	.098		4.88	2.35	1.41	8.64	10.80
1540	4" diameter, SDR 11		352	.114		11.15	2.73	1.63	15.51	18.60
1600	6" diameter, SDR 11		328	.122		34.50	2.93	1.75	39.18	45
1640	8" diameter, SDR 11	↓	272	.147	↓	47.50	3.53	2.11	53.14	60

33 51 13.20 Piping, Gas Service and Distribution, Steel

		Crew	Daily Output	Labor-Hours	Unit	Material	Labor	Equipment	Total	Total Incl O&P
0010	**PIPING, GAS SERVICE & DISTRIBUTION, STEEL**									
0020	not including excavation or backfill, tar coated and wrapped									
4000	Schedule 40, plain end									
4040	1" diameter	Q-4	300	.107	L.F.	5.70	3	.20	8.90	11.45
4080	2" diameter	Q-4	280	.114	L.F.	8.95	3.21	.22	12.38	15.40

Reference Section

All the reference information is in one section, making it easy to find what you need to know... and easy to use the book on a daily basis. This section is visually identified by a vertical gray bar on the page edges.

In this Reference Section, we've included Equipment Rental Costs, a listing of rental and operating costs; Crew Listings, a full listing of all crews, equipment, and their costs; Location Factors for adjusting costs to the region you are in; Reference Tables, where you will find explanations, estimating information and procedures, or technical data; and an explanation of all the Abbreviations in the book.

Table of Contents

Construction Equipment Rental Costs	293
Crew Listings	305
Location Factors	332
Reference Tables	337
R01 General Requirements	337
R02 Existing Conditions	344
R04 Masonry	345
R05 Metals	347
R06 Wood, Plastics & Composites	348
R07 Thermal & Moisture Protect.	348
R08 Openings	349
R09 Finishes	351
R13 Special Construction	353

Reference Tables (cont.)	
R22 Plumbing	354
R32 Exterior Improvements	354
R33 Utilities	355
Abbreviations	356

Equipment Rental Costs

01 54 | Construction Aids

01 54 33 | Equipment Rental

			UNIT	HOURLY OPER. COST	RENT PER DAY	RENT PER WEEK	RENT PER MONTH	CREW EQUIPMENT COST/DAY
10	0010	**CONCRETE EQUIPMENT RENTAL**, without operators						
	0150	For batch plant, see div. 01 54 33.50	R015433-10					
	0200	Bucket, concrete lightweight, 1/2 C.Y.	Ea.	.60	17.35	52	156	15.20
	0300	1 C.Y.	R033105-70	.65	21	63	189	17.80
	0400	1-1/2 C.Y.		.80	28.50	85	255	23.40
	0500	2 C.Y.		.90	33.50	100	300	27.20
	0580	8 C.Y.		4.85	223	670	2,000	172.80
	0600	Cart, concrete, self propelled, operator walking, 10 C.F.		2.30	58.50	175	525	53.40
	0700	Operator riding, 18 C.F.		3.60	83.50	250	750	78.80
	0800	Conveyer for concrete, portable, gas, 16" wide, 26' long		8.05	118	355	1,075	135.40
	0900	46' long		8.40	143	430	1,300	153.20
	1000	56' long		8.55	153	460	1,375	160.40
	1100	Core drill, electric, 2-1/2 H.P., 1" to 8" bit diameter		2.20	89.50	268	805	71.20
	1150	11 HP, 8" to 18" cores		4.75	112	335	1,000	105
	1200	Finisher, concrete floor, gas, riding trowel, 48" diameter		5.55	91.50	275	825	99.40
	1300	Gas, manual, 3 blade, 36" trowel	Ea.	1.15	16.65	50	150	19.20
	1400	4 blade, 48" trowel		1.65	21	63	189	25.80
	1500	Float, hand-operated (Bull float) 48" wide		.08	13	39	117	8.45
	1570	Curb builder, 14 H.P., gas, single screw		10.45	228	685	2,050	220.60
	1590	Double screw		11.10	270	810	2,425	250.80
	1600	Grinder, concrete and terrazzo, electric, floor		2.92	127	380	1,150	99.35
	1700	Wall grinder		1.46	63.50	190	570	49.70
	1800	Mixer, powered, mortar and concrete, gas, 6 C.F., 18 H.P.		5.70	113	340	1,025	113.60
	1900	10 C.F., 25 H.P.		6.95	138	415	1,250	138.60
	2000	16 C.F.		7.30	162	485	1,450	155.40
	2100	Concrete, stationary, tilt drum, 2 C.Y.		5.75	223	670	2,000	180
	2120	Pump, concrete, truck mounted 4" line 80' boom		21.95	890	2,665	8,000	708.60
	2140	5" line, 110' boom		28.70	1,175	3,545	10,600	938.60
	2160	Mud jack, 50 C.F. per hr.		5.75	127	380	1,150	122
	2180	225 C.F. per hr.		7.50	143	430	1,300	146
	2190	Shotcrete pump rig, 12 CY/hr		13.70	235	705	2,125	250.60
	2600	Saw, concrete, manual, gas, 18 H.P.		4.00	36.50	110	330	54
	2650	Self-propelled, gas, 30 H.P.		7.80	93.50	280	840	118.40
	2700	Vibrators, concrete, electric, 60 cycle, 2 H.P.		.42	7.65	23	69	7.95
	2800	3 H.P.		.58	9	27	81	10.05
	2900	Gas engine, 5 H.P.		1.05	13.65	41	123	16.60
	3000	8 H.P.		1.55	15	45	135	21.40
	3050	Vibrating screed, gas engine, 8 H.P.		2.71	73.50	221	665	65.90
	3100	Concrete transit mixer, hydraulic drive						
	3120	6 x 4, 250 H.P., 8 C.Y., rear discharge		39.50	550	1,645	4,925	645
	3200	Front discharge		46.50	680	2,040	6,125	780
	3300	6 x 6, 285 H.P., 12 C.Y., rear discharge		45.60	640	1,920	5,750	748.80
	3400	Front discharge		47.65	690	2,065	6,200	794.20
20	0010	**EARTHWORK EQUIPMENT RENTAL**, without operators	R015433-10					
	0040	Aggregate spreader, push type 8' to 12' wide	Ea.	2.00	24.50	73	219	30.60
	0045	Tailgate type, 8' wide	"	1.90	32.50	97	291	34.60
	0050	Augers for vertical drilling	R312323-30					
	0055	Earth auger, truck-mounted, for fence & sign posts	Ea.	10.40	485	1,460	4,375	375.20
	0060	For borings and monitoring wells	R312316-40	35.95	645	1,935	5,800	674.60
	0070	Earth auger, portable, trailer mounted		2.25	23	69	207	31.80
	0075	Earth auger, truck-mounted, for caissons, water wells, utility poles	R312316-45	174.00	3,500	10,535	31,600	3,499
	0080	Auger, horizontal boring machine, 12" to 36" diameter, 45 H.P.		19.00	192	575	1,725	267
	0090	12" to 48" diameter, 65 H.P.		27.05	340	1,025	3,075	421.40
	0095	Auger, for fence posts, gas engine, hand held		.40	4.67	14	42	6
	0100	Excavator, diesel hydraulic, crawler mounted, 1/2 C.Y. cap.		17.75	360	1,080	3,250	358
	0120	5/8 C.Y. capacity		21.40	485	1,455	4,375	462.20
	0140	3/4 C.Y. capacity		24.80	530	1,590	4,775	516.40

01 54 | Construction Aids

01 54 33 | Equipment Rental

			UNIT	HOURLY OPER. COST	RENT PER DAY	RENT PER WEEK	RENT PER MONTH	CREW EQUIPMENT COST/DAY
0150	1 C.Y. capacity		Ea.	30.85	590	1,775	5,325	601.80
0200	1-1/2 C.Y. capacity	R015433-10		37.95	785	2,360	7,075	775.60
0300	2 C.Y. capacity			49.45	1,000	2,995	8,975	994.60
0320	2-1/2 C.Y. capacity	R312316-30		65.60	1,350	4,040	12,100	1,333
0340	3-1/2 C.Y. capacity			106.15	2,200	6,585	19,800	2,166
0341	Attachments	R312316-40						
0342	Bucket thumbs			2.55	212	635	1,900	147.40
0345	Grapples	R312316-45		2.35	182	545	1,625	127.80
0350	Gradall type, truck mounted, 3 ton @ 15' radius, 5/8 C.Y.			42.35	945	2,835	8,500	905.80
0370	1 C.Y. capacity			47.80	1,075	3,220	9,650	1,026
0400	Backhoe-loader, 40 to 45 H.P., 5/8 C.Y. capacity			9.85	197	590	1,775	196.80
0450	45 H.P. to 60 H.P., 3/4 C.Y. capacity			12.30	242	725	2,175	243.40
0460	80 H.P., 1-1/4 C.Y. capacity			15.05	275	825	2,475	285.40
0470	112 H.P., 1-1/2 C.Y. capacity			20.75	405	1,220	3,650	410
0480	Attachments							
0482	Compactor, 20,000 lb			4.45	120	360	1,075	107.60
0485	Hydraulic hammer, 750 ft-lbs			2.05	70	210	630	58.40
0486	Hydraulic hammer, 1200 ft-lbs			4.30	135	405	1,225	115.40
0500	Brush chipper, gas engine, 6" cutter head, 35 H.P.			7.25	102	305	915	119
0550	12" cutter head, 130 H.P.			11.45	153	460	1,375	183.60
0600	15" cutter head, 165 H.P.			16.30	165	495	1,475	229.40
0750	Bucket, clamshell, general purpose, 3/8 C.Y.			1.05	35	105	315	29.40
0800	1/2 C.Y.			1.15	41.50	125	375	34.20
0850	3/4 C.Y.			1.30	51.50	155	465	41.40
0900	1 C.Y.			1.35	56.50	170	510	44.80
0950	1-1/2 C.Y.			2.10	76.50	230	690	62.80
1000	2 C.Y.			2.25	86.50	260	780	70
1010	Bucket, dragline, medium duty, 1/2 C.Y.			.60	22.50	67	201	18.20
1020	3/4 C.Y.			.65	23.50	71	213	19.40
1030	1 C.Y.			.65	25.50	77	231	20.60
1040	1-1/2 C.Y.			1.00	38.50	115	345	31
1050	2 C.Y.			1.05	43.50	130	390	34.40
1070	3 C.Y.			1.60	60	180	540	48.80
1200	Compactor, manually guided 2-drum vibratory smooth roller, 7.5 H.P.			5.35	152	455	1,375	133.80
1250	Rammer compactor, gas, 1000 lb. blow			2.05	36.50	110	330	38.40
1300	Vibratory plate, gas, 18" plate, 3000 lb. blow			2.00	22	66	198	29.20
1350	21" plate, 5000 lb. blow			2.45	27.50	83	249	36.20
1370	Curb builder/extruder, 14 H.P., gas, single screw			10.45	228	685	2,050	220.60
1390	Double screw			11.10	270	810	2,425	250.80
1500	Disc harrow attachment, for tractor			.38	63	189	565	40.85
1750	Extractor, piling, see lines 2500 to 2750							
1810	Feller buncher, shearing & accumulating trees, 100 H.P.		Ea.	24.80	525	1,575	4,725	513.40
1860	Grader, self-propelled, 25,000 lb.			20.85	425	1,270	3,800	420.80
1910	30,000 lb.			25.05	510	1,525	4,575	505.40
1920	40,000 lb.			34.40	765	2,300	6,900	735.20
1930	55,000 lb.			45.75	1,100	3,275	9,825	1,021
1950	Hammer, pavement demo., hyd., gas, self-prop., 1000 to 1250 lb.			20.15	310	935	2,800	348.20
2000	Diesel 1300 to 1500 lb.			31.20	615	1,850	5,550	619.60
2050	Pile driving hammer, steam or air, 4150 ft.-lb. @ 225 BPM			6.75	275	825	2,475	219
2100	8750 ft.-lb. @ 145 BPM			8.65	445	1,340	4,025	337.20
2150	15,000 ft.-lb. @ 60 BPM			9.00	480	1,440	4,325	360
2200	24,450 ft.-lb. @ 111 BPM			11.95	530	1,590	4,775	413.60
2250	Leads, 15,000 ft.-lb. hammers		L.F.	.03	1.28	3.83	11.50	1
2300	24,450 ft.-lb. hammers and heavier		"	.05	2.33	7	21	1.80
2350	Diesel type hammer, 22,400 ft.-lb.		Ea.	28.40	615	1,840	5,525	595.20
2400	41,300 ft.-lb.			39.70	665	1,990	5,975	715.60
2450	141,000 ft.-lb.			64.00	1,150	3,415	10,200	1,195
2500	Vib. elec. hammer/extractor, 200 KW diesel generator, 34 H.P.			33.55	640	1,925	5,775	653.40

01 54 | Construction Aids

01 54 33 | Equipment Rental

			UNIT	HOURLY OPER. COST	RENT PER DAY	RENT PER WEEK	RENT PER MONTH	CREW EQUIPMENT COST/DAY
2550	80 H.P.		Ea.	59.10	935	2,810	8,425	1,035
2600	150 H.P.	R015433-10		110.75	1,825	5,485	16,500	1,983
2700	Extractor, steam or air, 700 ft.-lb.	R312323-30		15.05	470	1,405	4,225	401.40
2750	1000 ft.-lb.			17.15	570	1,715	5,150	480.20
2800	Log chipper, up to 22" diam, 600 H.P.	R312316-40		32.75	430	1,290	3,875	520
2850	Logger, for skidding & stacking logs, 150 H.P.			39.55	835	2,500	7,500	816.40
2900	Rake, spring tooth, with tractor	R312316-45		8.04	219	658	1,975	195.90
3000	Roller, vibratory, tandem, smooth drum, 20 H.P.			6.15	122	365	1,100	122.20
3050	35 H.P.			8.75	235	705	2,125	211
3100	Towed type vibratory compactor, smooth drum, 50 H.P.			20.25	315	940	2,825	350
3150	Sheepsfoot, 50 H.P.			21.25	345	1,030	3,100	376
3170	Landfill compactor, 220 HP			62.00	1,225	3,650	11,000	1,226
3200	Pneumatic tire roller, 80 H.P.			12.20	330	990	2,975	295.60
3250	120 H.P.			18.75	565	1,690	5,075	488
3300	Sheepsfoot vibratory roller, 200 H.P.			50.15	925	2,775	8,325	956.20
3320	340 H.P.			69.85	1,375	4,090	12,300	1,377
3350	Smooth drum vibratory roller, 75 H.P.			18.30	505	1,510	4,525	448.40
3400	125 H.P.			23.95	625	1,870	5,600	565.60
3410	Rotary mower, brush, 60", with tractor			11.50	238	715	2,150	235
3420	Rototiller, 5 HP, walk-behind			1.89	55.50	166	500	48.30
3450	Scrapers, towed type, 9 to 12 C.Y. capacity			5.65	163	490	1,475	143.20
3500	12 to 17 C.Y. capacity			6.25	178	535	1,600	157
3550	Scrapers, self-propelled, 4 x 4 drive, 2 engine, 14 C.Y. capacity			98.80	1,550	4,635	13,900	1,717
3600	2 engine, 24 C.Y. capacity			137.50	2,425	7,245	21,700	2,549
3640	32 - 44 C.Y. capacity			163.05	2,825	8,465	25,400	2,997
3650	Self-loading, 11 C.Y. capacity			44.85	835	2,505	7,525	859.80
3700	22 C.Y. capacity			85.05	1,775	5,350	16,100	1,750
3710	Screening plant 110 H.P. w/ 5' x 10' screen			28.65	390	1,165	3,500	462.20
3720	5' x 16' screen			30.75	490	1,465	4,400	539
3850	Shovels, see Cranes division 01590-600							
3860	Shovel/backhoe bucket, 1/2 C.Y.		Ea.	1.90	56.50	170	510	49.20
3870	3/4 C.Y.			1.95	63.50	190	570	53.60
3880	1 C.Y.			2.05	73.50	220	660	60.40
3890	1-1/2 C.Y.			2.20	86.50	260	780	69.60
3910	3 C.Y.			2.50	122	365	1,100	93
3950	Stump chipper, 18" deep, 30 H.P.			5.54	103	310	930	106.30
4110	Tractor, crawler, with bulldozer, torque converter, diesel 80 H.P.			18.60	345	1,030	3,100	354.80
4150	105 H.P.			25.60	485	1,450	4,350	494.80
4200	140 H.P.			30.55	610	1,835	5,500	611.40
4260	200 H.P.			46.30	1,025	3,090	9,275	988.40
4310	300 H.P.			59.75	1,375	4,115	12,300	1,301
4360	410 H.P.			80.30	1,725	5,175	15,500	1,677
4370	500 H.P.			106.80	2,350	7,030	21,100	2,260
4380	700 H.P.			158.50	3,575	10,690	32,100	3,406
4400	Loader, crawler, torque conv., diesel, 1-1/2 C.Y., 80 H.P.			16.75	335	1,000	3,000	334
4450	1-1/2 to 1-3/4 C.Y., 95 H.P.			19.65	405	1,210	3,625	399.20
4510	1-3/4 to 2-1/4 C.Y., 130 H.P.			26.90	625	1,870	5,600	589.20
4530	2-1/2 to 3-1/4 C.Y., 190 H.P.			39.05	870	2,610	7,825	834.40
4560	3-1/2 to 5 C.Y., 275 H.P.			53.35	1,200	3,590	10,800	1,145
4610	Tractor loader, wheel, torque conv., 4 x 4, 1 to 1-1/4 C.Y., 65 H.P.			11.70	190	570	1,700	207.60
4620	1-1/2 to 1-3/4 C.Y., 80 H.P.			15.20	260	780	2,350	277.60
4650	1-3/4 to 2 C.Y., 100 H.P.			17.20	300	900	2,700	317.60
4710	2-1/2 to 3-1/2 C.Y., 130 H.P.			18.75	330	995	2,975	349
4730	3 to 4-1/2 C.Y., 170 H.P.			24.25	490	1,470	4,400	488
4760	5-1/4 to 5-3/4 C.Y., 270 H.P.			39.60	730	2,195	6,575	755.80
4810	7 to 8 C.Y., 375 H.P.			67.35	1,325	3,990	12,000	1,337
4870	12-1/2 C.Y., 690 H.P.			93.10	2,050	6,175	18,500	1,980
4880	Wheeled, skid steer, 10 C.F., 30 H.P. gas			6.50	117	350	1,050	122

01 54 | Construction Aids

01 54 33 | Equipment Rental

			UNIT	HOURLY OPER. COST	RENT PER DAY	RENT PER WEEK	RENT PER MONTH	CREW EQUIPMENT COST/DAY	
20	4890	1 C.Y., 78 H.P., diesel		12.15	208	625	1,875	222.20	20
	4891	Attachments for all skid steer loaders							
	4892	Auger	Ea.	.48	80.50	241	725	52.05	
	4893	Backhoe		.69	114	343	1,025	74.10	
	4894	Broom		.63	105	314	940	67.85	
	4895	Forks		.23	37.50	113	340	24.45	
	4896	Grapple		.56	93.50	281	845	60.70	
	4897	Concrete hammer		1.01	168	504	1,500	108.90	
	4898	Tree spade		1.04	173	519	1,550	112.10	
	4899	Trencher		.70	117	352	1,050	76	
	4900	Trencher, chain, boom type, gas, operator walking, 12 H.P.		3.00	45	135	405	51	
	4910	Operator riding, 40 H.P.		9.70	248	745	2,225	226.60	
	5000	Wheel type, diesel, 4' deep, 12" wide		54.15	765	2,290	6,875	891.20	
	5100	Diesel, 6' deep, 20" wide		68.95	1,800	5,370	16,100	1,626	
	5150	Ladder type, diesel, 5' deep, 8" wide		37.95	860	2,585	7,750	820.60	
	5200	Diesel, 8' deep, 16" wide		63.00	1,825	5,510	16,500	1,606	
	5210	Tree spade, self-propelled	Ea.	12.85	267	800	2,400	262.80	
	5250	Truck, dump, tandem, 12 ton payload		24.80	283	850	2,550	368.40	
	5300	Three axle dump, 16 ton payload		33.50	435	1,310	3,925	530	
	5350	Dump trailer only, rear dump, 16-1/2 C.Y.	Ea.	4.35	122	365	1,100	107.80	
	5400	20 C.Y.		4.75	137	410	1,225	120	
	5450	Flatbed, single axle, 1-1/2 ton rating		13.30	58.50	175	525	141.40	
	5500	3 ton rating		16.90	83.50	250	750	185.20	
	5550	Off highway rear dump, 25 ton capacity		46.35	1,025	3,110	9,325	992.80	
	5600	35 ton capacity		47.25	1,050	3,180	9,550	1,014	
	5610	50 ton capacity		61.20	1,375	4,090	12,300	1,308	
	5620	65 ton capacity		65.65	1,450	4,340	13,000	1,393	
	5630	100 ton capacity		84.30	1,875	5,590	16,800	1,792	
	6000	Vibratory plow, 25 H.P., walking		5.20	58.50	175	525	76.60	
40	0010	**GENERAL EQUIPMENT RENTAL**, without operators							40
	0150	Aerial lift, scissor type, to 15' high, 1000 lb. cap., electric	Ea.	2.40	45	135	405	46.20	
	0160	To 25' high, 2000 lb. capacity		2.80	63.50	190	570	60.40	
	0170	Telescoping boom to 40' high, 500 lb. capacity, gas		14.85	285	855	2,575	289.80	
	0180	To 45' high, 500 lb. capacity		15.80	325	980	2,950	322.40	
	0190	To 60' high, 600 lb. capacity		17.80	435	1,305	3,925	403.40	
	0195	Air compressor, portable, 6.5 CFM, electric		.42	10.35	31	93	9.55	
	0196	Gasoline		.67	15.65	47	141	14.75	
	0200	Air compressor, portable, gas engine, 60 C.F.M.		10.10	46.50	140	420	108.80	
	0300	160 C.F.M.		11.70	48.50	145	435	122.60	
	0400	Diesel engine, rotary screw, 250 C.F.M.		11.80	95	285	855	151.40	
	0500	365 C.F.M.		15.85	115	345	1,025	195.80	
	0550	450 C.F.M.		20.15	147	440	1,325	249.20	
	0600	600 C.F.M.		34.90	195	585	1,750	396.20	
	0700	750 C.F.M.		35.35	212	635	1,900	409.80	
	0800	For silenced models, small sizes, add							
	0900	Large sizes, add							
	0920	Air tools and accessories							
	0930	Breaker, pavement, 60 lb.	Ea.	.40	8.65	26	78	8.40	
	0940	80 lb.		.40	9	27	81	8.60	
	0950	Drills, hand (jackhammer) 65 lb.		.50	15.35	46	138	13.20	
	0960	Track or wagon, swing boom, 4" drifter		46.05	700	2,095	6,275	787.40	
	0970	5" drifter		56.05	760	2,285	6,850	905.40	
	0975	Track mounted quarry drill, 6" diameter drill		77.10	1,150	3,445	10,300	1,306	
	0980	Dust control per drill		.91	17.35	52	156	17.70	
	0990	Hammer, chipping, 12 lb.		.40	20.50	62	186	15.60	
	1000	Hose, air with couplings, 50' long, 3/4" diameter		.04	6.65	20	60	4.30	
	1100	1" diameter		.04	6.35	19	57	4.10	

01 54 | Construction Aids

01 54 33 | Equipment Rental

			UNIT	HOURLY OPER. COST	RENT PER DAY	RENT PER WEEK	RENT PER MONTH	CREW EQUIPMENT COST/DAY	
40	1200	1-1/2" diameter	Ea.	.05	9	27	81	5.80	40
	1300	2" diameter		.10	17	51	153	11	
	1400	2-1/2" diameter		.12	19.35	58	174	12.55	
	1410	3" diameter		.18	29.50	88	264	19.05	
	1450	Drill, steel, 7/8" x 2'		.07	12.35	37	111	7.95	
	1460	7/8" x 6'		.08	12.65	38	114	8.25	
	1520	Moil points		.03	4.33	13	39	2.85	
	1525	Pneumatic nailer w/accessories		.43	28.50	86	258	20.65	
	1530	Sheeting driver for 60 lb. breaker		.04	6	18	54	3.90	
	1540	For 90 lb. breaker		.12	8	24	72	5.75	
	1550	Spade, 25 lb.		.35	6.65	20	60	6.80	
	1560	Tamper, single, 35 lb.		.54	36	108	325	25.90	
	1570	Triple, 140 lb.		.81	54	162	485	38.90	
	1580	Wrenches, impact, air powered, up to 3/4" bolt		.30	8	24	72	7.20	
	1590	Up to 1-1/4" bolt		.35	16.65	50	150	12.80	
	1600	Barricades, barrels, reflectorized, 1 to 50 barrels		.02	3.33	10	30	2.15	
	1610	100 to 200 barrels		.02	2.53	7.60	23	1.70	
	1620	Barrels with flashers, 1 to 50 barrels		.02	4	12	36	2.55	
	1630	100 to 200 barrels		.02	3.20	9.60	29	2.10	
	1640	Barrels with steady burn type C lights		.03	5.35	16	48	3.45	
	1650	Illuminated board, trailer mounted, with generator		.65	117	350	1,050	75.20	
	1670	Portable barricade, stock, with flashers, 1 to 6 units		.02	4	12	36	2.55	
	1680	25 to 50 units		.02	3.73	11.20	33.50	2.40	
	1690	Butt fusion machine, electric		19.15	238	715	2,150	296.20	
	1695	Electro fusion machine		14.15	102	305	915	174.20	
	1700	Carts, brick, hand powered, 1000 lb. capacity		.25	41.50	124	370	26.80	
	1800	Gas engine, 1500 lb., 7-1/2' lift		3.59	102	306	920	89.90	
	1822	Dehumidifier, medium, 6 lb/hr, 150 CFM		.79	48	144	430	35.10	
	1824	Large, 18 lb/hr, 600 CFM		1.56	95.50	286	860	69.70	
	1830	Distributor, asphalt, trailer mtd, 2000 gal., 38 H.P. diesel		8.30	300	905	2,725	247.40	
	1840	3000 gal., 38 H.P. diesel		9.50	330	985	2,950	273	
	1850	Drill, rotary hammer, electric, 1-1/2" diameter		.79	25.50	76	228	21.50	
	1860	Carbide bit for above		.04	6.65	20	60	4.30	
	1865	Rotary, crawler, 250 H.P.		115.55	1,950	5,830	17,500	2,090	
	1870	Emulsion sprayer, 65 gal., 5 H.P. gas engine		2.44	88	264	790	72.30	
	1880	200 gal., 5 H.P. engine		6.10	147	440	1,325	136.80	
	1920	Floodlight, mercury vapor, or quartz, on tripod							
	1930	1000 watt	Ea.	.31	12	36	108	9.70	
	1940	2000 watt		.54	22	66	198	17.50	
	1950	Floodlights, trailer mounted with generator, 1 - 300 watt light		2.70	66.50	200	600	61.60	
	1960	2 - 1000 watt lights		3.80	112	335	1,000	97.40	
	2000	4 - 300 watt lights		3.15	78.50	235	705	72.20	
	2020	Forklift, wheeled, for brick, 18', 3000 lb., 2 wheel drive, gas		19.30	190	570	1,700	268.40	
	2040	28', 4000 lb., 4 wheel drive, diesel		15.80	243	730	2,200	272.40	
	2050	For rough terrain, 8000 lb., 16' lift, 68 H.P.		20.35	385	1,150	3,450	392.80	
	2060	For plant, 4 T. capacity, 80 H.P., 2 wheel drive, gas		11.85	88.50	265	795	147.80	
	2080	10 T. capacity, 120 H.P., 2 wheel drive, diesel		17.35	153	460	1,375	230.80	
	2100	Generator, electric, gas engine, 1.5 KW to 3 KW		2.65	9.35	28	84	26.80	
	2200	5 KW		3.40	13.65	41	123	35.40	
	2300	10 KW		6.35	24.50	73	219	65.40	
	2400	25 KW		8.25	55	165	495	99	
	2500	Diesel engine, 20 KW		8.40	63.50	190	570	105.20	
	2600	50 KW		15.20	86.50	260	780	173.60	
	2700	100 KW		29.45	117	350	1,050	305.60	
	2800	250 KW		58.15	220	660	1,975	597.20	
	2850	Hammer, hydraulic, for mounting on boom, to 500 ft.-lb.		1.90	65	195	585	54.20	
	2860	1000 ft.-lb.		3.35	103	310	930	88.80	
	2900	Heaters, space, oil or electric, 50 MBH		1.64	7.65	23	69	17.70	

01 54 | Construction Aids

01 54 33 | Equipment Rental

			UNIT	HOURLY OPER. COST	RENT PER DAY	RENT PER WEEK	RENT PER MONTH	CREW EQUIPMENT COST/DAY
3000	100 MBH	R312323-30	Ea.	2.96	10.65	32	96	30.10
3100	300 MBH			9.48	33.50	100	300	95.85
3150	500 MBH	R314116-40		19.15	50	150	450	183.20
3200	Hose, water, suction with coupling, 20' long, 2" diameter			.02	3	9	27	1.95
3210	3" diameter	R314116-45		.03	4.67	14	42	3.05
3220	4" diameter			.03	5	15	45	3.25
3230	6" diameter			.11	17.65	53	159	11.50
3240	8" diameter			.26	43.50	131	395	28.30
3250	Discharge hose with coupling, 50' long, 2" diameter			.01	1.33	4	12	.90
3260	3" diameter			.02	2.67	8	24	1.75
3270	4" diameter			.02	3.67	11	33	2.35
3280	6" diameter			.06	9.65	29	87	6.30
3290	8" diameter			.31	51.50	154	460	33.30
3295	Insulation blower			.25	7.35	22	66	6.40
3300	Ladders, extension type, 16' to 36' long			.19	31.50	95	285	20.50
3400	40' to 60' long			.28	47.50	142	425	30.65
3405	Lance for cutting concrete			2.50	79.50	239	715	67.80
3407	Lawn mower, rotary, 22", 5HP			1.63	43	129	385	38.85
3408	48" self propelled			3.92	131	394	1,175	110.15
3410	Level, laser type, for pipe and sewer leveling			1.21	80.50	242	725	58.10
3430	Electronic			.76	50.50	152	455	36.50
3440	Laser type, rotating beam for grade control			1.26	83.50	251	755	60.30
3460	Builders level with tripod and rod			.08	12.65	38	114	8.25
3500	Light towers, towable, with diesel generator, 2000 watt			3.15	78.50	235	705	72.20
3600	4000 watt			3.80	112	335	1,000	97.40
3700	Mixer, powered, plaster and mortar, 6 C.F., 7 H.P.			1.45	17.65	53	159	22.20
3800	10 C.F., 9 H.P.			1.75	29	87	261	31.40
3850	Nailer, pneumatic			.43	28.50	86	258	20.65
3900	Paint sprayers complete, 8 CFM			.72	48	144	430	34.55
4000	17 CFM			1.13	75.50	226	680	54.25
4020	Pavers, bituminous, rubber tires, 8' wide, 50 H.P., diesel			33.35	965	2,900	8,700	846.80
4030	10' wide, 150 H.P.			70.60	1,550	4,685	14,100	1,502
4050	Crawler, 8' wide, 100 H.P., diesel			70.50	1,625	4,905	14,700	1,545
4060	10' wide, 150 H.P.			79.35	1,975	5,900	17,700	1,815
4070	Concrete paver, 12' to 24' wide, 250 H.P.			70.40	1,525	4,555	13,700	1,474
4080	Placer-spreader-trimmer, 24' wide, 300 H.P.			101.85	2,525	7,545	22,600	2,324
4100	Pump, centrifugal gas pump, 1-1/2", 4 MGPH			3.05	40	120	360	48.40
4200	2", 8 MGPH			4.05	45	135	405	59.40
4300	3", 15 MGPH			4.30	48.50	145	435	63.40
4400	6", 90 MGPH			22.40	165	495	1,475	278.20
4500	Submersible electric pump, 1-1/4", 55 GPM			.37	15.35	46	138	12.15
4600	1-1/2", 83 GPM			.41	17.65	53	159	13.90
4700	2", 120 GPM			1.10	22	66	198	22
4800	3", 300 GPM			1.80	36.50	110	330	36.40
4900	4", 560 GPM			8.00	158	475	1,425	159
5000	6", 1590 GPM			11.75	215	645	1,925	223
5100	Diaphragm pump, gas, single, 1-1/2" diameter		Ea.	.98	41.50	124	370	32.65
5200	2" diameter			3.30	51.50	155	465	57.40
5300	3" diameter			3.30	51.50	155	465	57.40
5400	Double, 4" diameter			4.40	71.50	215	645	78.20
5500	Trash pump, self-priming, gas, 2" diameter			3.40	20.50	62	186	39.60
5600	Diesel, 4" diameter			8.85	56.50	170	510	104.80
5650	Diesel, 6" diameter			29.10	123	370	1,100	306.80
5655	Grout Pump			10.15	83.50	250	750	131.20
5660	Rollers, see 01 54 33 20							
5700	Salamanders, L.P. gas fired, 100,000 BTU		Ea.	2.97	11.35	34	102	30.55
5705	50,000 BTU			2.22	8	24	72	22.55
5720	Sandblaster, portable, open top, 3 C.F. capacity			.40	20.50	62	186	15.60

01 54 | Construction Aids

01 54 33 | Equipment Rental

			UNIT	HOURLY OPER. COST	RENT PER DAY	RENT PER WEEK	RENT PER MONTH	CREW EQUIPMENT COST/DAY	
40	5730	6 C.F. capacity	Ea.	.70	30	90	270	23.60	40
	5740	Accessories for above		.11	19	57	171	12.30	
	5750	Sander, floor		.79	19	57	171	17.70	
	5760	Edger		.56	17.35	52	156	14.90	
	5800	Saw, chain, gas engine, 18" long		1.65	16.35	49	147	23	
	5900	36" long		.55	50	150	450	34.40	
	5950	60" long		.55	50	150	450	34.40	
	6000	Masonry, table mounted, 14" diameter, 5 H.P.		1.28	54.50	164	490	43.05	
	6050	Portable cut-off, 8 H.P.		1.80	29.50	88	264	32	
	6100	Circular, hand held, electric, 7-1/4" diameter		.20	5	15	45	4.60	
	6200	12" diameter		.28	8.65	26	78	7.45	
	6250	Wall saw, w/hydraulic power, 10 H.P		5.65	55	165	495	78.20	
	6275	Shot blaster, walk behind, 20" wide		6.65	415	1,245	3,725	302.20	
	6280	Sidewalk broom, walk-behind		2.03	62	186	560	53.45	
	6300	Steam cleaner, 100 gallons per hour		2.80	63.50	190	570	60.40	
	6310	200 gallons per hour		3.90	78.50	235	705	78.20	
	6340	Tar Kettle/Pot, 400 gallon		3.99	51.50	155	465	62.90	
	6350	Torch, cutting, acetylene-oxygen, 150' hose		.50	21.50	65	195	17	
	6360	Hourly operating cost includes tips and gas		8.10				64.80	
	6410	Toilet, portable chemical		.11	18.35	55	165	11.90	
	6420	Recycle flush type		.13	22.50	67	201	14.45	
	6430	Toilet, fresh water flush, garden hose,		.15	25	75	225	16.20	
	6440	Hoisted, non-flush, for high rise		.13	22	66	198	14.25	
	6450	Toilet, trailers, minimum		.23	38	114	340	24.65	
	6460	Maximum		.69	115	344	1,025	74.30	
	6465	Tractor, farm with attachment		10.30	223	670	2,000	216.40	
	6470	Trailer, office, see division 01520-500							
	6500	Trailers, platform, flush deck, 2 axle, 25 ton capacity	Ea.	4.35	96.50	290	870	92.80	
	6600	40 ton capacity		5.60	135	405	1,225	125.80	
	6700	3 axle, 50 ton capacity		6.05	148	445	1,325	137.40	
	6800	75 ton capacity		7.55	195	585	1,750	177.40	
	6810	Trailer mounted cable reel for H.V. line work		4.66	222	665	2,000	170.30	
	6820	Trailer mounted cable tensioning rig		9.17	435	1,310	3,925	335.35	
	6830	Cable pulling rig		62.02	2,475	7,410	22,200	1,978	
	6850	Trailer, storage, see division 01520-500							
	6900	Water tank, engine driven discharge, 5000 gallons	Ea.	5.65	127	380	1,150	121.20	
	6925	10,000 gallons		7.80	177	530	1,600	168.40	
	6950	Water truck, off highway, 6000 gallons		56.50	735	2,210	6,625	894	
	7010	Tram car for H.V. line work, powered, 2 conductor		6.10	120	361	1,075	121	
	7020	Transit (builder's level) with tripod		.08	12.65	38	114	8.25	
	7030	Trench box, 3000 lbs. 6'x8'		.60	100	300	900	64.80	
	7040	7200 lbs. 6'x20'		.73	122	367	1,100	79.25	
	7050	8000 lbs., 8' x 16'		.93	154	463	1,400	100.05	
	7060	9500 lbs., 8'x20'		1.26	210	630	1,900	136.10	
	7065	11,000 lbs., 8'x24'		1.35	225	676	2,025	146	
	7070	12,000 lbs., 10' x 20'		1.50	250	750	2,250	162	
	7100	Truck, pickup, 3/4 ton, 2 wheel drive		6.35	56.50	170	510	84.80	
	7200	4 wheel drive		6.50	65	195	585	91	
	7250	Crew carrier, 9 passenger		8.65	78.50	235	705	116.20	
	7290	Tool van, 24,000 G.V.W.		10.85	108	325	975	151.80	
	7300	Tractor, 4 x 2, 30 ton capacity, 195 H.P.	Ea.	16.70	170	510	1,525	235.60	
	7410	250 H.P.		21.65	232	695	2,075	312.20	
	7500	6 x 2, 40 ton capacity, 240 H.P.		20.90	282	845	2,525	336.20	
	7600	6 x 4, 45 ton capacity, 240 H.P.		26.80	300	905	2,725	395.40	
	7620	Vacuum truck, hazardous material, 2500 gallon		10.15	295	885	2,650	258.20	
	7625	5,000 gallon		13.03	415	1,240	3,725	352.25	
	7640	Tractor, with A frame, boom and winch, 225 H.P.		19.10	242	725	2,175	297.80	
	7650	Vacuum, H.E.P.A., 16 gal., wet/dry	Ea.	.92	22	66	198	20.55	

01 54 | Construction Aids

01 54 33 | Equipment Rental

			UNIT	HOURLY OPER. COST	RENT PER DAY	RENT PER WEEK	RENT PER MONTH	CREW EQUIPMENT COST/DAY	
40	7655	55 gal, wet/dry	Ea.	.89	33	99	297	26.90	40
	7660	Water tank, portable		.15	25.50	75.90	228	16.40	
	7690	Large production vacuum loader, 3150 CFM		17.37	620	1,860	5,575	510.95	
	7700	Welder, electric, 200 amp	Ea.	3.37	31.50	95	285	45.95	
	7800	300 amp		4.98	36	108	325	61.45	
	7900	Gas engine, 200 amp		10.90	23.50	71	213	101.40	
	8000	300 amp		12.50	25.50	76	228	115.20	
	8100	Wheelbarrow, any size		.07	11.65	35	105	7.55	
	8200	Wrecking ball, 4000 lb.		2.00	70	210	630	58	
50	0010	**HIGHWAY EQUIPMENT RENTAL**, without operators							50
	0050	Asphalt batch plant, portable drum mixer, 100 ton/hr.	Ea.	57.45	1,350	4,025	12,100	1,265	
	0060	200 ton/hr.		63.80	1,400	4,230	12,700	1,356	
	0070	300 ton/hr.		74.25	1,675	5,005	15,000	1,595	
	0100	Backhoe attachment, long stick, up to 185 HP, 10.5' long		.31	20.50	62	186	14.90	
	0140	Up to 250 HP, 12' long		.34	22.50	67	201	16.10	
	0180	Over 250 HP, 15' long		.45	29.50	89	267	21.40	
	0200	Special dipper arm, up to 100 HP, 32' long		.92	61	183	550	43.95	
	0240	Over 100 HP, 33' long		1.14	76	228	685	54.70	
	0300	Concrete batch plant, portable, electric, 200 CY/Hr		10.79	505	1,515	4,550	389.30	
	0500	Grader attachment, ripper/scarifier, rear mounted							
	0520	Up to 135 HP	Ea.	2.95	61.50	185	555	60.60	
	0540	Up to 180 HP		3.55	80	240	720	76.40	
	0580	Up to 250 HP		3.90	90	270	810	85.20	
	0700	Pvmt. removal bucket, for hyd. excavator, up to 90 HP		1.50	45	135	405	39	
	0740	Up to 200 HP		1.70	66.50	200	600	53.60	
	0780	Over 200 HP		1.80	80	240	720	62.40	
	0900	Aggregate spreader, self-propelled, 187 HP		42.35	800	2,400	7,200	818.80	
	1000	Chemical spreader, 3 C.Y.		2.40	41.50	125	375	44.20	
	1900	Hammermill, traveling, 250 HP		55.98	1,775	5,320	16,000	1,512	
	2000	Horizontal borer, 3" diam, 13 HP gas driven		4.90	53.50	160	480	71.20	
	2200	Hydromulchers, gas power, 3000 gal., for truck mounting		12.90	188	565	1,700	216.20	
	2400	Joint & crack cleaner, walk behind, 25 HP		2.40	48.50	145	435	48.20	
	2500	Filler, trailer mounted, 400 gal., 20 HP		6.55	183	550	1,650	162.40	
	3000	Paint striper, self propelled, double line, 30 HP		5.50	157	470	1,400	138	
	3200	Post drivers, 6" I-Beam frame, for truck mounting		10.30	420	1,255	3,775	333.40	
	3400	Road sweeper, self propelled, 8' wide, 90 HP		27.75	550	1,645	4,925	551	
	4000	Road mixer, self-propelled, 130 HP		35.40	685	2,050	6,150	693.20	
	4100	310 HP		67.30	2,125	6,405	19,200	1,819	
	4200	Cold mix paver, incl pug mill and bitumen tank,							
	4220	165 HP	Ea.	79.35	1,975	5,910	17,700	1,817	
	4250	Paver, asphalt, wheel or crawler, 130 H.P., diesel		78.05	1,875	5,630	16,900	1,750	
	4300	Paver, road widener, gas 1' to 6', 67 HP		37.40	780	2,345	7,025	768.20	
	4400	Diesel, 2' to 14', 88 HP		49.05	1,000	3,025	9,075	997.40	
	4600	Slipform pavers, curb and gutter, 2 track, 75 HP		31.95	720	2,160	6,475	687.60	
	4700	4 track, 165 HP		38.75	770	2,305	6,925	771	
	4800	Median barrier, 215 HP		39.35	800	2,395	7,175	793.80	
	4901	Trailer, low bed, 75 ton capacity		8.15	193	580	1,750	181.20	
	5000	Road planer, walk behind, 10" cutting width, 10 HP		2.30	28	84	252	35.20	
	5100	Self propelled, 12" cutting width, 64 HP		6.25	112	335	1,000	117	
	5200	Pavement profiler, 4' to 6' wide, 450 HP		197.95	3,250	9,760	29,300	3,536	
	5300	8' to 10' wide, 750 HP		313.05	4,450	13,360	40,100	5,176	
	5400	Roadway plate, steel, 1"x8'x20'		.06	10.65	32	96	6.90	
	5600	Stabilizer, self-propelled, 150 HP		36.55	580	1,735	5,200	639.40	
	5700	310 HP		62.05	1,275	3,845	11,500	1,265	
	5800	Striper, thermal, truck mounted 120 gal. paint, 150 H.P.		37.90	490	1,465	4,400	596.20	
	6000	Tar kettle, 330 gal., trailer mounted		3.67	36.50	110	330	51.35	
	7000	Tunnel locomotive, diesel, 8 to 12 ton		25.05	560	1,685	5,050	537.40	

01 54 | Construction Aids

01 54 33 | Equipment Rental

			UNIT	HOURLY OPER. COST	RENT PER DAY	RENT PER WEEK	RENT PER MONTH	CREW EQUIPMENT COST/DAY	
50	7005	Electric, 10 ton	Ea.	21.95	640	1,915	5,750	558.60	50
	7010	Muck cars, 1/2 C.Y. capacity		1.65	21.50	64	192	26	
	7020	1 C.Y. capacity		1.85	30	90	270	32.80	
	7030	2 C.Y. capacity		1.95	35	105	315	36.60	
	7040	Side dump, 2 C.Y. capacity		2.15	41.50	125	375	42.20	
	7050	3 C.Y. capacity		2.90	48.50	145	435	52.20	
	7060	5 C.Y. capacity		4.10	61.50	185	555	69.80	
	7100	Ventilating blower for tunnel, 7-1/2 H.P.		1.29	50	150	450	40.30	
	7110	10 H.P.		1.49	51.50	155	465	42.90	
	7120	20 H.P.		2.44	67.50	202	605	59.90	
	7140	40 H.P.		4.29	95	285	855	91.30	
	7160	60 H.P.		6.49	147	440	1,325	139.90	
	7175	75 H.P.		8.30	196	587	1,750	183.80	
	7180	200 H.P.		18.70	293	880	2,650	325.60	
	7800	Windrow loader, elevating		39.00	930	2,785	8,350	869	
60	0010	**LIFTING & HOISTING EQUIPMENT RENTAL**, without operators							60
	0120	Aerial lift truck, 2 person, to 80'	Ea.	22.35	615	1,840	5,525	546.80	
	0140	Boom work platform, 40' snorkel		13.25	223	670	2,000	240	
	0150	Crane, flatbed mntd, 3 ton cap.		17.70	217	650	1,950	271.60	
	0200	Crane, climbing, 106' jib, 6000 lb. capacity, 410 FPM		68.15	1,400	4,230	12,700	1,391	
	0300	101' jib, 10,250 lb. capacity, 270 FPM		73.80	1,775	5,360	16,100	1,662	
	0400	Tower, static, 130' high, 106' jib,							
	0500	6200 lb. capacity at 400 FPM	Ea.	71.45	1,625	4,890	14,700	1,550	
	0600	Crawler mounted, lattice boom, 1/2 C.Y., 15 tons at 12' radius		26.58	605	1,820	5,450	576.65	
	0700	3/4 C.Y., 20 tons at 12' radius		35.44	755	2,270	6,800	737.50	
	0800	1 C.Y., 25 tons at 12' radius		47.25	1,000	3,025	9,075	983	
	0900	1-1/2 C.Y., 40 tons at 12' radius		51.20	975	2,925	8,775	994.60	
	1000	2 C.Y., 50 tons at 12' radius		54.00	1,175	3,495	10,500	1,131	
	1100	3 C.Y., 75 tons at 12' radius		66.25	1,425	4,270	12,800	1,384	
	1200	100 ton capacity, 60' boom		76.70	1,850	5,515	16,500	1,717	
	1300	165 ton capacity, 60' boom		102.30	2,300	6,890	20,700	2,196	
	1400	200 ton capacity, 70' boom		130.85	2,675	8,030	24,100	2,653	
	1500	350 ton capacity, 80' boom		182.90	3,500	10,530	31,600	3,569	
	1600	Truck mounted, lattice boom, 6 x 4, 20 tons at 10' radius		33.59	1,175	3,540	10,600	976.70	
	1700	25 tons at 10' radius		36.28	1,275	3,840	11,500	1,058	
	1800	8 x 4, 30 tons at 10' radius		49.47	1,375	4,130	12,400	1,222	
	1900	40 tons at 12' radius		50.73	1,425	4,310	12,900	1,268	
	2000	8 x 4, 60 tons at 15' radius		52.34	1,525	4,540	13,600	1,327	
	2050	82 tons at 15' radius		54.44	1,625	4,840	14,500	1,404	
	2100	90 tons at 15' radius		60.67	1,775	5,310	15,900	1,547	
	2200	115 tons at 15' radius		74.05	1,975	5,900	17,700	1,772	
	2300	150 tons at 18' radius		74.51	2,100	6,280	18,800	1,852	
	2350	165 tons at 18' radius		87.35	2,200	6,615	19,800	2,022	
	2400	Truck mounted, hydraulic, 12 ton capacity		45.25	605	1,810	5,425	724	
	2500	25 ton capacity		45.45	625	1,880	5,650	739.60	
	2550	33 ton capacity		46.30	660	1,975	5,925	765.40	
	2560	40 ton capacity		44.70	630	1,895	5,675	736.60	
	2600	55 ton capacity		66.40	880	2,645	7,925	1,060	
	2700	80 ton capacity		78.75	975	2,920	8,750	1,214	
	2720	100 ton capacity		95.75	2,500	7,525	22,600	2,271	
	2740	120 ton capacity		93.21	2,750	8,280	24,800	2,402	
	2760	150 ton capacity		114.91	3,525	10,540	31,600	3,027	
	2800	Self-propelled, 4 x 4, with telescoping boom, 5 ton		19.10	290	870	2,600	326.80	
	2900	12-1/2 ton capacity		32.75	520	1,565	4,700	575	
	3000	15 ton capacity		35.00	615	1,850	5,550	650	
	3050	20 ton capacity		35.85	640	1,920	5,750	670.80	
	3100	25 ton capacity		47.95	820	2,460	7,375	875.60	

01 54 | Construction Aids

01 54 33 | Equipment Rental

			UNIT	HOURLY OPER. COST	RENT PER DAY	RENT PER WEEK	RENT PER MONTH	CREW EQUIPMENT COST/DAY	
60	3150	40 ton capacity	Ea.	61.00	925	2,775	8,325	1,043	60
	3200	Derricks, guy, 20 ton capacity, 60' boom, 75' mast		19.00	345	1,036	3,100	359.20	
	3300	100' boom, 115' mast		30.02	590	1,770	5,300	594.15	
	3400	Stiffleg, 20 ton capacity, 70' boom, 37' mast		21.13	445	1,340	4,025	437.05	
	3500	100' boom, 47' mast		32.75	720	2,160	6,475	694	
	3550	Helicopter, small, lift to 1250 lbs. maximum, w/pilot		82.09	2,800	8,370	25,100	2,331	
	3600	Hoists, chain type, overhead, manual, 3/4 ton		.10	.67	2	6	1.20	
	3900	10 ton		.65	7.35	22	66	9.60	
	4000	Hoist and tower, 5000 lb. cap., portable electric, 40' high		4.42	199	597	1,800	154.75	
	4100	For each added 10' section, add		.09	15.65	47	141	10.10	
	4200	Hoist and single tubular tower, 5000 lb. electric, 100' high		5.95	278	833	2,500	214.20	
	4300	For each added 6'-6" section, add		.16	26.50	79	237	17.10	
	4400	Hoist and double tubular tower, 5000 lb., 100' high		6.38	305	918	2,750	234.65	
	4500	For each added 6'-6" section, add		.18	29.50	89	267	19.25	
	4550	Hoist and tower, mast type, 6000 lb., 100' high		6.91	315	952	2,850	245.70	
	4570	For each added 10' section, add		.11	19	57	171	12.30	
	4600	Hoist and tower, personnel, electric, 2000 lb., 100' @ 125 FPM		14.13	845	2,540	7,625	621.05	
	4700	3000 lb., 100' @ 200 FPM		16.14	955	2,870	8,600	703.10	
	4800	3000 lb., 150' @ 300 FPM		17.84	1,075	3,210	9,625	784.70	
	4900	4000 lb., 100' @ 300 FPM		18.50	1,100	3,270	9,800	802	
	5000	6000 lb., 100' @ 275 FPM		20.06	1,150	3,440	10,300	848.50	
	5100	For added heights up to 500', add	L.F.	.01	1.67	5	15	1.10	
	5200	Jacks, hydraulic, 20 ton	Ea.	.05	3.33	10	30	2.40	
	5500	100 ton	"	.30	9.65	29	87	8.20	
	6000	Jacks, hydraulic, climbing with 50' jackrods							
	6010	and control consoles, minimum 3 mo. rental							
	6100	30 ton capacity	Ea.	1.72	114	343	1,025	82.35	
	6150	For each added 10' jackrod section, add		.05	3.33	10	30	2.40	
	6300	50 ton capacity		2.76	184	552	1,650	132.50	
	6350	For each added 10' jackrod section, add		.06	4	12	36	2.90	
	6500	125 ton capacity		7.25	485	1,450	4,350	348	
	6550	For each added 10' jackrod section, add		.50	33	99	297	23.80	
	6600	Cable jack, 10 ton capacity with 200' cable		1.44	95.50	287	860	68.90	
	6650	For each added 50' of cable, add		.16	10.65	32	96	7.70	
70	0010	**WELLPOINT EQUIPMENT RENTAL**, without operators							70
	0020	Based on 2 months rental							
	0100	Combination jetting & wellpoint pump, 60 H.P. diesel	Ea.	13.00	283	850	2,550	274	
	0200	High pressure gas jet pump, 200 H.P., 300 psi	"	28.58	242	726	2,175	373.85	
	0300	Discharge pipe, 8" diameter	L.F.	.01	.46	1.38	4.14	.35	
	0350	12" diameter		.01	.68	2.04	6.10	.50	
	0400	Header pipe, flows up to 150 G.P.M., 4" diameter		.01	.41	1.24	3.72	.35	
	0500	400 G.P.M., 6" diameter		.01	.49	1.48	4.44	.40	
	0600	800 G.P.M., 8" diameter		.01	.68	2.04	6.10	.50	
	0700	1500 G.P.M., 10" diameter		.01	.71	2.14	6.40	.50	
	0800	2500 G.P.M., 12" diameter		.02	1.35	4.05	12.15	.95	
	0900	4500 G.P.M., 16" diameter		.03	1.73	5.18	15.55	1.30	
	0950	For quick coupling aluminum and plastic pipe, add		.03	1.79	5.36	16.10	1.30	
	1100	Wellpoint, 25' long, with fittings & riser pipe, 1-1/2" or 2" diameter	Ea.	.05	3.57	10.70	32	2.55	
	1200	Wellpoint pump, diesel powered, 4" diameter, 20 H.P.		5.78	163	490	1,475	144.25	
	1300	6" diameter, 30 H.P.		7.78	203	608	1,825	183.85	
	1400	8" suction, 40 H.P.		10.53	278	833	2,500	250.85	
	1500	10" suction, 75 H.P.		15.63	325	974	2,925	319.85	
	1600	12" suction, 100 H.P.		22.67	520	1,560	4,675	493.35	
	1700	12" suction, 175 H.P.		32.53	570	1,710	5,125	602.25	

Reference codes: R015433-10, R015433-15, R312316-45, R312319-90

01 54 | Construction Aids

01 54 33 | Equipment Rental

			UNIT	HOURLY OPER. COST	RENT PER DAY	RENT PER WEEK	RENT PER MONTH	CREW EQUIPMENT COST/DAY	
80	0010	**MARINE EQUIPMENT RENTAL**, without operators							80
	0200	Barge, 400 Ton, 30' wide x 90' long	Ea.	17.65	255	765	2,300	294.20	
	0240	800 Ton, 45' wide x 90' long		29.15	365	1,090	3,275	451.20	
	2000	Tugboat, diesel, 100 HP		22.25	180	540	1,625	286	
	2040	250 HP		43.65	335	1,005	3,025	550.20	
	2080	380 HP	Ea.	93.95	990	2,970	8,900	1,346	

Crews

Crew No.	Bare Costs		Incl. Subs O & P		Cost Per Labor-Hour	
Crew A-1	Hr.	Daily	Hr.	Daily	Bare Costs	Incl. O&P
1 Building Laborer	$18.70	$149.60	$31.75	$254.00	$18.70	$31.75
1 Concrete saw, gas manual		54.00		59.40	6.75	7.42
8 L.H., Daily Totals		$203.60		$313.40	$25.45	$39.17
Crew A-1A	Hr.	Daily	Hr.	Daily	Bare Costs	Incl. O&P
1 Skilled Worker	$26.10	$208.80	$44.25	$354.00	$26.10	$44.25
1 Shot Blaster, 20"		302.20		332.42	37.77	41.55
8 L.H., Daily Totals		$511.00		$686.42	$63.88	$85.80
Crew A-1B	Hr.	Daily	Hr.	Daily	Bare Costs	Incl. O&P
1 Building Laborer	$18.70	$149.60	$31.75	$254.00	$18.70	$31.75
1 Concrete Saw		118.40		130.24	14.80	16.28
8 L.H., Daily Totals		$268.00		$384.24	$33.50	$48.03
Crew A-1C	Hr.	Daily	Hr.	Daily	Bare Costs	Incl. O&P
1 Building Laborer	$18.70	$149.60	$31.75	$254.00	$18.70	$31.75
1 Chain saw, gas, 18"		23.00		25.30	2.88	3.16
8 L.H., Daily Totals		$172.60		$279.30	$21.57	$34.91
Crew A-1D	Hr.	Daily	Hr.	Daily	Bare Costs	Incl. O&P
1 Building Laborer	$18.70	$149.60	$31.75	$254.00	$18.70	$31.75
1 Vibrating plate, gas, 18"		29.20		32.12	3.65	4.01
8 L.H., Daily Totals		$178.80		$286.12	$22.35	$35.77
Crew A-1E	Hr.	Daily	Hr.	Daily	Bare Costs	Incl. O&P
1 Building Laborer	$18.70	$149.60	$31.75	$254.00	$18.70	$31.75
1 Vibratory Plate, gas, 21"		36.20		39.82	4.53	4.98
8 L.H., Daily Totals		$185.80		$293.82	$23.23	$36.73
Crew A-1F	Hr.	Daily	Hr.	Daily	Bare Costs	Incl. O&P
1 Building Laborer	$18.70	$149.60	$31.75	$254.00	$18.70	$31.75
1 Rammer/tamper, gas, 8"		38.40		42.24	4.80	5.28
8 L.H., Daily Totals		$188.00		$296.24	$23.50	$37.03
Crew A-1G	Hr.	Daily	Hr.	Daily	Bare Costs	Incl. O&P
1 Building Laborer	$18.70	$149.60	$31.75	$254.00	$18.70	$31.75
1 Rammer/tamper, gas, 8"		38.40		42.24	4.80	5.28
8 L.H., Daily Totals		$188.00		$296.24	$23.50	$37.03
Crew A-1H	Hr.	Daily	Hr.	Daily	Bare Costs	Incl. O&P
1 Building Laborer	$18.70	$149.60	$31.75	$254.00	$18.70	$31.75
1 Pressure washer		60.40		66.44	7.55	8.30
8 L.H., Daily Totals		$210.00		$320.44	$26.25	$40.06
Crew A-1J	Hr.	Daily	Hr.	Daily	Bare Costs	Incl. O&P
1 Building Laborer	$18.70	$149.60	$31.75	$254.00	$18.70	$31.75
1 Cultivator, Walk-Behind		48.30		53.13	6.04	6.64
8 L.H., Daily Totals		$197.90		$307.13	$24.74	$38.39
Crew A-1K	Hr.	Daily	Hr.	Daily	Bare Costs	Incl. O&P
1 Building Laborer	$18.70	$149.60	$31.75	$254.00	$18.70	$31.75
1 Cultivator, Walk-Behind		48.30		53.13	6.04	6.64
8 L.H., Daily Totals		$197.90		$307.13	$24.74	$38.39
Crew A-1M	Hr.	Daily	Hr.	Daily	Bare Costs	Incl. O&P
1 Building Laborer	$18.70	$149.60	$31.75	$254.00	$18.70	$31.75
1 Snow Blower, Walk-Behind		53.45		58.80	6.68	7.35
8 L.H., Daily Totals		$203.05		$312.80	$25.38	$39.10
Crew A-2	Hr.	Daily	Hr.	Daily	Bare Costs	Incl. O&P
2 Laborers	$18.70	$299.20	$31.75	$508.00	$19.22	$32.53
1 Truck Driver (light)	20.25	162.00	34.10	272.80		
1 Flatbed Truck, gas, 1.5 Ton		141.40		155.54	5.89	6.48
24 L.H., Daily Totals		$602.60		$936.34	$25.11	$39.01
Crew A-2A	Hr.	Daily	Hr.	Daily	Bare Costs	Incl. O&P
2 Laborers	$18.70	$299.20	$31.75	$508.00	$19.22	$32.53
1 Truck Driver (light)	20.25	162.00	34.10	272.80		
1 Flatbed Truck, gas, 1.5 Ton		141.40		155.54		
1 Concrete Saw		118.40		130.24	10.82	11.91
24 L.H., Daily Totals		$721.00		$1066.58	$30.04	$44.44
Crew A-3	Hr.	Daily	Hr.	Daily	Bare Costs	Incl. O&P
1 Truck Driver (heavy)	$21.00	$168.00	$35.35	$282.80	$21.00	$35.35
1 Dump Truck, 12 Ton, 8 C.Y.		368.40		405.24	46.05	50.66
8 L.H., Daily Totals		$536.40		$688.04	$67.05	$86.00
Crew A-3A	Hr.	Daily	Hr.	Daily	Bare Costs	Incl. O&P
1 Truck Driver (light)	$20.25	$162.00	$34.10	$272.80	$20.25	$34.10
1 Pickup truck, 4 x 4, 3/4 ton		91.00		100.10	11.38	12.51
8 L.H., Daily Totals		$253.00		$372.90	$31.63	$46.61
Crew A-3B	Hr.	Daily	Hr.	Daily	Bare Costs	Incl. O&P
1 Equip. Oper. (medium)	$26.50	$212.00	$43.65	$349.20	$23.75	$39.50
1 Truck Driver (heavy)	21.00	168.00	35.35	282.80		
1 Dump Truck, 16 Ton, 12 C.Y.		530.00		583.00		
1 F.E. Loader, W.M.,2.5 C.Y.		349.00		383.90	54.94	60.43
16 L.H., Daily Totals		$1259.00		$1598.90	$78.69	$99.93
Crew A-3C	Hr.	Daily	Hr.	Daily	Bare Costs	Incl. O&P
1 Equip. Oper. (light)	$25.45	$203.60	$41.90	$335.20	$25.45	$41.90
1 Loader, Skid Steer, 78 HP		222.20		244.42	27.77	30.55
8 L.H., Daily Totals		$425.80		$579.62	$53.23	$72.45
Crew A-3D	Hr.	Daily	Hr.	Daily	Bare Costs	Incl. O&P
1 Truck Driver, Light	$20.25	$162.00	$34.10	$272.80	$20.25	$34.10
1 Pickup truck, 4 x 4, 3/4 ton		91.00		100.10		
1 Flatbed Trailer, 25 Ton		92.80		102.08	22.98	25.27
8 L.H., Daily Totals		$345.80		$474.98	$43.23	$59.37
Crew A-3E	Hr.	Daily	Hr.	Daily	Bare Costs	Incl. O&P
1 Equip. Oper. (crane)	$27.45	$219.60	$45.20	$361.60	$24.23	$40.27
1 Truck Driver (heavy)	21.00	168.00	35.35	282.80		
1 Pickup truck, 4 x 4, 3/4 ton		91.00		100.10	5.69	6.26
16 L.H., Daily Totals		$478.60		$744.50	$29.91	$46.53
Crew A-3F	Hr.	Daily	Hr.	Daily	Bare Costs	Incl. O&P
1 Equip. Oper. (crane)	$27.45	$219.60	$45.20	$361.60	$24.23	$40.27
1 Truck Driver (heavy)	21.00	168.00	35.35	282.80		
1 Pickup truck, 4 x 4, 3/4 ton		91.00		100.10		
1 Truck Tractor, 240 H.P.		336.20		369.82		
1 Lowbed Trailer, 75 Ton		181.20		199.32	38.02	41.83
16 L.H., Daily Totals		$996.00		$1313.64	$62.25	$82.10

Crews

Crew No.	Bare Costs		Incl. Subs O & P		Cost Per Labor-Hour	
Crew A-3G	Hr.	Daily	Hr.	Daily	Bare Costs	Incl. O&P
1 Equip. Oper. (crane)	$27.45	$219.60	$45.20	$361.60	$24.23	$40.27
1 Truck Driver (heavy)	21.00	168.00	35.35	282.80		
1 Pickup truck, 4 x 4, 3/4 ton		91.00		100.10		
1 Truck Tractor, 240 H.P.		395.40		434.94		
1 Lowbed Trailer, 75 Ton		181.20		199.32	41.73	45.90
16 L.H., Daily Totals		$1055.20		$1378.76	$65.95	$86.17
Crew A-4	Hr.	Daily	Hr.	Daily	Bare Costs	Incl. O&P
2 Carpenters	$25.70	$411.20	$43.60	$697.60	$24.87	$41.78
1 Painter, Ordinary	23.20	185.60	38.15	305.20		
24 L.H., Daily Totals		$596.80		$1002.80	$24.87	$41.78
Crew A-5	Hr.	Daily	Hr.	Daily	Bare Costs	Incl. O&P
2 Laborers	$18.70	$299.20	$31.75	$508.00	$18.87	$32.01
.25 Truck Driver (light)	20.25	40.50	34.10	68.20		
.25 Flatbed Truck, gas, 1.5 Ton		35.35		38.88	1.96	2.16
18 L.H., Daily Totals		$375.05		$615.09	$20.84	$34.17
Crew A-6	Hr.	Daily	Hr.	Daily	Bare Costs	Incl. O&P
1 Instrument Man	$26.10	$208.80	$44.25	$354.00	$25.85	$43.30
1 Rodman/Chainman	25.60	204.80	42.35	338.80		
1 Laser Transit/Level		58.10		63.91	3.63	3.99
16 L.H., Daily Totals		$471.70		$756.71	$29.48	$47.29
Crew A-7	Hr.	Daily	Hr.	Daily	Bare Costs	Incl. O&P
1 Chief Of Party	$31.10	$248.80	$52.55	$420.40	$27.60	$46.38
1 Instrument Man	26.10	208.80	44.25	354.00		
1 Rodman/Chainman	25.60	204.80	42.35	338.80		
1 Laser Transit/Level		58.10		63.91	2.42	2.66
24 L.H., Daily Totals		$720.50		$1177.11	$30.02	$49.05
Crew A-8	Hr.	Daily	Hr.	Daily	Bare Costs	Incl. O&P
1 Chief of Party	$31.10	$248.80	$52.55	$420.40	$27.10	$45.38
1 Instrument Man	26.10	208.80	44.25	354.00		
2 Rodmen/Chainmen	25.60	409.60	42.35	677.60		
1 Laser Transit/Level		58.10		63.91	1.82	2.00
32 L.H., Daily Totals		$925.30		$1515.91	$28.92	$47.37
Crew A-9	Hr.	Daily	Hr.	Daily	Bare Costs	Incl. O&P
1 Asbestos Foreman	$27.90	$223.20	$48.00	$384.00	$27.46	$47.26
7 Asbestos Workers	27.40	1534.40	47.15	2640.40		
64 L.H., Daily Totals		$1757.60		$3024.40	$27.46	$47.26
Crew A-10	Hr.	Daily	Hr.	Daily	Bare Costs	Incl. O&P
1 Asbestos Foreman	$27.90	$223.20	$48.00	$384.00	$27.46	$47.26
7 Asbestos Workers	27.40	1534.40	47.15	2640.40		
64 L.H., Daily Totals		$1757.60		$3024.40	$27.46	$47.26
Crew A-10A	Hr.	Daily	Hr.	Daily	Bare Costs	Incl. O&P
1 Asbestos Foreman	$27.90	$223.20	$48.00	$384.00	$27.57	$47.43
2 Asbestos Workers	27.40	438.40	47.15	754.40		
24 L.H., Daily Totals		$661.60		$1138.40	$27.57	$47.43
Crew A-10B	Hr.	Daily	Hr.	Daily	Bare Costs	Incl. O&P
1 Asbestos Foreman	$27.90	$223.20	$48.00	$384.00	$27.52	$47.36
3 Asbestos Workers	27.40	657.60	47.15	1131.60		
32 L.H., Daily Totals		$880.80		$1515.60	$27.52	$47.36

Crew No.	Bare Costs		Incl. Subs O & P		Cost Per Labor-Hour	
Crew A-10C	Hr.	Daily	Hr.	Daily	Bare Costs	Incl. O&P
3 Asbestos Workers	$27.40	$657.60	$47.15	$1131.60	$27.40	$47.15
1 Flatbed Truck, gas, 1.5 Ton		141.40		155.54	5.89	6.48
24 L.H., Daily Totals		$799.00		$1287.14	$33.29	$53.63
Crew A-10D	Hr.	Daily	Hr.	Daily	Bare Costs	Incl. O&P
2 Asbestos Workers	$27.40	$438.40	$47.15	$754.40	$26.41	$44.51
1 Equip. Oper. (crane)	27.45	219.60	45.20	361.60		
1 Equip. Oper. Oiler	23.40	187.20	38.55	308.40		
1 Hydraulic Crane, 33 Ton		765.40		841.94	23.92	26.31
32 L.H., Daily Totals		$1610.60		$2266.34	$50.33	$70.82
Crew A-11	Hr.	Daily	Hr.	Daily	Bare Costs	Incl. O&P
1 Asbestos Foreman	$27.90	$223.20	$48.00	$384.00	$27.46	$47.26
7 Asbestos Workers	27.40	1534.40	47.15	2640.40		
2 Chipping Hammer, 12 Lb., Elec.		31.20		34.32	0.49	0.54
64 L.H., Daily Totals		$1788.80		$3058.72	$27.95	$47.79
Crew A-12	Hr.	Daily	Hr.	Daily	Bare Costs	Incl. O&P
1 Asbestos Foreman	$27.90	$223.20	$48.00	$384.00	$27.46	$47.26
7 Asbestos Workers	27.40	1534.40	47.15	2640.40		
1 Trk-mtd vac, 14 CY, 1500 Gal		510.95		562.04		
1 Flatbed Truck, 20,000 GVW		151.80		166.98	10.36	11.39
64 L.H., Daily Totals		$2420.35		$3753.43	$37.82	$58.65
Crew A-13	Hr.	Daily	Hr.	Daily	Bare Costs	Incl. O&P
1 Equip. Oper. (light)	$25.45	$203.60	$41.90	$335.20	$25.45	$41.90
1 Trk-mtd vac, 14 CY, 1500 Gal		510.95		562.04		
1 Flatbed Truck, 20,000 GVW		151.80		166.98	82.84	91.13
8 L.H., Daily Totals		$866.35		$1064.22	$108.29	$133.03
Crew B-1	Hr.	Daily	Hr.	Daily	Bare Costs	Incl. O&P
1 Labor Foreman (outside)	$20.70	$165.60	$35.15	$281.20	$19.37	$32.88
2 Laborers	18.70	299.20	31.75	508.00		
24 L.H., Daily Totals		$464.80		$789.20	$19.37	$32.88
Crew B-1A	Hr.	Daily	Hr.	Daily	Bare Costs	Incl. O&P
1 Laborer Foreman	$20.70	$165.60	$35.15	$281.20	$19.37	$32.88
2 Laborers	18.70	299.20	31.75	508.00		
2 Cutting Torches		34.00		37.40		
2 Gases		129.60		142.56	6.82	7.50
24 L.H., Daily Totals		$628.40		$969.16	$26.18	$40.38
Crew B-1B	Hr.	Daily	Hr.	Daily	Bare Costs	Incl. O&P
1 Laborer Foreman	$20.70	$165.60	$35.15	$281.20	$21.39	$35.96
2 Laborers	18.70	299.20	31.75	508.00		
1 Equip. Oper. (crane)	27.45	219.60	45.20	361.60		
2 Cutting Torches		34.00		37.40		
2 Gases		129.60		142.56		
1 Hyd. Crane, 12 Ton		724.00		796.40	27.74	30.51
32 L.H., Daily Totals		$1572.00		$2127.16	$49.13	$66.47
Crew B-2	Hr.	Daily	Hr.	Daily	Bare Costs	Incl. O&P
1 Labor Foreman (outside)	$20.70	$165.60	$35.15	$281.20	$19.10	$32.43
4 Laborers	18.70	598.40	31.75	1016.00		
40 L.H., Daily Totals		$764.00		$1297.20	$19.10	$32.43

Crews

Crew No.	Bare Costs		Incl. Subs O & P		Cost Per Labor-Hour	
Crew B-3	Hr.	Daily	Hr.	Daily	Bare Costs	Incl. O&P
1 Labor Foreman (outside)	$20.70	$165.60	$35.15	$281.20	$21.10	$35.50
2 Laborers	18.70	299.20	31.75	508.00		
1 Equip. Oper. (med.)	26.50	212.00	43.65	349.20		
2 Truck Drivers (heavy)	21.00	336.00	35.35	565.60		
1 Crawler Loader, 3 C.Y.		834.40		917.84		
2 Dump Trucks, 16 Ton, 12 C.Y.		1060.00		1166.00	39.47	43.41
48 L.H., Daily Totals		$2907.20		$3787.84	$60.57	$78.91
Crew B-3A	Hr.	Daily	Hr.	Daily	Bare Costs	Incl. O&P
4 Laborers	$18.70	$598.40	$31.75	$1016.00	$20.26	$34.13
1 Equip. Oper. (med.)	26.50	212.00	43.65	349.20		
1 Hyd. Excavator, 1.5 C.Y.		775.60		853.16	19.39	21.33
40 L.H., Daily Totals		$1586.00		$2218.36	$39.65	$55.46
Crew B-3B	Hr.	Daily	Hr.	Daily	Bare Costs	Incl. O&P
2 Laborers	$18.70	$299.20	$31.75	$508.00	$21.23	$35.63
1 Equip. Oper. (med.)	26.50	212.00	43.65	349.20		
1 Truck Driver (heavy)	21.00	168.00	35.35	282.80		
1 Backhoe Loader, 80 H.P.		285.40		313.94		
1 Dump Truck, 16 Ton, 12 C.Y.		530.00		583.00	25.48	28.03
32 L.H., Daily Totals		$1494.60		$2036.94	$46.71	$63.65
Crew B-3C	Hr.	Daily	Hr.	Daily	Bare Costs	Incl. O&P
3 Laborers	$18.70	$448.80	$31.75	$762.00	$20.65	$34.73
1 Equip. Oper. (med.)	26.50	212.00	43.65	349.20		
1 Crawler Loader, 4 C.Y.		1145.00		1259.50	35.78	39.36
32 L.H., Daily Totals		$1805.80		$2370.70	$56.43	$74.08
Crew B-4	Hr.	Daily	Hr.	Daily	Bare Costs	Incl. O&P
1 Labor Foreman (outside)	$20.70	$165.60	$35.15	$281.20	$19.42	$32.92
4 Laborers	18.70	598.40	31.75	1016.00		
1 Truck Driver (heavy)	21.00	168.00	35.35	282.80		
1 Truck Tractor, 195 H.P.		235.60		259.16		
1 Flatbed Trailer, 40 Ton		125.80		138.38	7.53	8.28
48 L.H., Daily Totals		$1293.40		$1977.54	$26.95	$41.20
Crew B-5	Hr.	Daily	Hr.	Daily	Bare Costs	Incl. O&P
1 Labor Foreman (outside)	$20.70	$165.60	$35.15	$281.20	$20.66	$34.81
3 Laborers	18.70	448.80	31.75	762.00		
1 Equip. Oper. (med.)	26.50	212.00	43.65	349.20		
1 Air Compressor, 250 C.F.M.		151.40		166.54		
2 Breaker, Pavement, 60 lb.		16.80		18.48		
2 -50' Air Hose, 1.5"		11.60		12.76		
1 Crawler Loader, 3 C.Y.		834.40		917.84	25.36	27.89
40 L.H., Daily Totals		$1840.60		$2508.02	$46.02	$62.70
Crew B-5A	Hr.	Daily	Hr.	Daily	Bare Costs	Incl. O&P
1 Foreman	$20.70	$165.60	$35.15	$281.20	$21.11	$35.46
6 Laborers	18.70	897.60	31.75	1524.00		
2 Equip. Oper. (med.)	26.50	424.00	43.65	698.40		
1 Equip. Oper. (light)	25.45	203.60	41.90	335.20		
2 Truck Drivers (heavy)	21.00	336.00	35.35	565.60		
1 Air Compressor, 365 C.F.M.		195.80		215.38		
2 Breaker, Pavement, 60 lb.		16.80		18.48		
8 -50' Air Hose, 1"		32.80		36.08		
2 Dump Trucks, 12 Ton, 8 C.Y.		736.80		810.48	10.23	11.25
96 L.H., Daily Totals		$3009.00		$4484.82	$31.34	$46.72

Crew No.	Bare Costs		Incl. Subs O & P		Cost Per Labor-Hour	
Crew B-5B	Hr.	Daily	Hr.	Daily	Bare Costs	Incl. O&P
1 Powderman	$26.10	$208.80	$44.25	$354.00	$23.68	$39.60
2 Equip. Oper. (med.)	26.50	424.00	43.65	698.40		
3 Truck Drivers (heavy)	21.00	504.00	35.35	848.40		
1 F.E. Loader, W.M., 2.5 C.Y.		349.00		383.90		
3 Dump Trucks, 16 Ton, 12 C.Y.		1590.00		1749.00		
1 Air Compressor, 365 C.F.M.		195.80		215.38	44.48	48.92
48 L.H., Daily Totals		$3271.60		$4249.08	$68.16	$88.52
Crew B-5C	Hr.	Daily	Hr.	Daily	Bare Costs	Incl. O&P
3 Laborers	$18.70	$448.80	$31.75	$762.00	$21.93	$36.67
1 Equip. Oper. (medium)	26.50	212.00	43.65	349.20		
2 Truck Drivers (heavy)	21.00	336.00	35.35	565.60		
1 Equip. Oper. (crane)	27.45	219.60	45.20	361.60		
1 Equip. Oper. Oiler	23.40	187.20	38.55	308.40		
2 Dump Trucks, 16 Ton, 12 C.Y.		1060.00		1166.00		
1 Crawler Loader, 4 C.Y.		1145.00		1259.50		
1 S.P. Crane, 4x4, 25 Ton		875.60		963.16	48.13	52.95
64 L.H., Daily Totals		$4484.20		$5735.46	$70.07	$89.62
Crew B-6	Hr.	Daily	Hr.	Daily	Bare Costs	Incl. O&P
2 Laborers	$18.70	$299.20	$31.75	$508.00	$20.95	$35.13
1 Equip. Oper. (light)	25.45	203.60	41.90	335.20		
1 Backhoe Loader, 48 H.P.		243.40		267.74	10.14	11.16
24 L.H., Daily Totals		$746.20		$1110.94	$31.09	$46.29
Crew B-6B	Hr.	Daily	Hr.	Daily	Bare Costs	Incl. O&P
2 Labor Foremen (out)	$20.70	$331.20	$35.15	$562.40	$19.37	$32.88
4 Laborers	18.70	598.40	31.75	1016.00		
1 S.P. Crane, 4x4, 5 Ton		326.80		359.48		
1 Flatbed Truck, Gas, 1.5 Ton		141.40		155.54		
1 Butt Fusion Machine		296.20		325.82	15.93	17.52
48 L.H., Daily Totals		$1694.00		$2419.24	$35.29	$50.40
Crew B-7	Hr.	Daily	Hr.	Daily	Bare Costs	Incl. O&P
1 Labor Foreman (outside)	$20.70	$165.60	$35.15	$281.20	$20.33	$34.30
4 Laborers	18.70	598.40	31.75	1016.00		
1 Equip. Oper. (med.)	26.50	212.00	43.65	349.20		
1 Brush Chipper, 12", 130 H.P.		183.60		201.96		
1 Crawler Loader, 3 C.Y.		834.40		917.84		
2 Chain saws gas, 36" Long		68.80		75.68	22.64	24.91
48 L.H., Daily Totals		$2062.80		$2841.88	$42.98	$59.21
Crew B-7A	Hr.	Daily	Hr.	Daily	Bare Costs	Incl. O&P
2 Laborers	$18.70	$299.20	$31.75	$508.00	$20.95	$35.13
1 Equip. Oper. (light)	25.45	203.60	41.90	335.20		
1 Rake w/Tractor		195.90		215.49		
2 Chain saws, gas, 18"		46.00		50.60	10.08	11.09
24 L.H., Daily Totals		$744.70		$1109.29	$31.03	$46.22
Crew B-8	Hr.	Daily	Hr.	Daily	Bare Costs	Incl. O&P
1 Labor Foreman (outside)	$20.70	$165.60	$35.15	$281.20	$21.87	$36.66
2 Laborers	18.70	299.20	31.75	508.00		
2 Equip. Oper. (med.)	26.50	424.00	43.65	698.40		
2 Truck Drivers (heavy)	21.00	336.00	35.35	565.60		
1 Hyd. Crane, 25 Ton		739.60		813.56		
1 Crawler Loader, 3 C.Y.		834.40		917.84		
2 Dump Trucks, 16 Ton, 12 C.Y.		1060.00		1166.00	47.04	51.74
56 L.H., Daily Totals		$3858.80		$4950.60	$68.91	$88.40

Crews

Crew No.	Bare Costs		Incl. Subs O & P		Cost Per Labor-Hour	
Crew B-9	Hr.	Daily	Hr.	Daily	Bare Costs	Incl. O&P
1 Labor Foreman (outside)	$20.70	$165.60	$35.15	$281.20	$19.10	$32.43
4 Laborers	18.70	598.40	31.75	1016.00		
1 Air Compressor, 250 C.F.M.		151.40		166.54		
2 Breaker, Pavement, 60 lb.		16.80		18.48		
2 -50' Air Hose, 1.5"		11.60		12.76	4.50	4.94
40 L.H., Daily Totals		$943.80		$1494.98	$23.59	$37.37
Crew B-9A	Hr.	Daily	Hr.	Daily	Bare Costs	Incl. O&P
2 Laborers	$18.70	$299.20	$31.75	$508.00	$19.47	$32.95
1 Truck Driver (heavy)	21.00	168.00	35.35	282.80		
1 Water Tanker, 5000 Gal.		121.20		133.32		
1 Truck Tractor, 195 H.P.		235.60		259.16		
2 -50' Discharge Hoses, 3"		3.50		3.85	15.01	16.51
24 L.H., Daily Totals		$827.50		$1187.13	$34.48	$49.46
Crew B-9B	Hr.	Daily	Hr.	Daily	Bare Costs	Incl. O&P
2 Laborers	$18.70	$299.20	$31.75	$508.00	$19.47	$32.95
1 Truck Driver (heavy)	21.00	168.00	35.35	282.80		
2 -50' Discharge Hoses, 3"		3.50		3.85		
1 Water Tanker, 5000 Gal.		121.20		133.32		
1 Truck Tractor, 195 H.P.		235.60		259.16		
1 Pressure Washer		55.80		61.38	17.34	19.07
24 L.H., Daily Totals		$883.30		$1248.51	$36.80	$52.02
Crew B-9C	Hr.	Daily	Hr.	Daily	Bare Costs	Incl. O&P
1 Labor Foreman (outside)	$20.70	$165.60	$35.15	$281.20	$19.10	$32.43
4 Laborers	18.70	598.40	31.75	1016.00		
1 Air Compressor, 250 C.F.M.		151.40		166.54		
2 -50' Air Hoses, 1.5"		11.60		12.76		
2 Breakers, Pavement, 60 lb.		16.80		18.48	4.50	4.94
40 L.H., Daily Totals		$943.80		$1494.98	$23.59	$37.37
Crew B-9D	Hr.	Daily	Hr.	Daily	Bare Costs	Incl. O&P
1 Labor Foreman (Outside)	$20.70	$165.60	$35.15	$281.20	$19.10	$32.43
4 Common Laborers	18.70	598.40	31.75	1016.00		
1 Air Compressor, 250 C.F.M.		151.40		166.54		
2 -50' Air Hoses, 1.5"		11.60		12.76		
2 Air Powered Tampers		51.80		56.98	5.37	5.91
40 L.H., Daily Totals		$978.80		$1533.48	$24.47	$38.34
Crew B-10	Hr.	Daily	Hr.	Daily	Bare Costs	Incl. O&P
1 Equip. Oper. (med.)	$26.50	$212.00	$43.65	$349.20	$26.50	$43.65
8 L.H., Daily Totals		$212.00		$349.20	$26.50	$43.65
Crew B-10A	Hr.	Daily	Hr.	Daily	Bare Costs	Incl. O&P
1 Equip. Oper. (med.)	$26.50	$212.00	$43.65	$349.20	$26.50	$43.65
1 Roller, 2-Drum, W.B., 7.5 HP		133.80		147.18	16.73	18.40
8 L.H., Daily Totals		$345.80		$496.38	$43.23	$62.05
Crew B-10B	Hr.	Daily	Hr.	Daily	Bare Costs	Incl. O&P
1 Equip. Oper. (med.)	$26.50	$212.00	$43.65	$349.20	$26.50	$43.65
1 Dozer, 200 H.P.		988.40		1087.24	123.55	135.91
8 L.H., Daily Totals		$1200.40		$1436.44	$150.05	$179.56
Crew B-10C	Hr.	Daily	Hr.	Daily	Bare Costs	Incl. O&P
1 Equip. Oper. (med.)	$26.50	$212.00	$43.65	$349.20	$26.50	$43.65
1 Dozer, 200 H.P.		988.40		1087.24		
1 Vibratory Roller, Towed, 23 Ton		350.00		385.00	167.30	184.03
8 L.H., Daily Totals		$1550.40		$1821.44	$193.80	$227.68
Crew B-10D	Hr.	Daily	Hr.	Daily	Bare Costs	Incl. O&P
1 Equip. Oper. (med.)	$26.50	$212.00	$43.65	$349.20	$26.50	$43.65
1 Dozer, 200 H.P.		988.40		1087.24		
1 Sheepsft. Roller, Towed		376.00		413.60	170.55	187.60
8 L.H., Daily Totals		$1576.40		$1850.04	$197.05	$231.26
Crew B-10E	Hr.	Daily	Hr.	Daily	Bare Costs	Incl. O&P
1 Equip. Oper. (med.)	$26.50	$212.00	$43.65	$349.20	$26.50	$43.65
1 Tandem Roller, 5 Ton		122.20		134.42	15.28	16.80
8 L.H., Daily Totals		$334.20		$483.62	$41.77	$60.45
Crew B-10F	Hr.	Daily	Hr.	Daily	Bare Costs	Incl. O&P
1 Equip. Oper. (med.)	$26.50	$212.00	$43.65	$349.20	$26.50	$43.65
1 Tandem Roller, 10 Ton		211.00		232.10	26.38	29.01
8 L.H., Daily Totals		$423.00		$581.30	$52.88	$72.66
Crew B-10G	Hr.	Daily	Hr.	Daily	Bare Costs	Incl. O&P
1 Equip. Oper. (med.)	$26.50	$212.00	$43.65	$349.20	$26.50	$43.65
1 Sheepsft. Roll., 130 H.P.		956.20		1051.82	119.53	131.48
8 L.H., Daily Totals		$1168.20		$1401.02	$146.03	$175.13
Crew B-10H	Hr.	Daily	Hr.	Daily	Bare Costs	Incl. O&P
1 Equip. Oper. (med.)	$26.50	$212.00	$43.65	$349.20	$26.50	$43.65
1 Diaphragm Water Pump, 2"		57.40		63.14		
1 -20' Suction Hose, 2"		1.95		2.15		
2 -50' Discharge Hoses, 2"		1.80		1.98	7.64	8.41
8 L.H., Daily Totals		$273.15		$416.46	$34.14	$52.06
Crew B-10I	Hr.	Daily	Hr.	Daily	Bare Costs	Incl. O&P
1 Equip. Oper. (med.)	$26.50	$212.00	$43.65	$349.20	$26.50	$43.65
1 Diaphragm Water Pump, 4"		78.20		86.02		
1 -20' Suction Hose, 4"		3.25		3.58		
2 -50' Discharge Hoses, 4"		4.70		5.17	10.77	11.85
8 L.H., Daily Totals		$298.15		$443.96	$37.27	$55.50
Crew B-10J	Hr.	Daily	Hr.	Daily	Bare Costs	Incl. O&P
1 Equip. Oper. (med.)	$26.50	$212.00	$43.65	$349.20	$26.50	$43.65
1 Centrifugal Water Pump, 3"		63.40		69.74		
1 -20' Suction Hose, 3"		3.05		3.36		
2 -50' Discharge Hoses, 3"		3.50		3.85	8.74	9.62
8 L.H., Daily Totals		$281.95		$426.14	$35.24	$53.27
Crew B-10K	Hr.	Daily	Hr.	Daily	Bare Costs	Incl. O&P
1 Equip. Oper. (med.)	$26.50	$212.00	$43.65	$349.20	$26.50	$43.65
1 Centr. Water Pump, 6"		278.20		306.02		
1 -20' Suction Hose, 6"		11.50		12.65		
2 -50' Discharge Hoses, 6"		12.60		13.86	37.79	41.57
8 L.H., Daily Totals		$514.30		$681.73	$64.29	$85.22
Crew B-10L	Hr.	Daily	Hr.	Daily	Bare Costs	Incl. O&P
1 Equip. Oper. (med.)	$26.50	$212.00	$43.65	$349.20	$26.50	$43.65
1 Dozer, 80 H.P.		354.80		390.28	44.35	48.78
8 L.H., Daily Totals		$566.80		$739.48	$70.85	$92.44
Crew B-10M	Hr.	Daily	Hr.	Daily	Bare Costs	Incl. O&P
1 Equip. Oper. (med.)	$26.50	$212.00	$43.65	$349.20	$26.50	$43.65
1 Dozer, 300 H.P.		1301.00		1431.10	162.63	178.89
8 L.H., Daily Totals		$1513.00		$1780.30	$189.13	$222.54

Crews

Crew No.	Bare Costs		Incl. Subs O & P		Cost Per Labor-Hour	
Crew B-10N	Hr.	Daily	Hr.	Daily	Bare Costs	Incl. O&P
1 Equip. Oper. (med.)	$26.50	$212.00	$43.65	$349.20	$26.50	$43.65
1 F.E. Loader, T.M., 1.5 C.Y.		334.00		367.40	41.75	45.92
8 L.H., Daily Totals		$546.00		$716.60	$68.25	$89.58
Crew B-10O	Hr.	Daily	Hr.	Daily	Bare Costs	Incl. O&P
1 Equip. Oper. (med.)	$26.50	$212.00	$43.65	$349.20	$26.50	$43.65
1 F.E. Loader, T.M., 2.25 C.Y.		589.20		648.12	73.65	81.02
8 L.H., Daily Totals		$801.20		$997.32	$100.15	$124.67
Crew B-10P	Hr.	Daily	Hr.	Daily	Bare Costs	Incl. O&P
1 Equip. Oper. (med.)	$26.50	$212.00	$43.65	$349.20	$26.50	$43.65
1 Crawler Loader, 3 C.Y.		834.40		917.84	104.30	114.73
8 L.H., Daily Totals		$1046.40		$1267.04	$130.80	$158.38
Crew B-10Q	Hr.	Daily	Hr.	Daily	Bare Costs	Incl. O&P
1 Equip. Oper. (med.)	$26.50	$212.00	$43.65	$349.20	$26.50	$43.65
1 Crawler Loader, 4 C.Y.		1145.00		1259.50	143.13	157.44
8 L.H., Daily Totals		$1357.00		$1608.70	$169.63	$201.09
Crew B-10R	Hr.	Daily	Hr.	Daily	Bare Costs	Incl. O&P
1 Equip. Oper. (med.)	$26.50	$212.00	$43.65	$349.20	$26.50	$43.65
1 F.E. Loader, W.M., 1 C.Y.		207.60		228.36	25.95	28.55
8 L.H., Daily Totals		$419.60		$577.56	$52.45	$72.19
Crew B-10S	Hr.	Daily	Hr.	Daily	Bare Costs	Incl. O&P
1 Equip. Oper. (med.)	$26.50	$212.00	$43.65	$349.20	$26.50	$43.65
1 F.E. Loader, W.M., 1.5 C.Y.		277.60		305.36	34.70	38.17
8 L.H., Daily Totals		$489.60		$654.56	$61.20	$81.82
Crew B-10T	Hr.	Daily	Hr.	Daily	Bare Costs	Incl. O&P
1 Equip. Oper. (med.)	$26.50	$212.00	$43.65	$349.20	$26.50	$43.65
1 F.E. Loader, W.M., 2.5 C.Y.		349.00		383.90	43.63	47.99
8 L.H., Daily Totals		$561.00		$733.10	$70.13	$91.64
Crew B-10U	Hr.	Daily	Hr.	Daily	Bare Costs	Incl. O&P
1 Equip. Oper. (med.)	$26.50	$212.00	$43.65	$349.20	$26.50	$43.65
1 F.E. Loader, W.M., 5.5 C.Y.		755.80		831.38	94.47	103.92
8 L.H., Daily Totals		$967.80		$1180.58	$120.97	$147.57
Crew B-10V	Hr.	Daily	Hr.	Daily	Bare Costs	Incl. O&P
1 Equip. Oper. (med.)	$26.50	$212.00	$43.65	$349.20	$26.50	$43.65
1 Dozer, 700 H.P.		3406.00		3746.60	425.75	468.32
8 L.H., Daily Totals		$3618.00		$4095.80	$452.25	$511.98
Crew B-10W	Hr.	Daily	Hr.	Daily	Bare Costs	Incl. O&P
1 Equip. Oper. (med.)	$26.50	$212.00	$43.65	$349.20	$26.50	$43.65
1 Dozer, 105 H.P.		494.80		544.28	61.85	68.03
8 L.H., Daily Totals		$706.80		$893.48	$88.35	$111.69
Crew B-10X	Hr.	Daily	Hr.	Daily	Bare Costs	Incl. O&P
1 Equip. Oper. (med.)	$26.50	$212.00	$43.65	$349.20	$26.50	$43.65
1 Dozer, 410 H.P.		1677.00		1844.70	209.63	230.59
8 L.H., Daily Totals		$1889.00		$2193.90	$236.13	$274.24
Crew B-10Y	Hr.	Daily	Hr.	Daily	Bare Costs	Incl. O&P
1 Equip. Oper. (med.)	$26.50	$212.00	$43.65	$349.20	$26.50	$43.65
1 Vibr. Roller, Towed, 12 Ton		448.40		493.24	56.05	61.66
8 L.H., Daily Totals		$660.40		$842.44	$82.55	$105.31

Crew No.	Bare Costs		Incl. Subs O & P		Cost Per Labor-Hour	
Crew B-11A	Hr.	Daily	Hr.	Daily	Bare Costs	Incl. O&P
1 Equipment Oper. (med.)	$26.50	$212.00	$43.65	$349.20	$22.60	$37.70
1 Laborer	18.70	149.60	31.75	254.00		
1 Dozer, 200 H.P.		988.40		1087.24	61.77	67.95
16 L.H., Daily Totals		$1350.00		$1690.44	$84.38	$105.65
Crew B-11B	Hr.	Daily	Hr.	Daily	Bare Costs	Incl. O&P
1 Equipment Oper. (light)	$25.45	$203.60	$41.90	$335.20	$22.07	$36.83
1 Laborer	18.70	149.60	31.75	254.00		
1 Air Powered Tamper		25.90		28.49		
1 Air Compressor, 365 C.F.M.		195.80		215.38		
2 -50' Air Hoses, 1.5"		11.60		12.76	14.58	16.04
16 L.H., Daily Totals		$586.50		$845.83	$36.66	$52.86
Crew B-11C	Hr.	Daily	Hr.	Daily	Bare Costs	Incl. O&P
1 Equipment Oper. (med.)	$26.50	$212.00	$43.65	$349.20	$22.60	$37.70
1 Laborer	18.70	149.60	31.75	254.00		
1 Backhoe Loader, 48 H.P.		243.40		267.74	15.21	16.73
16 L.H., Daily Totals		$605.00		$870.94	$37.81	$54.43
Crew B-11K	Hr.	Daily	Hr.	Daily	Bare Costs	Incl. O&P
1 Equipment Oper. (med.)	$26.50	$212.00	$43.65	$349.20	$22.60	$37.70
1 Laborer	18.70	149.60	31.75	254.00		
1 Trencher, Chain Type, 8' D		1606.00		1766.60	100.38	110.41
16 L.H., Daily Totals		$1967.60		$2369.80	$122.97	$148.11
Crew B-11L	Hr.	Daily	Hr.	Daily	Bare Costs	Incl. O&P
1 Equipment Oper. (med.)	$26.50	$212.00	$43.65	$349.20	$22.60	$37.70
1 Laborer	18.70	149.60	31.75	254.00		
1 Grader, 30,000 Lbs.		505.40		555.94	31.59	34.75
16 L.H., Daily Totals		$867.00		$1159.14	$54.19	$72.45
Crew B-11M	Hr.	Daily	Hr.	Daily	Bare Costs	Incl. O&P
1 Equipment Oper. (med.)	$26.50	$212.00	$43.65	$349.20	$22.60	$37.70
1 Laborer	18.70	149.60	31.75	254.00		
1 Backhoe Loader, 80 H.P.		285.40		313.94	17.84	19.62
16 L.H., Daily Totals		$647.00		$917.14	$40.44	$57.32
Crew B-11W	Hr.	Daily	Hr.	Daily	Bare Costs	Incl. O&P
1 Equipment Operator (med.)	$26.50	$212.00	$43.65	$349.20	$21.27	$35.74
1 Common Laborer	18.70	149.60	31.75	254.00		
10 Truck Drivers (hvy.)	21.00	1680.00	35.35	2828.00		
1 Dozer, 200 H.P.		988.40		1087.24		
1 Vibratory Roller, Towed, 23 Ton		350.00		385.00		
10 Dump Trucks, 12 Ton, 8 C.Y.		3684.00		4052.40	52.32	57.55
96 L.H., Daily Totals		$7064.00		$8955.84	$73.58	$93.29
Crew B-11Y	Hr.	Daily	Hr.	Daily	Bare Costs	Incl. O&P
1 Labor Foreman (Outside)	$20.70	$165.60	$35.15	$281.20	$21.52	$36.09
5 Common Laborers	18.70	748.00	31.75	1270.00		
3 Equipment Operators (med.)	26.50	636.00	43.65	1047.60		
1 Dozer, 80 H.P.		354.80		390.28		
2 Roller, 2-Drum, W.B., 7.5 HP		267.60		294.36		
4 Vibratory Plates, Gas, 21"		144.80		159.28	10.66	11.72
72 L.H., Daily Totals		$2316.80		$3442.72	$32.18	$47.82
Crew B-12A	Hr.	Daily	Hr.	Daily	Bare Costs	Incl. O&P
1 Equip. Oper. (crane)	$27.45	$219.60	$45.20	$361.60	$23.07	$38.48
1 Laborer	18.70	149.60	31.75	254.00		
1 Hyd. Excavator, 1 C.Y.		601.80		661.98	37.61	41.37
16 L.H., Daily Totals		$971.00		$1277.58	$60.69	$79.85

Crews

Crew No.	Bare Costs		Incl. Subs O & P		Cost Per Labor-Hour	
Crew B-12B	Hr.	Daily	Hr.	Daily	Bare Costs	Incl. O&P
1 Equip. Oper. (crane)	$27.45	$219.60	$45.20	$361.60	$23.07	$38.48
1 Laborer	18.70	149.60	31.75	254.00		
1 Hyd. Excavator, 1.5 C.Y.		775.60		853.16	48.48	53.32
16 L.H., Daily Totals		$1144.80		$1468.76	$71.55	$91.80
Crew B-12C	Hr.	Daily	Hr.	Daily	Bare Costs	Incl. O&P
1 Equip. Oper. (crane)	$27.45	$219.60	$45.20	$361.60	$23.07	$38.48
1 Laborer	18.70	149.60	31.75	254.00		
1 Hyd. Excavator, 2 C.Y.		994.60		1094.06	62.16	68.38
16 L.H., Daily Totals		$1363.80		$1709.66	$85.24	$106.85
Crew B-12D	Hr.	Daily	Hr.	Daily	Bare Costs	Incl. O&P
1 Equip. Oper. (crane)	$27.45	$219.60	$45.20	$361.60	$23.07	$38.48
1 Laborer	18.70	149.60	31.75	254.00		
1 Hyd. Excavator, 3.5 C.Y.		2166.00		2382.60	135.38	148.91
16 L.H., Daily Totals		$2535.20		$2998.20	$158.45	$187.39
Crew B-12E	Hr.	Daily	Hr.	Daily	Bare Costs	Incl. O&P
1 Equip. Oper. (crane)	$27.45	$219.60	$45.20	$361.60	$23.07	$38.48
1 Laborer	18.70	149.60	31.75	254.00		
1 Hyd. Excavator, .5 C.Y.		358.00		393.80	22.38	24.61
16 L.H., Daily Totals		$727.20		$1009.40	$45.45	$63.09
Crew B-12F	Hr.	Daily	Hr.	Daily	Bare Costs	Incl. O&P
1 Equip. Oper. (crane)	$27.45	$219.60	$45.20	$361.60	$23.07	$38.48
1 Laborer	18.70	149.60	31.75	254.00		
1 Hyd. Excavator, .75 C.Y.		516.40		568.04	32.27	35.50
16 L.H., Daily Totals		$885.60		$1183.64	$55.35	$73.98
Crew B-12G	Hr.	Daily	Hr.	Daily	Bare Costs	Incl. O&P
1 Equip. Oper. (crane)	$27.45	$219.60	$45.20	$361.60	$23.07	$38.48
1 Laborer	18.70	149.60	31.75	254.00		
1 Crawler Crane, 15 Ton		576.65		634.32		
1 Clamshell Bucket, .5 C.Y.		34.20		37.62	38.18	42.00
16 L.H., Daily Totals		$980.05		$1287.54	$61.25	$80.47
Crew B-12H	Hr.	Daily	Hr.	Daily	Bare Costs	Incl. O&P
1 Equip. Oper. (crane)	$27.45	$219.60	$45.20	$361.60	$23.07	$38.48
1 Laborer	18.70	149.60	31.75	254.00		
1 Crawler Crane, 25 Ton		983.00		1081.30		
1 Clamshell Bucket, 1 C.Y.		44.80		49.28	64.24	70.66
16 L.H., Daily Totals		$1397.00		$1746.18	$87.31	$109.14
Crew B-12I	Hr.	Daily	Hr.	Daily	Bare Costs	Incl. O&P
1 Equip. Oper. (crane)	$27.45	$219.60	$45.20	$361.60	$23.07	$38.48
1 Laborer	18.70	149.60	31.75	254.00		
1 Crawler Crane, 20 Ton		737.50		811.25		
1 Dragline Bucket, .75 C.Y.		19.40		21.34	47.31	52.04
16 L.H., Daily Totals		$1126.10		$1448.19	$70.38	$90.51
Crew B-12J	Hr.	Daily	Hr.	Daily	Bare Costs	Incl. O&P
1 Equip. Oper. (crane)	$27.45	$219.60	$45.20	$361.60	$23.07	$38.48
1 Laborer	18.70	149.60	31.75	254.00		
1 Gradall, 5/8 C.Y.		905.80		996.38	56.61	62.27
16 L.H., Daily Totals		$1275.00		$1611.98	$79.69	$100.75

Crew No.	Bare Costs		Incl. Subs O & P		Cost Per Labor-Hour	
Crew B-12K	Hr.	Daily	Hr.	Daily	Bare Costs	Incl. O&P
1 Equip. Oper. (crane)	$27.45	$219.60	$45.20	$361.60	$23.07	$38.48
1 Laborer	18.70	149.60	31.75	254.00		
1 Gradall, 3 Ton, 1 C.Y.		1026.00		1128.60	64.13	70.54
16 L.H., Daily Totals		$1395.20		$1744.20	$87.20	$109.01
Crew B-12L	Hr.	Daily	Hr.	Daily	Bare Costs	Incl. O&P
1 Equip. Oper. (crane)	$27.45	$219.60	$45.20	$361.60	$23.07	$38.48
1 Laborer	18.70	149.60	31.75	254.00		
1 Crawler Crane, 15 Ton		576.65		634.32		
1 F.E. Attachment, .5 C.Y.		49.20		54.12	39.12	43.03
16 L.H., Daily Totals		$995.05		$1304.04	$62.19	$81.50
Crew B-12M	Hr.	Daily	Hr.	Daily	Bare Costs	Incl. O&P
1 Equip. Oper. (crane)	$27.45	$219.60	$45.20	$361.60	$23.07	$38.48
1 Laborer	18.70	149.60	31.75	254.00		
1 Crawler Crane, 20 Ton		737.50		811.25		
1 F.E. Attachment, .75 C.Y.		53.60		58.96	49.44	54.39
16 L.H., Daily Totals		$1160.30		$1485.81	$72.52	$92.86
Crew B-12N	Hr.	Daily	Hr.	Daily	Bare Costs	Incl. O&P
1 Equip. Oper. (crane)	$27.45	$219.60	$45.20	$361.60	$23.07	$38.48
1 Laborer	18.70	149.60	31.75	254.00		
1 Crawler Crane, 25 Ton		983.00		1081.30		
1 F.E. Attachment, 1 C.Y.		60.40		66.44	65.21	71.73
16 L.H., Daily Totals		$1412.60		$1763.34	$88.29	$110.21
Crew B-12O	Hr.	Daily	Hr.	Daily	Bare Costs	Incl. O&P
1 Equip. Oper. (crane)	$27.45	$219.60	$45.20	$361.60	$23.07	$38.48
1 Laborer	18.70	149.60	31.75	254.00		
1 Crawler Crane, 40 Ton		994.60		1094.06		
1 F.E. Attachment, 1.5 C.Y.		69.60		76.56	66.51	73.16
16 L.H., Daily Totals		$1433.40		$1786.22	$89.59	$111.64
Crew B-12P	Hr.	Daily	Hr.	Daily	Bare Costs	Incl. O&P
1 Equip. Oper. (crane)	$27.45	$219.60	$45.20	$361.60	$23.07	$38.48
1 Laborer	18.70	149.60	31.75	254.00		
1 Crawler Crane, 40 Ton		994.60		1094.06		
1 Dragline Bucket, 1.5 C.Y.		31.00		34.10	64.10	70.51
16 L.H., Daily Totals		$1394.80		$1743.76	$87.17	$108.99
Crew B-12Q	Hr.	Daily	Hr.	Daily	Bare Costs	Incl. O&P
1 Equip. Oper. (crane)	$27.45	$219.60	$45.20	$361.60	$23.07	$38.48
1 Laborer	18.70	149.60	31.75	254.00		
1 Hyd. Excavator, 5/8 C.Y.		462.20		508.42	28.89	31.78
16 L.H., Daily Totals		$831.40		$1124.02	$51.96	$70.25
Crew B-12R	Hr.	Daily	Hr.	Daily	Bare Costs	Incl. O&P
1 Equip. Oper. (crane)	$27.45	$219.60	$45.20	$361.60	$23.07	$38.48
1 Laborer	18.70	149.60	31.75	254.00		
1 Hyd. Excavator, 1.5 C.Y.		775.60		853.16	48.48	53.32
16 L.H., Daily Totals		$1144.80		$1468.76	$71.55	$91.80
Crew B-12S	Hr.	Daily	Hr.	Daily	Bare Costs	Incl. O&P
1 Equip. Oper. (crane)	$27.45	$219.60	$45.20	$361.60	$23.07	$38.48
1 Laborer	18.70	149.60	31.75	254.00		
1 Hyd. Excavator, 2.5 C.Y.		1333.00		1466.30	83.31	91.64
16 L.H., Daily Totals		$1702.20		$2081.90	$106.39	$130.12

Crews

Crew No.	Bare Costs		Incl. Subs O & P		Cost Per Labor-Hour	
Crew B-12T	Hr.	Daily	Hr.	Daily	Bare Costs	Incl. O&P
1 Equip. Oper. (crane)	$27.45	$219.60	$45.20	$361.60	$23.07	$38.48
1 Laborer	18.70	149.60	31.75	254.00		
1 Crawler Crane, 75 Ton		1384.00		1522.40		
1 F.E. Attachment, 3 C.Y.		93.00		102.30	92.31	101.54
16 L.H., Daily Totals		$1846.20		$2240.30	$115.39	$140.02
Crew B-12V	Hr.	Daily	Hr.	Daily	Bare Costs	Incl. O&P
1 Equip. Oper. (crane)	$27.45	$219.60	$45.20	$361.60	$23.07	$38.48
1 Laborer	18.70	149.60	31.75	254.00		
1 Crawler Crane, 75 Ton		1384.00		1522.40		
1 Dragline Bucket, 3 C.Y.		48.80		53.68	89.55	98.50
16 L.H., Daily Totals		$1802.00		$2191.68	$112.63	$136.98
Crew B-13	Hr.	Daily	Hr.	Daily	Bare Costs	Incl. O&P
1 Labor Foreman (outside)	$20.70	$165.60	$35.15	$281.20	$20.49	$34.56
4 Laborers	18.70	598.40	31.75	1016.00		
1 Equip. Oper. (crane)	27.45	219.60	45.20	361.60		
1 Hyd. Crane, 25 Ton		739.60		813.56	15.41	16.95
48 L.H., Daily Totals		$1723.20		$2472.36	$35.90	$51.51
Crew B-13A	Hr.	Daily	Hr.	Daily	Bare Costs	Incl. O&P
1 Foreman	$20.70	$165.60	$35.15	$281.20	$21.87	$36.66
2 Laborers	18.70	299.20	31.75	508.00		
2 Equipment Operators	26.50	424.00	43.65	698.40		
2 Truck Drivers (heavy)	21.00	336.00	35.35	565.60		
1 Crawler Crane, 75 Ton		1384.00		1522.40		
1 Crawler Loader, 4 C.Y.		1145.00		1259.50		
2 Dump Trucks, 12 Ton, 8 C.Y.		736.80		810.48	58.32	64.15
56 L.H., Daily Totals		$4490.60		$5645.58	$80.19	$100.81
Crew B-13B	Hr.	Daily	Hr.	Daily	Bare Costs	Incl. O&P
1 Labor Foreman (outside)	$20.70	$165.60	$35.15	$281.20	$20.91	$35.13
4 Laborers	18.70	598.40	31.75	1016.00		
1 Equip. Oper. (crane)	27.45	219.60	45.20	361.60		
1 Equip. Oper. Oiler	23.40	187.20	38.55	308.40		
1 Hyd. Crane, 55 Ton		1060.00		1166.00	18.93	20.82
56 L.H., Daily Totals		$2230.80		$3133.20	$39.84	$55.95
Crew B-13C	Hr.	Daily	Hr.	Daily	Bare Costs	Incl. O&P
1 Labor Foreman (outside)	$20.70	$165.60	$35.15	$281.20	$20.91	$35.13
4 Laborers	18.70	598.40	31.75	1016.00		
1 Equip. Oper. (crane)	27.45	219.60	45.20	361.60		
1 Equip. Oper. Oiler	23.40	187.20	38.55	308.40		
1 Crawler Crane, 100 Ton		1717.00		1888.70	30.66	33.73
56 L.H., Daily Totals		$2887.80		$3855.90	$51.57	$68.86
Crew B-13D	Hr.	Daily	Hr.	Daily	Bare Costs	Incl. O&P
1 Laborer	$18.70	$149.60	$31.75	$254.00	$23.07	$38.48
1 Equip. Oper. (crane)	27.45	219.60	45.20	361.60		
1 Hyd. Excavator, 1 C.Y.		601.80		661.98		
1 Trench Box		79.25		87.17	42.57	46.82
16 L.H., Daily Totals		$1050.25		$1364.76	$65.64	$85.30
Crew B-13E	Hr.	Daily	Hr.	Daily	Bare Costs	Incl. O&P
1 Laborer	$18.70	$149.60	$31.75	$254.00	$23.07	$38.48
1 Equip. Oper. (crane)	27.45	219.60	45.20	361.60		
1 Hyd. Excavator, 1.5 C.Y.		775.60		853.16		
1 Trench Box		79.25		87.17	53.43	58.77
16 L.H., Daily Totals		$1224.05		$1555.93	$76.50	$97.25

Crew No.	Bare Costs		Incl. Subs O & P		Cost Per Labor-Hour	
Crew B-13F	Hr.	Daily	Hr.	Daily	Bare Costs	Incl. O&P
1 Laborer	$18.70	$149.60	$31.75	$254.00	$23.07	$38.48
1 Equip. Oper. (crane)	27.45	219.60	45.20	361.60		
1 Hyd. Excavator, 3.5 C.Y.		2166.00		2382.60		
1 Trench Box		79.25		87.17	140.33	154.36
16 L.H., Daily Totals		$2614.45		$3085.38	$163.40	$192.84
Crew B-13G	Hr.	Daily	Hr.	Daily	Bare Costs	Incl. O&P
1 Laborer	$18.70	$149.60	$31.75	$254.00	$23.07	$38.48
1 Equip. Oper. (crane)	27.45	219.60	45.20	361.60		
1 Hyd. Excavator, .75 C.Y.		516.40		568.04		
1 Trench Box		79.25		87.17	37.23	40.95
16 L.H., Daily Totals		$964.85		$1270.82	$60.30	$79.43
Crew B-13H	Hr.	Daily	Hr.	Daily	Bare Costs	Incl. O&P
1 Laborer	$18.70	$149.60	$31.75	$254.00	$23.07	$38.48
1 Equip. Oper. (crane)	27.45	219.60	45.20	361.60		
1 Gradall, 5/8 C.Y.		905.80		996.38		
1 Trench Box		79.25		87.17	61.57	67.72
16 L.H., Daily Totals		$1354.25		$1699.16	$84.64	$106.20
Crew B-13I	Hr.	Daily	Hr.	Daily	Bare Costs	Incl. O&P
1 Laborer	$18.70	$149.60	$31.75	$254.00	$23.07	$38.48
1 Equip. Oper. (crane)	27.45	219.60	45.20	361.60		
1 Gradall, 3 Ton, 1 C.Y.		1026.00		1128.60		
1 Trench Box		79.25		87.17	69.08	75.99
16 L.H., Daily Totals		$1474.45		$1831.38	$92.15	$114.46
Crew B-13J	Hr.	Daily	Hr.	Daily	Bare Costs	Incl. O&P
1 Laborer	$18.70	$149.60	$31.75	$254.00	$23.07	$38.48
1 Equip. Oper. (crane)	27.45	219.60	45.20	361.60		
1 Hyd. Excavator, 2.5 C.Y.		1333.00		1466.30		
1 Trench Box		79.25		87.17	88.27	97.09
16 L.H., Daily Totals		$1781.45		$2169.07	$111.34	$135.57
Crew B-14	Hr.	Daily	Hr.	Daily	Bare Costs	Incl. O&P
1 Labor Foreman (outside)	$20.70	$165.60	$35.15	$281.20	$20.16	$34.01
4 Laborers	18.70	598.40	31.75	1016.00		
1 Equip. Oper. (light)	25.45	203.60	41.90	335.20		
1 Backhoe Loader, 48 H.P.		243.40		267.74	5.07	5.58
48 L.H., Daily Totals		$1211.00		$1900.14	$25.23	$39.59
Crew B-15	Hr.	Daily	Hr.	Daily	Bare Costs	Incl. O&P
1 Equipment Oper. (med)	$26.50	$212.00	$43.65	$349.20	$22.24	$37.21
.5 Laborer	18.70	74.80	31.75	127.00		
2 Truck Drivers (heavy)	21.00	336.00	35.35	565.60		
2 Dump Trucks, 16 Ton, 12 C.Y.		1060.00		1166.00		
1 Dozer, 200 H.P.		988.40		1087.24	73.16	80.47
28 L.H., Daily Totals		$2671.20		$3295.04	$95.40	$117.68
Crew B-16	Hr.	Daily	Hr.	Daily	Bare Costs	Incl. O&P
1 Labor Foreman (outside)	$20.70	$165.60	$35.15	$281.20	$19.77	$33.50
2 Laborers	18.70	299.20	31.75	508.00		
1 Truck Driver (heavy)	21.00	168.00	35.35	282.80		
1 Dump Truck, 16 Ton, 12 C.Y.		530.00		583.00	16.56	18.22
32 L.H., Daily Totals		$1162.80		$1655.00	$36.34	$51.72

Crews

Crew No.	Bare Costs		Incl. Subs O & P		Cost Per Labor-Hour	
Crew B-17	Hr.	Daily	Hr.	Daily	Bare Costs	Incl. O&P
2 Laborers	$18.70	$299.20	$31.75	$508.00	$20.96	$35.19
1 Equip. Oper. (light)	25.45	203.60	41.90	335.20		
1 Truck Driver (heavy)	21.00	168.00	35.35	282.80		
1 Backhoe Loader, 48 H.P.		243.40		267.74		
1 Dump Truck, 12 Ton, 8 C.Y.		368.40		405.24	19.12	21.03
32 L.H., Daily Totals		$1282.60		$1798.98	$40.08	$56.22
Crew B-17A	Hr.	Daily	Hr.	Daily	Bare Costs	Incl. O&P
2 Laborer Foremen	$20.70	$331.20	$35.15	$562.40	$20.78	$35.27
6 Laborers	18.70	897.60	31.75	1524.00		
1 Skilled Worker Foreman	28.10	224.80	47.65	381.20		
1 Skilled Worker	26.10	208.80	44.25	354.00		
80 L.H., Daily Totals		$1662.40		$2821.60	$20.78	$35.27
Crew B-18	Hr.	Daily	Hr.	Daily	Bare Costs	Incl. O&P
1 Labor Foreman (outside)	$20.70	$165.60	$35.15	$281.20	$19.37	$32.88
2 Laborers	18.70	299.20	31.75	508.00		
1 Vibratory Plate, Gas, 21"		36.20		39.82	1.51	1.66
24 L.H., Daily Totals		$501.00		$829.02	$20.88	$34.54
Crew B-19	Hr.	Daily	Hr.	Daily	Bare Costs	Incl. O&P
1 Pile Driver Foreman	$27.15	$217.20	$48.25	$386.00	$24.84	$43.43
4 Pile Drivers	25.15	804.80	44.70	1430.40		
1 Equip. Oper. (crane)	27.45	219.60	45.20	361.60		
1 Building Laborer	18.70	149.60	31.75	254.00		
1 Crawler Crane, 40 Ton		994.60		1094.06		
60 L.F. of Leads, 15K Ft. Lbs.		60.00		66.00		
1 Hammer, Diesel, 22k Ft-Lb		595.20		654.72	29.46	32.41
56 L.H., Daily Totals		$3041.00		$4246.78	$54.30	$75.84
Crew B-19A	Hr.	Daily	Hr.	Daily	Bare Costs	Incl. O&P
1 Pile Driver Foreman	$27.15	$217.20	$48.25	$386.00	$25.76	$44.50
4 Pile Drivers	25.15	804.80	44.70	1430.40		
2 Equip. Oper. (crane)	27.45	439.20	45.20	723.20		
1 Equip. Oper. Oiler	23.40	187.20	38.55	308.40		
1 Crawler Crane, 75 Ton		1384.00		1522.40		
60 L.F. Leads, 25K Ft. Lbs.		108.00		118.80		
1 Hammer, Diesel, 41k Ft-Lb		715.60		787.16	34.49	37.94
64 L.H., Daily Totals		$3856.00		$5276.36	$60.25	$82.44
Crew B-20	Hr.	Daily	Hr.	Daily	Bare Costs	Incl. O&P
1 Labor Foreman (out)	$20.70	$165.60	$35.15	$281.20	$19.37	$32.88
2 Laborers	18.70	299.20	31.75	508.00		
24 L.H., Daily Totals		$464.80		$789.20	$19.37	$32.88
Crew B-20A	Hr.	Daily	Hr.	Daily	Bare Costs	Incl. O&P
1 Labor Foreman	$20.70	$165.60	$35.15	$281.20	$23.15	$38.60
1 Laborer	18.70	149.60	31.75	254.00		
1 Plumber	29.55	236.40	48.60	388.80		
1 Plumber Apprentice	23.65	189.20	38.90	311.20		
32 L.H., Daily Totals		$740.80		$1235.20	$23.15	$38.60
Crew B-21	Hr.	Daily	Hr.	Daily	Bare Costs	Incl. O&P
1 Labor Foreman (out)	$20.70	$165.60	$35.15	$281.20	$20.52	$34.64
2 Laborers	18.70	299.20	31.75	508.00		
.5 Equip. Oper. (crane)	27.45	109.80	45.20	180.80		
.5 S.P. Crane, 4x4, 5 Ton		163.40		179.74	5.84	6.42
28 L.H., Daily Totals		$738.00		$1149.74	$26.36	$41.06
Crew B-21A	Hr.	Daily	Hr.	Daily	Bare Costs	Incl. O&P
1 Labor Foreman	$20.70	$165.60	$35.15	$281.20	$24.01	$39.92
1 Laborer	18.70	149.60	31.75	254.00		
1 Plumber	29.55	236.40	48.60	388.80		
1 Plumber Apprentice	23.65	189.20	38.90	311.20		
1 Equip. Oper. (crane)	27.45	219.60	45.20	361.60		
1 S.P. Crane, 4x4, 12 Ton		575.00		632.50	14.38	15.81
40 L.H., Daily Totals		$1535.40		$2229.30	$38.38	$55.73
Crew B-21B	Hr.	Daily	Hr.	Daily	Bare Costs	Incl. O&P
1 Laborer Foreman	$20.70	$165.60	$35.15	$281.20	$20.85	$35.12
3 Laborers	18.70	448.80	31.75	762.00		
1 Equip. Oper. (crane)	27.45	219.60	45.20	361.60		
1 Hyd. Crane, 12 Ton		724.00		796.40	18.10	19.91
40 L.H., Daily Totals		$1558.00		$2201.20	$38.95	$55.03
Crew B-21C	Hr.	Daily	Hr.	Daily	Bare Costs	Incl. O&P
1 Laborer Foreman	$20.70	$165.60	$35.15	$281.20	$20.91	$35.13
4 Laborers	18.70	598.40	31.75	1016.00		
1 Equip. Oper. (crane)	27.45	219.60	45.20	361.60		
1 Equip. Oper. Oiler	23.40	187.20	38.55	308.40		
2 Cutting Torches		34.00		37.40		
2 Gases		129.60		142.56		
1 Lattice Boom Crane, 90 Ton		1547.00		1701.70	30.55	33.60
56 L.H., Daily Totals		$2881.40		$3848.86	$51.45	$68.73
Crew B-22	Hr.	Daily	Hr.	Daily	Bare Costs	Incl. O&P
1 Labor Foreman (out)	$20.70	$165.60	$35.15	$281.20	$20.98	$35.35
2 Laborers	18.70	299.20	31.75	508.00		
.75 Equip. Oper. (crane)	27.45	164.70	45.20	271.20		
.75 S.P. Crane, 4x4, 5 Ton		245.10		269.61	8.17	8.99
30 L.H., Daily Totals		$874.60		$1330.01	$29.15	$44.33
Crew B-22A	Hr.	Daily	Hr.	Daily	Bare Costs	Incl. O&P
1 Labor Foreman (out)	$20.70	$165.60	$35.15	$281.20	$22.06	$37.22
1 Skilled Worker	26.10	208.80	44.25	354.00		
2 Laborers	18.70	299.20	31.75	508.00		
.75 Equipment Oper. (crane)	27.45	164.70	45.20	271.20		
.75 S.P. Crane, 4x4, 5 Ton		245.10		269.61		
1 Generator, 5 KW		35.40		38.94		
1 Butt Fusion Machine		296.20		325.82	15.18	16.69
38 L.H., Daily Totals		$1415.00		$2048.77	$37.24	$53.91
Crew B-22B	Hr.	Daily	Hr.	Daily	Bare Costs	Incl. O&P
1 Skilled Worker	$26.10	$208.80	$44.25	$354.00	$22.40	$38.00
1 Laborer	18.70	149.60	31.75	254.00		
1 Electro Fusion Machine		174.20		191.62	10.89	11.98
16 L.H., Daily Totals		$532.60		$799.62	$33.29	$49.98
Crew B-23	Hr.	Daily	Hr.	Daily	Bare Costs	Incl. O&P
1 Labor Foreman (outside)	$20.70	$165.60	$35.15	$281.20	$19.10	$32.43
4 Laborers	18.70	598.40	31.75	1016.00		
1 Drill Rig, Truck-Mounted		3499.00		3848.90		
1 Flatbed Truck, gas, 3 Ton		185.20		203.72	92.11	101.32
40 L.H., Daily Totals		$4448.20		$5349.82	$111.21	$133.75

Crews

Crew No.	Bare Costs		Incl. Subs O & P		Cost Per Labor-Hour	
Crew B-23A	Hr.	Daily	Hr.	Daily	Bare Costs	Incl. O&P
1 Labor Foreman (outside)	$20.70	$165.60	$35.15	$281.20	$21.97	$36.85
1 Laborer	18.70	149.60	31.75	254.00		
1 Equip. Operator (medium)	26.50	212.00	43.65	349.20		
1 Drill Rig, Truck-Mounted		3499.00		3848.90		
1 Pickup Truck, 3/4 Ton		84.80		93.28	149.32	164.26
24 L.H., Daily Totals		$4111.00		$4826.58	$171.29	$201.11
Crew B-23B	Hr.	Daily	Hr.	Daily	Bare Costs	Incl. O&P
1 Labor Foreman (outside)	$20.70	$165.60	$35.15	$281.20	$21.97	$36.85
1 Laborer	18.70	149.60	31.75	254.00		
1 Equip. Operator (medium)	26.50	212.00	43.65	349.20		
1 Drill Rig, Truck-Mounted		3499.00		3848.90		
1 Pickup Truck, 3/4 Ton		84.80		93.28		
1 Centr. Water Pump, 6"		278.20		306.02	160.92	177.01
24 L.H., Daily Totals		$4389.20		$5132.60	$182.88	$213.86
Crew B-24	Hr.	Daily	Hr.	Daily	Bare Costs	Incl. O&P
1 Cement Finisher	$24.90	$199.20	$40.05	$320.40	$23.10	$38.47
1 Laborer	18.70	149.60	31.75	254.00		
1 Carpenter	25.70	205.60	43.60	348.80		
24 L.H., Daily Totals		$554.40		$923.20	$23.10	$38.47
Crew B-25	Hr.	Daily	Hr.	Daily	Bare Costs	Incl. O&P
1 Labor Foreman	$20.70	$165.60	$35.15	$281.20	$21.01	$35.30
7 Laborers	18.70	1047.20	31.75	1778.00		
3 Equip. Oper. (med.)	26.50	636.00	43.65	1047.60		
1 Asphalt Paver, 130 H.P.		1750.00		1925.00		
1 Tandem Roller, 10 Ton		211.00		232.10		
1 Roller, Pneum. Whl, 12 Ton		295.60		325.16	25.64	28.21
88 L.H., Daily Totals		$4105.40		$5589.06	$46.65	$63.51
Crew B-25B	Hr.	Daily	Hr.	Daily	Bare Costs	Incl. O&P
1 Labor Foreman	$20.70	$165.60	$35.15	$281.20	$21.47	$36.00
7 Laborers	18.70	1047.20	31.75	1778.00		
4 Equip. Oper. (medium)	26.50	848.00	43.65	1396.80		
1 Asphalt Paver, 130 H.P.		1750.00		1925.00		
2 Tandem Rollers, 10 Ton		422.00		464.20		
1 Roller, Pneum. Whl, 12 Ton		295.60		325.16	25.70	28.27
96 L.H., Daily Totals		$4528.40		$6170.36	$47.17	$64.27
Crew B-25C	Hr.	Daily	Hr.	Daily	Bare Costs	Incl. O&P
1 Labor Foreman	$20.70	$165.60	$35.15	$281.20	$21.63	$36.28
3 Laborers	18.70	448.80	31.75	762.00		
2 Equip. Oper. (medium)	26.50	424.00	43.65	698.40		
1 Asphalt Paver, 130 H.P.		1750.00		1925.00		
1 Tandem Roller, 10 Ton		211.00		232.10	40.85	44.94
48 L.H., Daily Totals		$2999.40		$3898.70	$62.49	$81.22
Crew B-26	Hr.	Daily	Hr.	Daily	Bare Costs	Incl. O&P
1 Labor Foreman (outside)	$20.70	$165.60	$35.15	$281.20	$21.68	$36.57
6 Laborers	18.70	897.60	31.75	1524.00		
2 Equip. Oper. (med.)	26.50	424.00	43.65	698.40		
1 Rodman (reinf.)	27.65	221.20	49.25	394.00		
1 Cement Finisher	24.90	199.20	40.05	320.40		
1 Grader, 30,000 Lbs.		505.40		555.94		
1 Paving Mach. & Equip.		2324.00		2556.40	32.15	35.37
88 L.H., Daily Totals		$4737.00		$6330.34	$53.83	$71.94

Crew No.	Bare Costs		Incl. Subs O & P		Cost Per Labor-Hour	
Crew B-26A	Hr.	Daily	Hr.	Daily	Bare Costs	Incl. O&P
1 Labor Foreman (outside)	$20.70	$165.60	$35.15	$281.20	$21.68	$36.57
6 Laborers	18.70	897.60	31.75	1524.00		
2 Equip. Oper. (med.)	26.50	424.00	43.65	698.40		
1 Rodman (reinf.)	27.65	221.20	49.25	394.00		
1 Cement Finisher	24.90	199.20	40.05	320.40		
1 Grader, 30,000 Lbs.		505.40		555.94		
1 Paving Mach. & Equip.		2324.00		2556.40		
1 Concrete Saw		118.40		130.24	33.50	36.85
88 L.H., Daily Totals		$4855.40		$6460.58	$55.17	$73.42
Crew B-26B	Hr.	Daily	Hr.	Daily	Bare Costs	Incl. O&P
1 Labor Foreman (outside)	$20.70	$165.60	$35.15	$281.20	$22.08	$37.16
6 Laborers	18.70	897.60	31.75	1524.00		
3 Equip. Oper. (med.)	26.50	636.00	43.65	1047.60		
1 Rodman (reinf.)	27.65	221.20	49.25	394.00		
1 Cement Finisher	24.90	199.20	40.05	320.40		
1 Grader, 30,000 Lbs.		505.40		555.94		
1 Paving Mach. & Equip.		2324.00		2556.40		
1 Concrete Pump, 110' Boom		938.60		1032.46	39.25	43.17
96 L.H., Daily Totals		$5887.60		$7712.00	$61.33	$80.33
Crew B-27	Hr.	Daily	Hr.	Daily	Bare Costs	Incl. O&P
1 Labor Foreman (outside)	$20.70	$165.60	$35.15	$281.20	$19.20	$32.60
3 Laborers	18.70	448.80	31.75	762.00		
1 Berm Machine		250.80		275.88	7.84	8.62
32 L.H., Daily Totals		$865.20		$1319.08	$27.04	$41.22
Crew B-28	Hr.	Daily	Hr.	Daily	Bare Costs	Incl. O&P
2 Carpenters	$25.70	$411.20	$43.60	$697.60	$23.37	$39.65
1 Laborer	18.70	149.60	31.75	254.00		
24 L.H., Daily Totals		$560.80		$951.60	$23.37	$39.65
Crew B-29	Hr.	Daily	Hr.	Daily	Bare Costs	Incl. O&P
1 Labor Foreman (outside)	$20.70	$165.60	$35.15	$281.20	$20.49	$34.56
4 Laborers	18.70	598.40	31.75	1016.00		
1 Equip. Oper. (crane)	27.45	219.60	45.20	361.60		
1 Gradall, 5/8 C.Y.		905.80		996.38	18.87	20.76
48 L.H., Daily Totals		$1889.40		$2655.18	$39.36	$55.32
Crew B-30	Hr.	Daily	Hr.	Daily	Bare Costs	Incl. O&P
1 Equip. Oper. (med.)	$26.50	$212.00	$43.65	$349.20	$22.83	$38.12
2 Truck Drivers (heavy)	21.00	336.00	35.35	565.60		
1 Hyd. Excavator, 1.5 C.Y.		775.60		853.16		
2 Dump Trucks, 16 Ton, 12 C.Y.		1060.00		1166.00	76.48	84.13
24 L.H., Daily Totals		$2383.60		$2933.96	$99.32	$122.25
Crew B-31	Hr.	Daily	Hr.	Daily	Bare Costs	Incl. O&P
1 Labor Foreman (outside)	$20.70	$165.60	$35.15	$281.20	$19.10	$32.43
4 Laborers	18.70	598.40	31.75	1016.00		
1 Air Compressor, 250 C.F.M.		151.40		166.54		
1 Sheeting Driver		5.75		6.33		
2 -50' Air Hoses, 1.5"		11.60		12.76	4.22	4.64
40 L.H., Daily Totals		$932.75		$1482.83	$23.32	$37.07
Crew B-32	Hr.	Daily	Hr.	Daily	Bare Costs	Incl. O&P
1 Laborer	$18.70	$149.60	$31.75	$254.00	$24.55	$40.67
3 Equip. Oper. (med.)	26.50	636.00	43.65	1047.60		
1 Grader, 30,000 Lbs.		505.40		555.94		
1 Tandem Roller, 10 Ton		211.00		232.10		
1 Dozer, 200 H.P.		988.40		1087.24	53.27	58.60
32 L.H., Daily Totals		$2490.40		$3176.88	$77.83	$99.28

Crews

Crew No.	Bare Costs		Incl. Subs O & P		Cost Per Labor-Hour	
Crew B-32A	Hr.	Daily	Hr.	Daily	Bare Costs	Incl. O&P
1 Laborer	$18.70	$149.60	$31.75	$254.00	$23.90	$39.68
2 Equip. Oper. (medium)	26.50	424.00	43.65	698.40		
1 Grader, 30,000 Lbs.		505.40		555.94		
1 Roller, Vibratory, 25 Ton		565.60		622.16	44.63	49.09
24 L.H., Daily Totals		$1644.60		$2130.50	$68.53	$88.77
Crew B-32B	Hr.	Daily	Hr.	Daily	Bare Costs	Incl. O&P
1 Laborer	$18.70	$149.60	$31.75	$254.00	$23.90	$39.68
2 Equip. Oper. (medium)	26.50	424.00	43.65	698.40		
1 Dozer, 200 H.P.		988.40		1087.24		
1 Roller, Vibratory, 25 Ton		565.60		622.16	64.75	71.22
24 L.H., Daily Totals		$2127.60		$2661.80	$88.65	$110.91
Crew B-32C	Hr.	Daily	Hr.	Daily	Bare Costs	Incl. O&P
1 Labor Foreman	$20.70	$165.60	$35.15	$281.20	$22.93	$38.27
2 Laborers	18.70	299.20	31.75	508.00		
3 Equip. Oper. (medium)	26.50	636.00	43.65	1047.60		
1 Grader, 30,000 Lbs.		505.40		555.94		
1 Tandem Roller, 10 Ton		211.00		232.10		
1 Dozer, 200 H.P.		988.40		1087.24	35.52	39.07
48 L.H., Daily Totals		$2805.60		$3712.08	$58.45	$77.33
Crew B-33A	Hr.	Daily	Hr.	Daily	Bare Costs	Incl. O&P
1 Equip. Oper. (med.)	$26.50	$212.00	$43.65	$349.20	$26.50	$43.65
.25 Equip. Oper. (med.)	26.50	53.00	43.65	87.30		
1 Scraper, Towed, 7 C.Y.		143.20		157.52		
1.250 Dozer, 300 H.P.		1626.25		1788.88	176.94	194.64
10 L.H., Daily Totals		$2034.45		$2382.90	$203.44	$238.29
Crew B-33B	Hr.	Daily	Hr.	Daily	Bare Costs	Incl. O&P
1 Equip. Oper. (med.)	$26.50	$212.00	$43.65	$349.20	$26.50	$43.65
.25 Equip. Oper. (med.)	26.50	53.00	43.65	87.30		
1 Scraper, Towed, 10 C.Y.		157.00		172.70		
1.250 Dozer, 300 H.P.		1626.25		1788.88	178.32	196.16
10 L.H., Daily Totals		$2048.25		$2398.07	$204.82	$239.81
Crew B-33C	Hr.	Daily	Hr.	Daily	Bare Costs	Incl. O&P
1 Equip. Oper. (med.)	$26.50	$212.00	$43.65	$349.20	$26.50	$43.65
.25 Equip. Oper. (med.)	26.50	53.00	43.65	87.30		
1 Scraper, Towed, 15 C.Y.		157.00		172.70		
1.250 Dozer, 300 H.P.		1626.25		1788.88	178.32	196.16
10 L.H., Daily Totals		$2048.25		$2398.07	$204.82	$239.81
Crew B-33D	Hr.	Daily	Hr.	Daily	Bare Costs	Incl. O&P
1 Equip. Oper. (med.)	$26.50	$212.00	$43.65	$349.20	$26.50	$43.65
.25 Equip. Oper. (med.)	26.50	53.00	43.65	87.30		
1 S.P. Scraper, 14 C.Y.		1717.00		1888.70		
.25 Dozer, 300 H.P.		325.25		357.77	204.22	224.65
10 L.H., Daily Totals		$2307.25		$2682.97	$230.72	$268.30
Crew B-33E	Hr.	Daily	Hr.	Daily	Bare Costs	Incl. O&P
1 Equip. Oper. (med.)	$26.50	$212.00	$43.65	$349.20	$26.50	$43.65
.25 Equip. Oper. (med.)	26.50	53.00	43.65	87.30		
1 S.P. Scraper, 24 C.Y.		2549.00		2803.90		
.25 Dozer, 300 H.P.		325.25		357.77	287.43	316.17
10 L.H., Daily Totals		$3139.25		$3598.18	$313.93	$359.82
Crew B-33F	Hr.	Daily	Hr.	Daily	Bare Costs	Incl. O&P
1 Equip. Oper. (med.)	$26.50	$212.00	$43.65	$349.20	$26.50	$43.65
.25 Equip. Oper. (med.)	26.50	53.00	43.65	87.30		
1 Elev. Scraper, 11 C.Y.		859.80		945.78		
.25 Dozer, 300 H.P.		325.25		357.77	118.51	130.36
10 L.H., Daily Totals		$1450.05		$1740.06	$145.01	$174.01
Crew B-33G	Hr.	Daily	Hr.	Daily	Bare Costs	Incl. O&P
1 Equip. Oper. (med.)	$26.50	$212.00	$43.65	$349.20	$26.50	$43.65
.25 Equip. Oper. (med.)	26.50	53.00	43.65	87.30		
1 Elev. Scraper, 20 C.Y.		1750.00		1925.00		
.25 Dozer, 300 H.P.		325.25		357.77	207.53	228.28
10 L.H., Daily Totals		$2340.25		$2719.28	$234.03	$271.93
Crew B-34A	Hr.	Daily	Hr.	Daily	Bare Costs	Incl. O&P
1 Truck Driver (heavy)	$21.00	$168.00	$35.35	$282.80	$21.00	$35.35
1 Dump Truck, 12 Ton, 8 C.Y.		368.40		405.24	46.05	50.66
8 L.H., Daily Totals		$536.40		$688.04	$67.05	$86.00
Crew B-34B	Hr.	Daily	Hr.	Daily	Bare Costs	Incl. O&P
1 Truck Driver (heavy)	$21.00	$168.00	$35.35	$282.80	$21.00	$35.35
1 Dump Truck, 16 Ton, 12 C.Y.		530.00		583.00	66.25	72.88
8 L.H., Daily Totals		$698.00		$865.80	$87.25	$108.22
Crew B-34C	Hr.	Daily	Hr.	Daily	Bare Costs	Incl. O&P
1 Truck Driver (heavy)	$21.00	$168.00	$35.35	$282.80	$21.00	$35.35
1 Truck Tractor, 240 H.P.		336.20		369.82		
1 Dump Trailer, 16.5 C.Y.		107.80		118.58	55.50	61.05
8 L.H., Daily Totals		$612.00		$771.20	$76.50	$96.40
Crew B-34D	Hr.	Daily	Hr.	Daily	Bare Costs	Incl. O&P
1 Truck Driver (heavy)	$21.00	$168.00	$35.35	$282.80	$21.00	$35.35
1 Truck Tractor, 240 H.P.		336.20		369.82		
1 Dump Trailer, 20 C.Y.		120.00		132.00	57.02	62.73
8 L.H., Daily Totals		$624.20		$784.62	$78.03	$98.08
Crew B-34E	Hr.	Daily	Hr.	Daily	Bare Costs	Incl. O&P
1 Truck Driver (heavy)	$21.00	$168.00	$35.35	$282.80	$21.00	$35.35
1 Dump Truck, Off Hwy., 25 Ton		992.80		1092.08	124.10	136.51
8 L.H., Daily Totals		$1160.80		$1374.88	$145.10	$171.86
Crew B-34F	Hr.	Daily	Hr.	Daily	Bare Costs	Incl. O&P
1 Truck Driver (heavy)	$21.00	$168.00	$35.35	$282.80	$21.00	$35.35
1 Dump Truck, Off Hwy., 35 Ton		1014.00		1115.40	126.75	139.43
8 L.H., Daily Totals		$1182.00		$1398.20	$147.75	$174.78
Crew B-34G	Hr.	Daily	Hr.	Daily	Bare Costs	Incl. O&P
1 Truck Driver (heavy)	$21.00	$168.00	$35.35	$282.80	$21.00	$35.35
1 Dump Truck, Off Hwy., 50 Ton		1308.00		1438.80	163.50	179.85
8 L.H., Daily Totals		$1476.00		$1721.60	$184.50	$215.20
Crew B-34H	Hr.	Daily	Hr.	Daily	Bare Costs	Incl. O&P
1 Truck Driver (heavy)	$21.00	$168.00	$35.35	$282.80	$21.00	$35.35
1 Dump Truck, Off Hwy., 65 Ton		1393.00		1532.30	174.13	191.54
8 L.H., Daily Totals		$1561.00		$1815.10	$195.13	$226.89
Crew B-34J	Hr.	Daily	Hr.	Daily	Bare Costs	Incl. O&P
1 Truck Driver (heavy)	$21.00	$168.00	$35.35	$282.80	$21.00	$35.35
1 Dump Truck, Off Hwy., 100 Ton		1792.00		1971.20	224.00	246.40
8 L.H., Daily Totals		$1960.00		$2254.00	$245.00	$281.75

Crews

Crew No.	Bare Costs		Incl. Subs O & P		Cost Per Labor-Hour	
Crew B-34K	**Hr.**	**Daily**	**Hr.**	**Daily**	**Bare Costs**	**Incl. O&P**
1 Truck Driver (heavy)	$21.00	$168.00	$35.35	$282.80	$21.00	$35.35
1 Truck Tractor, 240 H.P.		395.40		434.94		
1 Lowbed Trailer, 75 Ton		181.20		199.32	72.08	79.28
8 L.H., Daily Totals		$744.60		$917.06	$93.08	$114.63
Crew B-34N	**Hr.**	**Daily**	**Hr.**	**Daily**	**Bare Costs**	**Incl. O&P**
1 Truck Driver (heavy)	$21.00	$168.00	$35.35	$282.80	$21.00	$35.35
1 Dump Truck, 12 Ton, 8 C.Y.		368.40		405.24		
1 Flatbed Trailer, 40 Ton		125.80		138.38	61.77	67.95
8 L.H., Daily Totals		$662.20		$826.42	$82.78	$103.30
Crew B-34P	**Hr.**	**Daily**	**Hr.**	**Daily**	**Bare Costs**	**Incl. O&P**
1 Pipe Fitter	$29.85	$238.80	$49.05	$392.40	$25.53	$42.27
1 Truck Driver (light)	20.25	162.00	34.10	272.80		
1 Equip. Oper. (medium)	26.50	212.00	43.65	349.20		
1 Flatbed Truck, Gas, 3 Ton		185.20		203.72		
1 Backhoe Loader, 48 H.P.		243.40		267.74	17.86	19.64
24 L.H., Daily Totals		$1041.40		$1485.86	$43.39	$61.91
Crew B-34Q	**Hr.**	**Daily**	**Hr.**	**Daily**	**Bare Costs**	**Incl. O&P**
1 Pipe Fitter	$29.85	$238.80	$49.05	$392.40	$25.85	$42.78
1 Truck Driver (light)	20.25	162.00	34.10	272.80		
1 Eqip. Oper. (crane)	27.45	219.60	45.20	361.60		
1 Flatbed Trailer, 25 Ton		92.80		102.08		
1 Dump Truck, 12 Ton, 8 C.Y.		368.40		405.24		
1 Hyd. Crane, 25 Ton		739.60		813.56	50.03	55.04
24 L.H., Daily Totals		$1821.20		$2347.68	$75.88	$97.82
Crew B-34R	**Hr.**	**Daily**	**Hr.**	**Daily**	**Bare Costs**	**Incl. O&P**
1 Pipe Fitter	$29.85	$238.80	$49.05	$392.40	$25.85	$42.78
1 Truck Driver (light)	20.25	162.00	34.10	272.80		
1 Eqip. Oper. (crane)	27.45	219.60	45.20	361.60		
1 Flatbed Trailer, 25 Ton		92.80		102.08		
1 Dump Truck, 12 Ton, 8 C.Y.		368.40		405.24		
1 Hyd. Crane, 25 Ton		739.60		813.56		
1 Hyd. Excavator, 1 C.Y.		601.80		661.98	75.11	82.62
24 L.H., Daily Totals		$2423.00		$3009.66	$100.96	$125.40
Crew B-34S	**Hr.**	**Daily**	**Hr.**	**Daily**	**Bare Costs**	**Incl. O&P**
2 Pipe Fitters	$29.85	$477.60	$49.05	$784.80	$27.04	$44.66
1 Truck Driver (heavy)	21.00	168.00	35.35	282.80		
1 Eqip. Oper. (crane)	27.45	219.60	45.20	361.60		
1 Flatbed Trailer, 40 Ton		125.80		138.38		
1 Truck Tractor, 240 H.P.		336.20		369.82		
1 Hyd. Crane, 80 Ton		1214.00		1335.40		
1 Hyd. Excavator, 2 C.Y.		994.60		1094.06	83.46	91.80
32 L.H., Daily Totals		$3535.80		$4366.86	$110.49	$136.46
Crew B-34T	**Hr.**	**Daily**	**Hr.**	**Daily**	**Bare Costs**	**Incl. O&P**
2 Pipe Fitters	$29.85	$477.60	$49.05	$784.80	$27.04	$44.66
1 Truck Driver (heavy)	21.00	168.00	35.35	282.80		
1 Eqip. Oper. (crane)	27.45	219.60	45.20	361.60		
1 Flatbed Trailer, 40 Ton		125.80		138.38		
1 Truck Tractor, 240 H.P.		336.20		369.82		
1 Hyd. Crane, 80 Ton		1214.00		1335.40	52.38	57.61
32 L.H., Daily Totals		$2541.20		$3272.80	$79.41	$102.28

Crew No.	Bare Costs		Incl. Subs O & P		Cost Per Labor-Hour	
Crew B-35	**Hr.**	**Daily**	**Hr.**	**Daily**	**Bare Costs**	**Incl. O&P**
1 Laborer Foreman (out)	$20.70	$165.60	$35.15	$281.20	$24.50	$40.99
1 Skilled Worker	26.10	208.80	44.25	354.00		
1 Welder (plumber)	29.55	236.40	48.60	388.80		
1 Laborer	18.70	149.60	31.75	254.00		
1 Equip. Oper. (crane)	27.45	219.60	45.20	361.60		
1 Welder, electric, 300 amp		61.45		67.59		
1 Hyd. Excavator, .75 C.Y.		516.40		568.04	14.45	15.89
40 L.H., Daily Totals		$1557.85		$2275.24	$38.95	$56.88
Crew B-35A	**Hr.**	**Daily**	**Hr.**	**Daily**	**Bare Costs**	**Incl. O&P**
1 Laborer Foreman (out)	$20.70	$165.60	$35.15	$281.20	$23.51	$39.32
2 Laborers	18.70	299.20	31.75	508.00		
1 Skilled Worker	26.10	208.80	44.25	354.00		
1 Welder (plumber)	29.55	236.40	48.60	388.80		
1 Equip. Oper. (crane)	27.45	219.60	45.20	361.60		
1 Equip. Oper. Oiler	23.40	187.20	38.55	308.40		
1 Welder, gas engine, 300 amp		115.20		126.72		
1 Crawler Crane, 75 Ton		1384.00		1522.40	26.77	29.45
56 L.H., Daily Totals		$2816.00		$3851.12	$50.29	$68.77
Crew B-36	**Hr.**	**Daily**	**Hr.**	**Daily**	**Bare Costs**	**Incl. O&P**
1 Labor Foreman (outside)	$20.70	$165.60	$35.15	$281.20	$22.22	$37.19
2 Laborers	18.70	299.20	31.75	508.00		
2 Equip. Oper. (med.)	26.50	424.00	43.65	698.40		
1 Dozer, 200 H.P.		988.40		1087.24		
1 Aggregate Spreader		30.60		33.66		
1 Tandem Roller, 10 Ton		211.00		232.10	30.75	33.83
40 L.H., Daily Totals		$2118.80		$2840.60	$52.97	$71.02
Crew B-36A	**Hr.**	**Daily**	**Hr.**	**Daily**	**Bare Costs**	**Incl. O&P**
1 Labor Foreman (outside)	$20.70	$165.60	$35.15	$281.20	$23.44	$39.04
2 Laborers	18.70	299.20	31.75	508.00		
4 Equip. Oper. (med.)	26.50	848.00	43.65	1396.80		
1 Dozer, 200 H.P.		988.40		1087.24		
1 Aggregate Spreader		30.60		33.66		
1 Tandem Roller, 10 Ton		211.00		232.10		
1 Roller, Pneum. Whl, 12 Ton		295.60		325.16	27.24	29.97
56 L.H., Daily Totals		$2838.40		$3864.16	$50.69	$69.00
Crew B-36B	**Hr.**	**Daily**	**Hr.**	**Daily**	**Bare Costs**	**Incl. O&P**
1 Labor Foreman (outside)	$20.70	$165.60	$35.15	$281.20	$23.14	$38.58
2 Laborers	18.70	299.20	31.75	508.00		
4 Equip. Oper. (medium)	26.50	848.00	43.65	1396.80		
1 Truck Driver, Heavy	21.00	168.00	35.35	282.80		
1 Grader, 30,000 Lbs.		505.40		555.94		
1 F.E. Loader, crl, 1.5 C.Y.		399.20		439.12		
1 Dozer, 300 H.P.		1301.00		1431.10		
1 Roller, Vibratory, 25 Ton		565.60		622.16		
1 Truck Tractor, 240 H.P.		395.40		434.94		
1 Water Tanker, 5000 Gal.		121.20		133.32	51.37	56.51
64 L.H., Daily Totals		$4768.60		$6085.38	$74.51	$95.08
Crew B-36C	**Hr.**	**Daily**	**Hr.**	**Daily**	**Bare Costs**	**Incl. O&P**
1 Labor Foreman (outside)	$20.70	$165.60	$35.15	$281.20	$24.24	$40.29
3 Equip. Oper. (medium)	26.50	636.00	43.65	1047.60		
1 Truck Driver, Heavy	21.00	168.00	35.35	282.80		
1 Grader, 30,000 Lbs.		505.40		555.94		
1 Dozer, 300 H.P.		1301.00		1431.10		
1 Roller, Vibratory, 25 Ton		565.60		622.16		
1 Truck Tractor, 240 H.P.		395.40		434.94		
1 Water Tanker, 5000 Gal.		121.20		133.32	72.22	79.44
40 L.H., Daily Totals		$3858.20		$4789.06	$96.45	$119.73

Crews

Crew No.	Bare Costs		Incl. Subs O & P		Cost Per Labor-Hour	
Crew B-36E	**Hr.**	**Daily**	**Hr.**	**Daily**	**Bare Costs**	**Incl. O&P**
1 Labor Foreman (outside)	$20.70	$165.60	$35.15	$281.20	$24.62	$40.85
4 Equip. Oper. (medium)	26.50	848.00	43.65	1396.80		
1 Truck Driver, Heavy	21.00	168.00	35.35	282.80		
1 Grader, 30,000 Lbs.		505.40		555.94		
1 Dozer, 300 H.P.		1301.00		1431.10		
1 Roller, Vibratory, 25 Ton		565.60		622.16		
1 Truck Tractor, 240 H.P.		395.40		434.94		
1 Dist. Tanker, 3000 Gallon		273.00		300.30	63.34	69.68
48 L.H., Daily Totals		$4222.00		$5305.24	$87.96	$110.53
Crew B-37	**Hr.**	**Daily**	**Hr.**	**Daily**	**Bare Costs**	**Incl. O&P**
1 Labor Foreman (outside)	$20.70	$165.60	$35.15	$281.20	$20.16	$34.01
4 Laborers	18.70	598.40	31.75	1016.00		
1 Equip. Oper. (light)	25.45	203.60	41.90	335.20		
1 Tandem Roller, 5 Ton		122.20		134.42	2.55	2.80
48 L.H., Daily Totals		$1089.80		$1766.82	$22.70	$36.81
Crew B-38	**Hr.**	**Daily**	**Hr.**	**Daily**	**Bare Costs**	**Incl. O&P**
2 Laborers	$18.70	$299.20	$31.75	$508.00	$20.95	$35.13
1 Equip. Oper. (light)	25.45	203.60	41.90	335.20		
1 Backhoe Loader, 48 H.P.		243.40		267.74		
1 Hyd.Hammer, (1200 lb)		115.40		126.94	14.95	16.45
24 L.H., Daily Totals		$861.60		$1237.88	$35.90	$51.58
Crew B-39	**Hr.**	**Daily**	**Hr.**	**Daily**	**Bare Costs**	**Incl. O&P**
1 Labor Foreman (outside)	$20.70	$165.60	$35.15	$281.20	$19.03	$32.32
5 Laborers	18.70	748.00	31.75	1270.00		
1 Air Compressor, 250 C.F.M.		151.40		166.54		
2 Breakers, Pavement, 60 lb.		16.80		18.48		
2 -50' Air Hoses, 1.5"		11.60		12.76	3.75	4.12
48 L.H., Daily Totals		$1093.40		$1748.98	$22.78	$36.44
Crew B-40	**Hr.**	**Daily**	**Hr.**	**Daily**	**Bare Costs**	**Incl. O&P**
1 Pile Driver Foreman (out)	$27.15	$217.20	$48.25	$386.00	$24.84	$43.43
4 Pile Drivers	25.15	804.80	44.70	1430.40		
1 Building Laborer	18.70	149.60	31.75	254.00		
1 Equip. Oper. (crane)	27.45	219.60	45.20	361.60		
1 Crawler Crane, 40 Ton		994.60		1094.06		
1 Vibratory Hammer & Gen.		1983.00		2181.30	53.17	58.49
56 L.H., Daily Totals		$4368.80		$5707.36	$78.01	$101.92
Crew B-40B	**Hr.**	**Daily**	**Hr.**	**Daily**	**Bare Costs**	**Incl. O&P**
1 Laborer Foreman	$20.70	$165.60	$35.15	$281.20	$21.27	$35.69
3 Laborers	18.70	448.80	31.75	762.00		
1 Equip. Oper. (crane)	27.45	219.60	45.20	361.60		
1 Equip. Oper. Oiler	23.40	187.20	38.55	308.40		
1 Lattice Boom Crane, 40 Ton		1268.00		1394.80	26.42	29.06
48 L.H., Daily Totals		$2289.20		$3108.00	$47.69	$64.75
Crew B-41	**Hr.**	**Daily**	**Hr.**	**Daily**	**Bare Costs**	**Incl. O&P**
1 Labor Foreman (outside)	$20.70	$165.60	$35.15	$281.20	$19.68	$33.29
4 Laborers	18.70	598.40	31.75	1016.00		
.25 Equip. Oper. (crane)	27.45	54.90	45.20	90.40		
.25 Equip. Oper. Oiler	23.40	46.80	38.55	77.10		
.25 Crawler Crane, 40 Ton		248.65		273.51	5.65	6.22
44 L.H., Daily Totals		$1114.35		$1738.21	$25.33	$39.50

Crew No.	Bare Costs		Incl. Subs O & P		Cost Per Labor-Hour	
Crew B-42	**Hr.**	**Daily**	**Hr.**	**Daily**	**Bare Costs**	**Incl. O&P**
1 Labor Foreman (outside)	$20.70	$165.60	$35.15	$281.20	$21.79	$36.56
4 Laborers	18.70	598.40	31.75	1016.00		
1 Equip. Oper. (crane)	27.45	219.60	45.20	361.60		
1 Welder	29.55	236.40	48.60	388.80		
1 Hyd. Crane, 25 Ton		739.60		813.56		
1 Welder, gas engine, 300 amp		115.20		126.72		
1 Horz. Boring Csg. Mch.		421.40		463.54	22.79	25.07
56 L.H., Daily Totals		$2496.20		$3451.42	$44.58	$61.63
Crew B-43	**Hr.**	**Daily**	**Hr.**	**Daily**	**Bare Costs**	**Incl. O&P**
1 Labor Foreman (outside)	$20.70	$165.60	$35.15	$281.20	$19.10	$32.43
4 Laborers	18.70	598.40	31.75	1016.00		
1 Drill Rig, Truck-Mounted		3499.00		3848.90	87.47	96.22
40 L.H., Daily Totals		$4263.00		$5146.10	$106.58	$128.65
Crew B-44	**Hr.**	**Daily**	**Hr.**	**Daily**	**Bare Costs**	**Incl. O&P**
1 Pile Driver Foreman	$27.15	$217.20	$48.25	$386.00	$24.07	$41.97
4 Pile Drivers	25.15	804.80	44.70	1430.40		
1 Equip. Oper. (crane)	27.45	219.60	45.20	361.60		
2 Laborers	18.70	299.20	31.75	508.00		
1 Crawler Crane, 40 Ton		994.60		1094.06		
45 L.F. of Leads, 15K Ft. Lbs.		45.00		49.50	16.24	17.87
64 L.H., Daily Totals		$2580.40		$3829.56	$40.32	$59.84
Crew B-45	**Hr.**	**Daily**	**Hr.**	**Daily**	**Bare Costs**	**Incl. O&P**
1 Building Laborer	$18.70	$149.60	$31.75	$254.00	$19.85	$33.55
1 Truck Driver (heavy)	21.00	168.00	35.35	282.80		
1 Dist. Tanker, 3000 Gallon		273.00		300.30	17.06	18.77
16 L.H., Daily Totals		$590.60		$837.10	$36.91	$52.32
Crew B-46	**Hr.**	**Daily**	**Hr.**	**Daily**	**Bare Costs**	**Incl. O&P**
1 Pile Driver Foreman	$27.15	$217.20	$48.25	$386.00	$22.26	$38.82
2 Pile Drivers	25.15	402.40	44.70	715.20		
3 Laborers	18.70	448.80	31.75	762.00		
1 Chain saw, gas, 36" Long		34.40		37.84	0.72	0.79
48 L.H., Daily Totals		$1102.80		$1901.04	$22.98	$39.60
Crew B-47	**Hr.**	**Daily**	**Hr.**	**Daily**	**Bare Costs**	**Incl. O&P**
1 Blast Foreman	$20.70	$165.60	$35.15	$281.20	$19.70	$33.45
1 Driller	18.70	149.60	31.75	254.00		
1 Air Track Drill, 4"		787.40		866.14		
1 Air Compressor, 600 C.F.M.		396.20		435.82		
2 -50' Air Hoses, 3"		38.10		41.91	76.36	83.99
16 L.H., Daily Totals		$1536.90		$1879.07	$96.06	$117.44
Crew B-47A	**Hr.**	**Daily**	**Hr.**	**Daily**	**Bare Costs**	**Incl. O&P**
1 Drilling Foreman	$20.70	$165.60	$35.15	$281.20	$23.85	$39.63
1 Equip. Oper. (heavy)	27.45	219.60	45.20	361.60		
1 Oiler	23.40	187.20	38.55	308.40		
1 Air Track Drill, 5"		905.40		995.94	37.73	41.50
24 L.H., Daily Totals		$1477.80		$1947.14	$61.58	$81.13
Crew B-47C	**Hr.**	**Daily**	**Hr.**	**Daily**	**Bare Costs**	**Incl. O&P**
1 Laborer	$18.70	$149.60	$31.75	$254.00	$22.07	$36.83
1 Equip. Oper. (light)	25.45	203.60	41.90	335.20		
1 Air Compressor, 750 C.F.M.		409.80		450.78		
2 -50' Air Hoses, 3"		38.10		41.91		
1 Air Track Drill, 4"		787.40		866.14	77.21	84.93
16 L.H., Daily Totals		$1588.50		$1948.03	$99.28	$121.75

Crews

Crew No.	Bare Costs		Incl. Subs O & P		Cost Per Labor-Hour	
Crew B-47E	Hr.	Daily	Hr.	Daily	Bare Costs	Incl. O&P
1 Laborer Foreman	$20.70	$165.60	$35.15	$281.20	$19.20	$32.60
3 Laborers	18.70	448.80	31.75	762.00		
1 Flatbed Truck, gas, 3 Ton		185.20		203.72	5.79	6.37
32 L.H., Daily Totals		$799.60		$1246.92	$24.99	$38.97
Crew B-47G	Hr.	Daily	Hr.	Daily	Bare Costs	Incl. O&P
1 Laborer Foreman	$20.70	$165.60	$35.15	$281.20	$19.37	$32.88
2 Laborers	18.70	299.20	31.75	508.00		
1 Air Track Drill, 4"		787.40		866.14		
1 Air Compressor, 600 C.F.M.		396.20		435.82		
2 -50' Air Hoses, 3"		38.10		41.91		
1 Grout Pump		131.20		144.32	56.37	62.01
24 L.H., Daily Totals		$1817.70		$2277.39	$75.74	$94.89
Crew B-48	Hr.	Daily	Hr.	Daily	Bare Costs	Incl. O&P
1 Labor Foreman (outside)	$20.70	$165.60	$35.15	$281.20	$20.49	$34.56
4 Laborers	18.70	598.40	31.75	1016.00		
1 Equip. Oper. (crane)	27.45	219.60	45.20	361.60		
1 Centr. Water Pump, 6"		278.20		306.02		
1 -20' Suction Hose, 6"		11.50		12.65		
1 -50' Discharge Hose, 6"		6.30		6.93		
1 Drill Rig, Truck-Mounted		3499.00		3848.90	79.06	86.97
48 L.H., Daily Totals		$4778.60		$5833.30	$99.55	$121.53
Crew B-49	Hr.	Daily	Hr.	Daily	Bare Costs	Incl. O&P
1 Labor Foreman (outside)	$20.70	$165.60	$35.15	$281.20	$21.33	$36.50
5 Laborers	18.70	748.00	31.75	1270.00		
1 Equip. Oper. (crane)	27.45	219.60	45.20	361.60		
2 Pile Drivers	25.15	402.40	44.70	715.20		
1 Hyd. Crane, 25 Ton		739.60		813.56		
1 Centr. Water Pump, 6"		278.20		306.02		
1 -20' Suction Hose, 6"		11.50		12.65		
1 -50' Discharge Hose, 6"		6.30		6.93		
1 Drill Rig, Truck-Mounted		3499.00		3848.90	62.98	69.28
72 L.H., Daily Totals		$6070.20		$7616.06	$84.31	$105.78
Crew B-50	Hr.	Daily	Hr.	Daily	Bare Costs	Incl. O&P
1 Pile Driver Foreman	$27.15	$217.20	$48.25	$386.00	$23.00	$40.03
6 Pile Drivers	25.15	1207.20	44.70	2145.60		
1 Equip. Oper. (crane)	27.45	219.60	45.20	361.60		
5 Laborers	18.70	748.00	31.75	1270.00		
1 Crawler Crane, 40 Ton		994.60		1094.06		
60 L.F. of Leads, 15K Ft. Lbs.		60.00		66.00		
1 Hammer, 15K Ft. Lbs.		360.00		396.00		
1 Air Compressor, 600 C.F.M.		396.20		435.82		
2 -50' Air Hoses, 3"		38.10		41.91		
1 Chain saw, gas, 36" Long		34.40		37.84	18.11	19.92
104 L.H., Daily Totals		$4275.30		$6234.83	$41.11	$59.95
Crew B-51	Hr.	Daily	Hr.	Daily	Bare Costs	Incl. O&P
1 Labor Foreman (outside)	$20.70	$165.60	$35.15	$281.20	$19.29	$32.71
4 Laborers	18.70	598.40	31.75	1016.00		
1 Truck Driver (light)	20.25	162.00	34.10	272.80		
1 Flatbed Truck, Gas, 1.5 Ton		141.40		155.54	2.95	3.24
48 L.H., Daily Totals		$1067.40		$1725.54	$22.24	$35.95

Crew No.	Bare Costs		Incl. Subs O & P		Cost Per Labor-Hour	
Crew B-52	Hr.	Daily	Hr.	Daily	Bare Costs	Incl. O&P
1 Labor Foreman	$20.70	$165.60	$35.15	$281.20	$21.18	$36.03
1 Carpenter	25.70	205.60	43.60	348.80		
4 Laborers	18.70	598.40	31.75	1016.00		
.5 Rodman (reinf.)	27.65	110.60	49.25	197.00		
.5 Equip. Oper. (med.)	26.50	106.00	43.65	174.60		
.5 Crawler Loader, 3 C.Y.		417.20		458.92	7.45	8.20
56 L.H., Daily Totals		$1603.40		$2476.52	$28.63	$44.22
Crew B-53	Hr.	Daily	Hr.	Daily	Bare Costs	Incl. O&P
1 Building Laborer	$18.70	$149.60	$31.75	$254.00	$18.70	$31.75
1 Trencher, Chain, 12 H.P.		51.00		56.10	6.38	7.01
8 L.H., Daily Totals		$200.60		$310.10	$25.07	$38.76
Crew B-54	Hr.	Daily	Hr.	Daily	Bare Costs	Incl. O&P
1 Equip. Oper. (light)	$25.45	$203.60	$41.90	$335.20	$25.45	$41.90
1 Trencher, Chain, 40 H.P.		226.60		249.26	28.32	31.16
8 L.H., Daily Totals		$430.20		$584.46	$53.77	$73.06
Crew B-54A	Hr.	Daily	Hr.	Daily	Bare Costs	Incl. O&P
.17 Labor Foreman (outside)	$20.70	$28.15	$35.15	$47.80	$25.66	$42.41
1 Equipment Operator (med.)	26.50	212.00	43.65	349.20		
1 Wheel Trencher, 67 H.P.		891.20		980.32	95.21	104.74
9.36 L.H., Daily Totals		$1131.35		$1377.32	$120.87	$147.15
Crew B-54B	Hr.	Daily	Hr.	Daily	Bare Costs	Incl. O&P
.25 Labor Foreman (outside)	$20.70	$41.40	$35.15	$70.30	$25.34	$41.95
1 Equipment Operator (med.)	26.50	212.00	43.65	349.20		
1 Wheel Trencher, 150 H.P.		1626.00		1788.60	162.60	178.86
10 L.H., Daily Totals		$1879.40		$2208.10	$187.94	$220.81
Crew B-55	Hr.	Daily	Hr.	Daily	Bare Costs	Incl. O&P
1 Laborer	$18.70	$149.60	$31.75	$254.00	$19.48	$32.92
1 Truck Driver (light)	20.25	162.00	34.10	272.80		
1 Truck-mounted earth auger		674.60		742.06		
1 Flatbed Truck, gas, 3 Ton		185.20		203.72	53.74	59.11
16 L.H., Daily Totals		$1171.40		$1472.58	$73.21	$92.04
Crew B-56	Hr.	Daily	Hr.	Daily	Bare Costs	Incl. O&P
2 Laborers	$18.70	$299.20	$31.75	$508.00	$18.70	$31.75
1 Air Track Drill, 4"		787.40		866.14		
1 Air Compressor, 600 C.F.M.		396.20		435.82		
1 -50' Air Hose, 3"		19.05		20.95	75.17	82.68
16 L.H., Daily Totals		$1501.85		$1830.92	$93.87	$114.43
Crew B-57	Hr.	Daily	Hr.	Daily	Bare Costs	Incl. O&P
1 Labor Foreman (outside)	$20.70	$165.60	$35.15	$281.20	$20.85	$35.12
3 Laborers	18.70	448.80	31.75	762.00		
1 Equip. Oper. (crane)	27.45	219.60	45.20	361.60		
1 Barge, 400 Ton		294.20		323.62		
1 Crawler Crane, 25 Ton		983.00		1081.30		
1 Clamshell Bucket, 1 C.Y.		44.80		49.28		
1 Centr. Water Pump, 6"		278.20		306.02		
1 -20' Suction Hose, 6"		11.50		12.65		
20 -50' Discharge Hose, 6"		126.00		138.60	43.44	47.79
40 L.H., Daily Totals		$2571.70		$3316.27	$64.29	$82.91

Crews

Crew No.	Bare Costs		Incl. Subs O & P		Cost Per Labor-Hour	
Crew B-58	Hr.	Daily	Hr.	Daily	Bare Costs	Incl. O&P
2 Laborers	$18.70	$299.20	$31.75	$508.00	$20.95	$35.13
1 Equip. Oper. (light)	25.45	203.60	41.90	335.20		
1 Backhoe Loader, 48 H.P.		243.40		267.74		
1 Small Helicopter, w/pilot		2331.00		2564.10	107.27	117.99
24 L.H., Daily Totals		$3077.20		$3675.04	$128.22	$153.13
Crew B-59	Hr.	Daily	Hr.	Daily	Bare Costs	Incl. O&P
1 Truck Driver (heavy)	$21.00	$168.00	$35.35	$282.80	$21.00	$35.35
1 Truck Tractor, 195 H.P.		235.60		259.16		
1 Water Tanker, 5000 Gal.		121.20		133.32	44.60	49.06
8 L.H., Daily Totals		$524.80		$675.28	$65.60	$84.41
Crew B-60	Hr.	Daily	Hr.	Daily	Bare Costs	Incl. O&P
1 Labor Foreman (outside)	$20.70	$165.60	$35.15	$281.20	$21.62	$36.25
3 Laborers	18.70	448.80	31.75	762.00		
1 Equip. Oper. (crane)	27.45	219.60	45.20	361.60		
1 Equip. Oper. (light)	25.45	203.60	41.90	335.20		
1 Crawler Crane, 40 Ton		994.60		1094.06		
45 L.F. of Leads, 15K Ft. Lbs.		45.00		49.50		
1 Backhoe Loader, 48 H.P.		243.40		267.74	26.73	29.40
48 L.H., Daily Totals		$2320.60		$3151.30	$48.35	$65.65
Crew B-61	Hr.	Daily	Hr.	Daily	Bare Costs	Incl. O&P
1 Labor Foreman (outside)	$20.70	$165.60	$35.15	$281.20	$19.10	$32.43
4 Laborers	18.70	598.40	31.75	1016.00		
1 Cement Mixer, 2 C.Y.		180.00		198.00		
1 Air Compressor, 160 C.F.M.		122.60		134.86	7.57	8.32
40 L.H., Daily Totals		$1066.60		$1630.06	$26.66	$40.75
Crew B-62	Hr.	Daily	Hr.	Daily	Bare Costs	Incl. O&P
2 Laborers	$18.70	$299.20	$31.75	$508.00	$20.95	$35.13
1 Equip. Oper. (light)	25.45	203.60	41.90	335.20		
1 Loader, Skid Steer, 30 HP, gas		122.00		134.20	5.08	5.59
24 L.H., Daily Totals		$624.80		$977.40	$26.03	$40.73
Crew B-63	Hr.	Daily	Hr.	Daily	Bare Costs	Incl. O&P
5 Laborers	$18.70	$748.00	$31.75	$1270.00	$18.70	$31.75
1 Loader, Skid Steer, 30 HP, gas		122.00		134.20	3.05	3.36
40 L.H., Daily Totals		$870.00		$1404.20	$21.75	$35.10
Crew B-64	Hr.	Daily	Hr.	Daily	Bare Costs	Incl. O&P
1 Laborer	$18.70	$149.60	$31.75	$254.00	$19.48	$32.92
1 Truck Driver (light)	20.25	162.00	34.10	272.80		
1 Power Mulcher (small)		119.00		130.90		
1 Flatbed Truck, Gas, 1.5 Ton		141.40		155.54	16.27	17.90
16 L.H., Daily Totals		$572.00		$813.24	$35.75	$50.83
Crew B-65	Hr.	Daily	Hr.	Daily	Bare Costs	Incl. O&P
1 Laborer	$18.70	$149.60	$31.75	$254.00	$19.48	$32.92
1 Truck Driver (light)	20.25	162.00	34.10	272.80		
1 Power Mulcher (large)		204.20		224.62		
1 Flatbed Truck, Gas, 1.5 Ton		141.40		155.54	21.60	23.76
16 L.H., Daily Totals		$657.20		$906.96	$41.08	$56.69
Crew B-66	Hr.	Daily	Hr.	Daily	Bare Costs	Incl. O&P
1 Equip. Oper. (light)	$25.45	$203.60	$41.90	$335.20	$25.45	$41.90
1 Loader-Backhoe		196.80		216.48	24.60	27.06
8 L.H., Daily Totals		$400.40		$551.68	$50.05	$68.96

Crew No.	Bare Costs		Incl. Subs O & P		Cost Per Labor-Hour	
Crew B-67	Hr.	Daily	Hr.	Daily	Bare Costs	Incl. O&P
1 Millwright	$26.75	$214.00	$43.20	$345.60	$26.10	$42.55
1 Equip. Oper. (light)	25.45	203.60	41.90	335.20		
1 Forklift, R/T, 4,000 Lb.		272.40		299.64	17.02	18.73
16 L.H., Daily Totals		$690.00		$980.44	$43.13	$61.28
Crew B-68	Hr.	Daily	Hr.	Daily	Bare Costs	Incl. O&P
2 Millwrights	$26.75	$428.00	$43.20	$691.20	$26.32	$42.77
1 Equip. Oper. (light)	25.45	203.60	41.90	335.20		
1 Forklift, R/T, 4,000 Lb.		272.40		299.64	11.35	12.48
24 L.H., Daily Totals		$904.00		$1326.04	$37.67	$55.25
Crew B-69	Hr.	Daily	Hr.	Daily	Bare Costs	Incl. O&P
1 Labor Foreman (outside)	$20.70	$165.60	$35.15	$281.20	$21.27	$35.69
3 Laborers	18.70	448.80	31.75	762.00		
1 Equip Oper. (crane)	27.45	219.60	45.20	361.60		
1 Equip Oper. Oiler	23.40	187.20	38.55	308.40		
1 Hyd. Crane, 80 Ton		1214.00		1335.40	25.29	27.82
48 L.H., Daily Totals		$2235.20		$3048.60	$46.57	$63.51
Crew B-69A	Hr.	Daily	Hr.	Daily	Bare Costs	Incl. O&P
1 Labor Foreman	$20.70	$165.60	$35.15	$281.20	$21.37	$35.68
3 Laborers	18.70	448.80	31.75	762.00		
1 Equip. Oper. (medium)	26.50	212.00	43.65	349.20		
1 Concrete Finisher	24.90	199.20	40.05	320.40		
1 Curb/Gutter Paver, 2-Track		687.60		756.36	14.32	15.76
48 L.H., Daily Totals		$1713.20		$2469.16	$35.69	$51.44
Crew B-69B	Hr.	Daily	Hr.	Daily	Bare Costs	Incl. O&P
1 Labor Foreman	$20.70	$165.60	$35.15	$281.20	$21.37	$35.68
3 Laborers	18.70	448.80	31.75	762.00		
1 Equip. Oper. (medium)	26.50	212.00	43.65	349.20		
1 Cement Finisher	24.90	199.20	40.05	320.40		
1 Curb/Gutter Paver, 4-Track		771.00		848.10	16.06	17.67
48 L.H., Daily Totals		$1796.60		$2560.90	$37.43	$53.35
Crew B-70	Hr.	Daily	Hr.	Daily	Bare Costs	Incl. O&P
1 Labor Foreman (outside)	$20.70	$165.60	$35.15	$281.20	$22.33	$37.34
3 Laborers	18.70	448.80	31.75	762.00		
3 Equip. Oper. (med.)	26.50	636.00	43.65	1047.60		
1 Grader, 30,000 Lbs.		505.40		555.94		
1 Ripper, beam & 1 shank		76.40		84.04		
1 Road Sweeper, SP, 8' wide		551.00		606.10		
1 F.E. Loader, W.M., 1.5 C.Y.		277.60		305.36	25.19	27.70
56 L.H., Daily Totals		$2660.80		$3642.24	$47.51	$65.04
Crew B-71	Hr.	Daily	Hr.	Daily	Bare Costs	Incl. O&P
1 Labor Foreman (outside)	$20.70	$165.60	$35.15	$281.20	$22.33	$37.34
3 Laborers	18.70	448.80	31.75	762.00		
3 Equip. Oper. (med.)	26.50	636.00	43.65	1047.60		
1 Pvmt. Profiler, 750 H.P.		5176.00		5693.60		
1 Road Sweeper, SP, 8' wide		551.00		606.10		
1 F.E. Loader, W.M., 1.5 C.Y.		277.60		305.36	107.22	117.95
56 L.H., Daily Totals		$7255.00		$8695.86	$129.55	$155.28

Crews

Crew No.	Bare Costs		Incl. Subs O & P		Cost Per Labor-Hour	
Crew B-72	Hr.	Daily	Hr.	Daily	Bare Costs	Incl. O&P
1 Labor Foreman (outside)	$20.70	$165.60	$35.15	$281.20	$22.85	$38.13
3 Laborers	18.70	448.80	31.75	762.00		
4 Equip. Oper. (med.)	26.50	848.00	43.65	1396.80		
1 Pvmt. Profiler, 750 H.P.		5176.00		5693.60		
1 Hammermill, 250 H.P.		1512.00		1663.20		
1 Windrow Loader		869.00		955.90		
1 Mix Paver 165 H.P.		1817.00		1998.70		
1 Roller, Pneum. Whl, 12 Ton		295.60		325.16	151.09	166.20
64 L.H., Daily Totals		$11132.00		$13076.56	$173.94	$204.32

Crew B-73	Hr.	Daily	Hr.	Daily	Bare Costs	Incl. O&P
1 Labor Foreman (outside)	$20.70	$165.60	$35.15	$281.20	$23.82	$39.61
2 Laborers	18.70	299.20	31.75	508.00		
5 Equip. Oper. (med.)	26.50	1060.00	43.65	1746.00		
1 Road Mixer, 310 H.P.		1819.00		2000.90		
1 Tandem Roller, 10 Ton		211.00		232.10		
1 Hammermill, 250 H.P.		1512.00		1663.20		
1 Grader, 30,000 Lbs.		505.40		555.94		
.5 F.E. Loader, W.M., 1.5 C.Y.		138.80		152.68		
.5 Truck Tractor, 195 H.P.		117.80		129.58		
.5 Water Tanker, 5000 Gal.		60.60		66.66	68.20	75.02
64 L.H., Daily Totals		$5889.40		$7336.26	$92.02	$114.63

Crew B-74	Hr.	Daily	Hr.	Daily	Bare Costs	Incl. O&P
1 Labor Foreman (outside)	$20.70	$165.60	$35.15	$281.20	$23.43	$39.02
1 Laborer	18.70	149.60	31.75	254.00		
4 Equip. Oper. (med.)	26.50	848.00	43.65	1396.80		
2 Truck Drivers (heavy)	21.00	336.00	35.35	565.60		
1 Grader, 30,000 Lbs.		505.40		555.94		
1 Ripper, beam & 1 shank		76.40		84.04		
2 Stabilizers, 310 H.P.		2530.00		2783.00		
1 Flatbed Truck, Gas, 3 Ton		185.20		203.72		
1 Chem. Spreader, Towed		44.20		48.62		
1 Roller, Vibratory, 25 Ton		565.60		622.16		
1 Water Tanker, 5000 Gal.		121.20		133.32		
1 Truck Tractor, 195 H.P.		235.60		259.16	66.62	73.28
64 L.H., Daily Totals		$5762.80		$7187.56	$90.04	$112.31

Crew B-75	Hr.	Daily	Hr.	Daily	Bare Costs	Incl. O&P
1 Labor Foreman (outside)	$20.70	$165.60	$35.15	$281.20	$23.77	$39.55
1 Laborer	18.70	149.60	31.75	254.00		
4 Equip. Oper. (med.)	26.50	848.00	43.65	1396.80		
1 Truck Driver (heavy)	21.00	168.00	35.35	282.80		
1 Grader, 30,000 Lbs.		505.40		555.94		
1 Ripper, beam & 1 shank		76.40		84.04		
2 Stabilizers, 310 H.P.		2530.00		2783.00		
1 Dist. Tanker, 3000 Gallon		273.00		300.30		
1 Truck Tractor, 240 H.P.		273.00		300.30		
1 Roller, Vibratory, 25 Ton		565.60		622.16	75.42	82.96
56 L.H., Daily Totals		$5554.60		$6860.54	$99.19	$122.51

Crew B-76	Hr.	Daily	Hr.	Daily	Bare Costs	Incl. O&P
1 Dock Builder Foreman	$27.15	$217.20	$48.25	$386.00	$25.69	$44.52
5 Dock Builders	25.15	1006.00	44.70	1788.00		
2 Equip. Oper. (crane)	27.45	439.20	45.20	723.20		
1 Equip. Oper. Oiler	23.40	187.20	38.55	308.40		
1 Crawler Crane, 50 Ton		1131.00		1244.10		
1 Barge, 400 Ton		294.20		323.62		
1 Hammer, 15K Ft. Lbs.		360.00		396.00		
60 L.F. of Leads, 15K Ft. Lbs.		60.00		66.00		
1 Air Compressor, 600 C.F.M.		396.20		435.82		
2 -50' Air Hoses, 3"		38.10		41.91	31.66	34.83
72 L.H., Daily Totals		$4129.10		$5713.05	$57.35	$79.35

Crew B-76A	Hr.	Daily	Hr.	Daily	Bare Costs	Incl. O&P
1 Laborer Foreman	$20.70	$165.60	$35.15	$281.20	$20.63	$34.71
5 Laborers	18.70	748.00	31.75	1270.00		
1 Equip. Oper. (crane)	27.45	219.60	45.20	361.60		
1 Equip. Oper. Oiler	23.40	187.20	38.55	308.40		
1 Crawler Crane, 50 Ton		1131.00		1244.10		
1 Barge, 400 Ton		294.20		323.62	22.27	24.50
64 L.H., Daily Totals		$2745.60		$3788.92	$42.90	$59.20

Crew B-77	Hr.	Daily	Hr.	Daily	Bare Costs	Incl. O&P
1 Labor Foreman	$20.70	$165.60	$35.15	$281.20	$19.41	$32.90
3 Laborers	18.70	448.80	31.75	762.00		
1 Truck Driver (light)	20.25	162.00	34.10	272.80		
1 Crack Cleaner, 25 H.P.		48.20		53.02		
1 Crack Filler, Trailer Mtd.		162.40		178.64		
1 Flatbed Truck, gas, 3 Ton		185.20		203.72	9.89	10.88
40 L.H., Daily Totals		$1172.20		$1751.38	$29.31	$43.78

Crew B-78	Hr.	Daily	Hr.	Daily	Bare Costs	Incl. O&P
1 Labor Foreman	$20.70	$165.60	$35.15	$281.20	$19.10	$32.43
4 Laborers	18.70	598.40	31.75	1016.00		
1 Paint Striper, S.P.		138.00		151.80		
1 Flatbed Truck, Gas, 3 Ton		185.20		203.72		
1 Pickup Truck, 3/4 Ton		84.80		93.28	10.20	11.22
40 L.H., Daily Totals		$1172.00		$1746.00	$29.30	$43.65

Crew B-79	Hr.	Daily	Hr.	Daily	Bare Costs	Incl. O&P
1 Labor Foreman	$20.70	$165.60	$35.15	$281.20	$19.20	$32.60
3 Laborers	18.70	448.80	31.75	762.00		
1 Thermo. Striper, T.M.		596.20		655.82		
1 Flatbed Truck, Gas, 3 Ton		185.20		203.72		
2 Pickup Trucks, 3/4 Ton		169.60		186.56	29.72	32.69
32 L.H., Daily Totals		$1565.40		$2089.30	$48.92	$65.29

Crew B-80	Hr.	Daily	Hr.	Daily	Bare Costs	Incl. O&P
1 Labor Foreman	$20.70	$165.60	$35.15	$281.20	$19.37	$32.88
2 Laborers	18.70	299.20	31.75	508.00		
1 Flatbed Truck, gas, 3 Ton		185.20		203.72		
1 Fence Post Auger, T.M.		375.20		412.72	23.35	25.68
24 L.H., Daily Totals		$1025.20		$1405.64	$42.72	$58.57

Crew B-80A	Hr.	Daily	Hr.	Daily	Bare Costs	Incl. O&P
3 Laborers	$18.70	$448.80	$31.75	$762.00	$18.70	$31.75
1 Flatbed Truck, gas, 3 Ton		185.20		203.72	7.72	8.49
24 L.H., Daily Totals		$634.00		$965.72	$26.42	$40.24

Crews

Crew No.	Bare Costs		Incl. Subs O & P		Cost Per Labor-Hour	
Crew B-80B	Hr.	Daily	Hr.	Daily	Bare Costs	Incl. O&P
3 Laborers	$18.70	$448.80	$31.75	$762.00	$20.39	$34.29
1 Equip. Oper. (light)	25.45	203.60	41.90	335.20		
1 Crane, Flatbed Mounted, 3 Ton		271.60		298.76	8.49	9.34
32 L.H., Daily Totals		$924.00		$1395.96	$28.88	$43.62
Crew B-80C	Hr.	Daily	Hr.	Daily	Bare Costs	Incl. O&P
2 Laborers	$18.70	$299.20	$31.75	$508.00	$19.22	$32.53
1 Truck Driver (light)	20.25	162.00	34.10	272.80		
1 Flatbed Truck, Gas, 1.5 Ton		141.40		155.54		
1 Manual fence post auger, gas		6.00		6.60	6.14	6.76
24 L.H., Daily Totals		$608.60		$942.94	$25.36	$39.29
Crew B-81	Hr.	Daily	Hr.	Daily	Bare Costs	Incl. O&P
1 Laborer	$18.70	$149.60	$31.75	$254.00	$19.85	$33.55
1 Truck Driver (heavy)	21.00	168.00	35.35	282.80		
1 Hydromulcher, T.M.		216.20		237.82		
1 Truck Tractor, 195 H.P.		235.60		259.16	28.24	31.06
16 L.H., Daily Totals		$769.40		$1033.78	$48.09	$64.61
Crew B-82	Hr.	Daily	Hr.	Daily	Bare Costs	Incl. O&P
1 Laborer	$18.70	$149.60	$31.75	$254.00	$22.07	$36.83
1 Equip. Oper. (light)	25.45	203.60	41.90	335.20		
1 Horiz. Borer, 6 H.P.		71.20		78.32	4.45	4.89
16 L.H., Daily Totals		$424.40		$667.52	$26.52	$41.72
Crew B-83	Hr.	Daily	Hr.	Daily	Bare Costs	Incl. O&P
1 Tugboat Captain	$26.50	$212.00	$43.65	$349.20	$22.60	$37.70
1 Tugboat Hand	18.70	149.60	31.75	254.00		
1 Tugboat, 250 H.P.		550.20		605.22	34.39	37.83
16 L.H., Daily Totals		$911.80		$1208.42	$56.99	$75.53
Crew B-84	Hr.	Daily	Hr.	Daily	Bare Costs	Incl. O&P
1 Equip. Oper. (med.)	$26.50	$212.00	$43.65	$349.20	$26.50	$43.65
1 Rotary Mower/Tractor		235.00		258.50	29.38	32.31
8 L.H., Daily Totals		$447.00		$607.70	$55.88	$75.96
Crew B-85	Hr.	Daily	Hr.	Daily	Bare Costs	Incl. O&P
3 Laborers	$18.70	$448.80	$31.75	$762.00	$20.72	$34.85
1 Equip. Oper. (med.)	26.50	212.00	43.65	349.20		
1 Truck Driver (heavy)	21.00	168.00	35.35	282.80		
1 Aerial Lift Truck, 80'		546.80		601.48		
1 Brush Chipper, 12", 130 H.P.		183.60		201.96		
1 Pruning Saw, Rotary		7.45		8.20	18.45	20.29
40 L.H., Daily Totals		$1566.65		$2205.64	$39.17	$55.14
Crew B-86	Hr.	Daily	Hr.	Daily	Bare Costs	Incl. O&P
1 Equip. Oper. (med.)	$26.50	$212.00	$43.65	$349.20	$26.50	$43.65
1 Stump Chipper, S.P.		106.30		116.93	13.29	14.62
8 L.H., Daily Totals		$318.30		$466.13	$39.79	$58.27
Crew B-86A	Hr.	Daily	Hr.	Daily	Bare Costs	Incl. O&P
1 Equip. Oper. (medium)	$26.50	$212.00	$43.65	$349.20	$26.50	$43.65
1 Grader, 30,000 Lbs.		505.40		555.94	63.17	69.49
8 L.H., Daily Totals		$717.40		$905.14	$89.67	$113.14
Crew B-86B	Hr.	Daily	Hr.	Daily	Bare Costs	Incl. O&P
1 Equip. Oper. (medium)	$26.50	$212.00	$43.65	$349.20	$26.50	$43.65
1 Dozer, 200 H.P.		988.40		1087.24	123.55	135.91
8 L.H., Daily Totals		$1200.40		$1436.44	$150.05	$179.56

Crew No.	Bare Costs		Incl. Subs O & P		Cost Per Labor-Hour	
Crew B-87	Hr.	Daily	Hr.	Daily	Bare Costs	Incl. O&P
1 Laborer	$18.70	$149.60	$31.75	$254.00	$24.94	$41.27
4 Equip. Oper. (med.)	26.50	848.00	43.65	1396.80		
2 Feller Buncher, 100 H.P.		1026.80		1129.48		
1 Log Chipper, 22" Tree		520.00		572.00		
1 Dozer, 105 H.P.		494.80		544.28		
1 Chain saw, gas, 36" Long		34.40		37.84	51.90	57.09
40 L.H., Daily Totals		$3073.60		$3934.40	$76.84	$98.36
Crew B-88	Hr.	Daily	Hr.	Daily	Bare Costs	Incl. O&P
1 Laborer	$18.70	$149.60	$31.75	$254.00	$25.39	$41.95
6 Equip. Oper. (med.)	26.50	1272.00	43.65	2095.20		
2 Feller Bunchers, 100 H.P.		1026.80		1129.48		
1 Log Chipper, 22" Tree		520.00		572.00		
2 Log Skidders, 50 H.P.		1632.80		1796.08		
1 Dozer, 105 H.P.		494.80		544.28		
1 Chain saw, gas, 36" Long		34.40		37.84	66.23	72.85
56 L.H., Daily Totals		$5130.40		$6428.88	$91.61	$114.80
Crew B-89	Hr.	Daily	Hr.	Daily	Bare Costs	Incl. O&P
1 Skilled Worker	$26.10	$208.80	$44.25	$354.00	$22.40	$38.00
1 Building Laborer	18.70	149.60	31.75	254.00		
1 Flatbed Truck, gas, 3 Ton		185.20		203.72		
1 Concrete Saw		118.40		130.24		
1 Water Tank, 65 Gal.		16.40		18.04	20.00	22.00
16 L.H., Daily Totals		$678.40		$960.00	$42.40	$60.00
Crew B-89A	Hr.	Daily	Hr.	Daily	Bare Costs	Incl. O&P
1 Skilled Worker	$26.10	$208.80	$44.25	$354.00	$22.40	$38.00
1 Laborer	18.70	149.60	31.75	254.00		
1 Core Drill (large)		105.00		115.50	6.56	7.22
16 L.H., Daily Totals		$463.40		$723.50	$28.96	$45.22
Crew B-89B	Hr.	Daily	Hr.	Daily	Bare Costs	Incl. O&P
1 Equip. Oper. (light)	$25.45	$203.60	$41.90	$335.20	$22.85	$38.00
1 Truck Driver, Light	20.25	162.00	34.10	272.80		
1 Wall Saw, Hydraulic, 10 H.P.		78.20		86.02		
1 Generator, Diesel, 100 KW		305.60		336.16		
1 Water Tank, 65 Gal.		16.40		18.04		
1 Flatbed Truck, Gas, 3 Ton		185.20		203.72	36.59	40.25
16 L.H., Daily Totals		$951.00		$1251.94	$59.44	$78.25
Crew B-90	Hr.	Daily	Hr.	Daily	Bare Costs	Incl. O&P
1 Labor Foreman (outside)	$20.70	$165.60	$35.15	$281.20	$21.21	$35.61
3 Laborers	18.70	448.80	31.75	762.00		
2 Equip. Oper. (light)	25.45	407.20	41.90	670.40		
2 Truck Drivers (heavy)	21.00	336.00	35.35	565.60		
1 Road Mixer, 310 H.P.		1819.00		2000.90		
1 Dist. Truck, 2000 Gal.		247.40		272.14	32.29	35.52
64 L.H., Daily Totals		$3424.00		$4552.24	$53.50	$71.13
Crew B-90A	Hr.	Daily	Hr.	Daily	Bare Costs	Incl. O&P
1 Labor Foreman	$20.70	$165.60	$35.15	$281.20	$23.44	$39.04
2 Laborers	18.70	299.20	31.75	508.00		
4 Equip. Oper. (medium)	26.50	848.00	43.65	1396.80		
2 Grader, 30,000 Lbs.		1010.80		1111.88		
1 Tandem Roller, 10 Ton		211.00		232.10		
1 Roller, Pneum. Whl, 12 Ton		295.60		325.16	27.10	29.81
56 L.H., Daily Totals		$2830.20		$3855.14	$50.54	$68.84

Crews

Crew No.	Bare Costs		Incl. Subs O & P		Cost Per Labor-Hour	
Crew B-90B	Hr.	Daily	Hr.	Daily	Bare Costs	Incl. O&P
1 Labor Foreman	$20.70	$165.60	$35.15	$281.20	$22.93	$38.27
2 Laborers	18.70	299.20	31.75	508.00		
3 Equip. Oper. (medium)	26.50	636.00	43.65	1047.60		
1 Tandem Roller, 10 Ton		211.00		232.10		
1 Roller, Pneum. Whl, 12 Ton		295.60		325.16		
1 Road Mixer, 310 H.P.		1819.00		2000.90	48.45	53.30
48 L.H., Daily Totals		$3426.40		$4394.96	$71.38	$91.56
Crew B-91	Hr.	Daily	Hr.	Daily	Bare Costs	Incl. O&P
1 Labor Foreman (outside)	$20.70	$165.60	$35.15	$281.20	$23.14	$38.58
2 Laborers	18.70	299.20	31.75	508.00		
4 Equip. Oper. (med.)	26.50	848.00	43.65	1396.80		
1 Truck Driver (heavy)	21.00	168.00	35.35	282.80		
1 Dist. Tanker, 3000 Gallon		273.00		300.30		
1 Truck Tractor, 240 H.P.		273.00		300.30		
1 Aggreg. Spreader, S.P.		818.80		900.68		
1 Roller, Pneum. Whl, 12 Ton		295.60		325.16		
1 Tandem Roller, 10 Ton		211.00		232.10	29.24	32.16
64 L.H., Daily Totals		$3352.20		$4527.34	$52.38	$70.74
Crew B-92	Hr.	Daily	Hr.	Daily	Bare Costs	Incl. O&P
1 Labor Foreman (outside)	$20.70	$165.60	$35.15	$281.20	$19.20	$32.60
3 Laborers	18.70	448.80	31.75	762.00		
1 Crack Cleaner, 25 H.P.		48.20		53.02		
1 Air Compressor, 60 C.F.M.		108.80		119.68		
1 Tar Kettle, T.M.		51.35		56.48		
1 Flatbed Truck, gas, 3 Ton		185.20		203.72	12.30	13.53
32 L.H., Daily Totals		$1007.95		$1476.11	$31.50	$46.13
Crew B-93	Hr.	Daily	Hr.	Daily	Bare Costs	Incl. O&P
1 Equip. Oper. (med.)	$26.50	$212.00	$43.65	$349.20	$26.50	$43.65
1 Feller Buncher, 100 H.P.		513.40		564.74	64.17	70.59
8 L.H., Daily Totals		$725.40		$913.94	$90.67	$114.24
Crew B-94A	Hr.	Daily	Hr.	Daily	Bare Costs	Incl. O&P
1 Laborer	$18.70	$149.60	$31.75	$254.00	$18.70	$31.75
1 Diaphragm Water Pump, 2"		57.40		63.14		
1 -20' Suction Hose, 2"		1.95		2.15		
2 -50' Discharge Hoses, 2"		1.80		1.98	7.64	8.41
8 L.H., Daily Totals		$210.75		$321.26	$26.34	$40.16
Crew B-94B	Hr.	Daily	Hr.	Daily	Bare Costs	Incl. O&P
1 Laborer	$18.70	$149.60	$31.75	$254.00	$18.70	$31.75
1 Diaphragm Water Pump, 4"		78.20		86.02		
1 -20' Suction Hose, 4"		3.25		3.58		
2 -50' Discharge Hoses, 4"		4.70		5.17	10.77	11.85
8 L.H., Daily Totals		$235.75		$348.76	$29.47	$43.60
Crew B-94C	Hr.	Daily	Hr.	Daily	Bare Costs	Incl. O&P
1 Laborer	$18.70	$149.60	$31.75	$254.00	$18.70	$31.75
1 Centrifugal Water Pump, 3"		63.40		69.74		
1 -20' Suction Hose, 3"		3.05		3.36		
2 -50' Discharge Hoses, 3"		3.50		3.85	8.74	9.62
8 L.H., Daily Totals		$219.55		$330.94	$27.44	$41.37
Crew B-94D	Hr.	Daily	Hr.	Daily	Bare Costs	Incl. O&P
1 Laborer	$18.70	$149.60	$31.75	$254.00	$18.70	$31.75
1 Centr. Water Pump, 6"		278.20		306.02		
1 -20' Suction Hose, 6"		11.50		12.65		
2 -50' Discharge Hoses, 6"		12.60		13.86	37.79	41.57
8 L.H., Daily Totals		$451.90		$586.53	$56.49	$73.32

Crew No.	Bare Costs		Incl. Subs O & P		Cost Per Labor-Hour	
Crew B-95A	Hr.	Daily	Hr.	Daily	Bare Costs	Incl. O&P
1 Equip. Oper. (crane)	$27.45	$219.60	$45.20	$361.60	$23.07	$38.48
1 Laborer	18.70	149.60	31.75	254.00		
1 Hyd. Excavator, 5/8 C.Y.		462.20		508.42	28.89	31.78
16 L.H., Daily Totals		$831.40		$1124.02	$51.96	$70.25
Crew B-95B	Hr.	Daily	Hr.	Daily	Bare Costs	Incl. O&P
1 Equip. Oper. (crane)	$27.45	$219.60	$45.20	$361.60	$23.07	$38.48
1 Laborer	18.70	149.60	31.75	254.00		
1 Hyd. Excavator, 1.5 C.Y.		775.60		853.16	48.48	53.32
16 L.H., Daily Totals		$1144.80		$1468.76	$71.55	$91.80
Crew B-95C	Hr.	Daily	Hr.	Daily	Bare Costs	Incl. O&P
1 Equip. Oper. (crane)	$27.45	$219.60	$45.20	$361.60	$23.07	$38.48
1 Laborer	18.70	149.60	31.75	254.00		
1 Hyd. Excavator, 2.5 C.Y.		1333.00		1466.30	83.31	91.64
16 L.H., Daily Totals		$1702.20		$2081.90	$106.39	$130.12
Crew C-1	Hr.	Daily	Hr.	Daily	Bare Costs	Incl. O&P
2 Carpenters	$25.70	$411.20	$43.60	$697.60	$22.23	$37.69
1 Carpenter Helper	18.80	150.40	31.80	254.40		
1 Laborer	18.70	149.60	31.75	254.00		
32 L.H., Daily Totals		$711.20		$1206.00	$22.23	$37.69
Crew C-2	Hr.	Daily	Hr.	Daily	Bare Costs	Incl. O&P
1 Carpenter Foreman (out)	$27.70	$221.60	$47.00	$376.00	$22.57	$38.26
2 Carpenters	25.70	411.20	43.60	697.60		
2 Carpenter Helpers	18.80	300.80	31.80	508.80		
1 Laborer	18.70	149.60	31.75	254.00		
48 L.H., Daily Totals		$1083.20		$1836.40	$22.57	$38.26
Crew C-2A	Hr.	Daily	Hr.	Daily	Bare Costs	Incl. O&P
1 Carpenter Foreman (out)	$27.70	$221.60	$47.00	$376.00	$24.73	$41.60
3 Carpenters	25.70	616.80	43.60	1046.40		
1 Cement Finisher	24.90	199.20	40.05	320.40		
1 Laborer	18.70	149.60	31.75	254.00		
48 L.H., Daily Totals		$1187.20		$1996.80	$24.73	$41.60
Crew C-3	Hr.	Daily	Hr.	Daily	Bare Costs	Incl. O&P
1 Rodman Foreman	$29.65	$237.20	$52.80	$422.40	$24.27	$42.21
3 Rodmen (reinf.)	27.65	663.60	49.25	1182.00		
1 Equip. Oper. (light)	25.45	203.60	41.90	335.20		
3 Laborers	18.70	448.80	31.75	762.00		
3 Stressing Equipment		24.60		27.06		
.5 Grouting Equipment		73.00		80.30	1.52	1.68
64 L.H., Daily Totals		$1650.80		$2808.96	$25.79	$43.89
Crew C-4	Hr.	Daily	Hr.	Daily	Bare Costs	Incl. O&P
1 Rodman Foreman	$29.65	$237.20	$52.80	$422.40	$25.91	$45.76
2 Rodmen (reinf.)	27.65	442.40	49.25	788.00		
1 Building Laborer	18.70	149.60	31.75	254.00		
3 Stressing Equipment		24.60		27.06	0.77	0.85
32 L.H., Daily Totals		$853.80		$1491.46	$26.68	$46.61
Crew C-5	Hr.	Daily	Hr.	Daily	Bare Costs	Incl. O&P
1 Rodman Foreman	$29.65	$237.20	$52.80	$422.40	$24.97	$43.33
2 Rodmen (reinf.)	27.65	442.40	49.25	788.00		
1 Equip. Oper. (crane)	27.45	219.60	45.20	361.60		
2 Building Laborers	18.70	299.20	31.75	508.00		
1 Hyd. Crane, 25 Ton		739.60		813.56	15.41	16.95
48 L.H., Daily Totals		$1938.00		$2893.56	$40.38	$60.28

Crews

Crew No.	Bare Costs		Incl. Subs O & P		Cost Per Labor-Hour	
Crew C-6	Hr.	Daily	Hr.	Daily	Bare Costs	Incl. O&P
1 Labor Foreman (outside)	$20.70	$165.60	$35.15	$281.20	$20.07	$33.70
4 Laborers	18.70	598.40	31.75	1016.00		
1 Cement Finisher	24.90	199.20	40.05	320.40		
2 Gas Engine Vibrators		42.80		47.08	.89	.98
48 L.H., Daily Totals		$1006.00		$1664.68	$20.96	$34.68
Crew C-7	Hr.	Daily	Hr.	Daily	Bare Costs	Incl. O&P
1 Labor Foreman (outside)	$20.70	$165.60	$35.15	$281.20	$21.00	$35.13
5 Laborers	18.70	748.00	31.75	1270.00		
1 Cement Finisher	24.90	199.20	40.05	320.40		
1 Equip. Oper. (med.)	26.50	212.00	43.65	349.20		
1 Equip. Oper. (oiler)	23.40	187.20	38.55	308.40		
2 Gas Engine Vibrator		42.80		47.08		
1 Concrete Bucket, 1 C.Y.		17.80		19.58		
1 Hyd. Crane, 55 Ton		1060.00		1166.00	15.56	17.12
72 L.H., Daily Totals		$2632.60		$3761.86	$36.56	$52.25
Crew C-8	Hr.	Daily	Hr.	Daily	Bare Costs	Incl. O&P
1 Labor Foreman (outside)	$20.70	$165.60	$35.15	$281.20	$21.87	$36.31
3 Laborers	18.70	448.80	31.75	762.00		
2 Cement Finishers	24.90	398.40	40.05	640.80		
1 Equip. Oper. (med.)	26.50	212.00	43.65	349.20		
1 Concrete Pump (small)		708.60		779.46	12.65	13.92
56 L.H., Daily Totals		$1933.40		$2812.66	$34.52	$50.23
Crew C-8A	Hr.	Daily	Hr.	Daily	Bare Costs	Incl. O&P
1 Labor Foreman (outside)	$20.70	$165.60	$35.15	$281.20	$21.10	$35.08
3 Laborers	18.70	448.80	31.75	762.00		
2 Cement Finishers	24.90	398.40	40.05	640.80		
48 L.H., Daily Totals		$1012.80		$1684.00	$21.10	$35.08
Crew C-8B	Hr.	Daily	Hr.	Daily	Bare Costs	Incl. O&P
1 Labor Foreman (outside)	$20.70	$165.60	$35.15	$281.20	$20.66	$34.81
3 Laborers	18.70	448.80	31.75	762.00		
1 Equipment Operator	26.50	212.00	43.65	349.20		
1 Vibrating Screed		65.90		72.49		
1 Roller, Vibratory, 25 Ton		565.60		622.16		
1 Dozer, 200 H.P.		988.40		1087.24	40.50	44.55
40 L.H., Daily Totals		$2446.30		$3174.29	$61.16	$79.36
Crew C-8C	Hr.	Daily	Hr.	Daily	Bare Costs	Incl. O&P
1 Labor Foreman (outside)	$20.70	$165.60	$35.15	$281.20	$21.37	$35.68
3 Laborers	18.70	448.80	31.75	762.00		
1 Cement Finisher	24.90	199.20	40.05	320.40		
1 Equipment Operator (med.)	26.50	212.00	43.65	349.20		
1 Shotcrete Rig, 12 CY/hr		250.60		275.66	5.22	5.74
48 L.H., Daily Totals		$1276.20		$1988.46	$26.59	$41.43
Crew C-8D	Hr.	Daily	Hr.	Daily	Bare Costs	Incl. O&P
1 Labor Foreman (outside)	$20.70	$165.60	$35.15	$281.20	$22.44	$37.21
1 Laborer	18.70	149.60	31.75	254.00		
1 Cement Finisher	24.90	199.20	40.05	320.40		
1 Equipment Operator (light)	25.45	203.60	41.90	335.20		
1 Air Compressor, 250 C.F.M.		151.40		166.54		
2 -50' Air Hoses, 1"		8.20		9.02	4.99	5.49
32 L.H., Daily Totals		$877.60		$1366.36	$27.43	$42.70

Crew No.	Bare Costs		Incl. Subs O & P		Cost Per Labor-Hour	
Crew C-8E	Hr.	Daily	Hr.	Daily	Bare Costs	Incl. O&P
1 Labor Foreman (outside)	$20.70	$165.60	$35.15	$281.20	$22.44	$37.21
1 Laborer	18.70	149.60	31.75	254.00		
1 Cement Finisher	24.90	199.20	40.05	320.40		
1 Equipment Operator (light)	25.45	203.60	41.90	335.20		
1 Air Compressor, 250 C.F.M.		151.40		166.54		
2 -50' Air Hoses, 1"		8.20		9.02		
1 Concrete Pump (small)		708.60		779.46	27.13	29.84
32 L.H., Daily Totals		$1586.20		$2145.82	$49.57	$67.06
Crew C-10	Hr.	Daily	Hr.	Daily	Bare Costs	Incl. O&P
1 Laborer	$18.70	$149.60	$31.75	$254.00	$22.83	$37.28
2 Cement Finishers	24.90	398.40	40.05	640.80		
24 L.H., Daily Totals		$548.00		$894.80	$22.83	$37.28
Crew C-10B	Hr.	Daily	Hr.	Daily	Bare Costs	Incl. O&P
3 Laborers	$18.70	$448.80	$31.75	$762.00	$21.18	$35.07
2 Cement Finishers	24.90	398.40	40.05	640.80		
1 Concrete Mixer, 10 CF		138.60		152.46		
2 Conc. Finisher, 46" Walk-Behind		51.60		56.76	4.75	5.23
40 L.H., Daily Totals		$1037.40		$1612.02	$25.93	$40.30
Crew C-11	Hr.	Daily	Hr.	Daily	Bare Costs	Incl. O&P
1 Skilled Worker Foreman	$28.10	$224.80	$47.65	$381.20	$26.58	$44.87
5 Skilled Workers	26.10	1044.00	44.25	1770.00		
1 Equip. Oper. (crane)	27.45	219.60	45.20	361.60		
1 Lattice Boom Crane, 150 Ton		1852.00		2037.20	33.07	36.38
56 L.H., Daily Totals		$3340.40		$4550.00	$59.65	$81.25
Crew C-12	Hr.	Daily	Hr.	Daily	Bare Costs	Incl. O&P
1 Carpenter Foreman (out)	$27.70	$221.60	$47.00	$376.00	$25.16	$42.46
3 Carpenters	25.70	616.80	43.60	1046.40		
1 Laborer	18.70	149.60	31.75	254.00		
1 Equip. Oper. (crane)	27.45	219.60	45.20	361.60		
1 Hyd. Crane, 12 Ton		724.00		796.40	15.08	16.59
48 L.H., Daily Totals		$1931.60		$2834.40	$40.24	$59.05
Crew C-13	Hr.	Daily	Hr.	Daily	Bare Costs	Incl. O&P
2 Struc. Steel Workers	$27.70	$443.20	$54.05	$864.80	$27.03	$50.57
1 Carpenter	25.70	205.60	43.60	348.80		
1 Welder, gas engine, 300 amp		115.20		126.72	4.80	5.28
24 L.H., Daily Totals		$764.00		$1340.32	$31.83	$55.85
Crew C-14	Hr.	Daily	Hr.	Daily	Bare Costs	Incl. O&P
1 Carpenter Foreman (out)	$27.70	$221.60	$47.00	$376.00	$22.79	$38.58
3 Carpenters	25.70	616.80	43.60	1046.40		
2 Carpenter Helpers	18.80	300.80	31.80	508.80		
4 Laborers	18.70	598.40	31.75	1016.00		
2 Rodmen (reinf.)	27.65	442.40	49.25	788.00		
2 Rodman Helpers	18.80	300.80	31.80	508.80		
2 Cement Finishers	24.90	398.40	40.05	640.80		
1 Equip. Oper. (crane)	27.45	219.60	45.20	361.60		
1 Hyd. Crane, 80 Ton		1214.00		1335.40	8.93	9.82
136 L.H., Daily Totals		$4312.80		$6581.80	$31.71	$48.40

Crews

Crew No.	Bare Costs		Incl. Subs O & P		Cost Per Labor-Hour	
Crew C-14A	Hr.	Daily	Hr.	Daily	Bare Costs	Incl. O&P
1 Carpenter Foreman (out)	$27.70	$221.60	$47.00	$376.00	$25.53	$43.55
16 Carpenters	25.70	3289.60	43.60	5580.80		
4 Rodmen (reinf.)	27.65	884.80	49.25	1576.00		
2 Laborers	18.70	299.20	31.75	508.00		
1 Cement Finisher	24.90	199.20	40.05	320.40		
1 Equip. Oper. (med.)	26.50	212.00	43.65	349.20		
1 Gas Engine Vibrator		21.40		23.54		
1 Concrete Pump (small)		708.60		779.46	3.65	4.01
200 L.H., Daily Totals		$5836.40		$9513.40	$29.18	$47.57
Crew C-14B	Hr.	Daily	Hr.	Daily	Bare Costs	Incl. O&P
1 Carpenter Foreman (out)	$27.70	$221.60	$47.00	$376.00	$25.51	$43.42
16 Carpenters	25.70	3289.60	43.60	5580.80		
4 Rodmen (reinf.)	27.65	884.80	49.25	1576.00		
2 Laborers	18.70	299.20	31.75	508.00		
2 Cement Finishers	24.90	398.40	40.05	640.80		
1 Equip. Oper. (med.)	26.50	212.00	43.65	349.20		
1 Gas Engine Vibrator		21.40		23.54		
1 Concrete Pump (small)		708.60		779.46	3.51	3.86
208 L.H., Daily Totals		$6035.60		$9833.80	$29.02	$47.28
Crew C-14C	Hr.	Daily	Hr.	Daily	Bare Costs	Incl. O&P
1 Carpenter Foreman (out)	$27.70	$221.60	$47.00	$376.00	$24.06	$41.01
6 Carpenters	25.70	1233.60	43.60	2092.80		
2 Rodmen (reinf.)	27.65	442.40	49.25	788.00		
4 Laborers	18.70	598.40	31.75	1016.00		
1 Cement Finisher	24.90	199.20	40.05	320.40		
1 Gas Engine Vibrator		21.40		23.54	0.19	0.21
112 L.H., Daily Totals		$2716.60		$4616.74	$24.26	$41.22
Crew C-14D	Hr.	Daily	Hr.	Daily	Bare Costs	Incl. O&P
1 Carpenter Foreman (out)	$27.70	$221.60	$47.00	$376.00	$25.38	$43.10
18 Carpenters	25.70	3700.80	43.60	6278.40		
2 Rodmen (reinf.)	27.65	442.40	49.25	788.00		
2 Laborers	18.70	299.20	31.75	508.00		
1 Cement Finisher	24.90	199.20	40.05	320.40		
1 Equip. Oper. (med.)	26.50	212.00	43.65	349.20		
1 Gas Engine Vibrator		21.40		23.54		
1 Concrete Pump (small)		708.60		779.46	3.65	4.01
200 L.H., Daily Totals		$5805.20		$9423.00	$29.03	$47.12
Crew C-14E	Hr.	Daily	Hr.	Daily	Bare Costs	Incl. O&P
1 Carpenter Foreman (out)	$27.70	$221.60	$47.00	$376.00	$24.61	$42.41
2 Carpenters	25.70	411.20	43.60	697.60		
4 Rodmen (reinf.)	27.65	884.80	49.25	1576.00		
3 Laborers	18.70	448.80	31.75	762.00		
1 Cement Finisher	24.90	199.20	40.05	320.40		
1 Gas Engine Vibrator		21.40		23.54	0.24	0.27
88 L.H., Daily Totals		$2187.00		$3755.54	$24.85	$42.68
Crew C-14F	Hr.	Daily	Hr.	Daily	Bare Costs	Incl. O&P
1 Laborer Foreman (out)	$20.70	$165.60	$35.15	$281.20	$23.06	$37.66
2 Laborers	18.70	299.20	31.75	508.00		
6 Cement Finishers	24.90	1195.20	40.05	1922.40		
1 Gas Engine Vibrator		21.40		23.54	0.30	0.33
72 L.H., Daily Totals		$1681.40		$2735.14	$23.35	$37.99

Crew No.	Bare Costs		Incl. Subs O & P		Cost Per Labor-Hour	
Crew C-14G	Hr.	Daily	Hr.	Daily	Bare Costs	Incl. O&P
1 Laborer Foreman (out)	$20.70	$165.60	$35.15	$281.20	$22.53	$36.98
2 Laborers	18.70	299.20	31.75	508.00		
4 Cement Finishers	24.90	796.80	40.05	1281.60		
1 Gas Engine Vibrator		21.40		23.54	0.38	0.42
56 L.H., Daily Totals		$1283.00		$2094.34	$22.91	$37.40
Crew C-14H	Hr.	Daily	Hr.	Daily	Bare Costs	Incl. O&P
1 Carpenter Foreman (out)	$27.70	$221.60	$47.00	$376.00	$25.06	$42.54
2 Carpenters	25.70	411.20	43.60	697.60		
1 Rodman (reinf.)	27.65	221.20	49.25	394.00		
1 Laborer	18.70	149.60	31.75	254.00		
1 Cement Finisher	24.90	199.20	40.05	320.40		
1 Gas Engine Vibrator		21.40		23.54	0.45	0.49
48 L.H., Daily Totals		$1224.20		$2065.54	$25.50	$43.03
Crew C-15	Hr.	Daily	Hr.	Daily	Bare Costs	Incl. O&P
1 Carpenter Foreman (out)	$27.70	$221.60	$47.00	$376.00	$23.63	$39.87
2 Carpenters	25.70	411.20	43.60	697.60		
3 Laborers	18.70	448.80	31.75	762.00		
2 Cement Finishers	24.90	398.40	40.05	640.80		
1 Rodman (reinf.)	27.65	221.20	49.25	394.00		
72 L.H., Daily Totals		$1701.20		$2870.40	$23.63	$39.87
Crew C-16	Hr.	Daily	Hr.	Daily	Bare Costs	Incl. O&P
1 Labor Foreman (outside)	$20.70	$165.60	$35.15	$281.20	$23.16	$39.18
3 Laborers	18.70	448.80	31.75	762.00		
2 Cement Finishers	24.90	398.40	40.05	640.80		
1 Equip. Oper. (med.)	26.50	212.00	43.65	349.20		
2 Rodmen (reinf.)	27.65	442.40	49.25	788.00		
1 Concrete Pump (small)		708.60		779.46	9.84	10.83
72 L.H., Daily Totals		$2375.80		$3600.66	$33.00	$50.01
Crew C-17	Hr.	Daily	Hr.	Daily	Bare Costs	Incl. O&P
2 Skilled Worker Foremen	$28.10	$449.60	$47.65	$762.40	$26.50	$44.93
8 Skilled Workers	26.10	1670.40	44.25	2832.00		
80 L.H., Daily Totals		$2120.00		$3594.40	$26.50	$44.93
Crew C-17A	Hr.	Daily	Hr.	Daily	Bare Costs	Incl. O&P
2 Skilled Worker Foremen	$28.10	$449.60	$47.65	$762.40	$26.51	$44.93
8 Skilled Workers	26.10	1670.40	44.25	2832.00		
.125 Equip. Oper. (crane)	27.45	27.45	45.20	45.20		
.125 Hyd. Crane, 80 Ton		151.75		166.93	1.87	2.06
81 L.H., Daily Totals		$2299.20		$3806.53	$28.39	$46.99
Crew C-17B	Hr.	Daily	Hr.	Daily	Bare Costs	Incl. O&P
2 Skilled Worker Foremen	$28.10	$449.60	$47.65	$762.40	$26.52	$44.94
8 Skilled Workers	26.10	1670.40	44.25	2832.00		
.25 Equip. Oper. (crane)	27.45	54.90	45.20	90.40		
.25 Hyd. Crane, 80 Ton		303.50		333.85		
.25 Conc. Finish., 46" Wlk-Behind		6.45		7.09	3.78	4.16
82 L.H., Daily Totals		$2484.85		$4025.74	$30.30	$49.09
Crew C-17C	Hr.	Daily	Hr.	Daily	Bare Costs	Incl. O&P
2 Skilled Worker Foremen	$28.10	$449.60	$47.65	$762.40	$26.53	$44.94
8 Skilled Workers	26.10	1670.40	44.25	2832.00		
.375 Equip. Oper. (crane)	27.45	82.35	45.20	135.60		
.375 Hyd. Crane, 80 Ton		455.25		500.77	5.48	6.03
83 L.H., Daily Totals		$2657.60		$4230.77	$32.02	$50.97

Crews

Crew No.	Bare Costs		Incl. Subs O & P		Cost Per Labor-Hour	
Crew C-17D	Hr.	Daily	Hr.	Daily	Bare Costs	Incl. O&P
2 Skilled Worker Foremen	$28.10	$449.60	$47.65	$762.40	$26.55	$44.94
8 Skilled Workers	26.10	1670.40	44.25	2832.00		
.5 Equip. Oper. (crane)	27.45	109.80	45.20	180.80		
.5 Hyd. Crane, 80 Ton		607.00		667.70	7.23	7.95
84 L.H., Daily Totals		$2836.80		$4442.90	$33.77	$52.89
Crew C-17E	Hr.	Daily	Hr.	Daily	Bare Costs	Incl. O&P
2 Skilled Worker Foremen	$28.10	$449.60	$47.65	$762.40	$26.50	$44.93
8 Skilled Workers	26.10	1670.40	44.25	2832.00		
1 Hyd. Jack with Rods		82.35		90.58	1.03	1.13
80 L.H., Daily Totals		$2202.35		$3684.99	$27.53	$46.06
Crew C-18	Hr.	Daily	Hr.	Daily	Bare Costs	Incl. O&P
.125 Labor Foreman (out)	$20.70	$20.70	$35.15	$35.15	$18.92	$32.13
1 Laborer	18.70	149.60	31.75	254.00		
1 Concrete Cart, 10 C.F.		53.40		58.74	5.93	6.53
9 L.H., Daily Totals		$223.70		$347.89	$24.86	$38.65
Crew C-19	Hr.	Daily	Hr.	Daily	Bare Costs	Incl. O&P
.125 Labor Foreman (out)	$20.70	$20.70	$35.15	$35.15	$18.92	$32.13
1 Laborer	18.70	149.60	31.75	254.00		
1 Concrete Cart, 18 C.F.		78.80		86.68	8.76	9.63
9 L.H., Daily Totals		$249.10		$375.83	$27.68	$41.76
Crew C-20	Hr.	Daily	Hr.	Daily	Bare Costs	Incl. O&P
1 Labor Foreman (outside)	$20.70	$165.60	$35.15	$281.20	$20.70	$34.70
5 Laborers	18.70	748.00	31.75	1270.00		
1 Cement Finisher	24.90	199.20	40.05	320.40		
1 Equip. Oper. (med.)	26.50	212.00	43.65	349.20		
2 Gas Engine Vibrator		42.80		47.08		
1 Concrete Pump (small)		708.60		779.46	11.74	12.91
64 L.H., Daily Totals		$2076.20		$3047.34	$32.44	$47.61
Crew C-21	Hr.	Daily	Hr.	Daily	Bare Costs	Incl. O&P
1 Labor Foreman (outside)	$20.70	$165.60	$35.15	$281.20	$20.70	$34.70
5 Laborers	18.70	748.00	31.75	1270.00		
1 Cement Finisher	24.90	199.20	40.05	320.40		
1 Equip. Oper. (med.)	26.50	212.00	43.65	349.20		
2 Gas Engine Vibrator		42.80		47.08		
1 Concrete Conveyer		160.40		176.44	3.17	3.49
64 L.H., Daily Totals		$1528.00		$2444.32	$23.88	$38.19
Crew C-22	Hr.	Daily	Hr.	Daily	Bare Costs	Incl. O&P
1 Rodman Foreman	$29.65	$237.20	$52.80	$422.40	$27.93	$49.58
4 Rodmen (reinf.)	27.65	884.80	49.25	1576.00		
.125 Equip. Oper. (crane)	27.45	27.45	45.20	45.20		
.125 Equip. Oper. Oiler	23.40	23.40	38.55	38.55		
.125 Hyd. Crane, 25 Ton		92.45		101.69	2.20	2.42
42 L.H., Daily Totals		$1265.30		$2183.84	$30.13	$52.00
Crew C-23	Hr.	Daily	Hr.	Daily	Bare Costs	Incl. O&P
2 Skilled Worker Foremen	$28.10	$449.60	$47.65	$762.40	$26.36	$44.45
6 Skilled Workers	26.10	1252.80	44.25	2124.00		
1 Equip. Oper. (crane)	27.45	219.60	45.20	361.60		
1 Equip. Oper. Oiler	23.40	187.20	38.55	308.40		
1 Lattice Boom Crane, 90 Ton		1547.00		1701.70	19.34	21.27
80 L.H., Daily Totals		$3656.20		$5258.10	$45.70	$65.73

Crew No.	Bare Costs		Incl. Subs O & P		Cost Per Labor-Hour	
Crew C-24	Hr.	Daily	Hr.	Daily	Bare Costs	Incl. O&P
2 Skilled Worker Foremen	$28.10	$449.60	$47.65	$762.40	$26.36	$44.45
6 Skilled Workers	26.10	1252.80	44.25	2124.00		
1 Equip. Oper. (crane)	27.45	219.60	45.20	361.60		
1 Equip. Oper. Oiler	23.40	187.20	38.55	308.40		
1 Lattice Boom Crane, 150 Ton		1852.00		2037.20	23.15	25.47
80 L.H., Daily Totals		$3961.20		$5593.60	$49.52	$69.92
Crew C-25	Hr.	Daily	Hr.	Daily	Bare Costs	Incl. O&P
2 Rodmen (reinf.)	$27.65	$442.40	$49.25	$788.00	$21.88	$39.40
2 Rodmen Helpers	16.10	257.60	29.55	472.80		
32 L.H., Daily Totals		$700.00		$1260.80	$21.88	$39.40
Crew C-27	Hr.	Daily	Hr.	Daily	Bare Costs	Incl. O&P
2 Cement Finishers	$24.90	$398.40	$40.05	$640.80	$24.90	$40.05
1 Concrete Saw		118.40		130.24	7.40	8.14
16 L.H., Daily Totals		$516.80		$771.04	$32.30	$48.19
Crew C-28	Hr.	Daily	Hr.	Daily	Bare Costs	Incl. O&P
1 Cement Finisher	$24.90	$199.20	$40.05	$320.40	$24.90	$40.05
1 Portable Air Compressor, gas		14.75		16.23	1.84	2.03
8 L.H., Daily Totals		$213.95		$336.63	$26.74	$42.08
Crew D-1	Hr.	Daily	Hr.	Daily	Bare Costs	Incl. O&P
1 Bricklayer	$26.65	$213.20	$44.30	$354.40	$23.35	$38.80
1 Bricklayer Helper	20.05	160.40	33.30	266.40		
16 L.H., Daily Totals		$373.60		$620.80	$23.35	$38.80
Crew D-2	Hr.	Daily	Hr.	Daily	Bare Costs	Incl. O&P
3 Bricklayers	$26.65	$639.60	$44.30	$1063.20	$24.01	$39.90
2 Bricklayer Helpers	20.05	320.80	33.30	532.80		
40 L.H., Daily Totals		$960.40		$1596.00	$24.01	$39.90
Crew D-3	Hr.	Daily	Hr.	Daily	Bare Costs	Incl. O&P
3 Bricklayers	$26.65	$639.60	$44.30	$1063.20	$24.09	$40.08
2 Bricklayer Helpers	20.05	320.80	33.30	532.80		
.25 Carpenter	25.70	51.40	43.60	87.20		
42 L.H., Daily Totals		$1011.80		$1683.20	$24.09	$40.08
Crew D-4	Hr.	Daily	Hr.	Daily	Bare Costs	Incl. O&P
1 Bricklayer	$26.65	$213.20	$44.30	$354.40	$21.10	$35.19
3 Bricklayer Helpers	20.05	481.20	33.30	799.20		
1 Building Laborer	18.70	149.60	31.75	254.00		
1 Grout Pump, 50 C.F./hr		122.00		134.20	3.05	3.36
40 L.H., Daily Totals		$966.00		$1541.80	$24.15	$38.55
Crew D-5	Hr.	Daily	Hr.	Daily	Bare Costs	Incl. O&P
1 Block Mason Helper	$20.05	$160.40	$33.30	$266.40	$20.05	$33.30
8 L.H., Daily Totals		$160.40		$266.40	$20.05	$33.30
Crew D-6	Hr.	Daily	Hr.	Daily	Bare Costs	Incl. O&P
3 Bricklayers	$26.65	$639.60	$44.30	$1063.20	$23.35	$38.80
3 Bricklayer Helpers	20.05	481.20	33.30	799.20		
48 L.H., Daily Totals		$1120.80		$1862.40	$23.35	$38.80
Crew D-7	Hr.	Daily	Hr.	Daily	Bare Costs	Incl. O&P
1 Tile Layer	$24.85	$198.80	$39.95	$319.60	$21.98	$35.33
1 Tile Layer Helper	19.10	152.80	30.70	245.60		
16 L.H., Daily Totals		$351.60		$565.20	$21.98	$35.33

Crews

Crew No.	Bare Costs		Incl. Subs O & P		Cost Per Labor-Hour	
Crew D-8	Hr.	Daily	Hr.	Daily	Bare Costs	Incl. O&P
3 Bricklayers	$26.65	$639.60	$44.30	$1063.20	$24.01	$39.90
2 Bricklayer Helpers	20.05	320.80	33.30	532.80		
40 L.H., Daily Totals		$960.40		$1596.00	$24.01	$39.90

Crew D-9	Hr.	Daily	Hr.	Daily	Bare Costs	Incl. O&P
3 Bricklayers	$26.65	$639.60	$44.30	$1063.20	$23.35	$38.80
3 Bricklayer Helpers	20.05	481.20	33.30	799.20		
48 L.H., Daily Totals		$1120.80		$1862.40	$23.35	$38.80

Crew D-10	Hr.	Daily	Hr.	Daily	Bare Costs	Incl. O&P
1 Bricklayer Foreman	$28.65	$229.20	$47.60	$380.80	$24.57	$40.74
1 Bricklayer	26.65	213.20	44.30	354.40		
2 Bricklayer Helpers	20.05	320.80	33.30	532.80		
1 Equip. Oper. (crane)	27.45	219.60	45.20	361.60		
1 S.P. Crane, 4x4, 12 Ton		575.00		632.50	14.38	15.81
40 L.H., Daily Totals		$1557.80		$2262.10	$38.95	$56.55

Crew D-11	Hr.	Daily	Hr.	Daily	Bare Costs	Incl. O&P
2 Bricklayers	$26.65	$426.40	$44.30	$708.80	$24.45	$40.63
1 Bricklayer Helper	20.05	160.40	33.30	266.40		
24 L.H., Daily Totals		$586.80		$975.20	$24.45	$40.63

Crew D-12	Hr.	Daily	Hr.	Daily	Bare Costs	Incl. O&P
2 Bricklayers	$26.65	$426.40	$44.30	$708.80	$23.35	$38.80
2 Bricklayer Helpers	20.05	320.80	33.30	532.80		
32 L.H., Daily Totals		$747.20		$1241.60	$23.35	$38.80

Crew D-13	Hr.	Daily	Hr.	Daily	Bare Costs	Incl. O&P
1 Bricklayer Foreman	$28.65	$229.20	$47.60	$380.80	$24.92	$41.33
2 Bricklayers	26.65	426.40	44.30	708.80		
2 Bricklayer Helpers	20.05	320.80	33.30	532.80		
1 Equip. Oper. (crane)	27.45	219.60	45.20	361.60		
1 S.P. Crane, 4x4, 12 Ton		575.00		632.50	11.98	13.18
48 L.H., Daily Totals		$1771.00		$2616.50	$36.90	$54.51

Crew E-1	Hr.	Daily	Hr.	Daily	Bare Costs	Incl. O&P
2 Struc. Steel Workers	$27.70	$443.20	$54.05	$864.80	$27.70	$54.05
1 Welder, gas engine, 300 amp		115.20		126.72	7.20	7.92
16 L.H., Daily Totals		$558.40		$991.52	$34.90	$61.97

Crew E-2	Hr.	Daily	Hr.	Daily	Bare Costs	Incl. O&P
1 Struc. Steel Foreman	$29.70	$237.60	$57.95	$463.60	$27.99	$53.23
4 Struc. Steel Workers	27.70	886.40	54.05	1729.60		
1 Equip. Oper. (crane)	27.45	219.60	45.20	361.60		
1 Lattice Boom Crane, 90 Ton		1547.00		1701.70	32.23	35.45
48 L.H., Daily Totals		$2890.60		$4256.50	$60.22	$88.68

Crew E-3	Hr.	Daily	Hr.	Daily	Bare Costs	Incl. O&P
1 Struc. Steel Foreman	$29.70	$237.60	$57.95	$463.60	$28.37	$55.35
2 Struc. Steel Workers	27.70	443.20	54.05	864.80		
1 Welder, gas engine, 300 amp		115.20		126.72	4.80	5.28
24 L.H., Daily Totals		$796.00		$1455.12	$33.17	$60.63

Crew E-4	Hr.	Daily	Hr.	Daily	Bare Costs	Incl. O&P
1 Struc. Steel Foreman	$29.70	$237.60	$57.95	$463.60	$28.20	$55.02
3 Struc. Steel Workers	27.70	664.80	54.05	1297.20		
1 Welder, gas engine, 300 amp		115.20		126.72	3.60	3.96
32 L.H., Daily Totals		$1017.60		$1887.52	$31.80	$58.98

Crew E-5	Hr.	Daily	Hr.	Daily	Bare Costs	Incl. O&P
1 Struc. Steel Foreman	$29.70	$237.60	$57.95	$463.60	$27.89	$53.50
7 Struc. Steel Workers	27.70	1551.20	54.05	3026.80		
1 Equip. Oper. (crane)	27.45	219.60	45.20	361.60		
1 Lattice Boom Crane, 90 Ton		1547.00		1701.70		
1 Welder, gas engine, 300 amp		115.20		126.72	23.09	25.39
72 L.H., Daily Totals		$3670.60		$5680.42	$50.98	$78.89

Crew E-6	Hr.	Daily	Hr.	Daily	Bare Costs	Incl. O&P
1 Struc. Steel Foreman	$29.70	$237.60	$57.95	$463.60	$27.67	$52.91
12 Struc. Steel Workers	27.70	2659.20	54.05	5188.80		
1 Equip. Oper. (crane)	27.45	219.60	45.20	361.60		
1 Equip. Oper. (light)	25.45	203.60	41.90	335.20		
1 Lattice Boom Crane, 90 Ton		1547.00		1701.70		
1 Welder, gas engine, 300 amp		115.20		126.72		
1 Air Compressor, 160 C.F.M.		122.60		134.86		
2 Impact Wrenches		25.60		28.16	15.09	16.60
120 L.H., Daily Totals		$5130.40		$8340.64	$42.75	$69.51

Crew E-7	Hr.	Daily	Hr.	Daily	Bare Costs	Incl. O&P
1 Struc. Steel Foreman	$29.70	$237.60	$57.95	$463.60	$27.89	$53.50
7 Struc. Steel Workers	27.70	1551.20	54.05	3026.80		
1 Equip. Oper. (crane)	27.45	219.60	45.20	361.60		
1 Lattice Boom Crane, 90 Ton		1547.00		1701.70		
2 Welders, gas engine, 300 amp		230.40		253.44	24.69	27.15
72 L.H., Daily Totals		$3785.80		$5807.14	$52.58	$80.65

Crew E-8	Hr.	Daily	Hr.	Daily	Bare Costs	Incl. O&P
1 Struc. Steel Foreman	$29.70	$237.60	$57.95	$463.60	$27.86	$53.60
9 Struc. Steel Workers	27.70	1994.40	54.05	3891.60		
1 Equip. Oper. (crane)	27.45	219.60	45.20	361.60		
1 Lattice Boom Crane, 90 Ton		1547.00		1701.70		
4 Welders, gas engine, 300 amp		460.80		506.88	22.82	25.10
88 L.H., Daily Totals		$4459.40		$6925.38	$50.67	$78.70

Crew E-9	Hr.	Daily	Hr.	Daily	Bare Costs	Incl. O&P
2 Struc. Steel Foremen	$29.70	$475.20	$57.95	$927.20	$27.65	$52.50
5 Struc. Steel Workers	27.70	1108.00	54.05	2162.00		
1 Welder Foreman	29.70	237.60	57.95	463.60		
5 Welders	27.70	1108.00	54.05	2162.00		
1 Equip. Oper. (crane)	27.45	219.60	45.20	361.60		
1 Equip. Oper. Oiler	23.40	187.20	38.55	308.40		
1 Equip. Oper. (light)	25.45	203.60	41.90	335.20		
1 Lattice Boom Crane, 90 Ton		1547.00		1701.70		
5 Welders, gas engine, 300 amp		576.00		633.60	16.59	18.24
128 L.H., Daily Totals		$5662.20		$9055.30	$44.24	$70.74

Crew E-10	Hr.	Daily	Hr.	Daily	Bare Costs	Incl. O&P
1 Struc. Steel Foreman	$29.70	$237.60	$57.95	$463.60	$28.37	$55.35
2 Struc. Steel Workers	27.70	443.20	54.05	864.80		
1 Welder, gas engine, 300 amp		115.20		126.72		
1 Flatbed Truck, Gas, 3 Ton		185.20		203.72	12.52	13.77
24 L.H., Daily Totals		$981.20		$1658.84	$40.88	$69.12

Crew E-11	Hr.	Daily	Hr.	Daily	Bare Costs	Incl. O&P
2 Painters, Struc. Steel	$23.80	$380.80	$46.80	$748.80	$22.10	$41.78
1 Building Laborer	18.70	149.60	31.75	254.00		
1 Air Compressor, 250 C.F.M.		151.40		166.54		
1 Sandblaster, portable, 3 C.F.		15.60		17.16		
1 Sand Blasting Accessories		12.30		13.53	7.47	8.22
24 L.H., Daily Totals		$709.70		$1200.03	$29.57	$50.00

Crews

Crew No.	Bare Costs		Incl. Subs O & P		Cost Per Labor-Hour	
Crew E-12	Hr.	Daily	Hr.	Daily	Bare Costs	Incl. O&P
1 Welder Foreman	$29.70	$237.60	$57.95	$463.60	$27.57	$49.92
1 Equip. Oper. (light)	25.45	203.60	41.90	335.20		
1 Welder, gas engine, 300 amp		115.20		126.72	7.20	7.92
16 L.H., Daily Totals		$556.40		$925.52	$34.77	$57.84
Crew E-13	Hr.	Daily	Hr.	Daily	Bare Costs	Incl. O&P
1 Welder Foreman	$29.70	$237.60	$57.95	$463.60	$28.28	$52.60
.5 Equip. Oper. (light)	25.45	101.80	41.90	167.60		
1 Welder, gas engine, 300 amp		115.20		126.72	9.60	10.56
12 L.H., Daily Totals		$454.60		$757.92	$37.88	$63.16
Crew E-14	Hr.	Daily	Hr.	Daily	Bare Costs	Incl. O&P
1 Struc. Steel Worker	$27.70	$221.60	$54.05	$432.40	$27.70	$54.05
1 Welder, gas engine, 300 amp		115.20		126.72	14.40	15.84
8 L.H., Daily Totals		$336.80		$559.12	$42.10	$69.89
Crew E-16	Hr.	Daily	Hr.	Daily	Bare Costs	Incl. O&P
1 Welder Foreman	$29.70	$237.60	$57.95	$463.60	$28.70	$56.00
1 Welder	27.70	221.60	54.05	432.40		
1 Welder, gas engine, 300 amp		115.20		126.72	7.20	7.92
16 L.H., Daily Totals		$574.40		$1022.72	$35.90	$63.92
Crew E-17	Hr.	Daily	Hr.	Daily	Bare Costs	Incl. O&P
1 Structural Steel Foreman	$29.70	$237.60	$57.95	$463.60	$28.70	$56.00
1 Structural Steel Worker	27.70	221.60	54.05	432.40		
16 L.H., Daily Totals		$459.20		$896.00	$28.70	$56.00
Crew E-18	Hr.	Daily	Hr.	Daily	Bare Costs	Incl. O&P
1 Structural Steel Foreman	$29.70	$237.60	$57.95	$463.60	$27.86	$52.75
3 Structural Steel Workers	27.70	664.80	54.05	1297.20		
1 Equipment Operator (med.)	26.50	212.00	43.65	349.20		
1 Lattice Boom Crane, 20 Ton		976.70		1074.37	24.42	26.86
40 L.H., Daily Totals		$2091.10		$3184.37	$52.28	$79.61
Crew E-19	Hr.	Daily	Hr.	Daily	Bare Costs	Incl. O&P
1 Structural Steel Worker	$27.70	$221.60	$54.05	$432.40	$27.62	$51.30
1 Structural Steel Foreman	29.70	237.60	57.95	463.60		
1 Equip. Oper. (light)	25.45	203.60	41.90	335.20		
1 Lattice Boom Crane, 20 Ton		976.70		1074.37	40.70	44.77
24 L.H., Daily Totals		$1639.50		$2305.57	$68.31	$96.07
Crew E-20	Hr.	Daily	Hr.	Daily	Bare Costs	Incl. O&P
1 Structural Steel Foreman	$29.70	$237.60	$57.95	$463.60	$27.38	$51.49
5 Structural Steel Workers	27.70	1108.00	54.05	2162.00		
1 Equip. Oper. (crane)	27.45	219.60	45.20	361.60		
1 Oiler	23.40	187.20	38.55	308.40		
1 Lattice Boom Crane, 40 Ton		1268.00		1394.80	19.81	21.79
64 L.H., Daily Totals		$3020.40		$4690.40	$47.19	$73.29
Crew E-22	Hr.	Daily	Hr.	Daily	Bare Costs	Incl. O&P
1 Skilled Worker Foreman	$28.10	$224.80	$47.65	$381.20	$26.77	$45.38
2 Skilled Workers	26.10	417.60	44.25	708.00		
24 L.H., Daily Totals		$642.40		$1089.20	$26.77	$45.38
Crew E-24	Hr.	Daily	Hr.	Daily	Bare Costs	Incl. O&P
3 Structural Steel Workers	$27.70	$664.80	$54.05	$1297.20	$27.40	$51.45
1 Equipment Operator (medium)	26.50	212.00	43.65	349.20		
1 -25 Ton Crane		739.60		813.56	23.11	25.42
32 L.H., Daily Totals		$1616.40		$2459.96	$50.51	$76.87

Crew No.	Bare Costs		Incl. Subs O & P		Cost Per Labor-Hour	
Crew E-25	Hr.	Daily	Hr.	Daily	Bare Costs	Incl. O&P
1 Welder	$27.70	$221.60	$54.05	$432.40	$27.70	$54.05
1 Cutting Torch		17.00		18.70		
1 Gas		64.80		71.28	10.23	11.25
8 L.H., Daily Totals		$303.40		$522.38	$37.92	$65.30
Crew F-3	Hr.	Daily	Hr.	Daily	Bare Costs	Incl. O&P
2 Carpenters	$25.70	$411.20	$43.60	$697.60	$23.29	$39.20
2 Carpenter Helpers	18.80	300.80	31.80	508.80		
1 Equip. Oper. (crane)	27.45	219.60	45.20	361.60		
1 Hyd. Crane, 12 Ton		724.00		796.40	18.10	19.91
40 L.H., Daily Totals		$1655.60		$2364.40	$41.39	$59.11
Crew F-4	Hr.	Daily	Hr.	Daily	Bare Costs	Incl. O&P
2 Carpenters	$25.70	$411.20	$43.60	$697.60	$23.29	$39.20
2 Carpenter Helpers	18.80	300.80	31.80	508.80		
1 Equip. Oper. (crane)	27.45	219.60	45.20	361.60		
1 Hyd. Crane, 55 Ton		1060.00		1166.00	26.50	29.15
40 L.H., Daily Totals		$1991.60		$2734.00	$49.79	$68.35
Crew F-5	Hr.	Daily	Hr.	Daily	Bare Costs	Incl. O&P
2 Carpenters	$25.70	$411.20	$43.60	$697.60	$22.25	$37.70
2 Carpenter Helpers	18.80	300.80	31.80	508.80		
32 L.H., Daily Totals		$712.00		$1206.40	$22.25	$37.70
Crew F-6	Hr.	Daily	Hr.	Daily	Bare Costs	Incl. O&P
2 Carpenters	$25.70	$411.20	$43.60	$697.60	$23.25	$39.18
2 Building Laborers	18.70	299.20	31.75	508.00		
1 Equip. Oper. (crane)	27.45	219.60	45.20	361.60		
1 Hyd. Crane, 12 Ton		724.00		796.40	18.10	19.91
40 L.H., Daily Totals		$1654.00		$2363.60	$41.35	$59.09
Crew F-7	Hr.	Daily	Hr.	Daily	Bare Costs	Incl. O&P
2 Carpenters	$25.70	$411.20	$43.60	$697.60	$22.20	$37.67
2 Building Laborers	18.70	299.20	31.75	508.00		
32 L.H., Daily Totals		$710.40		$1205.60	$22.20	$37.67
Crew G-1	Hr.	Daily	Hr.	Daily	Bare Costs	Incl. O&P
1 Roofer Foreman	$23.95	$191.60	$43.95	$351.60	$20.56	$37.75
4 Roofers, Composition	21.95	702.40	40.30	1289.60		
2 Roofer Helpers	16.10	257.60	29.55	472.80		
1 Application Equipment		153.20		168.52		
1 Tar Kettle/Pot		62.90		69.19		
1 Crew Truck		116.20		127.82	5.93	6.53
56 L.H., Daily Totals		$1483.90		$2479.53	$26.50	$44.28
Crew G-2	Hr.	Daily	Hr.	Daily	Bare Costs	Incl. O&P
1 Plasterer	$23.50	$188.00	$38.85	$310.80	$20.78	$34.63
1 Plasterer Helper	20.15	161.20	33.30	266.40		
1 Building Laborer	18.70	149.60	31.75	254.00		
1 Grout Pump, 50 C.F./hr		122.00		134.20	5.08	5.59
24 L.H., Daily Totals		$620.80		$965.40	$25.87	$40.23
Crew G-2A	Hr.	Daily	Hr.	Daily	Bare Costs	Incl. O&P
1 Roofer, composition	$21.95	$175.60	$40.30	$322.40	$18.92	$33.87
1 Roofer Helper	16.10	128.80	29.55	236.40		
1 Building Laborer	18.70	149.60	31.75	254.00		
1 Grout Pump, 50 C.F./hr		122.00		134.20	5.08	5.59
24 L.H., Daily Totals		$576.00		$947.00	$24.00	$39.46

Crews

Crew No.	Bare Costs Hr.	Bare Costs Daily	Incl. Subs O & P Hr.	Incl. Subs O & P Daily	Cost Per Labor-Hour Bare Costs	Cost Per Labor-Hour Incl. O&P
Crew G-3						
2 Sheet Metal Workers	$28.75	$460.00	$48.35	$773.60	$23.73	$40.05
2 Building Laborers	18.70	299.20	31.75	508.00		
32 L.H., Daily Totals		$759.20		$1281.60	$23.73	$40.05
Crew G-4						
1 Labor Foreman (outside)	$20.70	$165.60	$35.15	$281.20	$19.37	$32.88
2 Building Laborers	18.70	299.20	31.75	508.00		
1 Flatbed Truck, Gas, 1.5 Ton		141.40		155.54		
1 Air Compressor, 160 C.F.M.		122.60		134.86	11.00	12.10
24 L.H., Daily Totals		$728.80		$1079.60	$30.37	$44.98
Crew G-5						
1 Roofer Foreman	$23.95	$191.60	$43.95	$351.60	$20.01	$36.73
2 Roofers, Composition	21.95	351.20	40.30	644.80		
2 Roofer Helpers	16.10	257.60	29.55	472.80		
1 Application Equipment		153.20		168.52	3.83	4.21
40 L.H., Daily Totals		$953.60		$1637.72	$23.84	$40.94
Crew G-6A						
2 Roofers Composition	$21.95	$351.20	$40.30	$644.80	$21.95	$40.30
1 Small Compressor, Electric		9.55		10.51		
2 Pneumatic Nailers		41.30		45.43	3.18	3.50
16 L.H., Daily Totals		$402.05		$700.74	$25.13	$43.80
Crew G-7						
1 Carpenter	$25.70	$205.60	$43.60	$348.80	$25.70	$43.60
1 Small Compressor, Electric		9.55		10.51		
1 Pneumatic Nailer		20.65		22.72	3.77	4.15
8 L.H., Daily Totals		$235.80		$382.02	$29.48	$47.75
Crew H-1						
2 Glaziers	$25.60	$409.60	$42.35	$677.60	$26.65	$48.20
2 Struc. Steel Workers	27.70	443.20	54.05	864.80		
32 L.H., Daily Totals		$852.80		$1542.40	$26.65	$48.20
Crew H-2						
2 Glaziers	$25.60	$409.60	$42.35	$677.60	$23.30	$38.82
1 Building Laborer	18.70	149.60	31.75	254.00		
24 L.H., Daily Totals		$559.20		$931.60	$23.30	$38.82
Crew H-3						
1 Glazier	$25.60	$204.80	$42.35	$338.80	$22.20	$37.08
1 Helper	18.80	150.40	31.80	254.40		
16 L.H., Daily Totals		$355.20		$593.20	$22.20	$37.08
Crew J-1						
3 Plasterers	$23.50	$564.00	$38.85	$932.40	$22.16	$36.63
2 Plasterer Helpers	20.15	322.40	33.30	532.80		
1 Mixing Machine, 6 C.F.		113.60		124.96	2.84	3.12
40 L.H., Daily Totals		$1000.00		$1590.16	$25.00	$39.75
Crew J-2						
3 Plasterers	$23.50	$564.00	$38.85	$932.40	$22.40	$36.93
2 Plasterer Helpers	20.15	322.40	33.30	532.80		
1 Lather	23.60	188.80	38.45	307.60		
1 Mixing Machine, 6 C.F.		113.60		124.96	2.37	2.60
48 L.H., Daily Totals		$1188.80		$1897.76	$24.77	$39.54

Crew No.	Bare Costs Hr.	Bare Costs Daily	Incl. Subs O & P Hr.	Incl. Subs O & P Daily	Cost Per Labor-Hour Bare Costs	Cost Per Labor-Hour Incl. O&P
Crew J-3						
1 Terrazzo Worker	$24.60	$196.80	$39.55	$316.40	$22.20	$35.70
1 Terrazzo Helper	19.80	158.40	31.85	254.80		
1 Terrazzo Grinder, Electric		99.35		109.29		
1 Terrazzo Mixer		155.40		170.94	15.92	17.51
16 L.H., Daily Totals		$609.95		$851.42	$38.12	$53.21
Crew J-4						
1 Tile Layer	$24.85	$198.80	$39.95	$319.60	$21.98	$35.33
1 Tile Layer Helper	19.10	152.80	30.70	245.60		
16 L.H., Daily Totals		$351.60		$565.20	$21.98	$35.33
Crew K-1						
1 Carpenter	$25.70	$205.60	$43.60	$348.80	$22.98	$38.85
1 Truck Driver (light)	20.25	162.00	34.10	272.80		
1 Flatbed Truck, gas, 3 Ton		185.20		203.72	11.57	12.73
16 L.H., Daily Totals		$552.80		$825.32	$34.55	$51.58
Crew K-2						
1 Struc. Steel Foreman	$29.70	$237.60	$57.95	$463.60	$25.88	$48.70
1 Struc. Steel Worker	27.70	221.60	54.05	432.40		
1 Truck Driver (light)	20.25	162.00	34.10	272.80		
1 Flatbed Truck, gas, 3 Ton		185.20		203.72	7.72	8.49
24 L.H., Daily Totals		$806.40		$1372.52	$33.60	$57.19
Crew L-1						
.25 Electrician	$29.40	$58.80	$47.90	$95.80	$29.52	$48.46
1 Plumber	29.55	236.40	48.60	388.80		
10 L.H., Daily Totals		$295.20		$484.60	$29.52	$48.46
Crew L-2						
1 Carpenter	$25.70	$205.60	$43.60	$348.80	$22.25	$37.70
1 Carpenter Helper	18.80	150.40	31.80	254.40		
16 L.H., Daily Totals		$356.00		$603.20	$22.25	$37.70
Crew L-3						
1 Carpenter	$25.70	$205.60	$43.60	$348.80	$26.44	$44.46
.25 Electrician	29.40	58.80	47.90	95.80		
10 L.H., Daily Totals		$264.40		$444.60	$26.44	$44.46
Crew L-3A						
1 Carpenter Foreman (outside)	$27.70	$221.60	$47.00	$376.00	$28.05	$47.45
.5 Sheet Metal Worker	28.75	115.00	48.35	193.40		
12 L.H., Daily Totals		$336.60		$569.40	$28.05	$47.45
Crew L-4						
1 Skilled Worker	$26.10	$208.80	$44.25	$354.00	$22.45	$38.02
1 Helper	18.80	150.40	31.80	254.40		
16 L.H., Daily Totals		$359.20		$608.40	$22.45	$38.02
Crew L-5						
1 Struc. Steel Foreman	$29.70	$237.60	$57.95	$463.60	$27.95	$53.34
5 Struc. Steel Workers	27.70	1108.00	54.05	2162.00		
1 Equip. Oper. (crane)	27.45	219.60	45.20	361.60		
1 Hyd. Crane, 25 Ton		739.60		813.56	13.21	14.53
56 L.H., Daily Totals		$2304.80		$3800.76	$41.16	$67.87

Crews

Crew No.	Bare Costs		Incl. Subs O & P		Cost Per Labor-Hour	
Crew L-5A	**Hr.**	**Daily**	**Hr.**	**Daily**	**Bare Costs**	**Incl. O&P**
1 Structural Steel Foreman	$29.70	$237.60	$57.95	$463.60	$28.14	$52.81
2 Structural Steel Workers	27.70	443.20	54.05	864.80		
1 Equip. Oper. (crane)	27.45	219.60	45.20	361.60		
1 S.P. Crane, 4x4, 25 Ton		875.60		963.16	27.36	30.10
32 L.H., Daily Totals		$1776.00		$2653.16	$55.50	$82.91
Crew L-6	**Hr.**	**Daily**	**Hr.**	**Daily**	**Bare Costs**	**Incl. O&P**
1 Plumber	$29.55	$236.40	$48.60	$388.80	$29.50	$48.37
.5 Electrician	29.40	117.60	47.90	191.60		
12 L.H., Daily Totals		$354.00		$580.40	$29.50	$48.37
Crew L-7	**Hr.**	**Daily**	**Hr.**	**Daily**	**Bare Costs**	**Incl. O&P**
1 Carpenter	$25.70	$205.60	$43.60	$348.80	$21.74	$36.67
2 Carpenter Helpers	18.80	300.80	31.80	508.80		
.25 Electrician	29.40	58.80	47.90	95.80		
26 L.H., Daily Totals		$565.20		$953.40	$21.74	$36.67
Crew L-8	**Hr.**	**Daily**	**Hr.**	**Daily**	**Bare Costs**	**Incl. O&P**
1 Carpenter	$25.70	$205.60	$43.60	$348.80	$23.71	$39.88
1 Carpenter Helper	18.80	150.40	31.80	254.40		
.5 Plumber	29.55	118.20	48.60	194.40		
20 L.H., Daily Totals		$474.20		$797.60	$23.71	$39.88
Crew L-9	**Hr.**	**Daily**	**Hr.**	**Daily**	**Bare Costs**	**Incl. O&P**
1 Skilled Worker Foreman	$28.10	$224.80	$47.65	$381.20	$23.67	$39.88
1 Skilled Worker	26.10	208.80	44.25	354.00		
2 Helpers	18.80	300.80	31.80	508.80		
.5 Electrician	29.40	117.60	47.90	191.60		
36 L.H., Daily Totals		$852.00		$1435.60	$23.67	$39.88
Crew L-10	**Hr.**	**Daily**	**Hr.**	**Daily**	**Bare Costs**	**Incl. O&P**
1 Structural Steel Foreman	$29.70	$237.60	$57.95	$463.60	$28.28	$52.40
1 Structural Steel Worker	27.70	221.60	54.05	432.40		
1 Equip. Oper. (crane)	27.45	219.60	45.20	361.60		
1 Hyd. Crane, 12 Ton		724.00		796.40	30.17	33.18
24 L.H., Daily Totals		$1402.80		$2054.00	$58.45	$85.58
Crew L-11	**Hr.**	**Daily**	**Hr.**	**Daily**	**Bare Costs**	**Incl. O&P**
2 Wreckers	$19.25	$308.00	$36.65	$586.40	$22.85	$40.10
1 Equip. Oper. (crane)	27.45	219.60	45.20	361.60		
1 Equip. Oper. (light)	25.45	203.60	41.90	335.20		
1 Hyd. Excavator, 2.5 C.Y.		1333.00		1466.30		
1 Loader, Skid Steer, 78 HP		222.20		244.42	48.60	53.46
32 L.H., Daily Totals		$2286.40		$2993.92	$71.45	$93.56
Crew M-1	**Hr.**	**Daily**	**Hr.**	**Daily**	**Bare Costs**	**Incl. O&P**
3 Elevator Constructors	$35.80	$859.20	$58.45	$1402.80	$34.01	$55.52
1 Elevator Apprentice	28.65	229.20	46.75	374.00		
5 Hand Tools		48.00		52.80	1.50	1.65
32 L.H., Daily Totals		$1136.40		$1829.60	$35.51	$57.17
Crew M-3	**Hr.**	**Daily**	**Hr.**	**Daily**	**Bare Costs**	**Incl. O&P**
1 Electrician Foreman (out)	$31.40	$251.20	$51.15	$409.20	$28.51	$46.83
1 Common Laborer	18.70	149.60	31.75	254.00		
.25 Equipment Operator, Medium	26.50	53.00	43.65	87.30		
1 Elevator Constructor	35.80	286.40	58.45	467.60		
1 Elevator Apprentice	28.65	229.20	46.75	374.00		
.25 S.P. Crane, 4x4, 20 Ton		167.70		184.47	4.93	5.43
34 L.H., Daily Totals		$1137.10		$1776.57	$33.44	$52.25

Crew No.	Bare Costs		Incl. Subs O & P		Cost Per Labor-Hour	
Crew M-4	**Hr.**	**Daily**	**Hr.**	**Daily**	**Bare Costs**	**Incl. O&P**
1 Electrician Foreman (out)	$31.40	$251.20	$51.15	$409.20	$28.28	$46.45
1 Common Laborer	18.70	149.60	31.75	254.00		
.25 Equipment Operator, Crane	27.45	54.90	45.20	90.40		
.25 Equipment Operator, Oiler	23.40	46.80	38.55	77.10		
1 Elevator Constructor	35.80	286.40	58.45	467.60		
1 Elevator Apprentice	28.65	229.20	46.75	374.00		
.25 S.P. Crane, 4x4, 40 Ton		260.75		286.82	7.24	7.97
36 L.H., Daily Totals		$1278.85		$1959.13	$35.52	$54.42
Crew Q-1	**Hr.**	**Daily**	**Hr.**	**Daily**	**Bare Costs**	**Incl. O&P**
1 Plumber	$29.55	$236.40	$48.60	$388.80	$26.60	$43.75
1 Plumber Apprentice	23.65	189.20	38.90	311.20		
16 L.H., Daily Totals		$425.60		$700.00	$26.60	$43.75
Crew Q-1C	**Hr.**	**Daily**	**Hr.**	**Daily**	**Bare Costs**	**Incl. O&P**
1 Plumber	$29.55	$236.40	$48.60	$388.80	$26.57	$43.72
1 Plumber Apprentice	23.65	189.20	38.90	311.20		
1 Equip. Oper. (medium)	26.50	212.00	43.65	349.20		
1 Trencher, Chain Type, 8' D		1606.00		1766.60	66.92	73.61
24 L.H., Daily Totals		$2243.60		$2815.80	$93.48	$117.33
Crew Q-2	**Hr.**	**Daily**	**Hr.**	**Daily**	**Bare Costs**	**Incl. O&P**
1 Plumber	$29.55	$236.40	$48.60	$388.80	$25.62	$42.13
2 Plumber Apprentices	23.65	378.40	38.90	622.40		
24 L.H., Daily Totals		$614.80		$1011.20	$25.62	$42.13
Crew Q-3	**Hr.**	**Daily**	**Hr.**	**Daily**	**Bare Costs**	**Incl. O&P**
2 Plumbers	$29.55	$472.80	$48.60	$777.60	$26.60	$43.75
2 Plumber Apprentices	23.65	378.40	38.90	622.40		
32 L.H., Daily Totals		$851.20		$1400.00	$26.60	$43.75
Crew Q-4	**Hr.**	**Daily**	**Hr.**	**Daily**	**Bare Costs**	**Incl. O&P**
2 Plumbers	$29.55	$472.80	$48.60	$777.60	$28.07	$46.17
1 Welder (plumber)	29.55	236.40	48.60	388.80		
1 Plumber Apprentice	23.65	189.20	38.90	311.20		
1 Welder, electric, 300 amp		61.45		67.59	1.92	2.11
32 L.H., Daily Totals		$959.85		$1545.19	$30.00	$48.29
Crew Q-5	**Hr.**	**Daily**	**Hr.**	**Daily**	**Bare Costs**	**Incl. O&P**
1 Steamfitter	$29.85	$238.80	$49.05	$392.40	$26.88	$44.17
1 Steamfitter Apprentice	23.90	191.20	39.30	314.40		
16 L.H., Daily Totals		$430.00		$706.80	$26.88	$44.17
Crew Q-6	**Hr.**	**Daily**	**Hr.**	**Daily**	**Bare Costs**	**Incl. O&P**
1 Steamfitter	$29.85	$238.80	$49.05	$392.40	$25.88	$42.55
2 Steamfitter Apprentices	23.90	382.40	39.30	628.80		
24 L.H., Daily Totals		$621.20		$1021.20	$25.88	$42.55
Crew Q-7	**Hr.**	**Daily**	**Hr.**	**Daily**	**Bare Costs**	**Incl. O&P**
2 Steamfitters	$29.85	$477.60	$49.05	$784.80	$26.88	$44.17
2 Steamfitter Apprentices	23.90	382.40	39.30	628.80		
32 L.H., Daily Totals		$860.00		$1413.60	$26.88	$44.17
Crew Q-8	**Hr.**	**Daily**	**Hr.**	**Daily**	**Bare Costs**	**Incl. O&P**
2 Steamfitters	$29.85	$477.60	$49.05	$784.80	$28.36	$46.61
1 Welder (steamfitter)	29.85	238.80	49.05	392.40		
1 Steamfitter Apprentice	23.90	191.20	39.30	314.40		
1 Welder, electric, 300 amp		61.45		67.59	1.92	2.11
32 L.H., Daily Totals		$969.05		$1559.19	$30.28	$48.72

Crews

Crew No.	Bare Costs		Incl. Subs O & P		Cost Per Labor-Hour	
Crew Q-9	Hr.	Daily	Hr.	Daily	Bare Costs	Incl. O&P
1 Sheet Metal Worker	$28.75	$230.00	$48.35	$386.80	$25.88	$43.52
1 Sheet Metal Apprentice	23.00	184.00	38.70	309.60		
16 L.H., Daily Totals		$414.00		$696.40	$25.88	$43.52
Crew Q-10	Hr.	Daily	Hr.	Daily	Bare Costs	Incl. O&P
2 Sheet Metal Workers	$28.75	$460.00	$48.35	$773.60	$26.83	$45.13
1 Sheet Metal Apprentice	23.00	184.00	38.70	309.60		
24 L.H., Daily Totals		$644.00		$1083.20	$26.83	$45.13
Crew Q-11	Hr.	Daily	Hr.	Daily	Bare Costs	Incl. O&P
2 Sheet Metal Workers	$28.75	$460.00	$48.35	$773.60	$25.88	$43.52
2 Sheet Metal Apprentices	23.00	368.00	38.70	619.20		
32 L.H., Daily Totals		$828.00		$1392.80	$25.88	$43.52
Crew Q-12	Hr.	Daily	Hr.	Daily	Bare Costs	Incl. O&P
1 Sprinkler Installer	$29.25	$234.00	$48.15	$385.20	$26.32	$43.33
1 Sprinkler Apprentice	23.40	187.20	38.50	308.00		
16 L.H., Daily Totals		$421.20		$693.20	$26.32	$43.33
Crew Q-13	Hr.	Daily	Hr.	Daily	Bare Costs	Incl. O&P
2 Sprinkler Installers	$29.25	$468.00	$48.15	$770.40	$26.32	$43.33
2 Sprinkler Apprentices	23.40	374.40	38.50	616.00		
32 L.H., Daily Totals		$842.40		$1386.40	$26.32	$43.33
Crew Q-14	Hr.	Daily	Hr.	Daily	Bare Costs	Incl. O&P
1 Asbestos Worker	$27.40	$219.20	$47.15	$377.20	$24.65	$42.42
1 Asbestos Apprentice	21.90	175.20	37.70	301.60		
16 L.H., Daily Totals		$394.40		$678.80	$24.65	$42.42
Crew Q-15	Hr.	Daily	Hr.	Daily	Bare Costs	Incl. O&P
1 Plumber	$29.55	$236.40	$48.60	$388.80	$26.60	$43.75
1 Plumber Apprentice	23.65	189.20	38.90	311.20		
1 Welder, electric, 300 amp		61.45		67.59	3.84	4.22
16 L.H., Daily Totals		$487.05		$767.60	$30.44	$47.97
Crew Q-16	Hr.	Daily	Hr.	Daily	Bare Costs	Incl. O&P
2 Plumbers	$29.55	$472.80	$48.60	$777.60	$27.58	$45.37
1 Plumber Apprentice	23.65	189.20	38.90	311.20		
1 Welder, electric, 300 amp		61.45		67.59	2.56	2.82
24 L.H., Daily Totals		$723.45		$1156.40	$30.14	$48.18
Crew Q-17	Hr.	Daily	Hr.	Daily	Bare Costs	Incl. O&P
1 Steamfitter	$29.85	$238.80	$49.05	$392.40	$26.88	$44.17
1 Steamfitter Apprentice	23.90	191.20	39.30	314.40		
1 Welder, electric, 300 amp		61.45		67.59	3.84	4.22
16 L.H., Daily Totals		$491.45		$774.39	$30.72	$48.40
Crew Q-17A	Hr.	Daily	Hr.	Daily	Bare Costs	Incl. O&P
1 Steamfitter	$29.85	$238.80	$49.05	$392.40	$27.07	$44.52
1 Steamfitter Apprentice	23.90	191.20	39.30	314.40		
1 Equip. Oper. (crane)	27.45	219.60	45.20	361.60		
1 Hyd. Crane, 12 Ton		724.00		796.40		
1 Welder, electric, 300 amp		61.45		67.59	32.73	36.00
24 L.H., Daily Totals		$1435.05		$1932.40	$59.79	$80.52
Crew Q-18	Hr.	Daily	Hr.	Daily	Bare Costs	Incl. O&P
2 Steamfitters	$29.85	$477.60	$49.05	$784.80	$27.87	$45.80
1 Steamfitter Apprentice	23.90	191.20	39.30	314.40		
1 Welder, electric, 300 amp		61.45		67.59	2.56	2.82
24 L.H., Daily Totals		$730.25		$1166.80	$30.43	$48.62
Crew Q-19	Hr.	Daily	Hr.	Daily	Bare Costs	Incl. O&P
1 Steamfitter	$29.85	$238.80	$49.05	$392.40	$27.72	$45.42
1 Steamfitter Apprentice	23.90	191.20	39.30	314.40		
1 Electrician	29.40	235.20	47.90	383.20		
24 L.H., Daily Totals		$665.20		$1090.00	$27.72	$45.42
Crew Q-20	Hr.	Daily	Hr.	Daily	Bare Costs	Incl. O&P
1 Sheet Metal Worker	$28.75	$230.00	$48.35	$386.80	$26.58	$44.40
1 Sheet Metal Apprentice	23.00	184.00	38.70	309.60		
.5 Electrician	29.40	117.60	47.90	191.60		
20 L.H., Daily Totals		$531.60		$888.00	$26.58	$44.40
Crew Q-21	Hr.	Daily	Hr.	Daily	Bare Costs	Incl. O&P
2 Steamfitters	$29.85	$477.60	$49.05	$784.80	$28.25	$46.33
1 Steamfitter Apprentice	23.90	191.20	39.30	314.40		
1 Electrician	29.40	235.20	47.90	383.20		
32 L.H., Daily Totals		$904.00		$1482.40	$28.25	$46.33
Crew Q-22	Hr.	Daily	Hr.	Daily	Bare Costs	Incl. O&P
1 Plumber	$29.55	$236.40	$48.60	$388.80	$26.60	$43.75
1 Plumber Apprentice	23.65	189.20	38.90	311.20		
1 Hyd. Crane, 12 Ton		724.00		796.40	45.25	49.77
16 L.H., Daily Totals		$1149.60		$1496.40	$71.85	$93.53
Crew Q-22A	Hr.	Daily	Hr.	Daily	Bare Costs	Incl. O&P
1 Plumber	$29.55	$236.40	$48.60	$388.80	$24.84	$41.11
1 Plumber Apprentice	23.65	189.20	38.90	311.20		
1 Laborer	18.70	149.60	31.75	254.00		
1 Equip. Oper. (crane)	27.45	219.60	45.20	361.60		
1 Hyd. Crane, 12 Ton		724.00		796.40	22.63	24.89
32 L.H., Daily Totals		$1518.80		$2112.00	$47.46	$66.00
Crew Q-23	Hr.	Daily	Hr.	Daily	Bare Costs	Incl. O&P
1 Plumber Foreman	$31.55	$252.40	$51.85	$414.80	$29.20	$48.03
1 Plumber	29.55	236.40	48.60	388.80		
1 Equip. Oper. (medium)	26.50	212.00	43.65	349.20		
1 Lattice Boom Crane, 20 Ton		976.70		1074.37	40.70	44.77
24 L.H., Daily Totals		$1677.50		$2227.17	$69.90	$92.80
Crew R-1	Hr.	Daily	Hr.	Daily	Bare Costs	Incl. O&P
1 Electrician Foreman	$29.90	$239.20	$48.70	$389.60	$25.95	$42.67
3 Electricians	29.40	705.60	47.90	1149.60		
2 Helpers	18.80	300.80	31.80	508.80		
48 L.H., Daily Totals		$1245.60		$2048.00	$25.95	$42.67
Crew R-1A	Hr.	Daily	Hr.	Daily	Bare Costs	Incl. O&P
1 Electrician	$29.40	$235.20	$47.90	$383.20	$24.10	$39.85
1 Helper	18.80	150.40	31.80	254.40		
16 L.H., Daily Totals		$385.60		$637.60	$24.10	$39.85

Crews

Crew No.	Bare Costs		Incl. Subs O & P		Cost Per Labor-Hour	
Crew R-2	Hr.	Daily	Hr.	Daily	Bare Costs	Incl. O&P
1 Electrician Foreman	$29.90	$239.20	$48.70	$389.60	$26.16	$43.03
3 Electricians	29.40	705.60	47.90	1149.60		
2 Helpers	18.80	300.80	31.80	508.80		
1 Equip. Oper. (crane)	27.45	219.60	45.20	361.60		
1 S.P. Crane, 4x4, 5 Ton		326.80		359.48	5.84	6.42
56 L.H., Daily Totals		$1792.00		$2769.08	$32.00	$49.45
Crew R-3	Hr.	Daily	Hr.	Daily	Bare Costs	Incl. O&P
1 Electrician Foreman	$29.90	$239.20	$48.70	$389.60	$29.21	$47.68
1 Electrician	29.40	235.20	47.90	383.20		
.5 Equip. Oper. (crane)	27.45	109.80	45.20	180.80		
.5 S.P. Crane, 4x4, 5 Ton		163.40		179.74	8.17	8.99
20 L.H., Daily Totals		$747.60		$1133.34	$37.38	$56.67
Crew R-4	Hr.	Daily	Hr.	Daily	Bare Costs	Incl. O&P
1 Struc. Steel Foreman	$29.70	$237.60	$57.95	$463.60	$28.44	$53.60
3 Struc. Steel Workers	27.70	664.80	54.05	1297.20		
1 Electrician	29.40	235.20	47.90	383.20		
1 Welder, gas engine, 300 amp		115.20		126.72	2.88	3.17
40 L.H., Daily Totals		$1252.80		$2270.72	$31.32	$56.77
Crew R-5	Hr.	Daily	Hr.	Daily	Bare Costs	Incl. O&P
1 Electrician Foreman	$29.90	$239.20	$48.70	$389.60	$25.59	$42.12
4 Electrician Linemen	29.40	940.80	47.90	1532.80		
2 Electrician Operators	29.40	470.40	47.90	766.40		
4 Electrician Groundmen	18.80	601.60	31.80	1017.60		
1 Crew Truck		116.20		127.82		
1 Tool Van		151.80		166.98		
1 Pickup Truck, 3/4 Ton		84.80		93.28		
.2 Hyd. Crane, 55 Ton		212.00		233.20		
.2 Hyd. Crane, 12 Ton		144.80		159.28		
.2 Drill Rig, Truck-Mounted		699.80		769.78		
1 Tractor w/Winch		297.80		327.58	19.40	21.34
88 L.H., Daily Totals		$3959.20		$5584.32	$44.99	$63.46
Crew R-6	Hr.	Daily	Hr.	Daily	Bare Costs	Incl. O&P
1 Electrician Foreman	$29.90	$239.20	$48.70	$389.60	$25.59	$42.12
4 Electrician Linemen	29.40	940.80	47.90	1532.80		
2 Electrician Operators	29.40	470.40	47.90	766.40		
4 Electrician Groundmen	18.80	601.60	31.80	1017.60		
1 Crew Truck		116.20		127.82		
1 Tool Van		151.80		166.98		
1 Pickup Truck, 3/4 Ton		84.80		93.28		
.2 Hyd. Crane, 55 Ton		212.00		233.20		
.2 Hyd. Crane, 12 Ton		144.80		159.28		
.2 Drill Rig, Truck-Mounted		699.80		769.78		
1 Tractor w/Winch		297.80		327.58		
3 Cable Trailers		510.90		561.99		
.5 Tensioning Rig		167.68		184.44		
.5 Cable Pulling Rig		989.00		1087.90	38.35	42.18
88 L.H., Daily Totals		$5626.77		$7418.65	$63.94	$84.30
Crew R-7	Hr.	Daily	Hr.	Daily	Bare Costs	Incl. O&P
1 Electrician Foreman	$29.90	$239.20	$48.70	$389.60	$20.65	$34.62
5 Electrician Groundmen	18.80	752.00	31.80	1272.00		
1 Crew Truck		116.20		127.82	2.42	2.66
48 L.H., Daily Totals		$1107.40		$1789.42	$23.07	$37.28

Crew No.	Bare Costs		Incl. Subs O & P		Cost Per Labor-Hour	
Crew R-8	Hr.	Daily	Hr.	Daily	Bare Costs	Incl. O&P
1 Electrician Foreman	$29.90	$239.20	$48.70	$389.60	$25.95	$42.67
3 Electrician Linemen	29.40	705.60	47.90	1149.60		
2 Electrician Groundmen	18.80	300.80	31.80	508.80		
1 Pickup Truck, 3/4 Ton		84.80		93.28		
1 Crew Truck		116.20		127.82	4.19	4.61
48 L.H., Daily Totals		$1446.60		$2269.10	$30.14	$47.27
Crew R-9	Hr.	Daily	Hr.	Daily	Bare Costs	Incl. O&P
1 Electrician Foreman	$29.90	$239.20	$48.70	$389.60	$24.16	$39.95
1 Electrician Lineman	29.40	235.20	47.90	383.20		
2 Electrician Operators	29.40	470.40	47.90	766.40		
4 Electrician Groundmen	18.80	601.60	31.80	1017.60		
1 Pickup Truck, 3/4 Ton		84.80		93.28		
1 Crew Truck		116.20		127.82	3.14	3.45
64 L.H., Daily Totals		$1747.40		$2777.90	$27.30	$43.40
Crew R-10	Hr.	Daily	Hr.	Daily	Bare Costs	Incl. O&P
1 Electrician Foreman	$29.90	$239.20	$48.70	$389.60	$27.72	$45.35
4 Electrician Linemen	29.40	940.80	47.90	1532.80		
1 Electrician Groundman	18.80	150.40	31.80	254.40		
1 Crew Truck		116.20		127.82		
3 Tram Cars		363.00		399.30	9.98	10.98
48 L.H., Daily Totals		$1809.60		$2703.92	$37.70	$56.33
Crew R-11	Hr.	Daily	Hr.	Daily	Bare Costs	Incl. O&P
1 Electrician Foreman	$29.90	$239.20	$48.70	$389.60	$27.66	$45.32
4 Electricians	29.40	940.80	47.90	1532.80		
1 Equip. Oper. (crane)	27.45	219.60	45.20	361.60		
1 Common Laborer	18.70	149.60	31.75	254.00		
1 Crew Truck		116.20		127.82		
1 Hyd. Crane, 12 Ton		724.00		796.40	15.00	16.50
56 L.H., Daily Totals		$2389.40		$3462.22	$42.67	$61.83
Crew R-12	Hr.	Daily	Hr.	Daily	Bare Costs	Incl. O&P
1 Carpenter Foreman	$26.20	$209.60	$44.45	$355.60	$23.45	$40.32
4 Carpenters	25.70	822.40	43.60	1395.20		
4 Common Laborers	18.70	598.40	31.75	1016.00		
1 Equip. Oper. (med.)	26.50	212.00	43.65	349.20		
1 Steel Worker	27.70	221.60	54.05	432.40		
1 Dozer, 200 H.P.		988.40		1087.24		
1 Pickup Truck, 3/4 Ton		84.80		93.28	12.20	13.41
88 L.H., Daily Totals		$3137.20		$4728.92	$35.65	$53.74
Crew R-15	Hr.	Daily	Hr.	Daily	Bare Costs	Incl. O&P
1 Electrician Foreman	$29.90	$239.20	$48.70	$389.60	$28.82	$47.03
4 Electricians	29.40	940.80	47.90	1532.80		
1 Equipment Operator	25.45	203.60	41.90	335.20		
1 Aerial Lift Truck		289.80		318.78	6.04	6.64
48 L.H., Daily Totals		$1673.40		$2576.38	$34.86	$53.67
Crew R-15A	Hr.	Daily	Hr.	Daily	Bare Costs	Incl. O&P
1 Electrician Foreman	$29.90	$239.20	$48.70	$389.60	$25.26	$41.65
2 Electricians	29.40	470.40	47.90	766.40		
2 Common Laborers	18.70	299.20	31.75	508.00		
1 Equipment Operator	25.45	203.60	41.90	335.20		
1 Aerial Lift Truck		289.80		318.78	6.04	6.64
48 L.H., Daily Totals		$1502.20		$2317.98	$31.30	$48.29

Crews

Crew No.	Bare Costs		Incl. Sub O & P		Cost Per Labor-Hour	
Crew R-18	**Hr.**	**Daily**	**Hr.**	**Daily**	**Bare Costs**	**Incl. O&P**
.25 Electrician Foreman	$29.90	$59.80	$48.70	$97.40	$22.92	$38.05
1 Electrician	29.40	235.20	47.90	383.20		
2 Helpers	18.80	300.80	31.80	508.80		
26 L.H., Daily Totals		$595.80		$989.40	$22.92	$38.05
Crew R-19	**Hr.**	**Daily**	**Hr.**	**Daily**	**Bare Costs**	**Incl. O&P**
.5 Electrician Foreman	$29.90	$119.60	$48.70	$194.80	$29.50	$48.06
2 Electricians	29.40	470.40	47.90	766.40		
20 L.H., Daily Totals		$590.00		$961.20	$29.50	$48.06
Crew R-21	**Hr.**	**Daily**	**Hr.**	**Daily**	**Bare Costs**	**Incl. O&P**
1 Electrician Foreman	$29.90	$239.20	$48.70	$389.60	$29.45	$47.99
3 Electricians	29.40	705.60	47.90	1149.60		
.1 Equip. Oper. (med.)	26.50	21.20	43.65	34.92		
.1 S.P. Crane, 4x4, 25 Ton		87.56		96.32	2.67	2.94
32.8 L.H., Daily Totals		$1053.56		$1670.44	$32.12	$50.93
Crew R-22	**Hr.**	**Daily**	**Hr.**	**Daily**	**Bare Costs**	**Incl. O&P**
.66 Electrician Foreman	$29.90	$157.87	$48.70	$257.14	$24.92	$41.10
2 Helpers	18.80	300.80	31.80	508.80		
2 Electricians	29.40	470.40	47.90	766.40		
37.28 L.H., Daily Totals		$929.07		$1532.34	$24.92	$41.10
Crew R-30	**Hr.**	**Daily**	**Hr.**	**Daily**	**Bare Costs**	**Incl. O&P**
.25 Electrician Foreman (out)	$31.40	$62.80	$51.15	$102.30	$22.97	$38.21
1 Electrician	29.40	235.20	47.90	383.20		
2 Laborers, (Semi-Skilled)	18.70	299.20	31.75	508.00		
26 L.H., Daily Totals		$597.20		$993.50	$22.97	$38.21

Location Factors

Costs shown in *RSMeans cost data publications* are based on National Averages for materials and installation. To adjust these costs to a specific location, simply multiply the base cost by the factor for that city. The data is arranged alphabetically by state and postal zip code numbers. For a city not listed, use the factor for a nearby city with similar economic characteristics.

STATE	CITY	Residential
ALABAMA		
350-352	Birmingham	.86
354	Tuscaloosa	.73
355	Jasper	.71
356	Decatur	.76
357-358	Huntsville	.84
359	Gadsden	.73
360-361	Montgomery	.75
362	Anniston	.68
363	Dothan	.74
364	Evergreen	.70
365-366	Mobile	.79
367	Selma	.72
368	Phenix City	.73
369	Butler	.71
ALASKA		
995-996	Anchorage	1.27
997	Fairbanks	1.29
998	Juneau	1.27
999	Ketchikan	1.29
ARIZONA		
850,853	Phoenix	.86
852	Mesa/Tempe	.83
855	Globe	.79
856-857	Tucson	.84
859	Show Low	.81
860	Flagstaff	.86
863	Prescott	.81
864	Kingman	.83
865	Chambers	.80
ARKANSAS		
716	Pine Bluff	.81
717	Camden	.70
718	Texarkana	.75
719	Hot Springs	.70
720-722	Little Rock	.87
723	West Memphis	.81
724	Jonesboro	.79
725	Batesville	.76
726	Harrison	.78
727	Fayetteville	.72
728	Russellville	.77
729	Fort Smith	.79
CALIFORNIA		
900-902	Los Angeles	1.06
903-905	Inglewood	1.05
906-908	Long Beach	1.04
910-912	Pasadena	1.05
913-916	Van Nuys	1.08
917-918	Alhambra	1.09
919-921	San Diego	1.04
922	Palm Springs	1.04
923-924	San Bernardino	1.05
925	Riverside	1.05
926-927	Santa Ana	1.06
928	Anaheim	1.05
930	Oxnard	1.07
931	Santa Barbara	1.06
932-933	Bakersfield	1.03
934	San Luis Obispo	1.08
935	Mojave	1.06
936-938	Fresno	1.09
939	Salinas	1.12
940-941	San Francisco	1.23
942,956-958	Sacramento	1.11
943	Palo Alto	1.18
944	San Mateo	1.22
945	Vallejo	1.15
946	Oakland	1.21
947	Berkeley	1.24
948	Richmond	1.24
949	San Rafael	1.22
950	Santa Cruz	1.15
951	San Jose	1.19
952	Stockton	1.09
953	Modesto	1.08
954	Santa Rosa	1.16

STATE	CITY	Residential
955	Eureka	1.12
959	Marysville	1.10
960	Redding	1.10
961	Susanville	1.09
COLORADO		
800-802	Denver	.94
803	Boulder	.93
804	Golden	.91
805	Fort Collins	.90
806	Greeley	.80
807	Fort Morgan	.93
808-809	Colorado Springs	.90
810	Pueblo	.91
811	Alamosa	.88
812	Salida	.90
813	Durango	.91
814	Montrose	.87
815	Grand Junction	.92
816	Glenwood Springs	.90
CONNECTICUT		
060	New Britain	1.08
061	Hartford	1.08
062	Willimantic	1.08
063	New London	1.08
064	Meriden	1.08
065	New Haven	1.08
066	Bridgeport	1.09
067	Waterbury	1.09
068	Norwalk	1.09
069	Stamford	1.10
D.C.		
200-205	Washington	.95
DELAWARE		
197	Newark	.99
198	Wilmington	1.00
199	Dover	.99
FLORIDA		
320,322	Jacksonville	.77
321	Daytona Beach	.84
323	Tallahassee	.73
324	Panama City	.67
325	Pensacola	.78
326,344	Gainesville	.77
327-328,347	Orlando	.84
329	Melbourne	.86
330-332,340	Miami	.85
333	Fort Lauderdale	.84
334,349	West Palm Beach	.84
335-336,346	Tampa	.86
337	St. Petersburg	.76
338	Lakeland	.83
339,341	Fort Myers	.80
342	Sarasota	.84
GEORGIA		
300-303,399	Atlanta	.90
304	Statesboro	.71
305	Gainesville	.79
306	Athens	.79
307	Dalton	.75
308-309	Augusta	.81
310-312	Macon	.82
313-314	Savannah	.82
315	Waycross	.76
316	Valdosta	.73
317,398	Albany	.79
318-319	Columbus	.83
HAWAII		
967	Hilo	1.21
968	Honolulu	1.23
STATES & POSS.		
969	Guam	.89

Location Factors

STATE	CITY	Residential
IDAHO		
832	Pocatello	.86
833	Twin Falls	.74
834	Idaho Falls	.75
835	Lewiston	.96
836-837	Boise	.87
838	Coeur d'Alene	.94
ILLINOIS		
600-603	North Suburban	1.10
604	Joliet	1.10
605	South Suburban	1.10
606-608	Chicago	1.16
609	Kankakee	1.00
610-611	Rockford	1.04
612	Rock Island	.96
613	La Salle	1.02
614	Galesburg	.99
615-616	Peoria	.98
617	Bloomington	.98
618-619	Champaign	.99
620-622	East St. Louis	1.00
623	Quincy	.98
624	Effingham	.96
625	Decatur	.97
626-627	Springfield	.97
628	Centralia	1.00
629	Carbondale	.95
INDIANA		
460	Anderson	.91
461-462	Indianapolis	.95
463-464	Gary	1.01
465-466	South Bend	.91
467-468	Fort Wayne	.91
469	Kokomo	.92
470	Lawrenceburg	.87
471	New Albany	.86
472	Columbus	.92
473	Muncie	.91
474	Bloomington	.94
475	Washington	.91
476-477	Evansville	.90
478	Terre Haute	.90
479	Lafayette	.91
IOWA		
500-503,509	Des Moines	.91
504	Mason City	.77
505	Fort Dodge	.76
506-507	Waterloo	.79
508	Creston	.81
510-511	Sioux City	.87
512	Sibley	.73
513	Spencer	.74
514	Carroll	.74
515	Council Bluffs	.81
516	Shenandoah	.75
520	Dubuque	.86
521	Decorah	.76
522-524	Cedar Rapids	.94
525	Ottumwa	.84
526	Burlington	.87
527-528	Davenport	.97
KANSAS		
660-662	Kansas City	.99
664-666	Topeka	.80
667	Fort Scott	.85
668	Emporia	.72
669	Belleville	.78
670-672	Wichita	.80
673	Independence	.85
674	Salina	.76
675	Hutchinson	.77
676	Hays	.82
677	Colby	.83
678	Dodge City	.82
679	Liberal	.79
KENTUCKY		
400-402	Louisville	.92
403-405	Lexington	.89
406	Frankfort	.89
407-409	Corbin	.78
410	Covington	1.00
411-412	Ashland	.98

STATE	CITY	Residential
413-414	Campton	.79
415-416	Pikeville	.86
417-418	Hazard	.73
420	Paducah	.90
421-422	Bowling Green	.90
423	Owensboro	.89
424	Henderson	.91
425-426	Somerset	.78
427	Elizabethtown	.87
LOUISIANA		
700-701	New Orleans	.86
703	Thibodaux	.84
704	Hammond	.79
705	Lafayette	.82
706	Lake Charles	.83
707-708	Baton Rouge	.82
710-711	Shreveport	.79
712	Monroe	.74
713-714	Alexandria	.74
MAINE		
039	Kittery	.79
040-041	Portland	.90
042	Lewiston	.89
043	Augusta	.82
044	Bangor	.88
045	Bath	.80
046	Machias	.81
047	Houlton	.85
048	Rockland	.81
049	Waterville	.80
MARYLAND		
206	Waldorf	.85
207-208	College Park	.87
209	Silver Spring	.86
210-212	Baltimore	.90
214	Annapolis	.85
215	Cumberland	.86
216	Easton	.68
217	Hagerstown	.86
218	Salisbury	.75
219	Elkton	.81
MASSACHUSETTS		
010-011	Springfield	1.04
012	Pittsfield	1.01
013	Greenfield	1.00
014	Fitchburg	1.12
015-016	Worcester	1.14
017	Framingham	1.12
018	Lowell	1.13
019	Lawrence	1.13
020-022, 024	Boston	1.19
023	Brockton	1.12
025	Buzzards Bay	1.10
026	Hyannis	1.09
027	New Bedford	1.12
MICHIGAN		
480,483	Royal Oak	1.03
481	Ann Arbor	1.05
482	Detroit	1.07
484-485	Flint	.97
486	Saginaw	.94
487	Bay City	.95
488-489	Lansing	.97
490	Battle Creek	.93
491	Kalamazoo	.92
492	Jackson	.95
493,495	Grand Rapids	.82
494	Muskegon	.89
496	Traverse City	.80
497	Gaylord	.83
498-499	Iron Mountain	.90
MINNESOTA		
550-551	Saint Paul	1.13
553-555	Minneapolis	1.17
556-558	Duluth	1.09
559	Rochester	1.05
560	Mankato	1.03
561	Windom	.83
562	Willmar	.85
563	St. Cloud	1.07
564	Brainerd	.98

Location Factors

STATE	CITY	Residential
565	Detroit Lakes	.96
566	Bemidji	.96
567	Thief River Falls	.95
MISSISSIPPI		
386	Clarksdale	.62
387	Greenville	.69
388	Tupelo	.64
389	Greenwood	.65
390-392	Jackson	.73
393	Meridian	.66
394	Laurel	.63
395	Biloxi	.75
396	McComb	.74
397	Columbus	.65
MISSOURI		
630-631	St. Louis	1.03
633	Bowling Green	.94
634	Hannibal	.87
635	Kirksville	.80
636	Flat River	.94
637	Cape Girardeau	.87
638	Sikeston	.82
639	Poplar Bluff	.82
640-641	Kansas City	1.03
644-645	St. Joseph	.95
646	Chillicothe	.84
647	Harrisonville	.94
648	Joplin	.85
650-651	Jefferson City	.88
652	Columbia	.88
653	Sedalia	.85
654-655	Rolla	.88
656-658	Springfield	.86
MONTANA		
590-591	Billings	.87
592	Wolf Point	.83
593	Miles City	.85
594	Great Falls	.88
595	Havre	.80
596	Helena	.87
597	Butte	.83
598	Missoula	.83
599	Kalispell	.81
NEBRASKA		
680-681	Omaha	.89
683-685	Lincoln	.78
686	Columbus	.70
687	Norfolk	.77
688	Grand Island	.77
689	Hastings	.76
690	Mccook	.70
691	North Platte	.75
692	Valentine	.66
693	Alliance	.66
NEVADA		
889-891	Las Vegas	1.00
893	Ely	.87
894-895	Reno	.94
897	Carson City	.95
898	Elko	.93
NEW HAMPSHIRE		
030	Nashua	.95
031	Manchester	.95
032-033	Concord	.92
034	Keene	.72
035	Littleton	.80
036	Charleston	.70
037	Claremont	.71
038	Portsmouth	.89
NEW JERSEY		
070-071	Newark	1.13
072	Elizabeth	1.15
073	Jersey City	1.11
074-075	Paterson	1.12
076	Hackensack	1.11
077	Long Branch	1.13
078	Dover	1.12
079	Summit	1.12
080,083	Vineland	1.09
081	Camden	1.10

STATE	CITY	Residential
082,084	Atlantic City	1.13
085-086	Trenton	1.11
087	Point Pleasant	1.10
088-089	New Brunswick	1.12
NEW MEXICO		
870-872	Albuquerque	.85
873	Gallup	.85
874	Farmington	.85
875	Santa Fe	.85
877	Las Vegas	.85
878	Socorro	.85
879	Truth/Consequences	.84
880	Las Cruces	.82
881	Clovis	.84
882	Roswell	.85
883	Carrizozo	.85
884	Tucumcari	.85
NEW YORK		
100-102	New York	1.34
103	Staten Island	1.25
104	Bronx	1.27
105	Mount Vernon	1.15
106	White Plains	1.18
107	Yonkers	1.19
108	New Rochelle	1.19
109	Suffern	1.12
110	Queens	1.26
111	Long Island City	1.29
112	Brooklyn	1.31
113	Flushing	1.28
114	Jamaica	1.28
115,117,118	Hicksville	1.19
116	Far Rockaway	1.27
119	Riverhead	1.21
120-122	Albany	.95
123	Schenectady	.96
124	Kingston	1.03
125-126	Poughkeepsie	1.04
127	Monticello	1.05
128	Glens Falls	.89
129	Plattsburgh	.93
130-132	Syracuse	.96
133-135	Utica	.93
136	Watertown	.90
137-139	Binghamton	.95
140-142	Buffalo	1.05
143	Niagara Falls	1.01
144-146	Rochester	.99
147	Jamestown	.89
148-149	Elmira	.86
NORTH CAROLINA		
270,272-274	Greensboro	.85
271	Winston-Salem	.85
275-276	Raleigh	.87
277	Durham	.85
278	Rocky Mount	.75
279	Elizabeth City	.75
280	Gastonia	.86
281-282	Charlotte	.88
283	Fayetteville	.84
284	Wilmington	.83
285	Kinston	.75
286	Hickory	.80
287-288	Asheville	.83
289	Murphy	.74
NORTH DAKOTA		
580-581	Fargo	.79
582	Grand Forks	.76
583	Devils Lake	.79
584	Jamestown	.74
585	Bismarck	.79
586	Dickinson	.77
587	Minot	.80
588	Williston	.77
OHIO		
430-432	Columbus	.94
433	Marion	.91
434-436	Toledo	.99
437-438	Zanesville	.90
439	Steubenville	.95
440	Lorain	.99
441	Cleveland	1.02

Location Factors

STATE	CITY	Residential
442-443	Akron	.98
444-445	Youngstown	.96
446-447	Canton	.94
448-449	Mansfield	.95
450	Hamilton	.93
451-452	Cincinnati	.93
453-454	Dayton	.91
455	Springfield	.93
456	Chillicothe	.96
457	Athens	.88
458	Lima	.90
OKLAHOMA		
730-731	Oklahoma City	.80
734	Ardmore	.79
735	Lawton	.82
736	Clinton	.78
737	Enid	.78
738	Woodward	.77
739	Guymon	.68
740-741	Tulsa	.79
743	Miami	.82
744	Muskogee	.73
745	Mcalester	.74
746	Ponca City	.78
747	Durant	.77
748	Shawnee	.76
749	Poteau	.78
OREGON		
970-972	Portland	1.02
973	Salem	1.01
974	Eugene	1.01
975	Medford	.99
976	Klamath Falls	1.00
977	Bend	1.02
978	Pendleton	.99
979	Vale	.98
PENNSYLVANIA		
150-152	Pittsburgh	.99
153	Washington	.94
154	Uniontown	.90
155	Bedford	.88
156	Greensburg	.94
157	Indiana	.91
158	Dubois	.90
159	Johnstown	.89
160	Butler	.92
161	New Castle	.92
162	Kittanning	.93
163	Oil City	.90
164-165	Erie	.95
166	Altoona	.88
167	Bradford	.90
168	State College	.91
169	Wellsboro	.90
170-171	Harrisburg	.94
172	Chambersburg	.89
173-174	York	.91
175-176	Lancaster	.91
177	Williamsport	.84
178	Sunbury	.91
179	Pottsville	.91
180	Lehigh Valley	1.01
181	Allentown	1.04
182	Hazleton	.91
183	Stroudsburg	.92
184-185	Scranton	.96
186-187	Wilkes-Barre	.93
188	Montrose	.90
189	Doylestown	1.04
190-191	Philadelphia	1.15
193	Westchester	1.09
194	Norristown	1.08
195-196	Reading	.97
PUERTO RICO		
009	San Juan	.74
RHODE ISLAND		
028	Newport	1.06
029	Providence	1.07
SOUTH CAROLINA		
290-292	Columbia	.82
293	Spartanburg	.81

STATE	CITY	Residential
294	Charleston	.82
295	Florence	.76
296	Greenville	.80
297	Rock Hill	.72
298	Aiken	.97
299	Beaufort	.75
SOUTH DAKOTA		
570-571	Sioux Falls	.76
572	Watertown	.72
573	Mitchell	.74
574	Aberdeen	.75
575	Pierre	.75
576	Mobridge	.73
577	Rapid City	.75
TENNESSEE		
370-372	Nashville	.85
373-374	Chattanooga	.77
375,380-381	Memphis	.83
376	Johnson City	.72
377-379	Knoxville	.75
382	Mckenzie	.73
383	Jackson	.71
384	Columbia	.73
385	Cookeville	.72
TEXAS		
750	Mckinney	.75
751	Waxahackie	.75
752-753	Dallas	.82
754	Greenville	.69
755	Texarkana	.73
756	Longview	.67
757	Tyler	.74
758	Palestine	.66
759	Lufkin	.71
760-761	Fort Worth	.82
762	Denton	.77
763	Wichita Falls	.79
764	Eastland	.73
765	Temple	.75
766-767	Waco	.78
768	Brownwood	.69
769	San Angelo	.72
770-772	Houston	.86
773	Huntsville	.70
774	Wharton	.71
775	Galveston	.84
776-777	Beaumont	.83
778	Bryan	.74
779	Victoria	.75
780	Laredo	.73
781-782	San Antonio	.81
783-784	Corpus Christi	.78
785	Mc Allen	.76
786-787	Austin	.80
788	Del Rio	.67
789	Giddings	.70
790-791	Amarillo	.78
792	Childress	.76
793-794	Lubbock	.76
795-796	Abilene	.75
797	Midland	.76
798-799,885	El Paso	.75
UTAH		
840-841	Salt Lake City	.81
842,844	Ogden	.79
843	Logan	.80
845	Price	.72
846-847	Provo	.81
VERMONT		
050	White River Jct.	.73
051	Bellows Falls	.75
052	Bennington	.82
053	Brattleboro	.78
054	Burlington	.80
056	Montpelier	.81
057	Rutland	.80
058	St. Johnsbury	.80
059	Guildhall	.79
VIRGINIA		
220-221	Fairfax	1.02
222	Arlington	1.04

Location Factors

STATE	CITY	Residential
223	Alexandria	1.06
224-225	Fredericksburg	.95
226	Winchester	.93
227	Culpeper	1.00
228	Harrisonburg	.90
229	Charlottesville	.92
230-232	Richmond	1.01
233-235	Norfolk	1.02
236	Newport News	1.01
237	Portsmouth	.92
238	Petersburg	.99
239	Farmville	.91
240-241	Roanoke	.99
242	Bristol	.86
243	Pulaski	.84
244	Staunton	.93
245	Lynchburg	.97
246	Grundy	.85
WASHINGTON		
980-981,987	Seattle	1.02
982	Everett	1.05
983-984	Tacoma	1.01
985	Olympia	1.01
986	Vancouver	.98
988	Wenatchee	.93
989	Yakima	.97
990-992	Spokane	.99
993	Richland	.97
994	Clarkston	.97
WEST VIRGINIA		
247-248	Bluefield	.88
249	Lewisburg	.89
250-253	Charleston	.97
254	Martinsburg	.86
255-257	Huntington	1.01
258-259	Beckley	.90
260	Wheeling	.93
261	Parkersburg	.92
262	Buckhannon	.92
263-264	Clarksburg	.92
265	Morgantown	.93
266	Gassaway	.92
267	Romney	.88
268	Petersburg	.90
WISCONSIN		
530,532	Milwaukee	1.07
531	Kenosha	1.04
534	Racine	1.02
535	Beloit	1.00
537	Madison	.99
538	Lancaster	.97
539	Portage	.96
540	New Richmond	1.00
541-543	Green Bay	1.01
544	Wausau	.95
545	Rhinelander	.95
546	La Crosse	.94
547	Eau Claire	.98
548	Superior	.99
549	Oshkosh	.95
WYOMING		
820	Cheyenne	.84
821	Yellowstone Nat. Pk.	.75
822	Wheatland	.75
823	Rawlins	.76
824	Worland	.75
825	Riverton	.74
826	Casper	.78
827	Newcastle	.74
828	Sheridan	.80
829-831	Rock Springs	.79

CANADIAN FACTORS (reflect Canadian Currency)

STATE	CITY	Residential
ALBERTA		
	Calgary	1.14
	Edmonton	1.13
	Fort McMurray	1.09
	Lethbridge	1.10
	Lloydminster	1.09
	Medicine Hat	1.10
	Red Deer	1.10
BRITISH COLUMBIA		
	Kamloops	1.08
	Prince George	1.08
	Vancouver	1.09
	Victoria	1.03
MANITOBA		
	Brandon	1.06
	Portage la Prairie	1.06
	Winnipeg	1.05
NEW BRUNSWICK		
	Bathurst	.97
	Dalhousie	.97
	Fredericton	1.05
	Moncton	.98
	Newcastle	.97
	Saint John	1.05
NEWFOUNDLAND		
	Corner Brook	.99
	St. John's	1.01
NORTHWEST TERRITORIES		
	Yellowknife	1.10
NOVA SCOTIA		
	Dartmouth	1.00
	Halifax	1.02
	New Glasgow	1.00
	Sydney	.99
	Yarmouth	1.00
ONTARIO		
	Barrie	1.17
	Brantford	1.19
	Cornwall	1.19
	Hamilton	1.19
	Kingston	1.19
	Kitchener	1.11
	London	1.17
	North Bay	1.15
	Oshawa	1.17
	Ottawa	1.19
	Owen Sound	1.15
	Peterborough	1.16
	Sarnia	1.19
	Sudbury	1.09
	Thunder Bay	1.15
	Toronto	1.20
	Windsor	1.14
PRINCE EDWARD ISLAND		
	Charlottetown	.95
	Summerside	.94
QUEBEC		
	Cap-de-la-Madeleine	1.18
	Charlesbourg	1.18
	Chicoutimi	1.20
	Gatineau	1.16
	Laval	1.17
	Montreal	1.21
	Quebec	1.22
	Sherbrooke	1.17
	Trois Rivieres	1.18
SASKATCHEWAN		
	Moose Jaw	.97
	Prince Albert	.96
	Regina	.99
	Saskatoon	.97
YUKON		
	Whitehorse	.96

General Requirements — R0111 Summary of Work

R011105-05 Tips for Accurate Estimating

1. Use pre-printed or columnar forms for orderly sequence of dimensions and locations and for recording telephone quotations.
2. Use only the front side of each paper or form except for certain pre-printed summary forms.
3. Be consistent in listing dimensions: For example, length x width x height. This helps in rechecking to ensure that, the total length of partitions is appropriate for the building area.
4. Use printed (rather than measured) dimensions where given.
5. Add up multiple printed dimensions for a single entry where possible.
6. Measure all other dimensions carefully.
7. Use each set of dimensions to calculate multiple related quantities.
8. Convert foot and inch measurements to decimal feet when listing. Memorize decimal equivalents to .01 parts of a foot (1/8" equals approximately .01').
9. Do not "round off" quantities until the final summary.
10. Mark drawings with different colors as items are taken off.
11. Keep similar items together, different items separate.
12. Identify location and drawing numbers to aid in future checking for completeness.
13. Measure or list everything on the drawings or mentioned in the specifications.
14. It may be necessary to list items not called for to make the job complete.
15. Be alert for: Notes on plans such as N.T.S. (not to scale); changes in scale throughout the drawings; reduced size drawings; discrepancies between the specifications and the drawings.
16. Develop a consistent pattern of performing an estimate. For example:
    a. Start the quantity takeoff at the lower floor and move to the next higher floor.
    b. Proceed from the main section of the building to the wings.
    c. Proceed from south to north or vice versa, clockwise or counterclockwise.
    d. Take off floor plan quantities first, elevations next, then detail drawings.
17. List all gross dimensions that can be either used again for different quantities, or used as a rough check of other quantities for verification (exterior perimeter, gross floor area, individual floor areas, etc.).
18. Utilize design symmetry or repetition (repetitive floors, repetitive wings, symmetrical design around a center line, similar room layouts, etc.). Note: Extreme caution is needed here so as not to omit or duplicate an area.
19. Do not convert units until the final total is obtained. For instance, when estimating concrete work, keep all units to the nearest cubic foot, then summarize and convert to cubic yards.
20. When figuring alternatives, it is best to total all items involved in the basic system, then total all items involved in the alternates. Therefore you work with positive numbers in all cases. When adds and deducts are used, it is often confusing whether to add or subtract a portion of an item; especially on a complicated or involved alternate.

R011110-10 Architectural Fees

Tabulated below are typical percentage fees by project size, for good professional architectural service. Fees may vary from those listed depending upon degree of design difficulty and economic conditions in any particular area.

Rates can be interpolated horizontally and vertically. Various portions of the same project requiring different rates should be adjusted proportionately. For alterations, add 50% to the fee for the first $500,000 of project cost and add 25% to the fee for project cost over $500,000.

Architectural fees tabulated below include Structural, Mechanical and Electrical Engineering Fees. They do not include the fees for special consultants such as kitchen planning, security, acoustical, interior design, etc.

Civil Engineering fees are included in the Architectural fee for project sites requiring minimal design such as city sites. However, separate Civil Engineering fees must be added when utility connections require design, drainage calculations are needed, stepped foundations are required, or provisions are required to protect adjacent wetlands.

Building Types	Total Project Size in Thousands of Dollars						
	100	250	500	1,000	5,000	10,000	50,000
Factories, garages, warehouses, repetitive housing	9.0%	8.0%	7.0%	6.2%	5.3%	4.9%	4.5%
Apartments, banks, schools, libraries, offices, municipal buildings	12.2	12.3	9.2	8.0	7.0	6.6	6.2
Churches, hospitals, homes, laboratories, museums, research	15.0	13.6	12.7	11.9	9.5	8.8	8.0
Memorials, monumental work, decorative furnishings	—	16.0	14.5	13.1	10.0	9.0	8.3

General Requirements — R0129 Payment Procedures

R012909-80 Sales Tax by State

State sales tax on materials is tabulated below (5 states have no sales tax). Many states allow local jurisdictions, such as a county or city, to levy additional sales tax.

Some projects may be sales tax exempt, particularly those constructed with public funds.

State	Tax (%)	State	Tax (%)	State	Tax (%)	State	Tax (%)
Alabama	4	Illinois	6.25	Montana	0	Rhode Island	7
Alaska	0	Indiana	6	Nebraska	5.5	South Carolina	5
Arizona	5.6	Iowa	5	Nevada	6.5	South Dakota	4
Arkansas	6	Kansas	5.3	New Hampshire	0	Tennessee	7
California	7.25	Kentucky	6	New Jersey	6	Texas	6.25
Colorado	2.9	Louisiana	4	New Mexico	5	Utah	4.75
Connecticut	6	Maine	5	New York	4	Vermont	6
Delaware	0	Maryland	5	North Carolina	4.5	Virginia	5
District of Columbia	5.75	Massachusetts	5	North Dakota	5	Washington	6.5
Florida	6	Michigan	6	Ohio	5.5	West Virginia	6
Georgia	4	Minnesota	6.5	Oklahoma	4.5	Wisconsin	5
Hawaii	4	Mississippi	7	Oregon	0	Wyoming	4
Idaho	5	Missouri	4.225	Pennsylvania	6	Average	4.84 %

Sales Tax by Province (Canada)

GST - a value-added tax, which the government imposes on most goods and services provided in or imported into Canada. PST - a retail sales tax, which five of the provinces impose on the price of most goods and some services. QST - a value-added tax, similar to the federal GST, which Quebec imposes. HST - Three provinces have combined their retail sales tax with the federal GST into one harmonized tax.

Province	PST (%)	QST (%)	GST (%)	HST (%)
Alberta	0	0	6	0
British Columbia	7	0	6	0
Manitoba	7	0	6	0
New Brunswick	0	0	0	14
Newfoundland	0	0	0	14
Northwest Territories	0	0	6	0
Nova Scotia	0	0	0	14
Ontario	8	0	6	0
Prince Edward Island	10	0	6	0
Quebec	0	7.5	6	0
Saskatchewan	7	0	6	0
Yukon	0	0	6	0

R012909-85 Unemployment Taxes and Social Security Taxes

State Unemployment Tax rates vary not only from state to state, but also with the experience rating of the contractor. The Federal Unemployment Tax rate is 6.2% of the first $7,000 of wages. This is reduced by a credit of up to 5.4% for timely payment to the state. The minimum Federal Unemployment Tax is 0.8% after all credits.

Social Security (FICA) for 2007 is estimated at time of publication to be 7.65% of wages up to $94,200.

General Requirements — R0131 Project Management & Coordination

R013113-40 Builder's Risk Insurance

Builder's Risk Insurance is insurance on a building during construction. Premiums are paid by the owner or the contractor. Blasting, collapse and underground insurance would raise total insurance costs above those listed. Floater policy for materials delivered to the job runs $.75 to $1.25 per $100 value. Contractor equipment insurance runs $.50 to $1.50 per $100 value. Insurance for miscellaneous tools to $1,500 value runs from $3.00 to $7.50 per $100 value.

Tabulated below are New England Builder's Risk insurance rates in dollars per $100 value for $1,000 deductible. For $25,000 deductible, rates can be reduced 13% to 34%. On contracts over $1,000,000, rates may be lower than those tabulated. Policies are written annually for the total completed value in place. For "all risk" insurance (excluding flood, earthquake and certain other perils) add $.025 to total rates below.

Coverage	Frame Construction (Class 1) Range	Frame Construction (Class 1) Average	Brick Construction (Class 4) Range	Brick Construction (Class 4) Average	Fire Resistive (Class 6) Range	Fire Resistive (Class 6) Average
Fire Insurance	$.350 to $.850	$.600	$.158 to $.189	$.174	$.052 to $.080	$.070
Extended Coverage	.115 to .200	.158	.080 to .105	.101	.081 to .105	.100
Vandalism	.012 to .016	.014	.008 to .011	.011	.008 to .011	.010
Total Annual Rate	$.477 to $1.066	$.772	$.246 to $.305	$.286	$.141 to $.196	$.180

R013113-50 General Contractor's Overhead

There are two distinct types of overhead on a construction project: Project Overhead and Main Office Overhead. Project Overhead includes those costs at a construction site not directly associated with the installation of construction materials. Examples of Project Overhead costs include the following:

1. Superintendent
2. Construction office and storage trailers
3. Temporary sanitary facilities
4. Temporary utilities
5. Security fencing
6. Photographs
7. Clean up
8. Performance and payment bonds

The above Project Overhead items are also referred to as General Requirements and therefore are estimated in Division 1. Division 1 is the first division listed in the CSI MasterFormat but it is usually the last division estimated. The sum of the costs in Divisions 1 through 49 is referred to as the sum of the direct costs.

All construction projects also include indirect costs. The primary components of indirect costs are the contractor's Main Office Overhead and profit. The amount of the Main Office Overhead expense varies depending on the the following:

1. Owner's compensation
2. Project managers and estimator's wages
3. Clerical support wages
4. Office rent and utilities
5. Corporate legal and accounting costs
6. Advertising
7. Automobile expenses
8. Association dues
9. Travel and entertainment expenses

These costs are usually calculated as a percentage of annual sales volume. This percentage can range from 35% for a small contractor doing less than $500,000 to 5% for a large contractor with sales in excess of $100 million.

General Requirements — R0131 Project Management & Coordination

R013113-60 Workers' Compensation Insurance Rates by Trade

The table below tabulates the national averages for Workers' Compensation insurance rates by trade and type of building. The average "Insurance Rate" is multiplied by the "% of Building Cost" for each trade. This produces the "Workers' Compensation Cost" by % of total labor cost, to be added for each trade by building type to determine the weighted average Workers' Compensation rate for the building types analyzed.

Trade	Insurance Rate (% Labor Cost) Range	Insurance Rate (% Labor Cost) Average	% of Building Cost Office Bldgs.	% of Building Cost Schools & Apts.	% of Building Cost Mfg.	Workers' Compensation Office Bldgs.	Workers' Compensation Schools & Apts.	Workers' Compensation Mfg.
Excavation, Grading, etc.	4.5% to 10.5%	10.4%	4.8%	4.9%	4.5%	.50	.51	.47
Piles & Foundations	8.9 to 42.9	21.4	7.1	5.2	8.7	1.52	1.11	1.86
Concrete	5.1 to 29.8	15.5	5.0	14.8	3.7	.78	2.29	.57
Masonry	5.4 to 34.5	14.9	6.9	7.5	1.9	1.03	1.12	.28
Structural Steel	8.6 to 104.1	40.8	10.7	3.9	17.6	4.37	1.59	7.18
Miscellaneous & Ornamental Metals	5.2 to 22.6	11.9	2.8	4.0	3.6	.33	.48	.43
Carpentry & Millwork	7.6 to 53.2	18.4	3.7	4.0	0.5	.68	.74	.09
Metal or Composition Siding	6.8 to 34.1	16.7	2.3	0.3	4.3	.38	.05	.72
Roofing	8.4 to 77.1	32.3	2.3	2.6	3.1	.74	.84	1.00
Doors & Hardware	5 to 24.9	11.1	0.9	1.4	0.4	.10	.16	.04
Sash & Glazing	5.6 to 28.9	14.0	3.5	4.0	1.0	.49	.56	.14
Lath & Plaster	3.1 to 35.4	14.0	3.3	6.9	0.8	.46	.97	.11
Tile, Marble & Floors	2 to 29.4	9.5	2.6	3.0	0.5	.25	.29	.05
Acoustical Ceilings	2.3 to 50.2	11.6	2.4	0.2	0.3	.28	.02	.03
Painting	4.9 to 29.6	13.2	1.5	1.6	1.6	.20	.21	.21
Interior Partitions	7.6 to 53.2	18.4	3.9	4.3	4.4	.72	.79	.81
Miscellaneous Items	2.3 to 197.9	16.8	5.2	3.7	9.7	.88	.62	1.63
Elevators	2.9 to 13.7	6.9	2.1	1.1	2.2	.14	.08	.15
Sprinklers	2.8 to 15.3	8.3	0.5	—	2.0	.04	—	.17
Plumbing	2.9 to 14.2	8.1	4.9	7.2	5.2	.40	.58	.42
Heat., Vent., Air Conditioning	4.6 to 24.8	11.9	13.5	11.0	12.9	1.61	1.31	1.54
Electrical	2.7 to 13.6	6.6	10.1	8.4	11.1	.67	.55	.73
Total	2% to 197.9%	—	100.0%	100.0%	100.0%	16.57%	14.87%	18.63%

Overall Weighted Average 16.69%

Workers' Compensation Insurance Rates by States

The table below lists the weighted average Workers' Compensation base rate for each state with a factor comparing this with the national average of 16.3%.

State	Weighted Average	Factor	State	Weighted Average	Factor	State	Weighted Average	Factor
Alabama	26.2%	161	Kentucky	19.6%	120	North Dakota	13.7%	84
Alaska	25.0	153	Louisiana	28.4	174	Ohio	14.1	87
Arizona	7.5	46	Maine	21.1	129	Oklahoma	14.6	90
Arkansas	14.0	86	Maryland	19.2	118	Oregon	13.3	82
California	18.9	116	Massachusetts	15.3	94	Pennsylvania	13.0	80
Colorado	13.3	82	Michigan	17.4	107	Rhode Island	21.3	131
Connecticut	21.3	131	Minnesota	25.7	158	South Carolina	17.0	104
Delaware	19.0	117	Mississippi	18.5	113	South Dakota	18.6	114
District of Columbia	15.9	98	Missouri	18.1	111	Tennessee	15.6	96
Florida	20.4	125	Montana	17.9	110	Texas	13.2	81
Georgia	22.8	140	Nebraska	24.4	150	Utah	12.4	76
Hawaii	17.0	104	Nevada	12.5	77	Vermont	24.3	149
Idaho	12.1	74	New Hampshire	21.5	132	Virginia	12.6	77
Illinois	18.8	115	New Jersey	12.9	79	Washington	11.0	67
Indiana	6.6	40	New Mexico	19.4	119	West Virginia	9.5	58
Iowa	12.0	74	New York	13.8	85	Wisconsin	14.8	91
Kansas	8.7	53	North Carolina	17.3	106	Wyoming	9.9	61

Weighted Average for U.S. is 16.7% of payroll = 100%

Rates in the following table are the base or manual costs per $100 of payroll for Workers' Compensation in each state. Rates are usually applied to straight time wages only and not to premium time wages and bonuses.

The weighted average skilled worker rate for 35 trades is 16.3%. For bidding purposes, apply the full value of Workers' Compensation directly to total labor costs, or if labor is 38%, materials 42% and overhead and profit 20% of total cost, carry 38/80 x 16.3% =7.7% of cost (before overhead and profit) into overhead. Rates vary not only from state to state but also with the experience rating of the contractor.

Rates are the most current available at the time of publication.

General Requirements — R0131 Project Management & Coordination

R013113-60 Workers' Compensation Insurance Rates by Trade and State (cont.)

State	Carpentry — 3 stories or less	Carpentry — interior cab. work	Carpentry — general	Concrete Work — NOC	Concrete Work — flat (flr., sdwk.)	Electrical Wiring — inside	Excavation — earth NOC	Excavation — rock	Glaziers	Insulation Work	Lathing	Masonry	Painting & Decorating	Pile Driving	Plastering	Plumbing	Roofing	Sheet Metal Work (HVAC)	Steel Erection — door & sash	Steel Erection — inter., ornam.	Steel Erection — structure	Steel Erection — NOC	Tile Work — (interior ceramic)	Waterproofing	Wrecking
	5651	5437	5403	5213	5221	5190	6217	6217	5462	5479	5443	5022	5474	6003	5480	5183	5551	5538	5102	5102	5040	5057	5348	9014	5701
AL	34.13	19.34	33.95	12.96	10.74	9.34	14.00	14.00	23.27	16.95	13.45	26.80	27.85	42.90	27.83	10.23	57.63	24.81	16.77	16.77	64.80	28.72	12.05	7.74	64.80
AK	20.75	14.96	16.46	15.76	14.11	13.57	20.48	20.48	28.87	34.86	12.30	34.35	24.83	41.80	33.32	11.83	46.92	12.89	13.20	13.20	53.01	25.10	9.07	8.50	53.01
AZ	6.88	5.05	11.62	5.56	3.49	3.85	4.74	4.74	7.40	7.69	4.27	5.37	4.87	9.83	5.20	3.90	12.19	5.65	9.50	9.50	18.98	8.11	2.04	2.34	50.31
AR	15.09	8.06	16.49	13.30	6.63	5.14	8.49	8.49	11.79	23.72	6.63	9.32	9.64	16.15	13.39	5.21	28.01	13.09	7.75	7.75	36.16	21.58	6.29	4.74	36.16
CA	28.28	9.07	28.28	13.18	13.18	9.93	7.88	7.88	16.37	23.34	10.90	14.55	19.45	21.18	18.12	11.59	39.79	15.33	13.55	13.55	23.93	21.91	7.74	19.45	21.91
CO	16.28	8.03	12.76	13.19	7.46	5.18	10.35	10.35	9.79	13.91	6.03	14.30	10.29	15.08	9.00	8.45	23.95	11.46	8.38	8.38	35.52	15.85	8.60	5.74	15.85
CT	18.32	16.06	25.93	24.48	12.29	8.04	11.27	11.27	19.79	16.24	22.62	25.73	16.24	24.89	18.63	10.52	37.59	15.83	16.42	16.42	50.14	26.51	9.01	6.09	56.71
DE	19.38	19.38	15.10	15.16	12.03	8.22	12.07	12.07	14.07	19.38	16.67	15.69	20.18	25.15	16.67	10.12	34.39	12.62	16.11	16.11	35.78	16.11	12.50	15.69	35.78
DC	10.42	10.52	11.61	12.06	13.54	6.02	10.60	10.60	20.19	9.97	11.19	17.94	9.87	13.76	12.24	14.19	18.73	8.23	13.03	13.03	45.04	17.90	29.37	4.20	45.04
FL	24.38	15.34	24.74	21.86	10.37	8.91	11.12	11.12	15.53	14.31	10.58	17.48	16.78	38.24	28.82	10.04	35.40	14.90	11.74	11.74	44.71	25.87	10.03	7.76	44.71
GA	32.85	16.50	24.47	14.19	10.63	9.63	16.68	16.68	15.64	22.01	14.48	19.28	18.57	26.27	18.16	10.69	43.72	20.35	15.97	15.97	51.69	40.46	9.92	8.47	51.69
HI	17.38	11.62	28.20	13.74	12.27	7.10	7.73	7.73	20.55	21.19	10.94	17.87	11.33	20.15	15.61	6.06	33.24	8.02	11.11	11.11	31.71	21.70	9.78	11.54	31.71
ID	11.92	6.31	13.47	10.64	5.99	5.25	6.90	6.90	9.58	8.14	6.31	9.91	8.88	15.41	9.45	5.99	29.54	9.11	7.94	7.94	33.81	14.13	13.67	4.09	33.81
IL	17.33	15.90	18.72	26.98	10.79	8.21	9.42	9.42	19.93	15.96	10.26	18.26	10.08	25.32	13.09	9.80	29.37	14.49	15.54	15.54	50.39	21.95	14.63	4.66	50.39
IN	6.77	5.04	7.58	5.08	3.27	2.66	4.64	4.64	5.63	5.55	2.25	5.64	5.54	9.05	4.04	2.94	11.82	4.60	5.20	5.20	21.57	8.02	3.19	2.68	21.57
IA	11.41	6.24	9.43	11.35	6.29	4.24	5.17	5.17	11.28	8.38	5.51	8.91	7.85	8.93	7.26	5.08	16.80	7.56	7.75	7.75	46.05	40.59	5.48	3.96	27.19
KS	8.44	8.42	10.36	7.95	5.18	3.44	4.51	4.51	8.63	8.03	4.28	7.83	6.15	11.55	6.35	5.42	8.44	7.02	7.09	7.09	25.33	12.78	5.36	3.20	27.57
KY	22.12	12.10	19.05	18.45	7.85	6.61	11.25	11.25	20.44	20.69	13.02	10.03	13.56	25.37	15.33	7.48	50.97	19.58	14.49	14.49	46.98	23.88	11.89	6.30	46.98
LA	23.14	24.93	53.17	26.43	15.61	9.88	17.49	17.49	20.14	21.43	24.51	28.33	29.58	31.25	22.37	8.64	77.12	21.20	18.98	18.98	51.76	24.21	13.77	13.78	66.41
ME	13.43	10.42	43.59	24.29	11.12	3.72	12.56	12.56	13.35	15.39	14.03	17.26	16.07	31.44	17.95	9.07	32.07	10.07	15.08	15.08	37.84	61.81	11.66	5.92	37.84
MD	13.71	5.95	10.55	18.82	8.70	5.15	12.91	12.91	25.52	20.07	11.37	13.31	9.37	26.22	14.21	8.58	48.72	12.74	13.66	13.66	56.25	37.81	8.03	7.14	26.80
MA	9.03	6.88	16.48	22.32	9.18	4.18	6.08	6.08	9.25	14.59	5.89	15.67	6.85	15.92	6.62	4.80	47.57	6.72	13.48	13.48	43.69	34.06	8.78	2.65	24.03
MI	18.40	11.39	20.64	19.45	8.95	4.72	10.20	10.20	12.41	14.75	13.59	15.54	13.64	40.27	15.18	6.84	30.67	10.00	10.16	10.16	40.27	22.21	11.11	4.87	39.06
MN	19.53	19.80	41.48	14.65	15.73	7.53	14.95	14.95	15.68	11.90	22.05	19.18	16.95	24.63	22.05	10.88	70.18	14.23	10.58	10.58	104.13	33.30	14.55	6.68	36.03
MS	15.56	12.84	23.21	11.49	8.98	6.57	12.44	12.44	15.13	18.03	7.40	9.64	14.52	24.87	35.36	8.27	35.09	19.67	13.57	13.57	35.55	29.83	10.22	5.63	35.55
MO	23.87	12.38	16.85	16.75	11.59	7.61	9.87	9.87	11.40	18.47	10.12	14.58	15.09	18.46	14.96	10.16	33.26	15.34	11.85	11.85	47.95	33.09	9.40	6.12	47.95
MT	16.07	12.88	21.13	12.84	11.28	7.97	14.85	14.85	11.94	30.09	15.86	13.71	9.70	33.39	11.78	10.03	39.71	11.20	9.70	9.70	36.35	18.79	7.76	6.04	36.35
NE	27.13	14.63	21.57	29.20	14.30	10.22	17.88	17.88	25.50	35.50	12.00	25.82	22.27	24.63	20.60	11.52	39.10	18.35	14.60	14.60	65.40	26.27	12.02	7.63	58.22
NV	16.61	7.10	11.57	9.28	6.51	5.39	9.50	9.50	12.80	10.21	8.27	8.90	9.60	14.19	8.05	6.78	14.14	19.04	10.96	10.96	23.34	26.59	6.18	5.18	33.18
NH	26.55	11.21	21.62	32.68	12.33	7.21	17.79	17.79	11.16	29.06	9.32	22.91	15.04	21.80	13.65	10.84	58.14	14.88	12.51	12.51	49.35	18.42	11.43	6.01	49.35
NJ	13.50	9.36	13.50	12.14	8.95	4.45	8.62	8.62	7.20	13.18	11.54	13.14	11.44	13.54	11.54	6.26	34.09	6.04	11.03	11.03	25.41	13.62	4.87	4.83	23.62
NM	24.50	8.22	21.85	20.69	11.35	7.41	11.51	11.51	19.30	14.97	10.20	16.38	16.09	24.75	12.34	9.96	33.44	12.58	18.13	18.13	52.69	29.16	7.79	7.63	52.69
NY	14.21	12.39	12.39	17.19	11.26	6.69	8.37	8.37	10.18	9.91	11.27	15.70	10.41	15.68	9.04	7.33	33.93	12.79	9.59	9.59	21.71	15.20	8.72	6.92	24.30
NC	15.87	13.33	16.07	15.61	7.64	10.21	10.51	10.51	11.41	13.79	14.12	11.11	12.95	17.63	13.75	9.20	26.70	13.53	9.22	9.22	79.26	20.69	6.50	5.68	79.26
ND	10.92	10.92	10.92	6.26	6.26	3.90	6.46	6.46	10.92	10.92	8.84	7.94	6.41	22.61	8.84	5.47	22.68	5.47	22.61	22.61	22.61	22.61	10.92	22.68	12.19
OH	7.70	7.61	9.64	11.57	10.29	5.45	8.10	8.10	7.09	17.31	50.15	11.49	11.69	28.24	3.08	6.58	24.34	9.30	8.14	8.14	26.41	10.66	9.21	5.70	26.41
OK	13.42	9.17	11.76	11.76	6.34	5.73	10.83	10.83	16.07	19.47	8.45	10.02	8.47	20.25	12.19	7.09	21.13	9.05	13.76	13.76	39.72	23.97	6.50	5.71	39.78
OR	16.00	8.81	15.78	13.64	8.79	4.26	9.88	9.88	14.41	10.15	7.41	14.86	10.88	15.81	11.82	5.87	23.65	9.60	8.95	8.95	28.75	14.60	11.37	4.49	28.75
PA	12.67	6.29	11.41	13.57	9.77	5.98	7.88	7.88	9.15	11.41	10.98	11.40	13.00	15.49	10.98	6.94	26.17	7.49	13.91	13.91	21.88	13.91	7.85	11.40	21.88
RI	19.53	11.65	18.07	18.23	16.24	4.43	10.38	10.38	12.85	22.78	11.97	25.11	24.13	37.66	17.25	8.52	33.92	10.29	14.07	14.07	59.49	37.50	14.36	7.70	78.79
SC	22.26	17.21	19.93	13.51	7.81	9.42	11.95	11.95	14.32	11.08	8.53	13.27	16.47	18.32	18.15	8.83	43.19	10.97	11.47	11.47	24.55	27.29	8.90	5.98	47.05
SD	22.94	7.74	24.42	28.24	8.33	5.60	12.27	12.27	11.43	12.46	8.11	9.46	10.34	21.36	12.74	12.97	19.88	12.48	8.97	8.97	84.39	33.82	8.80	4.25	84.39
TN	17.21	13.36	22.83	16.35	9.27	7.36	12.04	12.04	13.17	11.21	11.71	14.74	10.49	18.13	12.60	9.19	24.86	13.88	9.95	9.95	32.82	20.02	8.84	5.72	32.82
TX	13.44	9.34	13.44	11.97	7.27	7.70	8.65	8.65	9.83	12.10	7.83	12.38	9.08	20.20	9.08	6.92	19.68	16.57	12.11	12.11	30.40	12.18	6.82	6.89	13.00
UT	9.45	6.23	10.52	7.68	7.99	7.05	6.38	6.38	9.55	10.95	12.32	13.75	15.29	17.24	10.60	6.41	26.27	6.30	8.99	8.99	23.51	23.51	6.06	5.83	26.53
VT	23.50	10.31	19.77	29.84	12.44	8.34	14.86	14.86	24.74	25.93	10.12	20.66	14.15	21.89	15.99	10.47	45.95	14.75	14.56	14.56	74.50	62.00	10.22	10.24	74.50
VA	11.41	7.88	10.02	11.54	5.24	5.47	7.04	7.04	8.77	10.82	18.78	8.94	9.54	12.50	9.73	5.46	21.49	8.23	9.60	9.60	43.89	22.16	5.03	2.89	43.89
WA	9.30	9.30	9.30	8.12	8.12	3.35	8.04	8.04	12.92	10.56	9.30	10.21	9.88	17.82	11.50	4.64	18.57	4.61	8.65	8.65	8.65	8.65	10.40	18.57	7.87
WV	9.77	7.20	10.47	8.58	4.65	4.82	5.92	5.92	7.80	7.09	7.11	8.14	8.24	11.16	8.73	4.78	19.23	7.63	6.45	6.45	26.02	12.50	4.53	2.83	23.67
WI	12.40	11.36	16.86	12.29	9.90	4.24	7.44	7.44	13.31	11.62	8.09	18.99	14.32	13.64	12.11	5.42	34.72	7.16	9.18	9.18	37.88	24.06	13.99	5.52	37.88
WY	8.93	8.93	8.93	8.93	8.93	8.93	8.93	8.93	8.93	8.93	8.93	8.93	8.93	8.93	8.93	8.93	8.93	8.93	8.93	8.93	8.93	8.93	8.93	8.93	8.93
AVG.	16.75	11.08	18.39	15.53	9.55	6.59	10.39	10.39	14.05	15.81	11.61	14.92	13.19	21.39	14.04	8.10	32.30	11.89	11.86	11.86	40.80	23.82	9.53	7.16	39.10

General Requirements — R0131 Project Management & Coordination

R013113-60 Workers' Compensation (cont.) (Canada in Canadian dollars)

Province		Alberta	British Columbia	Manitoba	Ontario	New Brunswick	Newfndld. & Labrador	Northwest Territories	Nova Scotia	Prince Edward Island	Quebec	Saskatchewan	Yukon
Carpentry—3 stories or less	Rate	8.17	5.91	4.48	4.58	3.94	8.84	4.16	7.67	5.73	15.41	7.33	4.79
	Code	42143	721028	40102	723	4226	4226	4-41	4226	401	80110	B1317	202
Carpentry—interior cab. work	Rate	2.56	6.13	4.48	4.58	4.67	4.92	4.16	5.74	3.98	15.41	3.68	4.79
	Code	42133	721021	40102	723	4279	4270	4-41	4274	402	80110	B11-27	202
CARPENTRY—general	Rate	8.17	5.91	4.48	4.58	3.94	4.92	4.16	7.67	5.73	15.41	7.33	4.79
	Code	42143	721028	40102	723	4226	4299	4-41	4226	401	80110	B1317	202
CONCRETE WORK—NOC	Rate	5.12	6.02	7.42	15.40	3.94	8.84	4.16	4.83	5.73	16.51	7.33	3.86
	Code	42104	721010	40110	748	4224	4224	4-41	4224	401	80100	B13-14	203
CONCRETE WORK—flat (flr. sidewalk)	Rate	5.12	6.02	7.42	15.40	3.94	8.84	4.16	4.83	5.73	16.51	7.33	3.86
	Code	42104	721010	40110	748	4224	4224	4-41	4224	401	80100	B13-14	203
ELECTRICAL Wiring—inside	Rate	2.22	2.18	2.56	3.25	2.45	2.92	3.91	2.23	3.98	7.64	3.68	3.86
	Code	42124	721019	40203	704	4261	4261	4-46	4261	402	80170	B11-05	206
EXCAVATION—earth NOC	Rate	2.95	4.20	3.78	4.55	2.88	4.29	4.04	4.11	4.23	8.36	4.37	3.86
	Code	40604	721031	40706	711	4214	4214	4-43	4214	404	80030	R11-06	207
EXCAVATION—rock	Rate	2.95	4.20	3.78	4.55	2.88	4.29	4.04	4.11	4.23	8.36	4.37	3.86
	Code	40604	721031	40706	711	4214	4214	4-43	4214	404	80030	R11-06	207
GLAZIERS	Rate	3.66	4.07	4.48	8.90	5.20	6.52	4.16	7.67	3.98	14.36	7.33	3.86
	Code	42121	715020	40109	751	4233	4233	4-41	4233	402	80150	B13-04	212
INSULATION WORK	Rate	3.03	7.02	4.48	8.90	5.20	6.52	4.16	7.67	5.73	15.41	6.15	4.79
	Code	42184	721029	40102	751	4234	4234	4-41	4234	401	80110	B12-07	202
LATHING	Rate	6.18	8.48	4.48	4.58	4.67	4.92	4.16	5.74	3.98	15.41	7.33	4.79
	Code	42135	721033	40102	723	4273	4279	4-41	4271	402	80110	B13-16	202
MASONRY	Rate	5.12	8.48	4.48	11.79	5.20	6.52	4.16	7.67	5.73	16.51	7.33	4.79
	Code	42102	721037	40102	741	4231	4231	4-41	4231	401	80100	B13-18	202
PAINTING & DECORATING	Rate	5.08	5.21	36.17	6.75	4.67	4.92	4.16	5.74	3.98	15.41	6.15	4.79
	Code	42111	721041	40105	719	4275	4275	4-41	4275	402	80110	B12-01	202
PILE DRIVING	Rate	5.12	5.52	3.78	6.26	3.94	9.78	4.04	4.83	5.73	8.36	7.33	4.79
	Code	42159	722004	40706	732	4221	4221	4-43	4221	401	80030	B13-10	202
PLASTERING	Rate	6.18	8.48	5.38	6.75	4.67	4.92	4.16	5.74	3.98	15.41	6.15	4.79
	Code	42135	721042	40108	719	4271	4271	4-41	4271	402	80110	B12-21	202
PLUMBING	Rate	2.22	3.89	2.92	4.02	2.89	3.24	3.91	2.23	3.98	7.61	3.68	3.27
	Code	42122	721043	40204	707	4241	4241	4-46	4241	402	80160	B11-01	214
ROOFING	Rate	9.20	10.86	7.12	12.53	8.10	8.84	4.16	9.49	5.73	22.49	7.33	4.79
	Code	42118	721036	40403	728	4236	4236	4-41	4236	401	80130	B13-20	202
SHEET METAL WORK (HVAC)	Rate	2.22	3.89	7.12	4.02	2.89	3.24	3.91	3.23	3.98	7.61	3.68	4.35
	Code	42117	721043	40402	707	4244	4244	4-46	4244	402	80160	B11-07	208
STEEL ERECTION—door & sash	Rate	3.03	16.01	12.96	15.40	3.94	8.84	4.16	7.67	5.73	29.92	7.33	4.79
	Code	42106	722005	40502	748	4227	4227	4-41	4227	401	80080	B13-22	202
STEEL ERECTION—inter., ornam.	Rate	3.03	16.01	12.96	15.40	3.94	8.84	4.16	7.67	5.73	29.92	7.33	4.79
	Code	42106	722005	40502	748	4227	4227	4-41	4227	401	80080	B13-22	202
STEEL ERECTION—structure	Rate	3.03	16.01	12.96	15.47	3.94	8.84	4.16	7.67	5.73	29.92	7.33	4.79
	Code	42106	722005	40502	748	4227	4227	4-41	4227	401	80080	B13-22	202
STEEL ERECTION—NOC	Rate	3.03	16.01	12.96	15.40	3.94	8.84	4.16	7.67	5.73	29.92	7.33	4.79
	Code	42106	722005	40502	748	4227	4227	4-41	4227	401	80080	B13-22	202
TILE WORK—inter. (ceramic)	Rate	4.22	6.58	2.05	6.75	4.67	4.92	4.16	5.74	3.98	15.41	7.33	4.79
	Code	42113	721054	40103	719	4276	4276	4-41	4276	402	80110	B13-01	202
WATERPROOFING	Rate	5.08	5.19	4.48	4.58	5.20	4.92	4.16	7.67	3.98	22.49	6.15	4.79
	Code	42139	721016	40102	723	4239	4299	4-41	4239	402	80130	B12-17	202
WRECKING	Rate	2.95	5.74	7.07	15.40	2.88	4.29	4.04	4.11	5.73	15.41	7.33	4.79
	Code	40604	721005	40106	748	4211	4211	4-43	4211	401	80110	B13-09	202

General Requirements — R0154 Construction Aids

R015423-10 Steel Tubular Scaffolding

On new construction, tubular scaffolding is efficient up to 60' high or five stories. Above this it is usually better to use a hung scaffolding if construction permits. Swing scaffolding operations may interfere with tenants. In this case, the tubular is more practical at all heights.

In repairing or cleaning the front of an existing building the cost of tubular scaffolding per S.F. of building front increases as the height increases above the first tier. The first tier cost is relatively high due to leveling and alignment.

The minimum efficient crew for erecting and dismantling is three workers. They can set up and remove 18 frame sections per day up to 5 stories high. For 6 to 12 stories high, a crew of four is most efficient. Use two or more on top and two on the bottom for handing up or hoisting. They can also set up and remove 18 frame sections per day. At 7' horizontal spacing, this will run about 800 S.F. per day of erecting and dismantling. Time for placing and removing planks must be added to the above. A crew of three can place and remove 72 planks per day up to 5 stories. For over 5 stories, a crew of four can place and remove 80 planks per day.

The table below shows the number of pieces required to erect tubular steel scaffolding for 1000 S.F. of building frontage. This area is made up of a scaffolding system that is 12 frames (11 bays) long by 2 frames high.

For jobs under twenty-five frames, add 50% to rental cost. Rental rates will be lower for jobs over three months duration. Large quantities for long periods can reduce rental rates by 20%.

Description of Component	Number of Pieces for 1000 S.F. of Building Front	Unit
5' Wide Standard Frame, 6'-4" High	24	Ea.
Leveling Jack & Plate	24	
Cross Brace	44	
Side Arm Bracket, 21"	12	
Guardrail Post	12	
Guardrail, 7' section	22	
Stairway Section	2	
Stairway Starter Bar	1	
Stairway Inside Handrail	2	
Stairway Outside Handrail	2	
Walk-Thru Frame Guardrail	2	

Scaffolding is often used as falsework over 15' high during construction of cast-in-place concrete beams and slabs. Two foot wide scaffolding is generally used for heavy beam construction. The span between frames depends upon the load to be carried with a maximum span of 5'.

Heavy duty shoring frames with a capacity of 10,000#/leg can be spaced up to 10' O.C. depending upon form support design and loading.

Scaffolding used as horizontal shoring requires less than half the material required with conventional shoring.

On new construction, erection is done by carpenters.

Rolling towers supporting horizontal shores can reduce labor and speed the job. For maintenance work, catwalks with spans up to 70' can be supported by the rolling towers.

R015423-20 Pump Staging

Pump staging is generally not available for rent. The table below shows the number of pieces required to erect pump staging for 2400 S.F. of building frontage. This area is made up of a pump jack system that is 3 poles (2 bays) wide by 2 poles high.

Item	Number of Pieces for 2400 S.F. of Building Front	Unit
Aluminum pole section, 24' long	6	Ea.
Aluminum splice joint, 6' long	3	
Aluminum foldable brace	3	
Aluminum pump jack	3	
Aluminum support for workbench/back safety rail	3	
Aluminum scaffold plank/workbench, 14" wide x 24' long	4	
Safety net, 22' long	2	
Aluminum plank end safety rail	2	

The cost in place for this 2400 S.F. will depend on how many uses are realized during the life of the equipment.

Existing Conditions — R0241 Demolition

R024119-10 Demolition Defined

Whole Building Demolition - Demolition of the whole building with no concern for any particular building element, component, or material type being demolished. This type of demolition is accomplished with large pieces of construction equipment that break up the structure, load it into trucks and haul it to a disposal site, but disposal or dump fees are not included. Demolition of below-grade foundation elements, such as footings, foundation walls, grade beams, slabs on grade, etc., is not included. Certain mechanical equipment containing flammable liquids or ozone-depleting refrigerants, electric lighting elements, communication equipment components, and other building elements may contain hazardous waste, and must be removed, either selectively or carefully, as hazardous waste before the building can be demolished.

Foundation Demolition - Demolition of below-grade foundation footings, foundation walls, grade beams, and slabs on grade. This type of demolition is accomplished by hand or pneumatic hand tools, and does not include saw cutting, or handling, loading, hauling, or disposal of the debris.

Gutting - Removal of building interior finishes and electrical/mechanical systems down to the load-bearing and sub-floor elements of the rough building frame, with no concern for any particular building element, component, or material type being demolished. This type of demolition is accomplished by hand or pneumatic hand tools, and includes loading into trucks, but not hauling, disposal or dump fees, scaffolding, or shoring. Certain mechanical equipment containing flammable liquids or ozone-depleting refrigerants, electric lighting elements, communication equipment components, and other building elements may contain hazardous waste, and must be removed, either selectively or carefully, as hazardous waste, before the building is gutted.

Selective Demolition - Demolition of a selected building element, component, or finish, with some concern for surrounding or adjacent elements, components, or finishes (see the first Subdivision (s) at the beginning of appropriate Divisions). This type of demolition is accomplished by hand or pneumatic hand tools, and does not include handling, loading, storing, hauling, or disposal of the debris, scaffolding, or shoring. "Gutting" methods may be used in order to save time, but damage that is caused to surrounding or adjacent elements, components, or finishes may have to be repaired at a later time.

Careful Removal - Removal of a piece of service equipment, building element or component, or material type, with great concern for both the removed item and surrounding or adjacent elements, components or finishes. The purpose of careful removal may be to protect the removed item for later re-use, preserve a higher salvage value of the removed item, or replace an item while taking care to protect surrounding or adjacent elements, components, connections, or finishes from cosmetic and/or structural damage. An approximation of the time required to perform this type of removal is 1/3 to 1/2 the time it would take to install a new item of like kind. This type of removal is accomplished by hand or pneumatic hand tools, and does not include loading, hauling, or storing the removed item, scaffolding, shoring, or lifting equipment.

Cutout Demolition - Demolition of a small quantity of floor, wall, roof, or other assembly, with concern for the appearance and structural integrity of the surrounding materials. This type of demolition is accomplished by hand or pneumatic hand tools, and does not include saw cutting, handling, loading, hauling, or disposal of debris, scaffolding, or shoring.

Rubbish Handling - Work activities that involve handling, loading or hauling of debris. Generally, the cost of rubbish handling must be added to the cost of all types of demolition, with the exception of whole building demolition.

Minor Site Demolition - Demolition of site elements outside the footprint of a building. This type of demolition is accomplished by hand or pneumatic hand tools, or with larger pieces of construction equipment, and may include loading a removed item onto a truck (check the Crew for equipment used). It does not include saw cutting, hauling or disposal of debris, and, sometimes, handling or loading.

Masonry — R0401 Maintenance of Masonry

R040130-10 Cleaning Face Brick

On smooth brick a person can clean 70 S.F. an hour; on rough brick 50 S.F. per hour. Use one gallon muriatic acid to 20 gallons of water for 1000 S.F. Do not use acid solution until wall is at least seven days old, but a mild soap solution may be used after two days.

Time has been allowed for clean-up in brick prices.

Masonry — R0405 Common Work Results for Masonry

R040513-10 Cement Mortar (material only)

Type N - 1:1:6 mix by volume. Use everywhere above grade except as noted below. - 1:3 mix using conventional masonry cement which saves handling two separate bagged materials.

Type M - 1:1/4:3 mix by volume, or 1 part cement, 1/4 (10% by wt.) lime, 3 parts sand. Use for heavy loads and where earthquakes or hurricanes may occur. Also for reinforced brick, sewers, manholes and everywhere below grade.

Mix Proportions by Volume and Compressive Strength of Mortar

Where Used	Mortar Type	Allowable Proportions by Volume				Compressive Strength @ 28 days
		Portland Cement	Masonry Cement	Hydrated Lime	Masonry Sand	
Plain Masonry	M	1	1	—	6	
		1	—	1/4	3	2500 psi
	S	1/2	1	—	4	
		1	—	1/4 to 1/2	4	1800 psi
	N	—	1	—	3	
		1	—	1/2 to 1-1/4	6	750 psi
	O	—	1	—	3	
		1	—	1-1/4 to 2-1/2	9	350 psi
	K	1	—	2-1/2 to 4	12	75 psi
Reinforced Masonry	PM	1	1	—	6	2500 psi
	PL	1	—	1/4 to 1/2	4	2500 psi

Note: The total aggregate should be between 2.25 to 3 times the sum of the cement and lime used.

The labor cost to mix the mortar is included in the productivity and labor cost of unit price lines in unit cost sections for brickwork, blockwork and stonework.

The material cost of mixed mortar is included in the material cost of those same unit price lines and includes the cost of renting and operating a 10 C.F. mixer at the rate of 200 C.F. per day.

There are two types of mortar color used. One type is the inert additive type with about 100 lbs. per M brick as the typical quantity required. These colors are also available in smaller-batch-sized bags (1 lb. to 15 lb.) which can be placed directly into the mixer without measuring. The other type is premixed and replaces the masonry cement. Dark green color has the highest cost.

R040519-50 Masonry Reinforcing

Horizontal joint reinforcing helps prevent wall cracks where wall movement may occur and in many locations is required by code. Horizontal joint reinforcing is generally not considered to be structural reinforcing and an unreinforced wall may still contain joint reinforcing.

Reinforcing strips come in 10′ and 12′ lengths and in truss and ladder shapes, with and without drips. Field labor runs between 2.7 to 5.3 hours per 1000 L.F. for wall thicknesses up to 12″.

The wire meets ASTM A82 for cold drawn steel wire and the typical size is 9 ga. sides and ties with 3/16″ diameter also available. Typical finish is mill galvanized with zinc coating at .10 oz. per S.F. Class I (.40 oz. per S.F.) and Class III (.80 oz per S.F.) are also available, as is hot dipped galvanizing at 1.50 oz. per S.F.

Masonry — R0421 Clay Unit Masonry

R042110-10 Economy in Bricklaying

Have adequate supervision. Be sure bricklayers are always supplied with materials so there is no waiting. Place best bricklayers at corners and openings.

Use only screened sand for mortar. Otherwise, labor time will be wasted picking out pebbles. Use seamless metal tubs for mortar as they do not leak or catch the trowel. Locate stack and mortar for easy wheeling.

Have brick delivered for stacking. This makes for faster handling, reduces chipping and breakage, and requires less storage space. Many dealers will deliver select common in 2' x 3' x 4' pallets or face brick packaged. This affords quick handling with a crane or forklift and easy tonging in units of ten, which reduces waste.

Use wider bricks for one wythe wall construction. Keep scaffolding away from wall to allow mortar to fall clear and not stain wall.

On large jobs develop specialized crews for each type of masonry unit.

Consider designing for prefabricated panel construction on high rise projects.

Avoid excessive corners or openings. Each opening adds about 50% to labor cost for area of opening.

Bolting stone panels and using window frames as stops reduces labor costs and speeds up erection.

R042110-20 Common and Face Brick

Common building brick manufactured according to ASTM C62 and facing brick manufactured according to ASTM C216 are the two standard bricks available for general building use.

Building brick is made in three grades; SW, where high resistance to damage caused by cyclic freezing is required; MW, where moderate resistance to cyclic freezing is needed; and NW, where little resistance to cyclic freezing is needed. Facing brick is made in only the two grades SW and MW. Additionally, facing brick is available in three types; FBS, for general use; FBX, for general use where a higher degree of precision and lower permissible variation in size than FBS is needed; and FBA, for general use to produce characteristic architectural effects resulting from non-uniformity in size and texture of the units.

In figuring the material cost of brickwork, an allowance of 25% mortar waste and 3% brick breakage was included. If bricks are delivered palletized with 280 to 300 per pallet, or packaged, allow only 1-1/2% for breakage. Packaged or palletized delivery is practical when a job is big enough to have a crane or other equipment available to handle a package of brick. This is so on all industrial work but not always true on small commercial buildings.

The use of buff and gray face is increasing, and there is a continuing trend to the Norman, Roman, Jumbo and SCR brick.

Common red clay brick for backup is not used that often. Concrete block is the most usual backup material with occasional use of sand lime or cement brick. Building brick is commonly used in solid walls for strength and as a fire stop.

Brick panels built on the ground and then crane erected to the upper floors have proven to be economical. This allows the work to be done under cover and without scaffolding.

R042110-50 Brick, Block & Mortar Quantities

Type Brick	Nominal Size (incl. mortar) L H W	Modular Coursing	Number of Brick per S.F.	C.F. of Mortar per M Bricks, Waste Included 3/8" Joint	1/2" Joint	Bond Type	Description	Factor
Standard	8 x 2-2/3 x 4	3C=8"	6.75	10.3	12.9	Common	full header every fifth course	+20%
Economy	8 x 4 x 4	1C=4"	4.50	11.4	14.6		full header every sixth course	+16.7%
Engineer	8 x 3-1/5 x 4	5C=16"	5.63	10.6	13.6	English	full header every second course	+50%
Fire	9 x 2-1/2 x 4-1/2	2C=5"	6.40	550 # Fireclay	—	Flemish	alternate headers every course	+33.3%
Jumbo	12 x 4 x 6 or 8	1C=4"	3.00	23.8	30.8		every sixth course	+5.6%
Norman	12 x 2-2/3 x 4	3C=8"	4.50	14.0	17.9		Header = W x H exposed	+100%
Norwegian	12 x 3-1/5 x 4	5C=16"	3.75	14.6	18.6		Rowlock = H x W exposed	+100%
Roman	12 x 2 x 4	2C=4"	6.00	13.4	17.0		Rowlock stretcher = L x W exposed	+33.3%
SCR	12 x 2-2/3 x 6	3C=8"	4.50	21.8	28.0		Soldier = H x L exposed	—
Utility	12 x 4 x 4	1C=4"	3.00	15.4	19.6		Sailor = W x L exposed	-33.3%

Concrete Blocks Nominal Size	Approximate Weight per S.F. Standard	Lightweight	Blocks per 100 S.F.	Mortar per M block, waste included Partitions	Back up
2" x 8" x 16"	20 PSF	15 PSF	113	27 C.F.	36 C.F.
4"	30	20		41	51
6"	42	30		56	66
8"	55	38		72	82
10"	70	47		87	97
12"	85	55		102	112

Masonry — R0422 Concrete Unit Masonry

R042210-20 Concrete Block

The material cost of special block such as corner, jamb and head block can be figured at the same price as ordinary block of same size. Labor on specials is about the same as equal-sized regular block.

Bond beam and 16" high lintel blocks are more expensive than regular units of equal size. Lintel blocks are 8" long and either 8" or 16" high.

Use of motorized mortar spreader box will speed construction of continuous walls.

Hollow non-load-bearing units are made according to ASTM C129 and hollow load-bearing units according to ASTM C90.

Metals — R0531 Steel Decking

R053100-10 Decking Descriptions

General - All Deck Products

Steel deck is made by cold forming structural grade sheet steel into a repeating pattern of parallel ribs. The strength and stiffness of the panels are the result of the ribs and the material properties of the steel. Deck lengths can be varied to suit job conditions, but because of shipping considerations, are usually less than 40 feet. Standard deck width varies with the product used but full sheets are usually 12", 18", 24", 30", or 36". Deck is typically furnished in a standard width with the ends cut square. Any cutting for width, such as at openings or for angular fit, is done at the job site.

Deck is typically attached to the building frame with arc puddle welds, self-drilling screws, or powder or pneumatically driven pins. Sheet to sheet fastening is done with screws, button punching (crimping), or welds.

Composite Floor Deck

After installation and adequate fastening, floor deck serves several purposes. It (a) acts as a working platform, (b) stabilizes the frame, (c) serves as a concrete form for the slab, and (d) reinforces the slab to carry the design loads applied during the life of the building. Composite decks are distinguished by the presence of shear connector devices as part of the deck. These devices are designed to mechanically lock the concrete and deck together so that the concrete and the deck work together to carry subsequent floor loads. These shear connector devices can be rolled-in embossments, lugs, holes, or wires welded to the panels. The deck profile can also be used to interlock concrete and steel.

Composite deck finishes are either galvanized (zinc coated) or phosphatized/painted. Galvanized deck has a zinc coating on both the top and bottom surfaces. The phosphatized/painted deck has a bare (phosphatized) top surface that will come into contact with the concrete. This bare top surface can be expected to develop rust before the concrete is placed. The bottom side of the deck has a primer coat of paint.

Composite floor deck is normally installed so the panel ends do not overlap on the supporting beams. Shear lugs or panel profile shape often prevent a tight metal to metal fit if the panel ends overlap; the air gap caused by overlapping will prevent proper fusion with the structural steel supports when the panel end laps are shear stud welded.

Adequate end bearing of the deck must be obtained as shown on the drawings. If bearing is actually less in the field than shown on the drawings, further investigation is required.

Roof Deck

Roof deck is not designed to act compositely with other materials. Roof deck acts alone in transferring horizontal and vertical loads into the building frame. Roof deck rib openings are usually narrower than floor deck rib openings. This provides adequate support of rigid thermal insulation board.

Roof deck is typically installed to endlap approximately 2" over supports. However, it can be butted (or lapped more than 2") to solve field fit problems. Since designers frequently use the installed deck system as part of the horizontal bracing system (the deck as a diaphragm), any fastening substitution or change should be approved by the designer. Continuous perimeter support of the deck is necessary to limit edge deflection in the finished roof and may be required for diaphragm shear transfer.

Standard roof deck finishes are galvanized or primer painted. The standard factory applied paint for roof deck is a primer paint and is not intended to weather for extended periods of time. Field painting or touching up of abrasions and deterioration of the primer coat or other protective finishes is the responsibility of the contractor.

Cellular Deck

Cellular deck is made by attaching a bottom steel sheet to a roof deck or composite floor deck panel. Cellular deck can be used in the same manner as floor deck. Electrical, telephone, and data wires are easily run through the chase created between the deck panel and the bottom sheet.

When used as part of the electrical distribution system, the cellular deck must be installed so that the ribs line up and create a smooth cell transition at abutting ends. The joint that occurs at butting cell ends must be taped or otherwise sealed to prevent wet concrete from seeping into the cell. Cell interiors must be free of welding burrs, or other sharp intrusions, to prevent damage to wires.

When used as a roof deck, the bottom flat plate is usually left exposed to view. Care must be maintained during erection to keep good alignment and prevent damage.

Cellular deck is sometimes used with the flat plate on the top side to provide a flat working surface. Installation of the deck for this purpose requires special methods for attachment to the frame because the flat plate, now on the top, can prevent direct access to the deck material that is bearing on the structural steel. It may be advisable to treat the flat top surface to prevent slipping.

Cellular deck is always furnished galvanized or painted over galvanized.

Form Deck

Form deck can be any floor or roof deck product used as a concrete form. Connections to the frame are by the same methods used to anchor floor and roof deck. Welding washers are recommended when welding deck that is less than 20 gauge thickness.

Form deck is furnished galvanized, prime painted, or uncoated. Galvanized deck must be used for those roof deck systems where form deck is used to carry a lightweight insulating concrete fill.

Wood, Plastics & Comp. — R0611 Wood Framing

R061110-30 Lumber Product Material Prices

The price of forest products fluctuates widely from location to location and from season to season depending upon economic conditions. The bare material prices in the unit cost sections of the book show the National Average material prices in effect Jan. 1 of this book year. It must be noted that lumber prices in general may change significantly during the year.

Availability of certain items depends upon geographic location and must be checked prior to firm-price bidding.

Wood, Plastics & Comp. — R0616 Sheathing

R061636-20 Plywood

There are two types of plywood used in construction: interior, which is moisture-resistant but not waterproofed, and exterior, which is waterproofed.

The grade of the exterior surface of the plywood sheets is designated by the first letter: A, for smooth surface with patches allowed; B, for solid surface with patches and plugs allowed; C, which may be surface plugged or may have knot holes up to 1" wide; and D, which is used only for interior type plywood and may have knot holes up to 2-1/2" wide. "Structural Grade" is specifically designed for engineered applications such as box beams. All CC & DD grades have roof and floor spans marked on them.

Underlayment-grade plywood runs from 1/4" to 1-1/4" thick. Thicknesses 5/8" and over have optional tongue and groove joints which eliminate the need for blocking the edges. Underlayment 19/32" and over may be referred to as Sturd-i-Floor.

The price of plywood can fluctuate widely due to geographic and economic conditions.

Typical uses for various plywood grades are as follows:

AA-AD Interior — cupboards, shelving, paneling, furniture
BB Plyform — concrete form plywood
CDX — wall and roof sheathing
Structural — box beams, girders, stressed skin panels
AA-AC Exterior — fences, signs, siding, soffits, etc.
Underlayment — base for resilient floor coverings
Overlaid HDO — high density for concrete forms & highway signs
Overlaid MDO — medium density for painting, siding, soffits & signs
303 Siding — exterior siding, textured, striated, embossed, etc.

Thermal & Moist. Protec. — R0731 Shingles & Shakes

R073126-20 Roof Slate

16", 18" and 20" are standard lengths, and slate usually comes in random widths. For standard 3/16" thickness use 1-1/2" copper nails. Allow for 3% breakage.

Thermal & Moist. Protec. — R0752 Modified Bituminous Membrane Roofing

R075213-30 Modified Bitumen Roofing

The cost of modified bitumen roofing is highly dependent on the type of installation that is planned. Installation is based on the type of modifier used in the bitumen. The two most popular modifiers are atactic polypropylene (APP) and styrene butadiene styrene (SBS). The modifiers are added to heated bitumen during the manufacturing process to change its characteristics. A polyethylene, polyester or fiberglass reinforcing sheet is then sandwiched between layers of this bitumen. When completed, the result is a pre-assembled, built-up roof that has increased elasticity and weatherablility. Some manufacturers include a surfacing material such as ceramic or mineral granules, metal particles or sand.

The preferred method of adhering SBS-modified bitumen roofing to the substrate is with hot-mopped asphalt (much the same as built-up roofing). This installation method requires a tar kettle/pot to heat the asphalt, as well as the labor, tools and equipment necessary to distribute and spread the hot asphalt.

The alternative method for applying APP and SBS modified bitumen is as follows. A skilled installer uses a torch to melt a small pool of bitumen off the membrane. This pool must form across the entire roll for proper adhesion. The installer must unroll the roofing at a pace slow enough to melt the bitumen, but fast enough to prevent damage to the rest of the membrane.

Modified bitumen roofing provides the advantages of both built-up and single-ply roofing. Labor costs are reduced over those of built-up roofing because only a single ply is necessary. The elasticity of single-ply roofing is attained with the reinforcing sheet and polymer modifiers. Modifieds have some self-healing characteristics and because of their multi-layer construction, they offer the reliability and safety of built-up roofing.

Openings   R0813 Metal Doors

R081313-20 Steel Door Selection Guide

Standard steel doors are classified into four levels, as recommended by the Steel Door Institute in the chart below. Each of the four levels offers a range of construction models and designs, to meet architectural requirements for preference and appearance, including full flush, seamless, and stile & rail. Recommended minimum gauge requirements are also included.

For complete standard steel door construction specifications and available sizes, refer to the Steel Door Institute Technical Data Series, ANSI A250.8-98 (SDI-100), and ANSI A250.4-94 Test Procedure and Acceptance Criteria for Physical Endurance of Steel Door and Hardware Reinforcements.

Level		Model	Construction	For Full Flush or Seamless		
				Min. Gauge	Thickness (in)	Thickness (mm)
I	Standard Duty	1	Full Flush	20	0.032	0.8
		2	Seamless			
II	Heavy Duty	1	Full Flush	18	0.042	1.0
		2	Seamless			
III	Extra Heavy Duty	1	Full Flush	16	0.053	1.3
		2	Seamless			
		3	*Stile & Rail			
IV	Maximum Duty	1	Full Flush	14	0.067	1.6
		2	Seamless			

*Stiles & rails are 16 gauge; flush panels, when specified, are 18 gauge.

Openings   R0852 Wood Windows

R085216-10 Window Estimates

To ensure a complete window estimate, be sure to include the material and labor costs for each window, as well as the material and labor costs for an interior wood trim set.

Openings   R0853 Plastic Windows

R085313-20 Replacement Windows

Replacement windows are typically measured per United Inch.

United Inches are calculated by rounding the width and height of the window opening up to the nearest inch, then adding the two figures.

The labor cost for replacement windows includes removal of sash, existing sash balance or weights, parting bead where necessary and installation of new window.

Debris hauling and dump fees are not included.

Openings — R0871 Door Hardware

R087120-10 Hinges

All closer equipped doors should have ball bearing hinges. Lead lined or extremely heavy doors require special strength hinges. Usually 1-1/2 pair of hinges are used per door up to 7'-6" high openings. Table below shows typical hinge requirements.

Use Frequency	Type Hinge Required	Type of Opening	Type of Structure
High	Heavy weight	Entrances	Banks, Office buildings, Schools, Stores & Theaters
	ball bearing	Toilet Rooms	Office buildings and Schools
Average	Standard	Entrances	Dwellings
	weight	Corridors	Office buildings and Schools
	ball bearing	Toilet Rooms	Stores
Low	Plain bearing	Interior	Dwellings

Door Thickness	Weight of Doors in Pounds per Square Foot				
	White Pine	Oak	Hollow Core	Solid Core	Hollow Metal
1-3/8"	3psf	6psf	1-1/2psf	3-1/2 — 4psf	6-1/2psf
1-3/4"	3-1/2	7	2-1/2	4-1/2 — 5-1/4	6-1/2
2-1/4"	4-1/2	9	—	5-1/2 — 6-3/4	6-1/2

Finishes

R0920 Plaster & Gypsum Board

R092000-50 Lath, Plaster and Gypsum Board

Gypsum board lath is available in 3/8" thick x 16" wide x 4' long sheets as a base material for multi-layer plaster applications. It is also available as a base for either multi-layer or veneer plaster applications in 1/2" and 5/8" thick–4' wide x 8', 10' or 12' long sheets. Fasteners are screws or blued ring shank nails for wood framing and screws for metal framing.

Metal lath is available in diamond mesh pattern with flat or self-furring profiles. Paper backing is available for applications where excessive plaster waste needs to be avoided. A slotted mesh ribbed lath should be used in areas where the span between structural supports is greater than normal. Most metal lath comes in 27" x 96" sheets. Diamond mesh weighs 1.75, 2.5 or 3.4 pounds per square yard, slotted mesh lath weighs 2.75 or 3.4 pounds per square yard. Metal lath can be nailed, screwed or tied in place.

Many **accessories** are available. Corner beads, flat reinforcing strips, casing beads, control and expansion joints, furring brackets and channels are some examples. Note that accessories are not included in plaster or stucco line items.

Plaster is defined as a material or combination of materials that when mixed with a suitable amount of water, forms a plastic mass or paste. When applied to a surface, the paste adheres to it and subsequently hardens, preserving in a rigid state the form or texture imposed during the period of elasticity.

Gypsum plaster is made from ground calcined gypsum. It is mixed with aggregates and water for use as a base coat plaster.

Vermiculite plaster is a fire-retardant plaster covering used on steel beams, concrete slabs and other heavy construction materials. Vermiculite is a group name for certain clay minerals, hydrous silicates or aluminum, magnesium and iron that have been expanded by heat.

Perlite plaster is a plaster using perlite as an aggregate instead of sand. Perlite is a volcanic glass that has been expanded by heat.

Gauging plaster is a mix of gypsum plaster and lime putty that when applied produces a quick drying finish coat.

Veneer plaster is a one or two component gypsum plaster used as a thin finish coat over special gypsum board.

Keenes cement is a white cementitious material manufactured from gypsum that has been burned at a high temperature and ground to a fine powder. Alum is added to accelerate the set. The resulting plaster is hard and strong and accepts and maintains a high polish, hence it is used as a finishing plaster.

Stucco is a Portland cement based plaster used primarily as an exterior finish.

Plaster is used on both interior and exterior surfaces. Generally it is applied in multiple-coat systems. A three-coat system uses the terms scratch, brown and finish to identify each coat. A two-coat system uses base and finish to describe each coat. Each type of plaster and application system has attributes that are chosen by the designer to best fit the intended use.

Gypsum Plaster	2 Coat, 5/8" Thick		3 Coat, 3/4" Thick		
Quantities for 100 S.Y.	Base	Finish	Scratch	Brown	Finish
	1:3 Mix	2:1 Mix	1:2 Mix	1:3 Mix	2:1 Mix
Gypsum plaster	1,300 lb.		1,350 lb.	650 lb.	
Sand	1.75 C.Y.		1.85 C.Y.	1.35 C.Y.	
Finish hydrated lime		340 lb.			340 lb.
Gauging plaster		170 lb.			170 lb.

Vermiculite or Perlite Plaster	2 Coat, 5/8" Thick		3 Coat, 3/4" Thick		
Quantities for 100 S.Y.	Base	Finish	Scratch	Brown	Finish
Gypsum plaster	1,250 lb.		1,450 lb.	800 lb.	
Vermiculite or perlite	7.8 bags		8.0 bags	3.3 bags	
Finish hydrated lime		340 lb.			340 lb.
Gauging plaster		170 lb.			170 lb.

Stucco–Three-Coat System Quantities for 100 S.Y.	On Wood Frame	On Masonry
Portland cement	29 bags	21 bags
Sand	2.6 C.Y.	2.0 C.Y.
Hydrated lime	180 lb.	120 lb.

Finishes — R0929 Gypsum Board

R092910-10 Levels of Gypsum Drywall Finish

In the past, contract documents often used phrases such as "industry standard" and "workmanlike finish" to specify the expected quality of gypsum board wall and ceiling installations. The vagueness of these descriptions led to unacceptable work and disputes.

In order to resolve this problem, four major trade associations concerned with the manufacture, erection, finish and decoration of gypsum board wall and ceiling systems have developed an industry-wide *Recommended Levels of Gypsum Board Finish*.

The finish of gypsum board walls and ceilings for specific final decoration is dependent on a number of factors. A primary consideration is the location of the surface and the degree of decorative treatment desired. Painted and unpainted surfaces in warehouses and other areas where appearance is normally not critical may simply require the taping of wallboard joints and 'spotting' of fastener heads. Blemish-free, smooth, monolithic surfaces often intended for painted and decorated walls and ceilings in habitated structures, ranging from single-family dwellings through monumental buildings, require additional finishing prior to the application of the final decoration.

Other factors to be considered in determining the level of finish of the gypsum board surface are (1) the type of angle of surface illumination (both natural and artificial lighting), and (2) the paint and method of application or the type and finish of wallcovering specified as the final decoration. Critical lighting conditions, gloss paints, and thin wallcoverings require a higher level of gypsum board finish than do heavily textured surfaces which are subsequently painted or surfaces which are to be decorated with heavy grade wallcoverings.

The following descriptions were developed jointly by the Association of the Wall and Ceiling Industries-International (AWCI), Ceiling & Interior Systems Construction Association (CISCA), Gypsum Association (GA), and Painting and Decorating Contractors of America (PDCA) as a guide.

Level 0: No taping, finishing, or accessories required. This level of finish may be useful in temporary construction or whenever the final decoration has not been determined.

Level 1: All joints and interior angles shall have tape set in joint compound. Surface shall be free of excess joint compound. Tool marks and ridges are acceptable. Frequently specified in plenum areas above ceilings, in attics, in areas where the assembly would generally be concealed or in building service corridors, and other areas not normally open to public view.

Level 2: All joints and interior angles shall have tape embedded in joint compound and wiped with a joint knife leaving a thin coating of joint compound over all joints and interior angles. Fastener heads and accessories shall be covered with a coat of joint compound. Surface shall be free of excess joint compound. Tool marks and ridges are acceptable. Joint compound applied over the body of the tape at the time of tape embedment shall be considered a separate coat of joint compound and shall satisfy the conditions of this level. Specified where water-resistant gypsum backing board is used as a substrate for tile; may be specified in garages, warehouse storage, or other similar areas where surface appearance is not of primary concern.

Level 3: All joints and interior angles shall have tape embedded in joint compound and one additional coat of joint compound applied over all joints and interior angles. Fastener heads and accessories shall be covered with two separate coats of joint compound. All joint compound shall be smooth and free of tool marks and ridges. Typically specified in appearance areas which are to receive heavy- or medium-texture (spray or hand applied) finishes before final painting, or where heavy-grade wallcoverings are to be applied as the final decoration. This level of finish is not recommended where smooth painted surfaces or light to medium wallcoverings are specified.

Level 4: All joints and interior angles shall have tape embedded in joint compound and two separate coats of joint compound applied over all flat joints and one separate coat of joint compound applied over interior angles. Fastener heads and accessories shall be covered with three separate coats of joint compound. All joint compound shall be smooth and free of tool marks and ridges. This level should be specified where flat paints, light textures, or wallcoverings are to be applied. In critical lighting areas, flat paints applied over light textures tend to reduce joint photographing. Gloss, semi-gloss, and enamel paints are not recommended over this level of finish. The weight, texture, and sheen level of wallcoverings applied over this level of finish should be carefully evaluated. Joints and fasteners must be adequately concealed if the wallcovering material is lightweight, contains limited pattern, has a gloss finish, or any combination of these finishes is present. Unbacked vinyl wallcoverings are not recommended over this level of finish.

Level 5: All joints and interior angles shall have tape embedded in joint compound and two separate coats of joint compound applied over all flat joints and one separate coat of joint compound applied over interior angles. Fastener heads and accessories shall be covered with three separate coats of joint compound. A thin skim coat of joint compound or a material manufactured especially for this purpose, shall be applied to the entire surface. The surface shall be smooth and free of tool marks and ridges. This level of finish is highly recommended where gloss, semi-gloss, enamel, or nontextured flat paints are specified or where severe lighting conditions occur. This highest quality finish is the most effective method to provide a uniform surface and minimize the possibility of joint photographing and of fasteners showing through the final decoration.

Finishes — R0972 Wall Coverings

R097223-10 Wall Covering

The table below lists the quantities required for 100 S.F. of wall covering.

Description	Medium-Priced Paper	Expensive Paper
Paper	1.6 dbl. rolls	1.6 dbl. rolls
Wall sizing	0.25 gallon	0.25 gallon
Vinyl wall paste	0.6 gallon	0.6 gallon
Apply sizing	0.3 hour	0.3 hour
Apply paper	1.2 hours	1.5 hours

Most wallpapers now come in double rolls only.
To remove old paper, allow 1.3 hours per 100 S.F.

Finishes — R0991 Painting

R099100-20 Painting

Item	Coat	One Gallon Covers			In 8 Hours a Laborer Covers			Labor-Hours per 100 S.F.		
		Brush	Roller	Spray	Brush	Roller	Spray	Brush	Roller	Spray
Paint wood siding	prime	250 S.F.	225 S.F.	290 S.F.	1150 S.F.	1300 S.F.	2275 S.F.	.695	.615	.351
	others	270	250	290	1300	1625	2600	.615	.492	.307
Paint exterior trim	prime	400	—	—	650	—	—	1.230	—	—
	1st	475	—	—	800	—	—	1.000	—	—
	2nd	520	—	—	975	—	—	.820	—	—
Paint shingle siding	prime	270	255	300	650	975	1950	1.230	.820	.410
	others	360	340	380	800	1150	2275	1.000	.695	.351
Stain shingle siding	1st	180	170	200	750	1125	2250	1.068	.711	.355
	2nd	270	250	290	900	1325	2600	.888	.603	.307
Paint brick masonry	prime	180	135	160	750	800	1800	1.066	1.000	.444
	1st	270	225	290	815	975	2275	.981	.820	.351
	2nd	340	305	360	815	1150	2925	.981	.695	.273
Paint interior plaster or drywall	prime	400	380	495	1150	2000	3250	.695	.400	.246
	others	450	425	495	1300	2300	4000	.615	.347	.200
Paint interior doors and windows	prime	400	—	—	650	—	—	1.230	—	—
	1st	425	—	—	800	—	—	1.000	—	—
	2nd	450	—	—	975	—	—	.820	—	—

Special Construction — R1311 Swimming Pools

R131113-20 Swimming Pools

Pool prices given per square foot of surface area include pool structure, filter and chlorination equipment, pumps, related piping, ladders/steps, maintenance kit, skimmer and vacuum system. Decks and electrical service to equipment are not included.

Residential in-ground pool construction can be divided into two categories: vinyl lined and gunite. Vinyl lined pool walls are constructed of different materials including wood, concrete, plastic or metal. The bottom is often graded with sand over which the vinyl liner is installed. Vermiculite or soil cement bottoms may be substituted for an added cost.

Gunite pool construction is used both in residential and municipal installations. These structures are steel reinforced for strength and finished with a white cement limestone plaster.

Municipal pools will have a higher cost because plumbing codes require more expensive materials, chlorination equipment and higher filtration rates.

Municipal pools greater than 1,800 S.F. require gutter systems to control waves. This gutter may be formed into the concrete wall. Often a vinyl/stainless steel gutter or gutter/wall system is specified, which will raise the pool cost.

Competition pools usually require tile bottoms and sides with contrasting lane striping, which will also raise the pool cost.

Plumbing — R2211 Facility Water Distribution

R221113-50 Pipe Material Considerations

1. Malleable fittings should be used for gas service.
2. Malleable fittings are used where there are stresses/strains due to expansion and vibration.
3. Cast fittings may be broken as an aid to disassembling of heating lines frozen by long use, temperature and minerals.
4. Cast iron pipe is extensively used for underground and submerged service.
5. Type M (light wall) copper tubing is available in hard temper only and is used for nonpressure and less severe applications than K and L.
6. Type L (medium wall) copper tubing, available hard or soft for interior service.
7. Type K (heavy wall) copper tubing, available in hard or soft temper for use where conditions are severe. For underground and interior service.
8. Hard drawn tubing requires fewer hangers or supports but should not be bent. Silver brazed fittings are recommended, however soft solder is normally used.
9. Type DMV (very light wall) copper tubing designed for drainage, waste and vent plus other non-critical pressure services.

Domestic/Imported Pipe and Fittings Cost

The prices shown in this publication for steel/cast iron pipe and steel, cast iron, malleable iron fittings are based on domestic production sold at the normal trade discounts. The above listed items of foreign manufacture may be available at prices of 1/3 to 1/2 those shown. Some imported items after minor machining or finishing operations are being sold as domestic to further complicate the system.

Caution: Most pipe prices in this book also include a coupling and pipe hangers which for the larger sizes can add significantly to the per foot cost and should be taken into account when comparing "book cost" with quoted supplier's cost.

Plumbing — R2240 Plumbing Fixtures

R224000-40 Plumbing Fixture Installation Time

Item	Rough-In	Set	Total Hours	Item	Rough-In	Set	Total Hours
Bathtub	5	5	10	Shower head only	2	1	3
Bathtub and shower, cast iron	6	6	12	Shower drain	3	1	4
Fire hose reel and cabinet	4	2	6	Shower stall, slate		15	15
Floor drain to 4 inch diameter	3	1	4	Slop sink	5	3	8
Grease trap, single, cast iron	5	3	8	Test 6 fixtures			14
Kitchen gas range		4	4	Urinal, wall	6	2	8
Kitchen sink, single	4	4	8	Urinal, pedestal or floor	6	4	10
Kitchen sink, double	6	6	12	Water closet and tank	4	3	7
Laundry tubs	4	2	6	Water closet and tank, wall hung	5	3	8
Lavatory wall hung	5	3	8	Water heater, 45 gals. gas, automatic	5	2	7
Lavatory pedestal	5	3	8	Water heaters, 65 gals. gas, automatic	5	2	7
Shower and stall	6	4	10	Water heaters, electric, plumbing only	4	2	6

Fixture prices in front of book are based on the cost per fixture set in place. The rough-in cost, which must be added for each fixture, includes carrier, if required, some supply, waste and vent pipe connecting fittings and stops. The lengths of rough-in pipe are nominal runs which would connect to the larger runs and stacks. The supply runs and DWV runs and stacks must be accounted for in separate entries. In the eastern half of the United States it is common for the plumber to carry these to a point 5' outside the building.

Exterior Improvements — R3292 Turf & Grasses

R329219-50 Seeding

The type of grass is determined by light, shade and moisture content of soil plus intended use. Fertilizer should be disked 4" before seeding. For steep slopes disk five tons of mulch and lay two tons of hay or straw on surface per acre after seeding. Surface mulch can be staked, lightly disked or tar emulsion sprayed. Material for mulch can be wood chips, peat moss, partially rotted hay or straw, wood fibers and sprayed emulsions. Hemp seed blankets with fertilizer are also available. For spring seeding, watering is necessary. Late fall seeding may have to be reseeded in the spring. Hydraulic seeding, power mulching, and aerial seeding can be used on large areas.

Utilities — R3311 Water Utility Distribution Piping

R331113-80 Piping Designations

There are several systems currently in use to describe pipe and fittings. The following paragraphs will help to identify and clarify classifications of piping systems used for water distribution.

Piping may be classified by schedule. Piping schedules include 5S, 10S, 10, 20, 30, Standard, 40, 60, Extra Strong, 80, 100, 120, 140, 160 and Double Extra Strong. These schedules are dependent upon the pipe wall thickness. The wall thickness of a particular schedule may vary with pipe size.

Ductile iron pipe for water distribution is classified by Pressure Classes such as Class 150, 200, 250, 300 and 350. These classes are actually the rated water working pressure of the pipe in pounds per square inch (psi). The pipe in these pressure classes is designed to withstand the rated water working pressure plus a surge allowance of 100 psi.

The American Water Works Association (AWWA) provides standards for various types of **plastic pipe.** C-900 is the specification for polyvinyl chloride (PVC) piping used for water distribution in sizes ranging from 4″ through 12″. C-901 is the specification for polyethylene (PE) pressure pipe, tubing and fittings used for water distribution in sizes ranging from 1/2″ through 3″. C-905 is the specification for PVC piping sizes 14″ and greater.

PVC pressure-rated pipe is identified using the standard dimensional ratio (SDR) method. This method is defined by the American Society for Testing and Materials (ASTM) Standard D 2241. This pipe is available in SDR numbers 64, 41, 32.5, 26, 21, 17, and 13.5. Pipe with an SDR of 64 will have the thinnest wall while pipe with an SDR of 13.5 will have the thickest wall. When the pressure rating (PR) of a pipe is given in psi, it is based on a line supplying water at 73 degrees F.

The National Sanitation Foundation (NSF) seal of approval is applied to products that can be used with potable water. These products have been tested to ANSI/NSF Standard 14.

Valves and strainers are classified by American National Standards Institute (ANSI) Classes. These Classes are 125, 150, 200, 250, 300, 400, 600, 900, 1500 and 2500. Within each class there is an operating pressure range dependent upon temperature. Design parameters should be compared to the appropriate material dependent, pressure-temperature rating chart for accurate valve selection.

Abbreviations

A	Area Square Feet; Ampere	Cab.	Cabinet	Demob.	Demobilization
ABS	Acrylonitrile Butadiene Stryrene; Asbestos Bonded Steel	Cair.	Air Tool Laborer	d.f.u.	Drainage Fixture Units
		Calc	Calculated	D.H.	Double Hung
A.C.	Alternating Current; Air-Conditioning; Asbestos Cement; Plywood Grade A & C	Cap.	Capacity	DHW	Domestic Hot Water
		Carp.	Carpenter	Diag.	Diagonal
		C.B.	Circuit Breaker	Diam.	Diameter
		C.C.A.	Chromate Copper Arsenate	Distrib.	Distribution
A.C.I.	American Concrete Institute	C.C.F.	Hundred Cubic Feet	Dk.	Deck
AD	Plywood, Grade A & D	cd	Candela	D.L.	Dead Load; Diesel
Addit.	Additional	cd/sf	Candela per Square Foot	DLH	Deep Long Span Bar Joist
Adj.	Adjustable	CD	Grade of Plywood Face & Back	Do.	Ditto
af	Audio-frequency	CDX	Plywood, Grade C & D, exterior glue	Dp.	Depth
A.G.A.	American Gas Association			D.P.S.T.	Double Pole, Single Throw
Agg.	Aggregate	Cefi.	Cement Finisher	Dr.	Driver
A.H.	Ampere Hours	Cem.	Cement	Drink.	Drinking
A hr.	Ampere-hour	CF	Hundred Feet	D.S.	Double Strength
A.H.U.	Air Handling Unit	C.F.	Cubic Feet	D.S.A.	Double Strength A Grade
A.I.A.	American Institute of Architects	CFM	Cubic Feet per Minute	D.S.B.	Double Strength B Grade
AIC	Ampere Interrupting Capacity	c.g.	Center of Gravity	Dty.	Duty
Allow.	Allowance	CHW	Chilled Water; Commercial Hot Water	DWV	Drain Waste Vent
alt.	Altitude			DX	Deluxe White, Direct Expansion
Alum.	Aluminum	C.I.	Cast Iron	dyn	Dyne
a.m.	Ante Meridiem	C.I.P.	Cast in Place	e	Eccentricity
Amp.	Ampere	Circ.	Circuit	E	Equipment Only; East
Anod.	Anodized	C.L.	Carload Lot	Ea.	Each
Approx.	Approximate	Clab.	Common Laborer	E.B.	Encased Burial
Apt.	Apartment	Clam	Common maintenance laborer	Econ.	Economy
Asb.	Asbestos	C.L.F.	Hundred Linear Feet	E.C.Y	Embankment Cubic Yards
A.S.B.C.	American Standard Building Code	CLF	Current Limiting Fuse	EDP	Electronic Data Processing
Asbe.	Asbestos Worker	CLP	Cross Linked Polyethylene	EIFS	Exterior Insulation Finish System
A.S.H.R.A.E.	American Society of Heating, Refrig. & AC Engineers	cm	Centimeter	E.D.R.	Equiv. Direct Radiation
		CMP	Corr. Metal Pipe	Eq.	Equation
A.S.M.E.	American Society of Mechanical Engineers	C.M.U.	Concrete Masonry Unit	Elec.	Electrician; Electrical
		CN	Change Notice	Elev.	Elevator; Elevating
A.S.T.M.	American Society for Testing and Materials	Col.	Column	EMT	Electrical Metallic Conduit; Thin Wall Conduit
		CO_2	Carbon Dioxide		
Attchmt.	Attachment	Comb.	Combination	Eng.	Engine, Engineered
Avg.	Average	Compr.	Compressor	EPDM	Ethylene Propylene Diene Monomer
A.W.G.	American Wire Gauge	Conc.	Concrete		
AWWA	American Water Works Assoc.	Cont.	Continuous; Continued	EPS	Expanded Polystyrene
Bbl.	Barrel	Corr.	Corrugated	Eqhv.	Equip. Oper., Heavy
B&B	Grade B and Better; Balled & Burlapped	Cos	Cosine	Eqlt.	Equip. Oper., Light
		Cot	Cotangent	Eqmd.	Equip. Oper., Medium
B.&S.	Bell and Spigot	Cov.	Cover	Eqmm.	Equip. Oper., Master Mechanic
B.&W.	Black and White	C/P	Cedar on Paneling	Eqol.	Equip. Oper., Oilers
b.c.c.	Body-centered Cubic	CPA	Control Point Adjustment	Equip.	Equipment
B.C.Y.	Bank Cubic Yards	Cplg.	Coupling	ERW	Electric Resistance Welded
BE	Bevel End	C.P.M.	Critical Path Method	E.S.	Energy Saver
B.F.	Board Feet	CPVC	Chlorinated Polyvinyl Chloride	Est.	Estimated
Bg. cem.	Bag of Cement	C.Pr.	Hundred Pair	esu	Electrostatic Units
BHP	Boiler Horsepower; Brake Horsepower	CRC	Cold Rolled Channel	E.W.	Each Way
		Creos.	Creosote	EWT	Entering Water Temperature
B.I.	Black Iron	Crpt.	Carpet & Linoleum Layer	Excav.	Excavation
Bit.; Bitum.	Bituminous	CRT	Cathode-ray Tube	Exp.	Expansion, Exposure
Bk.	Backed	CS	Carbon Steel, Constant Shear Bar Joist	Ext.	Exterior
Bkrs.	Breakers			Extru.	Extrusion
Bldg.	Building	Csc	Cosecant	f.	Fiber stress
Blk.	Block	C.S.F.	Hundred Square Feet	F	Fahrenheit; Female; Fill
Bm.	Beam	CSI	Construction Specifications Institute	Fab.	Fabricated
Boil.	Boilermaker			FBGS	Fiberglass
B.P.M.	Blows per Minute	C.T.	Current Transformer	F.C.	Footcandles
BR	Bedroom	CTS	Copper Tube Size	f.c.c.	Face-centered Cubic
Brg.	Bearing	Cu	Copper, Cubic	f'c.	Compressive Stress in Concrete; Extreme Compressive Stress
Brhe.	Bricklayer Helper	Cu. Ft.	Cubic Foot		
Bric.	Bricklayer	cw	Continuous Wave	F.E.	Front End
Brk.	Brick	C.W.	Cool White; Cold Water	FEP	Fluorinated Ethylene Propylene (Teflon)
Brng.	Bearing	Cwt.	100 Pounds		
Brs.	Brass	C.W.X.	Cool White Deluxe	F.G.	Flat Grain
Brz.	Bronze	C.Y.	Cubic Yard (27 cubic feet)	F.H.A.	Federal Housing Administration
Bsn.	Basin	C.Y./Hr.	Cubic Yard per Hour	Fig.	Figure
Btr.	Better	Cyl.	Cylinder	Fin.	Finished
BTU	British Thermal Unit	d	Penny (nail size)	Fixt.	Fixture
BTUH	BTU per Hour	D	Deep; Depth; Discharge	Fl. Oz.	Fluid Ounces
B.U.R.	Built-up Roofing	Dis.;Disch.	Discharge	Flr.	Floor
BX	Interlocked Armored Cable	Db.	Decibel	F.M.	Frequency Modulation; Factory Mutual
c	Conductivity, Copper Sweat	Dbl.	Double		
C	Hundred; Centigrade	DC	Direct Current	Fmg.	Framing
C/C	Center to Center, Cedar on Cedar	DDC	Direct Digital Control	Fndtn.	Foundation

Abbreviations

Fori.	Foreman, Inside	I.W.	Indirect Waste	M.C.F.	Thousand Cubic Feet
Foro.	Foreman, Outside	J	Joule	M.C.F.M.	Thousand Cubic Feet per Minute
Fount.	Fountain	J.I.C.	Joint Industrial Council	M.C.M.	Thousand Circular Mils
FPM	Feet per Minute	K	Thousand; Thousand Pounds;	M.C.P.	Motor Circuit Protector
FPT	Female Pipe Thread		Heavy Wall Copper Tubing, Kelvin	MD	Medium Duty
Fr.	Frame	K.A.H.	Thousand Amp. Hours	M.D.O.	Medium Density Overlaid
F.R.	Fire Rating	KCMIL	Thousand Circular Mils	Med.	Medium
FRK	Foil Reinforced Kraft	KD	Knock Down	MF	Thousand Feet
FRP	Fiberglass Reinforced Plastic	K.D.A.T.	Kiln Dried After Treatment	M.F.B.M.	Thousand Feet Board Measure
FS	Forged Steel	kg	Kilogram	Mfg.	Manufacturing
FSC	Cast Body; Cast Switch Box	kG	Kilogauss	Mfrs.	Manufacturers
Ft.	Foot; Feet	kgf	Kilogram Force	mg	Milligram
Ftng.	Fitting	kHz	Kilohertz	MGD	Million Gallons per Day
Ftg.	Footing	Kip.	1000 Pounds	MGPH	Thousand Gallons per Hour
Ft. Lb.	Foot Pound	KJ	Kiljoule	MH, M.H.	Manhole; Metal Halide; Man-Hour
Furn.	Furniture	K.L.	Effective Length Factor	MHz	Megahertz
FVNR	Full Voltage Non-Reversing	K.L.F.	Kips per Linear Foot	Mi.	Mile
FXM	Female by Male	Km	Kilometer	MI	Malleable Iron; Mineral Insulated
Fy.	Minimum Yield Stress of Steel	K.S.F.	Kips per Square Foot	mm	Millimeter
g	Gram	K.S.I.	Kips per Square Inch	Mill.	Millwright
G	Gauss	kV	Kilovolt	Min., min.	Minimum, minute
Ga.	Gauge	kVA	Kilovolt Ampere	Misc.	Miscellaneous
Gal.	Gallon	K.V.A.R.	Kilovar (Reactance)	ml	Milliliter, Mainline
Gal./Min.	Gallon per Minute	KW	Kilowatt	M.L.F.	Thousand Linear Feet
Galv.	Galvanized	KWh	Kilowatt-hour	Mo.	Month
Gen.	General	L	Labor Only; Length; Long;	Mobil.	Mobilization
G.F.I.	Ground Fault Interrupter		Medium Wall Copper Tubing	Mog.	Mogul Base
Glaz.	Glazier	Lab.	Labor	MPH	Miles per Hour
GPD	Gallons per Day	lat	Latitude	MPT	Male Pipe Thread
GPH	Gallons per Hour	Lath.	Lather	MRT	Mile Round Trip
GPM	Gallons per Minute	Lav.	Lavatory	ms	Millisecond
GR	Grade	lb.; #	Pound	M.S.F.	Thousand Square Feet
Gran.	Granular	L.B.	Load Bearing; L Conduit Body	Mstz.	Mosaic & Terrazzo Worker
Grnd.	Ground	L. & E.	Labor & Equipment	M.S.Y.	Thousand Square Yards
H	High; High Strength Bar Joist;	lb./hr.	Pounds per Hour	Mtd.	Mounted
	Henry	lb./L.F.	Pounds per Linear Foot	Mthe.	Mosaic & Terrazzo Helper
H.C.	High Capacity	lbf/sq.in.	Pound-force per Square Inch	Mtng.	Mounting
H.D.	Heavy Duty; High Density	L.C.L.	Less than Carload Lot	Mult.	Multi; Multiply
H.D.O.	High Density Overlaid	L.C.Y.	Loose Cubic Yard	M.V.A.	Million Volt Amperes
Hdr.	Header	Ld.	Load	M.V.A.R.	Million Volt Amperes Reactance
Hdwe.	Hardware	LE	Lead Equivalent	MV	Megavolt
Help.	Helper Average	LED	Light Emitting Diode	MW	Megawatt
HEPA	High Efficiency Particulate Air	L.F.	Linear Foot	MXM	Male by Male
	Filter	Lg.	Long; Length; Large	MYD	Thousand Yards
Hg	Mercury	L & H	Light and Heat	N	Natural; North
HIC	High Interrupting Capacity	LH	Long Span Bar Joist	nA	Nanoampere
HM	Hollow Metal	L.H.	Labor Hours	NA	Not Available; Not Applicable
H.O.	High Output	L.L.	Live Load	N.B.C.	National Building Code
Horiz.	Horizontal	L.L.D.	Lamp Lumen Depreciation	NC	Normally Closed
H.P.	Horsepower; High Pressure	lm	Lumen	N.E.M.A.	National Electrical Manufacturers
H.P.F.	High Power Factor	lm/sf	Lumen per Square Foot		Assoc.
Hr.	Hour	lm/W	Lumen per Watt	NEHB	Bolted Circuit Breaker to 600V.
Hrs./Day	Hours per Day	L.O.A.	Length Over All	N.L.B.	Non-Load-Bearing
HSC	High Short Circuit	log	Logarithm	NM	Non-Metallic Cable
Ht.	Height	L-O-L	Lateralolet	nm	Nanometer
Htg.	Heating	L.P.	Liquefied Petroleum; Low Pressure	No.	Number
Htrs.	Heaters	L.P.F.	Low Power Factor	NO	Normally Open
HVAC	Heating, Ventilation & Air-	LR	Long Radius	N.O.C.	Not Otherwise Classified
	Conditioning	L.S.	Lump Sum	Nose.	Nosing
Hvy.	Heavy	Lt.	Light	N.P.T.	National Pipe Thread
HW	Hot Water	Lt. Ga.	Light Gauge	NQOD	Combination Plug-on/Bolt on
Hyd.;Hydr.	Hydraulic	L.T.L.	Less than Truckload Lot		Circuit Breaker to 240V.
Hz.	Hertz (cycles)	Lt. Wt.	Lightweight	N.R.C.	Noise Reduction Coefficient
I.	Moment of Inertia	L.V.	Low Voltage	N.R.S.	Non Rising Stem
I.C.	Interrupting Capacity	M	Thousand; Material; Male;	ns	Nanosecond
ID	Inside Diameter		Light Wall Copper Tubing	nW	Nanowatt
I.D.	Inside Dimension; Identification	M^2CA	Meters Squared Contact Area	OB	Opposing Blade
I.F.	Inside Frosted	m/hr; M.H.	Man-hour	OC	On Center
I.M.C.	Intermediate Metal Conduit	mA	Milliampere	OD	Outside Diameter
In.	Inch	Mach.	Machine	O.D.	Outside Dimension
Incan.	Incandescent	Mag. Str.	Magnetic Starter	ODS	Overhead Distribution System
Incl.	Included; Including	Maint.	Maintenance	O.G.	Ogee
Int.	Interior	Marb.	Marble Setter	O.H.	Overhead
Inst.	Installation	Mat; Mat'l.	Material	O&P	Overhead and Profit
Insul.	Insulation/Insulated	Max.	Maximum	Oper.	Operator
I.P.	Iron Pipe	MBF	Thousand Board Feet	Opng.	Opening
I.P.S.	Iron Pipe Size	MBH	Thousand BTU's per hr.	Orna.	Ornamental
I.P.T.	Iron Pipe Threaded	MC	Metal Clad Cable	OSB	Oriented Strand Board

Abbreviations

O.S.&Y.	Outside Screw and Yoke	Rsr	Riser	Th.; Thk.	Thick
Ovhd.	Overhead	RT	Round Trip	Thn.	Thin
OWG	Oil, Water or Gas	S.	Suction; Single Entrance; South	Thrded	Threaded
Oz.	Ounce	SC	Screw Cover	Tilf.	Tile Layer, Floor
P.	Pole; Applied Load; Projection	SCFM	Standard Cubic Feet per Minute	Tilh.	Tile Layer, Helper
p.	Page	Scaf.	Scaffold	THHN	Nylon Jacketed Wire
Pape.	Paperhanger	Sch.; Sched.	Schedule	THW.	Insulated Strand Wire
P.A.P.R.	Powered Air Purifying Respirator	S.C.R.	Modular Brick	THWN	Nylon Jacketed Wire
PAR	Parabolic Reflector	S.D.	Sound Deadening	T.L.	Truckload
Pc., Pcs.	Piece, Pieces	S.D.R.	Standard Dimension Ratio	T.M.	Track Mounted
P.C.	Portland Cement; Power Connector	S.E.	Surfaced Edge	Tot.	Total
P.C.F.	Pounds per Cubic Foot	Sel.	Select	T-O-L	Threadolet
P.C.M.	Phase Contrast Microscopy	S.E.R.; S.E.U.	Service Entrance Cable	T.S.	Trigger Start
P.E.	Professional Engineer; Porcelain Enamel; Polyethylene; Plain End	S.F.	Square Foot	Tr.	Trade
		S.F.C.A.	Square Foot Contact Area	Transf.	Transformer
		S.F. Flr.	Square Foot of Floor	Trhv.	Truck Driver, Heavy
Perf.	Perforated	S.F.G.	Square Foot of Ground	Trlr	Trailer
Ph.	Phase	S.F. Hor.	Square Foot Horizontal	Trlt.	Truck Driver, Light
P.I.	Pressure Injected	S.F.R.	Square Feet of Radiation	TTY	Teletypewriter
Pile.	Pile Driver	S.F. Shlf.	Square Foot of Shelf	TV	Television
Pkg.	Package	S4S	Surface 4 Sides	T.W.	Thermoplastic Water Resistant Wire
Pl.	Plate	Shee.	Sheet Metal Worker		
Plah.	Plasterer Helper	Sin.	Sine	UCI	Uniform Construction Index
Plas.	Plasterer	Skwk.	Skilled Worker	UF	Underground Feeder
Pluh.	Plumbers Helper	SL	Saran Lined	UGND	Underground Feeder
Plum.	Plumber	S.L.	Slimline	U.H.F.	Ultra High Frequency
Ply.	Plywood	Sldr.	Solder	U.L.	Underwriters Laboratory
p.m.	Post Meridiem	SLH	Super Long Span Bar Joist	Unfin.	Unfinished
Pntd.	Painted	S.N.	Solid Neutral	URD	Underground Residential Distribution
Pord.	Painter, Ordinary	S-O-L	Socketolet		
pp	Pages	sp	Standpipe	US	United States
PP; PPL	Polypropylene	S.P.	Static Pressure; Single Pole; Self-Propelled	USP	United States Primed
P.P.M.	Parts per Million			UTP	Unshielded Twisted Pair
Pr.	Pair	Spri.	Sprinkler Installer	V	Volt
P.E.S.B.	Pre-engineered Steel Building	spwg	Static Pressure Water Gauge	V.A.	Volt Amperes
Prefab.	Prefabricated	S.P.D.T.	Single Pole, Double Throw	V.C.T.	Vinyl Composition Tile
Prefin.	Prefinished	SPF	Spruce Pine Fir	VAV	Variable Air Volume
Prop.	Propelled	S.P.S.T.	Single Pole, Single Throw	VC	Veneer Core
PSF; psf	Pounds per Square Foot	SPT	Standard Pipe Thread	Vent.	Ventilation
PSI; psi	Pounds per Square Inch	Sq.	Square; 100 Square Feet	Vert.	Vertical
PSIG	Pounds per Square Inch Gauge	Sq. Hd.	Square Head	V.F.	Vinyl Faced
PSP	Plastic Sewer Pipe	Sq. In.	Square Inch	V.G.	Vertical Grain
Pspr.	Painter, Spray	S.S.	Single Strength; Stainless Steel	V.H.F.	Very High Frequency
Psst.	Painter, Structural Steel	S.S.B.	Single Strength B Grade	VHO	Very High Output
P.T.	Potential Transformer	sst	Stainless Steel	Vib.	Vibrating
P. & T.	Pressure & Temperature	Sswk.	Structural Steel Worker	V.L.F.	Vertical Linear Foot
Ptd.	Painted	Sswl.	Structural Steel Welder	Vol.	Volume
Ptns.	Partitions	St.; Stl.	Steel	VRP	Vinyl Reinforced Polyester
Pu	Ultimate Load	S.T.C.	Sound Transmission Coefficient	W	Wire; Watt; Wide; West
PVC	Polyvinyl Chloride	Std.	Standard	w/	With
Pvmt.	Pavement	STK	Select Tight Knot	W.C.	Water Column; Water Closet
Pwr.	Power	STP	Standard Temperature & Pressure	W.F.	Wide Flange
Q	Quantity Heat Flow	Stpi.	Steamfitter, Pipefitter	W.G.	Water Gauge
Quan.; Qty.	Quantity	Str.	Strength; Starter; Straight	Wldg.	Welding
Q.C.	Quick Coupling	Strd.	Stranded	W. Mile	Wire Mile
r	Radius of Gyration	Struct.	Structural	W-O-L	Weldolet
R	Resistance	Sty.	Story	W.R.	Water Resistant
R.C.P.	Reinforced Concrete Pipe	Subj.	Subject	Wrck.	Wrecker
Rect.	Rectangle	Subs.	Subcontractors	W.S.P.	Water, Steam, Petroleum
Reg.	Regular	Surf.	Surface	WT., Wt.	Weight
Reinf.	Reinforced	Sw.	Switch	WWF	Welded Wire Fabric
Req'd.	Required	Swbd.	Switchboard	XFER	Transfer
Res.	Resistant	S.Y.	Square Yard	XFMR	Transformer
Resi.	Residential	Syn.	Synthetic	XHD	Extra Heavy Duty
Rgh.	Rough	S.Y.P.	Southern Yellow Pine	XHHW; XLPE	Cross-Linked Polyethylene Wire Insulation
RGS	Rigid Galvanized Steel	Sys.	System		
R.H.W.	Rubber, Heat & Water Resistant; Residential Hot Water	t.	Thickness	XLP	Cross-linked Polyethylene
		T	Temperature; Ton	Y	Wye
rms	Root Mean Square	Tan	Tangent	yd	Yard
Rnd.	Round	T.C.	Terra Cotta	yr	Year
Rodm.	Rodman	T & C	Threaded and Coupled	Δ	Delta
Rofc.	Roofer, Composition	T.D.	Temperature Difference	%	Percent
Rofp.	Roofer, Precast	Tdd	Telecommunications Device for the Deaf	~	Approximately
Rohe.	Roofer Helpers (Composition)			Ø	Phase
Rots.	Roofer, Tile & Slate	T.E.M.	Transmission Electron Microscopy	@	At
R.O.W.	Right of Way	TFE	Tetrafluoroethylene (Teflon)	#	Pound; Number
RPM	Revolutions per Minute	T. & G.	Tongue & Groove; Tar & Gravel	<	Less Than
R.S.	Rapid Start			>	Greater Than

Index

A

ABC extinguisher	189
Abrasive floor tile	163
tread	163
ABS DWV pipe	220
Absorber shock	223
A/C coils furnace	240
packaged terminal	242
Access door and panels	131
door basement	27
door duct	234
Accessories bathroom	162, 187
column formwork	22
door	150
drywall	161
duct	234
plaster	157
toilet	187
wall formwork	23
Accessory drainage	117
fireplace	187
formwork	22, 23
masonry	30
Accordion door	129
Acid proof floor	165
Acoustic ceiling board	164
Acoustical ceiling	164
sealant	119, 161
wallboard	161
Acrylic ceiling	164
wall coating	184
wallcovering	170
wood block	166
Adhesive cement	167
roof	112
wallpaper	170
Adobe brick	36
Aggregate base course	272
exposed	26
stone	281
Aids staging	8
Air cleaner electronic	238
conditioner cooling & heating	242
conditioner packaged terminal	242
conditioner portable	242
conditioner receptacle	252
conditioner removal	232
conditioner rooftop	241
conditioner self-contained	242
conditioner thru-wall	242
conditioner window	242
conditioner wiring	255
conditioning ventilating	235, 242
cooled condensing unit	241
extractor	234
filter	237, 238
filter roll type	237
filters washable	237
handler heater	241
handler modular	240
return grille	236
supply register	236
Air-source heat pumps	243
Alarm burglar	264
residential	254
Alteration fee	6
Aluminum ceiling tile	165
chain link	275
column	93
coping	39
diffuser perforated	236
downspout	116
drip edge	118
ductwork	234
edging	282
flagpole	191
flashing	115
foil	103, 105
gravel stop	116
grille	236
gutter	117
louver	152
nail	66
plank scaffold	9
rivet	45
roof	108
service entrance cable	247
sheet metal	116
siding	109
siding accessories	109
siding paint	175
sliding door	132
storm door	124
tile	163
window	133
window demolition	123
Anchor bolt	22, 32
brick	32
buck	32
chemical	28, 44
epoxy	28, 44
expansion	44
framing	68
hollow wall	44
jute fiber	44
lead screw	45
masonry	32
metal nailing	44
nailing	44
nylon nailing	44
partition	32
plastic screw	45
rafter	68
rigid	32
roof safety	9
screw	44
sill	68
steel	32
stone	32
toggle-bolt	44
Antenna system	262
T.V.	262
Appliance	194
residential	194, 254
Apron wood	87
Arch laminated	86
radial	86
Architectural equipment	194
fee	6, 337
Armored cable	247
Arrester lightning	258
Ashlar stone	38
Asphalt base sheet	112
block	274
block floor	274
coating	101
curb	274, 275
felt	112
flood coat	112
paper	105
primer	167
sheathing	85
shingle	106
sidewalk	272
Asphaltic emulsion	279
Astragal	148
molding	89
one piece	148
Attic stair	196
ventilation fan	235
Auger hole	12
Autoclaved concrete block	34
Automatic transfer switch	258
washing machine	195
Awning canvas	191
window	135

B

Backerboard	159
Backfill	268
general	268
planting pit	280
trench	268
Backflow preventer	223
Backhoe	268
excavation	267
Backsplash countertop	205
Backup block	34
Baffle roof	119
Baked enamel frame	125
Baluster	92
Balustrade painting	181
Band joist framing	52
molding	87
Bankrun gravel	266
Bar grab	187
towel	187
Zee	162
Bark mulch	279
mulch redwood	279
Base cabinet	201, 203
carpet	170
ceramic tile	162
column	68
course	272, 273
course aggregate	272
cove	167
gravel	272
molding	87
quarry tile	163
resilient	167
sheet	112
sink	202
stone	38, 272
terrazzo	168
vanity	205
wood	87, 88
Baseboard demolition	66
heat	243
heat electric	243
heater electric	243
register	236
Basement stair	27
Basement/utility window	134
Baseplate scaffold	8
shoring	8
Basic finish materials	154
meter device	256
Basketweave fence	277
Basketweave fence vinyl	276
Bath communal	211
spa	211
steam	211
whirlpool	211
Bathroom	228
accessories	162, 187
faucet	229
fixture	227, 228
Bathtub	228
enclosure	186
removal	216
Batt insulation	102
Bay window	135
Bead casing	161
corner	157
parting	89
Beam & girder framing	71
bondcrete	158
ceiling	93
drywall	160
hanger	68
laminated	86
mantel	93
plaster	158
steel	47
wood	71, 82
Bed molding	88
Bedding brick	274
pipe	268
Beech tread	92
Belgian block	275
Bell & spigot pipe	224
Bench park	207
players	207
Benches	207
Berm pavement	275
road	275
Bevel siding	110
Bi-fold door	126, 128
Bin retaining walls metal	278
Bi-passing closet door	128
Birch door	128-130
molding	89
paneling	90
stair	92
wood frame	95
Bituminous block	274
coating	101
concrete curbs	274
Blanket insulation	102, 232
Blind exterior	95
venetian	200
window	200
Block asphalt	274
backup	34
belgian	275
bituminous	274
Block, brick and mortar	346
Block concrete	34-36, 347
concrete bond beam	35
concrete exterior	35
decorative concrete	35
filler	183
floor	166
glass	36
insulation	35
lightweight	36
lintel	35
partition	35, 36
profile	35
reflective	36
split rib	35
wall removal	14
Blocking	71
carpentry	71
wood	74
Blown in cellulose	103
in fiberglass	103
in insulation	103
Blueboard	159
Bluegrass sod	280
Bluestone sidewalk	272
sill	38
step	272
Board & batten siding	110
batten fence	278
ceiling	164
fence	277
gypsum	160
insulation	101
paneling	92

359

Index

sheathing 84
valance 202
verge 88
Boiler 238
   demolition 232
   electric 238
   electric steam 238
   gas fired 238
   gas/oil combination 238
   hot water 238, 239
   oil fired 239
   removal 232
   steam 238, 239
Bollard 46
   light 260
Bolt anchor 22, 32
   steel 68
Bondcrete 158
Bookcase 190, 203
Boring and exploratory drilling . . . 12
   cased 12
Borrow 266
Bow window 135, 139
Bowstring truss 86
Box distribution 289
   fence shadow 277
   pull 249
   stair 92
Boxed beam headers 49, 52
Boxes & wiring device 249
   utility 288
Bracing 71
   let-in 70
   metal joist 51
   roof rafter 55
   shoring 8
   stud wall 48
Bracket scaffold 8
   sidewall 8
Brass hinge 148
   screw 67
Breaker circuit 257
   vacuum 223
Brick adobe 36
   anchor 32
   bedding 274
Brick, block and mortar 346
Brick chimney simulated 187
   cleaning 30
   demolition 31, 154
   edging 282
   face 34, 346
   molding 88
   oversized 33
   paving 274
   removal 13
   sidewalk 274
   sill 38
   step 272
   veneer 33
   veneer demolition 31
Brick wall 37
Brick wash 30
Bricklaying 346
Bridging 71
   metal joist 51
   roof rafter 55
   stud wall 48
Bronze swing check valve 216
   valve 216
Broom cabinet 202
   finish concrete 26
Brownstone 38
Brush clearing 266
Buck anchor 32
   rough 73

Buggy concrete 26
Builder's risk insurance 339
Building demolition 13
   greenhouse 212
   hardware 147
   insulation 103
   paper 105, 106
   permit 7
   prefabricated 212
Built-in range 194
Built-up roof 112
Bulk bank measure excavating . . . 268
Bulkhead door 131
Bulldozer 268
Bumper door 147
   wall 147
Burglar alarm 264
Burlap curing 27
   rubbing 26
Burner gas conversion 239
   gun type 239
   residential 239
Bush hammer 26
   hammer finish concrete 26
Butterfly valve 287
Butyl caulking 119

C

Cabinet base 201, 203
   broom 202
   casework 203
   corner base 202
   corner wall 202
   demolition 66
   door 203, 204
   electrical 249
   hardboard 201
   hardware 205
   hinge 205
   kitchen 201
   medicine 187
   outdoor 205
   oven 202
   shower 186
   stain 178
   varnish 178
   wall 202
Cable armored 247
   electric 247, 248
   sheathed nonmetallic 247
   sheathed romex 247
Cafe door 130
Canopy door 190
   framing 77
Cant roof 77
Cantilever retaining wall 278
Canvas awning 191
Cap dowel 23
   post 68
   service entrance 247
Caps vent 225
Carbon dioxide extinguisher . . . 189
Carpet 169
   base 170
   felt pad 169
   floor 169
   nylon 169
   olefin 169
   pad 169
   padding 169
   removal 154
   stair 170
   tile 169
   urethane pad 169

wool 169
Carrier ceiling 157
   channel 165
   fixture 230
Cart concrete 26
Case work 202
Cased boring 12
   evaporator coils 240
Casement windows 135, 136, 139
   window vinyl 144
Casework cabinet 203
   custom 201
   ground 81
   painting 178
   stain 178
   varnish 178
Casing bead 161
   wood 87
Cast in place concrete 24
   in place retaining walls 278
   iron bench 207
   iron damper 188
   iron fitting 224
   iron pipe 224
   iron pipe fitting 224
   iron radiator 243
   iron trap 225
   iron vent caps 225
Caster scaffold 8
Catch basin 289
   basin precast 290
   basin removal 12
   door 205
Caulking 119
   polyurethane 119
   sealant 119
Cavity wall insulation 103
   wall reinforcing 32
Cedar closet 91
   fence 277
   paneling 92
   post 94
   roof deck 83
   roof plank 83
   shingle 107
   siding 110
Ceiling 164
   acoustical 164
   beam 93
   board 164
   board acoustic 164
   board fiberglass 164
   bondcrete 158
   carrier 157
   demolition 154
   diffuser 236
   drilling 28
   drywall 160
   eggcrate 164
   framing 72
   furring 81, 155, 157
   heater 196
   insulation 103
   lath 157
   luminous 164, 260
   molding 88
   painting 182, 183
   plaster 158
   stair 196
   support 46
   suspended 157, 164
   tile 164
Cellar door 131
   wine 197
Cellulose blown in 103
   insulation 103

Cement adhesive 167
   concrete curbs 274
   flashing 100
   masonry 31
   masonry unit 34-36
   mortar 163, 345
   parging 101
   terrazzo portland 168
Central vacuum 194
Ceramic mulch 279
   tile 162, 163
   tile countertop 206
   tile demolition 154
   tile floor 162
Chain link fence 275, 276
   link fence paint 173
   link fence removal 13
Chair lifts 214
   molding 89
   rail demolition 66
Chamfer strip 22
Channel carrier 165
   furring 155, 162
   siding 110
   steel 157
Charge powder 45
Charges dump 15
Chemical anchor 28, 44
   dry extinguisher 189
Chime door 257
Chimney 33
   accessories 187
   brick 33
   demolition 30, 31
   flue 39
   foundation 24
   metal 187
   screen 187
   simulated brick 187
   vent 238
Chips wood 279
Chlorination system 230
Chute rubbish 15
Circline fixture 259
Circuit-breaker 257
Circulating pump 234
Cladding sheet metal 114
Clamp water pipe ground 248
Clapboard painting 174
Clay roofing tile 108
   tile 108
Clean control joint 26
Cleaning brick 30
   face brick 345
   masonry 30
   up 10
Cleanout door 188
   floor 216
   pipe 216
   tee 216
Clear and grub 266
   grub 266
Clearing brush 266
   selective 266
   site 266
Clip plywood 68
Clock timer 254
Closet cedar 91
   door 126, 128, 129
   pole 89
   rod 190
   water 227
Closets water 230
Clothes dryer residential 195
CMU 34
Coal tar pitch 112

Index

Coat glaze 182
Coating bituminous 101
   roof 100
   rubber 101
   silicone 101
   spray 101
   trowel 101
   wall 184
   water repellent 101
   waterproofing 101
Coil tie formwork 23
Colonial door 128, 130
   wood frame 94, 95
Column 93
   base 68
   bondcrete 158
   brick 33
   demolition 31
   drywall 160
   formwork 20
   formwork accessories 22
   lally 46
   laminated wood 86
   lath 157
   plaster 158
   plywood formwork 20
   removal 31
   round fiber tube formwork 20
   steel 47
   steel framed formwork 20
   wood 72, 82, 94
Columns framing 72
   lightweight 46
   structural 46
Combination device 251
   storm door 128, 130
Commercial gutting 15
Common brick 34
   nail 66, 67
Communal bath 211
Compact fill 268
Compaction 269
   soil 268
   structural 269
Compactor residential 195
Compartments shower 186
Compensation workers' ... 6, 340-342
Components furnace 240
Composite decking 76
   insulation 105
   rafter 76
Concrete block 34-36, 347
   block autoclaved 34
   block back-up 34
   block bond beam 35
   block decorative 35
   block demolition 14
   block exterior 35
   block foundation 35
   block grout 32
   block insulation 103
   block painting 182
   broom finish 26
   buggy 26
   bush hammer finish 26
   cart 26
   cast in place 24, 25
   coping 39
   curb 274
   curing 27
   cutout 14
   demolition 14
   drilling 28
   finish 273
   float finish 26
   floor patch 20

footing 25
formwork insert 22
furring 81
hand trowel finish 26
machine trowel finish 26
membrane curing 27
monolithic finish 26
paving 273
pipe 288, 289
placing 25
ready mix 25
removal 12
retaining wall 278
septic tank 289
shingle 108
sidewalk 272
slab 26
stair 25
stamping 27
tile 108
utility vault 288
wall 26
wall finish 26
water curing 27
wheeling 26
Condensing unit air cooled 241
Conditioner portable air 242
Conditioning ventilating air .. 242
Conductive floor 168
Conductor 247
   & grounding 247
   wire 247
Conduit & fitting flexible 250
   electrical 248
   fitting 248
   flexible metallic 249
   greenfield 249
   in slab 249
   in slab PVC 249
   in trench electrical 250
   in trench steel 250
   PVC 248
   rigid in slab 249
   rigid steel 248
Connection motor 250
Connector joist 68
   timber 68
Construction management fee 6
   temporary 10
   time 6
Continuous footing formwork 20
Control component 233
   damper volume 234
   joint 26
   joint clean 26
   joint sawn 26
   joint sealant 26
Cooking equipment 194
   range 194
Cooling & heating A/C 242
Coping 37, 39
   aluminum 39
   concrete 39
   removal 31
   terra cotta 39
Copper cable 247
   downspout 116
   drum trap 225
   DWV tubing 218
   fitting 218
   flashing 115
   gutter 117
   pipe 218
   rivet 45
   roof 114
   wall covering 170

wire 247
Cork floor 167
   tile 167, 168
   wall tile 170
Corner base cabinet 202
   bead 157
   wall cabinet 202
Cornerboards PVC 96
Cornice 37
   molding 88
   painting 181
Corrugated metal pipe 289
   roof tile 108
   siding 108, 109
Counter top 205, 206
   top demolition 66
Countertop backsplash 205
   sink 228
Cove base 167
   base ceramic tile 162
   base terrazzo 168
   molding 88
   molding scotia 88
Cover ground 281
   pool 210
Covering wall 352
CPVC pipe 220
Crew survey 10
Crown molding 88
Crushed stone sidewalk 272
Cubicle shower 228
Cultured stone 39
Cupola 191
Curb 274
   asphalt 274, 275
   concrete 274
   edging 275
   granite 275
   inlet 275
   precast 274
   removal 12
   roof 77
   terrazzo 168
Curing concrete 27
   paper 105
Curtain damper fire 234
   rod 187
Custom casework 201
Cut stone trim 38
Cutout counter 206
   demolition 14
   slab 14
Cutting block 206
Cylinder lockset 147

D

Damper fire curtain 234
   fireplace 188
   multi-blade 234
Deadbolt 147
Deciduous shrub 281, 282
   tree 282
Deck framing 74, 75
   roof 83, 84
   slab form 47
   wood 83
Decking 347
   composite 76
   form 47
   roof 47
Decorator device 251
   switch 251
Deep freeze 194
Dehumidifier 196

Delivery charge 9
Demo thermal & moist. protect. . 100
Demolition 14, 31, 154, 216, 266, 344
   baseboard 66
   boiler 232
   brick 31, 154
   building 13
   cabinet 66
   ceiling 154
   ceramic tile 154
   chimney 30, 31
   column 31
   concrete 14
   concrete block 14
   cutout 14
   door 123
   ductwork 232
   electric 246
   fireplace 31
   flooring 154
   framing 63
   furnace 232
   glass 123
   granite 31
   gutter 100
   house 13
   HVAC 232
   joist 63
   masonry 14, 30, 31
   metal stud 154
   millwork 66
   paneling 66
   partition 154
   pavement 12
   plaster 154
   plenum 154
   plumbing 216
   plywood 100, 154, 155
   post 64
   rafter 64
   railing 66
   roofing 100
   selective 62, 123
   siding 100
   site 12
   tile 154
   truss 66
   walls and partitions 154
   window 123
   wood 154
Detection system 264
Detector infra-red 264
   motion 264
   smoke 264
Device combination 251
   decorator 251
   GFI 252
   receptacle 252
   residential 250
   wiring 256, 257
Diffuser ceiling 236
   linear 236
   opposed blade damper 236
   perforated aluminum 236
   rectangular 236
   steel 236
   T-bar mount 236
Dimmer switch 250, 257
Disappearing stairs 196
Disconnect safety switch 258
Dishwasher 195
Disposal 14
   field 289
   garbage 195
Distribution box 289
   pipe water 287

Index

system gas 290
Diving board 210
Door accessories 150
   accordion 129
   and panels access 131
   and window matls 123
   bell residential 253
   bell system 257
   bi-fold 126, 128
   bi-passing closet 128
   birch 129
   blind 96
   bulkhead 131
   bumper 147
   cabinet 203, 204
   cafe 130
   canopy 190
   casing PVC 96
   catch 205
   chime 257
   cleanout 188
   closet 126, 129
   combination storm 128
   demolition 123
   double 126
   dutch 128
   dutch oven 188
   entrance 128
   fiberglass 130
   fire 125
   flush 128
   folding accordion 129
   frame 94, 124
   frame interior 95
   french 130
   garage 132
   glass 132
   handle 205
   hardware 147
   jamb molding 89
   kick plate 148
   knocker 147
   labeled 125
   metal 124
   mirror 151
   molding 89
   moulded 127
   opener 133
   overhead 132
   panel 127
   paneled 125
   passage 129, 131
   pre-hung 130
   removal 123
   residential 94, 125, 126, 128
   residential garage 132
   rough buck 73
   sauna 211
   sectional overhead 132
   shower 186
   side light 126, 130
   silencer 147
   sill 90, 94, 147
   sliding 132
   special 131
   stain 178
   steel 125, 349
   stop 147
   storm 124
   switch burglar alarm 264
   threshold 95
   trim vinyl 111
   varnish 178
   weatherstrip 149
   weatherstrip garage 149
   wood 127, 130

Doors & windows exterior paint . 176
   & windows interior paint 179
Dormer gable 81
Double hung window ... 136, 138-140
Dowel cap 23
   reinforcing 23
   sleeve 23
Downspout 116, 117
   aluminum 116
   copper 116
   elbow 117
   lead coated copper 116
   painting 181
   steel 117
   strainer 116
Drain floor 226
Drainage accessory 117
   pipe 288, 289
   site 290
   trap 225
Drains roof 226
Drapery hardware 200
   rings 200
Drawer track 205
   wood 204
Drill main 288
   rig 12
   tap main 288
   wood 68
Drilling ceiling 28
   concrete 28
   plaster 155
   steel 43
Drinking fountain support 230
Drip edge 118
   edge aluminum 118
Driveway 272
   removal 12
Drop pan ceiling 164
Drum trap copper 225
Dry fall painting 183
Dryer clothes 195
   receptacle 252
   vent 195
Drywall 160
   accessories 161
   column 160
   cutout 15
   demolition 154
   finish 352
   frame 124
   gypsum 160
   nail 66
   painting 182
   prefinished 161
   screw 162
Duck tarpaulin 10
Duct access door 234
   accessories 234
   flexible insulated 235
   flexible noninsulated 235
   HVAC 234
   insulation 232
   liner 235
   mechanical 234
Ductile iron fitting 287
   iron fitting mech joint ... 287
   iron pipe 287
Ductless split system 242
Ducts flexible 235
Ductwork aluminum 234
   demolition 232
   fabric coated flexible 235
   fabricated 234
   galvanized 234
   metal 234
   rectangular 234
   rigid 234
Dump charges 15
Dumpster 15
Duplex receptacle 257
Dust partition 15
Dutch door 128
   oven door 188
DWV ABS pipe 220
   pipe ABS 220
   PVC pipe 220
   tubing copper 218

E

Earth vibrator 269
Earthwork 267
   equipment 268
Eave vent 118
Edge drip 118
Edging 282
   aluminum 282
   curb 275
Eggcrate ceiling 164
EIFS 105
Elbow downspout 117
Electric appliance 194
   baseboard heater 243
   boiler 238
   cable 247, 248
   demolition 246
   fixture 259, 260
   furnace 239
   generator set 258
   heater 196
   heating 243
   lamp 260
   log 188
   metallic tubing 248
   pool heater 239
   service 257
   stair 196
   switch 256-258
   water heater 226
   wire 247
Electrical cabinet 249
   conduit 248
   conduit greenfield 249
Electronic air cleaner 238
Embossed metal door 131
   print door 127
Employer liability 6
EMT 248
Emulsion adhesive 167
   asphaltic 279
   pavement 272
   penetration 273
Enclosure bathtub 186
   shower 186
   swimming pool 212
Engineered stone countertops .. 207
Entrance cable service 247
   door 128, 130
   frame 94
   lock 147
   weather cap 248
Epoxy anchor 28, 44
   grout 162, 163
   wall coating 184
Equipment earthwork 268
   insulation 232
   insurance 6
   pad 24
Erosion control synthetic ... 269
Estimate electrical heating ... 243

Estimates window 349
Estimating 337
Evaporator coils cased 240
Evergreen shrub 281
   tree 281
Excavating bulk bank measure .. 268
Excavation 267, 268
   backhoe 267
   hand 267, 268
   planting pit 280
   septic tank 289
   structural 268
   trench 268
Exhaust hood 195
Expansion anchor 44
   joint 22, 158
   shield 44
   tank 234
   tank steel 234
Exposed aggregate 26
   aggregate coating 184
Exterior blind 95
   concrete block 35
   door frame 94
   insulation 105
   insulation finish system ... 105
   lighting fixture 260
   molding 88
   plaster 158
   pre-hung door 130
   PVC molding 96
   residential door 128
   wood frame 94
Extinguisher carbon dioxide .. 189
   chemical dry 189
   fire 189
   standard 189
Extractor air 234

F

Fabric welded wire 24
Fabricated ductwork 234
Face brick 34, 346
   brick cleaning 345
Facing stone 38
Fan attic ventilation 235
   house ventilation 235
   paddle 254
   residential 254
   ventilation 254
   wiring 254
Fascia aluminum 109
   board 76
   board demolition 63
   metal 116
   PVC 96
   vinyl 111
   wood 88
Fastener timber 66
   wood 67
Faucet & fitting 229
   bathroom 229
   laundry 229
   lavatory 229
Fee architectural 6, 337
Felt asphalt 112
   carpet pad 169
Fence and gate 275
   basketweave 277
   board 277
   board batten 278
   cedar 277
   chain link 275
   metal 275, 276

362

Index

open rail	277
picket paint	173
removal	13
residential	275
vinyl	276
wire	276
wood	277
Fences	173
plastic	277
Fiber cement shingle	106
cement siding	111
Fiberboard insulation	104
Fiberglass blown in	103
ceiling board	164
door	130
insulation	101-103
panel	108
planter	207
shower stall	186
tank	233
wool	103
Field disposal	289
Fieldstone	37
Fill	268
gravel	266, 268
Filler block	183
Film polyethylene	279
security	151
Filter air	238
mechanical media	237
swimming pool	230
Filtration equipment	210
Fine grade	267, 280
Finish concrete	273
drywall	160
floor	166, 184
grading	267
nail	66
Finishes wall	170
Fir column	93
floor	166
molding	89
roof deck	83
roof plank	83
Fire call pullbox	264
damage repair	173
damper curtain type	234
door	125
door frame	125
escape disappearing	196
escape stairs	58
extinguisher	189
horn	264
resistant drywall	160, 161
retardant lumber	70
retardant plywood	70
Fireplace accessory	187
box	39
built-in	188
damper	188
demolition	31
form	188
free standing	188
mantel	93
masonry	39
prefabricated	187, 188
Firestop gypsum	157
wood	73
Fitting cast iron	224
cast iron pipe	224
conduit	248
copper	218
ductile iron	287
DWV pipe	221
malleable iron	220
plastic	221

plastic pipe	221, 223
PVC	221
steel	220
Fixture bathroom	227, 228
carrier	230
electric	259
fluorescent	259
incandescent	259
interior light	259
landscape	260
lantern	259
mirror light	259
plumbing	227, 234
removal	216
residential	253, 259
Flagging	165, 274
slate	274
Flagpole	191
aluminum	191
Flashing	114, 115
aluminum	115
cement	100
copper	115
lead coated	115
masonry	115
stainless	115
vent	225
Flexible conduit & fitting	250
ducts	235
ductwork fabric coated	235
insulated duct	235
metallic conduit	249
noninsulated duct	235
Float finish concrete	26
glass	150
Floater equipment	6
Floating floor	167
pin	148
Floor	165
acid proof	165
asphalt block	274
brick	165
carpet	169
ceramic tile	162
cleanout	216
conductive	168
cork	167
drain	226
finish	184
flagging	274
framing removal	15
insulation	102
marble	38
nail	67
oak	166
paint	180
parquet	166
plank	83
plywood	83
polyethylene	170
quarry tile	163
register	237
removal	13
rubber	167
stain	181
subfloor	83
tile terrazzo	168
underlayment	84
varnish	181
vinyl	168
wood	166
wood composition	166
Flooring	165
demolition	154
Flue chimney	39
chimney metal	238

liner	33, 39
screen	187
tile	39
Fluorescent fixture	259
Flush door	128
Foam glass insulation	102
insulation	103
Fog seal	272
Foil aluminum	103, 105
back insulation	103
Folding accordion door	129
door shower	186
Footing concrete	25
keyway	20
keyway formwork	20
reinforcing	23
removal	14
spread	24
Form decking	47
fireplace	188
Formwork accessory	22, 23
coil tie	23
column	20
column plywood	20
column round fiber tube	20
column steel framed	20
continuous footing	20
footing keyway	20
grade beam	20
insulating concrete	21
slab blockout	21
slab bulkhead	20
slab curb	21
slab edge	21
slab on grade	20
slab void	21
sleeve	23
snap tie	23
spread footing	20
stairs	22
wall	21
wall boxout	21
wall brick shelf	21
wall bulkhead	21
wall plywood	21
Foundation chimney	24
concrete block	35
mat	25, 26
removal	13
wall	35
Fountain outdoor	210
Fountains yard	210
Frame baked enamel	125
door	94, 124
drywall	124
entrance	94
fire door	125
labeled	125
metal	124
metal butt	125
scaffold	8
steel	124
welded	125
window	138
wood	94, 203
Framing anchor	68
band joist	52
beam & girder	71
canopy	77
columns	72
deck	74, 75
demolition	63
heavy	82
joist	72
laminated	86
metal joist	53

metal partition	49
metal stud	49
partition	156
porch	74
removal	62
roof metal rafter	55
roof metal truss	57
roof rafters	76
roof soffits	56
sill	77
sleepers	77
timber	82
treated lumber	78
wall	78
wood	71, 82, 83
Freeze deep	194
French door	130
Frieze PVC	96, 97
Front end loader	268
Furnace A/C coils	240
components	240
demolition	232
electric	239
gas fired	239
hot air	239
oil fired	240
Furnishings site	207
Furring and lathing	155
ceiling	81, 155, 157
channel	155, 162
metal	155
steel	155
wall	81, 155

G

Gable dormer	81
Galvanized ductwork	234
Garage door	132
door weatherstrip	149
Garbage disposal	195
Gas conversion burner	239
distribution system	290
fired boiler	238
fired furnace	239
generator	258
heat air conditioner	241
log	188
pipe	290
vent	238
Gas/oil combination boiler	238
Gasoline generator	258
Gate metal	275, 276
General contractor's overhead	339
Generator emergency	258
gas	258
set	258
GFI receptacle	252
Girder wood	71, 83
Glass	150, 151
block	36
demolition	123
door	132
door shower	186
float	150
heat reflective	151
lined water heater	195
low emissivity	150
mirror	151
plate	151
shower stall	186
tempered	150
tile	151
tinted	150
window	151

363

Index

Glaze coat 182
Glazed ceramic tile 162
    wall coating 184
Glazing 150
Glued laminated 86
Grab bar 187
Grade beam formwork 20
    fine 267, 280
Grading 266
    slope 267
Granite building 37
    chips 279
    curb 275
    demolition 31
    indian 275
    paver 37
    sidewalk 274
Grass cloth wallpaper 170
    lawn 280
    seed 280
    sprinkler 279
Gravel bankrun 266
    base 272
    fill 266, 268
    pea 279
    stop 116
Gravity retaining wall 278
Greenfield conduit 249
Greenhouse 212
Grid spike 68
Grille air return 236
    aluminum 236
    decorative wood 93
    painting 181
    window 141
Ground 81
    clamp water pipe 248
    cover 281
    rod 248
    wire 248
Grounding 248
    & conductor 247
Grout 32
    concrete block 32
    epoxy 162, 163
    tile 162
    wall 32
Guard gutter 118
    snow 119
Guardrail scaffold 8
Guards snow 119
Gun type burner 239
Gutter 118
    aluminum 117
    copper 117
    demolition 100
    guard 118
    lead coated copper 118
    stainless 118
    steel 118
    strainer 118
    vinyl 118
    wood 118
Gutting 15
Gypsum block demolition 14
    board 160
    board accessories 161
    drywall 160
    fabric wallcovering 170
    firestop 157
    lath 157
    lath nail 67
    plaster 158
    restoration 154
    sheathing 85
    weatherproof 85

H

Half round molding 89
    round window vinyl 145
Half-round window 140
Hammer bush 26
Hand carved door 127
    clearing 266
    excavation 267, 268
    hole 288
    split shake 107
    trowel finish concrete 26
Handicap lever 147
    ramp 25
Handle door 205
Handling rubbish 15
Handrail wood 89, 93
Hanger beam 68
    joist 68
Hardboard cabinet 201
    overhead door 132
    paneling 90
    siding 110
    tempered 90
    underlayment 84
Hardware 147
    cabinet 205
    door 147
    drapery 200
    finish 147
    rough 70
    window 147
Hardwood floor 166
    grille 93
Hauling 269
    truck 269
Hay 279
Header wood 78, 79
Headers boxed beam 49, 52
Hearth 39
Heat baseboard 243
    electric baseboard 243
    pump 243
    pump residential 255
    pumps air-source 243
    pumps water-source 243
    reflective glass 151
    temporary 20
Heated ready-mix concrete 20
Heater air handler 241
    electric 196
    infrared quartz 244
    quartz 244
    sauna 211
    swimming pool 239
    water 195, 227
Heating 217, 232, 233, 239
    electric 243
    estimate electrical 243
    hot air 239
    hydronic 239
    panel radiant 244
Heavy duty shoring 7, 8
    framing 82
    timber 82
Hemlock column 94
Hex bolt steel 43
Highway paver 274
Hinges 148, 350
    brass 148
    cabinet 205
    residential 148
    steel 147, 148
Hip rafter 76
Holdown 68
Hollow core door 129
metal 124
    wall anchor 44
Hood range 195
Hook robe 187
Horizontal aluminum siding 109
    vinyl siding 111
Horn fire 264
Hose bibb sillcock 229
Hospital tip pin 148
Hot air furnace 239
    air heating 239
    tub 211
    water boiler 238, 239
    water heating 238, 243
House demolition 13
    ventilation fan 235
Housewrap 105
Humidifier 196
Humus peat 279
HVAC demolition 232
    duct 234
Hydronic heating 239

I

Icemaker 194
In slab conduit 249
Incandescent fixture 259
Indian granite 275
Infra-red detector 264
Infrared quartz heater 244
Inlet curb 275
Insecticide 269
Insert concrete formwork 22
Insulated panels 105
    protectors ADA 217
Insulating concrete formwork .. 21
Insulation 101, 103
    batt 102
    blanket 102, 232
    board 101
    building 103
    cavity wall 103
    cellulose 103
    composite 105
    duct 232
    equipment 232
    exterior 105
    fiberglass 101, 103
    finish system exterior 105
    foam 103
    foam glass 102
    insert 35
    isocyanurate 102
    masonry 103
    mineral fiber 102
    pipe 217
    polystyrene 102, 103
    removal 100
    roof 104
    roof deck 104
    vapor barrier 106
    vermiculite 103
    wall 35, 102
    water heater 217
Insurance 6, 339-342
    builder risk 6, 339
    equipment 6, 339
    public liability 6, 339
Interior door frame 95
    light fixture 259
    pre-hung door 131
    residential door 128
    shutter 201
    wood frame 95
Interlocking retaining walls 278
Interval timer 251
Intrusion system 264
Ironing center 187
Ironspot brick 165
Irrigation system 279
Isocyanurate insulation 102

J

Jack rafter 76
Joint control 26
    expansion 22, 158
    push-on 287
    reinforcing 32
    sealer 119
    tyton 287
Joist connector 68
    demolition 63
    framing 72
    hanger 68
    metal framing 53
    removal 63
    sister 73
    web stiffeners 53
    wood 72, 85
Jute fiber anchor 44
    mesh 269

K

Kennel fence 276
Keyway footing 20
Kick plate 148
    plate door 148
Kiln dried lumber 70
Kitchen appliance 194
    cabinet 201
    sink 228
    sink faucet 229
    unit 194
Knocker door 147

L

Labeled door 125
    frame 125
Ladder swimming pool 210
Lag screw 45
    screw shield 44
Lally column 46
Laminated beam 86
    countertop 206
    countertop plastic 206
    framing 86
    glued 86
    roof deck 83
    veneer members 87
    wood 86
Lamp post 59
Lampholder 257
Landfill fees 15
Landing newel 92
Landscape fixture 260
    light 260
    surface 170
Lantern fixture 259
Latch set 147
Latex caulking 119
    underlayment 167
Lath gypsum 157
Lath, plaster and gypsum board . 351
Lattice molding 89

Index

Lauan door ... 129
Laundry faucet ... 229
   sink ... 228
   tray ... 228
Lavatory ... 227
   faucet ... 229
   removal ... 216
   sink ... 227
   support ... 230
   vanity top ... 227
   wall hung ... 227
Lawn grass ... 280
Lazy susan ... 202
Leaching pit ... 289
Lead coated copper downspout ... 116
   coated copper gutter ... 118
   coated downspout ... 116
   coated flashing ... 115
   flashing ... 115
   paint encapsulation ... 16
   paint removal ... 16
   screw anchor ... 45
Lean-to type greenhouse ... 212
Let-in bracing ... 70
Letter slot ... 189
Leveling jack shoring ... 8
Lever handicap ... 147
Liability employer ... 6
   insurance ... 339
Lift ... 214
Light bollard ... 260
   fixture interior ... 259
   landscape ... 260
   post ... 259
   support ... 46
   track ... 260
Lighting ... 259, 260
   fixture exterior ... 260
   incandescent ... 259
   outlet ... 253
   residential ... 253
   strip ... 259
   track ... 259
Lightning arrester ... 258
   protection ... 258
   suppressor ... 250
Lightweight block ... 36
   columns ... 46
Limestone ... 37, 280
   coping ... 39
Linear diffuser ... 236
Linen wallcovering ... 170
Liner duct ... 235
   flue ... 39
Lintel ... 37
   block ... 35
   precast concrete ... 27
   steel ... 46
Load center indoor ... 255
   center plug-in breaker ... 255
   center rainproof ... 255
   residential ... 250
Loader front end ... 268
Loam ... 266
Lock entrance ... 147
Locking receptacle ... 253
Lockset cylinder ... 147
Log electric ... 188
   gas ... 188
   structures ... 82
Louver ... 152
   aluminum ... 152
   midget ... 152
   redwood ... 152
   ventilation ... 152
   wall ... 152

   wood ... 93
Louvered blind ... 95
   door ... 129, 131
Low-voltage silicon rectifier ... 257
   switching ... 256
   switchplate ... 257
   transformer ... 257
Lumber ... 71
   core paneling ... 91
   kiln dried ... 70
   product prices ... 348
   treatment ... 70
Luminous ceiling ... 164, 260
   panel ... 164

M

Macadam ... 273
   penetration ... 273
Machine trowel finish concrete ... 26
Mahogany door ... 127
Malleable iron fitting ... 220
   iron pipe fitting ... 220
Management fee construction ... 6
Manhole precast ... 290
   removal ... 13
Mantel beam ... 93
   fireplace ... 93
Maple countertop ... 206
Marble ... 38
   chips ... 279
   coping ... 39
   countertop ... 206
   floor ... 38
   sill ... 38
   synthetic ... 166
   tile ... 166
Mason scaffold ... 7
Masonry accessory ... 30
   anchor ... 32
   brick ... 33, 165
   cement ... 31
   cleaning ... 30
   cornice ... 37
   demolition ... 14, 30, 31
   fireplace ... 39
   flashing ... 115
   furring ... 81
   insulation ... 103
   nail ... 67
   painting ... 182
   pointing ... 30
   reinforcing ... 32, 345
   removal ... 13, 31
   sill ... 38
   step ... 272
   toothing ... 14
   wall ... 34, 279
   wall tie ... 32
Mat foundation ... 25, 26
Mechanical duct ... 234
   media filter ... 237
Medicine cabinet ... 187
Membrane curing concrete ... 27
   roofing ... 112
Metal bin retaining walls ... 278
   butt frame ... 125
   chimney ... 187
   door ... 124
   door residential ... 126
   ductwork ... 234
   faced door ... 130
   fascia ... 116
   fence ... 275, 276
   flue chimney ... 238

   frame ... 124
   framing parapet ... 55
   furring ... 155
   gate ... 275, 276
   hollow ... 124
   joist bracing ... 51
   joist bridging ... 51
   joist framing ... 53
   nailing anchor ... 44
   overhead door ... 132
   partition framing ... 49
   pipe removal ... 216
   rafter framing ... 55
   roof ... 108
   roof parapet framing ... 55
   roof window ... 145
   sheet ... 114
   shelf ... 189
   soffit ... 112
   stud demolition ... 154
   stud framing ... 49
   studs ... 49, 156
   support assemblies ... 155
   threshold ... 147
   tile ... 163
   truss framing ... 57
   window ... 133, 134
Metallic foil ... 103
Meter center rainproof ... 256
   device basic ... 256
   socket ... 255
   water supply ... 223
   water supply domestic ... 223
Microtunneling ... 287
Microwave oven ... 194
Mill construction ... 82
Millwork ... 87, 203
   demolition ... 66
Mineral fiber ceiling ... 164
   fiber insulation ... 102
   roof ... 113
   wool blown in ... 103
Minor site demolition ... 12
Mirror ... 151
   ceiling board ... 164
   door ... 151
   glass ... 151
   light fixture ... 259
   wall ... 151
Miscellaneous painting ... 173, 178
Mix planting pit ... 280
Mixing valve ... 229
Mobilization or demob. ... 9
Modified bitumen roofing ... 348
Modular air handler ... 240
Module tub-shower ... 228
Molding ... 87
   base ... 87
   brick ... 88
   ceiling ... 88
   chair ... 89
   cornice ... 88
   exterior ... 88
   hardboard ... 90
   pine ... 88
   trim ... 89
   window and door ... 89
   wood ... 92
   wood transition ... 167
Monel rivet ... 45
Monitor support ... 46
Monolithic finish concrete ... 26
Monument survey ... 12
Mortar, brick and block ... 346
Mortar cement ... 163, 345
   thinset ... 163

Moss peat ... 279
Motion detector ... 264
Motor connection ... 250
   support ... 46
Moulded door ... 127
Mounting board plywood ... 84
Movable louver blind ... 201
Moving shrub ... 283
   tree ... 283
Mulch bark ... 279
   ceramic ... 279
   stone ... 279
Mulching ... 279
Multi-blade damper ... 234
Muntin ... 135
   window ... 141

N

Nail ... 66, 67
Nailer steel ... 73
   wood ... 73
Nailing anchor ... 44
Newel ... 92
No hub pipe ... 224
   hub pipe fitting ... 224
Non-removable pin ... 148
Nylon carpet ... 91, 169
   nailing anchor ... 44

O

Oak door frame ... 94
   floor ... 166
   molding ... 89
   paneling ... 91
   stair tread ... 92
   threshold ... 95
Oil fired boiler ... 239
   fired furnace ... 240
   water heater ... 227
Olefin carpet ... 169
One piece astragal ... 148
One-way vent ... 119
Onyx ... 168
Open rail fence ... 277
Opener door ... 133
Outdoor cabinet ... 205
   fountain ... 210
Outlet box plastic ... 249
   box steel ... 249
   lighting ... 253
Oven ... 194
   cabinet ... 202
   microwave ... 194
Overhaul ... 15
Overhead contractor ... 339
   door ... 132
Oversized brick ... 33

P

P trap ... 225
   trap running ... 225
Pad equipment ... 24
Padding carpet ... 169
Paddle fan ... 254
Paint aluminum siding ... 175
   chain link fence ... 173
   doors & windows exterior ... 176
   doors & windows interior ... 179
   encapsulation lead ... 16
   fence picket ... 173

365

Index

floor 180
floor concrete 180
floor wood 180
removal 16, 17
siding 175
trim exterior 177
walls masonry, exterior 178
Painting 353
  balustrade 181
  casework 178
  ceiling 182, 183
  clapboard 174
  concrete block 182
  cornice 181
  decking 174
  downspout 181
  drywall 182
  grille 181
  masonry 182
  miscellaneous 173, 178
  pipe 181
  plaster 182
  railing 174
  shutters 174
  siding 174
  stair stringers 174
  steel siding 174
  stucco 174
  swimming pool 210
  trellis/lattice 174
  trim 181
  truss 182
  wall 175, 182, 183
  window 179
Paints & coatings 171
Palladian windows 140
Pan shower 116
Panel door 127, 128
  fiberglass 108
  luminous 164
  radiant heat 244
  structural 81
Paneled door 125
  pine door 129
Paneling 90
  board 92
  cutout 15
  demolition 66
  hardboard 90
  plywood 90
  wood 90
Panels insulated 105
  prefabricated 105
Paper building 105
  curing 27
  sheathing 106
Paperhanging 170
Parapet metal framing 55
Parging cement 101
Park bench 207
Parquet floor 166
Particle board siding 110
  board underlayment 84
Parting bead 89
Partition anchor 32
  block 35, 36
  demolition 154
  dust 15
  framing 156
  shower 186
  support 46
  wood frame 73
Passage door 129, 131
Patch concrete floor 20
  roof 100
Patio door 132

Pavement berm 275
  demolition 12
  emulsion 272
Paver floor 165
  highway 274
Paving 273
  brick 274
  concrete 273
Pea gravel 279
Peastone 281
Peat humus 279
  moss 279
Pegboard 90
Penetration macadam 273
Perforated aluminum pipe 289
  ceiling 164
  pipe 290
Perlite insulation 102
  plaster 158
Permit building 7
Picket fence 277
  fence vinyl 276
  railing 59
Picture window 136, 138
Pier brick 33
Pilaster wood column 94
Pile sod 280
Pin floating 148
  non-removable 148
  powder 45
Pine door 128
  door frame 94
  fireplace mantel 93
  floor 166
  molding 88
  roof deck 83
  shelving 190
  siding 110
  stair 92
  stair tread 92
Pipe & fittings 216, 218, 220-222, 354
  bedding 268
  cast iron 224
  cleanout 216
  concrete 288, 289
  copper 218
  corrugated metal 289
  covering 217
  covering fiberglass 217
  CPVC 220
  drainage 288, 289
  ductile iron 287
  DWV ABS 220
  DWV PVC 220
  fitting cast iron 224
  fitting copper 218
  fitting DWV 221
  fitting no hub 224
  fitting plastic 221-223
  fitting soil 224
  gas 290
  insulation 217
  no hub 224
  painting 181
  perforated aluminum 289
  plastic 220, 221
  polyethylene 290
  PVC 221, 288
  rail aluminum 58
  rail galvanized 58
  rail stainless 58
  rail steel 58
  rail wall 58
  railing 58
  removal 13
  removal metal 216

residential PVC 221
sewage 288
single hub 224
soil 224
steel 290
subdrainage 289
supply register spiral 237
Piping designations 355
Piping subdrainage 289
Pit leaching 289
Pitch coal tar 112
  emulsion tar 272
Placing concrete 25
Plank floor 83
  roof 83
  scaffolding 8
  sheathing 84
Plant bed preparation 280
Planter 207
  fiberglass 207
Planting 283
Plaster accessories 157
  beam 158
  ceiling 158
  column 158
  cutout 15
  demolition 154
  drilling 155
  ground 81
  gypsum 158
  painting 182
  perlite 158
  thincoat 159
  wall 158
Plasterboard 160
Plastic faced hardboard 90
  fences 277
  fitting 221
  laminated countertop 206
  outlet box 249
  pipe 220, 221
  pipe fitting 222, 223
  screw anchor 45
  skylight 146
  trap 225
  vent caps 225
  window 142
Plate glass 151
  shear 68
  steel 46, 47
  stiffener 47
  wall switch 257
Plating zinc 67
Players bench 207
Plenum demolition 154
Plugmold raceway 256
Plumbing 223
  demolition 216
  fixture 227, 234, 354
  fixtures removal 216
Plywood 348
  clip 68
  demolition 100, 154, 155
  floor 83
  joist 85
  mounting board 84
  paneling 90
  sheathing roof & walls 84
  shelving 190
  siding 110
  soffit 90
  subfloor 83
  treatment 70
  underlayment 84
Pointing masonry 30
Pole closet 89

Police connect panel 264
Polyethylene film 279
  floor 170
  pipe 290
  pool cover 210
  septic tank 289
  tarpaulin 10
Polystyrene blind 96
  ceiling 164
  ceiling panel 164
  insulation 102, 103
Polysulfide caulking 119
Polyurethane caulking 119
Polyvinyl soffit 90, 112
Pool accessory 210
  cover 210
  cover polyethylene 210
  heater electric 239
  swimming 210
Porcelain tile 162
Porch framing 74
  molding 88
Portland cement terrazzo 168
Post cap 68
  cedar 94
  demolition 64
  lamp 59
  light 259
  wood 72
Powder actuated tool 45
  charge 45
  pin 45
Power wiring 257
Precast catch basin 290
  concrete lintel 27
  concrete stairs 27
  concrete window sill 27
  coping 39
  curb 274
  manhole 290
  receptor 186
  terrazzo 168
Prefabricated building 212
  fireplace 187
  panels 105
  stair 92
Prefinished drywall 161
  floor 166
  shelving 190
Preformed roof panel 108
  roofing & siding 108
Pre-hung door 130
Preparation plant bed 280
  surface 171, 172
Pressure valve relief 216
  wash 172
Pretreatment termite 269
Prices lumber products 348
Prime coat 273
Primer asphalt 167
Privacy fence vinyl 276
Profile block 35
Projected window 133
Property line survey 12
Protection lightning 258
  slope 269
  termite 269
  winter 20
Protectors ADA insulated 217
P&T relief valve 216
Pull box 249
  door 205
Pump circulating 234
  heat 243
  staging 7, 343
  submersible 226

Index

sump 195
   water 288
Purlin roof 83
Push button lock 147
Push-on joint 287
Puttying 171
PVC conduit 248
   conduit in slab 249
   cornerboards 96
   door casing 96
   DWV pipe 220, 221
   fascia 96
   fitting 221
   frieze 96, 97
   gravel stop 116
   molding exterior 96
   pipe 220, 221, 288
   pipe residential 221
   rake 97
   siding 111
   soffit 97
   waterstop 23

Q

Quarry tile 163
Quarter round molding 89
Quartz 279
   heater 244
Quoins 37, 38

R

Raceway plugmold 256
   surface 256
   wiremold 256
Radial arch 86
Radiant heating panel 244
Radiator cast iron 243
Rafter 76
   anchor 68
   composite 76
   demolition 64
   framing metal 55
   metal bracing 55
   metal bridging 55
   wood 76
Rail aluminum pipe 58
   galvanized pipe 58
   stainless pipe 58
   steel pipe 58
   wall pipe 58
Railing demolition 66
   picket 59
   pipe 58
   wood 89, 92, 93
Railroad tie 272, 282
   tie step 272
Rainproof meter center 256
Rake PVC 97
Ramp handicap 25
Ranch plank floor 166
Range cooking 194
   hood 195
   receptacle 253, 257
Ready mix concrete 25
Ready-mix concrete heated 20
Receptacle air conditioner 252
   device 252
   dryer 252
   duplex 257
   GFI 252
   locking 253
   range 253, 257

telephone 253
television 253
weatherproof 252
Receptor precast 186
   shower 186
   terrazzo 186
Rectangular diffuser 236
   ductwork 234
Rectifier low-voltage silicon 257
Redwood bark mulch 279
   cupola 191
   louver 152
   paneling 92
   siding 110
   trim 88
   tub 211
   wine cellar 197
Refinish floor 166
Reflective block 36
   insulation 103
Refrigerated wine cellar 197
Refrigeration residential 194
Register air supply 236
   baseboard 236
Reinforcing dowel 23
   footing 23
   joint 32
   masonry 32
   steel fiber 24
   synthetic fiber 24
   wall 23
Relief valve temperature 216
Removal air conditioner 232
   bathtub 216
   block wall 14
   boiler 232
   catch basin 12
   chain link fence 13
   concrete 12
   curb 12
   driveway 12
   fence 13
   fixture 216
   floor 13
   foundation 13
   lavatory 216
   masonry 13
   paint 16
   pipe 13
   plumbing fixtures 216
   shingle 100
   sidewalk 13
   sink 216
   sod 280
   stone 13
   tree 266
   water closet 216
   water heater 216
   window 123
Repair fire damage 173
Repellent water 183
Replacement sash 136
   sliding door 132
   windows 122, 349
Residential alarm 254
   appliance 194, 254
   burner 239
   closet door 126
   device 250
   door 94, 125, 126, 128
   door bell 253
   elevator 214
   fan 254
   fence 275
   fixture 253, 259
   garage door 132

greenhouse 212
gutting 15
heat pump 255
hinge 148
lighting 253
load center 250
lock 147
overhead door 132
PVC pipe 221
refrigeration 194
service 250
smoke detector 254
stair 92
storm door 124
switch 250
water heater 226, 255
wiring 250, 254
wood window 134
Resilient base 167
   floor 167
Resquared shingle 107
Restoration gypsum 154
   window 17
Retaining wall 279
   wall segmental 278
   walls interlocking 278
   walls stone 279
Retarder vapor 105
Ribbed waterstop 23
Ridge board 77
   shingle 107
   shingle clay 108
   shingle slate 107
   vent 118, 119
Rig drill 12
Rigid anchor 32
   conduit in-trench 250
   in slab conduit 249
   insulation 102
Ring split 68
   toothed 69
Rings drapery 200
Riser wood stair 92
Rivet 45
   aluminum 45
   copper 45
   monel 45
   stainless 45
   steel 45
   tool 45
Road berm 275
Robe hook 187
Rod closet 190
   curtain 187
   ground 248
Roll roof 112
   roofing 113
   type air filter 237
Romex copper 247
Roof adhesive 112
   aluminum 108
   baffle 119
   beam 86
   built-up 112
   cant 77, 112
   clay tile 108
   coating 100
   copper 114
   deck 83
   deck insulation 104
   deck laminated 83
   deck wood 83
   decking 47
   drains 226
   fiberglass 108
   framing removal 15

insulation 104
metal 108
metal rafter framing 55
metal truss framing 57
mineral 113
modified bitumen 113
nail 67
panel preformed 108
patch 100
purlin 83
rafter 76
rafter bracing 55
rafter bridging 55
rafters framing 76
roll 112
safety anchor 9
sheathing 84
skylight 146
slate 106, 348
soffits framing 56
specialties prefab 116
truss 83, 85, 86
window 146
window metal 145
zinc 114
Roofing & siding preformed 108
   demolition 100
   membrane 112
   modified bitumen 113
   roll 113
Rooftop air conditioner 241
Rope safety line 9
Rosewood door 127
Rosin paper 106
Rough buck 73
   hardware 70
   stone wall 37
Rough-in sink countertop 228
   sink raised deck 228
   sink service floor 230
   tub 228
Round diffuser 236
Rubber base 167
   coating 101
   floor 167
   floor tile 168
   sheet 167
   threshold 147
   tile 168
Rubbish chute 15
   handling 15

S

Safety line rope 9
   switch 257
   switch disconnect 258
Sales tax 6, 338
Salt treatment lumber 70
Sand fill 266
Sanding 171
   floor 166
Sandstone 38
   flagging 274
Sanitary base cove 162
Sash replacement 136
   steel 134
   wood 138
Sauna 211
   door 211
Sawn control joint 26
Scaffold aluminum plank 9
   baseplate 8
   bracket 8
   caster 8

Index

frame ... 8
guardrail ... 8
mason ... 7
stairway ... 8
wood plank ... 8
Scaffolding ... 343
  plank ... 8
  tubular ... 7
Scrape after damage ... 173
Screen chimney ... 187
  fence ... 278
  molding ... 88
  security ... 134
  squirrel and bird ... 187
  window ... 134, 141
  wood ... 141
Screw anchor ... 44
  brass ... 67
  drywall ... 162
  lag ... 45
  sheet metal ... 67
  steel ... 67
  wood ... 67
Seal fog ... 272
  pavement ... 272
Sealant ... 101
  acoustical ... 161
  caulking ... 119
  control joint ... 26
Sealcoat ... 272
Sealer joint ... 119
Sectional door ... 132
  overhead door ... 132
Security film ... 151
  screen ... 134
Seeding ... 280, 354
  general ... 280
Segmental retaining wall ... 278
Selective clearing ... 266
  demolition ... 14, 30, 62, 123
Self-contained air conditioner ... 242
Septic system ... 289
  tank concrete ... 289
  tanks ... 289
Service electric ... 257
  entrance cable aluminum ... 247
  entrance cap ... 247
  residential ... 250
  sink ... 230
  sink faucet ... 229
Sewage pipe ... 288
Shadow box fence ... 277
Shake wood ... 107
Shear plate ... 68
  wall ... 84
Sheathed nonmetallic cable ... 247
  romex cable ... 247
Sheathing ... 84
  asphalt ... 85
  gypsum ... 85
  paper ... 106
  roof ... 84
  roof & walls plywood ... 84
Sheet metal ... 114
  metal aluminum ... 116
  metal cladding ... 114
  metal screw ... 67
Shelf metal ... 189
Shellac door ... 178
Shelter temporary ... 20
Shelving ... 190
  pine ... 190
  plywood ... 190
  prefinished ... 190
  storage ... 189
  wood ... 190

Shield expansion ... 44
  lag screw ... 44
Shingle ... 106
  asphalt ... 106
  concrete ... 108
  removal ... 100
  ridge ... 107
  stain ... 175
  strip ... 106
  wood ... 107
Shock absorber ... 223
Shoring baseplate ... 8
  bracing ... 8
  heavy duty ... 7, 8
  leveling jack ... 8
Shower by-pass valve ... 229
  compartments ... 186
  cubicle ... 228
  door ... 186
  enclosure ... 186
  glass door ... 186
  pan ... 116
  partition ... 186
  receptor ... 186
  stall ... 228
  surround ... 186
Shower/tub control set ... 229
Shower-tub valve spout set ... 229
Shrub broadleaf evergreen ... 281
  deciduous ... 281, 282
  evergreen ... 281
  moving ... 283
Shutter ... 201
  interior ... 201
  wood ... 201
Side light ... 95
  door ... 126, 130
Sidewalk ... 165, 272
  asphalt ... 272
  brick ... 274
  concrete ... 272
  removal ... 13
Sidewall bracket ... 8
Siding aluminum ... 109
  bevel ... 110
  cedar ... 110
  demolition ... 100
  fiber cement ... 111
  fiberglass ... 108
  hardboard ... 110
  nail ... 67
  paint ... 175
  painting ... 174
  plywood ... 110
  PVC ... 111
  redwood ... 110
  removal ... 100
  stain ... 174
  steel ... 109
  vinyl ... 111
  wood ... 110
  wood product ... 110
Silencer door ... 147
Silicone caulking ... 120
  coating ... 101
  water repellent ... 101
Sill ... 37, 77
  anchor ... 68
  door ... 90, 94, 147
  framing ... 77
  masonry ... 38
  precast concrete window ... 27
  quarry tile ... 163
Sillcock hose bibb ... 229
Silt fence ... 269
Simulated stone ... 39

Single hub pipe ... 224
  hung window ... 133
  zone rooftop unit ... 241
Sink ... 227
  base ... 202
  countertop ... 228
  countertop rough-in ... 228
  kitchen ... 228
  laundry ... 228
  lavatory ... 227
  raised deck rough-in ... 228
  removal ... 216
  service ... 230
  service floor rough-in ... 230
  slop ... 230
Siren ... 264
Sister joist ... 73
Site clearing ... 266
  demolition ... 12
  demolition minor ... 12
  drainage ... 290
  furnishings ... 207
  improvement ... 207, 279
  preparation ... 12
Skirtboard ... 92
Skylight ... 146
  removal ... 100
  roof ... 146
Slab blockout formwork ... 21
  bulkhead formwork ... 20
  concrete ... 26
  curb formwork ... 21
  cutout ... 14
  edge formwork ... 21
  on grade ... 25
  on grade formwork ... 20
  on grade removal ... 13
  textured ... 25
  void formwork ... 21
Slate ... 38
  flagging ... 274
  removal ... 100
  roof ... 106, 348
  shingle ... 106
  sidewalk ... 274
  sill ... 38
  stair ... 38
  tile ... 166
Slatwall ... 91
Sleeper ... 77
Sleepers framing ... 77
Sleeve and tap ... 287
  dowel ... 23
  formwork ... 23
Sliding door ... 132
  glass door ... 132
  mirror ... 187
  window ... 134, 138, 139
Slop sink ... 228, 230
Slope grading ... 267
  protection ... 269
Slot letter ... 189
Slotted pipe ... 289
Small tools ... 9
Smoke detector ... 264
Snap tie formwork ... 23
Snow guards ... 119
Soap holder ... 187
Socket meter ... 255
Sod ... 280
Sodding ... 280
Sodium low pressure fixture ... 260
Soffit ... 78
  aluminum ... 109
  drywall ... 160
  metal ... 112

plaster ... 158
plywood ... 90
PVC ... 97
vent ... 118
vinyl ... 111
wood ... 90
Softener water ... 226
Soil compaction ... 268
  pipe ... 224
  tamping ... 268
  treatment ... 269
Solid surface countertops ... 206
  wood door ... 127
Spa bath ... 211
Spanish roof tile ... 108
Special door ... 131
  systems ... 264
Spike grid ... 68
Spiral pipe supply register ... 237
  stair ... 59, 92
Split rib block ... 35
  ring ... 68
  system ductless ... 242
Spray coating ... 101
Spread footing ... 24
  footing formwork ... 20
Sprinkler grass ... 279
Staging aids ... 8
  pump ... 7
Stain cabinet ... 178
  casework ... 178
  door ... 178
  floor ... 181
  lumber ... 86
  shingle ... 175
  siding ... 110, 174
  truss ... 182
Stainless flashing ... 115
  gutter ... 118
  rivet ... 45
  steel gravel stop ... 116
  steel hinge ... 148
Stair ... 92
  basement ... 27, 92
  carpet ... 170
  ceiling ... 196
  climber ... 214
  concrete ... 25
  electric ... 196
  prefabricated ... 92
  railroad tie ... 272
  removal ... 65
  residential ... 92
  slate ... 38
  spiral ... 59, 92
  stringer ... 73
  stringer wood ... 73
  tread tile ... 163
  tread wood ... 92
  wood ... 92
Stairs disappearing ... 196
  fire escape ... 58
  formwork ... 22
  precast concrete ... 27
Stairway scaffold ... 8
Stairwork and handrails ... 93
Stall shower ... 228
Stamping concrete ... 27
  texture ... 27
Standard extinguisher ... 189
Starting newel ... 92
Steam bath ... 211
  bath residential ... 211
  boiler ... 238, 239
  boiler electric ... 238
Steel anchor ... 32

Index

beam	47	wood burning	189
bin wall	278	Strainer downspout	116
bolt	68	gutter	118
bridging	71	roof	116
chain link	275	wire	117
channel	157	Strap tie	69
column	47	Straw	279
conduit in slab	249	Stringer stair	73
conduit in trench	250	Strip chamfer	22
conduit rigid	248	floor	166
diffuser	236	footing	24, 25
door	125, 349	lighting	259
downspout	117	shingle	106
drilling	43	soil	266
edging	272, 283	Stripping topsoil	266
expansion tank	234	Structural columns	46
fiber reinforcing	24	excavation	268
fitting	220	insulated panel	81
flashing	115	panel	81
frame	124	steel members	47
furring	155	Structures log	82
gravel stop	116	Stucco	158
gutter	118	painting	174
hex bolt	43	Stud demolition	65
hinge	147, 148	metal	156
lintel	46	partition	74
members structural	47	steel	156
nailer	73	wall	74, 156
pipe	290	wall bracing	48
plate	46, 47	wall bridging	48
rivet	45	Studs metal	49
sash	134	Subdrainage pipe	289
screw	67	piping	289
siding	109	system	289
stud	156	Subfloor	83
window	134	plywood	83
Steeple tip pin	148	Submersible pump	226, 288
Step bluestone	272	sump pump	226
brick	272	Sump pump	195
masonry	272	pump submersible	226
railroad tie	272	Support ceiling	46
stone	37	drinking fountain	230
Stiffener plate	47	lavatory	230
Stiffeners joist web	53	light	46
Stockade fence	277	monitor	46
Stockpiling of soil	266	motor	46
Stone aggregate	281	partition	46
anchor	32	X-ray	46
ashlar	38	Suppressor lightning	250
base	38, 272	Surface countertops solid	206
countertops engineered	207	landscape	170
cultured	39	preparation	171, 172
curbs	275	raceway	256
dust	272	Surfacing	272
fill	266	Surround shower	186
floor	38	tub	186
ground cover	281	Survey crew	10
mulch	279	monument	12
paver	37, 274	property line	12
removal	13	topographic	12
retaining walls	279	Suspended ceiling	157, 164
simulated	39	Suspension system ceiling	165
step	37	Swimming pools	210, 353
trim cut	38	pool enclosure	212
wall	279	pool heater	239
Stool cap	89	pool ladder	210
window	38	Swing check valve	216
Stop door	89	check valve bronze	216
gravel	116	Swing-up overhead door	133
Storage shelving	189	Switch box	249
tank	233	box plastic	249
tanks underground	233	decorator	251
Storm door	124	dimmer	250, 257
drainage manholes frames	290	electric	256-258
window	142	general duty	257
Stove	189	residential	250

safety	257	railroad	272, 282
time	258	strap	69
toggle	257	wall	32
Switching low-voltage	256	Tile	162, 167
Switchplate low-voltage	257	carpet	169
Synthetic erosion control	269	ceiling	164
fiber reinforcing	24	ceramic	162, 163
marble	166	clay	108
System antenna	262	concrete	108
ceiling suspension	165	cork	168
irrigation	279	cork wall	170
septic	289	demolition	154
subdrainage	289	flue	39
T.V.	262	glass	151
VHF	262	grout	162
		marble	166
T		metal	163
		porcelain	162
Tamping soil	268	quarry	163
Tank expansion	234	rubber	168
fiberglass	233	slate	166
septic	289	stainless steel	163
storage	233	stair tread	163
Tap and sleeve	287	vinyl	168
Tapping main	288	wall	162
Tar paper	279	window sill	163
pitch emulsion	272	Timber connector	68
Tarpaulin	10	fastener	66
duck	10	framing	82
polyethylene	10	heavy	82
Taxes	6, 338	laminated	86
sales	6, 338	Time switch	258
social security	6, 338	Timer clock	254
unemployment	6, 338	interval	251
T-bar mount diffuser	236	switch ventilator	236
Teak floor	166	Tinted glass	150
molding	88, 89	Toggle switch	257
paneling	91	Toggle-bolt anchor	44
Tee cleanout	216	Toilet accessories	187
Telephone receptacle	253	bowl	227
Television receptacle	253	Tool powder actuated	45
Temperature relief valve	216	rivet	45
Tempered glass	150	Tools small	9
hardboard	90	Toothed ring	69
Tempering valve	216	Toothing masonry	14
valve water	216	Top demolition counter	66
Temporary construction	10	dressing	280
heat	20	vanity	206
shelter	20	Topographic survey	12
Tennis court fence	276	Topsoil	266, 280
Terminal A/C packaged	242	stripping	266
Termite pretreatment	269	Towel bar	187
protection	269	Track drawer	205
Terne coated flashing	115	light	260
Terra cotta coping	39	lighting	259
cotta demolition	14	traverse	200
Terrazzo precast	168	Tractor	267
receptor	186	Transfer switch automatic	258
wainscot	169	Transformer low-voltage	257
Texture stamping	27	Transition molding wood	167
Textured slab	25	Transom lite frame	125
Thermal & moist. protect. demo	100	windows	140
Thermostat	233	Trap cast iron	225
integral	244	drainage	225
wire	255	P	225
Thincoat plaster	159	plastic	225
Thinset ceramic tile	162	Trapezoid windows	141
mortar	163	Traverse	200
Threshold	147	track	200
door	95	Travertine	38
stone	38	Tray laundry	228
wood	89	Tread abrasive	163
Thru-wall air conditioner	242	stone	38
Tie formwork snap	23	wood	92
rafter	77	Treated lumber framing	78

369

Index

Treatment lumber 70
   plywood 70
   wood 70
Tree 282
   deciduous 282
   evergreen 281
   guying 283
   moving 283
   removal 266
Trench backfill 268
   excavation 268
   utility 268
Trim exterior 88
   painting 181
   redwood 88
   tile 162
   wood 88
Trowel coating 101
Truck hauling 269
   loading 268
Truss bowstring 86
   demolition 66
   flat wood 85
   framing metal 57
   painting 182
   plate 69
   roof 83, 85, 86
   stain 182
   varnish 182
Tub hot 211
   redwood 211
   rough-in 228
   surround 186
Tubing copper 218
   electric metallic 248
Tub-shower module 228
Tubular scaffolding 7
   steel joist 85
Tumbler holder 187
Turned column 94
TV antenna 262
   system 262
Tyton joint 287

U

Undereave vent 152
Underground storage tanks .. 233
Underlayment 84
   latex 167
Unemployment taxes 6, 338
Unit kitchen 194
Utility boxes 288
   trench 268

V

Vacuum breaker 223
   central 194
   cleaning 194
Valance board 202
Valley rafter 76
Valve bronze 216
   mixing 229
   relief pressure 216
   shower by-pass 229
   spout set shower-tub ... 229
   swing check 216
   tempering 216
   water pressure 216
Vanity base 205
   top 206
   top lavatory 227
Vapor barrier 105

barrier sheathing 85
retarder 105
Varnish cabinet 178
   casework 178
   door 178
   floor 181, 184
   truss 182
VCT removal 154
Veneer brick 33
   core paneling 91
   members laminated 87
   removal 31
Venetian blind 200
Vent caps 225
   caps cast iron 225
   caps plastic 225
   chimney 238
   dryer 195
   eave 118
   flashing 225
   gas 238
   one-way 119
   ridge 118, 119
   ridge strip 152
   soffit 118
Ventilating air conditioning . 235, 242
Ventilation fan 254
   louver 152
Ventilator timer switch 236
Verge board 88
Vermiculite insulation 103
Vertical aluminum siding ... 109
   vinyl siding 111
VHF system 262
Vibrator earth 269
Vinyl blind 96
   casement window 144
   chain link 276
   composition floor 168
   door trim 111
   downspout 117
   faced wallboard 160
   fascia 111
   fence 276
   floor 168
   gutter 118
   half round window 145
   siding 111
   siding accessories 111
   soffit 111
   tile 168
   wallpaper 170
   window 143, 145
   window solid 142
   window trim 111
Volume control damper 234

W

Wainscot ceramic tile 162
   molding 89
   quarry tile 163
   terrazzo 169
Walk 272
Wall boxout formwork 21
Wall brick 37
Wall brick shelf formwork ... 21
   bulkhead formwork 21
   bumper 147
   cabinet 202
   ceramic tile 163
   coating 184
   concrete 26
   covering 352
   cutout 14, 15

drywall 160
finish concrete 26
finishes 170
formwork 21
formwork accessories 23
foundation 35
framing 78
framing removal 15
furring 81, 155
grout 32
heater 196
hung lavatory 227
insulation 35, 101-103
lath 157
louver 152
masonry 34, 279
mirror 151
painting 175, 182, 183
paneling 91
plaster 158
plywood formwork 21
reinforcing 23
removal 14
retaining 279
shear 84
sheathing 84, 85
siding 110
steel bin 278
stucco 158
stud 74, 156
switch plate 257
tie 32
tie masonry 32
tile 162
tile cork 170
Wallboard acoustical 161
Wallcovering 170
   acrylic 170
   gypsum fabric 170
Wallpaper 170, 352
   grass cloth 170
   vinyl 170
Walls and partitions demolition . 154
Walnut door frame 94
   floor 166
Wardrobe 190
   wood 203
Wash bowl 227
   brick 30
Washable air filters 237
Washer 69
   residential 195
Washing machine automatic .. 195
Water closets 227, 230
   closet removal 216
   closet support 230
   curing concrete 27
   distribution pipe 287
   hammer arrester 223
   heater 195, 227, 255
   heater electric 226
   heater insulation 217
   heater oil 227
   heater removal 216
   heater residential 226
   heater wrap kit 217
   heating hot 238, 243
   pipe ground clamp 248
   pressure relief valve .. 216
   pressure valve 216
   pump 195, 234, 288
   repellent 183
   repellent coating 101
   repellent silicone 101
   softener 226
   supply domestic meter .. 223

supply meter 223
tempering valve 216
well 288
Waterproofing coating 101
   demolition 100
Water-source heat pumps 243
Waterstop 23
   PVC 23
   ribbed 23
Weather cap entrance 248
Weatherproof receptacle 252
Weatherstrip 149
   door 149
Weatherstripping 148
Weathervanes 191
Welded frame 125
   wire fabric 24
Well 288
   water 288
Whirlpool bath 211
Window 133, 146
   air conditioner 242
   aluminum 133
   and door molding 89
   awning 135
   basement/utility 134
   bay 135
   blind 95, 200
   casement 136, 139
   casing 88
   demolition 123
   double hung 138, 140
   estimates 349
   frame 138
   glass 151
   grille 141
   half-round 140
   hardware 147
   metal 133, 134
   muntin 141
   painting 179
   picture 138
   plastic 142
   precast concrete sill .. 27
   removal 123
   restoration 17
   roof 146
   screen 134, 141
   sill 38
   sill marble 38
   sill tile 163
   sliding 138
   solid vinyl 142
   steel 134
   stool 38
   storm 142
   trim set 90
   trim vinyl 111
   vinyl 143, 145
   wood 134-136, 138, 139
Windows casement 136
   transom 140
   trapezoid 141
Wine cellar 197
Winter protection 20
Wire copper 247
   electric 247
   fence 276
   fences misc. metal 276
   ground 248
   mesh 158
   strainer 117
   thermostat 255
   THW 247
Wiremold raceway 256
Wiring air conditioner 255

Index

device 256, 257
  fan 254
  power 257
  residential 250, 254
Wood base 87
  beam 71, 82
  blind 95, 201
  block floor demolition 154
  blocking 74
  canopy 77
  casing 87
  chips 279
  column 72, 82, 94
  cupola 191
  deck 83
  demolition 154
  door 127, 128, 130
  drawer 204
  fascia 88
  fastener 67
  fence 277
  fiber ceiling 164
  fiber sheathing 85
  fiber underlayment 84
  floor 166
  floor demolition 154
  frame 94, 203
  framing 71
  furring 81
  girder 71
  gutter 118
  handrail 89
  joist 72, 85
  laminated 86
  louver 93
  molding 92
  nailer 73
  overhead door 132
  panel door 127
  paneling 90
  parquet 166
  partition 74
  plank scaffold 8
  planter 207
  product siding 110
  rafter 76
  railing 92, 93
  roof deck 83
  roof deck demolition 100
  sash 138
  screen 141
  screw 67
  shake 107
  sheathing 84
  shelving 190
  shingle 107
  sidewalk 272
  siding 110
  siding demolition 100
  sill 77
  soffit 90
  stair 92
  stair stringer 73
  storm door 130
  subfloor 84
  threshold 89
  tread 92
  treatment 70
  trim 88
  truss 85
  veneer wallpaper 170
  wardrobe 203
  window 134-136, 138, 139
  window demolition 123
Wool fiberglass 103
Workers' compensation . . 6, 340-342

X

X-ray support 46

Y

Yard fountains 210
Yellow pine floor 166

Z

Z bar suspension 165
Zee bar 162
Zinc plating 67
  roof 114
  weatherstrip 149

Contractor's Pricing Guides

For more information visit RSMeans Web site at www.rsmeans.com

Contractor's Pricing Guide: Residential Detailed Costs 2007

Every aspect of residential construction, from overhead costs to residential lighting and wiring, is in here. All the detail you need to accurately estimate the costs of your work with or without markups—labor-hours, typical crews, and equipment are included as well. When you need a detailed estimate, this publication has all the costs to help you come up with a complete, on the money, price you can rely on to win profitable work.

Unit Prices Now Updated to MasterFormat 2004!

$39.95 per copy
Available Nov. 2006
Catalog No. 60337

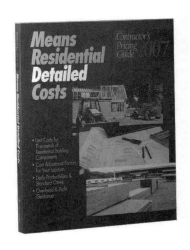

Contractor's Pricing Guide: Residential Repair & Remodeling Costs 2007

This book provides total unit price costs for every aspect of the most common repair and remodeling projects. Organized in the order of construction by component and activity, it includes demolition and installation, cleaning, painting, and more.

With simplified estimating methods, clear, concise descriptions, and technical specifications for each component, the book is a valuable tool for contractors who want to speed up their estimating time, while making sure their costs are on target.

$39.95 per copy
Available Nov. 2006
Catalog No. 60347

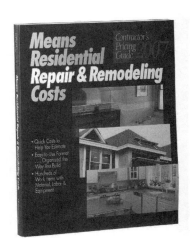

Contractor's Pricing Guide: Residential Square Foot Costs 2007

Now available in one concise volume, all you need to know to plan and budget the cost of new homes. If you are looking for a quick reference, the model home section contains costs for over 250 different sizes and types of residences, with hundreds of easily applied modifications. If you need even more detail, the Assemblies Section lets you build your own costs or modify the model costs further. Hundreds of graphics are provided, along with forms and procedures to help you get it right.

$39.95 per copy
Available Nov. 2006
Catalog No. 60327

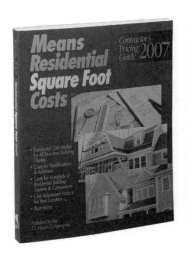

Annual Cost Guides

For more information visit RSMeans Web site at www.rsmeans.com

RSMeans Building Construction Cost Data 2007

Available in Both Softbound and Looseleaf Editions

Many customers enjoy the convenience and flexibility of the looseleaf binder, which increases the usefulness of *RSMeans Building Construction Cost Data 2007* by making it easy to add and remove pages. You can insert your own cost information pages, so everything is in one place. Copying pages for faxing is easier also. Whichever edition you prefer, softbound or the convenient looseleaf edition, you'll be eligible to receive *RSMeans Quarterly Update Service* FREE. Current subscribers can receive *RSMeans Quarterly Update Service* via e-mail.

Unit Prices Now Updated to MasterFormat 2004!

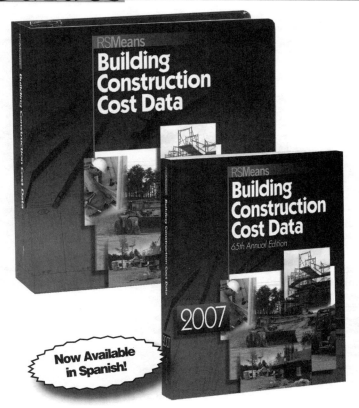

Now Available in Spanish!

$169.95 per copy, Looseleaf
Available Oct. 2006
Catalog No. 61017

RSMeans Building Construction Cost Data 2007

Offers you unchallenged unit price reliability in an easy-to-use arrangement. Whether used for complete, finished estimates or for periodic checks, it supplies more cost facts better and faster than any comparable source. Over 20,000 unit prices for 2007. The City Cost Indexes and Location Factors cover over 930 areas, for indexing to any project location in North America. Order and get *RSMeans Quarterly Update Service* FREE. You'll have year-long access to the RSMeans Estimating **HOTLINE** FREE with your subscription. Expert assistance when using RSMeans data is just a phone call away.

$136.95 per copy (English or Spanish)
Catalog No. 60017 (English) Available Oct. 2006
Catalog No. 60717 (Spanish) Available Jan. 2007

Now Available in Spanish!

Unit Prices Now Updated to MasterFormat 2004!

RSMeans Metric Construction Cost Data 2007

A massive compendium of all the data from both the 2007 *RSMeans Building Construction Cost Data* AND *Heavy Construction Cost Data*, in **metric** format! Access all of this vital information from one complete source. It contains more than 600 pages of unit costs and 40 pages of assemblies costs. The Reference Section contains over 200 pages of tables, charts and other estimating aids. A great way to stay in step with today's construction trends and rapidly changing costs.

$162.95 per copy
Available Dec. 2006
Catalog No. 63017

For more information visit RSMeans Web site at www.rsmeans.com

Annual Cost Guides

RSMeans Mechanical Cost Data 2007

- **HVAC**
- **Controls**

Total unit and systems price guidance for mechanical construction... materials, parts, fittings, and complete labor cost information. Includes prices for piping, heating, air conditioning, ventilation, and all related construction.

Plus new 2007 unit costs for:
- Over 2500 installed HVAC/controls assemblies
- "On Site" Location Factors for over 930 cities and towns in the U.S. and Canada
- Crews, labor, and equipment

$136.95 per copy
Available Oct. 2006
Catalog No. 60027

Unit Prices Now Updated to MasterFormat 2004!

RSMeans Plumbing Cost Data 2007

Comprehensive unit prices and assemblies for plumbing, irrigation systems, commercial and residential fire protection, point-of-use water heaters, and the latest approved materials. This publication and its companion, *RSMeans Mechanical Cost Data*, provide full-range cost estimating coverage for all the mechanical trades.

New for '07: More lines of no-hub CI soil pipe fittings, more flange-type escutcheons, fiberglass pipe insulation in a full range of sizes for 2-1/2" and 3" wall thicknesses, 220 lines of grease duct, and much more.

$136.95 per copy
Available Oct. 2006
Catalog No. 60217

RSMeans Electrical Cost Data 2007

Pricing information for every part of electrical cost planning. More than 13,000 unit and systems costs with design tables; clear specifications and drawings; engineering guides; illustrated estimating procedures; complete labor-hour and materials costs for better scheduling and procurement; and the latest electrical products and construction methods.
- A variety of special electrical systems including cathodic protection
- Costs for maintenance, demolition, HVAC/mechanical, specialties, equipment, and more

$136.95 per copy
Available Oct. 2006
Catalog No. 60037

Unit Prices Now Updated to MasterFormat 2004!

RSMeans Electrical Change Order Cost Data 2007

RSMeans Electrical Change Order Cost Data 2007 provides you with electrical unit prices exclusively for pricing change orders—based on the recent, direct experience of contractors and suppliers. Analyze and check your own change order estimates against the experience others have had doing the same work. It also covers productivity analysis and change order cost justifications. With useful information for calculating the effects of change orders and dealing with their administration.

$136.95 per copy
Available Nov. 2006
Catalog No. 60237

RSMeans Square Foot Costs 2007

It's Accurate and Easy To Use!

- **Updated 2007 price information**, based on nationwide figures from suppliers, estimators, labor experts, and contractors
- "How-to-Use" sections, with **clear examples** of commercial, residential, industrial, and institutional structures
- Realistic graphics, offering true-to-life illustrations of building projects
- Extensive information on using square foot cost data, including sample estimates and alternate pricing methods

$148.95 per copy
Over 450 pages, illustrated, available Nov. 2006
Catalog No. 60057

RSMeans Repair & Remodeling Cost Data 2007

Commercial/Residential

Use this valuable tool to estimate commercial and residential renovation and remodeling.

Includes: New costs for hundreds of unique methods, materials, and conditions that only come up in repair and remodeling, PLUS:
- Unit costs for over 15,000 construction components
- Installed costs for over 90 assemblies
- Over 930 "On-Site" localization factors for the U.S. and Canada.

Unit Prices Now Updated to MasterFormat 2004!
$116.95 per copy
Available Nov. 2006
Catalog No. 60047

Annual Cost Guides

For more information visit RSMeans Web site at www.rsmeans.com

RSMeans Facilities Construction Cost Data 2007

For the maintenance and construction of commercial, industrial, municipal, and institutional properties. Costs are shown for new and remodeling construction and are broken down into materials, labor, equipment, overhead, and profit. Special emphasis is given to sections on mechanical, electrical, furnishings, site work, building maintenance, finish work, and demolition.

More than 43,000 unit costs, plus assemblies costs and a comprehensive Reference Section are included.

$323.95 per copy
Available Dec. 2006
Catalog No. 60207

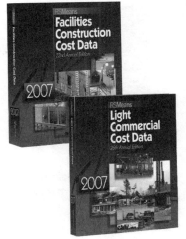

Unit Prices Now Updated to MasterFormat 2004!

RSMeans Light Commercial Cost Data 2007

Specifically addresses the light commercial market, which is a specialized niche in the construction industry. Aids you, the owner/designer/contractor, in preparing all types of estimates—from budgets to detailed bids. Includes new advances in methods and materials.

Assemblies Section allows you to evaluate alternatives in the early stages of design/planning.

Over 11,000 unit costs for 2007 ensure you have the prices you need... when you need them.

$116.95 per copy
Available Dec. 2006
Catalog No. 60187

RSMeans Residential Cost Data 2007

Contains square foot costs for 30 basic home models with the look of today, plus hundreds of custom additions and modifications you can quote right off the page. With costs for the 100 residential systems you're most likely to use in the year ahead. Complete with blank estimating forms, sample estimates, and step-by-step instructions.

Now contains line items for cultured stone and brick, PVC trim lumber, and TPO roofing.

$116.95 per copy
Available Oct. 2006
Catalog No. 60177

Unit Prices Now Updated to MasterFormat 2004!

RSMeans Site Work & Landscape Cost Data 2007

Includes unit and assemblies costs for earthwork, sewerage, piped utilities, site improvements, drainage, paving, trees & shrubs, street openings/repairs, underground tanks, and more. Contains 57 tables of Assemblies Costs for accurate conceptual estimates.

2007 update includes:
- Estimating for infrastructure improvements
- Environmentally-oriented construction
- ADA-mandated handicapped access
- Hazardous waste line items

$136.95 per copy
Available Nov. 2006
Catalog No. 60287

RSMeans Assemblies Cost Data 2007

RSMeans Assemblies Cost Data 2007 takes the guesswork out of preliminary or conceptual estimates. Now you don't have to try to calculate the assembled cost by working up individual component costs. We've done all the work for you.

Presents detailed illustrations, descriptions, specifications, and costs for every conceivable building assembly—240 types in all—arranged in the easy-to-use UNIFORMAT II system. Each illustrated "assembled" cost includes a complete grouping of materials and associated installation costs, including the installing contractor's overhead and profit.

$223.95 per copy
Available Oct. 2006
Catalog No. 60067

RSMeans Open Shop Building Construction Cost Data 2007

The latest costs for accurate budgeting and estimating of new commercial and residential construction... renovation work... change orders... cost engineering.

RSMeans Open Shop "BCCD" will assist you to:
- Develop benchmark prices for change orders
- Plug gaps in preliminary estimates and budgets
- Estimate complex projects
- Substantiate invoices on contracts
- Price ADA-related renovations

Unit Prices Now Updated to MasterFormat 2004!

$136.95 per copy
Available Dec. 2006
Catalog No. 60157

For more information
visit RSMeans Web site
at www.rsmeans.com

Annual Cost Guides

RSMeans Building Construction Cost Data 2007
Western Edition

This regional edition provides more precise cost information for western North America. Labor rates are based on union rates from 13 western states and western Canada. Included are western practices and materials not found in our national edition: tilt-up concrete walls, glu-lam structural systems, specialized timber construction, seismic restraints, and landscape and irrigation systems.

$136.95 per copy
Available Dec. 2006
Catalog No. 60227

Unit Prices Now Updated to MasterFormat 2004!

RSMeans Heavy Construction Cost Data 2007

A comprehensive guide to heavy construction costs. Includes costs for highly specialized projects such as tunnels, dams, highways, airports, and waterways. Information on labor rates, equipment, and material costs is included. Features unit price costs, systems costs, and numerous reference tables for costs and design.

$136.95 per copy
Available Dec. 2006
Catalog No. 60167

RSMeans Construction Cost Indexes 2007

Who knows what 2007 holds? What materials and labor costs will change unexpectedly? By how much?

- Breakdowns for 316 major cities
- National averages for 30 key cities
- Expanded five major city indexes
- Historical construction cost indexes

$294.00 per year (subscription)
$73.50 individual quarters
Catalog No. 60147 A,B,C,D

RSMeans Interior Cost Data 2007

Provides you with prices and guidance needed to make accurate interior work estimates. Contains costs on materials, equipment, hardware, custom installations, furnishings, and labor costs... for new and remodel commercial and industrial interior construction, including updated information on office furnishings, and reference information.

Unit Prices Now Updated to MasterFormat 2004!

$136.95 per copy
Available Nov. 2006
Catalog No. 60097

RSMeans Concrete & Masonry Cost Data 2007

Provides you with cost facts for virtually all concrete/masonry estimating needs, from complicated formwork to various sizes and face finishes of brick and block—all in great detail. The comprehensive unit cost section contains more than 7,500 selected entries. Also contains an Assemblies Cost section, and a detailed Reference section that supplements the cost data.

$124.95 per copy
Available Dec. 2006
Catalog No. 60117

Unit Prices Now Updated to MasterFormat 2004!

RSMeans Labor Rates for the Construction Industry 2007

Complete information for estimating labor costs, making comparisons, and negotiating wage rates by trade for over 300 U.S. and Canadian cities. With 46 construction trades listed by local union number in each city, and historical wage rates included for comparison. Each city chart lists the county and is alphabetically arranged with handy visual flip tabs for quick reference.

$296.95 per copy
Available Dec. 2006
Catalog No. 60127

RSMeans Facilities Maintenance & Repair Cost Data 2007

RSMeans Facilities Maintenance & Repair Cost Data gives you a complete system to manage and plan your facility repair and maintenance costs and budget efficiently. Guidelines for auditing a facility and developing an annual maintenance plan. Budgeting is included, along with reference tables on cost and management, and information on frequency and productivity of maintenance operations.

The only nationally recognized source of maintenance and repair costs. Developed in cooperation with the Civil Engineering Research Laboratory (CERL) of the Army Corps of Engineers.

$296.95 per copy
Available Dec. 2006
Catalog No. 60307

Reference Books

For more information visit RSMeans Web site at www.rsmeans.com

Home Addition & Renovation Project Costs

This essential home remodeling reference gives you 35 project estimates, and guidance for some of the most popular home renovation and addition projects... from opening up a simple interior wall to adding an entire second story. Each estimate includes a floor plan, color photos, and detailed costs. Use the project estimates as backup for pricing, to check your own estimates, or as a cost reference for preliminary discussion with homeowners.

Includes:

- Case studies—with creative solutions and design ideas.
- Alternate materials costs—so you can match the estimates to the particulars of your projects.
- Location Factors—easy multipliers to adjust the book's costs to your own location.

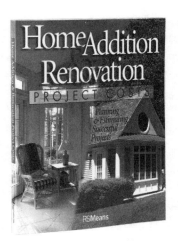

$29.95 per copy
Over 200 pages, illustrated, Softcover
Catalog No. 67349

Kitchen & Bath Project Costs:
Planning & Estimating Successful Projects

Project estimates for 35 of the most popular kitchen and bath renovations... from replacing a single fixture to whole-room remodels. Each estimate includes:

- All materials needed for the project
- Labor-hours to install (and demolish/remove) each item
- Subcontractor costs for certain trades and services
- An allocation for overhead and profit

PLUS! Takeoff and pricing worksheets—forms you can photocopy or access electronically from the book's Web site; alternate materials—unit costs for different finishes and fixtures; location factors—easy multipliers to adjust the costs to your location; and expert guidance on estimating methods, project design, contracts, marketing, working with homeowners, and tips for each of the estimated projects.

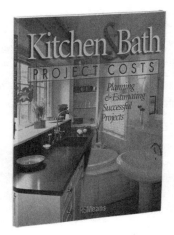

$29.95 per copy
Over 175 pages
Catalog No. 67347

Residential & Light Commercial Construction Standards 2nd Edition
By RSMeans and Contributing Authors

For contractors, subcontractors, owners, developers, architects, engineers, attorneys, and insurance personnel, this book provides authoritative requirements and recommendations compiled from the nation's leading professional associations, industry publications, and building code organizations.

It's an all-in-one reference for establishing a standard for workmanship, quickly resolving disputes, and avoiding defect claims. Includes practical guidance from professionals who are well-known in their respective fields for quality design and construction.

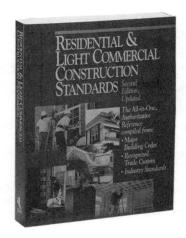

$59.95 per copy
600 pages, illustrated, Softcover
Catalog No. 67322A

For more information visit RSMeans Web site at www.rsmeans.com

Reference Books

Value Engineering: Practical Applications
By Alphonse Dell'Isola, PE
For Design, Construction, Maintenance & Operations

A tool for immediate application—for engineers, architects, facility managers, owners, and contractors. Includes: making the case for VE—the management briefing, integrating VE into planning, budgeting, and design, conducting life cycle costing, using VE methodology in design review and consultant selection, case studies, VE workbook, and a life cycle costing program on disk.

$79.95 per copy
Over 450 pages, illustrated, Softcover
Catalog No. 67319

Facilities Operations & Engineering Reference
By the Association for Facilities Engineering and RSMeans

An all-in-one technical reference for planning and managing facility projects and solving day-to-day operations problems. Selected as the official Certified Plant Engineer reference, this handbook covers financial analysis, maintenance, HVAC and energy efficiency, and more.

$54.98 per copy
Over 700 pages, illustrated, Hardcover
Catalog No. 67318

The Building Professional's Guide to Contract Documents
3rd Edition
By Waller S. Poage, AIA, CSI, CVS

A comprehensive reference for owners, design professionals, contractors, and students.

- Structure your documents for maximum efficiency.
- Effectively communicate construction requirements.
- Understand the roles and responsibilities of construction professionals.
- Improve methods of project delivery.

$32.48 per copy, 400 pages
Diagrams and construction forms, Hardcover
Catalog No. 67261A

Building Security: Strategies & Costs
By David Owen

This comprehensive resource will help you evaluate your facility's security needs, and design and budget for the materials and devices needed to fulfill them.

Includes over 130 pages of RSMeans cost data for installation of security systems and materials, plus a review of more than 50 security devices and construction solutions.

$44.98 per copy
350 pages, illustrated, Hardcover
Catalog No. 67339

Cost Planning & Estimating for Facilities Maintenance

In this unique book, a team of facilities management authorities shares their expertise on:

- Evaluating and budgeting maintenance operations
- Maintaining and repairing key building components
- Applying *RSMeans Facilities Maintenance & Repair Cost Data* to your estimating

Covers special maintenance requirements of the ten major building types.

$89.95 per copy
Over 475 pages, Hardcover
Catalog No. 67314

Life Cycle Costing for Facilities
By Alphonse Dell'Isola and Dr. Steven Kirk

Guidance for achieving higher quality design and construction projects at lower costs! Cost-cutting efforts often sacrifice quality to yield the cheapest product. Life cycle costing enables building designers and owners to achieve both. The authors of this book show how LCC can work for a variety of projects — from roads to HVAC upgrades to different types of buildings.

$99.95 per copy
450 pages, Hardcover
Catalog No. 67341

Planning & Managing Interior Projects 2nd Edition
By Carol E. Farren, CFM
Expert guidance on managing renovation & relocation projects.

This book guides you through every step in relocating to a new space or renovating an old one. From initial meeting through design and construction, to post-project administration, it helps you get the most for your company or client. Includes sample forms, spec lists, agreements, drawings, and much more!

$69.95 per copy
200 pages, Softcover
Catalog No. 67245A

Builder's Essentials: Best Business Practices for Builders & Remodelers
An Easy-to-Use Checklist System
By Thomas N. Frisby

A comprehensive guide covering all aspects of running a construction business, with more than 40 user-friendly checklists. Provides expert guidance on: increasing your revenue and keeping more of your profit, planning for long-term growth, keeping good employees, and managing subcontractors.

$29.95 per copy
Over 220 pages, Softcover
Catalog No. 67329

Reference Books

For more information visit RSMeans Web site at www.rsmeans.com

Interior Home Improvement Costs New 9th Edition

Updated estimates for the most popular remodeling and repair projects—from small, do-it-yourself jobs to major renovations and new construction. Includes: Kitchens & Baths; New Living Space from your Attic, Basement, or Garage; New Floors, Paint, and Wallpaper; Tearing Out or Building New Walls; Closets, Stairs, and Fireplaces; New Energy-Saving Improvements, Home Theaters, and More!

$24.95 per copy
250 pages, illustrated, Softcover
Catalog No. 67308E

Exterior Home Improvement Costs New 9th Edition

Updated estimates for the most popular remodeling and repair projects—from small, do-it-yourself jobs, to major renovations and new construction. Includes: Curb Appeal Projects—Landscaping, Patios, Porches, Driveways, and Walkways; New Windows and Doors; Decks, Greenhouses, and Sunrooms; Room Additions and Garages; Roofing, Siding, and Painting; "Green" Improvements to Save Energy & Water.

$24.95 per copy
Over 275 pages, illustrated, Softcover
Catalog No. 67309E

Builder's Essentials: Plan Reading & Material Takeoff
By Wayne J. DelPico
For Residential and Light Commercial Construction

A valuable tool for understanding plans and specs, and accurately calculating material quantities. Step-by-step instructions and takeoff procedures based on a full set of working drawings.

$35.95 per copy
Over 420 pages, Softcover
Catalog No. 67307

Builder's Essentials: Framing & Rough Carpentry 2nd Edition
By Scot Simpson

Develop and improve your skills with easy-to-follow instructions and illustrations. Learn proven techniques for framing walls, floors, roofs, stairs, doors, and windows. Updated guidance on standards, building codes, safety requirements, and more. Also available in Spanish!

$24.95 per copy
Over 150 pages, Softcover
Catalog No. 67298A
Spanish Catalog No. 67298AS

Concrete Repair and Maintenance Illustrated
By Peter Emmons

Hundreds of illustrations show users how to analyze, repair, clean, and maintain concrete structures for optimal performance and cost effectiveness. From parking garages to roads and bridges to structural concrete, this comprehensive book describes the causes, effects, and remedies for concrete wear and failure. Invaluable for planning jobs, selecting materials, and training employees, this book is a must-have for concrete specialists, general contractors, facility managers, civil and structural engineers, and architects.

$34.98 per copy
300 pages, illustrated, Softcover
Catalog No. 67146

Means Unit Price Estimating Methods
New 3rd Edition

This new edition includes up-to-date cost data and estimating examples, updated to reflect changes to the CSI numbering system and new features of RSMeans cost data. It describes the most productive, universally accepted ways to estimate, and uses checklists and forms to illustrate shortcuts and timesavers. A model estimate demonstrates procedures. A new chapter explores computer estimating alternatives.

$29.98 per copy
Over 350 pages, illustrated, Hardcover
Catalog No. 67303A

Total Productive Facilities Management
By Richard W. Sievert, Jr.

Today, facilities are viewed as strategic resources... elevating the facility manager to the role of asset manager supporting the organization's overall business goals. Now, Richard Sievert Jr., in this well-articulated guidebook, sets forth a new operational standard for the facility manager's emerging role... a comprehensive program for managing facilities as a true profit center.

$29.98 per copy
275 pages, Softcover
Catalog No. 67321

Means Environmental Remediation Estimating Methods 2nd Edition
By Richard R. Rast

Guidelines for estimating 50 standard remediation technologies. Use it to prepare preliminary budgets, develop estimates, compare costs and solutions, estimate liability, review quotes, negotiate settlements.

$49.98 per copy
Over 750 pages, illustrated, Hardcover
Catalog No. 64777A

For more information
visit RSMeans Web site
at www.rsmeans.com

Reference Books

Means Illustrated Construction Dictionary Condensed, 2nd Edition
Recognized in the industry as the best resource of its kind.

This essential tool has been further enhanced with updates to existing terms and the addition of hundreds of new terms and illustrations—in keeping with recent developments. For contractors, architects, insurance and real estate personnel, homeowners, and anyone who needs quick, clear definitions for construction terms.

$59.95 per copy
Over 500 pages, Softcover
Catalog No. 67282A

Means Repair & Remodeling Estimating New 4th Edition
By Edward B. Wetherill & RSMeans

This important reference focuses on the unique problems of estimating renovations of existing structures, and helps you determine the true costs of remodeling through careful evaluation of architectural details and a site visit.

New section on disaster restoration costs.

$69.95 per copy
Over 450 pages, illustrated, Hardcover
Catalog No. 67265B

Facilities Planning & Relocation
New, lower price and user-friendly format.
By David D. Owen

A complete system for planning space needs and managing relocations. Includes step-by-step manual, over 50 forms, and extensive reference section on materials and furnishings.

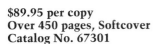

$89.95 per copy
Over 450 pages, Softcover
Catalog No. 67301

Means Square Foot & Assemblies Estimating Methods 3rd Edition

Develop realistic square foot and assemblies costs for budgeting and construction funding. The new edition features updated guidance on square foot and assemblies estimating using UNIFORMAT II. An essential reference for anyone who performs conceptual estimates.

$34.98 per copy
Over 300 pages, illustrated, Hardcover
Catalog No. 67145B

Means Electrical Estimating Methods 3rd Edition

Expanded edition includes sample estimates and cost information in keeping with the latest version of the CSI MasterFormat and UNIFORMAT II. Complete coverage of fiber optic and uninterruptible power supply electrical systems, broken down by components, and explained in detail. Includes a new chapter on computerized estimating methods. A practical companion to *RSMeans Electrical Cost Data*.

Means Mechanical Estimating Methods 3rd Edition

This guide assists you in making a review of plans, specs, and bid packages, with suggestions for takeoff procedures, listing substitutions, and pre-bid scheduling. Includes suggestions for budgeting labor and equipment usage. Compares material and construction methods to allow you to select the best option.

$64.95 per copy
Over 325 pages, Hardcover
Catalog No. 67230B

$64.95 per copy
Over 350 pages, illustrated, Hardcover
Catalog No. 67294A

Means ADA Compliance Pricing Guide New Second Edition
By Adaptive Environments and RSMeans

Completely updated and revised to the new 2004 *Americans with Disabilities Act Accessibility Guidelines*, this book features more than 70 of the most commonly needed modifications for ADA compliance. Projects range from installing ramps and walkways, widening doorways and entryways, and installing and refitting elevators, to relocating light switches and signage.

Project Scheduling & Management for Construction
New 3rd Edition
By David R. Pierce, Jr.

A comprehensive yet easy-to-follow guide to construction project scheduling and control—from vital project management principles through the latest scheduling, tracking, and controlling techniques. The author is a leading authority on scheduling, with years of field and teaching experience at leading academic institutions. Spend a few hours with this book and come away with a solid understanding of this essential management topic.

$79.95 per copy
Over 350 pages, illustrated, Softcover
Catalog No. 67310A

$64.95 per copy
Over 300 pages, illustrated, Hardcover
Catalog No. 67247B

Reference Books

For more information visit RSMeans Web site at www.rsmeans.com

The Practice of Cost Segregation Analysis
by Bruce A. Desrosiers and Wayne J. DelPico

This expert guide walks you through the practice of cost segregation analysis, which enables property owners to defer taxes and benefit from "accelerated cost recovery" through depreciation deductions on assets that are properly identified and classified.

With a glossary of terms, sample cost segregation estimates for various building types, key information resources, and updates via a dedicated Web site, this book is a critical resource for anyone involved in cost segregation analysis.

$99.95 per copy
Over 225 pages
Catalog No. 67345

Preventive Maintenance for Multi-Family Housing
by John C. Maciha

Prepared by one of the nation's leading experts on multi-family housing.

This complete PM system for apartment and condominium communities features expert guidance, checklists for buildings and grounds maintenance tasks and their frequencies, a reusable wall chart to track maintenance, and a dedicated Web site featuring customizable electronic forms. A must-have for anyone involved with multi-family housing maintenance and upkeep.

$89.95 per copy
225 pages
Catalog No. 67346

Means Landscape Estimating Methods
4th Edition
By Sylvia H. Fee

This revised edition offers expert guidance for preparing accurate estimates for new landscape construction and grounds maintenance. Includes a complete project estimate featuring the latest equipment and methods, and chapters on Life Cycle Costing and Landscape Maintenance Estimating.

$62.95 per copy
Over 300 pages, illustrated, Hardcover
Catalog No. 67295B

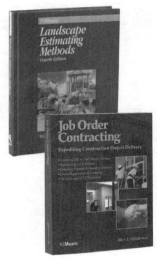

Job Order Contracting
Expediting Construction Project Delivery
by Allen Henderson

Expert guidance to help you implement JOC—fast becoming the preferred project delivery method for repair and renovation, minor new construction, and maintenance projects in the public sector and in many states and municipalities. The author, a leading JOC expert and practitioner, shows how to:

- Establish a JOC program
- Evaluate proposals and award contracts
- Handle general requirements and estimating
- Partner for maximum benefits

$89.95 per copy
192 pages, illustrated, Hardcover
Catalog No. 67348

Builder's Essentials: Estimating Building Costs
For the Residential & Light Commercial Contractor
By Wayne J. DelPico

Step-by-step estimating methods for residential and light commercial contractors. Includes a detailed look at every construction specialty—explaining all the components, takeoff units, and labor needed for well-organized, complete estimates. Covers correctly interpreting plans and specifications, and developing accurate and complete labor and material costs.

$29.95 per copy
Over 400 pages, illustrated, Softcover
Catalog No. 67343

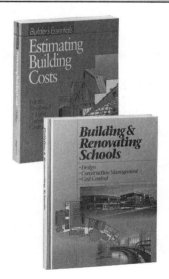

Building & Renovating Schools

This all-inclusive guide covers every step of the school construction process—from initial planning, needs assessment, and design, right through moving into the new facility. A must-have resource for anyone concerned with new school construction or renovation. With square foot cost models for elementary, middle, and high school facilities, and real-life case studies of recently completed school projects.

The contributors to this book—architects, construction project managers, contractors, and estimators who specialize in school construction—provide start-to-finish, expert guidance on the process.

$99.95 per copy
Over 425 pages, Hardcover
Catalog No. 67342

For more information
visit RSMeans Web site
at www.rsmeans.com

Reference Books

Historic Preservation: Project Planning & Estimating
By Swanke Hayden Connell Architects

Expert guidance on managing historic restoration, rehabilitation, and preservation building projects and determining and controlling their costs. Includes:
- How to determine whether a structure qualifies as historic
- Where to obtain funding and other assistance
- How to evaluate and repair more than 75 historic building materials

$49.98 per copy
Over 675 pages, Hardcover
Catalog No. 67323

Means Illustrated Construction Dictionary
Unabridged 3rd Edition, with CD-ROM

Long regarded as the industry's finest, *Means Illustrated Construction Dictionary* is now even better. With the addition of over 1,000 new terms and hundreds of new illustrations, it is the clear choice for the most comprehensive and current information. The companion CD-ROM that comes with this new edition adds many extra features: larger graphics, expanded definitions, and links to both CSI MasterFormat numbers and product information.

$99.95 per copy
Over 790 pages, illustrated, Hardcover
Catalog No. 67292A

Designing & Building with the IBC 2nd Edition
By Rolf Jensen & Associates, Inc.

This updated, comprehensive guide helps building professionals make the transition to the 2003 International Building Code®. Includes a side-by-side code comparison of the IBC 2003 to the IBC 2000 and the three primary model codes, a quick-find index, and professional code commentary. With illustrations, abbreviations key, and an extensive Resource section.

$99.95 per copy
Over 875 pages, Softcover
Catalog No. 67328A

Means Plumbing Estimating Methods 3rd Edition
By Joseph Galeno and Sheldon Greene

Updated and revised! This practical guide walks you through a plumbing estimate, from basic materials and installation methods through change order analysis. *Plumbing Estimating Methods* covers residential, commercial, industrial, and medical systems, and features sample takeoff and estimate forms and detailed illustrations of systems and components.

$29.98 per copy
330+ pages, Softcover
Catalog No. 67283B

Builder's Essentials: Advanced Framing Methods
By Scot Simpson

A highly illustrated, "framer-friendly" approach to advanced framing elements. Provides expert, but easy to interpret, instruction for laying out and framing complex walls, roofs, and stairs, and special requirements for earthquake and hurricane protection. Also helps bring framers up to date on the latest building code changes, and provides tips on the lead framer's role and responsibilities, how to prepare for a job, and how to get the crew started.

$24.95 per copy
250 pages, illustrated, Softcover
Catalog No. 67330

Means Estimating Handbook
2nd Edition

Updated Second Edition answers virtually any estimating technical question—all organized by CSI MasterFormat. This comprehensive reference covers the full spectrum of technical data required to estimate construction costs. The book includes information on sizing, productivity, equipment requirements, code-mandated specifications, design standards, and engineering factors.

$99.95 per copy
Over 900 pages, Hardcover
Catalog No. 67276A

Preventive Maintenance Guidelines for School Facilities
By John C. Maciha

A complete PM program for K-12 schools that ensures sustained security, safety, property integrity, user satisfaction, and reasonable ongoing expenditures.

Includes schedules for weekly, monthly, semiannual, and annual maintenance in hard copy and electronic format.

$149.95 per copy
Over 225 pages, Hardcover
Catalog No. 67326

Preventive Maintenance for Higher Education Facilities
By Applied Management Engineering, Inc.

An easy-to-use system to help facilities professionals establish the value of PM, and to develop and budget for an appropriate PM program for their college or university. Features interactive campus building models typical of those found in different-sized higher education facilities, and PM checklists linked to each piece of equipment or system in hard copy and electronic format.

$149.95 per copy
150 pages, Hardcover
Catalog No. 67337

For more information
visit RSMeans Web site
at www.rsmeans.com

Seminars

Means CostWorks® Training

This one-day seminar has been designed with the intention of assisting both new and existing users to become more familiar with the *Means CostWorks* program. The class is broken into two unique sections: (1) A one-half day presentation on the function of each icon; and each student will be shown how to use the software to develop a cost estimate. (2) Hands-on estimating exercises that will ensure that each student thoroughly understands how to use *CostWorks*. You must bring your own laptop computer to this course.

Means CostWorks Benefits/Features:
- Estimate in your own spreadsheet format
- Power of RSMeans National Database
- Database automatically regionalized
- Save time with keyword searches
- Save time by establishing common estimate items in "Bookmark" files
- Customize your spreadsheet template
- Hot Key to Product Manufacturers' listings and specs
- Merge capability for networking environments
- View crews and assembly components
- AutoSave capability
- Enhanced sorting capability

Unit Price Estimating

This interactive two-day seminar teaches attendees how to interpret project information and process it into final, detailed estimates with the greatest accuracy level.

The single most important credential an estimator can take to the job is the ability to visualize construction in the mind's eye, and thereby estimate accurately.

Some Of What You'll Learn:
- Interpreting the design in terms of cost
- The most detailed, time-tested methodology for accurate "pricing"
- Key cost drivers—material, labor, equipment, staging, and subcontracts
- Understanding direct and indirect costs for accurate job cost accounting and change order management

Who Should Attend: Corporate and government estimators and purchasers, architects, engineers… and others needing to produce accurate project estimates.

Square Foot and Assemblies Estimating

This two-day course teaches attendees how to quickly deliver accurate square foot estimates using limited budget and design information.

Some Of What You'll Learn:
- How square foot costing gets the estimate done faster
- Taking advantage of a "systems" or "assemblies" format
- The RSMeans "building assemblies/square foot cost approach"
- How to create a very reliable preliminary and systems estimate using bare-bones design information

Who Should Attend: Facilities managers, facilities engineers, estimators, planners, developers, construction finance professionals… and others needing to make quick, accurate construction cost estimates at commercial, government, educational, and medical facilities.

Repair and Remodeling Estimating

This two-day seminar emphasizes all the underlying considerations unique to repair/remodeling estimating and presents the correct methods for generating accurate, reliable R&R project costs using the unit price and assemblies methods.

Some Of What You'll Learn:
- Estimating considerations—like labor-hours, building code compliance, working within existing structures, purchasing materials in smaller quantities, unforeseen deficiencies
- Identifying problems and providing solutions to estimating building alterations
- Rules for factoring in minimum labor costs, accurate productivity estimates, and allowances for project contingencies
- R&R estimating examples calculated using unit price and assemblies data

Who Should Attend: Facilities managers, plant engineers, architects, contractors, estimators, builders… and others who are concerned with the proper preparation and/or evaluation of repair and remodeling estimates.

Mechanical and Electrical Estimating

This two-day course teaches attendees how to prepare more accurate and complete mechanical/electrical estimates, avoiding the pitfalls of omission and double-counting, while understanding the composition and rationale within the RSMeans Mechanical/Electrical database.

Some Of What You'll Learn:
- The unique way mechanical and electrical systems are interrelated
- M&E estimates–conceptual, planning, budgeting, and bidding stages
- Order of magnitude, square foot, assemblies, and unit price estimating
- Comparative cost analysis of equipment and design alternatives

Who Should Attend: Architects, engineers, facilities managers, mechanical and electrical contractors… and others needing a highly reliable method for developing, understanding, and evaluating mechanical and electrical contracts.

Plan Reading and Material Takeoff

This two-day program teaches attendees to read and understand construction documents and to use them in the preparation of material takeoffs.

Some of What You'll Learn:
- Skills necessary to read and understand typical contract documents—blueprints and specifications
- Details and symbols used by architects and engineers
- Construction specifications' importance in conjunction with blueprints
- Accurate takeoff of construction materials and industry-accepted takeoff methods

Who Should Attend: Facilities managers, construction supervisors, office managers… and others responsible for the execution and administration of a construction project, including government, medical, commercial, educational, or retail facilities.

Facilities Maintenance and Repair Estimating

This two-day course teaches attendees how to plan, budget, and estimate the cost of ongoing and preventive maintenance and repair for existing buildings and grounds.

Some Of What You'll Learn:
- The most financially favorable maintenance, repair, and replacement scheduling and estimating
- Auditing and value engineering facilities
- Preventive planning and facilities upgrading
- Determining both in-house and contract-out service costs
- Annual, asset-protecting M&R plan

Who Should Attend: Facility managers, maintenance supervisors, buildings and grounds superintendents, plant managers, planners, estimators… and others involved in facilities planning and budgeting.

Scheduling and Project Management

This two-day course teaches attendees the most current and proven scheduling and management techniques needed to bring projects in on time and on budget.

Some Of What You'll Learn:
- Crucial phases of planning and scheduling
- How to establish project priorities and develop realistic schedules and management techniques
- Critical Path and Precedence Methods
- Special emphasis on cost control

Who Should Attend: Construction project managers, supervisors, engineers, estimators, contractors… and others who want to improve their project planning, scheduling, and management skills.

Assessing Scope of Work for Facility Construction Estimating

This two-day course is a practical training program that addresses the vital importance of understanding the SCOPE of projects in order to produce accurate cost estimates in a facility repair and remodeling environment.

Some Of What You'll Learn:
- Discussions on site visits, plans/specs, record drawings of facilities, and site-specific lists
- Review of CSI divisions, including means, methods, materials, and the challenges of scoping each topic
- Exercises in SCOPE identification and SCOPE writing for accurate estimating of projects
- Hands-on exercises that require SCOPE, take-off, and pricing

Who Should Attend: Corporate and government estimators, planners, facility managers… and others needing to produce accurate project estimates.

Seminars

2007 RSMeans Seminar Schedule

For more information visit RSMeans Web site at www.rsmeans.com

Location	Dates
Las Vegas, NV	March
Washington, DC	April
Phoenix, AZ	April
Denver, CO	May
San Francisco, CA	June
Philadelphia, PA	June
Washington, DC	September
Dallas, TX	September
Las Vegas, NV	October
Orlando, FL	November
Atlantic City, NJ	November
San Diego, CA	December

Note: Call for exact dates and details.

Registration Information

Register Early... Save up to $100! Register 30 days before the start date of a seminar and save $100 off your total fee. *Note: This discount can be applied only once per order. It cannot be applied to team discount registrations or any other special offer.*

How to Register Register by phone today! RSMeans' toll-free number for making reservations is: **1-800-334-3509.**

Individual Seminar Registration Fee $935. *Means CostWorks*® Training Registration Fee $375. To register by mail, complete the registration form and return with your full fee to: Seminar Division, Reed Construction Data, RSMeans Seminars, 63 Smiths Lane, Kingston, MA 02364.

Federal Government Pricing All federal government employees save 25% off regular seminar price. Other promotional discounts cannot be combined with Federal Government discount.

Team Discount Program Two to four seminar registrations, call for pricing: 1-800-334-3509, Ext. 5115

Multiple Course Discounts When signing up for two or more courses, call for pricing.

Refund Policy Cancellations will be accepted up to ten days prior to the seminar start. There are no refunds for cancellations received later than ten working days prior to the first day of the seminar. A $150 processing fee will be applied for all cancellations. Written notice of cancellation is required. Substitutions can be made at any time before the session starts. **No-shows are subject to the full seminar fee.**

AACE Approved Courses Many seminars described and offered here have been approved for 14 hours (1.4 recertification credits) of credit by the AACE International Certification Board toward meeting the continuing education requirements for recertification as a Certified Cost Engineer/Certified Cost Consultant.

AIA Continuing Education We are registered with the AIA Continuing Education System (AIA/CES) and are committed to developing quality learning activities in accordance with the CES criteria. Many seminars meet the AIA/CES criteria for Quality Level 2. AIA members may receive (14) learning units (LUs) for each two-day RSMeans course.

NASBA CPE Sponsor Credits We are part of the National Registry of CPE Sponsors. Attendees may be eligible for (16) CPE credits.

Daily Course Schedule The first day of each seminar session begins at 8:30 A.M. and ends at 4:30 P.M. The second day is 8:00 A.M.–4:00 P.M. Participants are urged to bring a hand-held calculator since many actual problems will be worked out in each session.

Continental Breakfast Your registration includes the cost of a continental breakfast, a morning coffee break, and an afternoon break. These informal segments will allow you to discuss topics of mutual interest with other members of the seminar. (You are free to make your own lunch and dinner arrangements.)

Hotel/Transportation Arrangements RSMeans has arranged to hold a block of rooms at most host hotels. To take advantage of special group rates when making your reservation, be sure to mention that you are attending the RSMeans Seminar. You are, of course, free to stay at the lodging place of your choice. (**Hotel reservations and transportation arrangements should be made directly by seminar attendees.**)

Important Class sizes are limited, so please register as soon as possible.

Note: Pricing subject to change.

Registration Form
Call 1-800-334-3509 to register or FAX 1-800-632-6732. Visit our Web site: www.rsmeans.com

Please register the following people for the RSMeans Construction Seminars as shown here. We understand that we must make our own hotel reservations if overnight stays are necessary.

☐ Full payment of $_____ enclosed.

☐ Bill me

Name of Registrant(s)
(To appear on certificate of completion)

P.O. #:
GOVERNMENT AGENCIES MUST SUPPLY PURCHASE ORDER NUMBER OR TRAINING FORM.

Firm Name
Address
City/State/Zip
Telephone No.   Fax No.
E-mail Address
Charge our registration(s) to: ☐ MasterCard ☐ VISA ☐ American Express ☐ Discover
Account No.   Exp. Date
Cardholder's Signature
Seminar Name   City   Dates

Please mail check to: Seminar Division, Reed Construction Data, RSMeans Seminars, 63 Smiths Lane, P.O. Box 800, Kingston, MA 02364 USA

MeansData™

CONSTRUCTION COSTS FOR SOFTWARE APPLICATIONS
Your construction estimating software is only as good as your cost data.

A proven construction cost database is a mandatory part of any estimating package. The following list of software providers can offer you MeansData™ as an added feature for their estimating systems. See the table below for what types of products and services they offer (match their numbers). Visit online at **www.rsmeans.com/demosource/** for more information and free demos. Or call their numbers listed below.

1. **3D International**
   713-871-7000
   venegas@3di.com

2. **4Clicks-Solutions, LLC**
   719-574-7721
   mbrown@4clicks-solutions.com

3. **Aepco, Inc.**
   301-670-4642
   blueworks@aepco.com

4. **Applied Flow Technology**
   800-589-4943
   info@aft.com

5. **ArenaSoft Estimating**
   888-370-8806
   info@arenasoft.com

6. **Ares Corporation**
   925-299-6700
   sales@arescorporation.com

7. **Beck Technology**
   214-303-6293
   stewartcarroll@beckgroup.com

8. **BSD - Building Systems Design, Inc.**
   888-273-7638
   bsd@bsdsoftlink.com

9. **CMS - Computerized Micro Solutions**
   800-255-7407
   cms@proest.com

10. **Corecon Technologies, Inc.**
    714-895-7222
    sales@corecon.com

11. **CorVet Systems**
    301-622-9069
    sales@corvetsys.com

12. **Earth Tech**
    303-771-3103
    kyle.knudson@earthtech.com

13. **Estimating Systems, Inc.**
    800-967-8572
    esipulsar@adelphia.net

14. **HCSS**
    800-683-3196
    info@hcss.com

15. **MC² - Management Computer**
    800-225-5622
    vkeys@mc2-ice.com

16. **Maximus Asset Solutions**
    800-659-9001
    assetsolutions@maximus.com

17. **Sage Timberline Office**
    800-628-6583
    productinfo.timberline@sage.com

18. **Shaw Beneco Enterprises, Inc.**
    877-719-4748
    inquire@beneco.com

19. **US Cost, Inc.**
    800-372-4003
    sales@uscost.com

20. **Vanderweil Facility Advisors**
    617-451-5100
    info@VFA.com

21. **WinEstimator, Inc.**
    800-950-2374
    sales@winest.com

TYPE	1	2	3	4	5	6	7	8	9	10	11	12	13	14	15	16	17	18	19	20	21
BID					•			•	•		•			•	•		•				•
Estimating		•			•	•	•	•	•	•	•	•	•	•	•		•	•	•		•
DOC/JOC/SABER		•			•			•			•		•			•	•	•			•
ID/IQ		•									•		•			•	•	•			•
Asset Mgmt.																•				•	•
Facility Mgmt.	•		•													•				•	
Project Mgmt.	•	•					•				•			•			•	•			
TAKE-OFF					•					•	•	•		•	•		•		•		
EARTHWORK											•		•	•							
Pipe Flow				•										•							
HVAC/Plumbing					•				•		•										
Roofing					•				•		•										
Design	•				•		•										•	•			•
Other Offers/Links:																					
Accounting/HR		•			•										•		•	•			
Scheduling						•	•										•		•		•
CAD							•										•				•
PDA																	•		•		•
Lt. Versions		•							•								•				
Consulting	•				•		•		•		•	•			•	•			•	•	•
Training		•			•		•	•	•	•	•	•	•	•	•	•	•	•	•	•	•

Qualified re-seller applications now being accepted. Call Carol Polio, Ext. 5107.

FOR MORE INFORMATION
CALL 1-800-448-8182, EXT. 5107 OR FAX 1-800-632-6732

For more information
visit RSMeans Web site
at www.rsmeans.com

New Titles

Understanding & Negotiating Construction Contracts

By Kit Werremeyer

Take advantage of the author's 30 years' experience in small-to-large (including international) construction projects. Learn how to identify, understand, and evaluate high risk terms and conditions typically found in all construction contracts—then negotiate to lower or eliminate the risk, improve terms of payment, and reduce exposure to claims and disputes. The author avoids "legalese" and gives real-life examples from actual projects.

$69.95 per copy
300 pages, Softcover
Catalog No. 67350

Green Building: Project Planning & Cost Estimating, 2nd Edition

This new edition has been completely updated with the latest in green building technologies, design concepts, standards, and costs. Now includes a 2007 Green Building *CostWorks* CD with more than 300 green building assemblies and over 5,000 unit price line items for sustainable building. The new edition is also full-color with all new case studies—plus a new chapter on deconstruction, a key aspect of green building.

$129.95 per copy
350 pages, Softcover
Catalog No. 67338A

How to Estimate with Means Data & CostWorks
New 3rd Edition

By RSMeans and Saleh A. Mubarak, Ph.D.

New 3rd Edition—fully updated with new chapters, plus new CD with updated *CostWorks* cost data and MasterFormat organization. Includes all major construction items—with more than 300 exercises and two sets of plans that show how to estimate for a broad range of construction items and systems—including general conditions and equipment costs.

$59.95 per copy
272 pages, Softcover
Includes CostWorks CD
Catalog No. 67324B

Means Spanish/English Construction Dictionary
2nd Edition

By RSMeans and the International Code Council

This expanded edition features thousands of the most common words and useful phrases in the construction industry with easy-to-follow pronunciations in both Spanish and English. Over 800 new terms, phrases, and illustrations have been added. It also features a new stand-alone "Safety & Emergencies" section, with colored pages for quick access. Unique to this dictionary are the systems illustrations showing the relationship of components in the most common building systems for all major trades.

$23.95 per copy
Over 400 pages
Catalog No. 67327A

Construction Business Management

By Nick Ganaway

Only 43% of construction firms stay in business after four years. Make sure your company thrives with valuable guidance from a pro with 25 years of success as a commercial contractor. Find out what it takes to build all aspects of a business that is profitable, enjoyable, and enduring. With a bonus chapter on retail construction.

$49.95 per copy
200 pages, Softcover
Catalog No. 67352

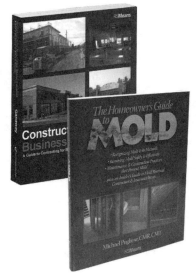

The Homeowner's Guide to Mold

Expert guidance to protect your health and your home.

Mold, whether caused by leaks, humidity or flooding, is a real health and financial issue—for homeowners and contractors. This full-color book explains:

- Construction and maintenance practices to prevent mold
- How to inspect for and remove mold
- Mold remediation procedures and costs
- What to do after a flood
- How to deal with insurance companies if you're thinking of submitting a mold damages claim

$21.95 per copy
144 pages, Softcover
Catalog No. 67344

2007 Order Form

ORDER TOLL FREE 1-800-334-3509
OR FAX 1-800-632-6732

Qty.	Book No.	COST ESTIMATING BOOKS	Unit Price	Total
	60067	Assemblies Cost Data 2007	$223.95	
	60017	Building Construction Cost Data 2007	136.95	
	61017	Building Const. Cost Data–Looseleaf Ed. 2007	169.95	
	60717	Building Const. Cost Data–Spanish 2007	136.95	
	60227	Building Const. Cost Data–Western Ed. 2007	136.95	
	60117	Concrete & Masonry Cost Data 2007	124.95	
	50147	Construction Cost Indexes 2007 (subscription)	294.00	
	60147A	Construction Cost Index–January 2007	73.50	
	60147B	Construction Cost Index–April 2007	73.50	
	60147C	Construction Cost Index–July 2007	73.50	
	60147D	Construction Cost Index–October 2007	73.50	
	60347	Contr. Pricing Guide: Resid. R & R Costs 2007	39.95	
	60337	Contr. Pricing Guide: Resid. Detailed 2007	39.95	
	60327	Contr. Pricing Guide: Resid. Sq. Ft. 2007	39.95	
	60237	Electrical Change Order Cost Data 2007	136.95	
	60037	Electrical Cost Data 2007	136.95	
	60207	Facilities Construction Cost Data 2007	323.95	
	60307	Facilities Maintenance & Repair Cost Data 2007	296.95	
	60167	Heavy Construction Cost Data 2007	136.95	
	60097	Interior Cost Data 2007	136.95	
	60127	Labor Rates for the Const. Industry 2007	296.95	
	60187	Light Commercial Cost Data 2007	116.95	
	60027	Mechanical Cost Data 2007	136.95	
	63017	Metric Construction Cost Data 2007	162.95	
	60157	Open Shop Building Const. Cost Data 2007	136.95	
	60217	Plumbing Cost Data 2007	136.95	
	60047	Repair and Remodeling Cost Data 2007	116.95	
	60177	Residential Cost Data 2007	116.95	
	60287	Site Work & Landscape Cost Data 2007	136.95	
	60057	Square Foot Costs 2007	148.95	
	62017	Yardsticks for Costing (2007)	136.95	
	62016	Yardsticks for Costing (2006)	126.95	
		REFERENCE BOOKS		
	67310A	ADA Compliance Pricing Guide, 2nd Ed.	79.95	
	67330	Bldrs Essentials: Adv. Framing Methods	24.95	
	67329	Bldrs Essentials: Best Bus. Practices for Bldrs	29.95	
	67298A	Bldrs Essentials: Framing/Carpentry 2nd Ed.	24.95	
	67298AS	Bldrs Essentials: Framing/Carpentry Spanish	24.95	
	67307	Bldrs Essentials: Plan Reading & Takeoff	35.95	
	67342	Building & Renovating Schools	99.95	
	67261A	Bldg. Prof. Guide to Contract Documents, 3rd Ed.	32.48	
	67339	Building Security: Strategies & Costs	44.98	
	67146	Concrete Repair & Maintenance Illustrated	34.98	
	67352	Construction Business Management	49.95	
	67314	Cost Planning & Est. for Facil. Maint.	89.95	
	67328A	Designing & Building with the IBC, 2nd Ed.	99.95	
	67230B	Electrical Estimating Methods, 3rd Ed.	64.95	
	64777A	Environmental Remediation Est. Methods, 2nd Ed.	49.98	
	67343	Estimating Bldg. Costs for Resi. & Lt. Comm.	29.95	

Qty.	Book No.	REFERENCE BOOKS (Cont.)	Unit Price	Total
	67276A	Estimating Handbook, 2nd Ed.	$99.95	
	67318	Facilities Operations & Engineering Reference	54.98	
	67301	Facilities Planning & Relocation	89.95	
	67338A	Green Building: Proj. Planning & Cost Est., 2nd Ed.	129.95	
	67323	Historic Preservation: Proj. Planning & Est.	49.98	
	67349	Home Addition & Renovation Project Costs	29.95	
	67308E	Home Improvement Costs–Int. Projects, 9th Ed.	24.95	
	67309E	Home Improvement Costs–Ext. Projects, 9th Ed.	24.95	
	67344	Homeowner's Guide to Mold	21.95	
	67324B	How to Est.w/Means Data & CostWorks, 3rd Ed.	59.95	
	67282A	Illustrated Const. Dictionary, Condensed, 2nd Ed.	59.95	
	67292A	Illustrated Const. Dictionary, w/CD-ROM, 3rd Ed.	99.95	
	67348	Job Order Contracting	89.95	
	67347	Kitchen & Bath Project Costs	29.95	
	67295B	Landscape Estimating Methods, 4th Ed.	62.95	
	67341	Life Cycle Costing for Facilities	99.95	
	67294A	Mechanical Estimating Methods, 3rd Ed.	64.95	
	67245A	Planning & Managing Interior Projects, 2nd Ed.	69.95	
	67283B	Plumbing Estimating Methods, 3rd Ed.	29.98	
	67345	Practice of Cost Segregation Analysis	99.95	
	67337	Preventive Maint. for Higher Education Facilities	149.95	
	67346	Preventive Maint. for Multi-Family Housing	89.95	
	67326	Preventive Maint. Guidelines for School Facil.	149.95	
	67247B	Project Scheduling & Management for Constr. 3rd Ed.	64.95	
	67265B	Repair & Remodeling Estimating Methods, 4th Ed.	69.95	
	67322A	Resi. & Light Commercial Const. Stds., 2nd Ed.	59.95	
	67327A	Spanish/English Construction Dictionary, 2nd Ed.	23.95	
	67145B	Sq. Ft. & Assem. Estimating Methods, 3rd Ed.	34.98	
	67321	Total Productive Facilities Management	29.98	
	67350	Understanding and Negotiating Const. Contracts	69.95	
	67303A	Unit Price Estimating Methods, 3rd Ed.	29.98	
	67319	Value Engineering: Practical Applications	79.95	

MA residents add 5% state sales tax

Shipping & Handling**

Total (U.S. Funds)*

Prices are subject to change and are for U.S. delivery only. *Canadian customers may call for current prices. **Shipping & handling charges: Add 7% of total order for check and credit card payments. Add 9% of total order for invoiced orders.

Send Order To: **ADDV-1000**

Name (Please Print) _____

Company _____

☐ Company
☐ Home Address _____

_____

City/State/Zip _____

Phone # _____ P.O. # _____

(Must accompany all orders being billed)

Mail To: RSMeans, P.O. Box 800, Kingston, MA 02364-0800

Reed Construction Data, Inc.

Reed Construction Data, Inc. is a leading worldwide provider of total construction information solutions. The company's portfolio of information products and services is designed specifically to help construction industry professionals advance their businesses with timely, accurate, and actionable project, product, and cost data. Each of these groups offers a variety of innovative products and services created for the full spectrum of design, construction, distribution, and manufacturing professionals. Reed Construction Data is a division of Reed Business Information, a member of the Reed Elsevier plc group of companies.

Cost Information
RSMeans, the undisputed market leader and authority on construction costs, publishes current cost and estimating information in annual cost books, on *Means CostWorks*® CD, and on the Web. RSMeans furnishes the construction industry with a rich library of corresponding reference books and a series of professional seminars that are designed to sharpen personal skills and maximize the effective use of cost estimating and management tools. RSMeans also provides construction cost consulting for owners, manufacturers, designers, and contractors.

Project Data
Reed Construction Data maintains a significant role in the overall construction process by facilitating the assembly of public and private project data and delivering this information to contractors, distributors, and building product manufacturers in a secure, accurate, and timely manner. Acting on behalf of architects, engineers, owner/developers, and government agencies, Reed Construction Data is the construction community's premier resource for project leads and bid documents. Our wide variety of products and services provides the most up-to-date, enhanced details on many of the country's largest public sector, commercial, industrial, and multi-family residential projects.

Reed Bulletin and Reed CONNECT™ provide project leads and project data through all stages of construction. Customers are supplied industry data through leads, project reports, contact lists, plans, specifications, and addenda—either online, via e-mail, or in paper format.

Building Product Information
The Reed First Source suite of products is the only integrated building product information system offered to the commercial construction industry for searching, selecting, and specifying building products. These online and print resources include the *First Source* catalog, SPEC-DATA™, MANU-SPEC ™, First Source CAD, and manufacturer's catalogs. Written by industry professionals and organized using CSI MasterFormat 2004 criteria, the design and construction community uses this information to make better, more informed design decisions.

Research & Analytics
Reed Construction Data's forecasting tools cover most aspects of the construction business. Our vast network of resources makes us uniquely qualified to give you the information you need to keep your business profitable.

Associated Construction Publications (ACP)
Reed Construction Data's regional construction magazines cover the nation through a network of 14 regional magazines. Serving the construction market for more than 100 years, our magazines are a trusted source of news and information in the local and national construction communities.

International
Reed Construction Data Canada is the "voice of construction" in Canada, and leading provider of project information and construction market intelligence. Products and services include: *Building Reports*, Canadian ICI project leads tracking construction projects through life cycles—from concept to construction. *Building Reports* can be customized and delivered through CONNECT™, RCD's web-based sales and contact management system, or by fax or email; *Daily Commercial News*, Ontario's leading construction industry daily newspaper, contains ICI project leads, Building Reports, Bidders' Register, news and trends, construction tenders, certificates of substantial performance, and CanaData Economic Snapshot; *Journal of Commerce*, Alberta and British Columbia's twice-weekly construction industry newspaper, is a one-stop source for private and public projects, tenders and expressions of interest, top industry news and construction trends, and more; *CanaData*, a variety of products tracking market performance, economic analysis, forecasts, and trends in the construction market; *Buildcore*, organized by MasterFormat™, is Canada's most comprehensive listing of commercial building products and manufacturers active and available in Canada—in print and online at www.buildcore.com.

Reed Business Information – Scandinavia, with offices in Denmark, Norway, Finland, and Sweden, is the construction industry's source for project information in the Nordic region.

Reed Construction Data—Australia (formerly Cordell Building Information Services) is Australia's leading provider and respected industry authority on building and construction information.

For more information, please visit our website at www.reedconstructiondata.com

Reed Construction Data, Inc.
30 Technology Parkway South
Norcross, GA 30092-2912
(800) 322-6996
(800) 895-8661 (fax)
Email: info@reedbusiness.com